INTERNATIONAL SERIES OF MONOGRAPHS ON PHYSICS

INTERNATIONAL SERIES OF MONOGRAPHS ON PHYSICS

166. N. Horing: *Quantum statistical field theory*
165. T.C. Choy: *Effective medium theory, Second edition*
164. L. Pitaevskii, S. Stringari: *Bose-Einstein condensation and superfluidity*
163. B.J. Dalton, J. Jeffers, S.M. Barnett: *Phase space methods for degenerate quantum gases*
162. W.D. McComb: *Homogeneous, isotropic turbulence - phenomenology, renormalization and statistical closures*
161. V.Z. Kresin, H. Morawitz, S.A. Wolf: *Superconducting state - mechanisms and properties*
160. C. Barrabès, P.A. Hogan: *Advanced general relativity - gravity waves, spinning particles, and black holes*
159. W. Barford: *Electronic and optical properties of conjugated polymers, Second edition*
158. F. Strocchi: *An introduction to non-perturbative foundations of quantum field theory*
157. K.H. Bennemann, J.B. Ketterson: *Novel superfluids, Volume 2*
156. K.H. Bennemann, J.B. Ketterson: *Novel superfluids, Volume 1*
155. C. Kiefer: *Quantum gravity, Third edition*
154. L. Mestel: *Stellar magnetism, Second edition*
153. R.A. Klemm: *Layered superconductors, Volume 1*
152. E.L. Wolf: *Principles of electron tunneling spectroscopy, Second edition*
151. R. Blinc: *Advanced ferroelectricity*
150. L. Berthier, G. Biroli, J.-P. Bouchaud, W. van Saarloos, L. Cipelletti: *Dynamical heterogeneities in glasses, colloids, and granular media*
149. J. Wesson: *Tokamaks, Fourth edition*
148. H. Asada, T. Futamase, P. Hogan: *Equations of motion in general relativity*
147. A. Yaouanc, P. Dalmas de Réotier: *Muon spin rotation, relaxation, and resonance*
146. B. McCoy: *Advanced statistical mechanics*
145. M. Bordag, G.L. Klimchitskaya, U. Mohideen, V.M. Mostepanenko: *Advances in the Casimir effect*
144. T.R. Field: *Electromagnetic scattering from random media*
143. W. Götze: *Complex dynamics of glass-forming liquids - a mode-coupling theory*
142. V.M. Agranovich: *Excitations in organic solids*
141. W.T. Grandy: *Entropy and the time evolution of macroscopic systems*
140. M. Alcubierre: *Introduction to 3+1 numerical relativity*
139. A.L. Ivanov, S.G. Tikhodeev: *Problems of condensed matter physics - quantum coherence phenomena in electron-hole and coupled matter-light systems*
138. I.M. Vardavas, F.W. Taylor: *Radiation and climate*
137. A.F. Borghesani: *Ions and electrons in liquid helium*
135. V. Fortov, I. Iakubov, A. Khrapak: *Physics of strongly coupled plasma*
134. G. Fredrickson: *The equilibrium theory of inhomogeneous polymers*
133. H. Suhl: *Relaxation processes in micromagnetics*
132. J. Terning: *Modern supersymmetry*
131. M. Mariño: *Chern-Simons theory, matrix models, and topological strings*
130. V. Gantmakher: *Electrons and disorder in solids*
129. W. Barford: *Electronic and optical properties of conjugated polymers*
128. R.E. Raab, O.L. de Lange: *Multipole theory in electromagnetism*
127. A. Larkin, A. Varlamov: *Theory of fluctuations in superconductors*
126. P. Goldbart, N. Goldenfeld, D. Sherrington: *Stealing the gold*
125. S. Atzeni, J. Meyer-ter-Vehn: *The physics of inertial fusion*
123. T. Fujimoto: *Plasma spectroscopy*
122. K. Fujikawa, H. Suzuki: *Path integrals and quantum anomalies*
121. T. Giamarchi: *Quantum physics in one dimension*
120. M. Warner, E. Terentjev: *Liquid crystal elastomers*
119. L. Jacak, P. Sitko, K. Wieczorek, A. Wojs: *Quantum hall systems*
117. G. Volovik: *The Universe in a helium droplet*
116. L. Pitaevskii, S. Stringari: *Bose-Einstein condensation*
115. G. Dissertori, I.G. Knowles, M. Schmelling: *Quantum chromodynamics*
114. B. DeWitt: *The global approach to quantum field theory*
113. J. Zinn-Justin: *Quantum field theory and critical phenomena, Fourth edition*
112. R.M. Mazo: *Brownian motion - fluctuations, dynamics, and applications*
111. H. Nishimori: *Statistical physics of spin glasses and information processing - an introduction*
110. N.B. Kopnin: *Theory of nonequilibrium superconductivity*
109. A. Aharoni: *Introduction to the theory of ferromagnetism, Second edition*
108. R. Dobbs: *Helium three*
107. R. Wigmans: *Calorimetry*
106. J. Kübler: *Theory of itinerant electron magnetism*
105. Y. Kuramoto, Y. Kitaoka: *Dynamics of heavy electrons*
104. D. Bardin, G. Passarino: *The Standard Model in the making*
103. G.C. Branco, L. Lavoura, J.P. Silva: *CP Violation*
101. H. Araki: *Mathematical theory of quantum fields*
100. L.M. Pismen: *Vortices in nonlinear fields*

Quantum Statistical Field Theory

An Introduction to Schwinger's Variational Method with Green's Function Nanoapplications, Graphene and Superconductivity

Norman J. Morgenstern Horing

Department of Physics and Engineering Physics, Stevens Institute of Technology

OXFORD
UNIVERSITY PRESS

OXFORD
UNIVERSITY PRESS

Great Clarendon Street, Oxford, OX2 6DP,
United Kingdom

Oxford University Press is a department of the University of Oxford.
It furthers the University's objective of excellence in research, scholarship,
and education by publishing worldwide. Oxford is a registered trade mark of
Oxford University Press in the UK and in certain other countries

First Edition published in 2017
Reprinted 2018

Published in the United States of America by Oxford University Press
198 Madison Avenue, New York, NY 10016, United States of America

British Library Cataloguing in Publication Data
Data available

Library of Congress Control Number: 2016952831

ISBN 978–0–19–879194–2

Rose Morgenstern Horing
1911–1999
Deeply loved mother and protector who, in the worst of times, braved severe adversity and privation to deliver me to manhood
Norman J. Morgenstern Horing

Contents

Prologue xiii
Acknowledgments xvii

1 Dirac Notation and Transformation Theory 1

1.1 Introductory Review 1
1.2 Dirac Notation 8
1.3 Complete Sets of Commuting Observables, Representations and Transformation Theory 13
1.4 Schwinger Measurement Symbols 15
1.5 Chapter 1 Problems 16
1.6 References for Chapter 1 19

2 Identical Particles and Second Quantization: Occupation Number Representation 20

2.1 Operators for Properties of Identical Particles 20
2.2 Second Quantization: Occupation Number Representation 24
2.3 Coherent States, Squeezed States 33
2.4 Chapter 2 Problems 38
2.5 References for Chapter 2 39

3 Q. M. Pictures; Heisenberg Equation; Linear Response; Superoperators and Non-Markovian Equations 40

3.1 Schrödinger Picture: Iteration Series for the Time-Development Operator 41
3.2 Heisenberg Picture and Equations of Motion 44
3.3 Interaction Picture 47
3.4 Linear Response 51
3.5 S-Matrix, Density Matrix, Superoperators and Non-Markovian Kinetic Equations 53
3.6 Chapter 3 Problems 59
3.7 References for Chapter 3 60

4 Schwinger Action Principle and Variational Calculus 61

4.1 Action Principle: Quantum Hamilton Equations 61
4.2 Variational Calculus and Sources 72
4.3 Particle Sources: General Considerations 81
4.4 Potential Source: General Considerations 87

4.5 An Identity 91
4.6 Chapter 4 Problems 93
4.7 References for Chapter 4 93

5 Retarded Green's Functions 95

5.1 Definition, Equation of Motion and Interpretation of Retarded Green's Functions 95
5.2 Partial Green's Functions, Applications to Localized States in a Continuum and Chemisorption 101
5.3 Dyson Integral Equation, T-Matrix for $\delta(\mathbf{x})$-Function Potential 105
5.4 Random Impurity Scattering: Born and Self-Consistent Born Approximations 107
5.5 Ando's Semi-Elliptic Density of States for the 2D Landau-Quantized Electron-Impurity System 113
5.6 Magnetic Field Green's Function for Electrons in Three and Two Dimensions 116
5.7 Green's Function Determination by Hamilton Equations: 2D Saddle Potential in $\mathbf{E}(t)$ Field 123
5.8 Product Green's Functions for Separable Hamiltonians: 1D Superlattice in Axial Fields 127
5.9 Electron Propagation in a Heterostructure with Spatially Variable Mass: Green's Function Matching 133
5.10 Green's Function Matching: Dynamic, Non-Local Electrostatic Screening at Interface; Surface Plasmons 138
5.11 Feynman Path Integrals 144
5.12 Chapter 5 Problems 145
5.13 References for Chapter 5 148

6 Quantum Mechanical Ensemble Averages and Statistical Thermodynamics 150

6.1 Quantum Mechanical Ensemble Averages and Thermodynamics 150
6.2 Boson and Fermion Statistics for Non-Interacting Systems: Bose Condensation 156
6.3 Helmholtz Free Energy and Magnetic Moment 159
6.4 Diamagnetic Susceptibility and the de Haas–van Alphen Effect 163
6.5 Chapter 6 Problems 166
6.6 References for Chapter 6 167

7 Thermodynamic Green's Functions and Spectral Structure 168

7.1 Definition and Interpretation of Thermal Green's Functions 168
7.2 Lehmann Spectral Representation of the One-Particle Thermal Green's Function 169
7.3 Other Spectral Considerations 173
7.4 Periodicity/Antiperiodicity in Imaginary Time, Matsubara Frequencies and Analytic Continuation 174

7.5 Commutator/Anticommutator Functions; Retarded/Advanced Green's Functions; Spectral Sum Rule 179
7.6 Statistical Thermodynamic Information in the One- and Two-Particle Green's Functions 182
7.7 Chapter 7 Problems 185
7.8 References for Chapter 7 186

8 Equations of Motion with Particle–Particle Interactions and Approximations 187

8.1 Multiparticle Green's Function Equations of Motion 188
8.2 Non-Correlation Decoupling Procedures for G_2: Hartree and Hartree–Fock Approximations 191
8.3 Non-Correlation Decoupling Procedures: Collisional Effects; *Sum of Ladder Diagrams* Integral Equation 195
8.4 Electron–Hole Interaction: Exciton States 198
8.5 Chapter 8 Problems 204
8.6 References for Chapter 8 205

9 Non-Equilibrium Green's Functions: Variational Relations and Approximations for Particle Interactions 206

9.1 Non-Equilibrium Green's Functions, Definitions and Variational Relations in Terms of Particle Sources 207
9.2 Exact Variational Relations in Terms of a Source Potential 216
9.3 Cumulants and the Linked Cluster Theorem 218
9.4 Variational Differential Formulation of the Random Phase Approximation 221
9.5 Variational Differential Formulation of Perturbation Theory 225
9.6 Self-Energy and Vertex Operator Perturbation Theory 227
9.7 Shielded Potential Perturbation Theory 229
9.8 Imaginary Time, Contour Time Ordering and the Langreth Algebra 231
9.9 Generalized Kadanoff–Baym Ansatz for Derivation of Quantum Transport Equations 234
9.10 Chapter 9 Problems 236
9.11 References for Chapter 9 237

10 Random Phase Approximation Plasma Phenomenology, Semiclassical and Hydrodynamic Models; Electrodynamics 238

10.1 Dielectric Response Functions for Inhomogeneous and Homogeneous Systems 239
10.2 RPA Plasmons and Shielding in 3D, 2D and 1D, Magnetoplasmas, Models and Surface Plasmons 244
10.3 Exchange and Correlation Energy 252

10.4 Atom–Surface van der Waals Attraction; Physisorption and the Plasma Image 256
10.5 Fast Particle Power Loss in Plasma 264
10.6 Electrodynamic Dielectric Response Functions for Quantum Plasmas: Dyadic Green's Functions 271
10.7 Dynamic, Non-Local Conductivity and Inverse Dielectric Tensors 280
10.8 Polaritons in Compound Nanostructures 290
10.9 Chapter 10 Problems 299
10.10 References for Chapter 10 300

11 Interacting Electron–Hole–Phonon System 302

11.1 Electron–Hole–Phonon Interaction Hamiltonian 303
11.2 Coupled Electron–Hole–Phonon Green's Functions and Equations 307
11.3 Electron–Phonon Interaction Phenomenology 314
11.4 Alternative Formulation of the Equations of Motion 323
11.5 Analysis of Electron–Hole Interaction Dynamics 327
11.6 Generalized Shielded Potential Approximation: Electron–Hole Interaction Operator 330
11.7 Dielectric Response of the Electron–Hole–Phonon System 333
11.8 Exciton-Plasmon Interaction 337
11.9 Chapter 11 Problems 338
11.10 References for Chapter 11 338

12 Graphene 340

12.1 Introduction 340
12.2 Graphene Hamiltonian, Green's Function and Dielectric Function 341
12.3 Graphene Green's Function in a Perpendicular Magnetic Field: Landau Quantization 346
12.4 Graphene Quantum Dot: Magnetic Field Green's Function 352
12.5 Graphene Antidot Lattice in a Magnetic Field 355
12.6 Graphene Statistical Thermodynamics and Magnetic Moment 358
12.7 Further Remarks and Developments 361
12.8 Chapter 12 Problems 363
12.9 References for Chapter 12 364

13 Superfluidity and Superconductivity 370

13.1 Superfluid Field Operators and the Gross–Pitaevskii Equations 371
13.2 Superfluid Green's Functions and Elementary Excitations 375

13.3 Superconductivity: BCS Theory; Anomalous Green's Functions and Gorkov Equations 382
13.4 Excitation Spectrum, Gap; Isotope Effect 386
13.5 Thermodynamics of Superconductors 389
13.6 Ginzburg–Landau Theory 391
13.7 London Theory; Meissner Effect 397
13.8 Electromagnetic Current Response of Superconductors: Gauge Invariance and Collective Modes 399
13.9 Magnetic Flux Quantization: Type I and Type II Superconductors 408
13.10 Josephson Tunneling 412
13.11 Further Remarks and Developments 415
13.12 Chapter 13 Problems 418
13.13 References for Chapter 13 418

Epilogue 421

Index 429

Prologue

Since the landmark breakthroughs in quantum electrodynamics, *c.*1950, the methods of advanced quantum mechanics, in particular Green's functions, have been the preferred and principal tools for theoretical research on the fundamental physics of particles and matter. Their use has facilitated the systematization of highly complicated analyses in the relativistic theory of interacting quantum fields, leading to profoundly important (indeed, Nobel prize-winning) developments, in the context of both variational differential methods and equivalent Feynman-diagrammatic methods. The transfer of these methods into the non-relativistic domain of condensed matter theory has also led to immensely improved understanding of the electronic properties of metals, semiconductors, superconductors, superfluids, magnetism, etc. Again, the use of Green's functions by physicists led to extraordinary scientific advances, not only in regard to the prediction and explanation of physical phenomena, but also in providing insight into the development of new devices. With this, physics is once again spawning new engineering disciplines, as it has done in the past with electronic, optical and nuclear engineering. It is now fairly common to find electrical engineering (as well as applied physics) departments around the world heavily staffed with faculty deeply involved in research in areas such as charge transport in semiconductors, nanostructures, quantum tunneling, superconductivity, etc.: areas of research that are well served by the use of Green's functions, to which engineering researchers increasingly turn to penetrate difficult problems.

This book has grown out of a two-semester course on the methods of advanced quantum mechanics that I have taught at Stevens Institute of Technology over many years. It is mainly based on the lectures of Professor Julian Schwinger at Harvard University (*c.*1956) and the seminal paper of P. C. Martin and J. Schwinger, followed by important contributions of their students and others. The book is eclectic in the material addressed from many points of view. It makes no reference to Feynman diagrammatics, as that subject already permeates the literature, but it does introduce superoperators; quantum-mechanical Green's functions are discussed in detail, but their applications are selectively limited while classical electromagnetic dyadic Green's functions are treated in this book on quantum theoretical methods; moreover, Landau diamagnetism is included, but ferromagnetism is not. In reality, a much fuller book of this type could and should have been written by Schwinger and/or Martin, but, unfortunately, that has not happened, although some of Schwinger's notes at the University of California at Los Angeles were published in his name in "Quantum Mechanics," edited by B. G. Englert, Springer (2001). If that book had gone further, it would have totally eclipsed this one. However, it did leave a bit of room for additional presentations (with some overlap).

This book has two objectives. The first is pedagogical, addressing students of science and engineering who have had just a first exposure to quantum mechanics in terms of the one-particle Schrödinger equation on the level of A. Beiser's book "Concepts of Modern Physics" or R. L. Liboff's "Introductory Quantum Mechanics." The present book is intended to introduce them to the advanced aspects of quantum mechanics that are important in current, mainly non-relativistic, research on interacting quantum systems. These include correlation phenomena due to electron–electron interactions, collective modes, electron–hole interactions, electron–phonon interactions, Graphene, superfluids, superconductors, magnetism, and semiconductor nanostructures that play a central role in the high-tech thrust for smaller, faster devices to facilitate the development of more powerful computers and other electronic devices. There are several mechanisms that I feel will help to rapidly advance the educational status of students of modest background in the arena of condensed matter quantum physics and engineering, to bring the reader to the point of being able to understand modern theoretical research papers and carry out such research themselves. One such mechanism is Schwinger's interpretation of Dirac ket-bra dyads in terms of *measurement symbols* that are factorized into annihilation and creation operators to introduce second quantization, which I consider to be more efficient educationally than other approaches to the subject. (Its effectiveness is undiminished by Schwinger's later adoption of a somewhat different notation.) Another mechanism is the use of variational differentiation jointly with the Schwinger Action Principle to mathematically describe interacting quantum systems and formulate various perturbation theories for them at arbitrary temperature and under non-equilibrium conditions, without explicit recourse to Wick's theorem. Providing the bit of introduction to variational calculus (in Chapter 4) that the reader needs is a small price to pay for the much greater efficiency of variational differential techniques. Such improved educational mechanisms have proved beneficial to rapidly advancing students' capabilities in the past. For example, the introduction of Dirac notation has certainly facilitated the study of quantum mechanics and the Dirac delta-function is employed universally because of its great utility. Vector calculus with its grad, div, curl makes it much easier to learn (as well as teach) potential theory and electromagnetics. Venerable older books like Kellog's "Potential Theory" and Sir James Jeans' "The Mathematical Theory of Electricity and Magnetism" present derivations that seem painfully lengthy in comparison with more modern and efficient presentations. Variational techniques offer such efficiencies of presentation, which we employ here for the reader's benefit. In some instances our explication/derivation of the material is detailed to the point of exhausting the patience of more knowledgeable readers, but it has been effective in promoting understanding of the subject in lower level students, including some mathematically strong undergraduates. While the material has traditionally been viewed as "advanced," the primary object here is to present it on an intermediate level using Schwinger's variational methods.

The second purpose of this book is to provide the reader with examples of applications to recent research on semiconductor nanostructures, including low-dimensional systems and Graphene. Introductions to the microscopic theories of superfluids and superconductors are included, as well as various facets of magnetic Landau quantization of electron orbits.

Following introductory chapters on Dirac notation, second quantization (including coherent states), Schrödinger, Heisenberg and Interaction Pictures and superoperators,

Chapter 4 presents the Schwinger Action Principle (encompassing all aspects of quantum theory) and variational calculus. An introduction to retarded Green's functions, the Dyson equation and T-matrix for a $\delta(\mathbf{x})$-potential, as well as for random impurity scattering, is presented in Chapter 5, where the self-consistent Born approximation is discussed. Partial Green's functions are also treated, with application to chemisorption, and the technique of Green's function matching is derived and applied to the Schrödinger equation and surface plasmons. Chapter 6 reviews some of the essentials of quantum statistical mechanics, ensemble theory and statistical thermodynamics, including the de Haas–van Alphen effect.

Thermal Green's functions and their spectral structure are presented in detail in Chapter 7, including the Lehmann spectral representation, periodicity/antiperiodicity in imaginary time, Matsubara frequencies, analytic continuation and the spectral sum rule. The equations of motion for multiparticle Green's functions with particle–particle interactions and non-correlation decoupling procedures, including the Ladder Diagram integral equation, are derived in Chapter 8, with application to electron–hole exciton states. Non-equilibrium Green's functions are introduced in Chapter 9 and exact variational relations in terms of particle and potential sources are presented. Cumulants and the Linked Cluster Theorem are discussed, as is a variational differential formulation of the random phase approximation (RPA). Variational differential formulations of perturbation theories for Green's functions, self-energy and vertex operators, as well as the shielded potential approximation, are derived, and the Langreth Algebra is discussed as well as the Generalized Kadanoff–Baym Ansatz. Chapter 10 is concerned with RPA plasma phenomenology and models, including dynamic, non-local dielectric response, plasmons, shielding, exchange and correlation energy, with an application to atom–surface van der Waals physisorption. Electromagnetic dyadic Green's functions are also discussed in Chapter 10, along with dynamic, non-local conductivity and dielectric tensors with applications to polaritons in compound nanostructures. A model of the interacting electron–hole–phonon system is treated in detail in Chapter 11, with an introductory discussion of the model Hamiltonian and definition and derivation of equations of motion for the coupled Green's functions of the system. An alternative formulation of the equations of motion with *five* interactions (electron–electron, hole–hole, electron–hole, electron–phonon and hole–phonon) is presented to facilitate the development of a generalized shielded potential approximation for this extremely complex system: A representation of the dynamics of such a system having five interactions would present a serious challenge to a Feynman-diagrammatic formulation. The subject of Graphene is introduced and discussed in Chapter 12. Finally, an extended introduction to the microscopic theories of superfluids and superconductors is presented in Chapter 13 and an Epilogue indicating directions for further reading is provided.

Acknowledgments

In appreciation of the fine education provided by Professors Mark Zemansky, Larry Wills, Roy Glauber, Paul Martin and Julian Schwinger, and the other denizens of the deep at the City College of New York (CUNY) and Harvard University.

With eternal appreciation for the steadfast support, encouragement,

and love of my wife

Jerilyn

in this long and difficult journey.

N.J.M.H.

1

Dirac Notation and Transformation Theory

Abstract
Chapter 1 opens with a brief review of some basic features of quantum mechanics, including the Schrödinger equation, linear and angular momentum and the theory of the hydrogenic atom. It also includes complete orthonormal sets of eigenfunctions, the translation operator, current, spin, equation of continuity, gauge transformation, determinant and permanent multiparticle energy eigenfunctions and the Pauli exclusion principle. Attention is then focused on Dirac bra-ket notation and complete sets of commuting observables. In this connection, representations and transformation among representations are discussed in detail for the Schrödinger system state vector and the eigenstates, as well as bra-ket matrix elements of operators. Finally, Schwinger's interpretation of ket-bra matrix operator structures in terms of annihilation and creation of systems in eigenstates is introduced.

1.1 Introductory Review

In a first exposure to quantum mechanics, the Hamiltonian, $H(\mathbf{x},\mathbf{p})$, for a single particle is usually written in position representation as $H(\mathbf{x},\frac{\hbar}{i}\nabla)$ and the quantum dynamical time development of the wave function $\Psi(\mathbf{x},t)$ is determined by the single-particle Schrödinger equation,[1–3]

$$H(\mathbf{x},\tfrac{\hbar}{i}\nabla)\Psi(\mathbf{x},t) = i\hbar\frac{\partial\Psi(\mathbf{x},t)}{\partial t}, \tag{1.1.1}$$

where $\mathbf{p} = \frac{\hbar}{i}\nabla$ is the momentum operator in position representation. The probability density for the particle to be found at $\mathbf{x}$ at time t is $|\Psi(\mathbf{x},t)|^2$. Solutions for $\Psi(\mathbf{x},t)$ are often sought by expansion in the complete orthonormal set of eigenfunctions $\phi_r(\mathbf{x})$ associated with a Hermitian operator $\Theta(\mathbf{x},\frac{\hbar}{i}\nabla)$ having eigenvalues λ_r:

$$\Theta(\mathbf{x},\tfrac{\hbar}{i}\nabla)\phi_r(\mathbf{x}) = \lambda_r\phi_r(\mathbf{x}), \tag{1.1.2}$$

and Θ is frequently (but not necessarily) chosen as a part of H that is relatively easy to analyze. The completeness and orthonormality of the eigenfunctions $\phi_r(\mathbf{x})$ may be expressed as follows ($\delta_{rs} = 1{,}0$ for $r = s$, $r \neq s$ is the Kronecker delta):

Quantum Statistical Field Theory: Schwinger's Variational Method. Norman J. Morgenstern Horing, Oxford University Press (2017). © Norman J. Morgenstern Horing. DOI 10.1093/oso/9780198791942.001.0001

$$\text{Completeness: } \Psi(\mathbf{x},t) = \sum_r C_r(t)\phi_r(\mathbf{x}), \tag{1.1.3}$$

$$\text{Orthonormality: } \int d^{(3)}\mathbf{x}\phi_r^*(\mathbf{x})\phi_s(\mathbf{x}) = \delta_{rs}, \tag{1.1.4}$$

which lead to C_r expressed as

$$C_r(t) = \int d^{(3)}\mathbf{x}\phi_r^*(\mathbf{x})\Psi(\mathbf{x},t). \tag{1.1.5}$$

An alternative expression of completeness is given by

$$\sum_r \phi_r^*(\mathbf{x}')\phi_r(\mathbf{x}) = \delta^{(3)}(\mathbf{x}-\mathbf{x}'), \tag{1.1.6}$$

where $\delta^{(3)}(\mathbf{x}-\mathbf{x}')$ is a three-dimensional (3D) Dirac delta function,[4,5] which plays the same role under integration as does the Kronecker delta, δ_{ij}, under summation.

A simple, but important, example is provided by choosing Θ to be the momentum operator, $\Theta = \mathbf{p} \Longrightarrow \frac{\hbar}{i}\nabla$. Its components, p_j, obey canonical commutation relations with coordinate components, $x_i : [x_i, p_j] \equiv (x_i p_j - p_j x_i) = i\hbar\delta_{ij}$. The eigenfunction equation, $\frac{\hbar}{i}\nabla\phi(\mathbf{x}) = \lambda\phi(\mathbf{x})$, has a continuum of infinite space eigenvalues, $\lambda \rightarrow \mathbf{p}'$ (all real 3-vector momenta), with the corresponding eigenfunctions $\phi_{p'}(\mathbf{x}) = \exp(\frac{i}{\hbar}\mathbf{p}'\cdot\mathbf{x})/(2\pi\hbar)^{3/2}$. The completeness relation, Eq. (1.1.3), is then just the Fourier transform, $\Psi(\mathbf{x},t) = \int \frac{d^{(3)}\mathbf{p}'}{(2\pi\hbar)^{3/2}} \exp\left(i\frac{\mathbf{p}'}{\hbar}\cdot\mathbf{x}\right) C_{p'}(t)$, which leads to Eq. (1.1.6) in the form $\int \frac{d^{(3)}\mathbf{p}'}{(2\pi\hbar)^3} \exp\left(i\frac{\mathbf{p}'}{\hbar}\cdot\mathbf{x}\right) \exp\left(-i\frac{\mathbf{p}'}{\hbar}\cdot\mathbf{x}'\right) = \delta^{(3)}(\mathbf{x}-\mathbf{x}')$ as a known identity, and the continuum form of the orthonormality relation, Eq. (1.1.4), is given by $\int \frac{d^{(3)}\mathbf{x}}{(2\pi\hbar)^3} \exp\left(i\frac{\mathbf{p}'}{\hbar}\cdot\mathbf{x}\right) \exp\left(-i\frac{\mathbf{p}''}{\hbar}\cdot\mathbf{x}\right) = \delta^{(3)}(\mathbf{p}'-\mathbf{p}'')$. It is important to note that the momentum operator, $\mathbf{p} = \frac{\hbar}{i}\nabla$, generates spatial translations since the 3D Taylor series for a positional displacement of $f(\mathbf{x})$ through $\mathbf{a}$ may be written as

$$f(\mathbf{x}+\mathbf{a}) = \sum_{n=0}^{\infty} \frac{(\mathbf{a}\cdot\nabla)^n}{n!} f(\mathbf{x}) = \exp(\mathbf{a}\cdot\nabla) f(\mathbf{x}) = \exp\left(\frac{i}{\hbar}\mathbf{a}\cdot\mathbf{p}\right) f(\mathbf{x}). \tag{1.1.7}$$

There are some systems of great intrinsic physical interest with correspondingly important Hamiltonians that can be treated exactly. In this regard, we briefly review the properties of H for a Coulombic force center, which is of special interest in connection with the theory of excitons to be discussed in a later chapter. The hydrogenic atom, which was of central significance in the historical development of quantum theory, involves the well-known Hamiltonian (Z is the charge number of the nucleus and r is the radial distance from it):

$$H = \frac{-\hbar^2}{2m}\nabla^2 - \frac{Ze^2}{r} = \frac{-\hbar^2}{2mr^2}\frac{\partial}{\partial r}\left(r^2\frac{\partial}{\partial r}\right) + \frac{\mathbf{L}^2}{2mr^2} - \frac{Ze^2}{r}. \tag{1.1.8}$$

The last term represents the Coulomb potential attracting the electron to the nuclear center. Here, **L** is the quantum mechanical orbital angular momentum operator for a single particle, defined by

$$\mathbf{L} = \mathbf{x} \times \mathbf{p} = \mathbf{x} \times \frac{\hbar}{i}\boldsymbol{\nabla}. \tag{1.1.9}$$

Its Cartesian components, L_i $(i = 1, 2, 3)$, obey the commutation relations

$$\left[L_i, L_j\right] = i\hbar \sum_{k=1}^{3} \varepsilon_{ijk} L_k, \tag{1.1.10}$$

where ε_{ijk} is $1, -1$ if i, j, k is an even, odd permutation of $1, 2, 3$ and it vanishes if any index is repeated. In spherical coordinates (θ is polar angle, ϕ is azimuthal angle), we have

$$\begin{aligned} L_1 &= \frac{\hbar}{i}\left(-\sin\phi\frac{\partial}{\partial\theta} - \cos\phi\cot\theta\frac{\partial}{\partial\phi}\right), \\ L_2 &= \frac{\hbar}{i}\left(\cos\phi\frac{\partial}{\partial\theta} - \sin\phi\cot\theta\frac{\partial}{\partial\phi}\right), \\ L_3 &= \frac{\hbar}{i}\frac{\partial}{\partial\phi}, \end{aligned} \tag{1.1.11}$$

and

$$\mathbf{L}^2 = \sum_{i=1}^{3} L_i^2 = -\hbar^2\left[\frac{1}{\sin^2\theta}\frac{\partial^2}{\partial\phi^2} + \frac{1}{\sin\theta}\frac{\partial}{\partial\theta}\left(\sin\theta\frac{\partial}{\partial\theta}\right)\right]. \tag{1.1.12}$$

Just as the momentum operator generates spatial translations, it is readily seen from Eq. (1.1.11) that L_3 generates spatial rotations about the 3-axis. The eigenfunctions and eigenvalue spectrum of $\mathbf{L}^2$ are given by

$$\mathbf{L}^2\varphi_{l,m} = \hbar^2 l(l+1)\varphi_{l,m}, \text{ where the orbital quantum number is } l = 0, 1, 2\ldots\infty, \tag{1.1.13}$$

$\varphi_{l,m} = Y_l^m(\theta, \phi)$ is a spherical harmonic with

$$L_3\varphi_{l,m} = \hbar m\varphi_{l,m} \text{ and the magnetic quantum number is } m = -l\ldots 0\ldots + l. \tag{1.1.14}$$

The separable energy eigenfunctions of H for a hydrogenic atom having the Hamiltonian of Eq. (1.1.8),

$$H\Phi_{n,l,m} = E_{n,l,m}\Phi_{n,l,m}, \tag{1.1.15}$$

which are simultaneously eigenfunctions of $\mathbf{L}^2$ and L_3 (Eqns. (1.1.13 and 1.1.14)) are given by (we define $R_n = 2Zr/a_0 n$, where a_0 is the Bohr radius):

$$\Phi_{n,l,m}(r, \theta, \phi) = \left(\frac{2Z}{a_0 n}\right)^{\frac{3}{2}} A_{nl}\,(R_n)^l e^{-\frac{R_n}{2}} F_{nl}(R_n)\,Y_l^m(\theta, \phi), \tag{1.1.16}$$

with energy eigenvalues

$$E_{n,l,m} = E_n = \frac{-Z^2 e^2}{2a_0 n^2} \tag{1.1.17}$$

(degenerate in l, m), where the principal quantum number n is an integer $n > l$, $F_{nl}(R_n)$ is an associated Laguerre polynomial, and normalization to unity is provided by

$$A_{nl} = \sqrt{\frac{(n-l-1)!}{2n\,[(n+l)!]^3}}\,. \tag{1.1.18}$$

The spatial localization of these energy eigenfunctions associated with the discrete character of the spectrum is evident from the factor $e^{-R_n/2}$.

The classical relation between the orbital angular momentum $\mathbf{L}$ of a particle of charge Q, mass m and its magnetic moment $\boldsymbol{\mu}$ also holds for the associated quantum mechanical operators,

$$\boldsymbol{\mu} = \frac{Q}{2mc}\mathbf{L}. \tag{1.1.19}$$

(Of course, there are anomalies due to interactions.) The analysis of atomic spectra made it clear that there is yet another internal coordinate related to an intrinsic magnetic moment of the electron and its associated intrinsic angular momentum, $\mathbf{S}$, which could take just two values, $S_z \to \pm\frac{\hbar}{2}$. As angular momenta (intrinsic as well as orbital), these operators obey the characteristic commutation relations, Eq. (1.1.10),

$$\left[S_i\,, S_j\right] = i\hbar \sum_{k=1}^{3} \varepsilon_{ijk} S_k. \tag{1.1.20}$$

Since they can only have two eigenvalues, they may be described by 2×2 Pauli spin matrices,

$$\boldsymbol{\sigma} = \frac{2}{\hbar}\mathbf{S}, \tag{1.1.21}$$

which obey similar relations:

$$\left[\sigma_i\,, \sigma_j\right] = 2i \sum_{k=1}^{3} \varepsilon_{ijk} \sigma_k. \tag{1.1.22}$$

A convenient representation is given by $\sigma_1 = \begin{pmatrix} 0 & 1 \\ 1 & 0 \end{pmatrix}$; $\sigma_2 = \begin{pmatrix} 0 & -i \\ i & 0 \end{pmatrix}$; $\sigma_3 = \begin{pmatrix} 1 & 0 \\ 0 & -1 \end{pmatrix}$. The Pauli matrices are Hermitian, idempotent and *anti*commutative in accordance with the relations (I is the unit 2×2 matrix, $i, j = 1, 2, 3$):

$$\sigma_i \sigma_j = I\delta_{ij} + i \sum_{k=1}^{3} \varepsilon_{ijk} \sigma_k \tag{1.1.23}$$

and

$$\sigma_i \sigma_j + \sigma_j \sigma_i = 2I\delta_{ij}. \tag{1.1.24}$$

The associated intrinsic magnetic moment of the electron is given by

$$\boldsymbol{\mu}_s = \frac{-e}{mc}\mathbf{S} \text{ corresponding to } \mu_{sz} \Longrightarrow \pm\frac{e\hbar}{2mc}, \tag{1.1.25}$$

since the gyromagnetic ratio for electron spin is approximately twice that of the orbital angular momentum. The total angular momentum operator for an electron is given by

$$\vec{\mathcal{J}} = \mathbf{L} + \mathbf{S}. \tag{1.1.26}$$

In the presence of an electromagnetic field having vector and scalar potentials, $\mathbf{A}(\mathbf{x}, t)$ and $\phi(\mathbf{x}, t)$, respectively, with $\mathbf{B} = \nabla \times \mathbf{A}$ and $\mathbf{E} = -\nabla\phi - \frac{1}{c}\partial\mathbf{A}/\partial t$, the nonrelativistic Hamiltonian of a particle of charge Q is given by

$$H = \frac{1}{2m}\left(\mathbf{p} - \frac{Q}{c}\mathbf{A}(\mathbf{x}, t)\right) \cdot \left(\mathbf{p} - \frac{Q}{c}\mathbf{A}(\mathbf{x}, t)\right) + Q\phi(\mathbf{x}, t). \tag{1.1.27}$$

Considered in conjunction with the Schrödinger equation, Eq. (1.1.1), the time rate of change of the charge density, $\rho(\mathbf{x}, t) = Q\Psi^*(\mathbf{x}, t)\Psi(\mathbf{x}, t)$, is given by

$$i\hbar\frac{\partial}{\partial t}\rho(\mathbf{x}, t) = Q\Psi^*(\mathbf{x}, t)\left[\frac{1}{2m}\left(\frac{\hbar}{i}\nabla - \frac{Q}{c}\mathbf{A}(\mathbf{x}, t)\right) \cdot \left(\frac{\hbar}{i}\nabla - \frac{Q}{c}\mathbf{A}(\mathbf{x}, t)\right) + Q\phi\right]\Psi(\mathbf{x}, t) - h.c., \tag{1.1.28}$$

where $h.c.$ denotes the Hermitian conjugate. This may be rewritten as

$$\begin{aligned}\frac{\partial\rho(\mathbf{x}, t)}{\partial t} = &-\nabla \cdot \left\{\frac{\hbar Q}{2m}\left[\left(i\nabla\Psi^*(\mathbf{x}, t)\right)\Psi(\mathbf{x}, t)) - \Psi^*(\mathbf{x}, t)\left(i\nabla\Psi(\mathbf{x}, t)\right)\right]\right.\\ &\left.-\frac{Q^2}{mc}\mathbf{A}(\mathbf{x}, t)\Psi^*(\mathbf{x}, t)\Psi(\mathbf{x}, t)\right\}.\end{aligned} \tag{1.1.29}$$

Considered with the equation of continuity for the conserved (normalized) charge-probability density, $\rho(\mathbf{x}, t)$,

$$\frac{\partial\rho(\mathbf{x}, t)}{\partial t} = -\nabla \cdot \mathbf{J}(\mathbf{x}, t), \tag{1.1.30}$$

we may identify the charge-current density, $\mathbf{J}(\mathbf{x}, t)$,

$$\mathbf{J}(\mathbf{x}, t) = \frac{\hbar Q}{2m}\left[\Psi^*(\mathbf{x}, t)\left(\frac{1}{i}\nabla\Psi(\mathbf{x}, t)\right) - \left(\frac{1}{i}\nabla\Psi^*(\mathbf{x}, t)\right)\Psi(\mathbf{x}, t)\right] - \frac{Q^2}{mc}\mathbf{A}(\mathbf{x}, t)\Psi^*(\mathbf{x}, t)\Psi(\mathbf{x}, t). \quad (1.1.31)$$

It is important to note that $\mathbf{J}$ is invariant under a gauge transformation, $\mathbf{A}(\mathbf{x}, t) \rightarrow \mathbf{A}'(\mathbf{x}, t) = \mathbf{A}(\mathbf{x}, t) + \nabla f(\mathbf{x}, t)$ and $\phi(\mathbf{x}, t) \rightarrow \phi'(\mathbf{x}, t) = \phi(\mathbf{x}, t) - \frac{1}{c}\frac{\partial}{\partial t}f(\mathbf{x}, t)$, taken jointly with the phase transformation,

$$\Psi(\mathbf{x}, t) \rightarrow e^{i\,Qf(\mathbf{x},t)/\hbar c}\Psi(\mathbf{x}, t). \quad (1.1.32)$$

This phase change describes the transformation that wave functions must undergo in conjunction with an electromagnetic gauge transformation to preserve the gauge invariance of the physical quantities in quantum mechanics.

In dealing with two particles, we have a Hamiltonian that depends on the coordinates and momenta of both particles,

$$H \longrightarrow H(\mathbf{x}_1, \tfrac{\hbar}{i}\nabla_1; \mathbf{x}_2, \tfrac{\hbar}{i}\nabla_2). \quad (1.1.33)$$

The wave function also depends on both coordinates,

$$\Psi \longrightarrow \Psi(\mathbf{x}_1, \mathbf{x}_2, t). \quad (1.1.34)$$

Correspondingly, Hermitian operators involving two particles are employed,

$$\Theta \longrightarrow \Theta(\mathbf{x}_1, \tfrac{\hbar}{i}\nabla_1; \mathbf{x}_2, \tfrac{\hbar}{i}\nabla_2), \quad (1.1.35)$$

and their eigenfunctions involve both coordinates,

$$\phi \longrightarrow \phi(\mathbf{x}_1, \mathbf{x}_2). \quad (1.1.36)$$

Similar extensions are made for N particles. Continuing for the moment with two particles, the quantum mechanical principle of indistinguishability for similar particles must be satisfied. This means that the state obtained by the interchange of the coordinates of any two identical particles must be physically indistinguishable from the original state and physically equivalent to it. Therefore, they can only differ by a constant phase factor since the probability density and current are unaffected:

$$\Psi(\mathbf{x}_1, \mathbf{x}_2, t) = e^{i\alpha}\Psi(\mathbf{x}_2, \mathbf{x}_1, t). \quad (1.1.37)$$

Interchanging the particle coordinates a second time,

$$\Psi(\mathbf{x}_2, \mathbf{x}_1, t) = e^{i\alpha}\Psi(\mathbf{x}_1, \mathbf{x}_2, t), \quad (1.1.38)$$

must yield the original wave function, so it is clear that $e^{2i\alpha} = 1$, whence

$$e^{i\alpha} = \pm 1 \qquad \text{and} \qquad \Psi(\mathbf{x}_1, \mathbf{x}_2, t) = \pm\Psi(\mathbf{x}_2, \mathbf{x}_1, t). \tag{1.1.39}$$

The upper sign (+) reflects a symmetric state and particles having such wave functions will be seen to obey Bose–Einstein statistics and are called "Bosons." The lower sign (–) reflects an antisymmetric state. The particles in this case will be seen to obey Fermi–Dirac statistics and are called "Fermions."

If the particles do not interact, it is possible to solve the time-independent Schrödinger energy eigenvalue equation easily by separation of variables, $H \longrightarrow h_1 + h_2$, since each part of the Hamiltonian depends only on one particle's coordinates:

$$\left[h_1\left(\mathbf{x}_1, \tfrac{\hbar}{i}\nabla_1\right) + h_2\left(\mathbf{x}_2, \tfrac{\hbar}{i}\nabla_2\right)\right]\phi_E(\mathbf{x}_1, \mathbf{x}_2) = E\phi_E(\mathbf{x}_1, \mathbf{x}_2), \tag{1.1.40}$$

with $\phi_E(\mathbf{x}_1, \mathbf{x}_2) = \phi_{E_1}(\mathbf{x}_1)\phi_{E_2}(\mathbf{x}_2)$ and $E_1 + E_2 = E$. The symmetry/antisymmetry requirements of the Pauli exclusion principle may be imposed on these separated solutions to obtain the properly normalized two-particle energy eigenfunctions in the symmetrical/antisymmetrical cases in the form

$$\phi_E(\mathbf{x}_1, \mathbf{x}_2) = \frac{[\phi_{E_1}(\mathbf{x}_1)\phi_{E_2}(\mathbf{x}_2) \pm \phi_{E_1}(\mathbf{x}_2)\phi_{E_2}(\mathbf{x}_1)]}{\sqrt{2}}. \tag{1.1.41}$$

Generalizing the non-interacting case to N particles, we similarly obtain non-interacting N-particle energy eigenfunctions as

$$\phi(\mathbf{x}_1, \mathbf{x}_2, \cdots, \mathbf{x}_\mathrm{N}) = \frac{1}{\sqrt{N!}}\begin{Bmatrix} det \\ perm \end{Bmatrix}\begin{vmatrix} \phi_{E_1}(\mathbf{x}_1)\phi_{E_1}(\mathbf{x}_2)\cdots\phi_{E_1}(\mathbf{x}_\mathrm{N}) \\ \phi_{E_2}(\mathbf{x}_1)\phi_{E_2}(\mathbf{x}_2)\cdots\phi_{E_2}(\mathbf{x}_\mathrm{N}) \\ \vdots \qquad\qquad \vdots \qquad\qquad \vdots \\ \phi_{E_n}(\mathbf{x}_1)\phi_{E_n}(\mathbf{x}_2)\cdots\phi_{E_n}(\mathbf{x}_\mathrm{N}) \end{vmatrix}, \tag{1.1.42}$$

with the energy eigenvalue $\sum_{i=1}^{N} E_i = E$. Here "det" and "perm" stand for determinant and permanent, respectively, and the permanent is the same linear combination of products of wave functions as the determinant, except that all signs are positive. In the antisymmetric case (Fermions) this is called the "Slater determinant," and it vanishes if any two "occupied states," ϕ_{E_i} and ϕ_{E_j}, are the same, so that no two (or more) Fermi particles can be in the same state. This is the Pauli exclusion principle for Fermions. Bosons are not subject to such a restriction on the occupation of states.

Note that if there is interaction, so that the eigenfunctions are not separable, we can still take a two-particle energy eigenfunction, $\phi_E(\mathbf{x}_1, \mathbf{x}_2)$, and symmetrize/antisymmetrize it as follows:

$$\frac{[\phi_E(\mathbf{x}_1, \mathbf{x}_2) \pm \phi_E(\mathbf{x}_2, \mathbf{x}_1)]}{\sqrt{2}}. \tag{1.1.43}$$

This is readily extended to N particles.

1.2 Dirac Notation

We will now redescribe wave functions in terms of abstract state vectors, introducing Dirac notation.[4–7] A single-particle wave function $\Psi(\mathbf{x})$ can be viewed as a column state vector or "function vector" (suppressing t-dependence),

$$\Psi(\mathbf{x}) \rightarrow \begin{pmatrix} \Psi(\mathbf{x_1}) \\ \Psi(\mathbf{x_2}) \\ \\ \vdots \\ \Psi(\mathbf{x_N}) \\ \vdots \end{pmatrix}, \tag{1.2.1}$$

in a space having a continuum of dimensions, one for each value of $\mathbf{x}$, with $\Psi(\mathbf{x_i})$ being the ith component of the vector. Denoting the corresponding abstract vector by $|\Psi\rangle$, termed a "ket" by Dirac,[4,5] and its adjoint $|\Psi\rangle^+ = \langle\Psi|$ (Dirac's "bra"), we have scalar (dot "inner") products of the type

$$\langle\Psi'| \cdot |\Psi\rangle \equiv \langle\Psi'|\Psi\rangle, \tag{1.2.2}$$

which form "bra·ket" structures, or "brackets." In these terms the ith component of $|\Psi\rangle$ can be written as

$$\langle\mathbf{x_i}|\boldsymbol{\Psi}\rangle \equiv \Psi(\mathbf{x_i}), \tag{1.2.3}$$

where the unit vector in the ith coordinate direction of our function vector space, the direction associated with $\mathbf{x_i}$, is denoted by $|\mathbf{x_i}\rangle$, an eigenvector of the position operator $\mathbf{X}$ with eigenvalue $\mathbf{x}_i$. We call this the "position" representation of the state vector $|\boldsymbol{\Psi}\rangle$. Since the unit vectors $|\mathbf{x_i}\rangle$ along the $\mathbf{x_i}$-coordinate directions represent mutually perpendicular unit vectors along the directions of mutually independent coordinate axes, the scalar product is given by

$$\langle\mathbf{x_i}|\mathbf{x_j}\rangle \Rightarrow \delta_{ij}, \tag{1.2.4}$$

and for the continuum of dimensions involved, this means that δ_{ij} is to be understood as a Dirac delta function,

$$\langle\mathbf{x_i}|\mathbf{x_j}\rangle = \delta^{(3)}(\mathbf{x_i} - \mathbf{x_j}). \tag{1.2.5}$$

Henceforth, we will drop superfluous subscripts on $\mathbf{x}$, understanding that each value of $\mathbf{x}$ indexes a separate orthogonal direction in the function vector space.

In terms of bras and kets, the completeness relation, $\Psi(\mathbf{x}) = \sum_r C_r\phi_r(\mathbf{x})$, becomes $\langle\mathbf{x}|\Psi\rangle = \sum_r C_r\langle\mathbf{x}|r\rangle$, so that $|\Psi\rangle = \sum_r C_r|r\rangle$. Alternatively, completeness can be expressed as (note that $\langle\mathbf{x}'|r\rangle^* = (\langle\mathbf{x}'|r\rangle)^+ = \langle r|\mathbf{x}'\rangle$ since the adjoint of a product is the product of the adjoints individually in reverse order, and $|r\rangle^+ \equiv \langle r|$, etc.):

$$\delta^{(3)}(\mathbf{x}-\mathbf{x}') = \sum_r \phi_r^*(\mathbf{x}')\phi_r(\mathbf{x}) = \sum_r \langle \mathbf{x}'|r\rangle^* \langle \mathbf{x}|r\rangle,$$
$$= \sum_r \langle \mathbf{x}|r\rangle \langle r|\mathbf{x}'\rangle, \tag{1.2.6}$$

and since

$$\langle \mathbf{x}|\mathbf{x}'\rangle = \delta^{(3)}(\mathbf{x}-\mathbf{x}') = \langle \mathbf{x}| \left(\sum_r |r\rangle\langle r| \right) |\mathbf{x}'\rangle, \tag{1.2.7}$$

we have the completeness relation as a sum of ket-bra "outer-product" structures as

$$\sum_r |r\rangle\langle r| = \mathbf{I}, \tag{1.2.8}$$

where **I** is the identity operator. In this form of the completeness relation, the sum is extended over all r-state vectors of the complete set, which means integration in the case of a continuum,

$$\int d^{(3)}\mathbf{x}|\mathbf{x}\rangle\langle \mathbf{x}| = \mathbf{I}. \tag{1.2.9}$$

The ket-bra outer-product structures involved here are *not* scalar-dot products, but are operators in the function vector space. They involve *no* multiplication between the ket and bra internally, but the ket can be dot multiplied to the left and the bra can be dot multiplied to the right.

Furthermore, the orthogonality relation is

$$\int d^{(3)}\mathbf{x}\phi_r^*(\mathbf{x})\phi_s(\mathbf{x}) = \int d^{(3)}\mathbf{x}\langle \mathbf{x}|r\rangle^* \langle \mathbf{x}|s\rangle,$$
$$= \langle r| \left[\int d^{(3)}\mathbf{x}|\mathbf{x}\rangle\langle \mathbf{x}| \right] |s\rangle,$$
$$= \langle r|\mathbf{I}|s\rangle = \langle r|s\rangle = \delta_{rs}. \tag{1.2.10}$$

(Again, in the case of a continuum, δ_{rs} is supplanted by a Dirac delta function.) Finally, we note that C_r is given by

$$C_r = \int d^{(3)}\mathbf{x}\phi_r^*(\mathbf{x})\Psi(\mathbf{x}),$$
$$= \int d^{(3)}\mathbf{x}\langle r|\mathbf{x}\rangle\langle \mathbf{x}|\Psi\rangle,$$
$$= \langle r| \left[\int d^{(3)}\mathbf{x}\,|\mathbf{x}\rangle\,\langle \mathbf{x}| \right] |\Psi\rangle\,, \tag{1.2.11}$$

or,

$$C_r = \langle r|\Psi\rangle, \tag{1.2.12}$$

which could have been obtained directly from the results above using

$$|\Psi\rangle = \sum_{r'} C_{r'}|r'\rangle \tag{1.2.13}$$

and

$$\langle r|r'\rangle = \delta_{rr'} = \langle r'|r\rangle, \tag{1.2.14}$$

since

$$\langle r|\Psi\rangle = \sum_{r'} C_{r'}\langle r|r'\rangle = \sum_{r'} C_{r'}\delta_{rr'} = C_r. \tag{1.2.15}$$

(Restoring time dependence, $|\Psi\rangle \to |\Psi(t)\rangle$ and $C_{r'} \to C_{r'}(t)$.)

Having identified the identity operator as a ket-bra structure, $\mathbf{I}=\sum_r |r\rangle\langle r|$, we now examine a more general operator structure corresponding to a matrix having matrix elements $M_{rr'}$:

$$\mathbf{M} = \sum_r \sum_{r'} |r\rangle M_{rr'}\langle r'|. \tag{1.2.16}$$

Again, there is no internal multiplication between the ket and bra of the outer product, but the ket can dot multiply to the left and the bra can dot multiply to the right. We will show below that with the above rules for multiplication, the operator $\mathbf{M}$ has operator multiplication properties that correspond exactly to those of the matrix with elements $M_{rr'}$. Consider first,

$$\mathbf{M}|r'\rangle = \sum_r \sum_{r''} |r\rangle M_{rr''}\langle r''|r'\rangle = \sum_r |r\rangle M_{rr'} \tag{1.2.17}$$

and

$$\langle r|\mathbf{M}|r'\rangle = \sum_{r'''} \langle r|r'''\rangle M_{r'''r'} = M_{rr'}, \tag{1.2.18}$$

where the orthogonality relation, $\langle r''|r'\rangle = \delta_{r''r'}$, has been used in both stages. This is a very convenient way of representing an operator and it justifies Dirac's bra-ket notation since the matrix element of $\mathbf{M}(M_{rr'})$ is here seen to be the bracket ("bra-ket") of the abstract operator $\mathbf{M}$ between the bra $\langle r|$ and the ket $|r'\rangle$. A final verification may be obtained by noting that we simply have an identity, since

$$\mathbf{M} = \sum_r \sum_{r'} |r\rangle\langle r|\mathbf{M}|r'\rangle\langle r'| = \left[\sum_r |r\rangle\langle r|\right]\mathbf{M}\left[\sum_{r'} |r'\rangle\langle r'|\right]$$
$$= \mathbf{I}\cdot\mathbf{M}\cdot\mathbf{I} = \mathbf{M}. \tag{1.2.19}$$

Dirac's ket-bra operators are just dyadic operators, such as those that have been used in electromagnetic theory for a long time.[8–10]

In regard to the implications of this for the position representation, where matrix elements are usually written as

$$M_{rr'} = \int d^{(3)}\mathbf{x}\phi_r^*(\mathbf{x})M(\mathbf{x}, \tfrac{\hbar}{i}\nabla)\phi_{r'}(\mathbf{x}), \tag{1.2.20}$$

or, alternatively,

$$\langle r|\mathbf{M}|r'\rangle = \langle r| \int d^{(3)}\mathbf{x}|\mathbf{x}\rangle M(\mathbf{x}, \tfrac{\hbar}{i}\nabla)\langle\mathbf{x}|r'\rangle, \tag{1.2.21}$$

we have

$$\mathbf{M} = \int d^{(3)}\mathbf{x}|\mathbf{x}\rangle M(\mathbf{x}, \tfrac{\hbar}{i}\nabla)\langle\mathbf{x}|. \tag{1.2.22}$$

However,

$$\begin{aligned}\langle\mathbf{x}'|\mathbf{M}|\mathbf{x}''\rangle &= \int d^{(3)}\mathbf{x}\langle\mathbf{x}'|\mathbf{x}\rangle M(\mathbf{x}, \tfrac{\hbar}{i}\nabla)\langle\mathbf{x}|\mathbf{x}''\rangle \\ &= \int d^{(3)}\mathbf{x}\delta^{(3)}(\mathbf{x}' - \mathbf{x})M(\mathbf{x}, \tfrac{\hbar}{i}\nabla)\delta^{(3)}(\mathbf{x} - \mathbf{x}''),\end{aligned} \tag{1.2.23}$$

so that

$$M_{x'x''} = \delta^{(3)}(\mathbf{x}' - \mathbf{x}'')M(\mathbf{x}', \tfrac{\hbar}{i}\nabla') \tag{1.2.24}$$

is diagonal in position representation. Applying this to the eigenfunction equation,

$$\Theta(\mathbf{x}, \tfrac{\hbar}{i}\nabla)\phi_r(\mathbf{x}) = \lambda_r\phi_r(\mathbf{x}), \tag{1.2.25}$$

we write:

$$\int d^{(3)}\mathbf{x}'\Theta(\mathbf{x}, \tfrac{\hbar}{i}\nabla)\delta^{(3)}(\mathbf{x}' - \mathbf{x})\phi_r(\mathbf{x}') = \lambda_r\langle\mathbf{x}|r\rangle. \tag{1.2.26}$$

Using the fact that $\Theta_{\mathbf{x}\mathbf{x}'} = \langle\mathbf{x}|\Theta|\mathbf{x}'\rangle$ is diagonal in position representation, we have

$$\int d^{(3)}\mathbf{x}'\langle\mathbf{x}|\Theta|\mathbf{x}'\rangle\langle\mathbf{x}'|r\rangle = \langle\mathbf{x}|\Theta|r\rangle = \lambda_r\langle\mathbf{x}|r\rangle, \tag{1.2.27}$$

since $\int d^{(3)}\mathbf{x}'|\mathbf{x}'\rangle\langle\mathbf{x}'| = \mathbf{I}$ is the identity operator. Thus, the eigenvector equation is just

$$\Theta|r\rangle = \lambda_r|r\rangle. \tag{1.2.28}$$

Similarly, for the Schrödinger equation,

$$H(\mathbf{x}, \tfrac{\hbar}{i}\nabla)\Psi(\mathbf{x}, t) = i\hbar\frac{\partial\Psi(\mathbf{x}, t)}{\partial t} \tag{1.2.29}$$

or,

$$\langle\mathbf{x}|H|\Psi\rangle = i\hbar\langle\mathbf{x}|\frac{\partial\Psi}{\partial t}\rangle, \tag{1.2.30}$$

from which

$$H|\Psi\rangle = i\hbar\frac{\partial|\Psi\rangle}{\partial t}. \tag{1.2.31}$$

It is, of course, important to verify that the multiplication properties of operators written as

$$\mathbf{M} = \sum_r\sum_{r'}|r\rangle M_{rr'}\langle r'| \tag{1.2.32}$$

just replicate the matrix multiplication properties. Consider a product of two such operators,

$$\mathbf{M}\cdot\mathbf{N} = \sum_r\sum_{r'}\sum_{r''}\sum_{r'''}|r\rangle M_{rr'}\langle r'|\cdot|r''\rangle N_{r''r'''}\langle r'''|. \tag{1.2.33}$$

Using the orthogonality relation, $\langle r'|r''\rangle = \delta_{r'r''}$, we have

$$\mathbf{M}\cdot\mathbf{N} = \sum_r\sum_{r'}\sum_{r'''}|r\rangle M_{rr'}N_{r'r'''}\langle r'''|. \tag{1.2.34}$$

Now (we write $r'''' \equiv r^{iv}$, etc.):

$$\begin{aligned}(\mathbf{M}\cdot\mathbf{N})_{r^{iv},r^{v}} &= \langle r^{iv}|\mathbf{M}\cdot\mathbf{N}|r^{v}\rangle,\\ &= \sum_r\sum_{r'}\sum_{r'''}\langle r^{iv}|\cdot|r\rangle M_{rr'}N_{r'r'''}\langle r'''|\cdot|r^{v}\rangle,\\ &= \sum_r\sum_{r'}\sum_{r'''}\delta_{r^{iv}r}M_{rr'}N_{r'r'''}\delta_{r'''r^{v}},\\ &= \sum_{r'}M_{r^{iv}r'}N_{r'r^{v}},\end{aligned} \tag{1.2.35}$$

which properly follows the rule for matrix multiplication. Dirac's introduction of dyadic notation facilitates many derivations and calculations in quantum theory, and greatly clarifies many aspects of quantum mechanics. As we will see, Schwinger has introduced further interpretation of the dyadic structures, which yield "second quantization" in a remarkably simple way.

1.3 Complete Sets of Commuting Observables, Representations and Transformation Theory

Taking a broader view of Eq. (1.2.28), we may understand Θ to stand for a complete set of commuting Hermitian operators, $A = \{A_1, A_2, \cdots, A_l\}$, corresponding to a complete set of compatible observables[2,4,7] (compatible in the sense that they can be measured precisely simultaneously). Such completeness means that all the members of the set have common eigenvectors, $|a'\rangle$, which are not degenerate, with a corresponding set of eigenvalues, $a' \to \{a'_i\}$, that uniquely define a state of maximum information permitted by the laws of quantum mechanics. Moreover, any other operator of the system that is compatible/commutative with $A = \{A_i\}$ is necessarily a function of the set $\{A_i\}$; and any other independent property of the system is incompatible/non-commutative with some member(s) of the set $\{A_i\}$. Thus, we understand A to represent a complete set of compatible observables/operators:

$$\Theta \to A = \{A_1, A_2, \cdots, A_l\}, \tag{1.3.1}$$

with $[A_i, A_j] = 0$, $(i = 1 \dots l; j = 1 \dots l)$, and

$$A\,|a'\rangle = a'\,|a'\rangle, \tag{1.3.2}$$

in the sense that

$$A_i\,|a'\rangle = a'_i\,|a'\rangle \qquad (i = 1, 2, \cdots, l) \tag{1.3.3}$$

describes the set of common eigenvectors of the complete set of commuting operators A, corresponding to the compatible observables that uniquely define a non-degenerate state $|a'\rangle$ of the system. The adjoint of such a ket $|a'\rangle^+ = \langle a'|$ is a bra-state, and the orthonormality of two such eigenstates of A, $|a'\rangle$ and $|a''\rangle$, is given by

$$\langle a'|\, a''\rangle = \delta_{a'a''}, \tag{1.3.4}$$

with completeness described by

$$\sum_{a'} |a'\rangle\langle a'| = \mathbf{I}. \tag{1.3.5}$$

Moreover, in summary, the Schrödinger equation is

$$H\,|\Psi(t)\rangle = i\hbar\frac{\partial\,|\Psi(t)\rangle}{\partial t} \tag{1.3.6}$$

and $|\Psi(t)\rangle$ may be expanded in the eigenvectors of A as

$$|\Psi(t)\rangle = \sum_{a'} C_{a'}(t)\,|a'\rangle, \tag{1.3.7}$$

where the coefficient $C_{a'}(t)$ is given by

$$C_{a'}(t) = \langle a' | \Psi(t) \rangle \equiv \Psi(a', t) \tag{1.3.8}$$

and is called "the state (or wave function) $|\Psi(t)\rangle$ in an A-representation," $\Psi(a', t)$. Also, an operator $\mathbf{M}$ having matrix elements $\mathbf{M}_{a'a''} = \langle a' | \mathbf{M} | a'' \rangle$ in an A-representation has the form

$$\mathbf{M} = \sum_{a'} \sum_{a''} |a'\rangle \mathbf{M}_{a'a''} \langle a''| . \tag{1.3.9}$$

This constitutes the description of an A-representation with basis vectors $|a'\rangle$.

We could equally well choose a different complete compatible and commuting set of observable Hermitian operators $B = \{B_1, B_2, \cdots, B_l\}$ having a complete set of common eigenvectors $|b'\rangle$,

$$B |b'\rangle = b' |b'\rangle, \tag{1.3.10}$$

and adjoints $|b'\rangle^+ = \langle b'|$, with orthogonality given by $\langle b' | b'' \rangle = \delta_{b'b''}$. In this case we would have the B-representation with basis vectors $|b'\rangle$, $\langle b' | \Psi(t) \rangle$ as the state $|\Psi(t)\rangle$ in the B-representation, $\Psi(b', t)$, and $\mathbf{M}_{b'b''} = \langle b' | \mathbf{M} | b'' \rangle$ as the matrix elements of $\mathbf{M}$ in the B-representation. It is straightforward to transform from one representation, or basis, to another. Consider $|\Psi\rangle$ in the A-representation,

$$\Psi(a') = \langle a' | \Psi \rangle = \langle a' | \mathbf{I} | \Psi \rangle, \tag{1.3.11}$$

and insert the identity in the B-representation,

$$\mathbf{I} = \sum_{b'} |b'\rangle\langle b'|, \tag{1.3.12}$$

to obtain

$$\Psi(a') = \sum_{b'} \langle a' | b' \rangle \langle b' | \Psi \rangle = \sum_{b'} \langle a' | b' \rangle \Psi(b'). \tag{1.3.13}$$

Furthermore,

$$\mathbf{M}_{a'a''} = \langle a' | \mathbf{M} | a'' \rangle = \langle a' | \mathbf{I} \cdot \mathbf{M} \cdot \mathbf{I} | a'' \rangle, \tag{1.3.14}$$

and again writing $\mathbf{I}$ in the B-representation, or basis, we have

$$\begin{aligned} \mathbf{M}_{a'a''} &= \sum_{b'} \sum_{b''} \langle a' | b' \rangle \langle b' | \mathbf{M} | b'' \rangle \langle b'' | a'' \rangle \\ &= \sum_{b',b''} \langle a' | b' \rangle \mathbf{M}_{b'b''} \langle b'' | a'' \rangle. \end{aligned} \tag{1.3.15}$$

Eqns. (1.3.13) and (1.3.15) are the fundamental relations of transformation theory. Note that while there is orthogonality among the eigenvectors $|a'\rangle$, and, separately, there is orthogonality among the eigenvectors $|b'\rangle$, they are *not* mutually orthogonal, $\langle a'|\, b'\rangle \neq 0$, because they are eigenvectors of different complete commuting operator sets. In fact, the $\langle a'|\, b'\rangle$ are the "transformation" functions that facilitate the change of basis described in Eqns. (1.3.13) and (1.3.15). Finally, it should be noted that any Hermitian operator **M** can be brought to diagonal form by transformation to the basis of its eigenvectors, $|m'\rangle|$, since the eigenvalue equation, $\mathbf{M}\,|m'\rangle = m'|\, m'\rangle$, from which

$$\mathbf{M}_{mm'} = \langle m\,|\mathbf{M}|\, m'\rangle = m'\langle m\,|m'\rangle = m\delta_{mm'}| \,. \tag{1.3.16}$$

1.4 Schwinger Measurement Symbols

It is sometimes useful to construct operators in a "mixed" representation by writing

$$\mathbf{M} = \mathbf{I}\cdot\mathbf{M}\cdot\mathbf{I}, \tag{1.4.1}$$

where the first identity is written in the A-representation and the second identity is written in the B-representation, obtaining

$$\begin{aligned}\mathbf{M} &= \sum_{a'} |a'\rangle\langle a'| \cdot\mathbf{M}\cdot \sum_{b'} |b'\rangle\langle b'| \\ &= \sum_{a',b'} |a'\rangle\,\mathbf{M}_{a'b'}\,\langle b'|\,,\end{aligned} \tag{1.4.2}$$

where

$$\mathbf{M}_{a'b'} = \langle a'|\,\mathbf{M}\,|b'\rangle \tag{1.4.3}$$

is the "mixed" matrix element of **M**. Also, Eq. (1.4.2) employs mixed dyadic operators, called "measurement symbols," which are defined as

$$\mathcal{M}(a',b') \equiv |a'\rangle\langle b'|\,, \tag{1.4.4}$$

such that

$$\mathbf{M} = \sum_{a',b'} \mathbf{M}_{a'b'}\mathcal{M}(a',b') = \sum_{a',b'} \langle a'|\,\mathbf{M}\,|b'\rangle\,\mathcal{M}(a',b'). \tag{1.4.5}$$

Considering the action of the measurement symbol $\mathcal{M}(a',b')$ on an arbitrary state,

$$|\Psi\rangle = \sum_{b''} |b''\rangle\langle b''|\,\Psi\rangle = \sum_{b''} |b''\rangle\,\Psi(b''), \tag{1.4.6}$$

we have

$$\mathcal{M}(a', b') \, |\Psi\rangle = \left|a'\right\rangle \sum_{b''} \langle b' \left| b'' \right\rangle \Psi(b'') = \left|a'\right\rangle \Psi(b'), \tag{1.4.7}$$

since $\langle b' \left| b'' \right\rangle = \delta_{b'b''}$. Thus, the action of $\mathcal{M}(a', b')$ on $|\Psi\rangle$ may be described as a two-step process:

1. It selects the component of $|\Psi\rangle$ associated with $\left|b'\right\rangle$, that is, it selects the probability amplitude for $|\Psi\rangle$ to be found in $\left|b'\right\rangle$ if a B-measurement were to be made.
2. It eliminates the state $\left|b'\right\rangle$, replacing it by $\left|a'\right\rangle$, carrying along the original amplitude $\Psi(b')$.

This has been interpreted by Schwinger in the following way:[11,12]

> "$\mathcal{M}(a', b')$ acts as an operator representing a selective measurement that changes the state upon which it acts. It selects only systems in the state $\left|b'\right\rangle$ out of the totality of impinging system states, rejecting all others; following this it emits these systems in the state $\left|a'\right\rangle$. The measurement itself entails an uncontrollable disturbance, which is beyond our scrutiny; it is an indivisible act about which we know only the initial and final states of the system, that is to say that the first stage selects systems in the state $\left|b'\right\rangle$ and that the last stage produces them in the state $\left|a'\right\rangle$. There is no physical meaning to any attempt to trace the history of the measurement process in detail, since no such information is available. However, so long as we do not interfere with the initial and final states, we may imagine this measurement process to take part in two stages associated with the interposition of a non-physical *null* state as an intermediary. We call this imagined *null* state 0 and write
>
> $$\mathcal{M}(a', b') = \mathcal{M}(a', 0)\mathcal{M}(0, b')." \tag{1.4.8}$$[11]

(Note that $\mathcal{M}(0, a') = \mathcal{M}^{+}(a', 0)$.) In accordance with this interpretation, $\mathcal{M}(0, b')$ represents the measurement process that selects a system in the state $\left|b'\right\rangle$, and emits it in the limbo state, called the *null* state (which is *not* any of the eigenstates $\left|a'\right\rangle$ or $\left|b'\right\rangle$ discussed above). This may be described as the *annihilation* of a system in the state $\left|b'\right\rangle$. Correspondingly, in this interpretive scheme, the ensuing action of $\mathcal{M}(a', 0)$ then *creates* (out of the limbo *null* state) and emits a system in the state $\left|a'\right\rangle$ (following its selection from the *null* state). This joint action of the operators $\mathcal{M}(a', 0)\mathcal{M}(0, b')$ thus has the same effect as $\mathcal{M}(a', b')$, which we can therefore describe as *annihilating* a system in $\left|b'\right\rangle$, along with the *creation* of a system in state $\left|a'\right\rangle$.[12–14] This will be considered further in a many-particle context.

1.5 Chapter 1 Problems

P.1.1. Eq. (1.1.10) says that $[L_1, L_2] \neq 0$. Prove that $\left[L_i, \mathbf{L}^2\right] = 0$.
P.1.2. Prove that L_3 generates infinitesimal rotations about the 3-axis.

P.1.3. Show that:

(a) $[A, BC] = B[A, C] + [A, B]C$ for quantum operators A,B,C, and that

(b) $[A, [B, C]] + [B, [C, A]] + [C, [A, B]] = 0$.

P.1.4. Consult the literature to discuss the relations between

(a) e^{A+B} and $e^A e^B$, also

(b) $\frac{de^{C(t)}}{dt}$ and $e^{C(t)} \frac{dC(t)}{dt}$.

P.1.5. Consider a one-dimensional (1D) simple harmonic oscillator with the Hamiltonian $H = \frac{p^2}{2m} + \frac{m\omega^2 q^2}{2}$, where q, p obey canonical commutation relations:

(a) Determine its eigenvalues, E_n, and eigenfunctions, ϕ_n, in position (q) representation.

(b) Defining operators

$$c = \sqrt{\frac{m\omega}{2\hbar}}(q + ip/m\omega),$$

$$c^+ = \sqrt{\frac{m\omega}{2\hbar}}(q - ip/m\omega),$$

write H in terms of c, c^+, and determine their commutation relations with each other. Show that c^+c commutes with H, and denoting their common eigenvectors by $\phi_n \to |n\rangle$, prove that

$$(c^+c)c^+ |n\rangle = (\lambda_n + 1)c^+ |n\rangle$$

$$(\lambda_n = n = 0, 1, 2 \ldots),$$

where $(c^+c) |n\rangle = \lambda_n |n\rangle$, so that c^+ has the interpretation of a "raising" operator. Show that c has the corresponding interpretation of a "lowering" operator.

(c) If we start with recognition that the lowest λ_n eigenvalue is $\lambda_0 = 0$ (and $c|0\rangle = 0$, correspondingly), show that the "ladder" procedure in the sequential application of c^+ operators yields the nth eigenstate as $|n\rangle = N_n(c^+)^n |0\rangle$. Determine the normalization constant N_n.

P.1.6. Verify Eqns. (1.1.23) and (1.1.24).

P.1.7. On the right-hand side of Eq. (1.1.28), why is (–*h.c.*; minus Hermitian conjugate) appropriate?

P.1.8. Verify Eq. (1.1.31).

P.1.9. Prove that Eq. (1.1.32) correctly describes the behavior of the wave function under a gauge transformation.

P.1.10. What is the relation between an eigenfunction of energy in a position representation and the corresponding position eigenfunction in an energy representation?

P.1.11. Verify that $M|\Psi\rangle$ and $\langle\Psi|M$ (where M is given by Eq. (1.2.16)). Follow the usual multiplication rules between a matrix and a state vector.

P.1.12. Prove that Dirac's representation of operators in terms of ket-bra ($|i\rangle\, \rangle j|$) structures is equivalent to the dyadic ($\hat{e}_i, \hat{e}_j^+$) notation of electromagnetic theory (see, for example, Reference 10 in Section 1.6, pp. 62–77), having the same multiplication and transformation properties under a change of basis.

P.1.13. A projection operator P has the property that $P^2 = P$. Show that for any subset m of dyadics of the type $|a_i'\rangle\langle a_i'|$, the sum $\sum_{i=1}^{m} |a_i'\rangle\langle a_i'|$ has this property.

P.1.14. Prove the identity for a variation δ of an operator X:

$$\delta(logdetX) = tr(X^{-1}\delta X).$$

P.1.15. Prove the following identity for Pauli spin matrices $\boldsymbol{\sigma} = \sigma_1\hat{\imath} + \sigma_2\hat{\jmath} + \sigma_3\hat{k}$ and two arbitrary vectors $\mathbf{a}, \mathbf{b}$:

$$(\boldsymbol{\sigma} \cdot \mathbf{a})(\boldsymbol{\sigma} \cdot \mathbf{b}) = \mathbf{a} \cdot \mathbf{b} + i\boldsymbol{\sigma} \cdot \mathbf{a} \times \mathbf{b},$$

and if $\hat{n}$ is an arbitrary unit vector, show that

$$(\boldsymbol{\sigma} \cdot \hat{n})^2 = 1.$$

P.1.16. Construct the unitary similarity transformation that diagonalizes the matrix $\boldsymbol{\sigma} \cdot \mathbf{H}$.

P.1.17. Hamilton's generalization of pure imaginaries $\underline{i}, \underline{j}, \underline{k}$ may be defined as:
$\underline{i}=-i\sigma_1, \underline{j}=-i\sigma_2, \underline{k}=-i\sigma_3$.
Show that $\underline{i}, \underline{j}, \underline{k}$ obey the following multiplication rules:
(a) $\underline{i}^2 = \underline{j}^2 = \underline{k}^2 = -I$
(b) $\underline{ij} = -\underline{ji} = \underline{k}$; $\underline{jk} = -\underline{kj} = \underline{i}$; and $\underline{ki} = -\underline{ik} = \underline{j}$.

P.1.18. A quaternion Q is defined as:
$Q \equiv t+\underline{i}x+\underline{j}y+\underline{k}z$,
where t, x, y and z are *real* numbers (notice that all real numbers are quaternions). Prove that:
(a) If Q_1 and Q_2 are quaternions, then $(Q_1 + Q_2)$ and $(Q_1 \cdot Q_2)$ are also quaternions.
(b) Q^+Q = real number, where Q^+ is called the "complex conjugate" of the quaternion Q and is given by $Q^+ = t-\underline{i}x-\underline{j}y-\underline{k}z$, with Q as above (notice that Q^+ is also a quaternion).
(c) For every quaternion Q there exists another quaternion Q^{-1} such that $Q^{-1}Q = QQ^{-1} = 1$.

P.1.19. The Hamiltonian of a charged particle in a magnetic field $\mathbf{B}$ given in Eq. (1.1.27) neglects a "Zeeman spin-splitting" term, $\overline{\mu}_0\boldsymbol{\sigma} \cdot \mathbf{B}$. Consider the case in which all *other* electromagnetic field effects are negligible and solve the following equation for the 2 × 2 matrix $G(\mathbf{p}, t-t')$ for $t > t'$, subject to the condition that $G(\mathbf{p}, t-t' < 0) \equiv 0$:

$$\left(\hat{I}i\frac{\partial}{\partial t} - \frac{\hat{I}p^2}{2m} - \overline{\mu}_0\boldsymbol{\sigma}\cdot\mathbf{B}\right)G(\mathbf{p}, t-t') = \hat{I}\delta(t-t').$$

($\hat{I}$ is the 2×2 unit matrix and $\overline{\mu}_0 = e\hbar/2mc$ is the Bohr magneton.)

1.6 References for Chapter 1

R.1.1. A. Beiser, "Concepts of Modern Physics," 5th Ed., McGraw-Hill, Inc. (1995). (Elementary introduction to the Schrödinger equation and atomic physics.)

R.1.2. R.L. Liboff, "Introductory Quantum Mechanics," Holden-Day, Inc. (1980). (Good introduction to quantum mechanics.)

R.1.3. E. Merzbacher, "Quantum Mechanics," 3rd Ed., John Wiley & Sons, Inc. (1998). (More advanced, fairly comprehensive presentation.)

R.1.4. P.A.M. Dirac, "Principles of Quantum Mechanics," 3rd Ed., Oxford/Clarendon Press (1949). (The master's voice.)

R.1.5. J.D. Jackson, "Mathematics for Quantum Mechanics," W.A. Benjamin, Inc. (1962).

R.1.6. J.J. Sakurai, "Modern Quantum Mechanics," Revised Edition, Editor: San Fu Tuan, Addison-Wesley Publishing Co. (1994).

R.1.7. R. Shankar, "Principles of Quantum Mechanics," Plenum Press (1980).

R.1.8. Chen-To Tai, "Dyadic Green's Functions in Electromagnetic Theory," InText Educational Publishers (1971).

R.1.9. P.M. Morse and H. Feshbach, "Methods of Theoretical Physics," Vol. I, McGraw-Hill Book Co., Inc. (1953).

R.1.10. L. Page and N.I. Adams, "Electrodynamics," D. van Nostrand Co., Inc. (1940).

R.1.11. Quoted from the lectures of J. Schwinger on Quantum Mechanics at Harvard University, unpublished (1956).

R.1.12. J. Schwinger, "The Geometry of Quantum States," Proc. Nat. Academy of Sciences, *46*, 257 (1960).

R.1.13. J. Schwinger, "Quantum Kinematics and Dynamics," W.A. Benjamin, Inc. (1970).

R.1.14. J. Schwinger, "Quantum Mechanics," Editor: B-G Englert, Springer (2001).

2

Identical Particles and Second Quantization: Occupation Number Representation

Abstract

Focusing on systems of many identical particles, Chapter 2 introduces appropriate operators to describe their properties in terms of Schwinger's "measurement symbols". The latter are then factorized into "creation" and "annihilation" operators, whose fundamental properties and commutation/anticommutation relations are derived in conjunction with the Pauli exclusion principle. This leads to "second quantization" with the Hamiltonian, number, linear and angular momentum operators expressed in terms of the annihilation and creation operators, as well as the occupation number representation. Finally, the concept of coherent states, as eigenstates of the annihilation operator, having minimum uncertainty, is introduced and discussed in detail.

2.1 Operators for Properties of Identical Particles

In accordance with the quantum mechanical principle of indistinguishability of identical particles, the operators representing their physical properties collectively must be constructed so as to ensure that they cannot be distinguished either kinematically or dynamically.[1] Kinematic identity requires that the particles are described by the same number and kind of dynamical variables, χ_k, such as position $\mathbf{r}_k$ and momentum $\mathbf{p}_k$ operators for each of the particles (indexed by k). Dynamic identity, which assures that the time development of the system does not cause a particular particle to be singled out from the others, requires that the Hamiltonian (which generates time development) must be symmetric with respect to the variables of the constituent particles no matter what perturbations may occur.

Since the constituent particles, which we will now term "subsystems," are therefore perpetually indistinguishable, the collective properties of the whole system must be composed symmetrically from subsystem contributions. For example, considering a property, $F^{(1)}$, of an n-particle system, which is composed of an analogous contribution, $f^{(1)}(\chi_k)$, from each subsystem singly (momentum, angular momentum, kinetic energy, etc.) involving only variables of that subsystem, we have

Quantum Statistical Field Theory: Schwinger's Variational Method. Norman J. Morgenstern Horing, Oxford University Press (2017). © Norman J. Morgenstern Horing. DOI 10.1093/oso/9780198791942.001.0001

$$F^{(1)} = \sum_{k=1}^{n} f^{(1)}(\chi_k) \tag{2.1.1}$$

(e.g., in the case of linear momentum $f^{(1)}(\chi_k) \to \mathbf{p}_k$ and the total momentum operator is $F^{(1)} \to \mathbf{P} = \sum_{k=1}^{n} \mathbf{p}_k$; for angular momentum $f^{(1)}(\chi_k) \to \mathbf{L}_k$ and $F^{(1)} \to \mathbf{L} = \sum_{k=1}^{n} \mathbf{L}_k$ is the total angular momentum operator; and for kinetic energy $f^{(1)}(\chi_k) \to p_k^2/2m$ and $F^{(1)} \to KE = \sum_{k=1}^{n} p_k^2/2m$ is the total kinetic energy operator).

For properties $F^{(2)}$ composed of analogous contributions from each pair of subsystems, such as particle–particle interaction energy, we have

$$F^{(2)} = \sum_{\substack{k=1 \\ (k \neq l)}}^{n} \sum_{l=1}^{n} \frac{1}{2!} f^{(2)}(\chi_k, \chi_l). \tag{2.1.2}$$

For example, in the case of the Coulomb interparticle potential, $f^{(2)}(\chi_k, \chi_l) = |\mathbf{r}_k - \mathbf{r}_l|^{-1}$. The exclusion of terms with $k = l$ from $F^{(2)}$ reflects the fact that these terms are of the $f^{(1)}$ type, and the $\frac{1}{2!}$-factor corrects for double counting in the double sum over k and l. Continuing along these lines, a property $F^{(3)}$ composed of analogous contributions from each triplet of subsystems has the form

$$F^{(3)} = \sum_{\substack{k=1 \\ (k \neq l \neq m)}}^{n} \sum_{l=1}^{n} \sum_{m=1}^{n} \frac{1}{3!} f^{(3)}(\chi_k, \chi_l, \chi_m). \tag{2.1.3}$$

Similar considerations apply to properties $F^{(m)}$ that are composed of analogous contributions involving the joint participation of m subsystems in a group, summed over all such groups of the n-particle system. These n-particle system operator forms ensure that no particle is distinguished from another.

Considering a subsystem k with a complete commuting set of Hermitian operators $A_k = \{A_{1k}, A_{2k}, \ldots, A_{lk}\}$ having the single-particle eigenvalue spectrum $\{a'_k\}$ and eigenvectors $|a'_k\rangle$, we may use Eq. (1.4.5) to write

$$f^{(1)}(\chi_k) = \sum_{a'_k, a''_k} \langle a'_k | f^{(1)}(\chi_k) | a''_k \rangle \mathcal{M}(a'_k, a''_k), \tag{2.1.4}$$

where the "measurement symbol," $\mathcal{M}(a'_k, a''_k)$, is (note that a'_k, a''_k both refer to the spectrum of A_k; a mixed representation will be discussed below):

$$\mathcal{M}(a'_k, a''_k) = |a'_k\rangle \langle a''_k| . \tag{2.1.5}$$

This corresponds to a measurement that accepts a subsystem in state a''_k and emits it in state a'_k. Because of identity, the spectra of all the subsystems are identical so the index k is irrelevant,

$$a'_k = a', \tag{2.1.6}$$

and it is clear that the matrix element is independent of k,

$$\langle a_k' | f^{(1)}(\chi_k) | a_k'' \rangle \equiv \langle a' | f^{(1)}(\chi) | a'' \rangle, \tag{2.1.7}$$

with the right-hand side being the same for each of the identical subsystems. Thus, from Eqns. (2.1.1) and (2.1.7), we have $\left(\sum_{a_k', a_k''} \to \sum_{a', a''}, \text{etc.}\right)$:

$$F^{(1)} = \sum_{a', a''} \langle a' | f^{(1)} | a'' \rangle \mathbf{M}(a', a''), \tag{2.1.8}$$

where

$$\mathbf{M}(a', a'') = \sum_{k=1}^{n} \mathcal{M}(a_k', a_k'') \tag{2.1.9}$$

is the *total* measurement symbol, which corresponds to actual observables of the system of identical particles, whereas $\mathcal{M}(a_k', a_k'')$ can have no direct meaning itself since it refers to a particular subsystem. It should be remarked that, as an operator, $\mathbf{M}(a', a'')$ acts in the product space of the n particles without distinguishing any of them. Clearly, we have the properties

$$\mathbf{M}^+(a', a'') = \mathbf{M}(a'', a'), \tag{2.1.10}$$

and

$$\sum_{a'} \mathbf{M}(a', a') = \sum_{k=1}^{n} \sum_{a_k'} \mathcal{M}(a_k', a_k') = \sum_{k=1}^{n} \mathbf{I}_k = \mathbf{N} \tag{2.1.11}$$

since

$$\sum_{a_k'} \mathcal{M}(a_k', a_k') = \sum_{a_k'} |a_k'\rangle \langle a_k'| = \mathbf{I}_k \tag{2.1.12}$$

is the identity operator for subsystem k, and summation over subsystems then yields the operator $\mathbf{N}$ for the total number of particles of the system, namely n.

Using the orthonormality of the eigenvectors and their adjoints constituting the dyadics $\mathcal{M}(a_k', a_k'')$, we have (note that $\delta(a', a'')$ is the Kronecker delta for a discrete spectrum and the Dirac delta function for a continuous spectrum):

$$\mathcal{M}(a_k', a_k'')\mathcal{M}(a_k''', a_k^{iv}) = \delta(a'', a''')\mathcal{M}(a_k', a_k^{iv}), \tag{2.1.13}$$

from which we can show that ($[A, B] = AB - BA$ is the commutator of A, B)

$$[\mathbf{M}(a', a''), \mathbf{M}(a''', a^{iv})] = \delta(a'', a''')\mathbf{M}(a', a^{iv}) - \delta(a', a^{iv})\mathbf{M}(a''', a''). \tag{2.1.14}$$

To prove this, we first note that kinematic independence of the subsystems implies that the operators of different subsystems commute with one another,

$$\left[\mathcal{M}(a'_k, a''_k), \mathcal{M}(a'''_l, a^{iv}_l)\right] = 0, \quad (k \neq l). \tag{2.1.15}$$

Furthermore (all sums over subsystems are from $1 \to n$),

$$\begin{aligned}\mathbf{M}(a', a'')\mathbf{M}(a''', a^{iv}) = &\sum_{\substack{k,l \\ (k\neq l)}} \mathcal{M}(a'_k, a''_k)\mathcal{M}(a'''_l, a^{iv}_l) \\ &+ \sum_k \delta(a'', a''')\mathcal{M}(a'_k, a^{iv}_k),\end{aligned} \tag{2.1.16}$$

or, transposing the last term,

$$\mathbf{M}(a', a'')\mathbf{M}(a''', a^{iv}) - \delta(a'', a''')\mathbf{M}(a', a^{iv}) = \sum_{\substack{k,l \\ (k\neq l)}} \mathcal{M}(a'_k, a''_k)\mathcal{M}(a'''_l, a^{iv}_l). \tag{2.1.17}$$

If we now interchange the roles $a', a'' \longleftrightarrow a''', a^{iv}$ and subtract the result from Eq. (2.1.17), the right-hand side vanishes by Eq. (2.1.15), and Eq. (2.1.14) is obtained.

Considering next a property $F^{(2)}$ that is composed of analogous contributions $f^{(2)}(\chi_k, \chi_l)$ from each pair of subsystems ($k \neq l$), a straightforward generalization involving direct product spaces for pairs of particles with basis vectors $|a'_k, a'''_l\rangle \equiv |a'_k\rangle\,|a'''_l\rangle$, etc., yields $f^{(2)}(\chi_k, \chi_l)$ in the form

$$f^{(2)}(\chi_k, \chi_l) = \sum_{a'_k, a''_k, a'''_l, a^{iv}_l} \langle a'_k, a'''_l |f^{(2)}(\chi_k, \chi_l)| a''_k, a^{iv}_l\rangle\, \mathcal{M}(a'_k, a'''_l; a''_k, a^{iv}_l), \tag{2.1.18}$$

where we have defined

$$\mathcal{M}(a'_k, a'''_l; a''_k, a^{iv}_l) \equiv \mathcal{M}(a'_k, a''_k)\mathcal{M}(a'''_l, a^{iv}_l), \tag{2.1.19}$$

which yields an identity when substituted into Eq. (2.1.18). In view of the fact that subsystems/particles are indistinguishable and the matrix elements are independent of particle indices, we may write

$$\begin{aligned}\langle a'_k, a'''_l |f^{(2)}(\chi_k, \chi_l)| a''_k, a^{iv}_l\rangle &= \langle a', a''' |f^{(2)}| a'', a^{iv}\rangle \\ &= \langle a''', a' |f^{(2)}| a^{iv}, a''\rangle,\end{aligned} \tag{2.1.20}$$

where the last equality reflects the indistinguishability of subsystems interchanged in pairs. Thus, $F^{(2)}$ takes the form

$$F^{(2)} = \sum_{a',a'',a''',a^{iv}} \left\langle a', a''' \left| f^{(2)} \right| a'', a^{iv} \right\rangle \frac{1}{2!} \sum_{(k \neq l)} \mathcal{M}(a'_k, a''_k) \mathcal{M}(a'''_l, a^{iv}_l). \tag{2.1.21}$$

Eq. (2.1.17) may be used to rewrite $F^{(2)}$ in terms of total measurement symbols, as

$$F^{(2)} = \sum_{a',a'',a''',a^{iv}} \left\langle a', a''' \left| f^{(2)} \right| a'', a^{iv} \right\rangle \times \frac{1}{2!} \left[\mathbf{M}(a', a'') \mathbf{M}(a''', a^{iv}) - \delta(a'', a''') \mathbf{M}(a', a^{iv}) \right]. \tag{2.1.22}$$

Similarly, $F^{(m)}$ can be written in terms of total measurement symbols alone.

2.2 Second Quantization: Occupation Number Representation

Recalling the interpretation of the single-particle measurement symbol $\mathcal{M}(a'_k, b'_k) = \mathcal{M}(a'_k, 0)\mathcal{M}(0, b'_k)$ as a sequence of operations in which the kth subsystem/particle is annihilated in state $|b'_k\rangle$, followed by the creation/emission of the kth subsystem in state $|a'_k\rangle$ with the original amplitude of the state $|b'_k\rangle$, it is clear that the total measurement symbol,

$$\mathbf{M}(a', b') = \sum_{k=1}^{n} \mathcal{M}(a'_k, b'_k), \tag{2.2.1}$$

has a similar interpretation. It operates on a system by removing/annihilating a subsystem in *mode* b', and creates a subsystem in *mode* a', without distinguishing which particular subsystem/particle is annihilated/created, since the contributions of all subsystems to $\mathbf{M}(a', b')$ in Eq. (2.2.1) are equally weighted and symmetrical. Moreover, the sum $\sum_{k=1}^{n}$ in Eq. (2.2.1) signifies the "either/or" logic of addition, so that just one subsystem undergoes the sequence of annihilation and creation operations, anonymously.

The interpretive scheme described previously can be given mathematical expression by introducing "second quantized" field operators $\psi^+(a')$, $\psi(b')$, such that $\mathbf{M}(a', b')$ is factored using the two operators as

$$\mathbf{M}(a', b') = \psi^+(a')\psi(b'), \tag{2.2.2}$$

where $\psi(b')$ is defined to annihilate a subsystem in mode b' and $\psi^+(a')$ is defined to create a subsystem in mode a'. This decomposition of $\mathbf{M}(a', b')$ is done in analogy to Eq. (1.4.8), with interpretations of $\psi(b')$ and $\psi^+(a')$ fully analogous to the interpretations of $\mathcal{M}(0, b')$ and $\mathcal{M}(a', 0) = \mathcal{M}^+(0, a')$, respectively, except that in the present context the indistinguishability of particles means that it is not possible to determine which particular subsystems are annihilated/created, since all of them are represented with equal weight in $\mathbf{M}(a', b')$. The transformation properties of $\mathbf{M}(a', b')$ under a change of representation are readily obtained as

$$\mathbf{M}(c', d') = \sum_{a', b'} \langle a' | c' \rangle \mathbf{M}(a', b') \langle d' | b' \rangle, \tag{2.2.3}$$

and they lead at once to the corresponding transformation properties for $\psi(a'), \psi^+(b')$, as

$$\psi(a') = \sum_{b'} \langle a' | b' \rangle \psi(b')$$
$$\psi^+(b') = \sum_{a'} \psi^+(a') \langle a' | b' \rangle, \tag{2.2.4}$$

since $\langle a' | b' \rangle = \langle b' | a' \rangle^*$. Moreover, since $F^{(1)}$ may be written in a mixed representation as

$$F^{(1)} = \sum_{a', b'} \langle a' | f^{(1)} | b' \rangle \mathbf{M}(a', b'), \tag{2.2.5}$$

we have

$$F^{(1)} = \sum_{a', b'} \psi^+(a') \langle a' | f^{(1)} | b' \rangle \psi(b'). \tag{2.2.6}$$

It is necessary to recognize that the mathematical implementation of this interpretive scheme requires the introduction of a Fock or a "number" space,[2] in which particle numbers may change with the annihilation or creation of subsystems. The description of a state, correspondingly, is given in terms of the occupation numbers of given subsystem modes and associated linear superpositions. For example, instead of describing a state of non-interacting particles by symmetrized/antisymmetrized products of the wave functions of n particles (Eq. (1.1.42)), the number space description is lodged merely in terms of the occupation numbers of the constituent subsystem modes.

The realization of the annihilation and creation properties of $\psi(a')$, $\psi^+(a')$ in terms of number space operations may be discussed by starting with Eq. (2.1.11),

$$\sum_{a'} \mathbf{M}(a', a') = \mathbf{N} = \sum_{a'} \psi^+(a') \psi(a'), \tag{2.2.7}$$

where $\mathbf{N}$ is the operator whose eigenvalues are the total (integer) number of particles, n, in the system, with eigenvector $|n\rangle$, $\mathbf{N}|n\rangle = n|n\rangle$. Moreover, the operator for the number of particles having eigenvalue a' of the A spectrum is $\psi^+(a')\psi(a')$. In position representation (and taking account of internal coordinates ξ, like spin),

$$\mathbf{N} = \int d^{(3)}\mathbf{x} \sum_{\xi} \psi^+(\mathbf{x}, \xi) \psi(\mathbf{x}, \xi). \tag{2.2.8}$$

Clearly, the operator for the particle density with spin ξ is just the integrand,

$$\rho(\mathbf{x}, \xi) = \psi^+(\mathbf{x}, \xi)\psi(\mathbf{x}, \xi). \tag{2.2.9}$$

Considering the annihilation properties of $\psi(a')$, the state $\psi(a')\,|n\rangle$ must have one less particle than does $|n\rangle$, so its number eigenvalue is reduced by one:

$$\mathbf{N}\psi(a')\,|n\rangle = (n-1)\psi(a')\,|n\rangle \tag{2.2.10}$$

(this state, $\psi(a')\,|n\rangle$, is *not* normalized to unity). Noting that $\mathbf{N}\,|n\rangle = n\,|n\rangle$, this relation may be alternatively written as

$$\mathbf{N}\psi(a')\,|n\rangle = \psi(a')(\mathbf{N}-\mathbf{I})\,|n\rangle\,. \tag{2.2.11}$$

Since this is true for every value of n, we have

$$\mathbf{N}\psi(a') = \psi(a')(\mathbf{N}-\mathbf{I}), \tag{2.2.12}$$

and, similarly, since the creation operator $\psi^+(a')$ increases by one the number of particles of the state $|n\rangle$ upon which it acts, we also have

$$\mathbf{N}\psi^+(a') = \psi^+(a')(\mathbf{N}+\mathbf{I}). \tag{2.2.13}$$

Eqns. (2.2.12) and (2.2.13) may be written as

$$\begin{aligned}\left[\psi(a'), \mathbf{N}\right] &= \psi(a'),\\ \left[\psi^+(a'), \mathbf{N}\right] &= -\psi^+(a').\end{aligned} \tag{2.2.14}$$

Since $\mathbf{N} = \mathbf{N}^+$ is Hermitian, the two equations in Eq. (2.2.14) are just Hermitian adjoints of each other. These considerations are easily generalized to arbitrary operator functions of $\mathbf{N}$, $f(\mathbf{N})$:

$$\begin{aligned}f(\mathbf{N})\psi(a') &= \psi(a')f(\mathbf{N}-\mathbf{I}),\\ f(\mathbf{N})\psi^+(a') &= \psi^+(a')f(\mathbf{N}+\mathbf{I}).\end{aligned} \tag{2.2.15}$$

An important special case is the unitary operator $e^{i\phi\mathbf{N}}$ (ϕ is a real phase):

$$e^{-i\phi\mathbf{N}}\psi(a') = \psi(a')e^{-i\phi(\mathbf{N}-\mathbf{I})}, \tag{2.2.16}$$

whence

$$e^{-i\phi\mathbf{N}}\psi(a')e^{i\phi\mathbf{N}} = e^{i\phi}\psi(a'), \tag{2.2.17}$$

and taking the adjoint,

$$e^{-i\phi \mathrm{N}} \psi^+(a') e^{i\phi \mathrm{N}} = e^{-i\phi} \psi^+(a'). \tag{2.2.18}$$

If we consider the addition of two particles to the system, one in state a' and one in state a'', we have

$$\psi^+(a')\psi^+(a'')\,|n\rangle = |n + 1_{a'} + 1_{a''}\rangle \text{ (unnormalized)} \tag{2.2.19}$$

(normalization will be discussed below). Adding the particles in the reverse order,

$$\psi^+(a'')\psi^+(a')\,|n\rangle = |n + 1_{a''} + 1_{a'}\rangle \text{ (unnormalized)} \tag{2.2.20}$$

must have the same physical result, so the states on the right-hand sides of the Eqns. (2.2.19) and (2.2.20) can differ only by a phase factor, $e^{i\theta}$. Correspondingly, the left-hand sides of these equations are related by

$$\psi^+(a')\psi^+(a'') = e^{i\theta} \psi^+(a'')\psi^+(a'). \tag{2.2.21}$$

(Note that any dependence of the phase factor on a', a'' is ruled out by the requirement that the result be consistent with the known transformation properties of ψ^+, ψ in Eq. (2.2.4).) Clearly, a second application of Eq. (2.2.21) to bring the order of $\psi^+(a'')\psi^+(a')$ on the right back to $\psi^+(a')\psi^+(a'')$ invokes a second phase factor, so

$$e^{2i\theta} = 1 \ ; \ e^{i\theta} = \pm 1 \tag{2.2.22}$$

is required for consistency. Therefore, there are two types of systems:

(a) First kind, $\pm \to +$,

$$\left[\psi(a'), \psi(a'')\right] = \left[\psi^+(a'), \psi^+(a'')\right] = 0, \tag{2.2.23}$$

where $[A, B] \equiv (AB - BA)$ is the commutator of A, B.

(b) Second kind, $\pm \to -$,

$$\left\{\psi(a'), \psi(a'')\right\} = \left\{\psi^+(a'), \psi^+(a'')\right\} = 0, \tag{2.2.24}$$

where $\{A, B\} \equiv (AB + BA)$ is the *anti*commutator of A, B.

Furthermore, considering Eqns. (2.2.14), we have

$$\left[\psi(a'), \sum_{a''} \psi^+(a'')\psi(a'')\right] = \psi(a'), \tag{2.2.25}$$

or

$$\sum_{a''} \left(\psi(a')\psi^+(a'')\psi(a'') - \psi^+(a'')\psi(a'')\psi(a')\right) = \psi(a'), \tag{2.2.26}$$

and using the commutation/anticommutation relations, Eqns. (2.2.23) and (2.2.24), we have

$$\sum_{a''} \{[\psi(a'), \psi^+(a'')]\}\, \psi(a'') = \psi(a'). \tag{2.2.27}$$

Since this is true for all a', the commutator $[\ldots]$ or anticommutator $\{\ldots\}$ in the summand on the left must be $\delta(a', a'')$:

$$\{[\psi(a'), \psi^+(a'')]\} = \delta(a', a''), \tag{2.2.28}$$

where $\delta(a', a'')$ is the Kronecker delta for a discrete spectrum and represents the Dirac δ-function for a continuum (corresponding to $\sum_{a''} \Rightarrow \int da''$ in Eq. (2.2.27)). This result shows that $\psi(a')$ and $\psi^+(a'')$ qualify as canonically paired *non-Hermitian* operator variables of both the first and second kinds. It will be seen in Chapter 6, Section 6.2, that particles of the first kind obey Bose–Einstein statistics and are called "Bosons," while those of the second kind obey Fermi–Dirac statistics and are called "Fermions."

Having expressed $F^{(1)}$ operators in terms of creation and annihilation operators (Eq. (2.2.6)), we need to do the same for $F^{(2)}$. Starting with Eq. (2.1.22), the factorization of Eq. (2.2.2) implies that

$$\begin{aligned} &\mathbf{M}(a', a'')\mathbf{M}(a''', a^{iv}) - \delta(a'', a''')\mathbf{M}(a', a^{iv}) \\ &= \psi^+(a')\psi(a'')\psi^+(a''')\psi(a^{iv}) - \delta(a'', a''')\psi^+(a')\psi(a^{iv}). \end{aligned} \tag{2.2.29}$$

Applying the canonical commutation/anticommutation relations, Eq. (2.2.28), to $\psi(a'')\psi^+(a''')$ results in

$$\mathbf{M}(a', a'')\mathbf{M}(a''', a^{iv}) - \delta(a'', a''')\mathbf{M}(a', a^{iv}) = \pm\psi^+(a')\psi^+(a''')\psi(a'')\psi(a^{iv}), \tag{2.2.30}$$

and the $\pm$ sign is absorbed using Eqns. (2.2.23) and (2.2.24) with the interchange of the last two factors, so that

$$\mathbf{M}(a', a'')\mathbf{M}(a''', a^{iv}) - \delta(a'', a''')\mathbf{M}(a', a^{iv}) = \psi^+(a')\psi^+(a''')\psi(a^{iv})\psi(a''). \tag{2.2.31}$$

Therefore,

$$F^{(2)} = \frac{1}{2!} \sum_{a', a'', a''', a^{iv}} \psi^+(a')\psi^+(a''') \left\langle a', a''' \right| f^{(2)} \left| a'', a^{iv} \right\rangle \psi(a^{iv})\psi(a'') \tag{2.2.32}$$

and similar results can be obtained for $F^{(m)}$.

Considering the number operator, Eq. (2.2.7), we write $\mathbf{N} = \sum_{a'} N(a')$ and interpret the summand

$$N(a') = \psi^+(a')\psi(a') \tag{2.2.33}$$

as the number operator for the number of subsystems/particles in mode a' of the A spectrum. Here, the commutation/anticommutation relations have important implications. For systems of the first kind (Bosons), $N(a')$ may assume eigenvalues $0, 1, 2, \ldots \infty$. However, for systems of the second kind (Fermions), $N(a')$ can only assume the eigenvalues 0 or 1, since in the latter case, the anticommutation relation (Eq. (2.2.28)) yields

$$\begin{aligned}\left[N(a')\right]^2 &= N(a')N(a') = \psi^+(a')\psi(a')\psi^+(a')\psi(a') \\ &= -\left(\psi^+(a')\right)^2\left(\psi(a')\right)^2 + N(a').\end{aligned} \tag{2.2.34}$$

However, Eq. (2.2.24) has the immediate implication that $(\psi^+(a'))^2 = (\psi(a'))^2 = 0$, so that

$$\left[N(a')\right]^2 = N(a'), \tag{2.2.35}$$

and the only possible eigenvalues of $N(a')$ are 0 and 1, which assures conformance with the Pauli exclusion principle for Fermions: That two electrons *cannot* occupy the same microscopic quantum state.

The description of a state in the number space under consideration (occupation number representation) involves specification of the occupation numbers, $\{n_{a'}\}$, of the subsystem modes $\{a'\}$, which are eigenvalues of $N(a') = \psi^+(a')\psi(a')$. For such states, we have

$$|\{n_{a'}\}\rangle = |n_{a'_1}, n_{a'_2} \ldots\ldots n_{a'_k} \ldots\ldots >, \tag{2.2.36}$$

with

$$N(a'_k)|\{n_{a'}\}\rangle = n_{a'_k}|\{n_{a'}\}\rangle, \tag{2.2.37}$$

and the total number is given by $n = \sum_k n_{a'_k}$:

$$\mathbf{N}|\{n_{a'}\}\rangle = \sum_k n_{a'_k}|\{n_{a'}\}\rangle = n|\{n_{a'}\}\rangle. \tag{2.2.38}$$

Defining the *vacuum* state as having null particle occupation,

$$|0\rangle = |\{0_{a'}\}\rangle; \qquad N(a'_k)|0\rangle = 0|0\rangle \text{ for all } k, \tag{2.2.39}$$

it has the property that the action of any annihilation operator on it yields a null result,

$$\psi(a'_k)|0\rangle \equiv 0, \text{ (for all } k), \tag{2.2.40}$$

since there is no state having a negative number of particles. The construction of a state with one particle may be undertaken as (C is a normalization constant):

$$|0 + 1_{a'_k}\rangle = C\psi^+(a'_k)|0\rangle, \tag{2.2.41}$$

and its single-particle occupation is readily verified by application of $N(a'_k)$:

$$\begin{aligned} N(a'_k)|0 + 1_{a'_k}\rangle &= C\psi^+(a'_k)\psi(a'_k)\psi^+(a'_k)|0\rangle \\ &= C\psi^+(a'_k)\left(1 \pm \psi^+(a'_k)\psi(a'_k)\right)|0\rangle, \end{aligned} \tag{2.2.42}$$

where we have used the commutation/anticommutation relations. The last term on the right vanishes since the number of particles in $|0\rangle$ vanishes, leaving

$$N(a'_k)|0 + 1_{a'_k}\rangle = C\psi^+(a'_k)|0\rangle = |0 + 1_{a'_k}\rangle, \tag{2.2.43}$$

with eigenvalue 1 for single-particle occupation. In regard to normalization of the state $|0 + 1_{a'_k}\rangle$, we again use the canonical commutation/anticommutation relations, verifying that $|C| = 1$ if $|0\rangle$ is normalized to unity. The construction of a state of n particles from one of $(n-1)$ particles may be accomplished as (K is a normalization constant):

$$|n\rangle = |1_{a'_k} + (n-1)\rangle = K\psi^+(a'_k)|n-1\rangle, \tag{2.2.44}$$

$$\mathbf{N}|n\rangle = \mathbf{N}K\psi^+(a'_k)|n-1\rangle = K\psi^+(a'_k)(\mathbf{N}+1)|n-1\rangle = n|n\rangle. \tag{2.2.45}$$

For the Boson case, n may be any integer, whereas for Fermions $n = 0, 1$ only. If the state $|n-1\rangle$ is normalized to unity and has $(n_k - 1) \leq (n-1)$ population in state a'_k (with the population difference $(n-n_k)$ in other states), then unit normalization for $|n\rangle$ requires that $1 = \langle n|n\rangle = |K|^2\langle n-1|\psi(a'_k)\psi^+(a'_k)|n-1\rangle$; and using the canonical commutation relation once more, we have

$$1 = \langle n|n\rangle = |K|^2 \left\langle n-1|(1 \pm N(a'_k))|n-1\right\rangle = |K|^2\,[1 \pm (n_k - 1)]\,. \tag{2.2.46}$$

For Bosons $|K| = \frac{1}{\sqrt{n_k}}$, so that the application of $\psi^+(a'_k)$ to add a particle carries a normalization factor $\frac{1}{\sqrt{n_k}}$,

$$|n\rangle = |1_{a'_k} + (n-1)\rangle = \frac{1}{\sqrt{n_k}}\psi^+(a'_k)|n-1\rangle. \tag{2.2.47}$$

In the case of Fermions, the population restriction of mode k in the lower state $|n-1\rangle$ to 0 or 1 means that the addition of one more particle can only result in mode k population of

1 in the upper state $|n\rangle$ (or nullification of the upper state if it would produce a prohibited double occupation). Thus for non-null Fermion states, $K = 1$.

Writing the $F^{(1)}$ and $F^{(2)}$ operators (Eqns. (2.2.6) and (2.2.32)) in terms of creation and annihilation operators, together with the commutation/anticommutation relations (Eqns. (2.2.23), (2.2.24) and (2.2.28)), constitutes a "second quantized" formulation of quantum mechanics. This is, in fact, a misnomer, as there is really only one stage of quantization, which is already implicit in the single-particle Schrödinger equation (Eq. (1.1.1)). However, the "second quantized" description offers great convenience in dealing with multiparticle and interacting quantum systems, and it will be employed extensively in this book. It should be noted that our formulation is in terms of an arbitrary one-particle representation, $A \to a'$, and that any other representation can be obtained using the transformation equations, Eq. (2.2.4), for a change of basis.

Finally, considering the Hamiltonian, $H = H^{(1)} + H^{(2)}$, which generates the time development of the quantum system, in position representation using Eqns. (2.2.6) and (2.2.32), the "one-particle" part is written as ($\sum_{a'} \to \int d^{(3)}\mathbf{x}'$, etc.):

$$F^{(1)} \to H^{(1)} = \int d^{(3)}\mathbf{x}' \int d^{(3)}\mathbf{x}'' \psi^+(\mathbf{x}') \left\langle \mathbf{x}' \right| h^{(1)} \left| \mathbf{x}'' \right\rangle \psi(\mathbf{x}''), \tag{2.2.48}$$

and with Eq. (1.2.24),

$$\left\langle \mathbf{x}' \right| h^{(1)} \left| \mathbf{x}'' \right\rangle = \delta^{(3)}(\mathbf{x}' - \mathbf{x}'') h^{(1)}(\mathbf{x}', \tfrac{\hbar}{i}\nabla'), \tag{2.2.49}$$

we have

$$H^{(1)} = \int d^{(3)}\mathbf{x}' \psi^+(\mathbf{x}') h^{(1)}(\mathbf{x}', \tfrac{\hbar}{i}\nabla') \psi(\mathbf{x}'). \tag{2.2.50}$$

Furthermore, for $H^{(2)}$ we have

$$\begin{aligned} H^{(2)} = \frac{1}{2!} \int d^{(3)}\mathbf{x}' \int d^{(3)}\mathbf{x}'' \int d^{(3)}\mathbf{x}''' \int d^{(3)}\mathbf{x}^{iv} \\ \psi^+(\mathbf{x}')\psi^+(\mathbf{x}''') \left\langle \mathbf{x}', \mathbf{x}''' \right| h^{(2)} \left| \mathbf{x}'', \mathbf{x}^{iv} \right\rangle \psi(\mathbf{x}^{iv})\psi(\mathbf{x}''), \end{aligned} \tag{2.2.51}$$

and noting that for a two-particle interaction potential $\mathcal{V}$

$$\left\langle \mathbf{x}', \mathbf{x}''' \right| h^{(2)} \left| \mathbf{x}'', \mathbf{x}^{iv} \right\rangle = \mathcal{V}(\mathbf{x}' - \mathbf{x}^{iv})\delta^{(3)}(\mathbf{x}''' - \mathbf{x}^{iv})\delta^{(3)}(\mathbf{x}' - \mathbf{x}''), \tag{2.2.52}$$

we have

$$H^{(2)} = \frac{1}{2!} \int d^{(3)}\mathbf{x}' \int d^{(3)}\mathbf{x}'' \psi^+(\mathbf{x}')\psi^+(\mathbf{x}'')\mathcal{V}(\mathbf{x}' - \mathbf{x}'')\psi(\mathbf{x}'')\psi(\mathbf{x}'). \tag{2.2.53}$$

(Here, we suppress notation indicating sums over internal variables ξ, such as spin.) Moreover, the number operator in position representation is

$$\mathbf{N} = \int d^{(3)}\mathbf{x}'\psi^+(\mathbf{x}')\psi(\mathbf{x}'). \tag{2.2.54}$$

Similarly, the many-particle momentum operator **P** may be written in position representation as (a $\frac{1}{2}$-factor and Hermitian conjugate are introduced to assure Hermiticity):

$$\mathbf{P} = \frac{1}{2}\left[\int d^{(3)}\mathbf{x}'\psi^+(\mathbf{x}')\frac{\hbar}{i}\nabla\,\psi(\mathbf{x}') + \text{Hermitian conjugate}\right] \tag{2.2.55}$$

and the multiparticle spatial translation operator for a displacement **a** is given by $\exp(\frac{i}{\hbar}\mathbf{a}\cdot\mathbf{P})$ (supplanting the single-particle translation operator $\exp(\frac{i}{\hbar}\mathbf{a}\cdot\ \mathbf{p})$ of Eq. (1.1.7)). Noting that its action on a multiparticle state $|n\rangle$, $e^{(\frac{i}{\hbar}\mathbf{a}\cdot\mathbf{P})}\,|n\rangle$, has the effect of translating the origin of spatial coordinates by **a**, we determine its action on an operator O by examining the matrix element $\langle n'\,|O|\,n\rangle$ as

$$\langle n'\,|O|\,n\rangle = \langle n'\,|IOI|\,n\rangle = \left\langle n'\left|e^{-\frac{i}{\hbar}\mathbf{a}\cdot\mathbf{P}}e^{\frac{i}{\hbar}\mathbf{a}\cdot\mathbf{P}}Oe^{-\frac{i}{\hbar}\mathbf{a}\cdot\mathbf{P}}e^{\frac{i}{\hbar}\mathbf{a}\cdot\mathbf{P}}\right|n\right\rangle, \tag{2.2.56}$$

finding that the role of O is played by $\widetilde{O}$,

$$\widetilde{O} = e^{\frac{i}{\hbar}\mathbf{a}\cdot\mathbf{P}}Oe^{-\frac{i}{\hbar}\mathbf{a}\cdot\mathbf{P}}, \tag{2.2.57}$$

with respect to the space-origin-translated states $|\widetilde{n}\rangle$,

$$|\widetilde{n}\rangle = e^{\frac{i}{\hbar}\mathbf{a}\cdot\mathbf{P}}\,|n\rangle \quad ; \quad \langle\widetilde{n'}| = \langle n'|e^{-\frac{i}{\hbar}\mathbf{a}\cdot\mathbf{P}}. \tag{2.2.58}$$

In particular, for the annihilation operator $\psi(\mathbf{x})$, we obtain

$$\widetilde{\psi}(\mathbf{x}) \equiv \psi(\mathbf{x}+\mathbf{a}) = e^{\frac{i}{\hbar}\mathbf{a}\cdot\mathbf{P}}\psi(\mathbf{x})e^{-\frac{i}{\hbar}\mathbf{a}\cdot\mathbf{P}}. \tag{2.2.59}$$

Moreover, for infinitesimal displacements, $\delta\mathbf{a}$, we have

$$\begin{aligned}\widetilde{O} - O \equiv \delta_{\mathbf{a}}\widetilde{O} &= \delta_{\mathbf{a}}(e^{\frac{i}{\hbar}\mathbf{a}\cdot\mathbf{P}}Oe^{-\frac{i}{\hbar}\mathbf{a}\cdot\mathbf{P}})\\ &= \delta_{\mathbf{a}}(e^{\frac{i}{\hbar}\mathbf{a}\cdot\mathbf{P}})Oe^{-\frac{i}{\hbar}\mathbf{a}\cdot\mathbf{P}} + e^{\frac{i}{\hbar}\mathbf{a}\cdot\mathbf{P}}O\delta_{\mathbf{a}}(e^{-\frac{i}{\hbar}\mathbf{a}\cdot\mathbf{P}})\\ &= \frac{i}{\hbar}\delta\mathbf{a}\cdot\mathbf{P}\widetilde{O} - \frac{i}{\hbar}\widetilde{O}\delta\mathbf{a}\cdot\mathbf{P} = -\frac{i}{\hbar}\left[\widetilde{O},\delta\mathbf{a}\cdot\mathbf{P}\right],\end{aligned} \tag{2.2.60}$$

from which the role of **P** as the generator of spatial translations in operators is clear. Moreover, the total many-particle angular momentum operator may be written in position representation as

$$\vec{\mathcal{J}} = \mathbf{L} + \mathbf{S} = \frac{1}{2}\int d^3\mathbf{x}'\psi^+(\mathbf{x}')\left[\mathbf{x}' \times \frac{\hbar}{i}\nabla' + \frac{\hbar}{2}\boldsymbol{\sigma}\right]\psi(\mathbf{x}') + \text{Hermitian conjugate}, \quad (2.2.61)$$

where $\psi(\mathbf{x}')$ is understood as a two-component column vector in spin-space and $\psi^+(\mathbf{x}')$ is its row-vector complex-conjugate adjoint. Of course, the first and second terms in the bracket on the right of Eq. (2.2.61) correspond to orbital and spin angular momentum operators, respectively, generalized to n-particle systems. The role of these many-particle angular momentum operators in generating spatial rotations in operators may be exhibited as above. Last, but not least, it is important to write the charge-current density operator corresponding to Eq. (1.1.31) for an n-particle system in second quantized form in position representation:

$$\mathbf{J}(\mathbf{x}) = \frac{\hbar Q}{2mi}\lim_{\mathbf{x}'\to\mathbf{x}}\left[(\nabla - \nabla')\,\psi^+(\mathbf{x}',t)\psi(\mathbf{x},t)\right] - \frac{Q^2}{mc}\mathbf{A}(\mathbf{x},t)\psi^+(\mathbf{x},t)\psi(\mathbf{x},t). \quad (2.2.62)$$

2.3 Coherent States, Squeezed States

While states involving small numbers of quanta of the electromagnetic field (photons) have been central to the successful development of quantum electrodynamics, Glauber[3–5] pointed out that superpositions of states having large numbers of photons are advantageous for the description of optical phenomena. In this regard, he defined coherent states as those that cause the correlation functions of the electromagnetic field to factorize. To explore this matter further, we start with an examination of the vacuum state $|0\rangle$ defined by $\psi(a_k')|0\rangle = 0$ for all k (Eq. (2.2.40)). As noted in connection with Eq. (2.2.28), $\psi(a')$ and $\psi^+(a'')$ are canonically paired *non*-Hermitian variables, but we can introduce corresponding Hermitian variables q and p that are canonically paired as (suppressing the index a', etc.):

$$\psi = \frac{1}{\sqrt{2}}(q + ip); \quad \psi^+ = \frac{1}{\sqrt{2}}(q - ip), \quad (2.3.1)$$

so that

$$q = \frac{1}{\sqrt{2}}(\psi + \psi^+) = q^+; \quad p = \frac{1}{i\sqrt{2}}(\psi - \psi^+) = p^+, \quad (2.3.2)$$

which obey the canonical commutation relations for bosons associated with photon quanta of the electromagnetic field,

$$[q(a'), q(a'')] = [p(a'), p(a'')] = 0; \quad [q(a'), p(a'')] = i\delta(a', a''). \quad (2.3.3)$$

Accordingly, the vacuum state is defined by

$$\psi\,|0\rangle = \frac{1}{\sqrt{2}}\,(q + ip)\,|0\rangle = 0. \tag{2.3.4}$$

Bearing in mind that in a q– representation we have $ip \rightarrow \partial/\partial q$, then the vacuum state in q– representation ($\langle q|0\rangle$) obeys the equation (restore index $a' \rightarrow k$):

$$q_k\,\langle q_k|0\rangle + \frac{\partial}{\partial q_k}\,\langle q_k|0\rangle = 0, \tag{2.3.5}$$

which has the obvious solution

$$\langle q_k|0\rangle = Aexp\Big[-\tfrac{1}{2}\sum_{k'} q_{k'}^2\Big], \tag{2.3.6}$$

where the normalization is given by $A = \pi^{-n/2}$, n being the number of modes.

Thus, the vacuum state is a non-trivial right eigenvector of the annihilation operator ψ having zero eigenvalue. On the other hand, the adjoint of Eq. (2.3.4),

$$\langle 0|\psi^+ = \langle 0|\frac{1}{\sqrt{2}}\,(q - ip) = 0, \tag{2.3.7}$$

confirms the existence of a left vacuum eigenvector with eigenvalue zero for the creation operator. However, it is readily seen that a corresponding right vacuum eigenvector with eigenvalue zero for $\psi^+|0\rangle$ does not exist, since

$$\psi^+|0\rangle = \frac{1}{\sqrt{2}}\,(q - ip)\,|0\rangle = 0, \text{or, equivalently,} \quad q_k\,\langle q_k|0\rangle = \frac{\partial}{\partial q_k}\,\langle q_k|0\rangle \tag{2.3.8}$$

would yield

$$\langle q_k|0\rangle = Aexp\Big[\tfrac{1}{2}\sum_{k} q_k^2\Big], \tag{2.3.9}$$

which *cannot* be normalized. Similarly, no left eigenvector of the annihilation operator with eigenvalue zero, $\langle 0|\psi$, exists. The non-existence of $\psi^+|0\rangle$ and $\langle 0|\psi$ can also be inferred from the Boson canonical commutation relations: The absence of such eigenstates is associated with the *non*-Hermiticity of the variables involved. It should be noted that the vacuum right eigenstate of the annihilation operator, Eq. (2.3.6), is just the ground state of decoupled simple harmonic oscillators.

Seeking other eigenvectors of the annihilation operator, we define "displaced" operators as

$$'\psi = \psi - \lambda \quad \text{and} \quad '\psi^+ = \psi^+ - \lambda^* \tag{2.3.10}$$

where λ is an arbitrary complex number. It is readily verified that $'\psi$ and $'\psi^+$ obey the same canonical commutation relations obeyed by ψ and ψ^+;

$$\left[{}'\psi\left(a'\right), {}'\psi^+\left(a''\right)\right] = \left[\psi\left(a'\right) - \lambda, \psi^+\left(a''\right) - \lambda^*\right] = \left[\psi\left(a'\right), \psi^+\left(a''\right)\right] = \delta\left(a', a'\right). \quad (2.3.11)$$

Following the same procedures as above, this implies the existence of a right eigenvector for the displaced annihilation operator $'\psi(a') \equiv {}'\psi(k)$:

$$'\psi(k)|\lambda_k\rangle = 0; \quad (\psi(k) - \lambda_k)\,|\lambda_k\rangle = 0 \qquad (2.3.12)$$

which assures us of the existence of a right eigenvector of ψ,

$$\psi(k)|\lambda_k\rangle = \lambda_k|\lambda_k\rangle, \qquad (2.3.13)$$

for *any* complex eigenvalue λ_k. In $q-$ representation, this takes the form

$$(q_k - \lambda_k)\,\langle q_n|\lambda_k\rangle + \frac{\partial}{\partial q_k}\langle q_k|\lambda_k\rangle = 0 \qquad (2.3.14)$$

corresponding to the ground state of decoupled displaced harmonic oscillators:

$$\langle q_k|\lambda_k\rangle = Aexp\Big[-\tfrac{1}{2}\sum_{k'}(q_{k'} - \lambda_{k'})^2\Big]. \qquad (2.3.15)$$

Similarly, we are assured of the existence of a left eigenvector of the creation operator

$$\langle\lambda_k|\psi^+(k) = \langle\lambda_k|\lambda_k^* \qquad (2.3.16)$$

for any complex number λ_k^*.

An alternative discussion of the displaced states is presented in Reference 3, Section 2.5, with the introduction of a displacement operator $D(\lambda_k)$ defined as

$$D(\lambda_k) = exp\left(\lambda_k\psi^+(k) - \lambda_k^*\psi(k)\right), \qquad (2.3.17)$$

which is shown to have the property

$$D(\lambda_k)|0\rangle_k = |\lambda_k\rangle_k. \qquad (2.3.18)$$

(Some insight into the structure of $D(\lambda_k)$ may be gained by observing that since the canonical momentum p corresponds to $-i\partial/\partial q$, the corresponding $q-$ translation operator (Eq. (1.1.7)) may be viewed as

$$exp(q\partial/\partial q) \rightarrow exp(i\lambda p) \rightarrow exp(\lambda_k\psi^+(k)), \qquad (2.3.19)$$

with a similar conjugate expression for the corresponding p– translation operator. The proof of Eq. (2.3.18) given in Reference 3, Section 2.5, is very clear and illuminating. The reader will also find it informative about operator algebra, particularly the Campbell–Baker–Hausdorff identity, which, for the case at hand, takes the form

$$e^A e^B = e^{A+B+[A,B]} \tag{2.3.20}$$

since A and B both commute with $[A, B]$ in the present context.[6,7]

Having analyzed the vacuum state, we can now construct the excited multiparticle number states in terms of it by repeated applications of $\psi^+(k)$– creation operators on the vacuum state using Eq. (2.2.47) with the normalized result:

$$|n_k\rangle_k = \left(\frac{(\psi^+(k))^{n_k}}{(n_k!)^{\frac{1}{2}}}\right)|0\rangle_k \qquad (n_k = 0, 1, 2, 3 \ldots). \tag{2.3.21}$$

Applying $\langle\lambda_k|$ on the left of Eq. (2.3.21), we have

$$\begin{aligned}\langle\lambda_k|n_k\rangle_k &= \langle\lambda_k|\left(\psi^+(k)\right)^{n_k}|0_k\rangle_k / (n_k!)^{\frac{1}{2}} \\ &= \left(\lambda_k^*\right)^{n_k}\langle\lambda_k|0_k\rangle_k / (n_k!)^{1/2},\end{aligned} \tag{2.3.22}$$

and taking the adjoint,

$$\langle n_k|\lambda_k\rangle_k = \lambda_k^{n_k}\langle 0_k|\lambda_k\rangle_k / (n_k!)^{\frac{1}{2}}. \tag{2.3.23}$$

With this we may identify $|\lambda_k\rangle_k$ as (using completeness of the number states):

$$|\lambda_k\rangle_k = \langle 0_k|\lambda_k\rangle_k \sum_{n_k} |n_k\rangle_k \frac{\lambda_k^{n_k}}{(n_k!)^{\frac{1}{2}}}. \tag{2.3.24}$$

Furthermore, due to the othonormality of the number states,

$$\langle\lambda_k|\lambda_k\rangle_k = |\langle 0_k|\lambda_k\rangle_k|^2 \sum_{n_k} \frac{|\lambda_k|^{2n_k}}{n_k!} = |\langle 0_k|\lambda_k\rangle|^2 e^{|\lambda_k|^2} \tag{2.3.25}$$

and with unit normalization of $|\lambda_k\rangle_k$ we can choose its phase so that

$$\langle 0|\lambda_k\rangle_k = e^{-|\lambda_k|^2/2}. \tag{2.3.26}$$

In consequence of this, we have

$$|\lambda_k\rangle_k = e^{-|\lambda_k|^2/2} \sum_{n_k} \frac{\lambda_k^{n_k}}{(n_k!)^{\frac{1}{2}}} |n_k\rangle_k \tag{2.3.27}$$

and

$$\langle\lambda_k|_k = e^{-|\lambda_k|^2/2}\sum_{n_k}\frac{\left(\lambda_k^*\right)^{n_k}}{(n_k!)^{\frac{1}{2}}}\langle n_k|_k. \tag{2.3.28}$$

The importance of these states lies in the fact that the electromagnetic field correlation functions do factorize for field eigenstates $|\ldots\rangle$ obeying the infinite succession of relations[3–5]:

$$\psi(k)|\ldots\rangle = \lambda_k|\ldots\rangle, \tag{2.3.29}$$

which were discussed above as right eigenstates of the annihilation operators. Correspondingly, the coherent states of the field are direct products of the decoupled individual states $|\lambda_k\rangle_k$,

$$|\ldots\rangle = \prod_k |\lambda_k\rangle_k. \tag{2.3.30}$$

However, these coherent states are *not* orthogonal. Forming a scalar product of two of them using Eqns. (2.3.27) and (2.3.28), we have

$$\begin{aligned}\langle\lambda_k|\beta_k\rangle &= e^{-\left(|\lambda_k|^2+|\beta_k|^2\right)/2}\sum_n\sum_m\frac{\left(\lambda_k^*\right)^n(\beta_k)^m}{(n!m!)^{\frac{1}{2}}}\langle n_k|m_k\rangle_k\\ &= e^{-\left(|\lambda_k|^2+|\beta_k|^2\right)/2}\sum_n\frac{\left(\lambda_k^*\beta_k\right)^n}{n!},\end{aligned} \tag{2.3.31}$$

where we have employed the orthonormality of the number states. Alternatively expressed, the overlap integral yields

$$|\langle\lambda_k|\beta_k\rangle|^2 = e^{-|\lambda_k-\beta_k|^2}, \tag{2.3.32}$$

clearly indicating the non-orthogonality of the coherent states. Nevertheless, these states form a *complete* set, and, in fact, are *over*complete in that an arbitrary state can be expanded in terms of them in (infinitely) many different ways. Although they are overcomplete, the coherent states satisfy the relation (I_k is the unit ket-bra dyadic for mode k):

$$\frac{1}{\pi}\int d^2\lambda_k|\lambda_k\rangle_k\langle\lambda_k|_k = \sum_{n_k}|n_k\rangle_k\langle n_k|_k = I_k \tag{2.3.33}$$

(where $\lambda_k = Re\lambda_k + iIm\lambda_k \equiv x + iy$ and $d^2\lambda_k \equiv dxdy$ with integration over the whole complex $\lambda_k(x,y)$ plane), which is in the nature of a completeness relation. This facilitates the expansion of arbitrary states and operators in terms of coherent states.[3–5]

Finally, it is important to recognize that the coherent states are states in which the variances of q_k and p_k, as defined by $(\langle \ldots \rangle \equiv \langle \lambda_k|_k \ldots |\lambda_k\rangle_k)$:

$$(\Delta q_k)^2 = \left\langle (q_k - \langle q_k \rangle)^2 \right\rangle_k = \left\langle q_k^2 \right\rangle - \langle q_k \rangle^2 ; \quad (\Delta p_k)^2 = \left\langle p_k^2 \right\rangle - \langle p_k \rangle^2 , \tag{2.3.34}$$

are equal and minimal in representing the closest approach to classical localization in accordance with the Heisenberg uncertainty principle:

$$\Delta q_k \Delta p_k = \frac{1}{2}. \tag{2.3.35}$$

This is readily verified using Eq. (2.3.15) for $\langle q_k|\lambda_k\rangle$ and its counterpart for $\langle p_k|\lambda_k\rangle$.

In comparing with the corresponding ground state result for a simple harmonic oscillator of mass m and frequency ω, it must be kept in mind that $m\omega \to 1$ and $\hbar \to 1$ here. In quantum optics[8,9] cases in which pertinent physical parameters involved permit manipulation of the states so that the q_k– variance Δq_k is reduced below $\frac{1}{\sqrt{2}}$, the state is said to be "squeezed:" Of course, the p_k– variance Δp_k is correspondingly increased to maintain the integrity of the minimum uncertainty relation, Eq. (2.3.35).

2.4 Chapter 2 Problems

P.2.1. Write the many-body Hamiltonian $H = H^{(1)} + H^{(2)}$ (Eqns. (2.2.50) and (2.2.51)) in momentum representation for $h^{(1)} \sim \nabla^2/2m$. Do the same for the number operator.

P.2.2. Prove that the number operator commutes with $H = H^{(1)} + H^{(2)}$ in position representation.

P.2.3. Verify Eqns. (2.2.3) and (2.2.4).

P.2.4. Determine the normalization constant C of Eq. (2.2.41) and also derive K of Eq. (2.2.44).

P.2.5. Write **P** (linear momentum operator), Eq. (2.2.55), and $\vec{\mathcal{J}}$ (angular momentum operator), Eq. (2.2.61), in momentum **p**-representation.

P.2.6. Consider a Fermi system having just two single-particle modes (±) and corresponding creation, annihilation operators $\psi_\pm^+$, $\psi_\pm$. Define Hermitian operator counterparts as

$$\Sigma_1^\pm = \psi_\pm + \psi_\pm^+,$$
$$\Sigma_2^\pm = i(\psi_\pm - \psi_\pm^+),$$

and also define

$$\Sigma_3^\pm = \frac{1}{i}\Sigma_1^\pm \Sigma_2^\pm .$$

Show that $\Sigma_3^{\pm} = \psi_{\pm}^{+}\psi_{\pm} - \psi_{\pm}\psi_{\pm}^{+}$ and that $\left(\Sigma_1^{\pm}\right)^2 = \left(\Sigma_2^{\pm}\right)^2 = \left(\Sigma_3^{\pm}\right)^2 = 1$. Also, verify that ($\{,\}$ denotes the anticommutator):

$$\left\{\Sigma_1^{\pm}, \Sigma_2^{\pm}\right\} = \left\{\Sigma_2^{\pm}, \Sigma_3^{\pm}\right\} = \left\{\Sigma_3^{\pm}, \Sigma_1^{\pm}\right\} = 0.$$

Prove that $N_{\pm} = \psi_{\pm}^{+}\psi_{\pm} = \frac{1}{2} + \frac{1}{2}\Sigma_3^{\pm}$ and that the only eigenvalues of $N_{\pm}$ are 1 and 0. Name a physical system to which this applies.

P.2.7. Verify Eqns. (2.2.61) and (2.2.62) and the comments between them.

P.2.8. Prove Eqns. (2.3.6, 2.3.14 and 2.3.15) including proper normalization.

P.2.9. Verify Eqns. (2.3.22–2.3.26).

P.2.10. Derive Eqns. (2.3.31–2.3.33).

P.2.11. Show that $\Delta q_k = \Delta p_k$ for the coherent states under consideration in Eqns. (2.3.15, 2.3.34 and 2.3.35).

P.2.12. Show that the field operator ψ must undergo the same phase transformation as the wave function Ψ in Eq. (1.1.32) to assure the invariance of the current density operator $\vec{J}$ (Eq. (2.2.62)) under a gauge transformation.

2.5 References for Chapter 2

R.2.1. Based on lectures of J. Schwinger on Quantum Mechanics at Harvard University, unpublished (1956).

R.2.2. An alternative discussion may be found in: A.L. Fetter and J.D. Walecka, "Quantum Theory of Many-Particle Systems," McGraw-Hill Book Co. (1971) as well as in many of the references cited in Chapter 1.

R.2.3. R.J. Glauber, Phys. Rev. *131*, 2766 (1963).

R.2.4. R.J. Glauber, Phys. Rev. Lett. *10*, 84 (1963).

R.2.5. R.J. Glauber in "Quantum Optics and Electronics," Editors: C. DeWitt, A. Blandin and C. Cohen-Tannoudji, Gordon and Breach Pub., p.63 (1965).

R.2.6. A. Messiah, "Quantum Mechanics," Vol. I, North-Holland Publishing Company, Amsterdam, p.442 (1961).

R.2.7. E. Merzbacher, "Quantum Mechanics," 3rd Edition, Wiley Pub. Co. p.40 (1998).

R.2.8. E. Mertzbacher, "Quantum Mechanics," 3rd Ed., Wiley (1998).

R.2.9. S.M. Barnett and P.M. Radmore, "Methods in Theoretical Quantum Optics," Oxford Science Pub. (1997).

3

Q. M. Pictures; Heisenberg Equation; Linear Response; Superoperators and Non-Markovian Equations

Abstract

In this chapter, three fundamental and equivalent mathematical frameworks (called "pictures") in which quantum theory can be lodged are exhibited, and their relations and relative advantages/disadvantages are discussed. First, and most familiar, is the Schrödinger picture, which considers the dynamical development of the overall system state vector as a function of time relative to a fixed complete set of time-independent basis eigenstates. On the other hand, the Heisenberg picture (which is convenient for the use of Green's functions) embeds the dynamical development of the system in a time-dependent counter-rotation of the complete set of basis eigenstates relative to the fixed, time-independent overall system state, so that the relation of the latter fixed system state to the counter-rotating basis eigenstates is indentically the same in the Heisenberg picture as it is in the Schrödinger picture. The Heisenberg equation of motion for operators is discussed in general, and is applied to annihilation and creation operators. Furthermore, the Interaction Picture addresses the situation in which a Hamiltonian, $H = H_0 + H_I$, involves a part, H_0, whose equations are relatively easy to solve, and a more complicated part, H_I. In this, the complete set of basis eigenstates are taken to have their counter-rotating time-development dynamics generated by H_0 while the overall system state dynamics are time-developed by H_I, which is convenient for the implementation of a perturbation theory in powers of H_I and is equivalent (if taken to infinite order) to the common results of the Schrödinger and Heisenberg pictures. The Interaction Picture is applied to the linear current–response to an external electromagnetic field, which is seen to involve the equilibrium current-current commutator correlation function. This chapter also addresses the S-matrix, density matrix and von Neumann equation, along with superoperators (also known as Liouville operators and tetradics) and their use in formulating non-Markovian kinetic equations for the dynamical development of a quantum system (with the help of projection operators), exhibiting phenomenology such as the intracollisional field effect.

Quantum Statistical Field Theory: Schwinger's Variational Method. Norman J. Morgenstern Horing, Oxford University Press (2017). © Norman J. Morgenstern Horing. DOI 10.1093/oso/9780198791942.001.0001

3.1 Schrödinger Picture: Iteration Series for the Time-Development Operator

There are three equivalent formats or "pictures" in which quantum mechanics is usually presented. The first to be discussed is the Schrödinger picture, which we have, in fact, dealt with from the outset in Chapter 1, Eq. (1.2.31). This picture describes the quantum dynamical temporal development of the state vector $|\Psi\rangle \equiv |\Psi_s\rangle$, the "official" state vector of the system, so to speak (as distinguished from basis eigenstates, $|a'\rangle$). In this picture, all operators of the system, **M** (as given by Eq. (1.2.32)), including the complete commuting set of compatible observable operators A and their common eigenvectors $|a'\rangle$, used as a basis, are all independent of time. To distinguish this Schrödinger picture of quantum mechanics from the others to be discussed below, we write the Schrödinger equation as

$$i\hbar\frac{\partial}{\partial t}\left|\Psi_s(t)\right\rangle = H\left|\Psi_s(t)\right\rangle. \tag{3.1.1}$$

If H is devoid of explicit time dependence, this first-order equation in time can be integrated, yielding

$$\left|\Psi_s(t)\right\rangle = e^{-iH(t-t_0)/\hbar}\left|\Psi_s(t_0)\right\rangle, \tag{3.1.2}$$

which provides a formal operator solution for the time dependence of $|\Psi_s(t)\rangle$ in terms of the initial state at t_0, $|\Psi_s(t_0)\rangle$. The operator,

$$U(t,t_0) = e^{-iH(t-t_0)/\hbar}, \tag{3.1.3}$$

may be understood as a time-development operator, since its action on $|\Psi_s(t_0)\rangle$ yields $|\Psi_s(t)\rangle$, through

$$\left|\Psi_s(t)\right\rangle = U(t,t_0)\left|\Psi_s(t_0)\right\rangle. \tag{3.1.4}$$

If the Hamiltonian, $H(t)$, has explicit time dependence, $U(t, t_0)$ is more complicated, obeying the equation

$$i\hbar\frac{\partial}{\partial t}U(t,t_0) = H(t)U(t,t_0), \tag{3.1.5}$$

and the adjoint equation

$$-i\hbar\frac{\partial}{\partial t}U^{+}(t,t_0) = U^{+}(t,t_0)H(t). \tag{3.1.6}$$

Since normalization is preserved under time development,

$$\langle\Psi_s(t)|\Psi_s(t)\rangle = \langle\Psi_s(t_0)|\Psi_s(t_0)\rangle = 1, \tag{3.1.7}$$

Eqns. (3.1.4 and 3.1.7) imply that $U(t, t_0)$ is unitary,

$$U(t,t_0)U^{+}(t,t_0) = \mathbf{I}, \tag{3.1.8}$$

with the obvious initial condition

$$U(t_0, t_0) = \mathbf{I}. \tag{3.1.9}$$

Its inverse is

$$U^{-1}(t, t_0) = U^{+}(t, t_0). \tag{3.1.10}$$

Moreover, from repeated application of the definition Eq. (3.1.4),

$$|\Psi_s(t)\rangle = U(t, t')\,|\Psi_s(t')\rangle = U(t, t')U(t', t_0)\,|\Psi_s(t_0)\rangle\,, \tag{3.1.11}$$

with the identification,

$$U(t, t_0) = U(t, t')U(t', t_0), \tag{3.1.12}$$

we see that the time-development operators form a group, and are unitary. Considering Eqns. (3.1.8) and (3.1.12), it is clear that

$$U^{-1}(t, t_0) = U(t_0, t). \tag{3.1.13}$$

Eq. (3.1.1) may be converted to an integral equation by integrating it with respect to time, with the result (also using Eq. (3.1.9)):

$$U(t, t_0) = \mathbf{I} - \frac{i}{\hbar}\int_{t_0}^{t} dt_1 H(t_1)U(t_1, t_0). \tag{3.1.14}$$

Iterating by substituting the whole right-hand side of Eq. (3.1.14) into the last $U(t_1, t_0)$-factor of the integrand yields yet another integral equation,

$$U(t, t_0) = \mathbf{I} - \frac{i}{\hbar}\int_{t_0}^{t} dt_1 H(t_1) + \left(\frac{-i}{\hbar}\right)^2 \int_{t_0}^{t} dt_1 H(t_1) \int_{t_0}^{t_1} dt_2 H(t_2)U(t_2, t_0). \tag{3.1.15}$$

Infinite repetition of this iteration process yields

$$\begin{aligned} U(t, t_0) = \mathbf{I} &- \frac{i}{\hbar}\int_{t_0}^{t} dt_1 H(t_1) \\ &+ \left(\frac{-i}{\hbar}\right)^2 \int_{t_0}^{t} dt_1 \int_{t_0}^{t_1} dt_2 H(t_1)H(t_2) + \cdots \\ &+ \left(\frac{-i}{\hbar}\right)^n \int_{t_0}^{t} dt_1 \int_{t_0}^{t_1} dt_2 \cdots \int_{t_0}^{t_{n-1}} dt_n H(t_1)H(t_2)\cdots H(t_n) + \cdots . \end{aligned} \tag{3.1.16}$$

To express the multiple integrals in a more convenient form, we start by exchanging dummy variables of integration and interchanging the order of integrations in the $n = 2$ term, with due attention to the variable limits, obtaining

$$\int_{t_0}^{t} dt_1 \int_{t_0}^{t_1} dt_2 H(t_1)H(t_2) = \int_{t_0}^{t} dt_2 \int_{t_0}^{t_2} dt_1 H(t_2)H(t_1)$$
$$= \int_{t_0}^{t} dt_1 \int_{t_1}^{t} dt_2 H(t_2)H(t_1). \qquad (3.1.17)$$

Averaging the two forms in Eq. (3.1.17), we have

$$\int_{t_0}^{t} dt_1 \int_{t_0}^{t_1} dt_2 H(t_1)H(t_2) = \frac{1}{2}\int_{t_0}^{t} dt_1 \int_{t_0}^{t_1} dt_2 H(t_1)H(t_2)$$
$$+ \frac{1}{2}\int_{t_0}^{t} dt_1 \int_{t_1}^{t} dt_2 H(t_2)H(t_1). \qquad (3.1.18)$$

Introducing the "time-ordering" symbol $(A(t_1)B(t_2))_+$, which is defined to automatically place the operator having the larger time argument on the left,

$$(A(t_1)B(t_2))_+ = \left\{ \begin{array}{l} A(t_1)B(t_2) \text{ for } t_1 > t_2 \\ B(t_2)A(t_1) \text{ for } t_2 > t_1 \end{array} \right\}, \qquad (3.1.19)$$

we may rewrite the double integral with the same limits on both integrals as

$$\int_{t_0}^{t} dt_1 \int_{t_0}^{t_1} dt_2 H(t_1)H(t_2) = \frac{1}{2}\int_{t_0}^{t} dt_1 \int_{t_0}^{t} dt_2 \,(H(t_1)H(t_2))_+ . \qquad (3.1.20)$$

Applying similar considerations in all subsequent terms we have (again, the time-ordering symbol, $(\ldots)_+$, places the operator with the largest time argument to the far left, and others to the right in descending order of their time arguments):

$$U(t, t_0) = \mathbf{I} + \sum_{n=1}^{\infty} \frac{1}{n!}\left(\frac{-i}{\hbar}\right)^n \int_{t_0}^{t} dt_1 \cdots \int_{t_0}^{t} dt_n \,(H(t_1)\cdots H(t_n))_+ , \qquad (3.1.21)$$

which has the form of a Taylor series defining the "time-ordered exponential" time-development operator,

$$U(t, t_0) \equiv \left(\exp\left[-\frac{i}{\hbar}\int_{t_0}^{t} dt_1 H(t_1) \right] \right)_+ . \qquad (3.1.22)$$

3.2 Heisenberg Picture and Equations of Motion

As the Schrödinger picture time evolution is governed by a unitary time-development operator $U(t, t_0)$, we may view the Schrödinger state $|\Psi_s(t)\rangle = U(t, t_0)|\Psi_s(t_0)\rangle$ as undergoing rotation in Hilbert space as a function of time relative to the fixed basis of eigenvectors $|a'\rangle$ and of the constant operators A. The Heisenberg picture is defined to have the "official" state vector $|\Psi_H(t)\rangle$ fixed in time and always coincident with $|\Psi_s(t_0)\rangle$, the Schrödinger state vector at initial time t_0,

$$|\Psi_H(t)\rangle \equiv |\Psi_s(t_0)\rangle \tag{3.2.1}$$

and

$$\frac{\partial}{\partial t}|\Psi_H(t)\rangle \equiv \frac{\partial}{\partial t}|\Psi_s(t_0)\rangle \equiv 0. \tag{3.2.2}$$

In order for the physically significant state vector projections onto the basis vectors to be the same in both pictures, it is clear that within the framework of the Heisenberg picture, the entire basis system, including the operators, must counter-rotate as a function of time. The requisite counter-rotation of the operators may be obtained from the requirement that operators of the Heisenberg picture, $O_H(t)$, have the same matrix elements with respect to Heisenberg picture states as Schrödinger picture operators, O_s, have with respect to Schrödinger picture states, i.e.:

$$\langle\Psi'_H|\, O_H(t)\,|\Psi_H\rangle \equiv \langle\Psi'_s(t)|\, O_s\,|\Psi_s(t)\rangle, \tag{3.2.3}$$

or, using Eq. (3.1.4) and its adjoint,

$$\langle\Psi'_s(t_0)|\, O_H(t)\,|\Psi_s(t_0)\rangle \equiv \langle\Psi'_s(t_0)|\, U^+(t, t_0) O_s U(t, t_0)\,|\Psi_s(t_0)\rangle. \tag{3.2.4}$$

Since this must be true for all possible states, the time-dependent Heisenberg operator $O_H(t)$ is related to its constant Schrödinger counterpart, O_s, as

$$O_H(t) = U^+(t, t_0) O_s U(t, t_0), \tag{3.2.5}$$

and the equation of motion describing the time dependence of its counter-rotation is given by time differentiation as

$$\frac{\partial O_H(t)}{\partial t} = \frac{\partial U^+(t, t_0)}{\partial t} O_s U(t, t_0) + U^+(t, t_0) O_s \frac{\partial U(t, t_0)}{\partial t}, \tag{3.2.6}$$

or, using Eqns. (3.1.5 and 3.1.6),

$$\frac{\partial O_H(t)}{\partial t} = \frac{i}{\hbar} U^+(t, t_0) H(t) O_s U(t, t_0) - \frac{i}{\hbar} U^+(t, t_0) O_s H(t) U(t, t_0). \tag{3.2.7}$$

Inserting the identity operator in the form $U(t,t_0)U^+(t,t_0) = \mathbf{I}$ (unitarity) between O_s and $H(t)$ in both terms on the right-hand side, and invoking Eq. 3.2.5, we have the Heisenberg equation of motion for the Heisenberg picture operators:

$$\frac{\partial O_H(t)}{\partial t} = \frac{i}{\hbar}\left[H_H(t), O_H(t)\right], \tag{3.2.8}$$

where

$$H_H(t) = U^+(t,t_0)H(t)U(t,t_0) \tag{3.2.9}$$

is the Hamiltonian operator in the Heisenberg picture. Based on Eq. (3.2.8), we can readily identify operators O_H, which are constants of the motion as those that commute with H_H. Since the Hamiltonian is composed of products of creation and annihilation operators, such as $\psi^+(\mathbf{x}), \psi(\mathbf{x}')$, in the Schrödinger picture, as discussed in Chapter 2,

$$H(t) \sim \psi^+(\mathbf{x})\cdots\psi(\mathbf{x}'), \tag{3.2.10}$$

the insertion of the identity operator in the form $U(t,t_0)U^+(t,t_0)$ always yields $H_H(t)$ in the form

$$H_H(t) \sim U^+(t,t_0)\psi^+(\mathbf{x})U(t,t_0)\cdots U^+(t,t_0)\psi(\mathbf{x}')U(t,t_0). \tag{3.2.11}$$

Identifying Heisenberg creation and annihilation operators as

$$\psi_H^+(\mathbf{x},t) = U^+(t,t_0)\psi^+(\mathbf{x})U(t,t_0), \tag{3.2.12}$$

$$\psi_H(\mathbf{x}',t) = U^+(t,t_0)\psi(\mathbf{x}')U(t,t_0), \tag{3.2.13}$$

the Heisenberg Hamiltonian,

$$H_H(t) \sim \psi_H^+(\mathbf{x},t)\cdots\psi_H(\mathbf{x}',t), \tag{3.2.14}$$

is composed of the same products of Heisenberg picture creation and annihilation operators. Thus,

$$H_H^{(1)} = \int d^{(3)}\mathbf{x}'\psi_H^+(\mathbf{x}',t)h^{(1)}(\mathbf{x}',\tfrac{\hbar}{i}\nabla',t)\psi_H(\mathbf{x}',t) \tag{3.2.15}$$

and

$$H_H^{(2)} = \frac{1}{2!}\int d^{(3)}\mathbf{x}'\int d^{(3)}\mathbf{x}''\psi_H^+(\mathbf{x}',t)\psi_H^+(\mathbf{x}'',t)\mathcal{V}(\mathbf{x}'-\mathbf{x}'')\psi_H(\mathbf{x}'',t)\psi_H(\mathbf{x}',t). \tag{3.2.16}$$

The canonical commutation relations are similarly converted to the Heisenberg picture as (Eqns. (2.2.23) and (2.2.24)):

$$\{[\psi_H(a',t),\psi_H(a'',t)]\} = \{[\psi_H^+(a',t),\psi_H^+(a'',t)]\} = 0 \tag{3.2.17}$$

and

$$\{[\psi_H(a',t),\psi_H^+(a'',t)]\} = \delta(a',a''), \tag{3.2.18}$$

which are in the form of "equal-time" commutation relations (it should be noted that the commutator/anticommutator of such operators at different times is generally *not* just a "c-number" (complex number)).

The equations of motion of the field operators $\psi_H(\mathbf{x},t), \psi_H^+(\mathbf{x},t)$ are readily obtained from the Heisenberg equation of motion, Eq. (3.2.8), and the Hamiltonian $H_H(t) = H_H^{(1)} + H_H^{(2)}$ (Eqns. (3.2.15) and (3.2.16)) as

$$\frac{\hbar}{i}\frac{\partial\psi_H(\mathbf{x},t)}{\partial t} = \int d^{(3)}\mathbf{x}' \left[\psi_H^+(\mathbf{x}',t)h^{(1)}(\mathbf{x}',\tfrac{\hbar}{i}\nabla',t)\psi_H(\mathbf{x}',t), \psi_H(\mathbf{x},t)\right]$$
$$+\frac{1}{2!}\int d^{(3)}\mathbf{x}'\int d^{(3)}\mathbf{x}''\ [\ \psi_H^+(\mathbf{x}',t)\psi_H^+(\mathbf{x}'',t)\mathcal{V}(\mathbf{x}'-\mathbf{x}'')$$
$$\times\ \psi_H(\mathbf{x}'',t)\psi_H(\mathbf{x}',t), \psi_H(\mathbf{x},t)]. \tag{3.2.19}$$

The commutator in the first integrand on the right may be evaluated using the canonical commutation relation as

$$\psi_H^+(\mathbf{x}',t)h^{(1)}(\mathbf{x}',\frac{\hbar}{i}\nabla',t)\psi_H(\mathbf{x}',t)\psi_H(\mathbf{x},t) - \psi_H(\mathbf{x},t)\psi_H^+(\mathbf{x}',t)h^{(1)}\left(\mathbf{x}',\frac{\hbar}{i}\nabla',t\right)\psi_H(\mathbf{x}',t)$$
$$= -\left\{\left[\psi_H(\mathbf{x},t), \psi_H^+(\mathbf{x}',t)\right]\right\} h^{(1)}\left(\mathbf{x}',\frac{\hbar}{i}\nabla',t\right)\psi_H(\mathbf{x}',t)$$
$$= -\delta^{(3)}(\mathbf{x}-\mathbf{x}')h^{(1)}\left(\mathbf{x}',\frac{\hbar}{i}\nabla',t\right)\psi_H(\mathbf{x}',t), \tag{3.2.20}$$

where we have commuted/anticommuted the last two factors of the first term and used the fact that $\left[h^{(1)}(\mathbf{x}',\frac{\hbar}{i}\nabla',t), \psi_H(\mathbf{x},t)\right] = 0$, which permits immediate execution of the integral. The second integrand on the right can be treated in a similar manner, and the resulting equation of motion for $\psi_H(\mathbf{x},t)$ is

$$\left(i\hbar\frac{\partial}{\partial t} - h^{(1)}(\mathbf{x},\tfrac{\hbar}{i}\nabla,t)\right)\psi_H(\mathbf{x},t)$$
$$-\int d^{(3)}\mathbf{x}'\mathcal{V}(\mathbf{x}-\mathbf{x}')\psi_H^+(\mathbf{x}',t)\psi_H(\mathbf{x}',t)\psi_H(\mathbf{x},t) = 0. \tag{3.2.21}$$

(An alternative derivation is presented in Chapter 4. See the discussion of Eqns. (4.1.64–4.1.70).) The adjoint equation may be written (interchanging primed and unprimed variables) as

$$\left(-i\hbar\frac{\partial}{\partial t'} - h^{(1)}(\mathbf{x}',\tfrac{\hbar}{i}\nabla',t')\right)\psi_H^+(\mathbf{x}',t')$$
$$-\int d^{(3)}\mathbf{x}\mathcal{V}(\mathbf{x}-\mathbf{x}')\psi_H^+(\mathbf{x}',t')\psi_H^+(\mathbf{x},t')\psi_H(\mathbf{x},t') = 0. \tag{3.2.22}$$

In the absence of interaction ($\mathcal{V}\to 0$), Eq. (3.2.21) is readily solved as

$$\psi_H(\mathbf{x},t) = e^{-\frac{it}{\hbar}h^{(1)}(\mathbf{x},\frac{\hbar}{i}\nabla)}\psi(\mathbf{x},0), \tag{3.2.23}$$

provided there is no explicit time dependence in $h^{(1)}$. Using Eq. (2.2.4) to transform $\psi(\mathbf{x},0) = \psi(\mathbf{x})$ to an energy representation (with energy eigenfunctions $\phi_{E\gamma}(\mathbf{x}), h^{(1)}(\mathbf{x}, \frac{\hbar}{i}\nabla)\phi_{E\gamma}(\mathbf{x}) = E_\gamma\phi_{E\gamma}(\mathbf{x})$), we have

$$\psi(\mathbf{x}) = \sum_{E_\gamma} \langle \mathbf{x} | E_\gamma \rangle \psi_{E_\gamma} = \sum_{E_\gamma} \phi_{E_\gamma}(\mathbf{x})\psi_{E_\gamma}, \tag{3.2.24}$$

and substitution into Eq. (3.2.23) yields

$$\psi_H(\mathbf{x}, t) = \sum_{E_\gamma} e^{-\frac{i}{\hbar}E_\gamma t}\phi_{E_\gamma}(\mathbf{x})\psi_{E_\gamma}, \tag{3.2.25}$$

where $\psi_{E_\gamma}, \psi^+_{E_\gamma}$ annihilates and creates a particle in energy state E_γ.

Much of the material in this book will be presented in the Heisenberg picture. In passing, it should be noted that with the Hamiltonian expressed in the Heisenberg picture (Eq. (3.2.9)) the equation for the unitary time-development operator $U(t, t_0)$ (Eq. (3.1.5)) may be rewritten as

$$\begin{aligned} i\hbar\frac{\partial U(t,t_0)}{\partial t} &= H(t)U(t,t_0) = U(t,t_0)\left(U^+(t,t_0)H(t)U(t,t_0)\right) \\ &= U(t,t_0)H_H(t) \end{aligned} \tag{3.2.26}$$

and the solution is given by the *anti*chronologically ordered exponential

$$U(t,t_0) = \left(\exp\left[-\frac{i}{\hbar}\int_{t_0}^{t} dt_1 H_H(t_1)\right]\right)_- = U_H(t,t_0), \tag{3.2.27}$$

whose Taylor-expansion expression has times ordered oppositely to the time order of Eq. (3.1.19), with operators having larger time arguments placed on the right.

3.3 Interaction Picture

The object of the Interaction Picture is to take advantage of a part, H_0, of the Hamiltonian,

$$H = H_0 + H', \tag{3.3.1}$$

which yields solutions relatively easily (for example, the kinetic energy), and treats the effects of the other, more difficult part, H', perturbatively, in an approximate manner. In this approach, the rotation of the state vector relative to basis vectors in Hilbert space is shared. An Interaction Picture state, $|\Psi_I(t)\rangle$, is defined to develop with time dependence generated by the "interaction" Hamiltonian H' alone, while the basis vectors and operators are counter-rotated using H_0 (so that the Interaction Picture becomes the Heisenberg picture if $H' = 0$). To start, we define the "official" Interaction Picture state vector $|\Psi_I(t)\rangle$ in terms of the Schrödinger state $|\Psi_s(t)\rangle$ at the same time, assuming that H_0 is independent of time (so all explicit time dependence is lodged in $H'(t)$), as

$$|\Psi_I(t)\rangle \equiv e^{iH_0t/\hbar}\,|\Psi_s(t)\rangle\,. \tag{3.3.2}$$

(We use the term "official" to indicate the actual state vector of the system to distinguish it from basis eigenvectors.) The equation of motion for $|\Psi_I(t)\rangle$ is given by

$$\begin{aligned} i\hbar\frac{\partial}{\partial t}\,|\Psi_I(t)\rangle &\equiv -H_0e^{iH_0t/\hbar}\,|\Psi_s(t)\rangle + i\hbar e^{iH_0t/\hbar}\frac{\partial}{\partial t}\,|\Psi_s(t)\rangle \\ &= e^{iH_0t/\hbar}(H-H_0)\,|\Psi_s(t)\rangle\,. \end{aligned} \tag{3.3.3}$$

Using Eqns. (3.3.1) and (3.3.2), we obtain

$$i\hbar\frac{\partial}{\partial t}\,|\Psi_I(t)\rangle = e^{iH_0t/\hbar}H'e^{-iH_0t/\hbar}\,|\Psi_I(t)\rangle. \tag{3.3.4}$$

In regard to equations of motion for operators, $O_I(t)$, in the interaction picture, we once again require equality of their matrix elements with those of the Schrödinger picture,

$$\begin{aligned} \langle\Psi_s'(t)|\,O_s\,|\Psi_s(t)\rangle &\equiv \langle\Psi_I'(t)|\,e^{iH_0t/\hbar}O_se^{-iH_0t/\hbar}\,|\Psi_I(t)\rangle \\ &= \langle\Psi_I'(t)|\,O_I(t)\,|\Psi_I(t)\rangle\,, \end{aligned} \tag{3.3.5}$$

from which we conclude that

$$O_I(t) = e^{iH_0t/\hbar}O_se^{-iH_0t/\hbar} \tag{3.3.6}$$

defines the counter-rotation of the operators generated by H_0. Applying this to Eq. (3.3.4) and introducing the interaction Hamiltonian in the interaction picture as

$$H_I'(t) = e^{iH_0t/\hbar}H'e^{-iH_0t/\hbar}, \tag{3.3.7}$$

we have

$$i\hbar\frac{\partial}{\partial t}\,|\Psi_I(t)\rangle = H_I'(t)\,|\Psi_I(t)\rangle\,, \tag{3.3.8}$$

which confirms that $|\Psi_I(t)\rangle$ has its time development generated by the interaction Hamiltonian, $H_I'(t)$, alone. On the other hand, the time dependence of the operators may be obtained by time differentiation of Eq. (3.3.6), with the result

$$i\hbar\frac{\partial}{\partial t}O_I(t) = e^{iH_0t/\hbar}\,(-H_0O_s + O_sH_0)\,e^{-iH_0t/\hbar}, \tag{3.3.9}$$

or

$$i\hbar\frac{\partial}{\partial t}O_I(t) = [O_I(t), H_0]\,, \tag{3.3.10}$$

(note that $H_{0I} = e^{iH_0 t/\hbar} H_0 e^{-iH_0 t/\hbar} = H_0$, so that H_0 is the same in the interaction picture as in the Schrödinger picture). This confirms that the time dependence of the interaction picture operators (and basis vectors) is generated by H_0 alone.

Focusing attention on the "more difficult" problem of Eq. (3.3.8), we define an interaction picture operator $V(t, t_0)$ in complete analogy to the Schrödinger time-development operator $U(t, t_0)$ of Eq. (3.1.4),

$$|\Psi_I(t)\rangle = V(t, t_0)\, |\Psi_I(t_0)\rangle\,. \tag{3.3.11}$$

By the usual arguments, it is readily shown that

$$i\hbar \frac{\partial}{\partial t} V(t, t_0) = H_I'(t)\, V(t, t_0), \tag{3.3.12}$$

$$-i\hbar \frac{\partial}{\partial t} V^+(t, t_0) = V^+(t, t_0) H_I'(t). \tag{3.3.13}$$

Again, $V(t, t_0)$ is unitary and satisfies the group properties. Moreover, the solution of Eq. (3.3.12) may be obtained in complete analogy to the solution of Eq. (3.1.5), with the result

$$V(t, t_0) = \left(\exp \left[-\frac{i}{\hbar} \int_{t_0}^{t} dt_1 H_I'(t_1) \right] \right)_+ , \tag{3.3.14}$$

with the role of $H(t)$ now played by $H_I'(t)$ in the interaction picture.

It is, of course, of interest to determine the precise relation of $U(t, t_0)$ to its analog $V(t, t_0)$. Using Eqns. (3.3.11) and (3.3.2), we have

$$|\Psi_I(t)\rangle = V(t, t_0)\, |\Psi_I(t_0)\rangle = V(t, t_0) e^{iH_0 t_0/\hbar}\, |\Psi_s(t_0)\rangle\,, \tag{3.3.15}$$

and, furthermore,

$$|\Psi_I(t)\rangle = e^{iH_0 t/\hbar}\, |\Psi_s(t)\rangle = e^{iH_0 t/\hbar} U(t, t_0)\, |\Psi_s(t_0)\rangle\,. \tag{3.3.16}$$

Equating the right-hand sides of these two equations, we find that

$$U(t, t_0) = U_H(t, t_0) = e^{-iH_{0I} t/\hbar} V(t, t_0) e^{iH_{0I} t_0/\hbar}, \tag{3.3.17}$$

where we have noted that $H_0 = H_{0I}$ (see discussion following Eq. (3.3.10)) and $U(t, t_0) = U_H(t, t_0)$ (Eq. (3.2.27)). This relation, Eq. (3.3.17), expresses the time-development operator on the left-hand side (written either in terms of Schrödinger or Heisenberg operators) entirely in terms of interaction picture operators on the right-hand side, and is an important point of departure for a Feynman-diagrammatic[1] formulation of perturbation expansions in powers of the strength of H_I'. Moreover, the relation of $V(t, t_0)$ to the Schrödinger and Heisenberg picture operators for time development may be determined by requiring that the Schrödinger picture time-development operator ($U(t,t')$) have the same matrix elements as its interaction picture counterpart ($U_I(t,t')$) between states at different times:

$$\left\langle \Psi_s'(t)|U(t,t')|\Psi_s(t')\right\rangle = \left\langle \Psi_I'(t)|U_I(t,t')|\Psi_I(t')\right\rangle. \tag{3.3.18}$$

Using Eq. (3.3.2), we have

$$\left\langle \Psi_I'(t)|e^{iH_0t/\hbar}U(t,t')e^{-iH_0t'/\hbar}|\Psi_I(t')\right\rangle = \left\langle \Psi_I'(t)|U_I(t,t')|\Psi_I(t')\right\rangle. \tag{3.3.19}$$

Since this is true for any state, we may make the identification

$$U_I(t,t') = e^{iH_0t/\hbar}U(t,t')e^{-iH_0t'/\hbar}, \tag{3.3.20}$$

and, employing Eq. (3.3.17), we obtain

$$U_I(t,t') \equiv V(t,t'), \tag{3.3.21}$$

which states that the interaction picture time-development operator, $V(t,t')$, is the same as that of the Schrödinger picture, but expressed in the Interaction Picture.

As indicated above, Eq. (3.3.14) is particularly useful in generating perturbation expansions in powers of the strength of H_I'. Moreover, the Hamiltonian in the interaction picture may be written, using Eqns. (3.3.6) and (3.3.7), in the form $\left(H_I = H_I^{(1)} + H_I^{(2)}\right)$:

$$H_I^{(1)} = \int d^{(3)}\mathbf{x}'\psi_I^+(\mathbf{x}',t)h^{(1)}(\mathbf{x}',\tfrac{\hbar}{i}\nabla',t)\psi_I(\mathbf{x}',t) \tag{3.3.22}$$

and

$$H_I^{(2)} = \frac{1}{2!}\int d^{(3)}\mathbf{x}'\int d^{(3)}\mathbf{x}''\psi_I^+(\mathbf{x}',t)\psi_I^+(\mathbf{x}'',t)V(\mathbf{x}'-\mathbf{x}'')\psi_I(\mathbf{x}'',t)\psi_I(\mathbf{x}',t), \tag{3.3.23}$$

where

$$\left(\psi_I \text{ or } \psi_I^+\right) = e^{iH_0t/\hbar}\left(\psi_S \text{ or } \psi_S^+\right)e^{-iH_0t/\hbar} \tag{3.3.24}$$

(using arguments similar to those involved in obtaining Eq. (3.3.6)). H_I' may be part of $H_I^{(1)}$, as in the case of electromagnetic scalar and vector potentials entering the one-particle term, and H_I' may also arise as all, or part of, $H_I^{(2)}$. As above, the canonical commutation relations are again equal-time commutation relations in the interaction picture:

$$\{[\psi_I(a',t),\psi_I(a'',t)]\} = \{[\psi_I^+(a',t),\psi_I^+(a'',t)]\} = 0, \tag{3.3.25}$$

$$\{[\psi_I(a',t),\psi_I^+(a'',t)]\} = \delta(a',a''). \tag{3.3.26}$$

The Interaction Picture is very well suited to describe scattering phenomena, with H' representing a mechanism of interaction that acts in a bounded spatial region on an "incoming" unperturbed wave-packet state developing in time under H_0 producing an "outgoing" packet state, with both such states subject to time development under H_0 *outside* the bounded region of interaction. Considering the time integral of Eq. (3.3.12) from t_0 to t (note that $V(t_0, t_0) = 1$):

$$V(t, t_0) = 1 - \frac{i}{\hbar}\int_{t_0}^{t} dt' H_I'(t') V(t', t_0) = 1 - \frac{i}{\hbar}\int_{t_0}^{t} dt' e^{iH_0 t'/\hbar} H'(t') e^{-iH_0 t/\hbar} V(t', t_0). \quad (3.3.27)$$

Employing a representation in terms of H_0-eigenstates, with eigenenergies E_n, E_r, we have (notation: $\omega_n = E_n/\hbar$ and $\omega_{nm} = \omega_n - \omega_m$):

$$\langle n| V(t, t_0)|m\rangle = \delta_{mn} - \frac{i}{\hbar}\sum_r \int_{t_0}^{t} dt' e^{i\omega_{nr} t} \langle n|H'(t')|r\rangle \langle r| V(t', t_0)|m\rangle . \quad (3.3.28)$$

Much of scattering theory rests on the assumption of the existence of a *transition matrix* T_{mn} such that, in the limit of the incident wave packet at $t_0 \to -\infty$ and the final packet-state at $t \to +\infty$, the right-hand side of Eq. (5.3.28) takes the form

$$\langle n| V(\infty, -\infty)|m\rangle = \delta_{mn} - \frac{i}{\hbar} T_{mn} \int_{-\infty}^{\infty} dt' e^{i\omega_{nm} t' + \alpha t'}, \quad (3.3.29)$$

where $0 << \alpha|t| << 1$. In the limit $\alpha \to 0$, this becomes

$$\langle n| V(\infty, -\infty)|m\rangle = \delta_{mn} - 2\pi i T_{mn} \delta(E_m - E_n). \quad (3.3.30)$$

Considering Eq. (3.3.21), $V(t, t') = U_I(t, t')$, the left-hand side of Eq. (3.30) represents the probability amplitude that state $|m >$ at $t' = -\infty$ evolves into state $< n|$ at $t' = +\infty$, defining the n, m-element of the *S-matrix*,

$$S_{mn} = \langle n| V(\infty, -\infty)|m\rangle = \langle n| U_I(\infty, -\infty)|m\rangle . \quad (3.3.31)$$

A careful discussion of the existence of the transition matrix T_{nm} is provided in Chapter 20 of Merzbacher's book "Quantum Mechanics,"[1] which also shows that the unitarity of the S-matrix elements leads to

$$2\pi \sum_n \delta(E_n - E_r) T_{nr}^* T_{nj} = i(T_{rj} - T_{jr}^*), \quad (3.3.32)$$

from which the optical theorem may be derived.

3.4 Linear Response

The interaction picture is frequently used to determine the linear response of a particular property represented by a quantum operator O, such as current density, $O \to \mathbf{J}$, when

H_I' is the part of the Hamiltonian due to the imposition of an electromagnetic field. Considering H_I' to be "turned on" at time t_0, Eqns. (3.3.5) and (3.3.11) yield

$$\left\langle \Psi_I'(t) \left| O_I(t) \right| \Psi_I(t) \right\rangle = \left\langle \Psi_I'(t_0) \left| V^+(t,t_0) O_I(t) V(t,t_0) \right| \Psi_I(t_0) \right\rangle, \tag{3.4.1}$$

which may be expanded to first order in H_I' using the linear-order term of Eq. (3.3.14) as follows:

$$\begin{aligned} &\left\langle \Psi_I'(t) | O_I(t) | \Psi_I(t) \right\rangle \\ &\cong \left\langle \Psi_I'(t_0) \left| \left(1 + \frac{i}{\hbar} \int_{t_0}^{t} dt' H_I'(t') \right) O_I(t) \left(1 - \frac{i}{\hbar} \int_{t_0}^{t} dt'' H_I'(t'') \right) \right| \Psi_I(t_0) \right\rangle \end{aligned} \tag{3.4.2}$$

$$\begin{aligned} &\cong \left\langle \Psi_I'(t_0) \left| O_I(t) \right| \Psi_I(t_0) \right\rangle + \frac{i}{\hbar} \left\langle \Psi_I'(t_0) \left| \int_{t_0}^{t} dt' \left[H_I'(t'), O_I(t) \right] \right| \Psi_I(t_0) \right\rangle \\ &\quad + O(H_I')^2 \quad \text{[terms of order } H_I'^2\text{]}, \end{aligned} \tag{3.4.3}$$

and by neglecting higher order terms starting with $O(H_I')^2$. The operator $O_I(t)$ is time-developed by H_0 alone but may contain terms referring to part of the external perturbation, as does the "diamagnetic" part of the current operator, resulting in a contribution from the first term, $\left\langle \Psi_I'(t_0) \left| O_I(t) \right| \Psi_I(t_0) \right\rangle$, in addition to the second term. All matrix elements of Eq. (3.4.3) are evaluated between states at time t_0, carrying no effect of H_I', so they correspond to the stationary Heisenberg state, as well as the initial Schrödinger state,

$$|\Psi_I(t_0)\rangle = |\Psi_H\rangle = |\Psi_s(t_0)\rangle \, . \tag{3.4.4}$$

An important example is that of linear current response, with O_I taken as the current density operator (Eq. (2.2.62), but in the Interaction Picture). Writing $\mathbf{J}_I(\mathbf{x},t) = \mathbf{J}_I^{(0)}(\mathbf{x},t) + \mathbf{J}_I^{(1)}(\mathbf{x},t)$, where

$$\mathbf{J}_I^{(0)}(\mathbf{x},t) = \frac{\hbar Q}{2mi} \lim_{\mathbf{x}' \to \mathbf{x}} \left[(\nabla - \nabla') \, \psi_I^+(\mathbf{x}',t) \, \psi_I(\mathbf{x},t) \right]$$

and

$$\mathbf{J}_I^{(1)}(\mathbf{x},t) = \frac{-Q^2}{mc} \mathbf{A}(\mathbf{x},t) \psi_I^+(\mathbf{x},t) \, \psi_I(\mathbf{x},t) \tag{3.4.5}$$

($\mathbf{J}_I^{(1)}(\mathbf{x},t)$ is the "diamagnetic" part of the current operator), we have the current density to linear order in $\mathbf{A}$ for the state $\Psi_I(t_0)$ given by Eq. (3.4.3) with $O_I(t) \to \mathbf{J}_I(\mathbf{x},t)$ as

$$\begin{aligned} \langle \Psi_I(t_0) | \mathbf{J}_I(\mathbf{x},t) | \Psi_I(t_0) \rangle &= \left\langle \Psi_I(t_0) | \mathbf{J}_I^{(1)}(\mathbf{x},t) | \Psi_I(t_0) \right\rangle \\ &\quad + \frac{i}{\hbar} \left\langle \Psi_I(t_0) | \int_{t_0}^{t} dt' \left[H_I'(t'), \mathbf{J}_I^{(0)}(\mathbf{x},t) \right] | \Psi_I(t_0) \right\rangle, \end{aligned} \tag{3.4.6}$$

where the part of the Hamiltonian of Eq. (1.1.27) involving the impressed electromagnetic field is given by

$$H_I'(t') = \int d^{(3)}\mathbf{x}'\psi_I^+(\mathbf{x}',t')\left[\frac{i\hbar Q}{2mc}\left(\nabla'\cdot\mathbf{A}(\mathbf{x}',t')+\mathbf{A}(\mathbf{x}',t')\cdot\nabla'\right)+Q\phi(\mathbf{x}',t')\right]\psi_I(\mathbf{x}',t')$$
$$+\frac{Q^2}{2mc^2}\int d^{(3)}\mathbf{x}'\psi_I^+(\mathbf{x}',t')\psi_I(\mathbf{x}',t')\mathbf{A}(\mathbf{x}',t')\cdot\mathbf{A}(\mathbf{x}',t'), \tag{3.4.7}$$

in second quantized form in position representation. Equivalently, in the gauge in which the scalar potential vanishes, we have

$$H_I'(t') = -\frac{1}{c}\int d^{(3)}\mathbf{x}'\,\mathbf{J}_I^{(0)}(\mathbf{x}',t')\cdot\mathbf{A}(\mathbf{x}',t') - \frac{1}{2c}\int d^{(3)}\mathbf{x}'\,\mathbf{J}_I^{(1)}(\mathbf{x}',t')\cdot\mathbf{A}(\mathbf{x}',t'). \tag{3.4.8}$$

As $H_I'(t')$ is at least of first order in the field variables, $\mathbf{A},\phi$, we have dropped the term involving $\left[H_I',\mathbf{J}^{(1)}\right]$ in Eq. (3.4.6) as being at least of second order. Also, we have assumed that the initial state $|\Psi_I(t_0)\rangle = |\Psi_S(t_0)\rangle = |\Psi_H\rangle$ is in the nature of an equilibrium state, for which the zero-order contribution $\langle\Psi_I(t_0)|\mathbf{J}_I^{(0)}(\mathbf{x},t)|\Psi_I(t_0)\rangle$ vanishes. Accordingly, we find from Eqns. (3.4.6 and 3.4.8) the result

$$\langle\Psi_I(t_0)|\mathbf{J}_I(\mathbf{x},t)|\Psi_I(t_0)\rangle$$
$$= -\frac{Q^2}{mc}\mathbf{A}(\mathbf{x},t)\left\langle\Psi_I(t_0)|\psi_I^+(\mathbf{x},t)\psi_I(\mathbf{x},t)|\Psi_I(t_0)\right\rangle$$
$$+\frac{i}{\hbar c}\int_{-\infty}^{t}dt'\int d^{(3)}\mathbf{x}'\left\langle\Psi_I(t_0)|\left[\mathbf{J}_I^{(0)}(\mathbf{x},t),\mathbf{J}_I^{(0)}(\mathbf{x}',t')\right]|\Psi_I(t_0)\right\rangle\cdot\mathbf{A}(\mathbf{x}',t'), \tag{3.4.9}$$

where we have taken the field to be turned on at $t_0 \to -\infty$ (and we have also dropped a term involving the second-order part of $H_I'(t') \sim \mathbf{J}_I^{(1)}\cdot\mathbf{A}$).

3.5 S-Matrix, Density Matrix, Superoperators and Non-Markovian Kinetic Equations

As briefly indicated at the end of Section 3.3, the S-matrix has the significance of an operator that time-develops the system state from initial time $t_i \to -\infty$ to the final value at $t_f \to \infty$. Thus, we may write

$$S = Lim_{\substack{t_f\to\infty \\ t_i\to-\infty}} U(t_f,t_i). \tag{3.5.1}$$

This operator is particularly useful in scattering situations in which there are large spatial separations of the scattering region from the incident particles (at $t_i \to -\infty$) and from the scattered particles (at $t_f \to +\infty$): Considered jointly with a slow, adiabatic turn-on of the scattering interaction, this may be taken to mean that the matrix elements of S

can be calculated using free (non-interacting) initial and final states. Such a situation does not commonly occur in many-body problems, since the particles are perpetually in interaction and in proximity to each other.

Another important concept frequently employed in the formulation of interaction problems having statistical features in both equlibrium and non-equilibrium conditions is that of the *statistical operator*, or *density matrix*, $\hat{\rho}$ (which we take to be normalized such that $Trace\,\hat{\rho} = 1$). It is introduced because statistical theoretical aspects require one to consider matrix elements of operators averaged over a spectrum of states to predict the probable value of the associated observables. In this sense, we write the average of a Heisenberg operator $O_H(t)$ as (Tr denotes trace):

$$\langle O(t)\rangle \equiv Tr(\hat{\rho}\, O_H(t)), \tag{3.5.2}$$

where $\hat{\rho}$ is a constant operator having the appropriate statistical features. Employing Eq. (3.2.5), this may be rewritten as

$$\langle O(t)\rangle = Tr(\hat{\rho}\, U^{+}(t,t_0)O_s U(t,t_0)), \tag{3.5.3}$$

and using the cyclic invariance of the trace we transfer the time dependence from the operator O to the density matrix,

$$\begin{aligned}\langle O(t)\rangle &= Tr(U(t,t_0)\hat{\rho}U^{+}(t,t_0)O_s)\\ &= Tr(\hat{\rho}(t)O_s),\end{aligned} \tag{3.5.4}$$

where $\hat{\rho}(t)$ is defined as

$$\hat{\rho}(t) \equiv U(t,t_0)\hat{\rho}U^{+}(t,t_0). \tag{3.5.5}$$

The time dependence of $\hat{\rho}(t)$ may be described in terms of its time derivative as

$$\frac{d\hat{\rho}(t)}{dt} = \frac{\partial U(t,t_0)}{\partial t}\hat{\rho}U^{+}(t,t_0) + U(t,t_0)\hat{\rho}\frac{\partial U^{+}(t,t_0)}{\partial t}, \tag{3.5.6}$$

or

$$\frac{d\hat{\rho}(t)}{dt} = -\frac{i}{\hbar}H(t)U(t,t_0)\hat{\rho}\,U^{+}(t,t_0) + \frac{i}{\hbar}U(t,t_0)\hat{\rho}\,U^{+}(t,t_0)H(t), \tag{3.5.7}$$

where we have used Eqns. (3.1.5 and 3.1.6). Again invoking Eq. (3.5.5), we have

$$\frac{d\hat{\rho}(t)}{dt} = -\frac{i}{\hbar}[H(t),\hat{\rho}(t)]. \tag{3.5.8}$$

This is the *von Neumann equation*, which is a quantum Liouville equation, and *differs* from the Heisenberg equation, Eq. (3.2.8), by a *negative* sign. Taken jointly with the initial condition,

$$\hat{\rho}(t=0) = \hat{\rho}, \tag{3.5.9}$$

Eq. (3.5.8) determines the time development of the density operator, which then provides $\langle O(t)\rangle$ using Eq. (3.5.4). If the system is started in equilibrium, the initial density matrix of Eq. (3.5.9) may represent a quantum mechanical canonical (or grand canonical) Gibbs ensemble[2] (which will be discussed in Chapter 6, along with its asymptotically equivalent microcanonical ensemble). Alternatively, $\hat{\rho}$ may be chosen to represent a "nearby" ensemble convenient for a perturbation-theoretical treatment, as has been done in the case of transport theory employing a *drifted*[3] equilibrium ensemble. Furthermore, $\hat{\rho}$ may represent a non-equilibrium ensemble,[3] such as a local equilibrium ensemble,[4] to study the phenomenology involved in relaxation processes, including diffusion, thermal conductivity, etc.

Yet another approach to the von Neumann equation may be taken writing Eq. (3.5.8) in the form

$$\frac{d\hat{\rho}(t)}{dt} = -\frac{i}{\hbar}\mathfrak{H}\hat{\rho}(t), \tag{3.5.10}$$

where $\mathfrak{H}$ is a *superoperator* or *Liouville operator* that acts in the space of operators (such as $\hat{\rho}(t)$), producing another operator by its action. Mathematically, it is a *tetradic* $\mathfrak{H}_{rs,\mu v}$, a four-index matrix that acts on an operator matrix having two indices, say $\hat{\rho}_{\mu v}$, in a matrix multiplication as

$$\frac{d\hat{\rho}_{rs}(t)}{dt} = -\frac{i}{\hbar}\sum_{\mu,v}\mathfrak{H}_{rs,\mu v}\hat{\rho}_{\mu v}(t). \tag{3.5.11}$$

The identity superoperator is

$$I_{rs,\mu v} = \delta_{r\mu}\delta_{sv}. \tag{3.5.12}$$

If the Hamiltonian has no explicit time dependence, Eq. (3.5.10) has the solution

$$\begin{aligned}\hat{\rho}(t) &= \exp\left(-\frac{i}{\hbar}\mathfrak{H}t\right)\hat{\rho},\\ &= e^{-\frac{i}{\hbar}Ht}\,\hat{\rho}\,e^{\frac{i}{\hbar}Ht}.\end{aligned} \tag{3.5.13}$$

The exponential of a tetradic superoperator may be understood in terms of its Taylor expansion,

$$\exp\left(-\frac{i}{\hbar}\mathfrak{H}t\right) \doteq \sum_n \frac{1}{n!}\left(\frac{-it}{\hbar}\right)^n (\mathfrak{H})^n, \tag{3.5.14}$$

with the action of powers of $\mathfrak{H}$ generating nested commutators, as

$$(\mathfrak{H})^2\hat{\rho} = [H,[H,\hat{\rho}]], \tag{3.5.15}$$

and for the nth power there are n commutators in the "nest,"

$$(\mathfrak{H})^n\hat{\rho} = [H\ldots,[H,[H,\hat{\rho}]]\ldots]. \tag{3.5.16}$$

Furthermore, if $F(\mathfrak{H})$ is an arbitrary function of $\mathfrak{H}$ and Q is an operator, then if $| E_\mu\rangle, | E_v\rangle$ are eigenstates of H, it may be readily shown that

$$\left\langle E_\mu \mid F(\mathfrak{H})Q \mid E_v\right\rangle = F(E_\mu - E_v) \left\langle E_\mu \mid Q \mid E_v\right\rangle. \tag{3.5.17}$$

Moreover, we note that the product of two arbitrary superoperators, $\mathfrak{H}_1, \mathfrak{H}_2$, is again a superoperator, since

$$(\mathfrak{H}_1\mathfrak{H}_2)_{mn,m'n'} = \sum_{k,l}(\mathfrak{H}_1)_{mn,kl}(\mathfrak{H}_2)_{kl,m'n'} \tag{3.5.18}$$

and their action on an operator A is associative in the sense that

$$(\mathfrak{H}_1\mathfrak{H}_2)A = \mathfrak{H}_1(\mathfrak{H}_2 A), \tag{3.5.19}$$

but the action of a superoperator on a product of operators AB is given by

$$\mathfrak{H}_1(AB) = (\mathfrak{H}_1 A)B + A(\mathfrak{H}_1 B). \tag{3.5.20}$$

Similarly, for a product of three operators, ABC, we have

$$\mathfrak{H}_1(ABC) = (\mathfrak{H}_1 A)BC + A(\mathfrak{H}_1 B)C + AB(\mathfrak{H}_1 C). \tag{3.5.21}$$

The density matrix approach, coupled with superoperators, rapidly produces deeply penetrating results relating to kinetic equations. Considering $\langle O(t)\rangle$ as given in Eq. (3.5.4), we note that a particular operator O_s (for example, the current operator) may have a subset of its matrix elements vanishing in a judiciously chosen representation. Thus, in this representation, we need not determine all of the elements of $\hat{\rho}(t)$ since the ones multiplying the vanishing elements of O_s do not contribute to $\langle O(t)\rangle$ and are irrelevant. Correspondingly, we introduce a superoperator *projection operator,* $\mathfrak{P}$, which eliminates the part of $\hat{\rho}(t)$, which is irrelevant in the chosen representation, yielding only that part of $\hat{\rho}(t)$ that has matrix elements that multiply non-zero matrix elements of O_s in the chosen representation, namely $\hat{f}(t)$,

$$\hat{f}(t) = \mathfrak{P}\,\hat{\rho}(t). \tag{3.5.22}$$

As a projection (super)operator, $\mathfrak{P}$ has the property

$$\mathfrak{P}^2 = \mathfrak{P}, \tag{3.5.23}$$

and, by definition,

$$\langle O(t)\rangle = Tr(\hat{f}(t)O_s). \tag{3.5.24}$$

Denoting the "irrelevant" part of $\hat{\rho}(t)$ by $\hat{g}(t)$, we have

$$\hat{\rho}(t) = \hat{f}(t) + \hat{g}(t), \tag{3.5.25}$$

and, correspondingly,

$$\hat{g}(t) = (1 - \mathfrak{P})\hat{\rho}(t) \equiv \overline{\mathfrak{P}}\,\hat{\rho}(t), \tag{3.5.26}$$

where we have defined

$$\overline{\mathfrak{P}} = 1 - \mathfrak{P} \tag{3.5.27}$$

to be a projection operator for the "irrelevant" part alone. It, too, obeys the characteristic projection relation,

$$(\overline{\mathfrak{P}})^2 = (1 - \mathfrak{P})^2 = 1 - 2\mathfrak{P} + \mathfrak{P}^2 = 1 - \mathfrak{P} = \overline{\mathfrak{P}}. \tag{3.5.28}$$

Clearly, the product $\mathfrak{P}\overline{\mathfrak{P}}$ eliminates everything,

$$\mathfrak{P}\overline{\mathfrak{P}} = 0. \tag{3.5.29}$$

Considering a Hamiltonian of the form

$$H = H_0 + V + F(t) \equiv \widetilde{H} + F(t), \tag{3.5.30}$$

where $F(t)$ is the operator for the energy of interaction with an external time-dependent field that is turned on at $t = 0$ and V represents a scattering interaction, we consider a density matrix $\hat{\rho}$ composed of a part $\hat{n}$ that describes an equilibrium governed by $\widetilde{H} = H_0 + V$, so that

$$\hat{n} = \hat{n}(\widetilde{H}) \quad \text{and} \quad \frac{d\hat{n}(t)}{dt} = -\frac{i}{\hbar}[\widetilde{H}, \hat{n}(\widetilde{H})] = 0, \tag{3.5.31}$$

and a part $\hat{R}(t)$ driven by $F(t)$,

$$\hat{\rho} = \hat{n} + \hat{R}(t) \quad \text{such that} \quad \hat{R}(0) = 0. \tag{3.5.32}$$

According to Eq. (3.5.10) taken to linear order in $F(t)$, we have

$$i\hbar\frac{d\,\hat{R}(t)}{dt} = \widetilde{\mathfrak{H}}\hat{R}(t) + \mathfrak{F}(t)\hat{n}(\widetilde{H}) \qquad (t > 0). \tag{3.5.33}$$

Here $\widetilde{\mathfrak{H}}, \mathfrak{H}_0, \mathfrak{V}, \mathfrak{F}(t)$ are the superoperators corresponding to $\widetilde{H}, H_0, V, F(t)$, respectively. Taking the representation basis of choice for projection $\mathfrak{P}$ to be that of the eigenvectors of H_0, we write $\left(\hat{R}(t) = \hat{f}(t) + \hat{g}(t)\right)$,

$$\hat{f}(t) = \mathfrak{P}\hat{R}(t) \quad \text{and} \quad \hat{g}(t) = \overline{\mathfrak{P}}\hat{R}(t). \tag{3.5.34}$$

Applying $\mathfrak{P}$ to Eq. (3.5.33), we obtain ($t > 0$):

$$i\hbar\frac{d\hat{f}(t)}{dt} = [\mathfrak{H}_0 + \mathfrak{P}\mathfrak{V}]\hat{f}(t) + \mathfrak{P}\mathfrak{V}\hat{g}(t) + \mathfrak{P}\mathfrak{F}(t)\hat{n}(H_0 + V), \tag{3.5.35}$$

where we have used the fact that $\mathfrak{P}\mathfrak{H}_0 = \mathfrak{H}_0\mathfrak{P}$ by virtue of our choice of basis. It is important to note that this equation for the dynamical development of the relevant $\hat{f}(t)$ couples it to the "irrelevant" $\hat{g}(t)$, so it is necessary to explore the dynamical development of $\hat{g}(t)$. This may be accomplished by applying $\overline{\mathfrak{P}}$ to Eq. (3.5.33), with the result

$$i\hbar\frac{d\hat{g}(t)}{dt} = [\mathfrak{H}_0 + \overline{\mathfrak{P}}\mathfrak{V}]\hat{g}(t) + \overline{\mathfrak{P}}\mathfrak{V}\hat{f}(t) + \overline{\mathfrak{P}}\mathfrak{F}(t)\hat{n}(H_0 + V). \tag{3.5.36}$$

Thus, $d\hat{g}(t)/dt$ involves $\hat{f}(t)$, but as a driving term, and this first-order ordinary differential equation may be solved subject to the initial condition, $\hat{g}(0) = 0$, by a change of dependent variable from $\hat{g}(t)$ to $\hat{y}(t)$:

$$\hat{g}(t) = \exp[-\tfrac{i}{\hbar}(\mathfrak{H}_0 + \overline{\mathfrak{P}}\mathfrak{V})t]\,\hat{y}(t), \tag{3.5.37}$$

which yields

$$i\hbar\frac{d\hat{y}(t)}{dt} = \exp\left[\tfrac{i}{\hbar}(\mathfrak{H}_0 + \overline{\mathfrak{P}}\mathfrak{V})t\right]\left\{\overline{\mathfrak{P}}\mathfrak{V}\hat{f}(t) + \overline{\mathfrak{P}}\mathfrak{F}(t)\hat{n}(H_0 + V)\right\}. \tag{3.5.38}$$

Integrating with $\hat{y}(0) = 0$, we obtain the solution for $\hat{g}(t)$ as

$$\hat{g}(t) = -\frac{i}{\hbar}\int_0^t d\tau \exp\left[-\tfrac{i}{\hbar}(\mathfrak{H}_0 + \overline{\mathfrak{P}}\mathfrak{V})\tau\right]\left\{\overline{\mathfrak{P}}\mathfrak{V}\hat{f}(t-\tau) + \overline{\mathfrak{P}}\mathfrak{F}(t-\tau)\hat{n}(H_0 + V)\right\}. \tag{3.5.39}$$

Substitution of Eq. (3.5.39) into Eq. (3.5.35) eliminates $\hat{g}(t)$, yielding an integro-differential equation for $\hat{f}(t)$ alone, as

$$\begin{aligned}\frac{d\hat{f}(t)}{dt} = &-\frac{i}{\hbar}[\mathfrak{H}_0 + \mathfrak{P}\mathfrak{V}]\hat{f}(t) - \frac{1}{\hbar^2}\mathfrak{P}\mathfrak{V}\int_0^t d\tau\,\mathfrak{E}(\tau)\overline{\mathfrak{P}}\mathfrak{V}\hat{f}(t-\tau)\\ &-\frac{i}{\hbar}\mathfrak{P}\mathfrak{F}(t)\hat{n}(H_0 + V) - \frac{1}{\hbar^2}\mathfrak{P}\mathfrak{V}\int_0^t d\tau\,\mathfrak{E}(\tau)\overline{\mathfrak{P}}\mathfrak{F}(t-\tau)\hat{n}(H_0 + V),\end{aligned} \tag{3.5.40}$$

where we have defined

$$\mathfrak{E}(t) = \exp\left[-\tfrac{i}{\hbar}(\mathfrak{H}_0 + \overline{\mathfrak{P}}\mathfrak{V})t\right]. \tag{3.5.41}$$

It is apparent that this kinetic equation is *non-Markovian* due to the second term on the right, which is to say that the determination of $\frac{d\hat{f}(t)}{dt}$ and $\hat{f}(t)$ depends not just upon

$\hat{f}(t)$ at the same time, but on the entire history, or memory, of $\hat{f}(t)$ at all prior times $t' < t$, with the scattering processes represented by V. The first term on the right of Eq. (3.5.40) describes the unperturbed kinetics of time development as modified by first-order scattering effects, which can be absorbed into H_0 shifting the unperturbed energies of the system. The third term on the right represents the driving action of the external field alone. The fourth and last term represents interference effects due to the action of the driving field during the scattering process, albeit in linear response to the external field. It is variously referred to as an *interference term* and an *intracollisional field effect* (in linear response). This type of superoperator analysis has been employed in the rigorous formulation of non-Markovian kinetic equations with applications to careful studies of transport, relaxation and irreversible processes by Zwanzig,[5] Fano[6] and Argyres,[7] and has recently been refined and vigorously developed by Buot and Rajagopal.[8]

3.6 Chapter 3 Problems

P.3.1. Prove the following relations:

(a) $U(t_0, t_0) = 1$
(b) $U^+(t, t_0)U(t, t_0) = 1$
(c) $U(t_1, t_2)U(t_2, t_3) = U(t_1, t_3)$
(d) $U(t, t_0)U(t_0, t) = 1$
(e) $U^+(t_0, t) = U(t, t_0)$

P.3.2. Determine the commutator of $\psi_H(\mathbf{x}, t)$ with $H_H^{(2)}$. Considering the factor 1/2! on the right of Eq. (3.2.19), explain why it does not appear on the right of Eq. (3.2.21).

P.3.3. Verify Eqns. (3.2.25 and 3.2.26).

P.3.4. Prove the equivalence of Eq. (3.1.22) for $U^+(t, t_0)$ as a chronologically (time)-ordered exponential with Eq. (3.2.27) for $U^+(t, t_0)$ as an antichronologically ordered exponential. Interpret the anti-chronological time order in Eq. (3.2.27) in terms of the counter-rotation of the basis vectors (and operators) in the Heisenberg picture discussed below Eq. (3.2.2), in contrast to the chronological time order of Eq. (3.1.22).

P.3.5. Consider a time-development operator from t_0 to a complex time τ, $U(\tau, t_0)$. Prove equivalence of the following trace (Tr) relation:

$$Tr[U(\tau + t_0, t_0)(\psi(1)\psi^+(2))_+] \equiv Tr[(U(\tau + t_0, t_0)\psi_I(1)\psi_I^+(2))_+],$$

where the time ordering on the right-hand side applies to $U(\tau, t_0)$ jointly with $\psi_I(t_1)$ and $\psi_I(t_2)$, whereas on the left, the time ordering of $U(\tau, t_0)$ is separate and distinct from the time ordering of $(\psi(t_1)\psi(t_2))_+$. Employing Eq. (3.2.27), further show that

$$Tr[U_H(\tau + t_0, t_0)(\psi_H(1)\psi_H^+(2))_+] = Tr[(U(\tau + t_0, t_0)\psi_I(1)\psi_I^+(2))_+].$$

This equation relates non-equilibrium Green's functions (to be defined later) in the full Heisenberg picture (left-hand side) with those in the interaction picture (right-hand side).

P.3.6. Reconsidering the derivation of Eq. (3.4.3), derive the terms of order $(H_I')^2$ that were omitted.

P.3.7. Although the approach taken in this book does not directly involve Feynman diagrams, it is appropriate for the reader to learn this widely used technique. Consult the literature on this matter, and write a report summarizing the derivation of the Feynman-diagrammatic approach (explaining in detail the roles of the GellMann-Low and Wick theorems and normal ordering vs. time ordering).

P.3.8. Determine the matrix elements of the Hamiltonian superoperator $\mathfrak{H}$ in Eq. (3.5.11) in terms of those for the Hamiltonian, H.

P.3.9. Verify the details of the derivation of Eqns. (3.5.17–3.5.21).

P.3.10. Verify the details of the derivation of Eqns. (3.5.33–3.5.41).

3.7 References for Chapter 3

R.3.1. See, for example, A.L. Fetter and J.D Walecka, "Quantum Theory of Many-Particle Systems," McGraw-Hill Book Co. (1971); A.A. Abrikosov, L.P. Gorkov and I.E. Dzyaloshinski, "Methods of Quantum Field Theory in Statistical Physics," Prentice-Hall Pub. (1963) and Pergamon Press (1965); E.M. Lifshitz and L.P. Pitaevskii, "Statistical Physics, Part 2 of Landau and Lifshitz 'Course of Theoretical Physics, Vol. 9'," 3rd Edition, Pergamon Press (1980); E. Merzbacher, "Quantum Mechanics," 3rd Edition, John Wiley & Sons (1998).

R.3.2. R. Pathria, "Statistical Mechanics," Pergamon Press (1972); K. Huang, "Statistical Mechanics," John Wiley & Sons (1987); D. ter Haar, "Elements of Statistical Mechanics," Rineheart, NY (1954).

R.3.3. D.N. Zubarev, "Nonequilibrium Statistical Thermodynamics," Consultants Bureau, NY (1974) and Plenum Press, NY (1974); D. Zubarev, V. Morozov and G. Röpke, "Statistical Mechanics of Nonequilibrium Processes," Vol. 2, Akademie Verlag (1997).

R.3.4. L.P. Kadanoff and P.C. Martin, Annals of Physics (NY) *24*, 419 (1963).

R.3.5. R. Zwanzig, in "Lectures in Theoretical Physics," Boulder, *3*, University of Colorado Press (1960).

R.3.6. U. Fano, Phys. Rev. *131*, 259 (1963); U. Fano, Reviews of Modern Physics *29*, 74 (1957).

R.3.7. P.N. Argyres, in "Lectures in Theoretical Physics," Boulder, Vol. VIII-A, University of Colorado Press (1966).

R.3.8. F.A. Buot, "General Theory of Quantum Distribution Function Transport Equations: Superfluid Systems and Ultrafast Dynamics of Optically Excited Semiconductors," Rivista del Nuovo Cimento, Vol. *20* No. 9, 1 (1997).

4

Schwinger Action Principle and Variational Calculus

Abstract

Chapter 4 introduces the Schwinger Action Principle, which is at the heart of the presentation of coupled quantum statistical field theory in this book. While the methods described here originally arose in the relativistic quantum field theory of elementary particle physics, they have also profoundly advanced our understanding of non-relativistic many-particle physics, some aspects of which will be elaborated in later chapters. The Schwinger action principle is a quantum mechanical variational principle that closely parallels the Hamilton Principle of Least Action of classical mechanics, generalizing it to include the role of quantum operators as generalized coordinates and momenta. As such, it unifies all aspects of quantum theory, incorporating Hamilton equations of motion for those operators and the Heisenberg equation, as well as producing the canonical equal-time commutation/anticommutation relations. It yields dynamical coupled field equations for the creation and annihilation operators of the interacting many-body system by variational differentiation of the Hamiltonian with respect to the field operators, and also yields equations for the development of matrix elements (underlying Green's functions) under variation of coupling strength. To facilitate understanding of the variational methods involved, Chapter 4 discusses the concepts of variational calculus involving impressed potentials, c-number coordinates and fields, and also quantum operator coordinates and fields. In particular, the variational development of bra-ket matrix elements of a time-dependent operator, $< a_1' t_1 | O_H(t) | a_2'' t_2 >_H$, are shown to devolve upon time-ordered operator matrix elements, to be identified later in terms of Green's functions. Considerable attention is also given to Fermion (electron, hole) and Boson (phonon), particle sources, as well as potential sources, with explicit solutions derived for the time-dependant bra-ket matrix elements $< a_1' t_1 | a_2'' t_2 >_{CS}$ driven by the full complement of interparticle coupling interactions (C) and all the various sources (S) (effectively re-driving the interaction picture, among other things). Finally, a useful identity relating to the variational derivative of a Green's function is derived.

4.1 Action Principle: Quantum Hamilton Equations

The *Schwinger Action Principle* is lodged in the Heisenberg picture, in which the "official" system state vector, $|\Psi_H\rangle$, is constant in time and all operators $O_H(t)$ evolve dynamically

Quantum Statistical Field Theory: Schwinger's Variational Method. Norman J. Morgenstern Horing, Oxford University Press (2017). © Norman J. Morgenstern Horing. DOI 10.1093/oso/9780198791942.001.0001

in accordance with the Heisenberg equation of motion Eq. (3.2.8). Accordingly, the basis states, as eigenvectors of such operators, evolve in time in a manner consistent with that of the operators. Starting with the Schrödinger picture,

$$A_s \left|a'\right\rangle_S = a' \left|a'\right\rangle_S, \tag{4.1.1}$$

Eq. (3.2.5) may be invoked to yield

$$U(t, t_0) A_H U^+(t, t_0) \left|a'\right\rangle_S = a' \left|a'\right\rangle_S, \tag{4.1.2}$$

or, by the unitarity of the time-development operator, we have

$$A_H \left[U^+(t, t_0) \left|a'\right\rangle_S\right] = a' \left[U^+(t, t_0) \left|a'\right\rangle_S\right], \tag{4.1.3}$$

from which we may identify the basis states in the Heisenberg picture, $\left|a'\right\rangle_H$, and see that they evolve in time according to

$$\left|a'\right\rangle_H = U^+(t, t_0) \left|a'\right\rangle_S . \tag{4.1.4}$$

Schwinger[1] focuses on the time development of the basis states as a fundamental element in formulating the Action Principle for quantum dynamics. Noting that an infinitesimal variation δ of a *transformation function*, $\delta\left(\left\langle a_1', t_1 | a_2'', t_2\right\rangle_H\right)$, can always be used to define a Hermitian infinitesimal operator $[\delta W_{12}]$ as (we take $\hbar \to 1$ henceforth, in most of this book):

$$\delta\left(\left\langle a_1', t_1 | a_2'', t_2\right\rangle_H\right) = i \left\langle a_1', t_1 \left| [\delta W_{12}] \right| a_2'', t_2\right\rangle_H, \tag{4.1.5}$$

consecutive variations can be accommodated in the form

$$[\delta W_{12}] + [\delta W_{23}] = [\delta W_{13}] . \tag{4.1.6}$$

Hermiticity of δW_{12} follows from the fact that an infinitesimal variation of a unitary operator (the transform in Eq. (4.1.5)) is skew-Hermitian, as one can readily see by writing the unitarity condition in the form

$$(U-1)^+(U-1) = -(U-1) - (U-1)^+$$

and noting that the left-hand side is negligibly (second-order) small in the infinitesimal $(U-1)$. (It should be noted that if $t_1 = t_2$, then $\left\langle a_1', t_1 | a_2'', t_1\right\rangle_H = \left\langle a_1' | a_2''\right\rangle_H$ merely involves non-dynamical transformation relations among the basis vectors, with a fixed numerical value and *null* dynamical variation.) The dynamical Action Principle postulates that there exists a special class of variations for which the associated operators δW_{12} are obtained by variation of a single finite operator W_{12}, the *action* operator,

$$[\delta W_{12}] = \delta\,[W_{12}]\,, \tag{4.1.7}$$

which is Hermitian, as above,

$$W_{12}^+ = W_{12}, \tag{4.1.8}$$

and also has the additivity property,

$$W_{12} + W_{23} = W_{13}. \tag{4.1.9}$$

This additivity property implies that a finite transformation from $a'_2 t_2$ to $a'_1 t_1$, corresponding to W_{12}, can be composed additively through a succession of infinitesimal transformations $W_{t+dt,t}$ as $(W_{t,t} = 0)$,

$$W_{12} = \sum_{t_2}^{t_1} W_{t+dt,t}, \tag{4.1.10}$$

or, defining the *Lagrangian* $L(t)$ through $W_{t+dt,t} = dtL(t)$, we have

$$W_{12} = \sum_{t_2}^{t_1} dtL \to \int_{t_2}^{t_1} dtL(t). \tag{4.1.11}$$

Since W_{12} is Hermitian, so is $L(t)$, which depends on dynamical variables of the system at hand, which may be Hermitian or non-Hermitian (which we will employ). $L(t)$ is composed of two parts:

$$L(t) = L_{kin} - H_H(t), \tag{4.1.12}$$

where L_{kin} is a quantum kinetic term analogous to the classical[2,3] expression $\sum_i p_i \dot{q}_i$ (p_i, q_i being canonical momentum, coordinate variables), which facilitates the Legendre transformation from the Lagrangian description in terms of coordinates (q_i) and velocities ($\dot{q}_i$) to the Hamiltonian description in terms of canonically paired coordinate and momentum variables. Bearing in mind that the canonically conjugate coordinate (q_i) and momentum (p_i) variables do not commute within the framework of quantum theory, it is convenient to symmetrize L_{kin} with respect to transference of the time derivative as ($\dot{q} = dq/dt$ and hc denotes the Hermitian conjugate):

$$L_{kin} = \frac{1}{2} \sum_i (p_i \cdot \dot{q}_i - \dot{p}_i \cdot q_i) + hc, \tag{4.1.13}$$

wherein the products $p_i \cdot \dot{q}_i$ and $\dot{p}_i \cdot q_i$ are also symmetrized/antisymmetrized in the Boson/Fermion cases, respectively, according to the definitions

$$p_i \cdot \dot{q}_i \equiv \tfrac{1}{2} (p_i \dot{q}_i \pm \dot{q}_i p_i) \text{ and } \dot{p}_i \cdot q_i \equiv \tfrac{1}{2} (\dot{p}_i q_i \pm q_i \dot{p}_i) . \tag{4.1.14}$$

(The difference between L_{kin} (Eq. (4.1.13)) and $\sum_i p_i \dot{q}_i$ is just a total time derivative, which is acceptable as a canonical transformation within the framework of the principle of stationary action and which maintains the Hamilton form for the equations of motion.) Since W_{12} is Hermitian, both L_{kin} and $H_H\ (t)$ must also be Hermitian.

There are some conditions on the special class of variations admissible for the Action Principle, Eq. (4.1.7). The variations of variables involved in the Lagrangian and Hamiltonian must be commutative for Bosons and anticommutative for Fermions,

$$\{[\delta q_i, q_j]\} = \{[\delta p_i, p_j]\} = \{[\delta q_i, p_j]\} = \{[q_i, \delta p_j]\} = 0, \tag{4.1.15}$$

and variations of variables of either species commute with the variables of the other species. These conditions are required for the consistency of the Action Principle formulation of quantum mechanics, as one may expect from the commutation/anticommutation relations of Chapter 2.

The Hamiltonian formulation provides quantum dynamical equations of motion analogous to those of classical mechanics. Considering the Hamiltonian as a function of canonically paired generalized coordinates q_i and momenta p_i, the Action Principle has the form

$$\delta\left(\langle a_1', t_1 | a_2'', t_2\rangle_H\right) = i\left\langle a_1', t_1 \left| \delta \left[\int_{t_2}^{t_1} dt \left(\frac{1}{2} \sum_i \left[(p_i \cdot \dot{q}_i - \dot{p}_i \cdot q_i) + hc \right] - H_H(t) \right) \right] \right| a_2'', t_2 \right\rangle_H . \tag{4.1.16}$$

The left-hand side may be written as

$$\delta\left(\langle a_1', t_1 | a_2'', t_2\rangle_H\right) = \left(\delta \langle a_1', t_1 |_H\right) | a_2'', t_2\rangle_H + \langle a_1', t_1 |_H \left(\delta | a_2'', t_2\rangle_H\right), \tag{4.1.17}$$

and anticipating deterministic equations of motion of finite order, the infinitesimal variations of an eigenstate on the right, say $\delta \langle a_1', t_1 |$, can be written as a unitary transformation of the system at the same time, t_1, generated by a Hermitian operator that depends only on the pertinent dynamical variables at the same time, i.e.:

$$\delta \langle a_1', t_1 |_H = i \langle a_1', t_1 |_H G_1, \tag{4.1.18}$$

$$\delta | a_2'', t_2\rangle_H = -iG_2 | a_2'', t_2\rangle_H , \tag{4.1.19}$$

where G_1 (G_2) depends only on the variables at t_1 (t_2), respectively, whence

$$\delta\left(\langle a_1', t_1 | a_2'', t_2\rangle_H\right) = i \langle a_1', t_1 | \left(G_1 - G_2\right) | a_2'', t_2\rangle_H . \tag{4.1.20}$$

Comparing Eqns. (4.1.16) and (4.1.20), we have the quantum principle of stationary action in the form

$$\begin{aligned} \delta\left[W_{12}\right] &= \delta \int_{t_2}^{t_1} dt L(t) \\ &= \delta \left[\int_{t_2}^{t_1} dt \left(\frac{1}{2} \sum_i \left[(p_i \cdot \dot{q}_i - \dot{p}_i \cdot q_i) + hc \right] - H_H(t) \right) \right] \\ &= G_1 - G_2. \end{aligned} \tag{4.1.21}$$

It is clear that variations involving intermediate times t, ($t_1 > t > t_2$), can have no effect since $\delta\left[W_{12}\right]$ depends only on the initial and final times in $(G_1 - G_2)$. Furthermore, we also have the familiar fact that the Lagrangian and the generators are not unique since we could equally well make the replacement,

$$L \to \bar{L} = L - \frac{dW}{dt}, \tag{4.1.22}$$

with the subtraction of a total time derivative. The stationarity of L implies the stationarity of $\bar{L}$, and the equations of motion derived from L and $\bar{L}$ will therefore be the same. The generators, however, do change:

$$G_1 \to \bar{G}_1 - \delta\left[W(t_1)\right]; \quad G_2 \to \bar{G}_2 - \delta\left[W(t_2)\right], \tag{4.1.23}$$

reflecting the change of actions

$$W_{12} \to \bar{W}_{12} = W_{12} - \left[W(t_1) - W(t_2)\right]. \tag{4.1.24}$$

These features of quantum canonical transformations are fully analogous to corresponding features of classical canonical transformations.[2]

To derive the equations of motion from the principle of stationary action, Eq. (4.1.21), we note the identity ($\dot{q}_i dt = dq_i$, $\dot{p}_i dt = dp_i$):

$$\begin{aligned} &\delta\left[\frac{1}{2}\sum_i \left[(p_i \cdot dq_i - dp_i \cdot q_i) + hc\right] - H_H(\{q_k\}, \{p_k\}, t)dt\right] \\ &= d\left[\frac{1}{2}\sum_i \left[(p_i \cdot \delta q_i - \delta p_i \cdot q_i) + hc\right] - H_H(\{q_k\}, \{p_k\}, t)\delta t\right] \\ &\quad + \sum_i (\delta p_i \cdot dq_i - dp_i \cdot \delta q_i) - \delta H_H dt + dH_H \delta t, \end{aligned} \tag{4.1.25}$$

which the reader may readily verify. The total differential on the right of Eq. (4.1.25) enables us to identify the generators at t_1, t_2 as

$$G_{1,2} = \frac{1}{2}\sum_i \left[\left(p_i(t_{1,2}) \cdot \delta q_i(t_{1,2}) - \delta p_i(t_{1,2}) \cdot q_i(t_{1,2})\right) + hc\right] - H_H\left(\left\{q_k(t_{1,2})\right\}, \left\{p_k(t_{1,2})\right\}\right)\delta t_{1,2}, \tag{4.1.26}$$

or (understanding the time of evaluation) we simply write

$$G = \frac{1}{2}\sum_i \left[(p_i \cdot \delta q_i - \delta p_i \cdot q_i) + hc\right] - H_H \delta t. \tag{4.1.27}$$

Since the variations at intermediate times cannot contribute, we may conclude from Eq. (4.1.25) that

$$\sum_i (\delta p_i \cdot dq_i - dp_i \cdot \delta q_i) - \delta H_H dt + dH_H \delta t = 0. \tag{4.1.28}$$

Separating the parts of the variation of H that come from its dependence on coordinates $q \leftrightarrow \{q_k\}$ and momenta $p \leftrightarrow \{p_k\}$ from its explicit dependence on t, we have

$$\delta H_H = (\delta H_H)_q + (\delta H_H)_p + \frac{\partial H_H}{\partial t}\delta t, \tag{4.1.29}$$

so that Eq. (4.1.28) may be rewritten as (dividing by dt):

$$\sum_i (\delta p_i \cdot \dot{q}_i - \dot{p}_i \cdot \delta q_i) - (\delta H_H)_q - (\delta H_H)_p - \frac{\partial H_H}{\partial t}\delta t + \frac{dH_H}{dt}\delta t = 0. \tag{4.1.30}$$

Since δt is arbitrary, it is clear that

$$\frac{dH_H}{dt} = \frac{\partial H_H}{\partial t}. \tag{4.1.31}$$

Moreover, the variation induced by a variation of the q-variables must be written with care concerning the anticommutative nature of variations of Fermi variables. Thus we define left- and right-handed derivatives (∂_l and ∂_r, respectively) in accordance with

$$(\delta H_H)_q \equiv \sum_i \delta q_i \frac{\partial_l H_H}{\partial q_i} = \sum_i \frac{\partial_r H_H}{\partial q_i}\delta q_i, \tag{4.1.32}$$

and, similarly,

$$(\delta H_H)_p \equiv \sum_i \delta p_i \frac{\partial_l H_H}{\partial p_i} = \sum_i \frac{\partial_r H_H}{\partial p_i}\delta p_i. \tag{4.1.33}$$

In the Fermion case of non-commutative variables, the right- and left-handed derivatives are not the same. In fact, consistency requires that H be an even function of Fermi variables, $\{q_k\}$, $\{p_k\}$, so that

$$\frac{\partial_l H_H}{\partial q_i} = -\frac{\partial_r H_H}{\partial q_i} \quad \text{and} \quad \frac{\partial_l H_H}{\partial p_i} = -\frac{\partial_r H_H}{\partial p_i}, \tag{4.1.34}$$

to assure that the two evaluations of $(\delta H_H)_q$ in Eq. (4.1.32) and $(\delta H_H)_p$ in Eq. (4.1.33) are the same. This issue is moot for Bosons, for which $\partial_l H_H \equiv \partial_r H_H$, since the variations of Boson variables are all commutative, and the subscripts l and r are superfluous and may be dropped (moreover, H need not be an even or odd function of Boson variables). Based on these considerations, Eq. (4.1.30) yields the quantum Hamilton equations as

$$\dot{q}_i = \frac{\partial_l H_H}{\partial p_i} \quad \text{and} \quad \dot{p}_i = -\frac{\partial_r H_H}{\partial q_i}, \tag{4.1.35}$$

in complete analogy to the classical Hamilton dynamical equations except for the anticommutation properties of Fermions. This form of the equations of motion is particularly convenient for dealing with complicated interaction Hamiltonians.

Returning to the generator, G, of Eq. (4.1.27), we write it in the form

$$G = G_q + G_p + G_t, \tag{4.1.36}$$

where

$$G_q = \frac{1}{2}\sum_i [p_i \cdot \delta q_i + hc] = \frac{1}{2}\sum_i [(p_i\delta q_i \pm \delta q_i p_i)/2 + hc] = \sum_i p_i \delta q_i \tag{4.1.37}$$

because of the commutation/anticommutation properties of δq_i, δp_i, and, furthermore,

$$G_p = -\frac{1}{2}\sum_i [\delta p_i \cdot q_i + hc] = -\sum_i \delta p_i q_i \tag{4.1.38}$$

and

$$G_t = -H_H \delta t. \tag{4.1.39}$$

Since G acts on states in accordance with Eqns. (4.1.18) and (4.1.19),

$$\delta \langle a', t|_H = i \langle a', t|_H G \quad \text{and} \quad \delta |a'', t\rangle_H = -iG |a'', t\rangle_H , \tag{4.1.40}$$

the variation of matrix elements of an arbitrary operator O between states at the same fixed time is given by

$$\begin{aligned} \delta \left(\langle a', t|O|a'', t\rangle\right) &= \delta \left(\langle a', t|\right) O |a'', t\rangle + \langle a', t| O \delta \left(|a'', t\rangle\right) \\ &= i \langle a', t| [G, O] |a'', t\rangle , \end{aligned} \tag{4.1.41}$$

which can be replicated by a variation of the operator, δO, between fixed states, i.e.:

$$\delta \left(\langle a', t|O|a'', t\rangle\right) = \langle a', t|\delta O|a'', t\rangle , \tag{4.1.42}$$

with the definition

$$\delta O = i [G, O] = \frac{1}{i} [O, G] . \tag{4.1.43}$$

Accordingly, we write in analogy to Eq. (4.1.36),

$$\delta O = \delta_q O + \delta_p O + \delta_t O, \tag{4.1.44}$$

with the associated variations through q's and p's identified as

$$\delta_q O = \frac{1}{i} [O, G_q]; \quad \delta_p O = \frac{1}{i} [O, G_p]; \quad \delta_t O = \frac{1}{i} [O, G_t] . \tag{4.1.45}$$

Considering changes through q's and p's relating to time, one must bear in mind that fixing the states in both Eq. (4.1.41) and Eq. (4.1.42) at time t when the variables advance as $q(t) \to q(t + \delta t) = q + \delta q$, $p(t) \to p(t + \delta t) = p + \delta p$ means that the operator δO in

Eq. (4.1.42) must be understood to have *counter-rotated* through δt, relative to the states fixed at the original time t. Thus we understand $\delta t \to -\delta t$ in G_t, with the result of:

$$\delta_t O = \frac{1}{i}\left[O, -H_H(-\delta t)\right], \tag{4.1.46}$$

whence

$$\frac{\delta_t O}{\delta t} = \frac{1}{i}\left[O, H_H\right] = i\left[H_H, O\right] \equiv \frac{dO}{dt} - \frac{\partial O}{\partial t},$$
$$\frac{dO}{dt} = \frac{\partial O}{\partial t} + i\left[H_H, O\right]. \tag{4.1.47}$$

Note that if O has no explicit time dependence, $\frac{\partial O}{\partial t} = 0$ and $\frac{dO}{dt}$ in Eq. (4.1.47) has the same meaning as does $\frac{\partial O_H}{\partial t}$ in Eq. (3.2.8). This result, which encompasses the Heisenberg equation of motion, clearly indicates that all of quantum mechanics follows directly from the Schwinger Action Principle. Its consistency with the quantum Hamilton equations, Eq. 4.1.35 is verified in Reference 1, Section 4.7.

Considering $\delta_q O$ and $\delta_p O$ in Eq. (4.1.45) further, we have

$$\delta_q O = \frac{1}{i}\left[O, \sum_i p_i \delta q_i\right]; \quad \delta_p O = i\left[O, \sum_i \delta p_i q_i\right]. \tag{4.1.48}$$

For Bosons, this yields

$$\frac{\partial O}{\partial q_i} = \frac{1}{i}\left[O, p_i\right]; \quad \frac{\partial O}{\partial p_i} = i\left[O, q_i\right], \tag{4.1.49}$$

which state that changes in q_i are generated by p_i, and vice versa, confirming the canonical nature of the variables. Further verification may be had by setting $O = q_k, p_k$ in succession, resulting in

$$\frac{\partial q_k}{\partial q_i} = \delta_{ki} = \frac{1}{i}\left[q_k, p_i\right]; \quad \frac{\partial q_k}{\partial p_i} = 0 = i\left[q_k, q_i\right] \tag{4.1.50}$$

$$\frac{\partial p_k}{\partial q_i} = 0 = \frac{1}{i}\left[p_k, p_i\right]; \quad \frac{\partial p_k}{\partial p_i} = \delta_{ki} = i\left[p_k, q_i\right], \tag{4.1.51}$$

which are the canonical commutation relations for Bosons. In the Fermion case, one must distinguish between operators O, which are even (O_e) and odd (O_o) in the Fermi variables p, q. For the even case, δq_i and δp_i commute with O_e, and we have

$$\frac{\partial_r O_e}{\partial q_i} = \frac{1}{i}\left[O_e, p_i\right] = -\frac{\partial_l O_e}{\partial q_i}; \quad \frac{\partial_l O_e}{\partial p_i} = i\left[O_e, q_i\right] = -\frac{\partial_r O_e}{\partial p_i}, \tag{4.1.52}$$

which again show that Fermi q_i and p_i variables generate changes in each other as canonical variables. For odd operators, on the other hand, we find anticommutators replacing commutators,

$$\frac{\partial_r O_o}{\partial q_i} = \frac{1}{i}\{O_o, p_i\} = \frac{\partial_l O_o}{\partial q_i}; \quad \frac{\partial_l O_o}{\partial p_i} = -i\{O_o, q_i\} = \frac{\partial_r O_o}{\partial p_i}. \tag{4.1.53}$$

In particular, choosing $O_o = q_k, p_k$ in succession, we have

$$\frac{\partial_r q_k}{\partial q_i} = \frac{\partial_l q_k}{\partial q_i} = \delta_{ki} = \frac{1}{i}\{q_k, p_i\}; \quad \frac{\partial_r q_k}{\partial p_i} = \frac{\partial_l q_k}{\partial p_i} = 0 = -i\{q_k, q_i\} \tag{4.1.54}$$

$$\frac{\partial_r p_k}{\partial q_i} = \frac{\partial_l p_k}{\partial q_i} = 0 = \frac{1}{i}\{p_k, p_i\}; \quad \frac{\partial_r p_k}{\partial p_i} = \frac{\partial_l p_k}{\partial p_i} = \delta_{ki} = -i\{p_k, q_i\}, \tag{4.1.55}$$

again obtaining canonical anticommutation relations for Fermions.

We may identify the non-Hermitian variables $\psi_H(a', t)$, $i\psi_H^+(a'', t)$ as being canonically paired in Eqns. (3.2.17) and (3.2.18), whence we associate

$$q_i \leftrightarrow \psi_H(a', t) \tag{4.1.56}$$

$$p_i \leftrightarrow i\psi_H^+(a', t). \tag{4.1.57}$$

Notwithstanding this non-Hermitian choice of variables, it is readily verified that L_{kin} in the form of Eq. (4.1.13) is Hermitian, as is $H_H = H_H^{(1)} + H_H^{(2)}$. Again, the Hamiltonian must be an even function of Fermion variables $\psi_H(a', t)$, $\psi_H^+(a'', t)$ in accordance with our earlier remarks. Moreover, $\delta\psi_H(a', t)$ and $\delta\psi_H^+(a'', t)$ have the same commutation/anticommutation properties as do δq_i and δp_i in the Boson/Fermion cases. Accordingly, if $\psi_H(a', t)$ is a Fermion variable, we have

$$\frac{\partial_l H_H}{\partial \psi_H(a', t)} = -\frac{\partial_r H_H}{\partial \psi_H(a', t)}, \text{ etc.} \tag{4.1.58}$$

However, if $\psi_H(a', t)$ is a Boson variable, we have

$$\frac{\partial_l H_H}{\partial \psi_H(a', t)} = \frac{\partial_r H_H}{\partial \psi_H(a', t)}, \text{ etc.} \tag{4.1.59}$$

with no restriction regarding the even/odd quality of H as a function of Boson variables since $\delta\psi_H$ is strictly commutative in this case.

This discussion of the Hamiltonian formulation of quantum mechanics has two important purposes: The first is to exhibit the quantum Hamilton equations of motion, Eq. (4.1.35), as an (equivalent) alternative to the Heisenberg equations of motion, since, as indicated above, the Hamilton equations are easier to use when dealing with complicated interaction Hamiltonians. For example, we can readily verify that the interacting electron field equation of motion for $\psi_H(\mathbf{x}, t)$, Eq. (3.2.21), follows immediately from the quantum Hamilton equation, Eq. (4.1.35), with $q_i \leftrightarrow \psi_H(\mathbf{x}, t)$ and $p_i \leftrightarrow i\psi_H^+(\mathbf{x}, t)$ using the Hamiltonian of Eqns. (3.2.15) and (3.2.16). In the course of this derivation, we associate

$$q_i \leftrightarrow \psi_H(\mathbf{x}, t), \tag{4.1.60}$$

$$p_i \leftrightarrow i\psi_H^+(\mathbf{x}, t), \tag{4.1.61}$$

where the spectrum of $\mathbf{x}$ values (three dimensional) is continuous. While the mutual independence of members of a discrete set of q_i, p_i variables is expressed as

$$\frac{\partial q_i}{\partial q_j} = \delta_{ij}; \quad \frac{\partial p_i}{\partial p_j} = \delta_{ij}; \quad \frac{\partial q_i}{\partial p_j} = \frac{\partial p_i}{\partial q_j} = 0, \tag{4.1.62}$$

and sums over them are denoted by $\sum_i$, the corresponding statements of mutual independence of the continuum of variables at all points $\mathbf{x}$ (for a fixed time t), namely $\psi_H(\mathbf{x},t)$, $\psi_H^+(\mathbf{x},t)$, are expressed by (δ symbolizes variation for members of a continuum of variables as does ∂ for a discrete set of variables):

$$\frac{\delta\psi_H(\mathbf{x},t)}{\delta\psi_H(\mathbf{x}',t)} = \delta^{(3)}(\mathbf{x}-\mathbf{x}'); \quad \frac{\delta\psi_H^+(\mathbf{x},t)}{\delta\psi_H^+(\mathbf{x}',t)} = \delta^{(3)}(\mathbf{x}-\mathbf{x}'); \quad \frac{\delta\psi_H(\mathbf{x},t)}{\delta\psi_H^+(\mathbf{x}',t)} = \frac{\delta\psi_H^+(\mathbf{x},t)}{\delta\psi_H(\mathbf{x}',t)} = 0. \tag{4.1.63}$$

Here, $\delta^{(3)}(\mathbf{x}-\mathbf{x}')$ plays the same role under integration over the continuum, $\int d^{(3)}\mathbf{x}$, as does δ_{ij} with respect to a discrete sum, $\sum_i$. The differentiations indicated by the variations in Eq. (4.1.63) are termed *variational derivatives*, and they follow all of the usual rules of calculus within the context of a continuum of variables. This will be discussed further in the next section.

The derivation of the equations of motion for $\psi_H(\mathbf{x},t)$, Eq. (3.2.21), can now proceed from the Hamilton equation, Eq. (4.1.35), as

$$i\frac{\partial\psi_H(\mathbf{x},t)}{\partial t} = \frac{\delta_l H_H^{(1)}}{\delta\psi_H^+(\mathbf{x},t)} + \frac{\delta_l H_H^{(2)}}{\delta\psi_H^+(\mathbf{x},t)}. \tag{4.1.64}$$

$\delta_l H_H^{(1)}$ in the first term on the right may be written as

$$\delta_l H_H^{(1)} = \int d^{(3)}\mathbf{x}'\, \delta_l\psi_H^+(\mathbf{x}',t)\, h^{(1)}(\mathbf{x}',\tfrac{\hbar}{i}\nabla',t)\psi_H(\mathbf{x}',t), \tag{4.1.65}$$

since the Hamilton formulation views $q_i \leftrightarrow \psi_H(\mathbf{x},t)$ and $p_i \leftrightarrow i\psi_H^+(\mathbf{x},t)$ as having mutually independent variations. Dividing Eq. (4.1.65) by $\delta_l\psi_H^+(\mathbf{x},t)$ and using Eq. (4.1.63), we obtain

$$\begin{aligned}\frac{\delta_l H_H^{(1)}}{\delta\psi_H^+(\mathbf{x},t)} &= \int d^{(3)}\mathbf{x}'\,\frac{\delta\psi_H^+(\mathbf{x}',t)}{\delta\psi_H^+(\mathbf{x},t)} h^{(1)}(\mathbf{x}',\tfrac{\hbar}{i}\nabla',t)\psi_H(\mathbf{x}',t)\\ &= \int d^{(3)}\mathbf{x}'\,\delta^{(3)}(\mathbf{x}'-\mathbf{x})h^{(1)}(\mathbf{x}',\tfrac{\hbar}{i}\nabla',t)\psi_H(\mathbf{x}',t)\\ &= h^{(1)}(\mathbf{x},\tfrac{\hbar}{i}\nabla,t)\psi_H(\mathbf{x},t).\end{aligned} \tag{4.1.66}$$

Considering $\delta_l H_H^{(2)}$ next, we have from Eq. (3.2.16),

$$\begin{aligned}\delta_l H_H^{(2)} = \frac{1}{2}\int d^{(3)}\mathbf{x}' \int d^{(3)}\mathbf{x}''\, &\{[\delta\psi_H^+(\mathbf{x}',t)]\,\psi_H^+(\mathbf{x}'',t)\\ &+\psi_H^+(\mathbf{x}',t)\,[\delta\psi_H^+(\mathbf{x}'',t)]\}\, V(\mathbf{x}'-\mathbf{x}'')\psi_H(\mathbf{x}'',t)\,\psi_H(\mathbf{x}',t).\end{aligned} \tag{4.1.67}$$

The case of Fermions will be examined first. In order to determine the left variation, the factor $\delta\psi_H^+(\mathbf{x}'',t)$ must be anticommuted to the left of $\psi_H^+(\mathbf{x}',t)$ in the second term of the curly bracket in the integrand on the right of Eq. (4.1.67), invoking a minus sign. Thus,

$$\delta_l H_H^{(2)} = \frac{1}{2}\int d^{(3)}\mathbf{x}'\int d^{(3)}\mathbf{x}''\,\left\{\left[\delta\psi_H^+(\mathbf{x}',t)\right]\psi_H^+(\mathbf{x}'',t)-\left[\delta\psi_H^+(\mathbf{x}'',t)\right]\psi_H^+(\mathbf{x}',t)\right\}$$
$$\times\mathcal{V}(\mathbf{x}'-\mathbf{x}'')\psi_H(\mathbf{x}'',t)\,\psi_H(\mathbf{x}',t)\,. \tag{4.1.68}$$

Proceeding as above,

$$\begin{aligned}\frac{\delta_l H_H^{(2)}}{\delta\psi_H^+(\mathbf{x},t)} &= \frac{1}{2}\int d^{(3)}\mathbf{x}'\int d^{(3)}\mathbf{x}''\,\left\{\delta^{(3)}(\mathbf{x}'-\mathbf{x})\psi_H^+(\mathbf{x}'',t)-\delta^{(3)}(\mathbf{x}''-\mathbf{x})\psi_H^+(\mathbf{x}',t)\right\}\\ &\quad\times\mathcal{V}(\mathbf{x}'-\mathbf{x}'')\psi_H(\mathbf{x}'',t)\,\psi_H(\mathbf{x}',t)\\ &= \frac{1}{2}\int d^{(3)}\mathbf{x}''\,\psi_H^+(\mathbf{x}'',t)\,\mathcal{V}(\mathbf{x}-\mathbf{x}'')\psi_H(\mathbf{x}'',t)\,\psi_H(\mathbf{x},t)\\ &\quad-\frac{1}{2}\int d^{(3)}\mathbf{x}'\psi_H^+(\mathbf{x}',t)\,\mathcal{V}(\mathbf{x}'-\mathbf{x})\psi_H(\mathbf{x},t)\,\psi_H(\mathbf{x}',t).\end{aligned} \tag{4.1.69}$$

In the last term on the right we employ the anticommutation relation $\psi_H(\mathbf{x},t)\,\psi_H(\mathbf{x}',t) = -\psi_H(\mathbf{x}',t)\,\psi_H(\mathbf{x},t)$, and in the first term, we change the dummy integration variable $\mathbf{x}''\to\mathbf{x}'$. The two terms on the right are thus seen to make equal contributions, canceling the prefactor 1/2, and yielding the equation of motion for $\psi_H(\mathbf{x},t)$ as (restore $\hbar$ temporarily):

$$i\hbar\frac{\partial\psi_H(\mathbf{x},t)}{\partial t} = h^{(1)}(\mathbf{x},\tfrac{\hbar}{i}\nabla,t)\psi_H(\mathbf{x},t) + \int d^{(3)}\mathbf{x}'\mathcal{V}(\mathbf{x}-\mathbf{x}')\psi_H^+(\mathbf{x}',t)\,\psi_H(\mathbf{x}',t)\,\psi_H(\mathbf{x},t)\,, \tag{4.1.70}$$

verifying Eq. (3.2.21) for Fermions. The reader should verify that the same result obtains for Bosons.

The second purpose of this discussion is rooted in the fact that the Schwinger Action Principle, lodged in the Hamilton formulation, provides an exceptionally convenient basis for the development of perturbation theories for the Green's function propagators determined by the equations of motion for interacting systems. These advantages will be evident in our treatment of the interacting electron–hole–phonon system and the interacting electron–electron system in later chapters. Moreover, if this short discussion arouses the reader's curiosity about the quantum mechanical canonical Hamilton formulation of the Action Principle sufficiently to seek out the fully detailed presentation in Schwinger's book *Quantum Kinematics and Dynamics*,[1] that could result in a deeper appreciation of the subject.

Our interest in the Action Principle is focused on a variety of variations in addition to those involving $q_i \leftrightarrow \psi_H(a',t)$ and $p_i \leftrightarrow i\psi_H^+(a'',t)$, since the latter variations $(\delta_{\psi_H},\delta_{\psi_H^+})$ are subsumed in the Hamilton equations of motion, as discussed above. In fact, the "special" class of variations includes most variations of strong interest, such as measures of coupling strength, applied fields at a given space-time point and "source" terms describing the creation/annihilation of Fermion and Boson particles, etc. The Action Principle provides an exceptionally valuable mechanism for the determination of such variations, and with this, it facilitates the formulation of various perturbation theories *without* explicit appeal to Wick's theorem.[4] For such variations (not involving $\delta_{\psi_H}, \delta_{\psi_H^+}$), the Action Principle does not involve L_{kin}, and we may write ($\hbar \to 1$):

$$\delta\left[W_{12}\right] = \int_{t_2}^{t_1} dt\, \delta L(t) = -\int_{t_2}^{t_1} dt\, \delta H_H(t) \tag{4.1.71}$$

and

$$\delta \left\langle a_1', t_1 | a_2'', t_2 \right\rangle_H = i \left\langle a_1', t_1 \left| \delta\left[W_{12}\right] \right| a_2'', t_2 \right\rangle_H = -i \left\langle a_1', t_1 \right| \int_{t_2}^{t_1} dt \delta H_H(t) \left| a_2'', t_2 \right\rangle_H . \tag{4.1.72}$$

As a simple example, suppose we attach to the interparticle potential a factor $\boldsymbol{\lambda}$ measuring its strength, $\mathcal{V} \to \boldsymbol{\lambda}\mathcal{V}$, with $\boldsymbol{\lambda} = 0$ or 1 corresponding to null or full strength. Correspondingly, the two-particle part of the Hamiltonian becomes $H_H^{(2)} \to \boldsymbol{\lambda} H_H^{(2)} = \boldsymbol{\lambda}(H_H - H_H^{(1)})$, and considering a variation $\delta \to \delta_\lambda$ in which $\boldsymbol{\lambda}$ is varied infinitesimally, we have

$$\delta_\lambda \left\langle a_1', t_1 | a_2'', t_2 \right\rangle_H = -i\delta\lambda \left\langle a_1', t_1 \left| \int_{t_2}^{t_1} dt\, H_H^{(2)}(t) \right| a_2'', t_2 \right\rangle_H . \tag{4.1.73}$$

If there is no explicit time dependence in the Hamiltonian, and anticipating that a subsequent trace will render the time dependence of the field operators in $H_H^{(2)}(t)$ inconsequential, we obtain

$$\begin{aligned} \frac{\partial}{\partial \lambda} \left\langle a_1', t_1 | a_2'', t_2 \right\rangle_H &= -\frac{i}{2}(t_1 - t_2) \int d^{(3)}\mathbf{x}' \int d^{(3)}\mathbf{x}'' \mathcal{V}(\mathbf{x}' - \mathbf{x}'') \\ &\times \left\langle a_1', t_1 \left| \psi_H^+(\mathbf{x}', t) \psi_H^+(\mathbf{x}'', t) \psi_H(\mathbf{x}'', t) \psi_H(\mathbf{x}', t) \right| a_2'', t_2 \right\rangle_H . \end{aligned} \tag{4.1.74}$$

This result will be useful at a later stage, in connection with Eq. (7.6.12) and the determination of electron correlation effects.

4.2 Variational Calculus and Sources

As briefly indicated above in connection with Eq. (4.1.63), the concepts of variational calculus relate to the passage from a discrete set of variables to a continuum. For example, the effective potential $V(1)$ at a space-time point $1 = \mathbf{x}_1, t_1$ in a dynamic, non-local polarizable medium is determined by the value of the impressed potential $U(2)$ at *all* space-time points 2 through a relation of the type

$$V(1) = \int d^4 2\, K(1, 2) U(2). \tag{4.2.1}$$

($d^4 2 \equiv d^{(3)}\mathbf{x}dt \equiv dx_1 dx_2 dx_3 dt$ will be written as $d^4 2 \to d2$ hereafter.) Assuming linearity of the medium in this relation, $K(1,2)$ is independent of $U(2)$, and we say that $V(1)$ is a *linear functional* of $U(2)$. In this, $U(2)$ is considered to be a distinct independent variable at each space-time point 2, whose independence of, say, $U(3)$ is expressed by

$$\frac{\delta U(2)}{\delta U(3)} = \delta^4(2-3), \tag{4.2.2}$$

($\delta^4(2\text{–}3) = \delta^{(3)}(\mathbf{x}_2\text{–}\mathbf{x}_3)\delta(t_2\text{–}t_3) \to \delta(2\text{–}3)$ hereafter) as a continuum analog of the mutual independence of variables $\{x_i\}$ in multivariable calculus expressed by $\partial x_i / \partial x_j = \delta_{ij}$. Given the linearity of Eq. (4.2.1), we may form a *variational/functional derivative* of $V(1)$ with respect to $U(1')$ in complete analogy to partial differentiation in multivariate calculus,

$$\begin{aligned}\frac{\delta V(1)}{\delta U(1')} &= \frac{\delta}{\delta U(1')}\int d^4 2\, K(1,2)U(2) \\ &= \int d^4 2\, K(1,2)\frac{\delta U(2)}{\delta U(1')} = \int d^4 2\, K(1,2)\delta^4(2-1')\end{aligned} \tag{4.2.3}$$

or

$$\frac{\delta V(1)}{\delta U(1')} = K(1,1'), \tag{4.2.4}$$

so that the linear coefficient of the integrand, $K(1,1')$, is just the variational derivative of $V(1)$ with respect to $U(1')$. Thus, Eq. (4.2.1) is the continuum analog of the multivariate calculus relation $y = \sum_i c_i x_i$ with Eq. (4.2.4) corresponding to $\partial y / \partial x_i = c_i$.

From these considerations, it is clear that other relations of multivariate calculus have analogous continuum relations in variational calculus, such as the chain rule of variational differentiation,

$$\frac{\delta V(1)}{\delta U(1')} = \int d^4 2\, \frac{\delta V(1)}{\delta Z(2)}\frac{\delta Z(2)}{\delta U(1')}, \tag{4.2.5}$$

as well as rules for variational derivatives of products and quotients, etc.

As a simple example of the use of variational differentiation, we can obtain the classical Euler–Lagrange equations[2,3] from the condition that the action integral be stationary (extremum) under fixed-endpoint variations $\delta q_i(t)$ of the action integral, i.e.:

$$\frac{\delta W}{\delta q_i(t)} = 0, \tag{4.2.6}$$

where $q_i(t)$ is a generalized coordinate of the classical mechanical system. Since the Lagrangian depends on all the generalized coordinates $\{q_k(t)\}$ and velocities $\{\dot{q}_k(t)\}$,

$$W = \int_{t_2}^{t_1} dt' L\left(\{q_k(t')\}, \{\dot{q}_k(t')\}, t'\right), \tag{4.2.7}$$

(we use t' for the dummy time integration variable) we have

$$\delta W = \int_{t_2}^{t_1} dt' \left(\sum_k \frac{\partial L}{\partial q_k} \delta q_k + \sum_k \frac{\partial L}{\partial \dot{q}_k} \delta \dot{q}_k \right).$$

Noting that $\delta \dot{q}_k = \frac{d}{dt'}(\delta q_k)$ and $\delta q_k(t_2) = \delta q_k(t_1) = 0$ for a fixed endpoint variation, integration by parts yields

$$\delta W = \int_{t_2}^{t_1} dt' \sum_k \left[\frac{\partial L}{\partial q_k} - \frac{d}{dt'} \left(\frac{\partial L}{\partial \dot{q}_k} \right) \right] \delta q_k \left(t'\right), \tag{4.2.8}$$

and, consequently,

$$\frac{\delta W}{\delta q_i(t)} = \int_{t_2}^{t_1} dt' \sum_k \left[\frac{\partial L}{\partial q_k} - \frac{d}{dt'} \left(\frac{\partial L}{\partial \dot{q}_k} \right) \right] \frac{\delta q_k(t')}{\delta q_i(t)}. \tag{4.2.9}$$

As the q_k's are mutually independent variables for different index-k-values, and as we regard $q_k(t')$ as a distinct independent variable for each value of t' in variational differentiation, we have

$$\frac{\delta q_k(t')}{\delta q_i(t)} = \delta_{ki} \delta(t' - t), \tag{4.2.10}$$

whence we immediately obtain the Euler–Lagrange equations from the extremum condition of Eq. (4.2.6):

$$\frac{\delta W}{\delta q_i(t)} = \frac{\partial L}{\partial q_i} - \frac{d}{dt} \left(\frac{\partial L}{\partial \dot{q}_i} \right) = 0, \quad i = 1 \ldots n. \tag{4.2.11}$$

Similar considerations apply if we require the extremum of a functional W of a multicomponent classical (c-number) field, ϕ_i $(i = 1, \ldots m)$, with the Lagrangian density $\mathscr{L}\left(\{\phi_i(\mathbf{x})\}, \{\partial \phi_i / \partial x_k\}, \mathbf{x}\right)$ that depends on several variables, $x_1, x_2, \ldots x_n = \mathbf{x}$, as

$$W = \int_V dx_1 \ldots \int_V dx_n \mathscr{L}\left(\{\phi_i(\mathbf{x})\}, \{\partial \phi_i / \partial x_k\}, \mathbf{x}\right), \tag{4.2.12}$$

where V is the volume of integration in the n-dimensional space of $x_1 \ldots x_n$, which is bounded by a closed $(n-1)$-dimensional surface S. Forming δW for a variation of ϕ_i, we have

$$\delta W = \int_V dx_1 \ldots \int_V dx_n \sum_{i=1}^{m} \left\{ \frac{\partial \mathscr{L}}{\partial \phi_i(\mathbf{x})} \delta\phi_i(\mathbf{x}) + \sum_{k=1}^{n} \frac{\partial \mathscr{L}}{\partial(\partial\phi_i/\partial x_k)} \delta\left(\frac{\partial\phi_i}{\partial x_k}\right) \right\}. \tag{4.2.13}$$

Noting that $\delta\left(\frac{\partial\phi_i}{\partial x_k}\right) = \frac{\partial(\delta\phi_i)}{\partial x_k}$, we integrate by parts with respect to x_k, obtaining

$$\delta W = \int_V dx_1 \ldots \int_V dx_n \sum_{i=1}^{m} \left[\frac{\partial \mathscr{L}}{\partial \phi_i(x)} - \sum_{k=1}^{n} \frac{\partial\left(\frac{\partial \mathscr{L}}{\partial(\partial\phi_i/\partial x_k)}\right)}{\partial x_k} \right] \delta\phi_i(\mathbf{x})$$

$$+ \int_V dx_1 \ldots \int_V dx_n \sum_{i=1}^{m} \sum_{k=1}^{n} \frac{\partial\left\{\frac{\partial \mathscr{L}}{\partial(\partial\phi_i/\partial x_k)}\delta\phi_i(\mathbf{x})\right\}}{\partial x_k}. \tag{4.2.14}$$

Introducing the notation $\phi_{i,k} \equiv \partial\phi_i/\partial x_k$, we observe that for a fixed value of i, $\partial\mathscr{L}/\partial\phi_{i,k}$ is an n-dimensional vector and $\sum_{k=1}^{n} \frac{\partial}{\partial x_k}\left(\partial\mathscr{L}/\partial\phi_{i,k}\right)$ is its n-dimensional divergence. A trivial extension of the divergence theorem to n-dimensions yields the last term on the right of Eq. (4.2.14) as the surface integral (over the closed $(n-1)$-dimensional surface S) of the normal component of $\left(\partial\mathscr{L}/\partial\phi_{i,k}\right)\delta\phi_i$, evaluated at the surface. For a fixed endpoint variation, $\delta\phi_i(\mathbf{x})$ is zero at the surface, so the last term on the right of Eq. (4.2.14) vanishes. Finally, we divide the rest of Eq. (4.2.14) by $\delta\phi_j(\mathbf{x}')$ and note that mutual independence of the variables $\phi_i(\mathbf{x})$ means that $\delta\phi_i(\mathbf{x})/\delta\phi_j(\mathbf{x}') = \delta_{ij}\delta^{(n)}(\mathbf{x}-\mathbf{x}')$, so the n-dimensional volume integral and m-dimensional sum, $\sum_{i=1}^{m}$, yield the result

$$\frac{\delta W}{\delta\phi_j(\mathbf{x}')} = \frac{\partial\mathscr{L}}{\delta\phi_j(\mathbf{x}')} - \sum_{k=1}^{n} \frac{\partial}{\partial x_k}\left(\partial\mathscr{L}/\partial\phi_{j,k}(\mathbf{x}')\right). \tag{4.2.15}$$

Moreover, the extremum condition $\delta W = 0$ then provides the classical Euler–Lagrange field equations as $(j = 1 \ldots m)$:

$$\frac{\partial\mathscr{L}}{\partial\phi_j(\mathbf{x})} - \sum_{k=1}^{n} \frac{\partial}{\partial x_k}\left(\partial\mathscr{L}/\partial\phi_{j,k}(\mathbf{x})\right) = 0. \tag{4.2.16}$$

The Action Principle in the form of Eq. (4.1.71) is a powerful vehicle for the determination of variational derivatives of matrix elements of operators that vary with time. Considering an operator $O_H(t)$, the variation

$$\delta\left\langle a_1', t_1 \left| O_H(t) \right| a_2'', t_2 \right\rangle_H \tag{4.2.17}$$

is most conveniently determined by referring the time dependence to the bra-kets on the right and left of $O_H(t)$ and using the Action Principle, Eqns. (4.1.5)–(4.1.72).

This may be accomplished by introducing unit operators $\mathbf{I} = \sum_{b'} |b', t\rangle_H \langle b', t|_H$ and $\mathbf{I} = \sum_{b''} |b'', t\rangle_H \langle b'', t|_H$ to the left and right of $O_H(t)$ in Eq. (4.2.17), respectively, obtaining

$$\delta \langle a_1', t_1 | O_H(t) | a_2'', t_2\rangle_H = \sum_{b',b''} \delta \left[\langle a_1', t_1 | b', t\rangle_H \langle b', t| \, O_H(t) \, |b'', t\rangle_H \langle b'', t | a_2'', t_2\rangle_H\right]. \quad (4.2.18)$$

Since $\langle b', t| \, O_H(t) \, |b'', t\rangle_H$ is independent of t, and noting that variational derivatives follow the usual rules for differentiation of products, we have

$$\begin{aligned}\delta \langle a_1', t_1 | O_H(t) | a_2'', t_2\rangle_H &= \sum_{b',b''} \delta \left(\langle a_1', t_1 | b', t\rangle_H\right) \langle b', t| \, O_H(t) \, |b'', t\rangle_H \langle b'', t | a_2'', t_2\rangle_H \\ &+ \sum_{b',b''} \langle a_1', t_1 | b', t\rangle_H \langle b', t| \, O_H(t) \, |b'', t\rangle_H \, \delta \left(\langle b'', t | a_2'', t_2\rangle_H\right).\end{aligned} \quad (4.2.19)$$

Restoring the identity $\mathbf{I} = \sum_{b''} |b'', t\rangle_H \langle b'', t|_H$ in the first term on the right of Eq. (4.2.19), and $\mathbf{I} = \sum_{b'} |b', t\rangle_H \langle b', t|_H$ in the second term, we have

$$\begin{aligned}\delta \langle a_1', t_1 | O_H(t) | a_2'', t_2\rangle_H &= \sum_{b'} \delta \left(\langle a_1', t_1 | b', t\rangle_H\right) \langle b', t | O_H(t) | a_2'', t_2\rangle_H \\ &+ \sum_{b'} \langle a_1', t_1 | O_H(t) | b', t\rangle_H \, \delta \left(\langle b', t | a_2'', t_2\rangle_H\right).\end{aligned} \quad (4.2.20)$$

The Action Principle (Eqns. (4.1.71) and (4.1.72)) is also directly applicable to the further variations of transformation functions *not* involving the field variables (coordinates and momenta),

$$\begin{aligned}\delta \left(\langle a', t_1 | b', t\rangle_H\right) &= -i \left\langle a', t_1 \left| \int_t^{t_1} dt' \delta H_H(t') \right| b', t \right\rangle_H, \\ \delta \left(\langle b', t | a'', t_2\rangle_H\right) &= -i \left\langle b', t \left| \int_{t_2}^{t} dt' \delta H_H(t') \right| a'', t_2 \right\rangle_H,\end{aligned} \quad (4.2.21)$$

and using the results in Eq. (4.2.20), again restoring the identity $\mathbf{I} = \sum_{b'} |b', t\rangle_H \langle b', t|_H$ we obtain

$$\delta \langle a_1', t_1 | O_H(t) | a_2'', t_2 \rangle_H = -i \left\langle a_1', t_1 \left| \int_t^{t_1} dt' \delta H_H(t') O_H(t) \right| a_2'', t_2 \right\rangle_H$$
$$- i \left\langle a_1', t_1 \left| \int_{t_2}^{t} dt' O_H(t) \delta H_H(t') \right| a_2'', t_2 \right\rangle_H . \qquad (4.2.22)$$

Noting that $t' > t$ in the first term on the right, and $t' < t$ in the second term, it is clear that the Action Principle automatically time-orders the operators $O_H(t)\delta H_H(t')$. Thus,

$$\delta \langle a_1', t_1 | O_H(t) | a_2'', t_2 \rangle_H = -i \left\langle a_1', t_1 \left| \int_{t_2}^{t_1} dt' \left(\delta H_H(t') O_H(t) \right)_+ \right| a_2'', t_2 \right\rangle_H \qquad (4.2.23)$$

(recall the time-ordering symbol $(\ldots)_+$ was defined in Eq. (3.1.19)). If the variation occurs in a space-time dependent c-number parameter $Z(\mathbf{x}, t)$, which enters part of the Hamiltonian density multiplying an operator $Q_H(\mathbf{x}, t)$ as $H_H(t) \to \int d^{(3)}\mathbf{x}\, Z(\mathbf{x}, t) Q_H(\mathbf{x}, t)$, then

$$\delta \langle a_1', t_1 | O_H(t) | a_2'', t_2 \rangle_H = -i \left\langle a_1', t_1 \left| \int_{t_2}^{t_1} dt' \int d^{(3)}\mathbf{x}' \delta Z(\mathbf{x}', t') \left(Q_H(\mathbf{x}', t') O_H(t) \right)_+ \right| a_2'', t_2 \right\rangle_H : \qquad (4.2.24)$$

and the variational derivative is given by (we assume that $\delta Z(\mathbf{x}, t)$ commutes with $O_H(t)$, which is always true for Bosons and is also true for Fermions if $O_H(t)$ is an even function of $\psi_H^+(1), \psi_H(1)$):

$$\frac{\delta \langle a_1', t_1 | O_H(t) | a_2'', t_2 \rangle_H}{\delta Z(\mathbf{x}', t')} = -i \langle a_1', t_1 | \left(Q_H(\mathbf{x}', t') O_H(t) \right)_+ | a_2'', t_2 \rangle_H \quad \text{for } t_2 \le t' \le t_1$$
$$= 0 \text{ otherwise,} \qquad (4.2.25)$$

since $\frac{\delta Z(\mathbf{x}', t')}{\delta Z(\mathbf{x}'', t'')} = \delta^{(3)}(\mathbf{x}' - \mathbf{x}'')\delta(t' - t'')$ contributes to the integrand $\int_{t_2}^{t_1} dt'$ only if t'' is in the integration range, $t_1 \ge t'' \ge t_2$. The only variations excluded from this result are those involving variations of the field variables $\psi_H^+(1), \psi_H(1)$, which are subsumed in the equations of motion in Hamilton form as discussed above (Eq. (4.1.35)).

It is often necessary to evaluate a variational derivative of the form

$$\frac{\delta \langle O_H(t) \rangle_H}{\delta Z(\mathbf{x}', t')}, \text{ where we define } \langle O_H(t) \rangle_H \equiv \frac{\langle a_1', t_1 | O_H(t) | a_2'', t_2 \rangle_H}{\langle a_1', t_1 | a_2'', t_2 \rangle_H}. \qquad (4.2.26)$$

Since variational derivatives follow similar rules for products and quotients as do ordinary derivatives, we have

$$\frac{\delta \langle O_H(t)\rangle_H}{\delta Z(\mathbf{x}',t')} = \frac{\left[\delta \langle a_1',t_1\,|O_H(t)|\,a_2'',t_2\rangle_H /\delta Z(\mathbf{x}',t')\right]}{\langle a_1',t_1|\,a_2'',t_2\rangle_H}$$

$$-\frac{\langle a_1',t_1\,|O_H(t)|\,a_2'',t_2\rangle_H}{\langle a_1',t_1|\,a_2'',t_2\rangle_H}\frac{\left[\delta \langle a_1',t_1|\,a_2'',t_2\rangle_H /\delta Z(\mathbf{x}',t')\right]}{\langle a_1',t_1|\,a_2'',t_2\rangle_H}. \quad (4.2.27)$$

For $t_2 \le t' \le t_1$, Eq. (4.2.27) yields

$$\frac{\delta \langle O_H(t)\rangle_H}{\delta Z(\mathbf{x}',t')} = \frac{-i\langle a_1',t_1\,|(Q_H(\mathbf{x}',t')O_H(t))_+|\,a_2'',t_2\rangle_H}{\langle a_1',t_1|\,a_2'',t_2\rangle_H} + i\,\langle O_H(t)\rangle_H \frac{\langle a_1',t_1\,|Q_H(\mathbf{x}',t')|\,a_2'',t_2\rangle_H}{\langle a_1',t_1|\,a_2'',t_2\rangle_H} \quad (4.2.28)$$

(where the last numerator factor on the right is obtained from Eq. (4.2.25) by setting $O_H(t) \to 1$). Thus we obtain

$$\frac{\delta \langle O_H(t)\rangle_H}{\delta Z(\mathbf{x}',t')} = -i\left[\langle (Q_H(\mathbf{x}',t')O_H(t))_+\rangle_H - \langle O_H(t)\rangle_H \langle Q_H(\mathbf{x}',t')\rangle_H\right], \quad (4.2.29)$$

for $t_2 \le t' \le t_1$; otherwise the right-hand side vanishes.

Our formulation of the theory of interacting systems will be based on the introduction of "sources" in the form of test fields that probe the system's response characteristics, paving the way for the development of perturbation theories. In this book we will employ the following sources and associated additional terms in the Hamiltonian density.

4.2.1 Potential Sources: $U(\mathrm{x}, t)$

A single-particle local potential $U(1)$ enters the one-particle part of the Hamiltonian density through a term of the form

$$H_H^U(\mathbf{x},t) \to eU(\mathbf{x},t)\,\psi_H^+(\mathbf{x},t)\psi_H(\mathbf{x},t). \quad (4.2.30)$$

The associated variational derivative is given by Eq. (4.2.25) as

$$\begin{aligned}\frac{\delta \langle a_1',t_1\,|O_H(t)|\,a_2'',t_2\rangle_H}{\delta U(\mathbf{x}',t')} &= -ie\langle a_1',t_1\,|(\psi_H^+(\mathbf{x}',t')\psi_H(\mathbf{x}',t')O_H(t))_+|\,a_2'',t_2\rangle_H \\ &= -ie\langle a_1',t_1\,|(\rho_H(\mathbf{x}',t')O_H(t))_+|\,a_2'',t_2\rangle_H, \quad (4.2.31)\end{aligned}$$

for $t_2 \le t' \le t_1$; otherwise it vanishes,

where $\rho_H(\mathbf{x}',t') = \psi_H^+(\mathbf{x}',t')\psi_H(\mathbf{x}',t')$ is the density operator.

4.2.2 Fermion/Boson Particle Sources: $\eta\,(\mathbf{x}, t)$, $\eta^+\,(\mathbf{x}, t)$

These particle source terms enter the Hamiltonian density through a term of the form (note that the sign of these source terms is the negative of those usually introduced in the Lagrangian):

$$H_H^{\eta}\,(\mathbf{x}, t) \rightarrow \left[\eta^+\,(\mathbf{x}, t)\,\psi_H(\mathbf{x}, t) + \psi_H^+(\mathbf{x}, t)\eta\,(\mathbf{x}, t)\right], \tag{4.2.32}$$

where $H_H^{\eta}\,(\mathbf{x}, t)$ is manifestly Hermitian and the requirement that it be an even function of Fermion variables means that $\eta\,(\mathbf{x}, t)$ and $\eta^+\,(\mathbf{x}, t)$ must be anticommutative, as well as $\delta\eta$ and $\delta\eta^+$, in the Fermion case. The introduction of such a particle source term, $H_H^{\eta}\,(\mathbf{x}, t)$, implies that the electron field Hamilton equation of motion for ψ_H (Eqns. (4.1.35), (4.1.64) and (4.1.70)) has an inhomogeneous driving term, as follows:

$$\left[i\hbar\frac{\partial}{\partial t} - h^{(1)}\left(\mathbf{x}, \tfrac{\hbar}{i}\nabla, t\right)\right]\psi_H(\mathbf{x}, t) - \int d^{(3)}\mathbf{x}'\mathcal{V}(\mathbf{x}-\mathbf{x}')\psi_H^+(\mathbf{x}', t)\psi_H(\mathbf{x}', t)\psi_H(\mathbf{x}, t) = \eta\,(\mathbf{x}, t)\,, \tag{4.2.33}$$

in contrast to Eq. (3.2.21), which lacks $\eta\,(\mathbf{x}, t)$. This leads to particle source dependence in associated Green's functions. In contrast to the Hamiltonian forms of $H_H^{(1)}$ (Eq. (3.2.15)), $H_H^{(2)}$ (Eq. (3.2.16)) H_H^U (Eq. (4.2.30)), which contain like numbers of ψ_H and ψ_H^+ operators and thereby conserve particle number, it should be noted that $\left[H_H^{\eta}, \mathbf{N}\right] \neq 0$, so that the presence of the particle sources in the Hamiltonian violates number conservation.

The associated variational derivatives with respect to particle sources are given by

$$\frac{\delta_l \langle a_1', t_1\,|O_H(t)|\,a_2'', t_2\rangle_H}{\delta\eta^+\,(\mathbf{x}', t')} = -i\langle a_1', t_1\,|\left(\psi_H(\mathbf{x}', t')O_H(t)\right)_+|\,a_2'', t_2\rangle_H \tag{4.2.34}$$

for Bosons with $O_H\,(t)$ either even or odd as a function of ψ_H, ψ_H^+; and also for Fermions if O_H is an even function of ψ_H, ψ_H^+, for both time orderings and $t_2 \leq t' \leq t_1$; otherwise it vanishes. Furthermore,

$$\frac{\delta_r \langle a_1', t_1\,|O_H(t)|\,a_2'', t_2\rangle_H}{\delta\eta\,(\mathbf{x}', t')} = -i\langle a_1', t_1\,|\left(\psi_H^+(\mathbf{x}', t')O_H(t)\right)_+|\,a_2'', t_2\rangle_H\,,$$

$$\text{for } t_2 \leq t' \leq t_1\text{; otherwise it vanishes,} \tag{4.2.35}$$

for Bosons with $O_H(t)$ either even or odd as a function of ψ_H, ψ_H^+; and also for Fermions if $O_H(t)$ is an *even* function of Fermion variables, $\psi_H^+(\mathbf{x}', t')$, $\psi_H(\mathbf{x}', t')$, for both time orderings. It is readily verified that $\delta_l/\delta\eta \equiv -\delta_r/\delta\eta$ in these relations for the Fermion case with O_H even, whereas $\delta_l/\delta\eta \equiv \delta_r/\delta\eta$ always for Bosons.

However, if $O_H(t)$ is an *odd* function of Fermion variables, then $\delta_l/\delta\eta \equiv \delta_r/\delta\eta$ for both time orderings for Fermions and Bosons. But the anticommutation property of $\delta\eta$ in the

Fermion case induces an additional sign change when O_H is an odd function for $t' > t$, so the result for odd O_H is given by

$$\frac{\delta_r \langle a_1', t_1 | O_H(t) | a_2'', t_2 \rangle_H}{\delta\eta(\mathbf{x}', t')} = \frac{\delta_l \langle a_1', t_1 | O_H(t) | a_2'', t_2 \rangle_H}{\delta\eta(\mathbf{x}', t')}$$
$$= -i\varepsilon \langle a_1', t_1 | \left(\psi_H^+(\mathbf{x}', t') O_H(t)\right)_+ | a_2'', t_2 \rangle_H \qquad (O_H \text{ odd})$$
$$\text{for } t_2 \le t' \le t_1; \text{ otherwise it vanishes,} \tag{4.2.36}$$

where $\varepsilon = +1$ for $t' < t$ and $\varepsilon = -1$ for $t' > t$ in the Fermion case (but $\varepsilon \equiv +1$ for Bosons). The particular case of $O_H(t) = \psi_H(\mathbf{x}, t)$ is of special interest, yielding

$$\frac{\delta_r \langle a_1', t_1 | \psi_H(\mathbf{x}, t) | a_2'', t_2 \rangle_H}{\delta\eta(\mathbf{x}', t')} = -i\varepsilon \langle a_1', t_1 | \left(\psi_H(\mathbf{x}, t) \psi_H^+(\mathbf{x}', t')\right)_+ | a_2'', t_2 \rangle_H$$
$$\text{for } t_2 \le t' \le t_1; \text{ otherwise it vanishes.} \tag{4.2.37}$$

Considering an average of an operator P defined as

$$\langle P \rangle_H \to \langle P \rangle = \frac{\langle a_1', t_1 | P | a_2'', t_2 \rangle_H}{\langle a_1', t_1 | a_2'', t_2 \rangle_H}, \tag{4.2.38}$$

we have, with $P \to \psi_H(\mathbf{x}, t)$,

$$\frac{\delta_r \langle \psi_H(\mathbf{x}, t) \rangle}{\delta\eta(\mathbf{x}', t')} = \frac{1}{\langle a_1', t_1 | a_2'', t_2 \rangle_H} \frac{\delta_r \langle a_1', t_1 | \psi_H(\mathbf{x}, t) | a_2'', t_2 \rangle_H}{\delta\eta(\mathbf{x}', t')}$$
$$- \frac{\langle a_1', t_1 | \psi_H(\mathbf{x}, t) | a_2'', t_2 \rangle_H}{\langle a_1', t_1 | a_2'', t_2 \rangle_H} \frac{1}{\langle a_1', t_1 | a_2'', t_2 \rangle_H} \frac{\delta_r \langle a_1', t_1 | a_2'', t_2 \rangle_H}{\delta\eta(\mathbf{x}', t')}$$
$$= -i\varepsilon \left\langle \left(\psi_H(\mathbf{x}, t) \psi_H^+(\mathbf{x}', t')\right)_+ \right\rangle + i \langle \psi_H(\mathbf{x}, t) \rangle \left\langle \psi_H^+(\mathbf{x}', t') \right\rangle. \tag{4.2.39}$$

(The last variational derivative, $\delta_r \langle a_1', t_1 | a_2'', t_2 \rangle_H / \delta\eta(\mathbf{x}', t')$, is evaluated using Eq. (4.2.35) with $O_H \to 1$.) In the limit of vanishing particle source ($\eta \to 0$), matrix elements of odd numbers of operators, such as $\langle \psi_H \rangle$ or $\langle \psi_H^+ \rangle$, vanish between states having the same particle numbers, since their operation involves changing the particle number of, say, the state on the right, which is then orthogonal to the state on the left. Thus, we obtain the limiting result for $\eta \to 0$ as

$$\frac{\delta_r \langle \psi_H(\mathbf{x}, t) \rangle}{\delta\eta(\mathbf{x}', t')} = \frac{\delta_l \langle \psi_H(\mathbf{x}, t) \rangle}{\delta\eta(\mathbf{x}', t')} = -i\varepsilon \left\langle \left(\psi_H(\mathbf{x}, t) \psi_H^+(\mathbf{x}', t')\right)_+ \right\rangle. \tag{4.2.40}$$

4.2.3 Boson–Phonon Particle Sources: $\mathcal{J}_{l,k}(t)$

Anticipating consideration of electron/hole interaction with phonons, which are quanta of lattice vibrations having the displacement operator ${}^{i}Q_{l,k}(t)$ for the kth ion in the lth

unit cell of a lattice (i^{th} component), the source ${}^{i}\mathcal{J}_{l,k}$ enters this part of the Hamiltonian (*not* Hamiltonian density) as

$$H_H^{\mathcal{J}}(t) \to \sum_{l,k,i} {}^{i}\mathcal{J}_{l,k}(t)\; {}^{i}Q_{l,k}(t)\,. \tag{4.2.41}$$

($H_H^{\mathcal{J}}$ does not conserve phonon number.) The associated variational derivative with respect to ${}^{i}\mathcal{J}_{lk}$ is given by

$$\frac{\delta \langle a_1', t_1 | O_H(t) | a_2'', t_2\rangle_H}{\delta\, {}^{i}\mathcal{J}_{l,k}(t')} = -i \langle a_1', t_1 | ({}^{i}Q_{l,k}(t')\, O_H(t))_+ | a_2'', t_2\rangle_H,$$
$$\text{for } t_2 \le t' \le t_1;\ \text{otherwise it vanishes.} \tag{4.2.42}$$

Other sources (such as an impressed electric field source, $\mathbf{E}_0$) to be employed will be explained as the need arises. Variational differentiation with respect to sources will be helpful in developing Green's function equations for interacting systems, including self-interactions such as electron–electron interactions, and in developing perturbation theories expeditiously.

4.3 Particle Sources: General Considerations

In the preceding section we saw that a variational derivative of an average with respect to particle source η (η^+) introduces a factor $\psi^+(\psi)$ into the average, subject to time ordering, along with a factor $-i$ (with due care for distinguishing between δ_r and δ_l in the Fermion case when O_H is an even function of Fermi variables). Moreover, a variational derivative with respect to the source ${}^{i}\mathcal{J}_{l,k}$ introduces a factor ${}^{i}Q_{l,k}$, subject to time ordering, with a factor $-i$. On the basis of these considerations one can systematically describe matrix elements of field operators and associated averages in terms of variational derivatives. Starting with $\langle a_1', t_1 | a_2'', t_2\rangle_H$, we may employ Eq. (4.2.34) to obtain (with $O_H \longrightarrow 1$):

$$\frac{\delta_l \langle a_1', t_1 | a_2'', t_2\rangle_H}{\delta\eta^+(\mathbf{x}, t)} = -i \langle a_1', t_1 | \psi_H(\mathbf{x}, t) | a_2'', t_2\rangle$$
$$\text{for } t_2 \le t \le t_1;\ \text{otherwise it vanishes.} \tag{4.3.1}$$

A second variational differentiation using Eq. (4.2.36) (with $O_H \longrightarrow \psi_H$) yields

$$\frac{\delta_r}{\delta\eta(\mathbf{x}', t')} \frac{\delta_l}{\delta\eta^+(\mathbf{x}, t)} \langle a_1', t_1 | a_2'', t_2\rangle_H = -i \frac{\delta_r}{\delta\eta(\mathbf{x}', t')} \langle a_1', t_1 | \psi_H(\mathbf{x}, t) | a_2'', t_2\rangle, \tag{4.3.2}$$

which was evaluated above in Eq. (4.2.40), with the result (noting that $\delta_r = \delta_l$ for O_H an odd function of Fermi variables):

$$\frac{\delta_l}{\delta\eta(\mathbf{x}', t')} \frac{\delta_l}{\delta\eta^+(\mathbf{x}, t)} \langle a_1', t_1 | a_2'', t_2\rangle_H = (-i)^2 \varepsilon \langle a_1', t_1 | (\psi_H(\mathbf{x}, t)\psi_H^+(\mathbf{x}', t'))_+ | a_2'', t_2\rangle, \tag{4.3.3}$$

for $t_2 \le t,\ t' \le t_1$; otherwise it vanishes. Similarly,

$$\frac{\delta}{\delta\, {}^{i}\mathcal{J}_{l,k}(t'')}\frac{\delta_l}{\delta\eta(\mathbf{x}',t')}\frac{\delta_l}{\delta\eta^{+}(\mathbf{x},t)}\left\langle a_1',t_1\left|a_2'',t_2\right.\right\rangle_H$$
$$= (-i)^3\varepsilon\left\langle a_1',t_1\left|(\psi_H(\mathbf{x},t)\psi_H^{+}(\mathbf{x}',t')\,{}^{i}Q_{l,k}(t''))_+\right|a_2'',t_2\right\rangle_H \tag{4.3.4}$$

for $t_2 \leq t,\ t',\ t'' \leq t_1$; otherwise it vanishes.

In this, we note that the order in which $\psi, \psi^+, {}^{i}Q_{l,k}$ are written within a time-ordered product is irrelevant, since the time ordering determines the actual order of such factors. Moreover, the order of variational derivatives can be interchanged even if the operators do not commute, and correspondingly, the order of the non-commuting operators is determined by their relative time arguments.

It is clear that we can generate the presence of M $-\ \psi$'s and N $-\ \psi^+$'s and P $-\ {}^{i}Q_{l,k}$'s in the time-ordered operator of the matrix element $\langle a_1', t_1|(\ldots)_+|a_2'', t_2\rangle_H$ with the application of M $-\ \delta_l/\delta\eta^+$ and N $-\ \delta_r/\delta\eta$ and P $-\ \delta/\delta\ {}^{i}\mathcal{J}_{l,k}$ variational derivatives (at differing space-time points) to the transformation function $\langle a_1', t_1|a_2'', t_2\rangle$ in the presence of sources, along with the various interactions in the Hamiltonian. Furthermore, considering the potential source of Eq. (4.2.30) for a single species, each variational derivative $\delta/\delta U(\mathbf{x}, t)$ generates the presence of a factor $-ie\psi_H^+(\mathbf{x}, t)\psi_H(\mathbf{x}, t)$ in the time-ordered operator of the matrix element.

It is convenient to define the closely related "normalized" transformation function,

$$e^{\mathcal{W}} = \frac{\left\langle a_1',t_1\left|a_2'',t_2\right.\right\rangle_H}{\left\langle a_1',t_1\left|a_2'',t_2\right.\right\rangle_0} \Longrightarrow \frac{\left\langle a_1',t_1\left|a_2'',t_2\right.\right\rangle_{CS}}{\left\langle a_1',t_1\left|a_2'',t_2\right.\right\rangle_C}, \tag{4.3.5}$$

where the denominator $\left\langle a_1',t_1\left|a_2'',t_2\right.\right\rangle_0 \equiv \left\langle a_1',t_1\left|a_2'',t_2\right.\right\rangle_C$ is the transformation function in the *absence of sources,* η, η^+, ${}^{i}\mathcal{J}_{l,k}$, $U \longrightarrow 0$, but which includes the effects of all other interaction couplings in the Hamiltonian, $H_H = H_H^{(1)} + H_H^{(2)}$, including particle–particle interactions, as in Eq. (3.2.16), as well as coupling with the phonon field written in the form

$$H_{ep} = -\sum_i\sum_{l,k}\int d^{(3)}\mathbf{x}_1\ {}^{i}\gamma_{l,k}^{e}(\mathbf{x}_1)\ {}^{i}Q_{l,k}(t_1)\ \psi_H^{+}(1^{+})\ \psi_H(1), \tag{4.3.6}$$

where ${}^{i}\gamma_{l,k}^{e}(\mathbf{x}_1)$ is a specified c-number function to be discussed in Chapter 11. It should be noted that H_{ep} does not conserve phonon number. In the numerator on the right of Eq. (4.3.5), we have written $\left\langle a_1',t_1\left|a_2'',t_2\right.\right\rangle_H \rightarrow \left\langle a_1',t_1\left|a_2'',t_2\right.\right\rangle_{CS}$ to emphasize that its time development takes place under the Hamiltonian $H_H + H_H^{\eta} + H_H^{\mathcal{J}} + H_H^{U}$, which includes all kinematic and interaction-coupling (subscript "C") parts of H_H *as well as* the source (subscript "S") terms, $H_H^{\eta} + H_H^{\mathcal{J}} + H_H^{U}$.

Since the denominator of Eq. (4.3.5), $\langle a_1', t_1 | a_2'', t_2 \rangle_C$, does not involve sources, it is inert under variational differentiations with respect to sources, acting as a constant. Consequently, we have

$$\prod_{m=1}^{M} \frac{\delta_l}{\delta \eta^+ (\mathbf{x}_m, t_m)} \prod_{n=1}^{N} \frac{\delta_r}{\delta \eta (\mathbf{x}_n', t_n')} \prod_{p=1}^{P} \frac{\delta}{\delta \, {}^i\mathcal{J}_{l,k} (t_p'')} e^{\mathcal{W}} = (-i)^{M+N+P} \varepsilon (t_1, \ldots, t_M, t_1', \ldots, t_N')$$

$$\times \frac{\left\langle a_1', t_1 \left| \left(\prod_{m=1}^{M} \psi_H (\mathbf{x}_m, t_m) \prod_{n=1}^{N} \psi_H^+ (\mathbf{x}_n', t_n') \prod_{p=1}^{P} {}^i Q_{l,k} (t_p'') \right)_+ \right| a_2'', t_2 \right\rangle}{\langle a_1', t_1 | a_2'', t_2 \rangle_C}, \tag{4.3.7}$$

where $\delta/\delta\eta^+$ and $\delta/\delta\eta$ are understood as taking place independently of each other and $\psi_H (\mathbf{x}_m, t_m)$, $\psi_H^+ (\mathbf{x}_n', t_n')$ denote creation, annihilation operators for Fermions at space-time coordinates $(\mathbf{x}_m, t_m)$, $(\mathbf{x}_n', t_n')$, etc. (the special cases M, N or $P = 0$ mean that there are no corresponding variational derivatives on the left and no corresponding ψ_H, ψ_H^+, ${}^iQ_{l,k}$ – operators on the right, with replacements of each by unity on both sides). $\varepsilon (t_1, \ldots t_M, t_1', \ldots t_N')$ is identically $+1$ for Bosons and it is $+1$ or -1 for Fermions if the actual time order of the ψ_H, ψ_H^+ operators is an even or odd permutation of the order shown above, respectively. In Eq. (4.3.7), we have held U fixed. However, as noted above, a variational derivative with respect to $U (\mathbf{x}', t')$ introduces the factor $-ie\psi_H^+ (\mathbf{x}', t') \psi_H (\mathbf{x}', t')$ into the average, subject to time ordering. In this context we may write (time order indicated by $t'^+ > t'$; set $e \to 1$):

$$-i \frac{\delta}{\delta U (\mathbf{x}', t')} \equiv \varepsilon \frac{\delta_l}{\delta \eta (\mathbf{x}', t'^+)} \frac{\delta_l}{\delta \eta^+ (\mathbf{x}', t')}. \tag{4.3.8}$$

One can discuss the role of sources from another point of view. If we assign an overall measure of source strength λ_S, such that

$$\begin{aligned} H_S(t) &= H_H^{\eta} (t) + H_H^U (t) + H_H^{\mathcal{J}}(t) \longrightarrow \lambda_S H_S (t) \\ &= \lambda_S \bigg[\int d^{(3)}\mathbf{x} \{ \eta^+ (\mathbf{x}, t) \psi_H (\mathbf{x}, t) + \psi_H^+ (\mathbf{x}, t) \eta (\mathbf{x}, t) \\ &\quad + eU(\mathbf{x}, t) \psi_H^+ (\mathbf{x}, t) \psi_H (\mathbf{x}, t) \} + \sum_{i,l,k} {}^i\mathcal{J}_{l,k} (t) \, {}^iQ_{l,k} (t) \bigg], \end{aligned} \tag{4.3.9}$$

then, by the Action Principle (Eq. (4.1.72)), we have (restricting our considerations to a single Fermion species for simplicity, but with obvious generalization to several species):

$$\frac{\partial}{\partial \lambda_S} e^{\mathcal{W}} = \frac{\delta e^{\mathcal{W}}}{\delta \lambda_S} = \frac{-i \langle a_1', t_1 | \int_{t_2}^{t_1} dt' H_S (t') | a_2'', t_2 \rangle_{CS}}{\langle a_1', t_1 | a_2'', t_2 \rangle_C}. \tag{4.3.10}$$

Repeatedly applying Eq. (4.1.72) n times, we obtain

$$\frac{\partial^n e^{\mathcal{W}}}{\partial \lambda_S^n} = (-i)^n \left[\langle a_1', t_1 | a_2'', t_2 \rangle_C\right]^{-1} \tag{4.3.11}$$

$$\times \left\langle a_1', t_1 \left| \int_{t_2}^{t_1} dt' \int dt'' \ldots \int_{t_2}^{t_1} dt^{(n)} \left(H_S(t') H_S(t'') \ldots H_S(t^{(n)}) \right)_+ \right| a_2'', t_2 \right\rangle_{CS},$$

since the denominator has no dependence on sources. Noting that

$$e^{\mathcal{W}} = \sum_{n=0}^{\infty} \frac{\lambda_S^n}{n!} \left[\frac{\partial^n e^{\mathcal{W}}}{\partial \lambda_S^n} \right]_{\lambda_S=0} = \frac{\left\langle a_1', t_1 \left| \left(\exp(-\lambda_S i \int_{t_2}^{t_1} dt' H_S(t') \right)_+ \right| a_2'', t_2 \right\rangle_{C(S\to 0)}}{\langle a_1', t_1 | a_2'', t_2 \rangle_C}, \tag{4.3.12}$$

we have effectively rederived the interaction picture operator for the source Hamiltonian H_S, corresponding to $V(t_1, t_2)$ of Eq. (3.3.14) directly from the Action Principle. The operators in the structure of H_S in Eq. (4.3.12) do *not* feel the sources, only the couplings.

In fact, a knowledge of the transformation function $\langle a_1', t_1 | a_2'', t_2 \rangle_{S(C=0)}$ in the presence of sources alone, but with *no coupling*, is sufficient to determine its value with coupling, $\langle a_1', t_1 | a_2'', t_2 \rangle_{CS}$, by variational differentiation. We will show this for the prototypical electron–phonon interaction/coupling Hamiltonian of Eq. (4.3.6), to which we attach a strength factor λ_{ep}, so that $H_{ep}(t) \longrightarrow \lambda_{ep} H_{ep}(t)$. Again applying the Action Principle, in the form of Eq. (4.1.72) to $\delta\lambda_{ep}$, we have

$$\frac{\delta \langle a_1', t_1 | a_2'', t_2 \rangle_{CS}}{\delta \lambda_{ep}} = \frac{\partial \langle a_1', t_1 | a_2'', t_2 \rangle_{CS}}{\partial \lambda_{ep}} = -i \langle a_1', t_1 | \int_{t_2}^{t_1} dt' H_{ep}(t') | a_2'', t_2 \rangle_{CS}$$

$$= -i \left\langle a_1', t_1 \left| \sum_i \sum_{l,k} \int d^{(3)}\mathbf{x}' \int_{t_2}^{t_1} dt' \, {}^i\gamma_{l,k}^e(\mathbf{x}') \, {}^iQ_{l,k}(t') \, \psi_H^+(1'^+) \, \psi_H(1') \right| a_2'', t_2 \right\rangle_{CS}. \tag{4.3.13}$$

However, since the integrand factors ${}^iQ_{l,k}(t')$, $\psi^+(1'^+)$, $\psi(1')$ can be generated by the variational derivatives $i\delta/\delta\, {}^i\mathcal{J}_{l,k}(t')$, $i\delta_l/\delta\eta(1'^+)$, $i\delta_r/\delta\eta^+(1')$, respectively, Eq. (4.3.13) may also be written as

$$\frac{\partial \langle a_1', t_1 | a_2'', t_2 \rangle_{CS}}{\partial \lambda_{ep}}$$

$$= -\sum_i \sum_{l,k} \int d^{(3)}\mathbf{x}' \int_{t_2}^{t_1} dt' \, {}^i\gamma_{l,k}^e(\mathbf{x}') \frac{\delta}{\delta \mathcal{J}_{l,k}(t')} \frac{\delta_l}{\delta\eta(1'^+)} \frac{\delta_r}{\delta\eta^+(1')} \langle a_1', t_1 | a_2'', t_2 \rangle_{CS}. \tag{4.3.14}$$

The solution of this differential equation with respect to λ_{ep}, with $\lambda_{ep} \longrightarrow 1$ at full coupling strength, is given by

$$\langle a_1', t_1 | a_2'', t_2 \rangle_{CS}$$
$$= \exp\left[-\sum_i \sum_{l,k} \int d^{(3)}\mathbf{x}' \int_{t_2}^{t_1} dt' \, {}^i\gamma_{l,k}^e (\mathbf{x}') \frac{\delta}{\delta \mathcal{J}_{l,k}(t')} \frac{\delta_l}{\delta \eta (1'^+)} \frac{\delta_r}{\delta \eta^+ (1')} \right] \langle a_1', t_1 | a_2'', t_2 \rangle_S , \tag{4.3.15}$$

where the final factor on the right, $\langle a_1', t_1 | a_2'', t_2 \rangle_S$, occurs as a constant of integration in the solution of Eq. (4.3.14) due to the fact that $\langle a_1', t_1 | a_2'', t_2 \rangle_{CS} \longrightarrow \langle a_1', t_1 | a_2'', t_2 \rangle_S$ is devoid of dependence on coupling strength as $\lambda_{ep} \longrightarrow 0$. Eq. (4.3.15) explicitly exhibits how the transformation function with coupling and sources, $\langle a_1', t_1 | a_2'', t_2 \rangle_{CS}$, may be constructed from its value at zero coupling through variational differentiation with respect to its source dependencies.

The next item to address is the determination of the transformation function $\langle a_1', t_1 | a_2'', t_2 \rangle_S$ in the presence of η, η^+ sources, but with no couplings. We will illustrate this for electrons, $\langle a_1', t_1 | a_2'', t_2 \rangle_S \longrightarrow \langle a_1', t_1 | a_2'', t_2 \rangle_{\eta, \eta^+}$, in the absence of electron–electron and electron–phonon interactions (also dropping the source potential U). Recalling that (Eq. (4.3.1)):

$$\frac{\delta_l \langle a_1', t_1 | a_2'', t_2 \rangle_{\eta,\eta^+}}{\delta \eta^+(\mathbf{x}, t)} = -i \langle a_1', t_1 | \psi_H(\mathbf{x}, t) | a_2'', t_2 \rangle_{\eta,\eta^+} , \tag{4.3.16}$$

we employ the equation of motion, Eq. (4.2.33) (in this case $\mathcal{V} \longrightarrow 0$ with no interactions; note the temporary restoration of $\hbar$):

$$\left[i\hbar \frac{\partial}{\partial t} - h^{(1)} \left(\mathbf{x}, \tfrac{\hbar}{i} \nabla, t \right) \right] \psi_H(\mathbf{x}, t) = \eta(\mathbf{x}, t) \tag{4.3.17}$$

and form the matrix element $\langle a_1', t_1 | \ldots | a_2'', t_2 \rangle_{\eta,\eta^+}$, resulting in the equation

$$\left[i\hbar \frac{\partial}{\partial t} - h^{(1)} \left(\mathbf{x}, \tfrac{\hbar}{i} \nabla, t \right) \right] \langle a_1', t_1 | \psi_H(\mathbf{x}, t) | a_2'', t_2 \rangle_{\eta,\eta^+} = \eta(\mathbf{x}, t) \langle a_1', t_1 | a_2'', t_2 \rangle_{\eta,\eta^+} . \tag{4.3.18}$$

This may be solved by introducing a Green's function G° which inverts the differential operator in brackets on the left of Eq. (4.3.18) as a matrix in space-time indices, defined by

$$\left[i\hbar \frac{\partial}{\partial t} - h^{(1)} \left(\mathbf{x}, \tfrac{\hbar}{i} \nabla, t \right) \right] G^\circ(\mathbf{x}, t; \mathbf{x}', t') = \delta^{(3)}(\mathbf{x} - \mathbf{x}') \delta(t - t'). \tag{4.3.19}$$

Using G°, the solution of Eq. (4.3.18) may be written as

$$\langle a_1', t_1 | \psi_H(\mathbf{x}, t) | a_2'', t_2 \rangle_{\eta,\eta^+} = \int d^{(3)}\mathbf{x}' \int_{t_2}^{t_1} dt' G^\circ(\mathbf{x}, t; \mathbf{x}', t') \eta(\mathbf{x}', t') \langle a_1', t_1 | a_2'', t_2 \rangle_{\eta,\eta^+} . \tag{4.3.20}$$

Comparing with Eq. (4.3.16), we have

$$\frac{\delta_l \langle a_1', t_1 | a_2'', t_2 \rangle_{\eta,\eta^+}}{\delta \eta^+(\mathbf{x}, t)} = -i \int d^{(3)}\mathbf{x}' \int_{t_2}^{t_1} dt' G^\circ(\mathbf{x}, t; \mathbf{x}', t') \eta(\mathbf{x}', t') \langle a_1', t_1 | a_2'', t_2 \rangle_{\eta,\eta^+} \tag{4.3.21}$$

or

$$\begin{aligned} \frac{\delta_l \langle a_1', t_1 | a_2'', t_2 \rangle_{\eta,\eta^+}}{\langle a_1', t_1 | a_2'', t_2 \rangle_{\eta,\eta^+}} &= \delta_l \left(\ln \langle a_1', t_1 | a_2'', t_2 \rangle_{\eta,\eta^+} \right) \\ &= -i \int d^{(3)}\mathbf{x} \int_{t_2}^{t_1} dt \delta\eta^+(\mathbf{x}, t) \int d^{(3)}\mathbf{x}' \int_{t_2}^{t_1} dt' G^\circ(\mathbf{x}, t; \mathbf{x}', t') \eta(\mathbf{x}', t'), \end{aligned} \tag{4.3.22}$$

with the result

$$\begin{aligned} \langle a_1', t_1 | a_2'', t_2 \rangle_{\eta,\eta^+} = \exp &\left[-i \int d^{(3)}\mathbf{x} \int_{t_2}^{t_1} dt \int d^{(3)}\mathbf{x}' \int_{t_2}^{t_1} dt' \, \eta^+(\mathbf{x}, t) G^\circ(\mathbf{x}, t, \mathbf{x}', t') \eta(\mathbf{x}', t') \right] \\ &\times \langle a_1', t_1 | a_2'', t_2 \rangle_{\eta,\eta^+=0} , \end{aligned} \tag{4.3.23}$$

where $\langle a_1', t_1 | a_2'', t_2 \rangle_{\eta,\eta^+=0}$ arises as a constant of integration. Again viewing $G^\circ(\mathbf{x}, t; \mathbf{x}', t')$ as a matrix in its space-time indices, and $\eta(\mathbf{x}', t')$ and $\eta^+(\mathbf{x}, t)$ as a vector and adjoint vector, respectively, in this context, we may alternatively write Eq. (4.3.23) as

$$\langle a_1', t_1 | a_2'', t_2 \rangle_{\eta,\eta^+} = \exp\left(-i\eta^+ G^\circ \eta\right) \langle a_1', t_1 | a_2'', t_2 \rangle_{\eta,\eta^+=0} . \tag{4.3.24}$$

It should be noted that the transformation function $\langle a_1', t_1 | a_2'', t_2 \rangle_S$ in the absence of coupling also involves a decoupled phonon transformation function $\langle a_1', t_1 | a_2'', t_2 \rangle_{\mathcal{J}}$ with phonon $\mathcal{J}$-source-dependent dynamics, which are independent of the electron dynamics and their η, η^+-dependence since there is no coupling. This independence of the phonon and electron dynamics is expressed in the product form of the relation

$$\langle a_1', t_1 | a_2'', t_2 \rangle_S = \langle a_1', t_1 | a_2'', t_2 \rangle_{\eta,\eta^+} \langle a_1', t_1 | a_2'', t_2 \rangle_{\mathcal{J}} . \tag{4.3.25}$$

We defer discussion of the phonon dynamics until Chapter 11 as we have not even described the phonon Hamiltonian yet, but we simply note that the decoupled phonon field obeys a linear differential equation with an associated Green's function D° analogous to G° of Eq. (4.3.19). Correspondingly, a similar derivation for phonons (entirely similar to the one for electrons leading to Eq. (4.3.24)) yields the phonon $\mathcal{J}$-source-dependent decoupled transformation function as

$$\langle a_1', t_1 | a_2'', t_2 \rangle_{\mathcal{J}} = \exp\left(-\frac{i}{2}\mathcal{J}D^\circ\mathcal{J}\right)\langle a_1', t_1 | a_2'', t_2 \rangle_{\mathcal{J}=0}\,. \tag{4.3.26}$$

(The exponent factor of 1/2 in the phonon case arises from the fact that the dependence on the $\mathcal{J}$-source in the exponent on the right of Eq. (4.3.26) is bilinear, as opposed to linear dependencies in the exponent on the right of Eq. (4.3.24) on each of the η, η^+ sources individually for electrons.)

The result for $e^{\mathcal{W}}$ in the presence of phonon interactions, according to Eqns. (4.3.5), (4.3.15) and ((4.3.24–4.3.26)), is given by

$$\begin{aligned} e^{\mathcal{W}} &= \exp\left[-\sum_i\sum_{l,k}\int d^{(3)}\mathbf{x}'\int_{t_2}^{t_1} dt'\, {}^i\gamma^e_{l,k}(\mathbf{x}')\frac{\delta}{\delta\mathcal{J}_{l,k}(t')}\frac{\delta_r}{\delta\eta^+(1')}\frac{\delta_l}{\delta\eta(1'^+)}\right] \\ &\times \exp\left(-i\eta^+G^\circ\eta\right)\exp\left(-\frac{i}{2}\mathcal{J}D^\circ\mathcal{J}\right)\left[\langle a_1', t_1 | a_2'', t_2\rangle_C\right]^{-1} \\ &\times \langle a_1', t_1 | a_2'', t_2\rangle_{\eta,\eta^+=0}\langle a_1', t_1 | a_2'', t_2\rangle_{\mathcal{J}=0}\,, \end{aligned} \tag{4.3.27}$$

where we have ignored electron–electron interactions (which will be treated later). The source-dependent transformation function, $\langle a_1', t_1 | a_2'', t_2\rangle_{CS}$, will eventually be seen to have the character of a non-equilibrium generalization of the grand partition function, and the source-independent normalizing denominator, $\langle a_1', t_1 | a_2'', t_2\rangle_C$, to have the character of a source-free grand partition function. The role of the latter is to remove "ground-state/vacuum polarization" effects from $e^{\mathcal{W}}$. In this context,

$$\begin{aligned} \mathcal{W} &= \ln\langle a_1', t_1 | a_2'', t_2\rangle_{CS} - \ln\langle a_1', t_1 | a_2'', t_2\rangle_C \\ &= -\beta(\Omega_{CS} - \Omega_C), \end{aligned} \tag{4.3.28}$$

where Ω_{CS} has the nature of a non-equilibrium generalization of the grand potential in the presence of sources as well as coupling, and Ω_C is its source-free counterpart, with β having the sense of an inverse thermal energy. It is often preferable to work with cumulants, which emerge naturally from an analysis of $\mathcal{W}$ itself.

4.4 Potential Source: General Considerations

To consider particle–particle interactions further, we focus attention on the source potential U, with the associated Hamiltonian part $H_H^U(\mathbf{x}, t)$ of Eq. (4.2.30), which we will discuss for the case of a single species (with $e \longrightarrow 1$),

$$H_H^U \longrightarrow \int d^3\mathbf{x}\, U(\mathbf{x}, t)\psi_H^+(\mathbf{x}, t)\psi_H(\mathbf{x}, t). \tag{4.4.1}$$

In the present discussion, we will ignore phonon interactions and $H_{\mathcal{J}}$. Using Eq. (4.2.31) with $e \longrightarrow 1$, we have

$$\frac{\delta e^{\mathcal{W}}}{\delta U(\mathbf{x},t)} = \frac{-i\langle a_1', t_1 | \psi_H^{+}(\mathbf{x},t)\psi_H(\mathbf{x},t)|a_2'', t_2\rangle_{CS}}{\langle a_1', t_1 | a_2'', t_2\rangle_C} . \tag{4.4.2}$$

Moreover, if we attach a strength parameter λ_{ee} to the particle–particle interaction $H^{(2)}$ (Eq. (2.2.53)),

$$H^{(2)} \longrightarrow \lambda_{ee} H^{(2)}, \tag{4.4.3}$$

then, by the Action Principle, we have (Eqns. (4.1.73) and (4.1.74)):

$$\begin{aligned}\frac{\partial \langle a_1', t_1 | a_2'', t_2\rangle_{CS}}{\partial \lambda_{ee}} &= \frac{-i}{2}\langle a_1', t_1 | \int_{t_2}^{t_1} dt \int d^{(3)}\mathbf{x}' \int d^{(3)}\mathbf{x}'' \psi_H^{+}(\mathbf{x}',t)\psi_H^{+}(\mathbf{x}'',t) \\ &\times \mathcal{V}(\mathbf{x}'-\mathbf{x}'')\psi_H(\mathbf{x}'',t)\psi_H(\mathbf{x}',t)|a_2'', t_2\rangle_{CS} \\ &= \frac{-i}{2}\int_{t_2}^{t_1} dt \int d^{(3)}\mathbf{x}' \int d^{(3)}\mathbf{x}'' \frac{\delta_l}{\delta\eta\,(\mathbf{x}',t^+)}\frac{\delta_l}{\delta\eta\,(\mathbf{x}'',t^+)}\mathcal{V}(\mathbf{x}'-\mathbf{x}'') \\ &\times \frac{\delta_l}{\delta\eta^+\,(\mathbf{x}'',t)}\frac{\delta_l}{\delta\eta^+\,(\mathbf{x}',t)}\langle a_1', t_1 | a_2'', t_2\rangle_{CS} .\end{aligned} \tag{4.4.4}$$

Following the same Taylor series development in powers of λ_{ee} that was used for λ_{ep} in Eq. ((4.3.10–4.3.15)), we obtain the result

$$\begin{aligned}\langle a_1', t_1 | a_2'', t_2\rangle_{CS} = \exp\Bigg[&\frac{-i}{2}\int_{t_2}^{t_1} dt \int d^{(3)}\mathbf{x}' \int d^{(3)}\mathbf{x}'' \frac{\delta_l}{\delta\eta\,(\mathbf{x}',t^+)}\frac{\delta_l}{\delta\eta\,(\mathbf{x}'',t^+)}\mathcal{V}(\mathbf{x}'-\mathbf{x}'') \\ &\times \frac{\delta_l}{\delta\eta^+\,(\mathbf{x}'',t)}\frac{\delta_l}{\delta\eta^+\,(\mathbf{x}',t)}\Bigg]\langle a_1', t_1 | a_2'', t_2\rangle_{S} ,\end{aligned} \tag{4.4.5}$$

where, as above (Eq. (4.3.15)), $\langle a_1', t_1 | a_2'', t_2\rangle_S$ depends on sources to the exclusion of couplings. Alternatively, using Eq. (4.3.8), it is clear that Eq. (4.4.4) may be rewritten as

$$\begin{aligned}\frac{\partial \langle a_1', t_1 | a_2'', t_2\rangle_{CS}}{\partial \lambda_{ee}} &= \frac{i}{2}\int_{t_2}^{t_1} dt \int d^{(3)}\mathbf{x}' \int d^{(3)}\mathbf{x}'' \frac{\delta}{\delta U\,(\mathbf{x}',t)}\mathcal{V}(\mathbf{x}'-\mathbf{x}'') \\ &\times \frac{\delta}{\delta U\,(\mathbf{x}'',t)}\langle a_1', t_1 | a_2'', t_2\rangle_{CS} ,\end{aligned} \tag{4.4.6}$$

where the sources involved include U as well as η, η^+. Again, a Taylor series in powers of λ_{ee} may be developed in full analogy with that for λ_{ep} (but bear in mind that phonon interactions and $H_{\mathcal{J}}$ are excluded from the present considerations), and the result again embeds the coupling dependence for λ_{ee} into variational derivatives, but this time with respect to the source potential U,

$$\langle a_1', t_1 | a_2'', t_2 \rangle_{CS} = \exp\left[\frac{i}{2}\int_{t_2}^{t_1} dt \int d^{(3)}\mathbf{x}' \int d^{(3)}\mathbf{x}'' \frac{\delta}{\delta U(\mathbf{x}', t)} \mathcal{V}(\mathbf{x}' - \mathbf{x}'') \times \frac{\delta}{\delta U(\mathbf{x}'', t)}\right] \langle a_1', t_1 | a_2'', t_2 \rangle_S , \tag{4.4.7}$$

where $\langle a_1', t_1 | a_2'', t_2 \rangle_S$, which is devoid of coupling dependence, now depends on the source potential U (which was ignored in the preceding section) as well as η, η^+. In the present case $\langle a_1', t_1 | a_2'', t_2 \rangle_S$ is determined by the equation (recall the discussion of Eqns. (4.3.16–4.3.23) and install the potential U):

$$\frac{\delta_l \langle a_1', t_1 | a_2'', t_2 \rangle_S}{\delta \eta^+(\mathbf{x}, t)} = -i \int d^{(3)}\mathbf{x}' \int_{t_2}^{t_1} dt' \, G_U^\circ(\mathbf{x}, t; \mathbf{x}', t') \eta(\mathbf{x}', t') \langle a_1', t_1 | a_2'', t_2 \rangle_S , \tag{4.4.8}$$

where $G_U^\circ(\mathbf{x}, t; \mathbf{x}', t')$ must now be modified to include U in its definition,

$$\left[i\hbar\frac{\partial}{\partial t} - h^{(1)}\left(\mathbf{x}, \tfrac{\hbar}{i}\nabla, t\right) - U(\mathbf{x}, t)\right] G_U^\circ(\mathbf{x}, t; \mathbf{x}', t') = \delta^{(3)}(\mathbf{x} - \mathbf{x}')\delta(t - t'). \tag{4.4.9}$$

Again employing space-time matrix notation, the solution of Eq. (4.4.8) yields

$$\langle a_1', t_1 | a_2'', t_2 \rangle_S = \left[\exp\left(-i\eta^+ G_U^\circ \eta\right)\right] \langle a_1', t_1 | a_2'', t_2 \rangle_U , \tag{4.4.10}$$

where, bearing in mind that the inverse of a product of matrices is the product of the individual inverse matrices in the reverse order, Eq. (4.4.9) yields

$$G_U^\circ = \left[(G^\circ)^{-1} - U\right]^{-1} = G^\circ \left[1 - UG^\circ\right]^{-1} = \left[1 - G^\circ U\right]^{-1} G^\circ . \tag{4.4.11}$$

Thus,

$$\langle a_1', t_1 | a_2'', t_2 \rangle_S = \left[\exp\left(-i\eta^+ \left[1 - G^\circ U\right]^{-1} G^\circ \eta\right)\right] \langle a_1', t_1 | a_2'', t_2 \rangle_U , \tag{4.4.12}$$

where $\langle a_1', t_1 | a_2'', t_2 \rangle_U$ appears as a constant of integration in the solution of Eq. (4.4.8) arising in connection with the fact that $\langle a_1', t_1 | a_2'', t_2 \rangle_C \longrightarrow \langle a_1', t_1 | a_2'', t_2 \rangle_U$ is devoid of dependence on η, η^+ (phonons and their sources ${}^i\mathcal{J}_{l,k}$ are ignored here).

To address the determination of the final transformation function $\langle a_1', t_1 | a_2'', t_2 \rangle_U$, it is helpful to note that for any diagonalizable operator X having a non-vanishing determinant,

$$Trace[(\delta X)X^{-1}] = \delta \log \det X \,. \tag{4.4.13}$$

This is readily verified in diagonal representation, since (eigenvalues of X are indicated by λ_i):

$$\begin{aligned} \delta \log \det X &= \delta \log \prod_i \lambda_i = \delta \sum_i \log \lambda_i \\ &= \sum_i (\delta \lambda_i) \lambda_i^{-1} = Trace[(\delta X)X^{-1}] \,. \end{aligned} \tag{4.4.14}$$

Considering $\delta \langle a_1', t_1 | a_2'', t_2 \rangle_U$, we have

$$\begin{aligned} \delta \langle a_1', t_1 | a_2'', t_2 \rangle_U &= -i \left\langle a_1', t_1 \left| \int_{t_2}^{t_1} dt \int d^{(3)}\mathbf{x} (\delta U(\mathbf{x}, t)) \psi_H^+(\mathbf{x}, t)\ \psi_H(\mathbf{x}, t) \right| a_2'', t_2 \right\rangle_U \\ &= -i \int_{t_2}^{t_1} dt \int d^{(3)}\mathbf{x}\, \delta U(\mathbf{x}, t) \langle a_1', t_1 \,|\, \psi_H^+(\mathbf{x}, t)\ \psi_H(\mathbf{x}, t) \,|\, a_2'', t_2 \rangle_U \,. \end{aligned} \tag{4.4.15}$$

From the fact that $\psi_H(\mathbf{x}, t)$ in the presence of only the source potential given by H_U satisfies the equation (in the absence of all other sources and couplings):

$$i\hbar \frac{\partial \psi_H(\mathbf{x}, t)}{\partial t} = h^{(1)} \psi_H(\mathbf{x}, t) + U(\mathbf{x}, t) \psi_H(\mathbf{x}, t) \tag{4.4.16}$$

(and $\psi_H^+(\mathbf{x}, t)$ satisfies the adjoint equation), one can readily show that Eq. (4.4.15) may be rewritten in terms of G_U° of Eq. (4.4.9) as (see Section 5.1):

$$\delta \langle a_1', t_1 | a_2'', t_2 \rangle_U = \mp \int_{t_2}^{t_1} dt \int d^{(3)}\mathbf{x}\, G_U^\circ(\mathbf{x}, t^+; \mathbf{x}, t) \delta U(\mathbf{x}, t) \langle a_1', t_1 | a_2'', t_2 \rangle_U \,, \tag{4.4.17}$$

where $\mp \to -$ for Bosons and $\mp \to +$ for Fermions (Chapters 5 and 7). Recalling Eq. (4.4.11) for G_U°, we may rewrite Eq. (4.4.17) in space-time matrix notation as

$$\begin{aligned} \frac{\delta \langle a_1', t_1 | a_2'', t_2 \rangle_U}{\langle a_1', t_1 | a_2'', t_2 \rangle_U} &\Longrightarrow \mp Trace(G_U^\circ \delta U) = \mp Trace[1 - G^\circ U]^{-1} \delta(G^\circ U) \\ &= \pm Trace[1 - G^\circ U]^{-1} \delta(1 - G^\circ U), \end{aligned} \tag{4.4.18}$$

where the *Trace* subsumes the space-time integrations in matrix notation. Applying the identity, Eq. (4.4.13), with $X = 1 - G^\circ U$, we have

$$\delta \log \langle a_1', t_1 | a_2'', t_2 \rangle_U = \mp \delta \log \det(1 - G^\circ U), \tag{4.4.19}$$

from which we may conclude that integration of the variational differential equation yields

$$\langle a_1', t_1 | a_2'', t_2 \rangle_U = \det{}^{\mp 1} (1 - G^\circ U) \langle a_1', t_1 | a_2'', t_2 \rangle_0 , \tag{4.4.20}$$

where $\langle a_1', t_1 | a_2'', t_2 \rangle_0$ is a constant of integration in the sense that it does not depend on U, η, η^+ or on particle–particle interactions. As we have explicitly extracted all effects of interparticle and phonon interactions as well as the effects of the external potential and sources, the time dependence of $\langle a_1', t_1 | a_2'', t_2 \rangle_{\eta, \eta^+ \mathcal{J}=0}$ is that generated by just the free single-particle Hamiltonian, $h^{(1)}$.

Combining the results above to include both particle–particle interactions and particle–phonon interactions, with particle sources η, η^+, phonon particle sources ${}^i\mathcal{J}_{l,k}$ and an external single-particle potential source U (single Fermion/Boson species thus far, but easy to generalize), we finally have the result:[5]

$$\begin{aligned}
\langle a_1', t_1 | a_2'', t_2 \rangle_{CS} &= \exp\left[-\sum_i \sum_{l,k} \int_{t_2}^{t_1} dt' \int d^{(3)}\mathbf{x}' \, {}^i\gamma_{l,k}^e (\mathbf{x}') \frac{\delta}{\delta^i \mathcal{J}_{l,k}(t')} \frac{\delta_r}{\delta \eta^+ (1')} \frac{\delta_l}{\delta \eta (1'^+)} \right] \\
&\times \left\{ \begin{array}{c} \exp\left[\frac{-i}{2} \int_{t_2}^{t_1} dt \int d^{(3)}\mathbf{x}' \int d^{(3)}\mathbf{x}'' \frac{\delta_l}{\delta\eta(\mathbf{x}',t^+)} \frac{\delta_l}{\delta\eta(\mathbf{x}'',t^+)} \mathcal{V}(\mathbf{x}' - \mathbf{x}'') \frac{\delta_l}{\delta\eta^+(\mathbf{x}'',t)} \frac{\delta_l}{\delta\eta^+(\mathbf{x}',t)} \right] \\ \text{OR} \\ \exp\left[\frac{i}{2} \int_{t_2}^{t_1} dt \int d^{(3)}\mathbf{x}' \int d^{(3)}\mathbf{x}'' \frac{\delta}{\delta U(\mathbf{x}',t)} \mathcal{V}(\mathbf{x}' - \mathbf{x}'') \frac{\delta}{\delta U(\mathbf{x}'',t)} \right] \end{array} \right\} \\
&\times \exp\left(-i\eta^+ [1 - G^\circ U]^{-1} G^\circ \eta \right) \det{}^{\pm 1} (1 - G^\circ U) \\
&\times \exp\left(\frac{-i}{2} \mathcal{J} D^\circ \mathcal{J} \right) \langle a_1', t_1 | a_2'', t_2 \rangle_0 \langle a_1', t_1 | a_2'', t_2 \rangle_{\mathcal{J}=0} .
\end{aligned} \tag{4.4.21}$$

Just as $\langle a_1', t_1 | a_2'', t_2 \rangle_0$ has its time dependence generated by $h^{(1)}$ alone, we also have the time dependence of $\langle a_1', t_1 | a_2'', t_2 \rangle_{\mathcal{J}=0}$ generated by the free phonon Hamiltonian alone (to be specified in Chapter 11, in the absence of all sources and couplings).

4.5 An Identity

We close this chapter with a discussion of a useful identity that will be employed later. Considering a function $G(1, 1') = -i\varepsilon \langle (\psi_H(1) \psi_H^+(1'))_+ \rangle$ which has an inverse $G^{-1}(1, 1')$ $[1 = \mathbf{x}_1, t_1, \text{and } 1' = \mathbf{x}_1', t_1', \text{etc.}]$, so that

$$\int d^4 2 G(1,2) G^{-1}(2,1') = \int d^4 2 G^{-1}(1,2) G(2,1')$$
$$= \delta^4(1-1') \equiv \delta^{(3)}(\mathbf{x}_1 - \mathbf{x}_1')\delta(t_1 - t_1'), \tag{4.5.1}$$

we may write this in space-time matrix notation as

$$GG^{-1} = I = G^{-1}G, \tag{4.5.2}$$

where the space-time integration $\int d^4 2 = \int d^{(3)}\mathbf{x}_2 \int dt_2$ is subsumed in the implicit summation of the matrix notation. Taking a variation δ of Eq. (4.5.2) and noting that $\delta I \equiv 0$, we have

$$(\delta G)G^{-1} + G\left(\delta\left[G^{-1}\right]\right) = 0 = \left(\delta\left[G^{-1}\right]\right)G + G^{-1}(\delta G), \tag{4.5.3}$$

whence

$$\delta G = -G(\delta\left[G^{-1}\right])G, \tag{4.5.4}$$

or

$$\delta G(1,1') = -\int d^4 3 \int d^4 4 G(1,3)\left(\delta G^{-1}(3,4)\right) G(4,1'). \tag{4.5.5}$$

In particular, for a variation with respect to a potential function $U(2)$, we obtain the identity

$$\frac{\delta G(1,1')}{\delta U(2)} = -\int d^4 3 \int d^4 4 G(1,3)\left(\frac{\delta G^{-1}(3,4)}{\delta U(2)}\right) G(4,1'). \tag{4.5.6}$$

It is important to note that the assumption that G^{-1} exists uniquely is by no means universally true, in particular for the case of Boson fields. Condensed Boson matter, specifically superfluids and superconductors, involve fields having non-vanishing averages such as

$$\langle\psi_H(1)\rangle \neq 0 \text{ or } \left\langle\psi_H(1)\psi_H(1')\right\rangle \neq 0,\ \left\langle\psi_H^+(1)\psi_H^+(1')\right\rangle \neq 0, \tag{4.5.7}$$

which will be seen to satisfy equations that are homogeneous counterparts of those satisfied by $G(1,1') = -i\varepsilon\left\langle\left(\psi_H(1)\psi_H^+(1')\right)_+\right\rangle$. When such non-trivial homogeneous solutions exist, the inverse G^{-1} does not exist.

4.6 Chapter 4 Problems

P.4.1. Consider a classical Lagrangian and its associated action integral,

$$\int_{t_1}^{t_2} dtL$$

with

$$L = \sum_i p_i \dot{q}_i - H.$$

[-3pt] Trace the classical analog of the variation of the action considered in Section 4.1, deriving the classical Hamilton equations and generators of translations in time, variations associated with δq_i, δp_i and the fundamental Poisson bracket relations. Determine the classical analog of the Heisenberg equation of motion. (Note that it is necessary to include "end-point" path variations as discussed on p.365, Eqns. 8–73, of Reference 2 in Section 4.7.)

P.4.2. Using the variational differential techniques of classical mechanics, determine the curve joining points (x_1, y_1) and (x_2, y_2) along which a particle released from rest subject to a uniform gravitational field (in the y-direction) travels (frictionlessly) in minimum time. (This is the celebrated brachistochrone problem.)

P.4.3. Verify Eqns. (4.1.25) to (4.1.35).

P.4.4. Carry out the derivation from Eq. (4.1.67) to Eq. (4.1.70) for Bosons.

P.4.5. Considering the additional particle-source term $H_H^{\eta}(\mathbf{x}, t)$ in the Hamiltonian density, derive the presence of the inhomogeneous driving term, $\eta(\mathbf{x}, t)$, on the right-hand side of Eq. (4.2.33).

P.4.6. Determine $\frac{dN}{dt}$ in terms of particle sources, $\eta(\mathbf{x}, t)$, $\eta^{+}(\mathbf{x}, t)$.

P.4.7. Verify Eqns. (4.3.7 and 4.3.8) and associated remarks.

P.4.8. Derive Eq. (4.4.10).

4.7 References for Chapter 4

R.4.1. J. Schwinger, "Quantum Kinematics and Dynamics" W.A. Benjamin Publishers (1970). (This book presents a comprehensive discussion of the canonical Hamiltonian formulation of quantum mechanics in terms of Schwinger's Action Principle.) Also see J. Schwinger, "Quantum Mechanics," Editor: B.G. Englert, Springer (2001; J. Schwinger, Proc. Nat. Acad. of Sciences *46*, 883 (1960); and J. Schwinger, *47*, 1075 (1961).

R.4.2. H. Goldstein, "Classical Mechanics" 2nd. Ed., Addison-Wesley Publ. Co. (1980). (Excellent introduction to classical Lagrangian and Hamiltonian dynamics and canonical theory.)

R.4.3. L.D. Landau and E.M. Lifshitz, "Mechanics," 3rd. Ed., Vol. 1, Pergamon Press (1976). (Landau's books are always reliable and instructive.)

R.4.4. A.L. Fetter and J.D. Walecka, "Quantum Theory of Many-Particle Systems," McGraw-Hill Book Co. (1971). (Chapter 3, Section 8.)

R.4.5. Much of this material is based on notes on lectures of J. Schwinger on quantum electrodynamics at Harvard University about 1953–1954. It has been adjusted here for application to non-relativistic many-particle systems. The notes came to this author without identification of the note-takers.

5

Retarded Green's Functions

Abstract

Chapter 5 introduces single-particle retarded Green's functions $G_1^{ret}(\mathbf{x}, t; \mathbf{x}', t')$, which are defined to vanish identically for $t < t'$, and trace Schrödinger particle dynamical time development for $t > t'$ by providing the probability amplitude that a particle created at $\mathbf{x}', t'$ will be annihilated at $\mathbf{x}, t$. This physical interpretation of G_1^{ret} is seen to devolve upon the action of the time-development operator $exp[-iH(t-t')]$ in the evolution of the creation/annihilation operator constituents of the Green's function, which is evident in the spectrum of its energy eigenfunction expansion. Partial Green's functions, which represent the time development of one (or a few) states that may be understood as localized but are in interaction with a continuum of states, are discussed with derivation of an explicit solution (which is applied to the phenomenon of chemisorption). We also briefly introduce the Dyson integral equation, T-matrix and the Dirac delta $\delta(x)$-function potential, with the latter applied to random impurity scattering. The retarded Green's function in the presence of random impurity scattering is exhibited in the Born and self-consistent Born approximations, with application to Ando's semi-elliptic density of states for the 2D Landau-quantized electron-impurity system. Furthermore, several important retarded Green's functions are discussed along with the methods of their derivation: These include Green's functions for electrons in magnetic fields in both three- and two dimensions, and also a Hamilton equation-of-motion method for the determination of Green's functions with application to a 2D saddle potential in a time-dependent electric field. Moreover, separable Hamiltonians and their product Green's functions are discussed with application to a one-dimensional superlattice in axial electric and magnetic fields. Finally, a Green's function matching method is exhibited, with applications to non-relativistic Schrödinger electrons in a heterostructure with spatially variable effective mass, and also to non-local electrostatics with concomitant surface plasmons. Although Feynman diagrammatics are not employed here (since they are available in most other treatments of the subject), the chapter is closed with a discussion of the equivalence of Green's functions with Feynman path integrals.

5.1 Definition, Equation of Motion and Interpretation of Retarded Green's Functions

The focus of this book is on non-relativistic many-particle Green's functions appropriate to condensed matter theory. However, before addressing this we will discuss a somewhat

Quantum Statistical Field Theory: Schwinger's Variational Method. Norman J. Morgenstern Horing,
Oxford University Press (2017). © Norman J. Morgenstern Horing. DOI 10.1093/oso/9780198791942.001.0001

more familiar and simpler type of Green's function, the retarded single-particle Green's function, to provide insight into its interpretation and phenomenology. The retarded Schrödinger Green's function, $G_1^{ret}(\mathbf{x}_1, t_1; \mathbf{x}_1', t_1')$, is defined to represent the response of the non-relativistic Schrödinger wave function at space-time point $\mathbf{x}_1, t_1$ to an impulsive $\delta^{(3)}(\mathbf{x}_1 - \mathbf{x}_1')\delta(t_1 - t_1')$-function driving term at $\mathbf{x}_1', t_1'$ (impulsive in both time and space variables) subject to the condition that the response vanish identically before the impulse occurs ($t_1 < t_1'$). This retardation condition is easy to incorporate by defining the Green's function in terms of a matrix element expectation value in a state of *no* particles, the vacuum state $|0\rangle$, which we take to be normalized to unity $\langle 0|0\rangle = 1$. This definition has the time-ordered (see Chapter 3) form,

$$G_1^{ret}(\mathbf{x}_1, t_1; \mathbf{x}_1', t_1') = -i\varepsilon \langle 0| \left(\psi_H(\mathbf{x}_1, t_1)\psi_H^+(\mathbf{x}_1', t_1')\right)_+ |0\rangle, \tag{5.1.1}$$

where ε is identically $+1$ for Bosons but it is $+1$ or -1 for Fermions for $t_1 > t_1'$ or $t_1 < t_1'$, respectively. The Schrödinger response character of G_1^{ret} as defined here is readily seen from its equation of motion, which may be derived using the equation of motion (Eq. (3.2.21)) for the Heisenberg field operator $\psi_H(\mathbf{x}, t) \to \psi(\mathbf{x}, t)$ (suppress the subscript H, also $\hbar \to 1$). At this stage, particle–particle interactions and sources are ignored, but an arbitrary local one-particle potential will be included in $h^{(1)}$ in the Schrödinger equation,

$$\left(i\frac{\partial}{\partial t_1} - h^{(1)}\left(\mathbf{x}_1, \tfrac{\hbar}{i}\nabla_1, t_1\right)\right)\psi(\mathbf{x}_1, t_1) = 0. \tag{5.1.2}$$

Multiplying Eq. (5.1.2) by $\psi^+(\mathbf{x}_1', t_1')$ from the left or right to form a time-ordered product for $t_1 \neq t_1'$ and taking the vacuum state average, we obtain the one-particle retarded Green's function equation of motion, which is homogeneous for all times except $t_1 = t_1'$ (we write $h^{(1)}\left(\mathbf{x}_1, \frac{\hbar}{i}\nabla_1, t_1\right) \to h^{(1)}$):

$$\left(i\frac{\partial}{\partial t_1} - h^{(1)}\right)G_1^{ret}\left(\mathbf{x}_1, t_1; \mathbf{x}_1', t_1'\right) = \delta^{(3)}\left(\mathbf{x}_1 - \mathbf{x}_1'\right)\delta\left(t_1 - t_1'\right). \tag{5.1.3}$$

It should be noted that if the system is translationally invariant, such that $h^{(1)}$ and an associated potential energy have *no* dependence on $\mathbf{x}_1$, then the only positional dependence on $\mathbf{x}_1$ and $\mathbf{x}_1'$ occurs through the combination $\mathbf{x}_1 - \mathbf{x}_1'$ in $\delta^{(3)}(\mathbf{x}_1 - \mathbf{x}_1')$ on the right of Eq. (5.1.3), and consequently in G_1^{ret} as well: $G_1^{ret}(\mathbf{x}_1, t_1; \mathbf{x}_1', t_1') \to G_1^{ret}(\mathbf{x}_1 - x_1', t_1, t_1')$. The Dirac delta function on the right-hand side vanishes for $t_1 \neq t_1'$, as it must due to the homogeneity of Eq. (5.1.2), and its correctness as an impulsive inhomogeneous driving term for $t_1 = t_1'$ may be verified by integrating Eq. (5.1.3) with respect to t_1 from $t_1 = t_1' - \delta$ to $t_1 = t_1' + \delta$ $(\delta \to 0^+)$, with the result

$$i\left(G_{1>}^{ret}\left(\mathbf{x}_1, t_1' + \delta; \mathbf{x}_1', t_1'\right) - G_{1<}^{ret}(\mathbf{x}_1, t_1' - \delta; \mathbf{x}_1', t_1')\right) = \delta^{(3)}\left(\mathbf{x}_1 - \mathbf{x}_1'\right). \tag{5.1.4}$$

(Here we have used the fact that $h_1^{(1)}G_1^{ret}$ is a finite quantity, which yields a negligibly small contribution when integrated over an infinitesimal time interval.) $G_{1>}^{ret}$, $G_{1<}^{ret}$ represent the functional forms of the time-ordered G_1^{ret} for $t_1 > t_1'$, $t_1 < t_1'$, respectively. Retardation is ensured by the fact that $G_{1<}^{ret}(t_1 < t_1') \equiv 0$ vanishes identically, since, for $t_1' > t_1$, the time ordering yields

$$G^{ret}_{1<}(\mathbf{x}_1, t_1; \mathbf{x}'_1, t'_1) = \mp i \langle 0 \left| \psi^+ \left(\mathbf{x}'_1, t'_1\right) \psi \left(\mathbf{x}_1, t_1\right) \right| 0 \rangle \equiv 0 \tag{5.1.5}$$

($\mp$ is $-$ or $+$ for Bosons or Fermions, respectively), where the action of $\psi(\mathbf{x}_1, t_1)$ on the vacuum calls for a state of -1 particles, whose non-existence is represented by 0 (see Chapter 2; this condition defines the vacuum state). It should be noted that when we later deal with many-particle systems and thermodynamic Green's functions, $G_{1<}(t_1 < t'_1)$ will *not* vanish. However, we consider only retarded Green's functions in this chapter. Considering next $t_1 > t'_1$,

$$G^{ret}_{1>}(\mathbf{x}_1, t_1; \mathbf{x}'_1, t'_1) = -i \langle 0 \left| \psi \left(\mathbf{x}_1, t_1\right) \psi^+ \left(\mathbf{x}'_1, t'_1\right) \right| 0 \rangle, \tag{5.1.6}$$

we may complete verification of the correctness of the δ-function driving term by using the equal-time canonical commutation/anticommutation relation, $\langle \{[\psi(\mathbf{x}_1, t'_1), \psi^+(\mathbf{x}'_1, t'_1)]\} \rangle = \delta^{(3)}(\mathbf{x}_1 - \mathbf{x}'_1)$, in Eqns. (5.1.6) and (5.1.4) jointly with Eq. (5.1.5).

To facilitate the physical interpretation of the retarded Green's function, we note that the matrix element on the right of Eq. (5.1.6) describes the state of a single particle created at point $\mathbf{x}'_1$ at time t'_1,

$$\left|\mathbf{x}'_1 t'_1\right\rangle = \psi^+ \left(\mathbf{x}'_1, t'_1\right) |0\rangle, \tag{5.1.7}$$

which is in a scalar product with a state describing the annihilation of the particle at point $\mathbf{x}_1$ at time t_1, the latter state being

$$\langle \mathbf{x}_1 t_1 | = \langle 0 | \psi (\mathbf{x}_1, t_1). \tag{5.1.8}$$

Thus, the probability amplitude for the creation of a particle at $\mathbf{x}'_1$, t'_1 which is subsequently annihilated after propagating to $\mathbf{x}_1$, t_1 is given by

$$\left\langle \mathbf{x}_1 t_1 | \mathbf{x}'_1 t'_1 \right\rangle = i G^{ret}_1(\mathbf{x}_1, t_1; \mathbf{x}'_1, t'_1). \tag{5.1.9}$$

Any question of normalization of the single-particle state, Eq. (5.1.7), and locality of creation (annihilation) is resolved by noting that

$$\left\langle \mathbf{x}_1 t'_1 | \mathbf{x}'_1 t'_1 \right\rangle = \langle 0 \left| \psi \left(\mathbf{x}_1, t'_1\right) \psi^+ \left(\mathbf{x}'_1, t'_1\right) \right| 0 \rangle, \tag{5.1.10}$$

and using the canonical commutation relations,

$$\left\langle \mathbf{x}_1 t'_1 | \mathbf{x}'_1 t'_1 \right\rangle = \langle 0 \left| \delta^{(3)} \left(\mathbf{x}_1 - \mathbf{x}'_1\right) \right| 0 \rangle = \delta^{(3)} \left(\mathbf{x}_1 - \mathbf{x}'_1\right), \tag{5.1.11}$$

since the commuted/anticommuted term involves $\langle 0 \left| \psi^+ \psi \right| 0 \rangle = 0$, by definition of the vacuum state, $\psi |0\rangle = 0$. This expresses the orthogonality of different positional eigenstates and their delta-function normalization to unity under integration. Since $G^{ret}_{1<} \equiv 0$, Eq. (5.1.11) may be understood as an initial condition on $G^{ret}_{1>}$,

$$i G^{ret}_{1>}(\mathbf{x}_1, t'_1 + \delta; \mathbf{x}'_1, t'_1) = \delta^{(3)} \left(\mathbf{x}_1 - \mathbf{x}'_1\right). \tag{5.1.12}$$

We note in passing that an initial single-particle state $\Phi(\mathbf{x}', t'_1)$ in position representation evolves into $\Phi(\mathbf{x}, t_1)$ for $t_1 > t'_1$ according to the relation $\Phi(\mathbf{x}_1, t_1) =$

$i\int d^3\mathbf{x}_1'\, G^{ret}_{1>}(\mathbf{x}_1,t_1;\mathbf{x}_1',t_1')\Phi(\mathbf{x}_1',t_1')$, since $(i\frac{\partial}{\partial t_1} - h^{(1)})G^{ret}_{1>}(\mathbf{x}_1,t_1;\mathbf{x}_1',t_1') = 0$ for $t_1 \neq t_1'$, and as $t_1 \to t_1' + \delta \to t_1' + 0^+$, $iG^{ret}_{1>} = \delta^{(3)}(\mathbf{x}_1 - \mathbf{x}_1')$, as one should expect for the solution of an initial value problem.

To explore the dynamical content of G_1^{ret} in describing particle propagation, we employ the simple exponential time-development operator for $H^{(1)}$ having no explicit time dependence (as opposed to the time-ordered exponential), to bring the time arguments of ψ and ψ^+ into coincidence:

$$\psi(\mathbf{x}_1,t_1) = e^{iH^{(1)}(t_1-t_1')}\psi(\mathbf{x}_1,t_1')\,e^{-iH^{(1)}(t_1-t_1')}. \tag{5.1.13}$$

Circumventing the issue of zero-point energy for the moment, we redefine the energy scale so that the vacuum state may be understood to have no energy as well as no particles,

$$\langle 0|\, e^{iH^{(1)}(t_1-t_1')} = \langle 0|\,, \tag{5.1.14}$$

and we define the *Heaviside unit step function* as

$$\eta_+(t) = 1 \text{ for } t > 0\,; 0 \text{ for } t < 0\ (1/2 \text{ for } t = 0). \tag{5.1.15}$$

The resulting retarded one-particle Green's function,

$$G_1^{ret}(\mathbf{x}_1,t_1;\mathbf{x}_1',t_1') = -i\eta_+(t_1-t_1')\left\langle 0\left|\psi(\mathbf{x}_1,t_1')\,e^{-iH^{(1)}(t_1-t_1')}\psi^+(\mathbf{x}_1',t_1')\right|0\right\rangle, \tag{5.1.16}$$

may now be expanded in single-particle energy eigenstates, $H^{(1)}\left|E_\gamma^{(1)}\right\rangle = E_\gamma^{(1)}\left|E_\gamma^{(1)}\right\rangle$, by inserting the unit operator $I = \sum_{E_\gamma}\left|E_\gamma^{(1)}\right\rangle\left\langle E_\gamma^{(1)}\right|$ (completeness) next to the time-development operator and noting that for an arbitrary function f, $f(H^{(1)})\left|E_\gamma^{(1)}\right\rangle = f\left(E_\gamma^{(1)}\right)\left|E_\gamma^{(1)}\right\rangle$. Thus, we obtain (for convenience we set $T = t_1 - t_1'$ and put $t_1' \to 0$):

$$G_1^{ret}(\mathbf{x}_1,t_1;\mathbf{x}_1',t_1') = -i\eta_+(T) \times \sum_{E_\gamma^{(1)}}\left\langle 0\left|\psi(\mathbf{x}_1,0)\left|E_\gamma^{(1)}\right\rangle e^{-iE_\gamma^{(1)}T}\left\langle E_\gamma^{(1)}\right|\psi^+(\mathbf{x}_1',0)\right|0\right\rangle. \tag{5.1.17}$$

The matrix element,

$$\left\langle 0\left|\psi(\mathbf{x}_1,0)\right|E_\gamma^{(1)}\right\rangle = \left\langle \mathbf{x}_1|E_\gamma^{(1)}\right\rangle = \phi_\gamma(\mathbf{x}_1), \tag{5.1.18}$$

is just the single-particle energy eigenfunction in position representation, $\phi_\gamma(\mathbf{x})$, and

$$\left\langle E_\gamma^{(1)}\left|\psi^+(\mathbf{x}_1',0)\right|0\right\rangle = \left\langle E_\gamma^{(1)}|\mathbf{x}_1'\right\rangle = \phi_\gamma^*(\mathbf{x}_1') \tag{5.1.19}$$

is its complex conjugate (at $\mathbf{x}_1'$ instead of $\mathbf{x}_1$). Thus,

$$G_1^{ret}(\mathbf{x}_1,t_1;\mathbf{x}_1',t_1') = -i\eta_+(T)\sum_{E_\gamma^{(1)}}\phi_\gamma(\mathbf{x}_1)\,\phi_\gamma^*(\mathbf{x}_1')\,e^{-iE_\gamma^{(1)}T}, \tag{5.1.20}$$

in position-time representation.

Our considerations here would be incomplete without some mention of the matrix operator form for the Green's function. One can rewrite Eq. (5.1.16) using Eqns. (5.1.7) and (5.1.8) as ($T = t_1 - t_1'; t_1' \to 0$):

$$G_1^{ret}(\mathbf{x}_1, t_1; \mathbf{x}_1', t_1') = -i\eta_+(T)\left\langle \mathbf{x}_1 \left| e^{-iH^{(1)}T} \right| \mathbf{x}_1' \right\rangle, \qquad (5.1.21)$$

and defining the matrix operator Green's function, $G_1^{ret}(T)$, in terms of its positional matrix elements as

$$G_1^{ret}(\mathbf{x}_1, t_1; \mathbf{x}_1', t_1') = \left\langle \mathbf{x}_1 \left| G_1^{ret}(T) \right| \mathbf{x}_1' \right\rangle, \qquad (5.1.22)$$

we may identify it as

$$G_1^{ret}(T) = -i\eta_+(T)e^{-iH^{(1)}T}. \qquad (5.1.23)$$

Fourier-transforming $T \to \omega$ (understanding ω to have a small imaginary part, $+i\delta$, to ensure convergence), we obtain

$$G_1^{ret}(\omega) = \frac{I}{\omega - H^{(1)} + i\delta}; \quad \left(\omega + i\delta - H^{(1)}\right) G_1^{ret}(\omega) = I, \qquad (5.1.24)$$

as an operator Green's function independent of representation. Using eigenvectors $|E\rangle$ of $H^{(1)}$, this may be written alternatively as ($I = \sum_E |E\rangle \langle E|$),

$$G_1^{ret}(\omega) = \sum_E |E\rangle \frac{1}{\omega - E + i\delta} \langle E| . \qquad (5.1.25)$$

It is immediately clear that the energy eigenvalues, E, are given by the frequency poles of $G_1^{ret}(\omega)$. Furthermore, forming $< \mathbf{x}_1 | G_1^{ret}(\omega) | \mathbf{x}_1' >$, the residues at such poles are proportional to $\phi_E(\mathbf{x}_1)\phi_E^*(\mathbf{x}_1')$, yielding information about the energy eigenfunctions. The trace of the imaginary part of $G_1^{ret}(\omega)$ (Dirac prescription[4], $[x \pm i\delta]^{-1} = P(\frac{1}{x}) \mp \pi i\delta(x)$, where P denotes subsequent principal part integration),

$$Tr\,\mathrm{Im}G_1^{ret}(\omega) = -\pi \sum_E \delta(\omega - E), \qquad (5.1.26)$$

is just proportional to the density $D(\omega)$ of single-particle energy eigenstates (per unit energy),

$$D(\omega) = \sum_E \delta(\omega - E) = -\frac{1}{\pi} Tr\,\mathrm{Im}G_1^{ret}(\omega). \qquad (5.1.27)$$

This is readily verified since the ω-integral of $D(\omega)$ has a unit contribution for every energy state.

It is important to note that our Green's function considerations are not at all limited to position representation as Eqns. (5.1.23–5.1.27) can be recast in *any* representation convenient to the problem at hand, e.g., momentum representation, angular momentum representation and Wannier representation, etc. For example, in the case of spatial translational invariance involving *no* positional dependence in the non-relativistic Hamiltonian

$h^{(1)} \sim -\nabla^2/2m$, the reader can readily verify that the retarded Green's function in momentum representation has the form

$$\langle \mathbf{p}|G_1^{ret}(\omega)|\mathbf{p}'\rangle = \sum_{\mathbf{x}}\sum_{\mathbf{x}'} \langle \mathbf{p}|\mathbf{x}\rangle \langle \mathbf{x}|G_1^{ret}(\omega)|\mathbf{x}'\rangle \langle \mathbf{x}'|\mathbf{p}'\rangle$$

$$= \int \frac{d^{(3)}\mathbf{x}}{(2\pi)^{3/2}} \int \frac{d^{(3)}\mathbf{x}'}{(2\pi)^{3/2}} e^{i\mathbf{p}\cdot\mathbf{x}} e^{-i\mathbf{p}'\cdot\mathbf{x}'} G_1^{ret}(\mathbf{x}-\mathbf{x}';\omega),$$

so that $(\mathbf{X} = (\mathbf{x}-\mathbf{x}'))$,

$$G_1^{ret}(\mathbf{p},\mathbf{p}';\omega) = \int \frac{d^{(3)}\mathbf{X}}{(2\pi)^3} e^{i\mathbf{p}\cdot\mathbf{X}} G_1^{ret}(\mathbf{X},\omega) \int d^{(3)}\mathbf{x}' e^{i(\mathbf{p}-\mathbf{p}')\cdot\mathbf{x}'} = \delta^{(3)}(\mathbf{p}-\mathbf{p}') G_1^{ret}(\mathbf{p},\omega),$$

where

$$G^{ret}(\mathbf{p},\omega) = \frac{1}{\omega - p^2/2m + i\delta}.$$

(Of course the transformation properties of ψ, ψ^+ in Eq. (2.2.4) can always be used to change representation as desired.)

Further to the case of spatial translational invariance, it is important to recognize that $G^{ret}(\mathbf{x},\mathbf{x}';\omega) \equiv G^{ret}(\mathbf{x}-\mathbf{x}';\omega)$ depends only on $(\mathbf{x}-\mathbf{x}')$ to the exclusion of $(\mathbf{x}+\mathbf{x}')$ in this case with *no* positional dependence in the Hamiltonian. If we use the spatial translation operator of Eq. (2.2.57) to refer both $\mathbf{x}$ and $\mathbf{x}'$ to a common origin 0, we have the Green's function as ($\mathbf{P}$ is the momentum operator, $\hbar = 1$):

$$G_1^{ret}(\mathbf{x}_1,t_1;\mathbf{x}_1',t_1') = -i\epsilon \langle 0|(\psi_H(\mathbf{x}_1,t_1)\psi_H^+(\mathbf{x}_1',t_1'))_+|0\rangle$$

$$= -i\epsilon \left\langle 0|(e^{i\mathbf{x}_1\cdot\mathbf{P}}\psi_H(0,t_1)e^{-i\mathbf{x}_1\cdot\mathbf{P}}e^{i\mathbf{x}_1'\cdot\mathbf{P}}\psi_H^+(0,t_1')e^{-i\mathbf{x}_1'\cdot\mathbf{P}})_+|0\right\rangle$$

$$= -i\epsilon \left\langle 0|(\psi_H(0,t_1)e^{-i(\mathbf{x}_1-\mathbf{x}_1')\cdot\mathbf{P}}\psi_H^+(0,t_1'))_+|0\right\rangle,$$

since the vacuum has no particles, and, correspondingly, no momentum, so $\mathbf{P}|0\rangle = 0$, etc.

In the case of three-dimensional (3D) infinite space with $H^{(1)} \to h^{(1)} = -\frac{\nabla^2}{2m}$, Eq. (5.1.24) yields the translationally invariant, field-free result

$$G_1^{ret}(\mathbf{x}_1-\mathbf{x}_1';\omega) = \frac{-m}{2\pi}\frac{e^{i\sqrt{2m\omega}|\mathbf{x}_1-\mathbf{x}_1'|}}{|\mathbf{x}_1-\mathbf{x}_1'|}; \quad [\omega > 0]. \tag{5.1.28}$$

For electrons confined to motion on a two-dimensional (2D) plane, the corresponding result is ($\overline{\mathbf{x}}$ is a 2D position vector):

$$G_1^{ret}(\overline{\mathbf{x}}_1-\overline{\mathbf{x}}_1';\omega) = \frac{-im}{2} H_0^{(1)}\left(\sqrt{2m\omega}|\overline{\mathbf{x}}_1-\overline{\mathbf{x}}_1'|\right); \quad [\omega > 0]. \tag{5.1.29}$$

($H_0^{(1)}$ is the Hankel function of the first kind of order zero.) In the case of one-dimensional (1D) electron motion, the Green's function is given by

$$G_1^{ret}(x_1 - x_1'; \omega) = \frac{-im}{\sqrt{2m\omega}} e^{i\sqrt{2m\omega}|x_1 - x_1'|}; \quad [\omega > 0]. \tag{5.1.30}$$

These results, Eqns. (5.1.28–5.1.30), are readily derived from Eq. (5.1.3) with ∇^2 written in 3D, 2D and 1D, respectively.

5.2 Partial Green's Functions, Applications to Localized States in a Continuum and Chemisorption

There is an interesting and important variant of Eq. (5.1.24) for situations in which the spectrum of energy states can be viewed as falling into two interacting subsets, wherein interest is focused on one of the subsets for which a partial Green's function is of special importance.[1] An interesting example is the Newns-Anderson theory[2] of chemisorption, in which an electron in a Fermi sea of occupied states behind the bounding surface of a metal interacts with a single isolated *resonant* state of an atom outside. Another important example is the theory of resonant tunneling,[3] in which a Fermi sea of occupied electron states behind a potential barrier interacts with a single isolated resonant state in a quantum well between the barrier and a second barrier a bit further away.

To analyze this type of problem, we assume that the two sets of states involved—eigenstates of subsystem 1, $|1i\rangle$ and eigenstates of subsystem 2, $|2j\rangle$—form a complete, orthonormal set,

$$\sum_i |1i\rangle\langle 1i| + \sum_j |2j\rangle\langle 2j| = I. \tag{5.2.1}$$

Due to interactions, the Hamiltonian has "transfer" matrix elements of the type

$$\langle 1i|H|2j\rangle \neq 0, \tag{5.2.2}$$

which connect the spaces of the two subsystems, as well as having diagonal elements in each of them. It is convenient to introduce measurement symbol projection operators,

$$P_1 = \sum_i |1i\rangle\langle 1i|, \tag{5.2.3}$$

$$P_2 = \sum_j |2j\rangle\langle 2j|, \tag{5.2.4}$$

which have the property $P_1^2 = P_1$, $P_2^2 = P_2$, since

$$\begin{aligned} P_1^2 &= \sum_{i,i'} |1i\rangle\langle 1i| \cdot |1i'\rangle\langle 1i'| \\ &= \sum_{i,i'} |1i\rangle\,\delta_{ii'}\langle 1i'| = \sum_i |1i\rangle\langle 1i| = P_1. \end{aligned} \tag{5.2.5}$$

Also, $P_1 P_2 = P_2 P_1 = 0$ since the subspaces are orthogonal and $\langle 1i|2j\rangle \equiv 0$. P_1 and P_2 project (accept) just the parts of a state lying in subspace 1 or 2, respectively. For example, if $|\Phi\rangle$ is given by

$$|\Phi\rangle = \sum_{i'} \Phi_{1i'} \left|1i'\right\rangle + \sum_{j'} \Phi_{2j'} \left|2j'\right\rangle, \tag{5.2.6}$$

then

$$\begin{aligned} P_1 |\Phi\rangle &= \sum_{i} |1i\rangle \langle 1i| \sum_{i'} \Phi_{1i'} \left|1i'\right\rangle + 0 \\ &= \sum_{i,i'} |1i\rangle \Phi_{1i'} \delta_{ii'} = \sum_{i} \Phi_{1i} |1i\rangle. \end{aligned} \tag{5.2.7}$$

The *partial* Green's operators for subspaces 1 and 2, denoted by G_{11} and G_{22}, respectively, are defined as

$$G_{11} \equiv P_1 G P_1, \tag{5.2.8}$$

$$G_{22} \equiv P_2 G P_2. \tag{5.2.9}$$

We will also encounter mixed elements that couple the subspaces defined by

$$G_{12} \equiv P_1 G P_2, \tag{5.2.10}$$

$$G_{21} \equiv P_2 G P_1. \tag{5.2.11}$$

Since, according to Eq. (5.2.1),

$$\sum_{m=1}^{2} P_m = P_1 + P_2 = I, \tag{5.2.12}$$

we may write Eq. (5.1.24) as (notation: $G_1^{ret}(\omega) \to G$, all sums are over $1, 2$; $\hbar \to 1$; suppress $i\delta$ so that $\omega + i\delta \to \omega$):

$$\left(\omega - H^{(1)}\right) \sum_{m} P_m \sum_{n} P_n G \sum_{l} P_l = I \tag{5.2.13}$$

and since $P_m P_n = P_m^2 \delta_{mn} = P_m \delta_{mn}$, we have

$$\left(\omega - H^{(1)}\right) \sum_{m,l} P_m G_{ml} = I. \tag{5.2.14}$$

Multiplying from the left by P_r and from the right by P_s (and noting that $P_r G_{rs} = P_r P_r G P_s = P_r G P_s = G_{rs}$, etc.), we have

$$\omega G_{rs} - \sum_{m} H_{rm}^{(1)} G_{ms} = \delta_{rs} P_r, \tag{5.2.15}$$

where $H_{rm}^{(1)} = P_r H^{(1)} P_m$ are the off-diagonal Hamiltonian elements responsible for the interaction of the two subspaces. (Although we use the notation $H^{(1)}$ of a single-particle

Hamiltonian, these considerations are more generally applicable; for example, to a system of interacting electrons coupled to phonons.) Thus,

$$\omega G_{11} - H_{11}^{(1)} G_{11} - H_{12}^{(1)} G_{21} = P_1, \tag{5.2.16}$$

$$\omega G_{21} - H_{21}^{(1)} G_{11} - H_{22}^{(1)} G_{21} = 0, \tag{5.2.17}$$

and two similar equations under the $(1 \leftrightarrow 2)$ interchange. Considering the partial Green's function for subspace 1, G_{11}, we may eliminate G_{21}, using the second equation,

$$G_{21} = \left(\omega - H_{22}^{(1)}\right)^{-1} H_{21}^{(1)} G_{11}, \tag{5.2.18}$$

which involves the partial operator Green's function of subspace 2, $\left(\omega - H_{22}^{(1)}\right)^{-1}$. The result for G_{11} is then given by the first equation as

$$\left(\omega - H_{11}^{(1)}\right) G_{11} - H_{12}^{(1)} \left(\omega - H_{22}^{(1)}\right)^{-1} H_{21}^{(1)} G_{11} = P_1. \tag{5.2.19}$$

A similar result obtains for the second set of equations,

$$\left(\omega - H_{22}^{(1)}\right) G_{22} - H_{21}^{(1)} \left(\omega - H_{11}^{(1)}\right)^{-1} H_{12}^{(1)} G_{22} = P_2. \tag{5.2.20}$$

This is particularly useful if subspace 1 involves one or just a few states. For example, if a system 1 involving a single localized state with energy ε_L is in interaction with a system 2 having a continuum of states ε_K, we may write $H_{12}^{(1)} \to \sum_K |L\rangle H_{LK}^{(1)} \langle K|$, and Eq. (5.2.19) becomes ($P_1 = |\varepsilon_L\rangle \langle \varepsilon_L| = |L\rangle\langle L|$; $P_2 = \sum_k |\varepsilon_K\rangle \langle \varepsilon_K| = \sum_K |K\rangle\langle K|$),

$$\left[P_1 \left(\omega - \varepsilon_L\right) P_1 - P_1 H_{12}^{(1)} P_2 \left(\omega - H_{22}^{(1)}\right)^{-1} P_2 H_{21}^{(1)} P_1\right] G_{11} = P_1, \tag{5.2.21}$$

or

$$|L\rangle \langle L| = \left[|L\rangle \left(\omega - \varepsilon_L\right) \langle L| \right.$$
$$\left. - \sum_K |L\rangle H_{LK}^{(1)} \langle K| \cdot \sum_{K'} \left|K'\right\rangle \frac{1}{\omega - \varepsilon_{K'}} \left\langle K'\right| \cdot \sum_{K''} \cdot \left|K''\right\rangle H_{K''L}^{(1)} \langle L| \right] |L\rangle\, G_{LL} \langle L|, \tag{5.2.22}$$

where we write $G_{11}=|L\rangle\, G_{LL}\, \langle L|$. Using orthonormality of the continuum states, $\left\langle K|K'\right\rangle = \delta_{KK'}$, we have

$$\left(\omega - \varepsilon_L - \sum_K \frac{H_{LK}^{(1)} H_{KL}^{(1)}}{\omega - \varepsilon_K}\right) G_{LL} = 1, \tag{5.2.23}$$

and the partial Green's operator for the localized state in interaction with the continuum is given in energy/frequency representation by

$$G_{LL} = \frac{1}{\omega + i\delta - \varepsilon_L - \sum_K \frac{H_{LK}H_{KL}}{\omega + i\delta - \varepsilon_K}}, \tag{5.2.24}$$

where we restore $\omega \rightarrow \omega + i\delta$ ($\delta \rightarrow 0^+$) to assure retardation under Fourier transformation to direct time representation. The result of the interaction with the continuum is to change the isolated localized state energy level ε_L by a "self-energy" $\Delta\varepsilon$ due to interaction, given by

$$\Delta\varepsilon = \sum_K \frac{H_{LK}H_{KL}}{\omega + i\delta - \varepsilon_K} = \int d^{(3)}K \frac{H_{LK}H_{KL}}{\omega + i\delta - \varepsilon_K}, \tag{5.2.25}$$

where we have suppressed a factor in the K-integrand describing the number density of energy states, ε_K, per unit K-space. Employing the Dirac prescription,

$$\Delta\varepsilon = \mathcal{P} \int d^{(3)}K \frac{H_{LK}H_{KL}}{\omega - \varepsilon_K} - \pi i \int d^{(3)}K H_{LK}H_{KL}\delta\left(\omega - \varepsilon_K\right), \tag{5.2.26}$$

we may identify the corrected (renormalized) energy E_L of the isolated localized state as the energy/frequency for which G_{LL} is singular, thus obtaining an equation of the Dyson form

$$E_L - \varepsilon_L - \mathcal{P} \int d^{(3)}K \frac{H_{LK}H_{KL}}{E_L - \varepsilon_K} + \pi i \int d^{(3)}K H_{LK}H_{KL}\delta\left(E_L - \varepsilon_K\right) = 0. \tag{5.2.27}$$

The principal part integral ($\mathcal{P}$) shifts the isolated localized energy level ε_L. (It can also result in additional roots, as in the theory of chemisorption.) The Dirac delta function term on the right of Eq. (5.2.27) contributes a finite imaginary part to E_L if $E_L \sim \varepsilon_L$ lies within the range of the K-integration for a band of states in the continuum under consideration (e.g., for a crystalline solid). If it lies outside the K-integration range, the imaginary part is zero. When the imaginary part is finite, it imparts a finite lifetime to the isolated localized energy level, signifying its decay into continuum states. In the context first considered by Anderson,[2] the localized state of an impurity atom in a crystalline solid would have a continuum electron hop onto the atom into the localized state and later hop off the atom back into a continuum state, the decay time involved being τ,

$$\tau \sim \frac{\hbar}{\mathrm{Im}E_L}, \tag{5.2.28}$$

and the state E_L functions as a localized scattering resonance. This interpretation in terms of a decay time for the localized state is readily verified if one considers the Fourier transform of G_{LL}, Eq. (5.2.24), from frequency to direct time representation, which yields

$$G_{LL}(T) = \int \frac{d\omega}{2\pi} e^{-i\omega T} \frac{1}{\omega - (\varepsilon_L + \text{Re}\Delta\varepsilon) + i\,(\delta + \text{Im}\Delta\varepsilon_L)}. \tag{5.2.29}$$

Assuming the perturbative coupling H_{KL} to be weak, we may replace the ω-dependence in the energy shift $\Delta\varepsilon$ by $\omega \to \varepsilon_L$, and with finite $\text{Im}\Delta\varepsilon_L$, we may drop $\delta \to 0^+$. The evaluation of the Fourier transform in Eq. (5.2.29) then yields

$$G_{LL}(T) = -i\eta_+(T) e^{-i(\varepsilon_L + \text{Re}\Delta\varepsilon)\,T} e^{-|\text{Im}\Delta\varepsilon|\,T}. \tag{5.2.30}$$

This clearly indicates decay of the state with decay time (restoring $\hbar$),

$$\tau \sim \frac{\hbar}{|\text{Im}\Delta\varepsilon|} = \frac{\hbar}{\text{Im}E_L}. \tag{5.2.31}$$

Another important application of the partial Green's function of Eq. (5.2.24) is to Newns' theory of hydrogen chemisorption,[2] in which the isolated localized state energy level is that of a single electronic energy level of an atom just outside the surface of a metal, and the continuum of states is that of a band of the metallic electrons. (We omit discussion of exchange energy, which was properly considered in Newns' work.) In this context, the bandwidth is not large and if the interaction is weak (atom far from the surface), one localized state exists, with the atomic orbital energy about ε_L with a small correction. However, if the interaction is stronger (atom close to the surface), two localized states emerge as roots of the Dyson equation for the level, approximating the bonding and antibonding states of the molecule consisting of the adatom and the particular metal electron states to which it directly couples, binding the atom to the metal strongly with a shared electron. Other important applications of the partial Green's function of Eq. (5.2.24) include double-barrier resonant tunneling, and charge transport in quantum wells, wires and dots.[3]

5.3 Dyson Integral Equation, T-Matrix for δ(x)-Function Potential

It is instructive and useful to consider some examples of retarded Green's functions as defined by Eq. (5.1.1) in terms of a vacuum state average. Since all Green's function equations of motion are based on Eq. (5.1.2) in the absence of particle–particle interactions, we expect that the equations have the same form (for example Eq. (5.1.3)) irrespective of the averaging process, which only determines side conditions (such as that of retardation as opposed to periodicity/antiperiodicity in imaginary time, which will be discussed in a later chapter). Dropping the particle–particle interaction potential $\mathcal{V}$ for the present, the retarded single-particle Green's function obeys the equation of motion:

$$\left(i\frac{\partial}{\partial t_1} - h_1^{(1)} \right) G_1^{ret}(1, 1') = \delta\left(1 - 1'\right). \tag{5.3.1}$$

In various problems it is of interest to include in $h_1^{(1)}$ a scattering potential $U(\mathbf{x}-\mathbf{x}_i)$ due to an impurity (or a quantum well) sited at $\mathbf{x}_i$. Thus, $h_1^{(1)} = h_1^0 + U(\mathbf{x}-\mathbf{x}_i)$, where h_1^0 includes the non-relativistic single-particle kinetic energy operator $-\frac{\nabla_1^2}{2m}$ (perhaps with external magnetic and/or electric fields), yielding

$$\left(i\frac{\partial}{\partial t_1} - h_1^0\right) G_1^{ret}(1,1') = U(\mathbf{x}_1-\mathbf{x}_i)\, G_1^{ret}(1,1') + \delta\left(1-1'\right). \tag{5.3.2}$$

Defining a simpler retarded Green's function $G_1^0(1,1')$ without the U-potential,

$$\left(i\frac{\partial}{\partial t_1} - h_1^0\right) G_1^0(1,1') = \delta\left(1-1'\right), \tag{5.3.3}$$

we may write the equation for $G_1^{ret}(1,1')$ in the four-dimensional integral form ($1 = \mathbf{x}_1, t_1$, etc.):

$$G_1^{ret}(1,1') = G_1^0(1,1') + \int d1'' G_1^0(1,1'') U\left(\mathbf{x}_1''-\mathbf{x}_i\right) G_1^{ret}(1'',1'). \tag{5.3.4}$$

Integral equations of this type occur quite generally in more complicated interaction problems and are said to be of the *Dyson* form, and the kernel $\left(\textit{inthiscase, trivially,} U\left(\mathbf{x}_1''-\mathbf{x}_i\right)\right)$ is the *self-energy*. Verification is immediate by application of the operator $\left(i\frac{\partial}{\partial t_1} - h_1^0\right)$ to both sides of Eq. (5.3.4) using Eq. (5.3.3). In the absence of explicit time dependence in $h_1^{(1)}$ we have time-translational invariance, and $G_1^{ret}(1,1')$ depends on $\left(t_1-t_1'\right)$ alone (to the exclusion of $\left(t_1+t_1'\right)$, see Eq. (5.1.22)), which implies that we may Fourier-transform the Green's function from time to frequency representation as

$$G_1^{ret}(1,1') = G_1^{ret}(\mathbf{x}_1,\mathbf{x}_1';t_1-t_1') \to G_1^{ret}(\mathbf{x}_1,\mathbf{x}_1';\omega),$$

with the result

$$G_1^{ret}(\mathbf{x}_1,\mathbf{x}_1';\omega) = G_1^0(\mathbf{x}_1,\mathbf{x}_1';\omega) + \int d^{(3)}\mathbf{x}_1'' G_1^0(\mathbf{x}_1,\mathbf{x}_1'';\omega) U\left(\mathbf{x}_1''-\mathbf{x}_i\right) G_1^{ret}(\mathbf{x}_1'',\mathbf{x}_1';\omega). \tag{5.3.5}$$

This integral equation describes a scattering process that can also be written in terms of a "T-matrix," $T\left(\mathbf{x}_1'',\mathbf{x}_1'''\right) = \left\langle \mathbf{x}_1''\,|T|\,\mathbf{x}_1'''\right\rangle$ defined by

$$G_1^{ret}(\mathbf{x}_1,\mathbf{x}_1';\omega) = G_1^0(\mathbf{x}_1,\mathbf{x}_1';\omega) + \int d^{(3)}\mathbf{x}_1''' \int d^{(3)}\mathbf{x}_1'' G_1^0(\mathbf{x}_1,\mathbf{x}_1'';\omega)\, T\left(\mathbf{x}_1'',\mathbf{x}_1'''\right) G_1^0(\mathbf{x}_1''',\mathbf{x}_1';\omega), \tag{5.3.6}$$

such that $T\left(\mathbf{x}_1'',\mathbf{x}_1'''\right)$ is in the nature of a resolvent kernel.

Modeling a short-range scattering potential by $U\left(\mathbf{x}_1''-\mathbf{x}_i\right) = U_0\delta^{(3)}(\mathbf{x}_1''-\mathbf{x}_i)$, Eq. (5.3.5) becomes

$$G_1^{ret}(\mathbf{x}_1,\mathbf{x}_1';\omega) = G_1^0(\mathbf{x}_1,\mathbf{x}_1';\omega) + U_0 G_1^0(\mathbf{x}_1,\mathbf{x}_i;\omega) G_1^{ret}(\mathbf{x}_i,\mathbf{x}_1';\omega). \tag{5.3.7}$$

Setting $\mathbf{x}_1 = \mathbf{x}_i$, we may solve Eq. (5.3.7), algebraically, as

$$G_1^{ret}(\mathbf{x}_i, \mathbf{x}_1'; \omega) = \left[1 - U_0 G_1^0(\mathbf{x}_i, \mathbf{x}_i; \omega)\right]^{-1} G_1^0(\mathbf{x}_i, \mathbf{x}_1'; \omega), \tag{5.3.8}$$

with the result given in frequency/energy representation by

$$G_1^{ret}(\mathbf{x}_1, \mathbf{x}_1'; \omega) = G_1^0(\mathbf{x}_1, \mathbf{x}_1'; \omega) + \frac{U_0 G_1^0(\mathbf{x}_1, \mathbf{x}_i; \omega) G_1^0(\mathbf{x}_i, \mathbf{x}_1'; \omega)}{1 - U_0 G_1^0(\mathbf{x}_i, \mathbf{x}_i; \omega)}. \tag{5.3.9}$$

In this case, the T-matrix is given in position representation by

$$T\left(\mathbf{x}_1'', \mathbf{x}_1'''\right) = U_0 \left[1 - U_0 G_1^0(\mathbf{x}_i, \mathbf{x}_i; \omega)\right]^{-1} \delta^{(3)}\left(\mathbf{x}_1'' - \mathbf{x}_i\right) \delta^{(3)}\left(\mathbf{x}_1''' - \mathbf{x}_i\right). \tag{5.3.10}$$

This exact result is discussed in the literature relating to impurity scattering in metals. When G_1^0 is translationally invariant, $G_1^0(\mathbf{x}_i, \mathbf{x}_i; \omega) = G_1^0(0, 0; \omega)$ does not depend on the impurity location $\mathbf{x}_i$. However, in the 3D case of Eq. (5.1.28), $G_1^0(\mathbf{x}_i, \mathbf{x}_i; \omega)$ on the right-hand side of Eq. (5.3.10) is divergent. To deal with this, we recognize that the single-point $U(\mathbf{x}_1'' - \mathbf{x}_i) = U_0 \delta^{(3)}(\mathbf{x}_1'' - \mathbf{x}_i)$-function potential is unrealistic and should be replaced by a scattering potential of small, finite radius, a. To estimate its effect without doing a serious recalculation of G_1^{ret} from the exact integral equation (Eq. (5.3.5)) involving the finite radius a, we employ Eq. (5.1.28) with the replacement $|\mathbf{x_i} - \mathbf{x_i}| \to a$ (instead of $|\mathbf{x_i} - \mathbf{x_i}| \to 0$) representing a "smear" of the $\delta^{(3)}$-function potential over a finite radius, such that $G_1^0(\mathbf{x}_i, \mathbf{x}_i; \omega) = \frac{-m}{2\pi} \frac{e^{-a\sqrt{2m|\omega|}}}{a}$. If the scattering potential is attractive, $U_0 = -\lambda$ $(\lambda > 0)$, there can also be a bound state ($\omega < 0$) characterized by the vanishing of the denominator on the right of Eqns. (5.3.9 and 5.3.10):

$$1 - U_0 G_1^0(\mathbf{x}_i, \mathbf{x}_i; \omega) = 1 - \frac{\lambda m}{2\pi} \frac{e^{-a\sqrt{2m|\omega|}}}{a} = 0, \tag{5.3.11}$$

from which the bound state frequency ω (energy) may be estimated from

$$\sqrt{2m\,|\omega|} = \frac{1}{a} \ln(\lambda m / 2\pi a), \tag{5.3.12}$$

provided that the condition $\lambda m / 2\pi a > 1$ is satisfied for the existence of the bound state. Corresponding results may be derived in 2D and 1D using $G_1^0(\mathbf{x}, \mathbf{x}'; \omega)$ as given by Eqns. (5.1.29 and 5.1.30), respectively. The divergence in the 2D case may be treated in a manner similar to that above. There is no divergence in the 1D case.

5.4 Random Impurity Scattering: Born and Self-Consistent Born Approximations

The Dyson integral equation approach embodies the closed form sum of an infinite series of selected Feynman diagrams. Moreover, randomization of impurity sites introduces an aspect of irreversibility into the analysis, which is manifested in an imaginary contribution to self-energy leading to decay and finite lifetime in the structure of $G_1^{ret}(T)$. To facilitate the analysis of a random array of impurity scatterers having more complicated potentials than the short-range delta-function model, we rewrite Eq. (5.3.4) in the form

$$G_1^{ret}(1,1') = G_1^0(1,1') + \int d1'' \int d1''' G_1^0(1,1'') U_i(1'',1''')\, G_1^{ret}(1''',1'), \qquad (5.4.1)$$

where

$$U_i(1'',1''') = U(\mathbf{x}_1'' - \mathbf{x}_i)\, \delta(1''-1''') . \qquad (5.4.2)$$

Introducing the space-time matrix notation $f(1',1'') = \langle 1' | f | 1'' \rangle$, we have the integral equation for G_1^{ret} as

$$G_1^{ret} = G_1^0 + G_1^0 U_i G_1^{ret}, \qquad (5.4.3)$$

where the integrals of Eq. (5.4.1) are subsumed in the rules of the space-time matrix multiplication. Considered jointly with the T-matrix equation in the form

$$G_1^{ret} = G_1^0 + G_1^0 T G_1^0, \qquad (5.4.4)$$

we can reformulate Eq. (5.3.4) in an advantageous manner for the introduction of fairly sophisticated approximations in dealing with difficult problems. To start, we note that the matrix/operator solution of Eq. (5.4.3) is

$$G_1^{ret} = \left[1 - G_1^0 U_i\right]^{-1} G_1^0, \qquad (5.4.5)$$

and, alternatively, if we solve Eq. (5.4.3) for G_1^0 in terms of G_1^{ret}, we have

$$G_1^0 = G_1^{ret} \left[1 + U_i G_1^{ret}\right]^{-1}. \qquad (5.4.6)$$

Moreover, comparing Eqns. (5.4.3) and (5.4.4), we find that

$$T G_1^0 = U_i G_1^{ret}, \qquad (5.4.7)$$

and, furthermore, using Eq. (5.4.3) jointly with Eq. (5.4.7), we may obtain a corresponding integral equation for the T-matrix using the fact that

$$U_i G_1^{ret} = U_i G_1^0 + U_i G_1^0 U_i G_1^{ret}, \qquad (5.4.8)$$

from which we find the *Lippman–Schwinger* equation,

$$T = U_i + U_i G_1^0 T. \qquad (5.4.9)$$

The formal solution for T,

$$T = \left[1 - U_i G_1^0\right]^{-1} U_i = \sum_{n=0}^{\infty} \left(U_i G_1^0\right)^n U_i, \tag{5.4.10}$$

can be expanded in powers of $U_i G_1^0$ as the *Born approximation series* for T,

$$T = U_i + U_i G_1^0 U_i + U_i G_1^0 U_i G_1^0 U_i + \cdots ,$$

and the corresponding Born series for G_1^{ret} is given by Eq. (5.4.5) as

$$G_1^{ret} = \sum_{n=0}^{\infty} \left(G_1^0 U_i\right)^n G_1^0 = G_1^0 + G_1^0 U_i G_1^0 + G_1^0 U_i G_1^0 U_i G_1^0 + \cdots , \tag{5.4.11}$$

which is popular but does not always have useful convergence properties. The first two terms on the right of Eq. (5.4.11) correspond to the first Born approximation. The terms in the series $\sum_n$ correspond to Feynman diagrams for the entire Born series.

To develop the *self-consistent Born approximation* for random impurity scattering, we rewrite Eq. (5.4.3) using Eq. (5.4.6), obtaining

$$G_1^{ret} = G_1^0 + G_1^{ret} \left[1 + U_i G_1^{ret}\right]^{-1} U_i G_1^{ret}, \tag{5.4.12}$$

and multiplication by U_i from the left yields

$$U_i G_1^{ret} = U_i G_1^0 + U_i G_1^{ret} [1 + U_i G_1^{ret}]^{-1} U_i G_1^{ret}. \tag{5.4.13}$$

Substituting this into the right-hand side of Eq. (5.4.3) we have

$$G_1^{ret} = G_1^0 + G_1^0 U_i G_1^0 + G_1^0 U_i G_1^{ret} [1 + U_i G_1^{ret}]^{-1} U_i G_1^{ret}, \tag{5.4.14}$$

and expanding the right-hand side to lowest order in a series in powers of $U_i G_1^{ret}$ by making the replacement $[1 + U_i G_1^{ret}]^{-1} \rightarrow 1$, we obtain the self-consistent Born approximation as

$$G_1^{ret} = G_1^0 + G_1^0 U_i G_1^0 + G_1^0 U_i G_1^{ret} U_i G_1^{ret}. \tag{5.4.15}$$

Installing a randomly sited array of $\mathfrak{N}$ impurities by the replacement $U_i \rightarrow \sum_{i=1}^{\mathfrak{N}} U_i$, we have

$$G_1^{ret}(1,1') = G_1^0(1,1') + \int d1'' G_1^0(1,1'') \sum_{i=1}^{\mathfrak{N}} U(\mathbf{x}'' - \mathbf{x}_i) G_1^0(1'',1')$$

$$+ \sum_{i=1}^{\mathfrak{N}} \sum_{j=1}^{\mathfrak{N}} \int d1'' G_1^0(1,1'') U(\mathbf{x}'' - \mathbf{x}_i) \int d1''' G_1^{ret}(1'',1''') U(\mathbf{x}''' - \mathbf{x}_j) G_1^{ret}(1''',1'). \tag{5.4.16}$$

A further approximation is introduced by assuming that the random impurity site averaging can be applied separately to G_1^{ret} wherever it appears (G_1^0 has no dependence on the impurity sites and is therefore unaffected by such averaging). Denoting impurity site averaging by $\langle\langle\cdots\rangle\rangle$, this means that

$$\langle\langle G_1^{ret}(1,1')\rangle\rangle = G_1^0(1,1') + \int d1'' G_1^0(1,1'') \sum_{i=1}^{\mathfrak{N}} \langle\langle U(\mathbf{x}''-\mathbf{x}_i)\rangle\rangle\, G_1^0(1'',1')$$
$$+ \sum_{i=1}^{\mathfrak{N}} \sum_{j=1}^{\mathfrak{N}} \int d1'' G_1^0(1,1'')$$
$$\times \left\langle\!\!\left\langle U(\mathbf{x}''-\mathbf{x}_i) \int d1''' G_1^{ret}(1'',1''') U(\mathbf{x}'''-\mathbf{x}_j) G_1^{ret}(1''',1') \right\rangle\!\!\right\rangle \qquad (5.4.17)$$

is approximately replaced by

$$\langle\langle G_1^{ret}(1,1')\rangle\rangle = G_1^0(1,1') + \int d1'' G_1^0(1,1'') \sum_{i=1}^{\mathfrak{N}} \langle\langle U(\mathbf{x}''-\mathbf{x}_i)\rangle\rangle\, G_1^0(1'',1')$$
$$+ \sum_{i=1}^{\mathfrak{N}} \sum_{j=1}^{\mathfrak{N}} \int d1'' G_1^0(1,1'') \int d1''' \langle\langle U(\mathbf{x}''-\mathbf{x}_i) U(\mathbf{x}'''-\mathbf{x}_j)\rangle\rangle$$
$$\times \langle\langle G_1^{ret}(1'',1''')\rangle\rangle \langle\langle G_1^{ret}(1''',1')\rangle\rangle, \qquad (5.4.18)$$

where the uniform random impurity site averaging is defined by

$$\langle\langle f(\{\mathbf{x}_i\})\rangle\rangle = \prod_{k=1}^{\mathfrak{N}} \frac{1}{vol} \int d^{(3)}\mathbf{x}_k f(\mathbf{x}_{i=1}, \mathbf{x}_{i=2}, \ldots \mathbf{x}_{i=\mathfrak{N}})$$
$$= \int \frac{d^{(3)}\mathbf{x}_1}{vol} \int \frac{d^{(3)}\mathbf{x}_2}{vol} \cdots \int \frac{d^{(3)}\mathbf{x}_n}{vol} f(\mathbf{x}_1, \mathbf{x}_2, \ldots \mathbf{x}_n). \qquad (5.4.19)$$

($\mathfrak{N}$ is the total number of impurities, *vol* denotes volume/area/length of the system depending on its dimensionality.) Considered jointly with the simplification of the second term on the right of Eq. (5.4.18) by the averaging process (see below), this leads to the Dyson equation for $\langle\langle G_1^{ret}(1,1')\rangle\rangle$ given by

$$\langle\langle G_1^{ret}(1,1')\rangle\rangle = G_1^0(1,1') + \sum_{i=1}^{\mathfrak{N}} \sum_{j=1}^{\mathfrak{N}} \int d1'' G_1^0(1,1'')$$
$$\times \int d1''' \langle\langle U(\mathbf{x}''-\mathbf{x}_i) U(\mathbf{x}'''-\mathbf{x}_j)\rangle\rangle \langle\langle G_1^{ret}(1'',1''')\rangle\rangle \langle\langle G_1^{ret}(1''',1')\rangle\rangle,$$
$$(5.4.20)$$

where we have ignored the first-order term

$$\int d1'' G_1^0(1,1'') G_1^0(1'',1') \sum_{i=1}^{\mathfrak{N}} \langle\langle U(\mathbf{x}''-\mathbf{x}_i)\rangle\rangle$$

on the following basis (note that $\langle\langle G_1^0\rangle\rangle = G_1^0$ since G_1^0 has no dependence on the impurity sites, $\mathbf{x}_i$):

$$\begin{aligned}
\sum_{i=1}^{\mathfrak{N}} \langle\langle U(\mathbf{x}''-\mathbf{x}_i)\rangle\rangle &= \sum_{i=1}^{\mathfrak{N}} \int \frac{d^{(3)}\mathbf{p}}{(2\pi)^3} \left\langle\left\langle e^{i\mathbf{p}\cdot(\mathbf{x}''-\mathbf{x}_i)}\right\rangle\right\rangle U(\mathbf{p}),\\
&= \int \frac{d^{(3)}\mathbf{p}}{(2\pi)^3} e^{i\mathbf{p}\cdot\mathbf{x}''} U(\mathbf{p}) \sum_{i=1}^{\mathfrak{N}} \prod_{j=1}^{\mathfrak{N}} \frac{1}{vol} \int d^3\mathbf{x}_j e^{-i\mathbf{p}\cdot\mathbf{x}_i},\\
&= \int d^{(3)}\mathbf{p}\, \delta^{(3)}(\mathbf{p}) U(\mathbf{p}) e^{i\mathbf{p}\cdot\mathbf{x}''} \sum_{i=1}^{\mathfrak{N}} \frac{1}{vol},\\
&= U(0)\frac{\mathfrak{N}}{vol} = n_i U(0),
\end{aligned} \tag{5.4.21}$$

where $n_i = \mathfrak{N}/vol$ is the impurity density. Accordingly, the first-order term,

$$\int d1'' G_1^0(1,1'') n_i U(0) G_1^0(1'',1'), \tag{5.4.22}$$

corresponds to a constant energy shift, $n_i U(0)$. Translating the energy scale through $n_i U(0)$, we may drop this term:

$$n_i U(0) \to 0. \tag{5.4.23}$$

Considering the last term on the right of Eq. (5.4.20), we identify a "self-energy," $\sum(1'',1''')$ (not to be confused with summation signs $\sum_{i,j}$) as:

$$\sum(1'',1''') = \sum_{i=1}^{\mathfrak{N}} \sum_{j=1}^{\mathfrak{N}} \langle\langle U(\mathbf{x}''-\mathbf{x}_i) U(\mathbf{x}'''-\mathbf{x}_j)\rangle\rangle \langle\langle G_1^{ret}(1'',1''')\rangle\rangle, \tag{5.4.24}$$

so that Eq. (5.4.20) has the standard Dyson form,

$$\langle\langle G_1^{ret}(1,1')\rangle\rangle = G_1^0(1,1') + \int d1'' \int d1''' G_1^0(1,1'') \sum(1'',1''') \langle\langle G_1^{ret}(1''',1')\rangle\rangle. \tag{5.4.25}$$

Fourier-analyzing $t_1 - t_1' \to \omega$, we evaluate $\sum(\mathbf{x}'', \mathbf{x}''', \omega)$ as

$$\sum(\mathbf{x}'', \mathbf{x}''', \omega) = \sum_{i=1}^{\mathfrak{N}} \sum_{j=1}^{\mathfrak{N}} \langle\langle G_1^{ret}(\mathbf{x}'', \mathbf{x}''', \omega)\rangle\rangle \prod_{k=1}^{\mathfrak{N}} \frac{1}{vol} \int d^{(3)}\mathbf{x}_k \langle\langle U(\mathbf{x}'' - \mathbf{x}_i) U(\mathbf{x}''' - \mathbf{x}_j)\rangle\rangle$$

$$= \langle\langle G_1^{ret}(\mathbf{x}'', \mathbf{x}''', \omega)\rangle\rangle \int \frac{d^{(3)}\mathbf{p}}{(2\pi)^3} \int \frac{d^{(3)}\mathbf{q}}{(2\pi)^3} e^{i(\mathbf{p}\cdot\mathbf{x}'' + \mathbf{q}\cdot\mathbf{x}''')}$$

$$\times U(\mathbf{p}) U(\mathbf{q}) \sum_{i=1}^{\mathfrak{N}} \sum_{j=1}^{\mathfrak{N}} \prod_{k=1}^{\mathfrak{N}} \frac{1}{vol} \int d^{(3)}\mathbf{x}_k e^{-i(\mathbf{p}\cdot\mathbf{x}_i + \mathbf{q}\cdot\mathbf{x}_j)}. \tag{5.4.26}$$

Considering the subsets of terms for which $i = j$ and those for which $i \neq j$ separately, we have

$$\sum_{i=1}^{\mathfrak{N}} \sum_{j=1}^{\mathfrak{N}} \prod_{k=1}^{\mathfrak{N}} \frac{1}{vol} \int d^{(3)}\mathbf{x}_k e^{-i(\mathbf{p}\cdot\mathbf{x}_i + \mathbf{q}\cdot\mathbf{x}_j)} = n_i (2\pi)^3 \delta^{(3)}(\mathbf{p} + \mathbf{q}) + n_i^2 (2\pi)^6 \delta^{(3)}(\mathbf{p}) \delta^{(3)}(\mathbf{q}). \tag{5.4.27}$$

This yields the self-energy as

$$\sum(\mathbf{x}'', \mathbf{x}''', \omega) = \langle\langle G_1^{ret}(\mathbf{x}'', \mathbf{x}''', \omega)\rangle\rangle n_i \int \frac{d^{(3)}\mathbf{p}}{(2\pi)^3} e^{i\mathbf{p}\cdot(\mathbf{x}'' - \mathbf{x}''')} U(\mathbf{p}) U(-\mathbf{p})$$

$$+ \langle\langle G_1^{ret}(\mathbf{x}'', \mathbf{x}''', \omega)\rangle\rangle n_i^2 U^2(0), \tag{5.4.28}$$

where the last term on the right may also be understood to be absorbed in another translation of the energy scale, as with Eq. (5.4.23). It is clear from the result,

$$\sum(\mathbf{x}'', \mathbf{x}''', \omega) = \langle\langle G_1^{ret}(\mathbf{x}'', \mathbf{x}''', \omega)\rangle\rangle n_i \int \frac{d^{(3)}\mathbf{p}}{(2\pi)^3} e^{i\mathbf{p}\cdot(\mathbf{x}'' - \mathbf{x}''')} U(\mathbf{p}) U(-\mathbf{p}), \tag{5.4.29}$$

that impurity site averaging leaves the self-energy, and hence $\langle\langle G_1^{ret}(\mathbf{x}, \mathbf{x}', \omega)\rangle\rangle$, translationally invariant if there is no other spatial inhomogeneity besides the randomly sited impurities (nor a magnetic field). In this case, we may Fourier-transform $\mathbf{x}'' - \mathbf{x}''' \to \mathbf{q}$, with the result

$$\sum(\mathbf{x}'' - \mathbf{x}''', \omega) \to \sum(\mathbf{q}, \omega) = n_i \int \frac{d^{(3)}\mathbf{p}}{(2\pi)^3} \langle\langle G_1^{ret}(\mathbf{q} - \mathbf{p}, \omega)\rangle\rangle U(\mathbf{p}) U(-\mathbf{p}). \tag{5.4.30}$$

The corresponding Dyson integral equation for $\langle\langle G_1^{ret}(\mathbf{x} - \mathbf{x}', \omega)\rangle\rangle \to \langle\langle G_1^{ret}(\mathbf{q}, \omega)\rangle\rangle$ is

$$\langle\langle G_1^{ret}(\mathbf{q}, \omega)\rangle\rangle = G_1^0(\mathbf{q}, \omega) + G_1^0(\mathbf{q}, \omega) \sum(\mathbf{q}, \omega) \langle\langle G_1^{ret}(\mathbf{q}, \omega)\rangle\rangle$$

or

$$\langle\langle G_1^{ret}(\mathbf{q},\omega)\rangle\rangle = G_1^0(\mathbf{q},\omega)+n_i G_1^0(\mathbf{q},\omega)\,\langle\langle G_1^{ret}(\mathbf{q},\omega)\rangle\rangle \int \frac{d^{(3)}\mathbf{p}}{(2\pi)^3}\,\langle\langle G_1^{ret}(\mathbf{q}-\mathbf{p},\omega)\rangle\rangle\, U(\mathbf{p})U(-\mathbf{p}). \tag{5.4.31}$$

For a 3D potential $U(\mathbf{x})$ that is slowly varying in space, we may expect $U(\mathbf{p})$ to be sharply peaked in the vicinity of $\mathbf{p}=\mathbf{0}$, so that the $\mathbf{p}$-integrand of Eq. (5.4.31) may be taken with $\langle\langle G_1^{ret}(\mathbf{q}-\mathbf{p},\omega)\rangle\rangle \to \langle\langle G_1^{ret}(\mathbf{q},\omega)\rangle\rangle$, which can be extracted from the integral, whence

$$\langle\langle G_1^{ret}(\mathbf{q},\omega)\rangle\rangle = G_1^0(\mathbf{q},\omega) + n_i G_1^0(\mathbf{q},\omega)\left[\langle\langle G_1^{ret}(\mathbf{q},\omega)\rangle\rangle\right]^2 \int \frac{d^{(3)}\mathbf{p}}{(2\pi)^3} U(\mathbf{p})U(-\mathbf{p}), \tag{5.4.32}$$

where $G_1^0(\mathbf{q},\omega) = (\omega - q^2/2m + i\delta)^{-1}$. This may be solved algebraically as a quadratic equation for $\langle\langle G_1^{ret}(\mathbf{q},\omega)\rangle\rangle$, yielding

$$\langle\langle G_1^{ret}(\mathbf{q},\omega)\rangle\rangle = \frac{1 \pm \sqrt{1-4n_i\left[G_1^0(\mathbf{q},\omega)\right]^2 I}}{2n_i G_1^0(\mathbf{q},\omega) I}; \quad I = \int \frac{d^{(3)}\mathbf{p}}{(2\pi)^3} U(\mathbf{p})U(-\mathbf{p}). \tag{5.4.33}$$

On the other hand, if $U(\mathbf{x})$ is a very short range rapidly varying potential in the nature of a Dirac delta function, $U(\mathbf{x}) \sim \delta^{(3)}(\mathbf{x})$, then $U(\mathbf{p})$ is essentially constant and the $\mathbf{p}$-integral of Eq. (5.4.31) can be seen (with a shift of the integration variable) to have only a very weak dependence on $\mathbf{q}$, and the solution may be analyzed accordingly. An example of this type is discussed in the next section.

5.5 Ando's Semi-Elliptic Density of States for the 2D Landau-Quantized Electron-Impurity System

The self-energy in the self-consistent Born approximation, Σ, due to random impurity scattering (Eq. (5.4.29)) has been employed by Ando[5] in the theory of two-dimensional (2D) Landau-quantized magnetoresistance. In this, the electron motions are confined to a deep narrow potential well constituting the x-y plane approximated by a planar sheet at $z = 0$, with a finite 2D sheet density of electrons (per unit area). Possible states of motion across this potential well (called "subband states") are strongly size-quantized by high, confining potential walls at $z = 0^-$ and $z = 0^+$ (hence the term "quantum well"), such that the only energetically accessible energy level for motion in the z-direction is the lowest one. This is to say that terms in Eq. (5.1.25) involving higher subband energies have such large energy denominators that they can be neglected, leaving only the lowest energy term for motion across the quantum well. Under a high magnetic field normal to the sheet, Landau quantization produces a discrete spectrum in which the energy eigenstates for motion on the sheet have energies $E_n = (n + \frac{1}{2})\hbar\omega_c$ (see Section 5.6, Eq. (5.6.31), $n = 0, 1, 2, \ldots$; ω_c is the cyclotron frequency). These levels are degenerate but we will ignore this degeneracy for illustrative purposes. Again, we assume that the energy separation $\hbar\omega_c$ between Landau eigenstates is so large compared to available excitation energies that transitions are not possible and only the lowest state is accessible for motion on the plane (as well as across it). In this case, which occurs in GaAs quantum wells and Si-based inversion layers, the pertinent part of G_1^0 in Eq. (5.1.25) is given in position-frequency representation by

$$G_1^0(\bar{\mathbf{x}}'', \bar{\mathbf{x}}''', \omega) = \frac{\phi_0(\bar{\mathbf{x}}'')\phi_0^*(\bar{\mathbf{x}}''')}{\omega - E_0 + i\delta}, \tag{5.5.1}$$

where $\phi_0(\bar{\mathbf{x}})$ is the lowest 2D energy eigenstate as a function of the 2D position vector $\bar{\mathbf{x}}$ in a high magnetic field corresponding to the lowest 2D value, $E_0 = \frac{1}{2}\hbar\omega_c$, on the plane (setting the energy scale to zero at the lowest energy for motion across the quantum well). For convenience, we assume the impurity scattering potentials to be randomly sited on the 2D electron sheet with 2D density $n_{i_{2D}}$, although it is not difficult to deal with them sited on a different "remote" sheet as well as distributed throughout 3D space, as has been done in many actual calculations for physically realizable impurity distributions. Adjusting Eq. (5.4.29) to this 2D context, we have

$$\sum(\bar{\mathbf{x}}'', \bar{\mathbf{x}}''', \omega) = \langle\langle G_1^{ret}(\bar{\mathbf{x}}'', \bar{\mathbf{x}}''', \omega)\rangle\rangle\, n_{i_{2D}} \int \frac{d^{(2)}\bar{\mathbf{p}}}{(2\pi)^2} e^{i\bar{\mathbf{p}}\cdot(\bar{\mathbf{x}}''-\bar{\mathbf{x}}''')}\, U(\bar{\mathbf{p}})U(-\bar{\mathbf{p}}). \tag{5.5.2}$$

For a short-range 2D delta-function impurity, $U(\bar{\mathbf{p}}) = U_0$, is constant and (Eq. (5.4.28)):

$$\sum(\bar{\mathbf{x}}'', \bar{\mathbf{x}}''', \omega) = \langle\langle G_1^{ret}(\bar{\mathbf{x}}'', \bar{\mathbf{x}}''', \omega)\rangle\rangle\, n_{i_{2D}} U_0^2 \delta^{(2)}(\bar{\mathbf{x}}'' - \bar{\mathbf{x}}'''). \tag{5.5.3}$$

The associated Dyson equation for $\langle\langle G_1^{ret}(\bar{\mathbf{x}}, \bar{\mathbf{x}}', \omega)\rangle\rangle$ is (Eq. (5.4.25)):

$$\begin{aligned}\langle\langle G_1^{ret}(\bar{\mathbf{x}}, \bar{\mathbf{x}}', \omega)\rangle\rangle &= G_1^0(\bar{\mathbf{x}}, \bar{\mathbf{x}}', \omega)\\ &+ n_{i_{2D}} U_0^2 \int d^{(2)}\bar{\mathbf{x}}''\, G_1^0(\bar{\mathbf{x}}, \bar{\mathbf{x}}'', \omega)\, \langle\langle G_1^{ret}(\bar{\mathbf{x}}'', \bar{\mathbf{x}}'', \omega)\rangle\rangle \langle\langle G_1^{ret}(\bar{\mathbf{x}}'', \bar{\mathbf{x}}', \omega)\rangle\rangle,\end{aligned} \tag{5.5.4}$$

and using Eq. (5.5.1) for G_1^0, we obtain the *separable* non-linear integral equation

$$\begin{aligned}\langle\langle G_1^{ret}(\bar{\mathbf{x}}, \bar{\mathbf{x}}', \omega)\rangle\rangle &= G_1^0(\bar{\mathbf{x}}, \bar{\mathbf{x}}', \omega)\\ &+ \frac{n_{i_{2D}} U_0^2 \phi_0(\bar{\mathbf{x}})}{\omega - E_0 + i\delta} \int d^{(2)}\bar{\mathbf{x}}''\phi^*(\bar{\mathbf{x}}'')\, \langle\langle G_1^{ret}(\bar{\mathbf{x}}'', \bar{\mathbf{x}}'', \omega)\rangle\rangle \langle\langle G_1^{ret}(\bar{\mathbf{x}}'', \bar{\mathbf{x}}', \omega)\rangle\rangle.\end{aligned} \tag{5.5.5}$$

Considering that both terms on the right of Eq. (5.5.5) are proportional to $\phi_0(\bar{\mathbf{x}})$, we attempt a solution of the form

$$\langle\langle G_1^{ret}(\bar{\mathbf{x}}, \bar{\mathbf{x}}', \omega)\rangle\rangle = Z(\omega)\phi_0(\bar{\mathbf{x}})\phi_0^*(\bar{\mathbf{x}}'), \tag{5.5.6}$$

which leads to the quadratic equation

$$Z(\omega) = \frac{1}{\omega - E_0 + i\delta} + \frac{n_{i_{2D}} U_0^2 Z^2(\omega)}{\omega - E_0 + i\delta} \int d^{(2)}\bar{\mathbf{x}}'' \left|\phi_0(\bar{\mathbf{x}}'')\right|^4. \tag{5.5.7}$$

Denoting the $\bar{\mathbf{x}}''$-integral by $\mathcal{I}$, we obtain

$$Z(\omega) = \frac{\omega - E_0}{2n_{i_{2D}} U_0^2 \mathcal{I}} \pm \frac{1}{2n_{i_{2D}} U_0^2 \mathcal{I}} \sqrt{(\omega - E_0)^2 - 4n_{i_{2D}} U_0^2 \mathcal{I}}. \tag{5.5.8}$$

In the case of a scattering potential $U(\mathbf{x})$ that is slowly varying in space, we again expect $U(\bar{\mathbf{p}})$ to contribute mainly at $\mathbf{p} = 0$, so Eq. (5.5.2) yields

$$\sum(\bar{\mathbf{x}}'', \bar{\mathbf{x}}''', \omega) = \left\langle\left\langle G_1^{ret}(\bar{\mathbf{x}}'', \bar{\mathbf{x}}''', \omega)\right\rangle\right\rangle n_{i_{2D}} \int \frac{d^{(2)}\bar{\mathbf{p}}}{(2\pi)^2} U(\bar{\mathbf{p}}) U(-\bar{\mathbf{p}}), \tag{5.5.9}$$

and the $\bar{\mathbf{p}}$-integral, denoted by $\mathcal{I}'$, is just a constant. In this case the Dyson equation has the form

$$\begin{aligned}\left\langle\left\langle G_1^{ret}(\bar{\mathbf{x}}, \bar{\mathbf{x}}', \omega)\right\rangle\right\rangle &= G_1^0(\bar{\mathbf{x}}, \bar{\mathbf{x}}', \omega) \\ &+ n_{i_{2D}} \mathcal{I}' \int d^{(2)}\mathbf{x}'' \int d^{(2)}\mathbf{x}''' G_1^0(\bar{\mathbf{x}}, \bar{\mathbf{x}}'', \omega) \left\langle\left\langle G_1^{ret}(\bar{\mathbf{x}}'', \bar{\mathbf{x}}''', \omega)\right\rangle\right\rangle \left\langle\left\langle G_1^{ret}(\bar{\mathbf{x}}''', \bar{\mathbf{x}}', \omega)\right\rangle\right\rangle.\end{aligned} \tag{5.5.10}$$

Eq. (5.5.1) again renders this non-linear integral equation separable, with both terms on the right proportional to $\phi_0(\bar{\mathbf{x}})$, so a solution of the form of Eq. (5.5.6) is to be expected, with the result (using the unit normalization of $\phi_0(\bar{\mathbf{x}})$),

$$Z(\omega) = \frac{\omega - E_0}{2n_{i_{2D}} \mathcal{I}'} \pm \frac{1}{2n_{i_{2D}} \mathcal{I}'} \sqrt{(\omega - E_0)^2 - 4n_{i_{2D}} \mathcal{I}'}. \tag{5.5.11}$$

These are just the types of results Ando obtained in the theory of impurity limited 2D magnetoconductivity using the self-consistent Born approximation, except for our neglect of the degeneracy of Landau-quantized electron energy eigenstates in a high magnetic field.

The results given by Eqns. (5.5.6), (5.5.8) and (5.5.11) yield a semi-elliptic density of one-electron energy eigenstates (per unit energy). Employing Eq. (5.1.27), we have

$$D(\omega) = -\frac{1}{\pi} Tr\, \mathrm{Im} G_1^{ret}(\omega) = -\frac{1}{\pi} \int d^{(2)}\mathbf{x}\, \mathrm{Im} \left\langle\left\langle G_1^{ret}(\bar{\mathbf{x}}, \bar{\mathbf{x}}, \omega)\right\rangle\right\rangle, \tag{5.5.12}$$

and using the unit normalization of $\phi_0(\bar{\mathbf{x}})$ we obtain ($\eta_+(x)$ is the Heaviside unit step function):

$$D(\omega) = \frac{-1}{\pi} \mathrm{Im} Z(\omega) = -\frac{2}{\pi\Gamma} \sqrt{1 - \left(\frac{\omega - E_0}{\Gamma}\right)^2}\, \eta_+ \left(1 - \left[\frac{\omega - E_0}{\Gamma}\right]^2\right). \tag{5.5.13}$$

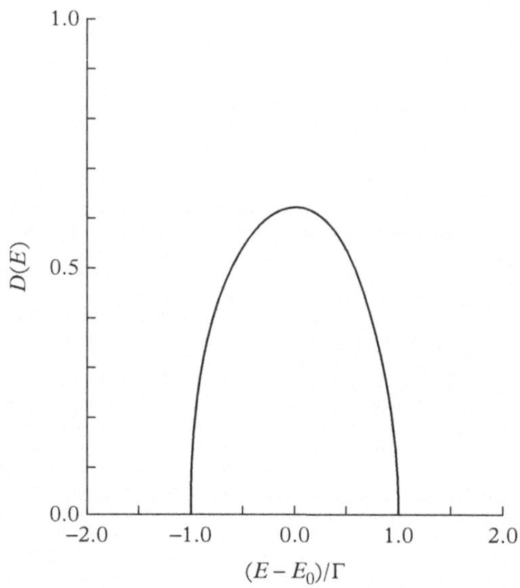

Figure 5.1: *Schematic of Ando's Semi-Elliptic Density of States for a 2D Landau Quantized Electron-Impurity system with Linewidth* Γ *as a function of Energy* $E - E_0 \leftrightarrow \omega$.

Here, Γ is the line width and $\Gamma^2 = 4n_{i_{2D}}U_0^2\mathcal{I}$ for short-range spatial delta-function impurities, whereas $\Gamma^2 = 4n_{i_{2D}}\mathcal{I}'$ for slowly varying scattering potentials. This semi-elliptic density of states of the self-consistent Born approximation for 2D Landau-quantized electrons in a high magnetic field is shown schematically in Figure (5.1). The unscattered delta-function density of states arising from $G_1^0(\omega)$ is represented by a vertical dotted line. Since the total number of states is unchanged, the ω-integral of the elliptic density matches that of the delta function. However, there is a redistribution of the state density weight due to scattering, which reflects the fact that the quantum state at E_0 no longer has an infinite lifetime. Random scattering induces a finite inverse lifetime determined by the half-width of the elliptic curve.

5.6 Magnetic Field Green's Function for Electrons in Three and Two Dimensions

The circularity of an electron orbit in a magnetic field reflects the fact that the direction of momentum is not conserved. Accordingly, the Green's function depends on $(\mathbf{x} + \mathbf{x}')$ as well as $(\mathbf{x} - \mathbf{x}')$. In connection with this, it is useful to consider a gauge transformation of the electromagnetic potentials, $\mathbf{A}(\mathbf{x}, t) \rightarrow \mathbf{A}(\mathbf{x}, t) + \nabla\lambda(\mathbf{x}, t)$ and $\phi(\mathbf{x}, t) \rightarrow \phi(\mathbf{x}, t) - \frac{1}{c}\partial\lambda(\mathbf{x}, t)/\partial t$. To maintain invariance of the Hamiltonian, the corresponding transformation of the field operator ψ is $\psi(\mathbf{x}, t) \rightarrow e^{\frac{ie\lambda(\mathbf{x},t)}{\hbar c}}\psi(\mathbf{x}, t)$ and due to its bilinear structure the associated Green's function transforms as (see discussion of Eq. (1.1.32)):

$$G_1^{ret}(\mathbf{x},t;\mathbf{x}',t') \to G_1'(\mathbf{x},t;\mathbf{x}',t') = \exp\left(\frac{ie}{\hbar c}[\lambda(\mathbf{x},t)-\lambda(\mathbf{x}',t')]\right) G_1^{ret}(\mathbf{x},t;\mathbf{x}',t'). \quad (5.6.1)$$

Denoting the Schrödinger operator as $\Lambda_1 = i\frac{\partial}{\partial t_1} - h_1^{(1)}$, the electron Green's function equation, including a constant magnetic field, has the structure,

$$\Lambda_1(\nabla_1, \mathbf{A}(\mathbf{x}_1), \mathbf{B}(\mathbf{x}_1)) G_1^{ret}(\mathbf{x}_1, t_1; \mathbf{x}_1', t_1') = \delta^{(3)}(\mathbf{x}_1 - \mathbf{x}_1')\delta(t_1 - t_1'), \quad (5.6.2)$$

or, equivalently,

$$\Lambda_1(\nabla_1, \mathbf{A}(\mathbf{x}_1) + \nabla\lambda(\mathbf{x}_1), \mathbf{B}(\mathbf{x}_1)) G_1'(\mathbf{x}_1, t_1; \mathbf{x}_1', t_1') = \delta^{(3)}(\mathbf{x}_1 - \mathbf{x}_1')\delta(t_1 - t_1'). \quad (5.6.3)$$

Considering application to a magnetic field that is uniform in space as well as constant in time, it is useful to choose λ to depend on *both* $\mathbf{x}_1$ *and* $\mathbf{x}_1'$, requiring that

$$\mathbf{A}(\mathbf{x}_1) + \nabla_1\lambda(\mathbf{x}_1, \mathbf{x}_1') = \frac{1}{2}\mathbf{B}(\mathbf{x}_1) \times (\mathbf{x}_1 - \mathbf{x}_1'). \quad (5.6.4)$$

The point of this requirement is that Eq. (5.6.3) becomes

$$\Lambda_1\left(\nabla_1, \tfrac{1}{2}\mathbf{B}(\mathbf{x}_1) \times (\mathbf{x}_1 - \mathbf{x}_1')\right) G_1'(\mathbf{x}_1, t_1; \mathbf{x}_1', t_1') = \delta^{(3)}(\mathbf{x}_1 - \mathbf{x}_1')\delta(t_1 - t_1'), \quad (5.6.5)$$

so that if $\mathbf{B}(\mathbf{x}_1) \to \mathbf{B}$ is uniform and constant (in both space and time) then Eq. (5.6.5) mandates that G_1' depends on $(\mathbf{x}_1 - \mathbf{x}_1')$ alone, with no dependence on $(\mathbf{x}_1 + \mathbf{x}_1')$, and we have

$$G_1'(\mathbf{x}_1, t_1; \mathbf{x}_1', t_1') = G_1'(\mathbf{x}_1 - \mathbf{x}_1'; t_1 - t_1') = \exp\left(\frac{ie}{\hbar c}[\lambda(\mathbf{x}_1, \mathbf{x}_1') - \lambda(\mathbf{x}_1', \mathbf{x}_1')]\right) G_1^{ret}(\mathbf{x}_1, t_1; \mathbf{x}_1', t_1'). \quad (5.6.6)$$

Notwithstanding the presence of the source point $\mathbf{x}_1'$, Eq. (5.6.4) emphasizes the fact that $\mathbf{A}(\mathbf{x}_1)$ can differ from $\frac{1}{2}\mathbf{B} \times \mathbf{x}_1$ only by a gauge transformation,

$$\mathbf{A}(\mathbf{x}_1) + \nabla_1\phi(\mathbf{x}_1) = \frac{1}{2}\mathbf{B} \times \mathbf{x}_1, \quad (5.6.7)$$

with $\phi(\mathbf{x}_1)$ as the gauge function. Subtracting Eq.(5.6.7) from Eq. (5.6.4), we obtain an equation for $\lambda(\mathbf{x}_1, \mathbf{x}_1')$:

$$\nabla_1[\lambda(\mathbf{x}_1, \mathbf{x}_1') - \phi(\mathbf{x}_1)] = -\frac{1}{2}\mathbf{B} \times \mathbf{x}_1', \quad (5.6.8)$$

and line-integration yields

$$\lambda(\mathbf{x}_1, \mathbf{x}_1') = -\frac{1}{2}\int d\mathbf{x}_1 \cdot \mathbf{B} \times \mathbf{x}_1' + \phi(\mathbf{x}_1) = -\frac{1}{2}\mathbf{x}_1 \cdot \mathbf{B} \times \mathbf{x}_1' + \phi(\mathbf{x}_1). \quad (5.6.9)$$

Carrying out the $\mathbf{x}_1$-integration, we readily find that for a uniform, constant magnetic field, both the $(\mathbf{x}_1 + \mathbf{x}_1')$ – and gauge-dependencies of G_1 are embedded in the factor $C(\mathbf{x}_1, \mathbf{x}_1')$, with

$$G_1^{ret}(\mathbf{x}_1, t_1; \mathbf{x}_1', t_1') = C(\mathbf{x}_1, \mathbf{x}_1') G_1'(\mathbf{x}_1 - \mathbf{x}_1'; t_1 - t_1'), \tag{5.6.10}$$

where the factor

$$C(\mathbf{x}_1, \mathbf{x}_1') = \exp\left[i\frac{e}{2\hbar c}\left(\mathbf{x}_1 \cdot \mathbf{B} \times \mathbf{x}_1' - \phi(\mathbf{x}_1) + \phi(\mathbf{x}_1')\right)\right] \tag{5.6.11}$$

accounts for the changing momentum direction associated with circularity of the electron path in a magnetic field (with the magnitude of momentum conserved).

Eq. (5.6.5) for $G_1'(\mathbf{x}_1 - \mathbf{x}_1', t_1 - t_1')$ is considerably simplified because, effectively, it has the source at the origin, with $\mathbf{A}$ replaced by $\frac{1}{2}\mathbf{B} \times (\mathbf{x}_1 - \mathbf{x}_1')$. The Schrödinger electron operator $\Lambda(\nabla, \mathbf{A}, \mathbf{B}) = i(\partial/\partial t) - h_1^{(1)}(\nabla, \mathbf{A}, \mathbf{B})$ with spin can be constructed from the field-free formula $\Lambda(\nabla) = i(\partial/\partial t) - h_1^{(1)}(\nabla)$ by appropriate adjustment of the canonical momentum operator $\frac{1}{i}\nabla \to \frac{1}{i}(\nabla - ie\mathbf{A}/c)$ and adding a Zeeman spin term $\overline{\mu}_0 \boldsymbol{\sigma} \cdot \mathbf{B}$. Taking the z-axis along the magnetic field and introducing the notation:

$$\mathbf{R} = \mathbf{x}_1 - \mathbf{x}_1' = \begin{pmatrix} X \\ Y \\ Z \end{pmatrix} = \begin{pmatrix} x_1 - x_1' \\ y_1 - y_1' \\ z_1 - z_1' \end{pmatrix}, \tag{5.6.12}$$

$$T = t_1 - t_1',$$
$$\boldsymbol{\sigma} = \sigma_1, \sigma_2, \sigma_3 \text{ are the Pauli spin matrices},$$
$$\omega_c = \text{cyclotron frequency} \to eB/mc,$$
$$\overline{\mu}_0 = \text{Bohr magneton} \to e\hbar/2mc,$$

Eq. (5.6.5) may be written as

$$\left(\frac{\nabla_R^2}{2m} - \frac{m\omega_c^2}{8}(X^2 + Y^2) + \frac{\omega_c}{4}L_Z - \overline{\mu}_0 B \sigma_3 + i\frac{\partial}{\partial T} \right) G_1'(\mathbf{R}, T) = \delta^{(3)}(\mathbf{R})\delta(T), \tag{5.6.13}$$

with

$$L_Z = \frac{1}{i}\left(X\frac{\partial}{\partial Y} - Y\frac{\partial}{\partial X} \right) = L_{z_1} + L_{z_1'}. \tag{5.6.14}$$

It is helpful to note that $L_Z G_1'(\mathbf{R}, T) = 0$. To verify this, we choose the gauge $\phi = 0$ so that $L_Z C(\mathbf{x}_1, \mathbf{x}_1') = 0$, and then $L_Z G_1'(\mathbf{R}, T) = C^{-1}(\mathbf{x}_1, \mathbf{x}_1') \cdot L_Z G_1^{ret}(\mathbf{x}_1, t_1; \mathbf{x}_1', t_1')$. The vanishing of

$$L_Z G_1^{ret}(\mathbf{x}_1, t_1; \mathbf{x}_1', t_1') = \left(L_{z_1} + L_{z_1'} \right) G_1^{ret}(\mathbf{x}_1, t_1; \mathbf{x}_1', t_1')$$

is a consequence of the bilinear dependence of $G_1^{ret}(\mathbf{x}_1, t_1; \mathbf{x}_1', t_1')$ on the field operators, $\psi(\mathbf{x}_1, t_1)$, $\psi^+(\mathbf{x}_1', t_1')$, and the associated interpretation of $G_1^{ret}(\mathbf{x}_1, t_1; \mathbf{x}_1', t_1')$ in terms of the creation of a particle at $(\mathbf{x}_1, t_1)$ which is subsequently annihilated at $(\mathbf{x}_1', t_1')$. The action of L_{z_1} on $G_1^{ret}(\mathbf{x}_1, t_1; \mathbf{x}_1', t_1')$ measures the component of orbital angular momentum parallel to the magnetic field at the point of creation, while the action of $L_{z_1'}$ measures the negative of it at the point of annihilation (bear in mind that $L_{z_1'}$ acts on the adjoint operator). Thus, $L_Z = L_{z_1} + L_{z_1'}$ measures the loss of orbital angular momentum parallel to the magnetic field during particle propagation from $(\mathbf{x}_1, t_1)$ to $(\mathbf{x}_1', t_1')$, which vanishes because it is a conserved quantity. (An alternate proof that $L_Z G_1^{ret}(\mathbf{x}_1, t_1; \mathbf{x}_1', t_1') = 0$ with $\phi \neq 0$ can be constructed by expanding the field operators in a series of angular momentum eigenfunctions, and applying the operator $L_Z = L_{z_1} + L_{z_1'}$.) These considerations reduce Eq. (5.6.13) to the simpler form

$$\left(\frac{\nabla_{\mathbf{R}}^2}{2m} - \frac{m\omega_c^2}{8}(X^2 + Y^2) - \overline{\mu}_0 B\sigma_3 + i\frac{\partial}{\partial T}\right) G_1'(\mathbf{R}, T) = \delta^{(3)}(\mathbf{R})\delta(T), \tag{5.6.15}$$

and for $T > 0$, G_1' satisfies the homogeneous counterpart of Eq. (5.6.15),

$$\left(\frac{\nabla_{\mathbf{R}}^2}{2m} - \frac{m\omega_c^2}{8}\left[X^2 + Y^2\right] - \overline{\mu}_0 B\sigma_3 + i\frac{\partial}{\partial T}\right) G_1'(\mathbf{R}, T) = 0. \tag{5.6.16}$$

Furthermore, for $T < 0$, $G_1' \equiv 0$, and T-integration of Eq. (5.6.15) from $T = 0^-$ to $T = 0^+$ yields the initial value of G_1' as

$$G_1'(\mathbf{R}, T = 0^+) = -i\delta^{(3)}(\mathbf{R}) = -i\delta^{(3)}(\mathbf{x}_1 - \mathbf{x}_1'). \tag{5.6.17}$$

Because of the presence of $\delta^{(3)}(\mathbf{x}_1 - \mathbf{x}_1')$ in Eq. (5.6.17) we may replace $C(\mathbf{x}_1, \mathbf{x}_1')$ by $C(\mathbf{x}_1, \mathbf{x}_1)$, which is identically unity, so we have the same initial value for G_1^{ret},

$$G_1^{ret}(\mathbf{x}_1, \mathbf{x}_1', T = 0^+) = G_1'(\mathbf{R}, T = 0^+) = -i\delta^{(3)}(\mathbf{x}_1 - \mathbf{x}_1'). \tag{5.6.18}$$

Considering Eq. (5.6.17) and Fourier-transforming $\mathbf{R} = \mathbf{x}_1 - \mathbf{x}_1' \rightarrow \mathbf{p} = (\overline{p}, p_z) = (p_x, p_y, p_z)$, and defining G_1'' as

$$G_1'(\mathbf{p}, T) = \eta_+(T) \exp\left[-i\left(\overline{\mu}_0 B\sigma_3 + \frac{p_z^2}{2m}\right) T\right] G_1''(\overline{p}, T) \tag{5.6.19}$$

(where $\eta_+(T) = 0$ for $T < 0$, and 1 for $T > 0$), we have

$$\left[-\frac{1}{2m}(p_x^2 + p_y^2) + \frac{m\omega_c^2}{8}\left(\frac{\partial^2}{\partial p_x^2} + \frac{\partial^2}{\partial p_y^2}\right) + i\frac{\partial}{\partial T}\right] G_1''(\overline{p}, T) = 0, \tag{5.6.20}$$

and Eq. (5.6.18) becomes

$$G_1''(\bar{p}, T = 0^+) = -i. \tag{5.6.21}$$

The solution of Eq. (5.6.20) is given by (K, C are constants):

$$G_1''(\bar{p}, T) = K \sec\left[\frac{\omega_c}{2}(T + C)\right] \exp\left(-i\frac{p_x^2 + p_y^2}{m\omega_c} \tan\left[\frac{\omega_c}{2}(T + C)\right]\right). \tag{5.6.22}$$

To satisfy Eq. (5.6.21) we set $C = 0$, $K = -i$, whence

$$G_1''(\bar{p}, T) = -i \sec\left(\frac{\omega_c}{2}T\right) \exp\left(-i\frac{p_x^2 + p_y^2}{m\omega_c} \tan\left(\frac{\omega_c}{2}T\right)\right). \tag{5.6.23}$$

Finally, we have the retarded 3D magnetic field Green's function G_1^{ret} as

$$G_1^{ret}(\mathbf{x}_1, \mathbf{x}_1', T) = -i\eta_+(T)C(\mathbf{x}_1, \mathbf{x}_1') \int \frac{d^{(3)}\mathbf{p}}{(2\pi)^3} e^{i\mathbf{p}\cdot\mathbf{R}} \exp\left[-i\left(\bar{\mu}_0 B\sigma_3 + \frac{p_z^2}{2m}\right)T\right] \cdot \sec\left(\frac{\omega_c}{2}T\right)$$
$$\exp\left(-i\frac{p_x^2 + p_y^2}{m\omega_c} \tan\left(\frac{\omega_c}{2}T\right)\right). \tag{5.6.24}$$

The $\mathbf{p}$-integrals of Eq. (5.6.24) are simply Fourier transforms of Gaussians, leading to the result for G_1^{ret} in position representation as

$$G_1^{ret}(\mathbf{x}_1, \mathbf{x}_1', T) = -i\eta_+(T)C(\mathbf{x}_1, \mathbf{x}_1') \exp\left[-i\left(\bar{\mu}_0 B\sigma_3 T\right)\right] \frac{m^{3/2}\omega_c}{2(2\pi i)^{3/2}} \frac{\exp\left[\frac{imZ^2}{2T} + \frac{im\omega_c(X^2+Y^2)}{4\tan(\omega_c T/2)}\right]}{T^{1/2}\sin(\omega_c T/2)}. \tag{5.6.25}$$

In the limit of zero magnetic field, we have ($C(\mathbf{x}, \mathbf{x}') \to 1$):

$$G_1^{ret}(\mathbf{x}_1, \mathbf{x}_1', T) = -i\left(\frac{m}{2\pi i T}\right)^{3/2} \exp\left[\frac{im}{2T}(X^2 + Y^2 + Z^2)\right]. \tag{5.6.26}$$

In the case of an electron confined to 2D motion on the $\bar{x} = (x, y)$ plane in a perpendicular magnetic field, $\mathbf{B} = B\hat{z}$, the result for G_1^{ret} is readily obtained by noting that the momentum component p_z along the normal magnetic field vanishes identically, so that $p_z \to 0$ in Eqns. (5.6.19–5.6.24) and the p_z-integral, $\int \frac{dp_z}{2\pi}$, is eliminated. Thus, we obtain the retarded 2D magnetic field Green's function as ($\bar{x}_1, \bar{x}_1', \bar{p}, \bar{R} = \bar{x}_1 - \bar{x}_1'$ are 2D vectors in the $\bar{x} = (x, y)$ plane and $X = x - x'$, $Y = y - y'$):

$$G_1^{ret}(\overline{x}_1, \overline{x}_1', T) =$$
$$- i\eta_+(T)C(\overline{x}_1, \overline{x}_1')\int \frac{d^{(2)}\overline{p}}{(2\pi)^2} e^{i\overline{p}\cdot\overline{R}} e^{-i(\overline{\mu}_0 B\sigma_3 T)} \sec\left(\frac{\omega_c}{2}T\right)\exp\left(-i\frac{\overline{p}^2 \tan(\omega_c T/2)}{m\omega_c}\right), \tag{5.6.27}$$

and, correspondingly,

$$G_1^{ret}(\overline{x}_1, \overline{x}_1, T) = -\eta_+(T)C(\overline{x}_1, \overline{x}_1')\frac{m\omega_c e^{-i\overline{\mu}_0 B\sigma_3 T}}{4\pi \sin(\omega_c T/2)} \exp\left(\frac{im\omega_c\left[X^2 + Y^2\right]}{4\tan(\omega_c T/2)}\right). \tag{5.6.28}$$

Employing the generating function for Laguerre polynomials, L_n, in the form

$$[\sin(\omega_c T/2)]^{-1} \exp\left[i\frac{x}{2}\frac{1}{\tan(\omega_c T/2)}\right] = 2ie^{-\frac{x}{2}}\sum_{n=0}^{\infty} L_n(x)e^{-i(n+\frac{1}{2})\omega_c T}, \tag{5.6.29}$$

we obtain G_1^{ret} as follows:

$$G_1^{ret}(\overline{x}_1, \overline{x}_1', T) = \eta_+(T)C(\overline{x}_1, \overline{x}_1')\frac{m\omega_c}{2\pi i} e^{-i\overline{\mu}_0 B\sigma_3 T} \exp\left(-m\omega_c\left[X^2 + Y^2\right]/4\right) \tag{5.6.30}$$
$$\times \sum_{n=0}^{\infty} L_n\left(m\omega_c\left[X^2 + Y^2\right]/2\right) e^{-i(n+1/2)\omega_c T}.$$

This may be re-expressed in terms of Hermite polynomials H_n using Feldheim's identity[6], $L_n(a^2 + b^2) = [(-1)^n/n!]\sum_{k=0}^{n}\binom{n}{k} H_{2k}(a)H_{2n-2k}(b)$, where $\binom{n}{k}$ is the binomial coefficient. Comparing Eq. (5.6.30) with Eq. (5.1.20), it is evident that the energy spectrum of an electron confined to motion in a plane normal to a constant, uniform magnetic field (ignoring spin, $\overline{\mu}_0 \to 0$) is given by

$$E_n^{(1)} = (n + 1/2)\hbar\omega_c, \tag{5.6.31}$$

so the 2D spectrum is discrete and *fully quantized*. Considering the Fourier time transform of Eq. (5.6.30) (with $\overline{\mu}_0 \to 0$), it is immediately apparent that it yields a discrete sum of terms of the "denominator" form $(\omega - [n + 1/2]\,\omega_c)^{-1}$, and if available excitations are not sufficiently energetic to cause transitions to higher n-levels (as represented by large denominators), then only the $n = 0$ level contributes, as in Eq. (5.5.1) with $E_0 = \hbar\omega_c/2$. The fully quantized nature of the discrete spectrum of states for motion in the plane transverse to the magnetic field, with uniform energy separation $\hbar\omega_c$, is said to be *Landau quantized.* Such magnetic field quantization is present in three dimensions as well as two dimensions for motion in the transverse plane: However, the 3D case also encompasses a continuum of states due to 1D motion along the magnetic field, which is unaffected by the field. For a uniform magnetic field, the part of the motion represented by everything in G_1' (Eqns. (5.6.10), (5.6.30) and (5.6.25)), *excluding* $C(\mathbf{x}_1, \mathbf{x}_1')$, depends only on $\mathbf{x}_1 - \mathbf{x}_1'$ and not at all on $\mathbf{x}_1 + \mathbf{x}_1'$, reflecting effective translational invariance associated with conservation of the *magnitude* of momentum (in part). In the limit of a weak magnetic field, when the discrete separation of Landau levels ($\hbar\omega_c$) may be considered negligible, G_1' may be approximated by the zero-field result. But the factor $C(\mathbf{x}_1, \mathbf{x}_1')$ (Eq. (5.6.11)) depends on both $\mathbf{x}_1 + \mathbf{x}_1'$ and $\mathbf{x}_1 - \mathbf{x}_1'$, lacking translational invariance and reflecting the non-conservation of the *direction* of momentum evident in the circularity of motion in a magnetic field; also embodying the gauge dependence of the Green's function.

Considering another point of view on the determination of $C(\mathbf{x}_1, \mathbf{x}_1')$ we note that the gauge change of vector potential is given by $\nabla\lambda(\mathbf{x})$, so that $\lambda(\mathbf{x}) \to \int_0^{\mathbf{x}} d\mathbf{x}_1 \cdot \nabla_1 \lambda(\mathbf{x}_1) \to \int_0^{\mathbf{x}} d\mathbf{x}_1 \cdot \mathbf{A}(\mathbf{x}_1)$ describes the corresponding phase change of $\psi(\mathbf{x}) \to e^{\frac{ie}{\hbar c}\lambda(\mathbf{x})}\psi(\mathbf{x}) \to \exp\left[\frac{ie}{\hbar c}\int_0^{\mathbf{x}} d\mathbf{x}_1 \cdot \mathbf{A}(\mathbf{x}_1)\right]\psi(\mathbf{x})$ (the lower limit, 0, is a reference point of line integration). In this context, $\lambda(\mathbf{x}') \to \int_0^{\mathbf{x}'} d\mathbf{x}_1 \cdot \mathbf{A}(\mathbf{x}_1)$ describes the negative phase change of the adjoint $\psi^+(\mathbf{x}') \to e^{-\frac{ie}{\hbar c}\lambda(\mathbf{x}')}\psi^+(\mathbf{x}') \to \exp\left[\frac{ie}{\hbar c}\int_0^{\mathbf{x}} d\mathbf{x}_1 \cdot \mathbf{A}(\mathbf{x}_1)\right]\psi^+(\mathbf{x}')$. Thus, the net phase change of the Green's function under the gauge transformation is given by a line integral that generally *depends upon path*:

$$\frac{ie}{\hbar c}\left[\lambda(\mathbf{x}) - \lambda(\mathbf{x}')\right] = \frac{ie}{\hbar c}\int_0^{\mathbf{x}} d\mathbf{x}_1 \cdot \mathbf{A}(\mathbf{x}_1) - \frac{ie}{\hbar c}\int_0^{\mathbf{x}'} d\mathbf{x}_1 \cdot \mathbf{A}(\mathbf{x}_1) = \frac{ie}{\hbar c}\int_{\mathbf{x}'}^{\mathbf{x}} d\mathbf{x}_1 \cdot \mathbf{A}(\mathbf{x}_1), \quad (5.6.32)$$

and this applies to inhomogeneous as well as homogeneous magnetic fields,

$$C(\mathbf{x}, \mathbf{x}') = e^{\frac{ie}{\hbar c}[\lambda(\mathbf{x}) - \lambda(\mathbf{x}')]} = e^{\frac{ie}{\hbar c}\int_{\mathbf{x}'}^{\mathbf{x}} d\mathbf{x}_1 \cdot \mathbf{A}(\mathbf{x}_1)}. \quad (5.6.33)$$

Examining such a line integral over a closed loop, $\oint_{\mathbf{x}}^{\mathbf{x}} d\mathbf{x}_1 \cdot \mathbf{A}(\mathbf{x}_1)$, yields the magnetic flux through the loop by a simple application of Stoke's theorem. (For a uniform magnetic field with gauge $\mathbf{A}(\mathbf{x}_1) = \frac{1}{2}\mathbf{B} \times \mathbf{x}_1$, integration over a straight line path $\mathbf{x}' \to \mathbf{x}$ in Eq. (5.6.33) yields $C(\mathbf{x}_1, \mathbf{x}_1') = \exp\left[\frac{ie}{2\hbar c}\mathbf{x} \cdot \mathbf{B} \times \mathbf{x}'\right]$, as in Eq. (5.6.11)). The Aharonov–Bohm effect nicely illustrates observable consequences of the dependence of phase on path: In it, an electron wave enters a two-slit device and moves from $\mathbf{x}'$ to $\mathbf{x}$ by different paths through the two slits, exhibiting constructive or destructive wave function interference when the two paths meet at a given point. This interference depends mainly on the path length difference experienced by the distinct parts of the electron wave passing through the two slits. It is interesting to note that the two parts of the electron wave can move through paths separated by a region of non-zero magnetic field between the slits without entering the finite field region, which, nevertheless, influences their constructive or destructive final interference by *variation of the field strength.* This is an elegant demonstration of the non-local character of quantum mechanics.

5.7 Green's Function Determination by Hamilton Equations: 2D Saddle Potential in E(t) Field

The determination of Green's functions can also be carried out employing Eq. (5.1.21) with the use of Hamilton equations of motion. In addressing this, we include the possibility of explicitly time-dependent Hamiltonians by the replacement (Eq. (3.1.22)):

$$e^{-iH_s^{(1)}(t_1-t_1')} \to U(t_1,t_1'), \tag{5.7.1}$$

so that Eq. (5.1.21) may be written in the Schrödinger picture as

$$iG_1^{ret}(\mathbf{x}_1,t_1;\mathbf{x}_1',t_1') = \eta_+(t_1-t_1')\langle\mathbf{x}_1|U(t_1,t_1')|\mathbf{x}_1'\rangle_s. \tag{5.7.2}$$

Differentiating with respect to time, we have

$$i\frac{\partial}{\partial t_1}G_1^{ret}(\mathbf{x}_1,t_1;\mathbf{x}_1',t_1') = \delta(t_1-t_1')\langle\mathbf{x}_1|U(t_1,t_1)|\mathbf{x}_1'\rangle_s + \eta_+(t_1-t_1')\langle\mathbf{x}_1|\frac{\partial}{\partial t_1}U(t_1,t_1')|\mathbf{x}_1'\rangle_s \tag{5.7.3}$$

and employing Eq. (3.1.5) yields ($U(t_1,t_1)=1$),

$$i\frac{\partial}{\partial t_1}G_1^{ret}(\mathbf{x}_1,t_1;\mathbf{x}_1',t_1') = \delta(t_1-t_1')\delta^{(3)}(\mathbf{x}_1-\mathbf{x}_1') + \frac{1}{i\hbar}\eta_+(t_1-t_1')\langle\mathbf{x}_1|H_s^{(1)}(t_1)U(t_1,t_1')|\mathbf{x}_1'\rangle_s. \tag{5.7.4}$$

The positional eigenstates $\langle\mathbf{x}_1|$ and $|x_1'\rangle$ are both evaluated at the "initial" time t_1'. Time-developing the state $\langle\mathbf{x}_1|$ frome time t_1' to time t_1 involves employing the adjoint of Eq. (4.1.4),

$$\langle\mathbf{x}_1|_sU(t_1,t_1') = \langle\mathbf{x}_1|_H \equiv \langle\mathbf{x}_1(t_1)|, \tag{5.7.5}$$

and

$$\langle\mathbf{x}_1(t_1')| = \langle\mathbf{x}_1(t_1)|U^+(t_1,t_1'). \tag{5.7.6}$$

Therefore, we have

$$\begin{aligned} i\frac{\partial G_1^{ret}(\mathbf{x}_1,t_1;\mathbf{x}_1',t_1')}{\partial t_1} &= \delta(t_1-t_1')\delta^{(3)}(\mathbf{x}_1-\mathbf{x}_1') \\ &+ \frac{1}{i\hbar}\eta_+(t_1-t_1')\langle\mathbf{x}_1(t_1)|U^+(t_1,t_1')H_s^{(1)}(t_1)U(t_1,t_1')|\mathbf{x}_1'(t_1')\rangle. \end{aligned} \tag{5.7.7}$$

Expressing this in terms of the Heisenberg picture Hamiltonian $H_H(t_1)$ (Eq. (3.2.9)) yields ($t_1 > t_1'$):

$$i\frac{\partial G_1^{ret}(\mathbf{x}_1,t_1;\mathbf{x}_1',t_1')}{\partial t_1} = -\frac{i}{\hbar}\langle\mathbf{x}_1(t_1)|H_H^{(1)}(t_1)|\mathbf{x}_1'(t_1')\rangle. \tag{5.7.8}$$

Division of Eq. (5.7.8) by G_1^{ret},

$$\frac{1}{G_1^{ret}(\mathbf{x}_1,t_1;\mathbf{x}_1',t_1')}\frac{\partial}{\partial t_1}G_1^{ret}(\mathbf{x}_1,t_1;\mathbf{x}_1',t_1') = -\frac{i}{\hbar}\frac{\langle \mathbf{x}_1(t_1)|H_H^{(1)}(t_1)|\mathbf{x}_1'(t_1')\rangle}{\langle \mathbf{x}_1(t_1)|\mathbf{x}_1'(t_1')\rangle}, \tag{5.7.9}$$

and time integration yields (notation $\mathbf{x}_1, t_1 \to \mathbf{x}, t$):

$$G_1^{ret}(\mathbf{x},t;\mathbf{x}',t') = \eta_+(t-t')K(\mathbf{x},\mathbf{x}';t')\exp\left[-\frac{i}{\hbar}\int_{t'}^{t} dt''\frac{\left\langle \mathbf{x}(t'')|H_H^{(1)}(t'')|\mathbf{x}'(t')\right\rangle}{\langle \mathbf{x}(t'')|\mathbf{x}'(t')\rangle}\right], \tag{5.7.10}$$

with $K(\mathbf{x},\mathbf{x}';t')$ being a constant of integration independent of t, which is determined by the initial condition of Eq. (5.1.12).

To determine G_1^{ret} using Eq. (5.7.10), it is necessary to express $H_H^{(1)}(t'')$ in terms of the Heisenberg position and momentum operators, $\widehat{\mathbf{x}}(t''),\widehat{\mathbf{p}}(t'')$, at time t' and derive their Hamilton equations of motion; and then solve for $\widehat{\mathbf{x}}(t''),\widehat{\mathbf{p}}(t'')$ explicitly in terms of $t'',\widehat{\mathbf{x}}(0)$ and $\widehat{\mathbf{p}}(0)$ (taking the initial time as $t_1' \equiv 0$). The resulting pair of solutions is then used to eliminate $\widehat{\mathbf{p}}(t'')$ and $\widehat{\mathbf{p}}(0)$ from $H_H(t'')$ in terms of $\widehat{\mathbf{x}}(t'')$ and $\widehat{\mathbf{x}}(0)$,while employing the equal-time canonical commutation relations to commute $\widehat{\mathbf{x}}(t'')$ to the far left where it acts as $\langle \mathbf{x}(t'')|\widehat{\mathbf{x}}(t'') = \langle \mathbf{x}(t'')|\mathbf{x}$, with $\widehat{\mathbf{x}}(0)$ on the right acting as $\widehat{\mathbf{x}}(0)|\mathbf{x}'(0)\rangle = \mathbf{x}'|\mathbf{x}'(0)\rangle$ (see discussion in Section 1.3). This procedure yields an explicit function of t'', $\mathbf{x}$ and $\mathbf{x}'$, which is to be integrated with respect to t'' as in Eq. (5.7.10) (the residual factors $\langle \mathbf{x}(t'')|\mathbf{x}'(t')\rangle$ appearing in both the numerator and denominator cancel).

As an illustration of this method, we discuss the retarded Green's function for an electron in a two-dimensional saddle potential with an arbitrary time-dependent electric field, $\mathbf{E}(t)$. The Hamiltonian involves different "spring" constants k_1,k_2 in the orthogonal $\overrightarrow{x_1}-$ and $\overrightarrow{x}_2-$ coordinate directions as follows:

$$H_H^{(1)}(t'') = \frac{\widehat{p}_1^2}{2m} + \frac{\widehat{p}_2^2}{2m} + \frac{1}{2}k_1\widehat{x}_1^2 - \frac{1}{2}k_2\widehat{x}_2^2 - e\mathbf{E}(t)\cdot\widehat{\mathbf{x}}. \tag{5.7.11}$$

Here, $(\widehat{p}_1,\widehat{x}_1)$, $(\widehat{p}_2,\widehat{x}_2)$ are canonically paired momentum and position operators for the 1, 2 directions on the plane, respectively. The "saddle" structure involves a normal parabolic potential with a minimum in the $\overrightarrow{x}_1-$ direction, and an inverted parabolic potential with a maximum in the $\overrightarrow{x}_2-$ direction (both k_2 and k_1 are positive). The Hamilton equations of motion of the position and momentum operators are

$$\frac{d\,\widehat{x}_1}{dt} = \frac{\widehat{p}_1}{m}, \tag{5.7.12}$$

$$\frac{d\,\widehat{p}_1}{dt} = eE_1 - k_1\widehat{x}_1, \tag{5.7.13}$$

and

$$\frac{d\widehat{x}_2}{dt} = \frac{\widehat{p}_2}{m}, \tag{5.7.14}$$

$$\frac{d\widehat{p}_2}{dt} = eE_2 + k_2\widehat{x}_2. \tag{5.7.15}$$

Forming the second time derivatives, $d^2\widehat{x}_{1,2}/dt^2$, we obtain

$$\frac{d^2\widehat{x}_1}{dt^2} + \frac{k_1}{m}\widehat{x}_1 = \frac{e}{m}E_1(t), \tag{5.7.16}$$

$$\frac{d^2\widehat{x}_2}{dt^2} - \frac{k_2}{m}\widehat{x}_2 = \frac{e}{m}E_2(t). \tag{5.7.17}$$

In the absence of an electric field ($E_{1,2} \to 0$) these equations have homogenous solutions of the form $\exp(\pm i\omega_{1,2}t)$ with $\omega_{1,2} = \sqrt{k_{1,2}/m}$. In the presence of an electric field, the driven solutions are given in terms of t'', $\widehat{x}_{1,2}(0)$ and $\widehat{p}_{1,2}(0)$ as (the initial time $t' \to 0$ here is not to be confused with the dummy variable of integration t' below):

$$\begin{aligned}\widehat{x}_1(t'') = &-\frac{e}{m\omega_1}\cos(\omega_1 t'')\int_0^{t''} dt' \sin(\omega_1 t')E_1(t') + \frac{e}{m\omega_1}\sin(\omega_1 t'') \\ &\times \int_0^{t''} dt' \cos(\omega_1 t')E_1(t') + \widehat{x}_1(0)\cos(\omega_1 t'') + \frac{\widehat{p}_1(0)}{m\omega_1}\sin(\omega_1 t''),\end{aligned} \tag{5.7.18}$$

and

$$\begin{aligned}\widehat{p}_1(t'') = &\left[\frac{e}{\omega_1}\int_0^{t''} dt' \sin(\omega_1 t')E_1(t') - m\widehat{x}_1(0)\right]\omega_1 \sin(\omega_1 t'') \\ &+ \left[e\int_0^{t''} dt' \cos(\omega_1 t')E_1(t')\right]\cos(\omega_1 t'') + \widehat{p}_1(0)\cos(\omega_1 t'').\end{aligned} \tag{5.7.19}$$

Eliminating $\widehat{p}_1(0)$ in terms of $\widehat{x}_1(t'')$ and $\widehat{x}_1(0)$, we have

$$\widehat{p}_1(t'') = \frac{e}{\sin\omega_1 t''}\int_0^{t''} dt' \sin(\omega_1 t')E_1(t') + m\omega_1 \cot(\omega_1 t'')\widehat{x}_1(t'') - \frac{m\omega_1}{\sin(\omega_1 t'')}\widehat{x}_1(0). \tag{5.7.20}$$

Similarly,

$$\widehat{x}_2(t'') = -\frac{e}{m\omega_2}\cosh(\omega_2 t'')\int_0^{t''} dt' \sinh(\omega_2 t')E_2(t')$$
$$+\frac{e}{m\omega_2}\sinh(\omega_2 t'')\int_0^{t''} dt' \cosh(\omega_2 t')E_2(t') + \widehat{x}_2(0)\cosh(\omega_2 t'') + \frac{\widehat{p}_2(0)}{m\omega_2}\sinh(\omega_2 t'') \tag{5.7.21}$$

and

$$\widehat{p}_2(t'') = \left[-e\int_0^{t''} dt' \sinh(\omega_2 t')E_2(t') + m\omega_2\widehat{x}_2(0)\right]\sinh(\omega_2 t'')$$
$$+\left[e\int_0^{t''} dt' \cosh(\omega_2 t')E_2(t') + \widehat{p}_2(0)\right]\cosh(\omega_2 t''). \tag{5.7.22}$$

Eliminating $\widehat{p}_2(0)$ in terms of $\widehat{x}_2(t'')$ and $\widehat{x}_2(0)$ yields

$$\widehat{p}_2(t'') = \frac{e}{\sinh(\omega_2 t'')}\int_0^{t''} dt' \sinh(\omega_2 t')E_1(t') + m\omega_2\coth(\omega_2 t'')\widehat{x}_2(t'') - \frac{m\omega_2}{\sinh(\omega_2 t'')}\widehat{x}_2(0). \tag{5.7.23}$$

The non-relativistic Hamiltonian $H_H^{(1)}(t'')$, Eq. (5.7.11), involves the operators $\widehat{p}_1^2, \widehat{p}_2^2, \widehat{x}_1^2, \widehat{x}_2^2$, which, in turn, involve products like $\widehat{p}_{1,2}(0)\widehat{x}_{1,2}(0)$ and $\widehat{x}_{1,2}(0)\widehat{p}_{1,2}(0)$. In such terms it is necessary to employ the equal-time canonical commutation relations to commute $\widehat{x}_{1,2}(0)$ to the right of $\widehat{p}_{1,2}(0)$, and then eliminate $\widehat{p}_{1,2}(0)$ in terms of $\widehat{x}_{1,2}(t'')$ and $\widehat{x}_{1,2}(0)$ using Eq. (5.7.18) and Eq. (5.7.21). With this, all squared operators in $H_H^{(1)}(t'')$ involve $\widehat{x}_{1,2}(t'')$ only to the left of $\widehat{x}_{1,2}(0)$, and the matrix element $\left\langle \mathbf{x}(t'')|H_H^{(1)}(t'')|\mathbf{x}'(0)\right\rangle$ (setting $t' \to 0$) may be evaluated to complete the determination of the Green's function using Eq. (5.7.10) (noting that $\widehat{\mathbf{x}}(0)|\mathbf{x}'(0)\rangle = \mathbf{x}'|\mathbf{x}'(0)\rangle$ and $\langle \mathbf{x}(t'')|\widehat{\mathbf{x}}(t'') = \langle \mathbf{x}(t'')|\mathbf{x})$. The resulting retarded Green's function for a non-relativistic electron in a saddle potential and a time-dependent electric field is given by

$$G_1^{ret}(\mathbf{x}, t; \mathbf{x}', t' = 0) = -\frac{\eta_+(t)m\sqrt{\omega_1\omega_2}}{2\pi}\sqrt{\frac{1}{\sin(\omega_1 t)\sinh(\omega_2 t)}} \tag{5.7.24}$$
$$\times \exp\left[-i\left(\begin{array}{l} -\frac{m\omega_1}{2}\cot(\omega_1 t)(x_1^2 + x_1'^2) + \frac{m\omega_1}{\sin(\omega_1 t)}x_1 x_1' \\ -\frac{e}{\sin(\omega_1 t)}\int_0^t dt' E_1(t')[x_1' \sin(\omega_1(t-t')) + x_1\sin(\omega_1 t')] \\ +\frac{e^2}{m\omega_1\sin(\omega_1 t)}\int_0^t dt'\int_0^{t'} dt'' E_1(t')\sin(\omega_1(t-t'))\sin(\omega_1 t'')E_1(t'') \\ -\frac{m\omega_2}{2}\coth(\omega_2 t)(x_2^2 + x_2'^2) + \frac{m\omega_2}{\sinh(\omega_2 t)}x_2' x_2 \\ -\frac{e}{\sinh(\omega_2 t)}\int_0^t dt' E_2(t')[x_2'\sinh(\omega_2(t-t')) - x_2\sinh(\omega_2 t')] \\ +\frac{e^2}{m\omega_2\sinh(\omega_2 t)}\int_0^t dt'\int_0^{t'} dt'' E_2(t')\cdot\sinh(\omega_2(t-t'))\sinh(\omega_2 t'')E_2(t'') \end{array}\right)\right].$$

In this, the prefactor $K(\mathbf{x}, \mathbf{x}'; t')$ of the exponential on the right of Eq. (5.7.10) was determined using the initial retardation condition, $G_1^{ret}(\mathbf{x}, t=0^+; \mathbf{x}', t'=0) = -i\delta^{(3)}(\mathbf{x}-\mathbf{x}')$.

The retarded Green's function for non-relativistic electrons in a parabolic potential with a constant uniform magnetic field and a time-dependent electric field was analyzed by Fessatidis, Horing and Sabeeh,[7] following an earlier derivation by Horing, Cui and Fiorenza.[8] The results are useful in quantum-coherent nanostructure transport calculations.

5.8 Product Green's Functions for Separable Hamiltonians: 1D Superlattice in Axial Fields

In this section, we address a composite system whose separable Hamiltonian $H^{(1)}$ is composed of two disjoint, commuting parts $h_A^{(1)}$ and $h_B^{(1)}$:

$$H^{(1)} = h_A^{(1)} + h_B^{(1)}; \quad \left[h_A^{(1)}, h_B^{(1)}\right] \equiv 0, \tag{5.8.1}$$

which involve mutually independent (commutative) sets of variables x_A and x_B, respectively. The retarded Green's function $G_1^{ret}(x_A, x_B, t; x_A', x_B', t')$ of the composite system is given by i times the product of the individual Green's functions associated with $h_A^{(1)}$ and $h_B^{(1)}$ individually as

$$G_1^{ret}(x_A, x_B, t_1; x_A', x_B', t_1') = iG_A^{ret}(x_A, t_1; x_A', t_1')G_B^{ret}(x_B, t_1; x_B', t_1'). \tag{5.8.2}$$

Here G_A^{ret} and G_B^{ret} satisfy the individual Green's function equations given by

$$\left(i\frac{\partial}{\partial t_1} - h_A^{(1)}(x_A, t)\right) G_A^{ret}(x_A, t_1; x_A', t_1') = \delta(x_A - x_A')\delta(t_1 - t_1') \to 0 \text{ for } t_1' > t_1, \tag{5.8.3}$$

and

$$\left(i\frac{\partial}{\partial t_1} - h_B^{(1)}(x_B, t)\right) G_B^{ret}(x_B, t_1; x_B', t_1') = \delta(x_B - x_B')\delta(t_1 - t_1') \to 0 \text{ for } t_1' > t_1. \tag{5.8.4}$$

The retardation condition mandates that both G_A^{ret} and G_B^{ret} vanish for $t_1 < t_1'$, resulting in the following initial conditions:

$$G_A^{ret}(x_A, t_1'^+; x_A', t_1') = -i\delta(x_A - x_A') \tag{5.8.5}$$

and

$$G_B^{ret}(x_B, t_1'^+; x_B', t_1') = -i\delta(x_B - x_B'). \tag{5.8.6}$$

To verify this, we form the time derivative $\frac{\partial}{\partial t_1} G_1^{ret}$ for $t_1 > t_1'$ using Eq. (5.8.2):

$$\frac{\partial}{\partial t_1} G_1^{ret} = i\left(\frac{\partial G_A^{ret}}{\partial t_1}\right) G_B^{ret} + iG_A^{ret}\left(\frac{\partial}{\partial t_1} G_B^{ret}\right). \tag{5.8.7}$$

Substitution of the homogeneous forms of Eq. (5.8.3) and Eq. (5.8.4) for $t_1 > t_1'$ yields

$$\frac{\partial}{\partial t_1} G_1^{ret} = [h_A^{(1)}(x_A, t) G_A^{ret}] G_B^{ret} + G_A^{ret} [h_B^{(1)}(x_B, t) G_B^{ret}] \tag{5.8.8}$$

or

$$\frac{\partial G_1^{ret}}{\partial t_1} - [h_A^{(1)}(x_A, t) + h_B^{(1)}(x_B, t)] G_A^{ret} G_B^{ret} = 0 \tag{5.8.9}$$

because of the mutual independence and commutativity of the variables in $h_A^{(1)}(x_A, t)$ and $h_B^{(1)}(x_B, t)$, and those in G_B^{ret} and G_A^{ret}, respectively. Therefore, $G_1^{ret} = iG_A^{ret} G_B^{ret}$ satisfies the homogeneous equation for $t_1 > t_1'$ with the separable Hamiltonian $h_A^{(1)} + h_B^{(1)}$, as expected,

$$i\frac{\partial G_1^{ret}}{\partial t_1} - [h_A^{(1)}(x_A, t) + h_B^{(1)}(x_B, t)] G_1^{ret} = 0. \tag{5.8.10}$$

Furthermore, evaluation of the retarded function $G_1^{ret}(t_1'^{+}; t_1')$ in accordance with Eq. (5.8.2) yields

$$G_1^{ret}(x_A, x_B, t_1'^{+}; x_A', x_B', t_1') = iG_A^{ret}(x_A, t_1'^{+}; x_A', t_1') G_B^{ret}(x_B, t_1'^{+}; x_B', t_1'). \tag{5.8.11}$$

On the other hand, $G_1^{ret}(t_1'^{+}; t_1')$ may be obtained directly by integrating $\partial G_1^{ret}/\partial t_1$ from $t_1'^{-}$ to $t_1'^{+}$ assuming the usual space-time impulsive driving term (in all variables, $x_A, x_B; t$) and using the retardation condition, with the result

$$G_1^{ret}(x_A, x_B, t_1'^{+}; x_A', x_B', t_1') = -i\delta(x_A - x_A')\delta(x_B - x_B'). \tag{5.8.12}$$

Equivalence of Eq. (5.8.12) and Eq. (5.8.11) is readily seen using Eqns. (5.8.5 and 5.8.6), verifying the correctness of the factor "i" on the right of Eq. (5.8.2), so that $G_1^{ret} = iG_A^{ret} G_B^{ret}$ is properly confirmed as the retarded composite system Green's function for $h^{(1)} = h_A^{(1)} + h_B^{(1)}$.

Considering a one-dimensional (1D) superlattice miniband in a constant, uniform magnetic field along the axis of the lattice (z-direction) and a uniform time-dependent parallel electric field, the 1D miniband part of the Hamiltonian in a tight-binding, nearest-neighbor approximation may be written in terms of variables referring only to the superlattice axis direction as[9]

$$h_A^{(1)} = \frac{\Delta}{2}(1 - \cos[\widehat{p}_z d/\hbar]) \tag{5.8.13}$$

(here $\widehat{p}_z$ is the crystal momentum operator in the z-axial direction and Δ and d are the miniband width and superlattice period, respectively). Furthermore, it is convenient to represent the axial time-dependent electric field, $E_z(t)$, by means of a vector potential using $E_z(t) = -(\frac{1}{c})\partial A_z(t)/\partial t$,

$$A_z(t) = -c\int_{t_0}^{t} dt' E_z(t')\ ;\ \widehat{p}_z \rightarrow \widehat{p}_z - \frac{e}{c}A_z(t). \tag{5.8.14}$$

In this way, the Hamiltonian $h_A^{(1)}$ remains spatially translationally invariant as

$$h_z^{(1)} = \frac{\Delta}{2}\left(1 - \cos\left[\frac{d}{\hbar}(\widehat{p}_z - \frac{e}{c}A_z(t))\right]\right), \tag{5.8.15}$$

still involving only variables referring to the z-axial direction. To obtain the associated Green's function G_A^{ret} we employ the method of Section 5.7, invoking the Hamilton operator equations for $\widehat{p}_z$ and $\widehat{z}$ as

$$\frac{d\,\widehat{p}_z}{dt} \equiv 0 \Longrightarrow \widehat{p}_z(t) = constant = \widehat{p}_z \tag{5.8.16}$$

and

$$\frac{d\,\widehat{z}}{dt} = \frac{\Delta d}{2\hbar}\sin\left[\frac{d}{\hbar}(\widehat{p}_z - \frac{e}{c}A_z(t))\right], \tag{5.8.17}$$

so that

$$\widehat{z}(t) - \widehat{z}(t') = \frac{\Delta d}{2\hbar}\int_{t'}^{t} dt''\sin\left[\frac{d}{\hbar}(\widehat{p}_z - \frac{e}{c}A_z(t''))\right]. \tag{5.8.18}$$

Taking advantage of the fact that $\widehat{p}_z$ is a constant operator for the miniband, we employ the $\widehat{p}_z$-representation. An analogous derivation to that of Eq. (5.7.10), but done in momentum representation (instead of position representation) leads to a similar result:

$$G_A^{ret}(p_z, t; p_z', t') = \eta_+(t-t')K(p_z, p_z'; t')\exp\left[-\frac{i}{\hbar}\int_{t'}^{t} dt''\frac{\langle p_z(t'')|h_A(t'')|p_z'(t')\rangle}{\langle p_z(t'')|p_z'(t')\rangle}\right], \tag{5.8.19}$$

where $p_z(t), p_z(t')$ are the eigenvalues of $\widehat{p}_z$ at times t, t', respectively, and K is devoid of t-dependence. Since $\widehat{p}_z$ is a constant of the motion, and $G_A^{ret}(p_z, t, p_z', t')$ is diagonal as a matrix in p_z, p_z' because $h_z^{(1)}$ is similarly diagonal, we have

$$G_1^{ret}(p_z, t; p_z', t') = \eta_+(t-t')\delta(p_z - p_z')K(p_z)\exp\left[-\frac{i}{\hbar}\int_{t'}^{t} dt''\frac{\langle p_z(t'')|h_A(t'')|p_z(t')\rangle}{\langle p_z(t'')|p_z(t')\rangle}\right] \tag{5.8.20}$$

and

$$\frac{\langle p_z(t'')|h_A(t'')|p_z(t')\rangle}{\langle p_z(t'')|p_z(t')\rangle} = \frac{\langle p_z(t'')|\frac{\Delta}{2}(1-\cos[\frac{d}{\hbar}(\widehat{p}_z - \frac{e}{c}A_z(t))])|p_z(t')\rangle}{\langle(p_z(t'')|p_z(t')\rangle}$$
$$= \frac{\Delta}{2}\left(1-\cos\left[\frac{d}{\hbar}(p_z - \frac{e}{c}A_z(t)\right]\right), \tag{5.8.21}$$

so that

$$G_A^{ret}(p_z, t; p_z', t') = \eta_+(t-t')\delta(p_z-p_z')K(p_z)\exp\left[-\frac{i\Delta}{2\hbar}\int_{t'}^{t} dt''\left(1-\cos\left[\frac{d}{\hbar}\left(p_z - \frac{e}{c}A_z(t'')\right)\right]\right)\right]. \tag{5.8.22}$$

$K(p_z)$ may be found using a Fourier transform to position representation,

$$G_A^{ret}(z, t; z', t') = \eta_+(t-t')\int_{-\infty}^{\infty}\frac{dp_z}{2\pi\hbar}e^{\frac{i}{\hbar}p_z(z-z')}K(p_z)$$
$$\times \exp\left[-\frac{i\Delta}{2\hbar}\int_{t'}^{t} dt''\left(1-\cos\left[\frac{d}{\hbar}(p_z - \frac{e}{c}A_z(t''))\right]\right)\right], \tag{5.8.23}$$

and introducing the initial condition in the form

$$G_1^{ret}(z, t'^+; z', t') = \int_{-\infty}^{\infty}\frac{dp_z}{2\pi\hbar}K(p_z)e^{\frac{i}{\hbar}p_z(z-z')} = -\frac{i}{\hbar}\delta(z-z'), \tag{5.8.24}$$

leads to the identification

$$K(p_z) = -\frac{i}{\hbar}. \tag{5.8.25}$$

Therefore, the 1D miniband Green's function is given by

$$G_A^{ret}(z, t; z', t') =$$
$$-\frac{i}{\hbar}\eta_+(t-t')\int_{-\infty}^{\infty}\frac{dp_z}{2\pi\hbar}e^{\frac{i}{\hbar}p_z(z-z')}\exp\left[-\frac{i\Delta}{2\hbar}\int_{t'}^{t} dt''\left(1-\cos\left[\frac{d}{\hbar}\left(p_z - \frac{e}{c}A_z(t'')\right)\right]\right)\right] \tag{5.8.26}$$

in vector potential representation. In the case of a constant electric field, $A_z(t) = -cE_zt$, Eq. (5.8.26) yields ($\omega_B = eE_zd/\hbar$ is the Bloch frequency):

$$G_A^{ret}(z, t; z', t') = -\frac{i}{\hbar}\eta_+(t-t')\int_{-\infty}^{\infty}\frac{dp_z}{2\pi\hbar}e^{\frac{i}{\hbar}p_z(z-z')} \tag{5.8.27}$$
$$\times \exp\left[-\frac{i\Delta}{2\hbar}\left(t-t'-\frac{2}{\omega_B}\sin(\omega_B(t-t')/2)\cos\left(\frac{p_zd}{\hbar}+\omega_B(t+t')/2\right)\right)\right].$$

Considering an additive Hamiltonian $h_B^{(1)}$ for a constant, uniform axial magnetic field, there is one part $\left(\partial^2/\partial z^2\right)$ returning to translational motion along the common axial direction of the electric and magnetic fields, which we take to be subsumed in $h_A^{(1)}$ (Eq. (5.8.13)). The rest of $h_B^{(1)}$ arises in connection with circular motion induced by the magnetic field in the lateral plane perpendicular to the axial fields and to the growth direction of the superlattice: This part of h_B involves the position and momentum operator components of the 2D lateral plane *only* and they are independent of, and commutative with, $h_A^{(1)}$ and its $\widehat{p}_z$ operators. Therefore, we can employ Eq. (5.8.11) using Eq. (5.6.28) for G_B^{ret} for the Green's function of the quantized circular motion on the lateral plane. The resulting Green's function of the composite system is given in terms of the product $iG_A^{ret}G_B^{ret}$ as (including spin):

$$G_1^{ret}(\mathbf{x},t;\mathbf{x}',t') = -\eta_+(T)\frac{m\omega_c}{4\pi\hbar^2}C(\overline{x},\overline{x'})\frac{e^{-\frac{i}{\hbar}\overline{\mu}_0 B\sigma_3 T}}{\sin(\omega_c T/2)}\exp\left(\frac{im\omega_c\left[X^2+Y^2\right]}{4\hbar\tan(\omega_c T/2)}\right) \tag{5.8.28}$$

$$\times\int_{-\infty}^{\infty}\frac{dp_z}{2\pi\hbar}e^{\frac{i}{\hbar}p_z Z}\exp\left\{-\frac{i\Delta}{2\hbar}\int_{t'}^{t}dt''\left(1-\cos\left(\frac{d}{\hbar}\left(p_z-\frac{e}{c}A_z(t'')\right)\right)\right)\right\}.$$

If the axial electric field E_z is constant as well as uniform, the p_z-integral becomes

$$\int_{-\infty}^{\infty}\frac{dp_z}{2\pi\hbar}\Longrightarrow$$

$$\int_{-\infty}^{\infty}\frac{dp_z}{2\pi\hbar}e^{\frac{i}{\hbar}p_z Z}\exp\left[-\frac{i\Delta}{2\hbar}\left(T-\frac{2}{\omega_B}\left(\sin\left(\frac{\omega_B}{2}(t-t')\right)\cos\left(\frac{p_z d}{\hbar}+\frac{\omega_B}{2}(t+t')\right)\right)\right)\right], \tag{5.8.29}$$

with $X = x-x'$, $Y = y-y'$, $Z = z-z'$, $T = t-t'$, $\overline{x} = (x,y)$, $\overline{x'} = (x',y')$. In these equations, $C(\overline{x},\overline{x'}) = \exp\left[\frac{i}{\hbar}\frac{e}{2c}\overline{x}\cdot\mathbf{B}\times\overline{x'}-\phi(\overline{x})+\phi(\overline{x'})\right]$, $\mathbf{B}= B\,\widehat{z}$ is the axial magnetic field and E_z is the axial electric field. Also, $\phi(\overline{x})$ is an arbitrary magnetic gauge function, $\overline{\mu}_0$ is the Bohr magneton, ω_c is the cyclotron frequency, d is the superlattice period, Δ is the miniband width and ω_B is the Bloch frequency; σ_3 is third Pauli spin matrix. In the case of vanishing electric field the p_z-integral of Eq. (5.8.28) is

$$\int_{-\infty}^{\infty}\frac{dp_z}{2\pi\hbar}\Longrightarrow\int_{-\infty}^{\infty}\frac{dp_z}{2\pi\hbar}e^{\frac{i}{\hbar}p_z Z}\exp\left\{-\frac{i\Delta}{2\hbar}\left(1-\cos\left(\frac{p_z d}{\hbar}\right)\right)T\right\}. \tag{5.8.30}$$

An advantage of the vector potential gauge for the electric field is that it maintains spatial-translation invariance if the field is uniform. However, the alternative scalar potential gauge offers the advantage of time-translation invariance provided that the electric field is independent of time (however, it is *not* lattice-translation invariant). The corresponding Green's function in the scalar potential gauge, $\tilde{G}_1$, may be determined from

G_1^{ret} in the vector potential gauge using the gauge transformation $\mathbf{A}(\mathbf{x},t) \to \mathbf{A}'(\mathbf{x},t) = \mathbf{A}(\mathbf{x},t) + \nabla\lambda(\mathbf{x},t)$ and $\varphi(\mathbf{x},t) \to \varphi'(\mathbf{x},t) = \varphi(\mathbf{x},t) - \frac{1}{c}\partial\lambda(\mathbf{x},t)/\partial t$: Accordingly, the Green's function transformation takes the form

$$G_1^{ret}(\mathbf{x},t;\mathbf{x}',t') \to \tilde{G}_1(\mathbf{x},t;\mathbf{x}',t') = \exp\left\{\frac{ie}{\hbar c}\left(\lambda(\mathbf{x},t) - \lambda(\mathbf{x}',t')\right)\right\} G_1^{ret}(\mathbf{x},t;\mathbf{x}',t'). \tag{5.8.31}$$

Bearing in mind that the vector potential gauge is given by $A_z(t) = -c\int_0^t dt' E_z(t')$, and that the scalar potential gauge is $\varphi'(\mathbf{x},t) = -E_z(t)z$, the gauge transformation between the two that assures the vanishing of $\varphi(\mathbf{x},t)$ and $A_z'(t)$ is $(t_0 \to 0)$:

$$\lambda(\mathbf{x},t) = cz\int_0^t dt' E_z(t'). \tag{5.8.32}$$

Therefore, the Green's function in the scalar potential gauge is given by

$$\tilde{G}_1(\mathbf{x},t;\mathbf{x}',t') = \exp\left[\frac{ie}{\hbar}\left(z\int_0^t dt'' E_z(t'') - z'\int_0^{t'} dt'' E_z(t'')\right)\right] G_1^{ret}(\mathbf{x},t;\mathbf{x}',t'), \tag{5.8.33}$$

with $G_1^{ret}(\mathbf{x},t;\mathbf{x}',t')$ of Eqns. (5.8.28) and (5.8.29). If the electric field is constant and uniform (Eq. (5.8.29)), the scalar potential gauge is devoid of time dependence, so there is time-translation invariance and $\tilde{G}_1(\mathbf{x},t;\mathbf{x}',t')$ is a function of $T = t-t'$, to the exclusion of $\tau = t + t'$. To see this explicitly, we write $t = (T+\tau)/2$ and $t' = (\tau - T)/2$ everywhere in Eq. (5.8.33) (with $G_1^{ret}(\mathbf{x},t;\mathbf{x}',t')$ given by Eqns. (5.8.28) and (5.8.29)): The fact that $\tilde{G}_1(\mathbf{x},t;\mathbf{x}',t')$ has *no* dependence on τ permits us to evaluate it at $\tau = 0$ to obtain the result for any τ. This means putting $t = T/2$ and $t' = -T/2$ throughout Eq. (5.8.33) and Eqns. (5.8.28) and (5.8.29), and it leads to the constant electric field result:

$$\tilde{G}_1(\mathbf{x},\mathbf{x}';T) = -\eta_+(T)\frac{m\omega_c}{4\pi\hbar^2}C(\overline{x},\overline{x}')e^{\frac{ie}{2\hbar}E_zT(z+z')}\frac{e^{-\frac{i}{\hbar}\overline{\mu}_0 B\sigma_3 T}}{\sin(\omega_c T/2)}\exp\left(\frac{im\omega_c\left[X^2+Y^2\right]}{4\hbar\tan(\omega_c T/2)}\right)$$
$$\times\int_{-\infty}^{\infty}\frac{dp_z}{2\pi\hbar}e^{\frac{i}{\hbar}p_z Z}\exp\left\{-\frac{i\Delta}{2\hbar}\left(T - \frac{2}{\omega_B}\cos(p_z d/\hbar)\sin(\omega_B T/2)\right)\right\}. \tag{5.8.34}$$

Employing the Jacobi–Anger formula for $e^{iz\cos\alpha}$ in terms of a Bessel function expansion, we have

$$\tilde{G}_1(\mathbf{x},\mathbf{x}';T) = -\eta_+(T)\frac{m\omega_c}{4\pi\hbar^2}C(\overline{x},\overline{x}')e^{\frac{ie}{2\hbar}E_zT(z+z')}\frac{e^{-\frac{i}{\hbar}\overline{\mu}_0 B\sigma_3 T}}{\sin(\omega_c T/2)}\exp\left(\frac{im\omega_c\left[X^2+Y^2\right]}{4\hbar\tan(\omega_c T/2)}\right)$$
$$\times\int_{-\infty}^{\infty}\frac{dp_z}{2\pi\hbar}e^{\frac{i}{\hbar}p_z Z}e^{-\frac{i\Delta T}{2\hbar}}\sum_{k=-\infty}^{\infty}i^k e^{\frac{i}{\hbar}kdp_z}\mathcal{J}_k\left(\frac{\Delta}{\hbar\omega_B}\sin(\omega_B T/2)\right), \tag{5.8.35}$$

where $\mathcal{J}_k(x)$ is the Bessel function of order k. The p_z-integral now yields $\delta(Z + kd)$-functions, leading to an alternative to Eq. (5.8.34):

$$\tilde{G}_1(\mathbf{x}, \mathbf{x}'; T) = -\eta_+(T)\frac{m\omega_c}{4\pi\hbar^2}C(\overline{x}, \overline{x}')e^{\frac{ie}{2\hbar}E_z T(z+z')}\frac{e^{-\frac{i}{\hbar}\overline{\mu}_0 B\sigma_3 T}}{\sin(\omega_c T/2)}\exp\left(\frac{im\omega_c\left[X^2 + Y^2\right]}{4\hbar\tan(\omega_c T/2)}\right)$$

$$\times e^{-\frac{i\Delta T}{2\hbar}}\sum_{k=-\infty}^{\infty} i^k \mathcal{J}_k\left(\frac{\Delta}{\hbar\omega_B}\sin(\omega_B T/2)\right)\delta(Z + kd), \qquad (5.8.36)$$

in scalar potential representation.

The above results for the Green's function in vector potential representation (Eq. (5.8.28)) and in scalar potential representation (Eq. (5.8.34)) have the advantage of being expressed in terms of *elementary functions* in direct time representation, although under a momentum Fourier transform. As such, they generate the Landau eigenfunctions associated with the magnetic field quantization in the lateral plane, and they also generate the miniband Wannier–Stark ladder Bloch states in scalar potential representation and Houston functions with accelerated Bloch states in vector potential representation.

5.9 Electron Propagation in a Heterostructure with Spatially Variable Mass: Green's Function Matching

Intensive study of surface states at the interface of two adjoining media having different masses and potentials has brought forth a technique for matching Green's functions rather than wave functions in the analysis of quantum electron dynamics in such systems. It must be borne in mind that for such heterostructures having spatially variable effective mass, the non-relativistic Hermitian kinetic energy operator has the form

$$h_1^{kin} = \nabla \cdot \frac{1}{2m(\mathbf{x})}\nabla, \qquad (5.9.1)$$

rather than $[2m(\mathbf{x})]^{-1}\,\nabla^2$, which is *not* Hermitian. Addressing the case of two adjoining media A, B separated by an interface S, this section will deal with the construction of the *global* Schrödinger Green's function, G, for the combined system A & B jointly in terms of the individual Green's function G_A of A by itself and G_B of B by itself, each determined separately in the absence of the other. In this, G_A and G_B are assumed to satisfy the same linear homogeneous boundary conditions as those required for G on the distant external closed surface bounding the combined regions of A and B together. However, *no* such requirement is imposed at the interface S, and G_A may be taken as if the properties of region A were continued throughout region B, and vice versa (Figure 5.2). Considering 3D Dirac δ-function source points $\mathbf{x}'$ and $\mathbf{x}''$ in A and assuming time-translational invariance involving a single Fourier time-transform variable, t–$t' \to \omega$, then for a field point $\mathbf{x}$ in A, we have (suppress ω):

$$\left(-\nabla \cdot \frac{1}{2m(\mathbf{x})}\nabla + V(\mathbf{x}) - \omega\right)G(\mathbf{x}, \mathbf{x}') = \delta^{(3)}(\mathbf{x} - \mathbf{x}'), \qquad (5.9.2)$$

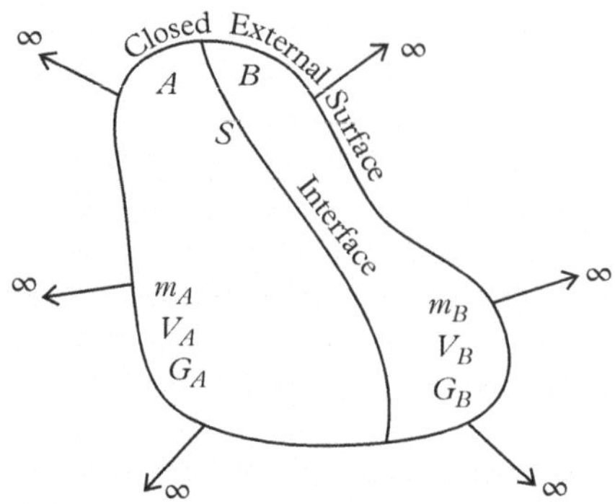

Figure 5.2: *Schematic of two regions A, B separated by an interface S, each having its own mass $m_{A,B}$ and potential $V_{A,B}$ and Green's function $G_{A,B}$ if infinitely extended. The outer closed boundary of the two-region system is understood to be infinitely distant in all directions.*

and complex conjugation with use of the reciprocity relation, $G^*(\mathbf{x},\mathbf{x}'') = G(\mathbf{x}'',\mathbf{x})$, yields

$$\left(-\nabla\cdot\frac{1}{2m(\mathbf{x})}\nabla + V(\mathbf{x}) - \omega\right) G_A(\mathbf{x}'',\mathbf{x}) = \delta^{(3)}(\mathbf{x}-\mathbf{x}''). \tag{5.9.3}$$

Multiplying Eq. (5.9.2) by $G_A(\mathbf{x}'',\mathbf{x})$ and Eq. (5.9.3) by $G(\mathbf{x},\mathbf{x}')$, subtracting and integrating over the volume $(d^{(3)}\mathbf{x})$ of A, the divergence/Gauss theorem yields a surface integral $\int_S dS$ over the interface between A and B:

$$G(\mathbf{x},\mathbf{x}') = G_A(\mathbf{x},\mathbf{x}') + \int_S dS \frac{1}{2m(\mathbf{x}_{S,A})}\left\{G_A(\mathbf{x},\mathbf{x}_S)\frac{\partial}{\partial n_{S,A}}G(\mathbf{x}_S,\mathbf{x}') - G(\mathbf{x}_S,\mathbf{x}')\frac{\partial}{\partial n_{S,A}}G_A(\mathbf{x},\mathbf{x}_S)\right\}. \tag{5.9.4}$$

Here, the mass on the A side of S is denoted by $m(\mathbf{x}_{S,A})$ (m may be discontinuous across S) and we have used the symmetry properties of Green's functions. Also, $\partial/\partial n_{S,A}$ is the outwardly directed normal derivative from A toward B. Surface integral contributions from the closed outer external boundaries of the combined regions $A\&B$ *vanish* since G_A and G satisfy the same linear homogeneous boundary conditions there.

In order to determine $G(\mathbf{x},\mathbf{x}')$ everywhere in A using Eq. (5.9.4), it is necessary to evaluate the boundary values of $G(\mathbf{x}_S,\mathbf{x}')$ and $\frac{\partial}{\partial n_{S,A}}G(\mathbf{x}_S,\mathbf{x}')$. To accomplish this, we consider a field point $\mathbf{x}$ in region B instead of A, but retain the source point $\mathbf{x}'$ of G in region A while taking a source point $\mathbf{x}''$ of G_B in region B, and we proceed as above: The only difference is that we integrate $\mathbf{x}$ over B where G satisfies the homogeneous counterpart of Eq. (5.9.2), with the result

$$G(\mathbf{x},\mathbf{x}') = \int_S dS \frac{1}{2m(\mathbf{x}_{S,B})}\left\{G_B(\mathbf{x},\mathbf{x}_S)\frac{\partial}{\partial n_{S,B}}G(\mathbf{x}_S,\mathbf{x}') - G(\mathbf{x}_S,\mathbf{x}')\frac{\partial}{\partial n_{S,B}}G_B(\mathbf{x},\mathbf{x}_S)\right\} \tag{5.9.5}$$

(the outward normal here, n_{SB}, is the negative of n_{SA}). The anticipated discontinuity of the effective mass across S does occur in semiconductor heterostructures, and we employ the *Ben Daniel-Duke* boundary condition at S to assure the continuity of probability current density across the interface,

$$\frac{1}{2m(\mathbf{x}_{S,A})}\frac{\partial}{\partial n_{S,A}}G(\mathbf{x}_S,\mathbf{x}') = \frac{-1}{2m(\mathbf{x}_{S,B})}\frac{\partial}{\partial n_{S,B}}G(\mathbf{x}_S,\mathbf{x}') \equiv \frac{1}{2m(\mathbf{x}_S)}\frac{\partial}{\partial n_S}G(\mathbf{x}_S,\mathbf{x}'), \tag{5.9.6}$$

as well as the continuity of $G(\mathbf{x}_S,\mathbf{x}')$ at the interface. In the definition on the far right of Eq. (5.9.6), we use the notation $n_S = n_{SA} = -n_{SB}$ for the outward normals. Considering an arbitrary point at the interface $\mathbf{x} = \mathbf{x}''_S$ in Eqns. (5.9.4) and (5.9.5), these equations constitute two coupled surface integral equations for the two unknowns $\{G(\mathbf{x}_S,\mathbf{x}')\}$ and $\left\{[2m(\mathbf{x}_S)]^{-1}\partial G(\mathbf{x}_S,\mathbf{x}')/\partial n_S\right\}$, as given by

$$\begin{aligned}G(\mathbf{x}''_S,\mathbf{x}') &= G_A(\mathbf{x}''_S,\mathbf{x}')\\ &+\int_S dS\left\{G_A(\mathbf{x}''_S,\mathbf{x}_S)\left[\frac{1}{2m(\mathbf{x}_S)}\frac{\partial}{\partial n_S}G(\mathbf{x}_S,\mathbf{x}')\right]-\frac{1}{2m(\mathbf{x}_{S,A})}G(\mathbf{x}_S,\mathbf{x}')\frac{\partial}{\partial n_S}G_A(\mathbf{x}''_S,\mathbf{x}_S)\right\}\end{aligned} \tag{5.9.7}$$

(note that $\frac{1}{m(\mathbf{x}_{S,A})} \neq \frac{1}{m(\mathbf{x}_{S,B})}$ in the last term), and

$$\begin{aligned}G(\mathbf{x}''_S,\mathbf{x}') = &-\int_S dS\left\{G_B(\mathbf{x}''_S,\mathbf{x}_S)\left[\frac{1}{2m(\mathbf{x}_S)}\frac{\partial}{\partial n_S}G(\mathbf{x}_S,\mathbf{x}')\right]\right.\\ &\left.-\frac{1}{2m(\mathbf{x}_{S,B})}G(\mathbf{x}_S,\mathbf{x}')\frac{\partial}{\partial n_S}G_B(\mathbf{x}''_S,\mathbf{x}_S)\right\}.\end{aligned} \tag{5.9.8}$$

As indicated above, the final determination of G at an arbitrary point in A using Eqns. (5.9.4 and 5.9.5) can be achieved by evaluating the quantities $\left[\frac{1}{2m(\mathbf{x}_S)}\frac{\partial}{\partial n_S}G(\mathbf{x}_S,\mathbf{x}')\right]$ and $G(\mathbf{x}_S,\mathbf{x}')$ from the coupled integral equations, Eqns. (5.9.7) and (5.9.8). Of course, G can be determined at an arbitrary point in B by interchanging the roles of A and B. Proceeding along the lines of Inglesfield,[10–12] we view $G(\mathbf{x}_S,\mathbf{x}')$; $\frac{1}{2m(\mathbf{x}_S)}\frac{\partial}{\partial n_S}G(\mathbf{x}_S,\mathbf{x}')$; $G_A(\mathbf{x}_S,\mathbf{x}')$; $G_B(\mathbf{x}_S,\mathbf{x}')$ as the components of vectors g; g'; g_A; g_B, respectively, in a space $\{\mathbf{x}_S\}$ in which both $G_{A,B}(\mathbf{x}''_S,\mathbf{x}_S)$ and $\frac{1}{2m(\mathbf{x}_{S,A,B})}\frac{\partial}{\partial n_S}G_{A,B}(\mathbf{x}''_S,\mathbf{x}_S)$ are the matrix elements in positional indices of operators denoted by $\mathcal{G}_{A,B}$; $\mathcal{G}'_{A,B}$. In terms of this notation, Eqns. (5.9.7) and (5.9.8) may be written in matrix form (the surface integrations of Eqns. (5.9.7 and 5.9.8) are subsumed in the indicated matrix multiplications):

$$g = g_A + \mathcal{G}_A g' - \mathcal{G}'_A g \tag{5.9.9}$$

and

$$g = -\mathcal{G}_B g' + \mathcal{G}'_B g. \tag{5.9.10}$$

These equations yield

$$g = \frac{1}{1 + \mathcal{G}'_A}\left(g_A + \mathcal{G}_A g'\right) = -\frac{1}{1 - \mathcal{G}'_B}\mathcal{G}_B g', \tag{5.9.11}$$

and they may be solved for g' as

$$g' = -\left[\left(1 + \mathcal{G}'_A\right)\left(1 - \mathcal{G}'_B\right)^{-1}\mathcal{G}_B + \mathcal{G}_A\right]^{-1} g_A. \tag{5.9.12}$$

Substituting this result for g' into Eq. (5.9.11), we obtain g as

$$g = \left(1 - \mathcal{G}'_B\right)^{-1}\mathcal{G}_B\left[\left(1 + \mathcal{G}'_A\right)\left(1 - \mathcal{G}'_B\right)^{-1}\mathcal{G}_B + \mathcal{G}_A\right]^{-1} g_A. \tag{5.9.13}$$

Eqns. (5.9.12) and (5.9.13) jointly provide all the information needed in Eqns. (5.9.4 and 5.9.5) to obtain explicit solutions for $G(\mathbf{x}, \mathbf{x}')$ in A and in B, respectively, expressed in terms of integrals of the known functions G_A, G_B, and their normal derivatives. The global composite heterostructure system Green's function G has eigenstate energies manifested in its new frequency poles involving $\mathcal{G}_A$ and $\mathcal{G}_B$ *jointly*. They are surface states with energies/frequencies that can be obtained from the vanishing of the denominator of Eq. (5.9.12) for g':

$$\det\left[\left(1 + \mathcal{G}'_A\right)\left(1 - \mathcal{G}'_B\right)^{-1}\mathcal{G}_B + \mathcal{G}_A\right] = 0. \tag{5.9.14}$$

It should be noted that there is no restriction on boundary conditions on G_A and G_B that can be imposed at the interface S *other than that they add only homogenous terms to* $G_{A,B}$ (as solutions of Eq. (5.9.3), etc., in the associated region of interest, A, B, respectively) to avoid the introduction of spurious source terms in the region of interest. In view of this, Eq. (5.9.14) may be simplified by choosing their derivatives to vanish at S, yielding $g' = -(\mathcal{G}_A + \mathcal{G}_B)^{-1} g_A$ so that Eq. (5.9.14) reads $\det(\mathcal{G}_A + \mathcal{G}_B) = 0$.

These operator manipulations are actually very simple to use in one-dimensional (1D) problems that arise in the case of planar heterostructures, in which the coupled surface integral equations reduce to simple algebraic equations. For example, consider the problem of two semi-infinite media, A and B, interfacing at the plane $z = 0$, taking the individual media components to have different constant, uniform single-electron potentials V_A, V_B and different uniform masses m_A, m_B, respectively (Figure 5.3). The overall 1D potential is

$$V(z) = (1 - \eta_+(z))\, V_A + \eta_+(z)\, V_B, \tag{5.9.15}$$

and, correspondingly, the mass is

$$m(z) = (1 - \eta_+(z))\, m_A + \eta_+(z) m_B. \tag{5.9.16}$$

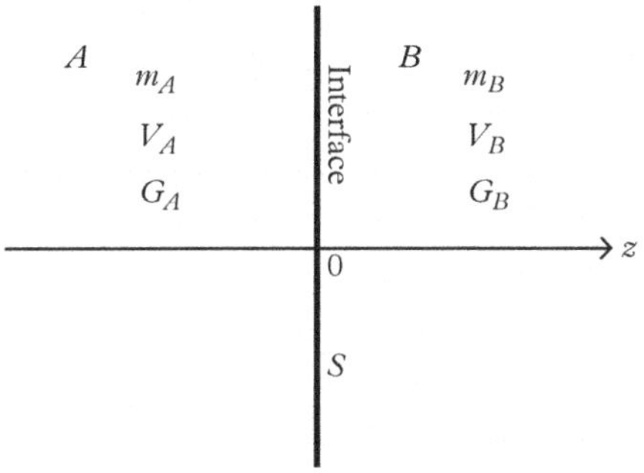

Figure 5.3: *Schematic of two semi-infinite regions A, B separated by a planar interface S at z = 0.*

To determine the global Green's function, G, we Fourier-transform in the translationally invariant lateral $\bar{\mathbf{x}} = (x, y)$-plane, $\bar{\mathbf{x}} - \bar{\mathbf{x}}' \to \bar{\mathbf{p}}$, as well as in time, $t - t' \to \omega$ (suppress $\bar{\mathbf{p}}$, ω), obtaining a 1D equation in the z-direction:

$$\left\{ -\frac{1}{2} \left[\frac{\partial}{\partial z} \left(\frac{1}{m(z)} \frac{\partial}{\partial z} \right) - \frac{\bar{\mathbf{p}}^2}{m(z)} \right] + V(z) - \omega \right\} G(z, z') = \delta(z - z'). \tag{5.9.17}$$

Doing the same for the *infinite*-space Green's functions G_A, G_B, we have

$$\left[-\frac{1}{2m_{A,B}} \left(\frac{\partial^2}{\partial z^2} - \bar{\mathbf{p}}^2 \right) + V_{A,B} - \omega \right] G_{A,B}(z, z') = \delta(z - z'), \tag{5.9.18}$$

which have the solutions (Eq. (5.1.30)):

$$G_{A,B}(z, z') = -i \frac{m_{A,B}}{k_{A,B}} e^{ik_{A,B}|z - z'|}, \tag{5.9.19}$$

where

$$k_{A,B} = \sqrt{2m_{A,B} \left(\omega - V_{A,B} \right) - \bar{\mathbf{p}}^2}. \tag{5.9.20}$$

Following the divergence/Gauss theorem considerations leading to Eqns. (5.9.4) and (5.9.5), we now obtain just boundary values at $z_S = 0$ instead of integrals at the interface $\mathbf{x}_S \to z_S = 0$, with results given by

Region A; $z < 0$

$$G(z, z') = G_A(z, z') + \frac{1}{2m_A} \left\{ G_A(z, 0) \left[\frac{\partial}{\partial z_S} G(z_S, z') \right]_{z_S = 0^-} - G(0, z') \left[\frac{\partial}{\partial z_S} G_A(z, z_S) \right]_{z_S = 0^-} \right\}, \tag{5.9.21}$$

Region B; $z > 0$

$$G(z, z') = \frac{1}{2m_B} \left\{ G_B(z, 0) \left[\frac{\partial}{\partial z_S} G(z_S, z') \right]_{z_S = 0^+} - G(0, z') \left[\frac{\partial}{\partial z_S} G_B(z, z_S) \right]_{z_S = 0^+} \right\}. \tag{5.9.22}$$

Evaluating Eqns. (5.9.21) and (5.9.22) at $z = 0$, the coupled surface integral equations, Eqns. (5.9.7) and (5.9.8), are seen to have become simple algebraic equations replacing Eqns. (5.9.9) and (5.9.10), and they are easily solved (with the Ben Daniel-Duke boundary condition) to determine g and g' (Eqns. (5.9.11 and 5.9.12)) leading to the global Green's function for all z, employing Eqns. (5.9.4 and 5.9.5).

5.10 Green's Function Matching: Dynamic, Non-Local Electrostatic Screening at Interface; Surface Plasmons

The prototypical Green's function of electrostatics is the Coulomb potential of a unit point charge in infinite space, which obeys the Poisson equation given by

$$\nabla \cdot \nabla \frac{1}{|\mathbf{x}-\mathbf{x}'|} = -4\pi \delta^{(3)}(\mathbf{x}-\mathbf{x}'), \tag{5.10.1}$$

and is denoted as

$$G_0^{\infty}(\mathbf{x}-\mathbf{x}') = \frac{1}{|\mathbf{x}-\mathbf{x}'|}. \tag{5.10.2}$$

If the infinite medium has a dielectric constant ε, the effective unit charge strength is reduced to $1/\varepsilon$ and $G_0^{\infty}(\mathbf{x}-\mathbf{x}')$ is correspondingly reduced. Furthermore, if the infinite medium is a homogeneous isotropic dynamic, non-local plasma with dielectric function $\epsilon\,(\mathbf{x}', \mathbf{x}''; t', t'') = \epsilon\,(\mathbf{x}'-\mathbf{x}''; t'-t'')$, the relation between electric displacement $\mathbf{D}(\mathbf{x})$ and electric field $\mathbf{E}(\mathbf{x})$ can be expressed in position-time representation as a Faltung-convolution integral:

$$\mathbf{D}(\mathbf{x}') = \int d^{(3)}\mathbf{x}'' \int dt'' \varepsilon(\mathbf{x}'-\mathbf{x}''; t'-t'')\mathbf{E}(\mathbf{x}'', t''). \tag{5.10.3}$$

Fourier-transforming $\mathbf{x}'-\mathbf{x}'' \to \mathbf{p}$, this takes the form

$$\mathbf{D}(\mathbf{p}, \omega) = \varepsilon(\mathbf{p}, \omega)\mathbf{E}(\mathbf{p}, \omega), \tag{5.10.4}$$

where we have also Fourier-transformed the time dependence to frequency representation $t' - t'' \to \omega$. Correspondingly, the shielded form of the electrostatic Coulomb Green's function for infinite space in $\mathbf{p}$-representation is given by

$$G_0^{\infty}(\mathbf{p}, \omega) = \frac{4\pi/p^2}{\varepsilon(\mathbf{p}, \omega)}, \tag{5.10.5}$$

in which the effective charge strength is reduced by the infinite-space dynamic, non-local dielectric function $\varepsilon(\mathbf{p}, \omega)$ of the plasma (time-translational invariance will be assumed below and ω will be suppressed).

The introduction of a spatial boundary or other inhomogeneity deprives the system of spatial translational invariance and the Green's function, $G(\mathbf{x}', \mathbf{x}'')$, depends on $\mathbf{x}'$ and

$\mathbf{x}''$ separately (no longer on the combination $\mathbf{x}'-\mathbf{x}''$ alone). Our focus here will be on the introduction of a planar interface at $z = 0$ separating two semi-infinite non-local plasma-like media A, B which are individually homogeneous (except for the interface), having bulk non-local dielectric functions $\varepsilon_A(\mathbf{x}' - \mathbf{x}'')$, $\varepsilon_B(\mathbf{x}' - \mathbf{x}'')$, respectively. These dielectric functions of the infinitely extended bulk media A, B are taken to represent their respective dielectric properties when restricted to their individual half-spaces by the interface. (In this regard, we ignore small quantum effects that change ε_A, ε_B over microscopic distances from the interface on the order of an inverse Fermi wave number.)

Our objective here is to apply the Green's function matching method of the preceding section to the determination of the global electrostatic Coulomb Green's function for the combined system $A\&B$: As the potential (G) of a unit point charge, this Green's function obeys the $\nabla \cdot \mathbf{D}$ equation in the form of a non-locality integral over the combined regions of A and B, as

$$\int\limits_{A\&B} d^{(3)}\mathbf{x}'''\nabla\cdot\left[\varepsilon(\mathbf{x},\mathbf{x}''')\nabla'''G(\mathbf{x}''',\mathbf{x}')\right] = -4\pi\delta^{(3)}(\mathbf{x}-\mathbf{x}'), \tag{5.10.6}$$

where $\varepsilon(\mathbf{x},\mathbf{x}') = \varepsilon_A(\mathbf{x}-\mathbf{x}')$ for $\mathbf{x}$, $\mathbf{x}'$ in A and $\varepsilon(\mathbf{x},\mathbf{x}') = \varepsilon_B(\mathbf{x}-\mathbf{x}')$ for $\mathbf{x}$, $\mathbf{x}'$ in B. However, if $\mathbf{x}$ is in $A(B)$ and $\mathbf{x}'$ is in $B(A)$, then $\varepsilon(\mathbf{x},\mathbf{x}') \equiv 0$. For the case in which both $\mathbf{x}$ and $\mathbf{x}'$ are in A, the global Green's function for all space, G, satisfies

$$\int\limits_A d^{(3)}\mathbf{x}G_A(\mathbf{x}'',\mathbf{x})\left[\int\limits_A d^{(3)}\mathbf{x}'''\nabla\cdot[\varepsilon_A(\mathbf{x},\mathbf{x}''')\nabla'''G(\mathbf{x}''',\mathbf{x}')] = -4\pi\delta^{(3)}(\mathbf{x}-\mathbf{x}')\right]$$
$$= -4\pi\, G_A(\mathbf{x}'';\mathbf{x}'), \tag{5.10.7}$$

where we have applied $\int_A d^3\mathbf{x}G_A(\mathbf{x}'',\mathbf{x})$ to Eq. (5.10.6), integrating over region A alone to the exclusion of region B on the other side of the bounding plane interface at $z = 0$. Furthermore, the Green's function for region A if it were infinitely extended, G_A, obeys a similar equation given by

$$\int\limits_A d^{(3)}\mathbf{x}'''G(\mathbf{x}''',\mathbf{x}')\left[\int\limits_A d^{(3)}\mathbf{x}\nabla'''\cdot[\varepsilon_A(\mathbf{x}''',\mathbf{x})\nabla G_A(\mathbf{x},\mathbf{x}'')] = -4\pi\delta^{(3)}(\mathbf{x}'''-\mathbf{x}'')\right]$$
$$= -4\pi\, G(\mathbf{x}'',\mathbf{x}'). \tag{5.10.8}$$

Application of integration by parts to Eqns. (5.10.7 and 5.10.8) yields

$$-4\pi\, G_A(\mathbf{x}'',\mathbf{x}') = \int\limits_A d^{(3)}\mathbf{x}\int\limits_A d^{(3)}\mathbf{x}'''\nabla\cdot\left\{G_A(\mathbf{x}'',\mathbf{x})\varepsilon_A(\mathbf{x},\mathbf{x}''')\nabla'''G(\mathbf{x}''',\mathbf{x}')\right\}$$
$$-\int\limits_A d^{(3)}\mathbf{x}\int\limits_A d^{(3)}\mathbf{x}'''\left(\nabla G_A(\mathbf{x}'',\mathbf{x})\right)\varepsilon_A(\mathbf{x},\mathbf{x}''')\cdot\nabla'''G(\mathbf{x}''',\mathbf{x}') \tag{5.10.9}$$

and

$$-4\pi G(\mathbf{x}'',\mathbf{x}') = \int_A d^{(3)}\mathbf{x}''' \int_A d^{(3)}\mathbf{x}\nabla''' \cdot \left\{G(\mathbf{x}''',\mathbf{x}')\varepsilon_A(\mathbf{x}''',\mathbf{x})\nabla G_A(\mathbf{x},\mathbf{x}'')\right\}$$
$$- \int_A d^{(3)}\mathbf{x}''' \int_A d^{(3)}\mathbf{x}\left(\nabla''' G(\mathbf{x}''',\mathbf{x}')\right)\varepsilon_A(\mathbf{x}''',\mathbf{x}) \cdot \nabla G_A(\mathbf{x},\mathbf{x}''). \tag{5.10.10}$$

Noting the reciprocity principle for (real) Green's functions ($G(\mathbf{x}''',\mathbf{x}') = G(\mathbf{x}',\mathbf{x}''')$ and $G_A(\mathbf{x},\mathbf{x}'') = G_A(\mathbf{x}'',\mathbf{x})$), we interchange the dummy integration variables in the last term on the right of Eq. (5.10.10) and subtract Eq. (5.10.10) from Eq. (5.10.9); this eliminates the last terms of these two equations, leaving only volume integrals that can be converted to surface integrals. Employing the divergence/Gauss theorem yields (with outwardly directed normal $\hat{n}$; notation – $\mathbf{x} = (\bar{\mathbf{x}}, z) = (x, y, z)$):

$$G(\mathbf{x}'',\mathbf{x}') - G_A(\mathbf{x}'',\mathbf{x}') = \frac{1}{4\pi}\int_S dS\hat{n} \cdot \int_A d^{(3)}\mathbf{x}''' G_A(\mathbf{x}'',\mathbf{x}_S)\varepsilon_A(\mathbf{x}_S,\mathbf{x}''')\nabla''' G(\mathbf{x}''',\mathbf{x}')$$
$$-\frac{1}{4\pi}\int_S dS\hat{n} \cdot \int_A d^{(3)}\mathbf{x} G(\mathbf{x}_S,\mathbf{x}')\varepsilon_A(\mathbf{x}_S,\mathbf{x})\nabla G_A(\mathbf{x},\mathbf{x}''). \tag{5.10.11}$$

Since the lateral plane has spatial translational symmetry involving dependence on $\bar{\mathbf{x}} - \bar{\mathbf{x}}'$ alone (to the exclusion of $\bar{\mathbf{x}} + \bar{\mathbf{x}}'$), we may Fourier-transform $\bar{\mathbf{x}} - \bar{\mathbf{x}}' \to \bar{\mathbf{p}}$, and then the planar surface integral reduces to the value of the integrand at the boundary $\mathbf{x}_S \to z_S = 0_A$ (here, $0_{\left\{A \atop B\right\}}$ indicates $z_S \to 0$ from the $\left\{A \atop B\right\}$ side). Thus, we obtain ($\bar{\mathbf{p}}$ and ω are suppressed):

$$G(z'',z') = G_A(z'',z') + \frac{1}{4\pi}\int_A dz''' \Big[G_A(z'',0_A)\varepsilon_A(0_A,z''')\frac{\partial}{\partial z'''}G(z''',z')$$
$$-G(0_A,z')\varepsilon_A(0_A,z''')\frac{\partial}{\partial z'''}G_A(z''',z'')\Big]. \tag{5.10.12}$$

Undertaking a similar procedure with the field point of G in B and its source point in A, but with the δ-function source point of G_B in B, a result similar to Eq. (5.10.11) is obtained with $G_A \to G_B$: However, the left-hand side of this counterpart of Eq. (5.10.11) has only G present—excluding a subtracted G_B term—since the δ-function, which could have produced such a G_B term on the left, does not contribute because its argument cannot vanish with a field point in B and a source point in A. Therefore, from the B side, we find (the outwardly directed normal reverses direction ($\hat{n} \to -\hat{n}$) from A to B):

$$G(\mathbf{x}'',\mathbf{x}') = \frac{1}{4\pi}\int_S dS(-\hat{n})\cdot\int_B d^{(3)}\mathbf{x}'''G_B(\mathbf{x}'',\mathbf{x}_S)\varepsilon_B(\mathbf{x}_S,\mathbf{x}''')\nabla'''G(\mathbf{x}''',\mathbf{x}')$$
$$-\frac{1}{4\pi}\int_S dS(-\hat{n})\cdot\int_B d^{(3)}\mathbf{x}G(\mathbf{x}_S,\mathbf{x}')\varepsilon_B(\mathbf{x}_S,\mathbf{x})\nabla G_B(\mathbf{x},\mathbf{x}''). \quad (5.10.13)$$

In the case of the plane boundary under consideration, Fourier transformation in the lateral plane $\bar{\mathbf{x}}-\bar{\mathbf{x}}'\to\bar{\mathbf{p}}$ (suppressed) again simplifies the result:

$$B\text{-Side:}\quad G(z'',z') = \frac{1}{4\pi}G_B(z'',0_B)D(0_B,z') - \frac{1}{4\pi}G(0_B,z')D_B(0_B,z''), \quad (5.10.14)$$

with the notation

$$D(0_{\left\{A \atop B\right\}},z) = -\int_{\left\{A \atop B\right\}} dz'''\varepsilon_{\left\{A \atop B\right\}}(0_{\left\{A \atop B\right\}},z''')\frac{\partial}{\partial z'''}G(z''',z) \quad (5.10.15)$$

and

$$D_{\left\{A \atop B\right\}}(0_{\left\{A \atop B\right\}},z) = -\int_{\left\{A \atop B\right\}} dz'''\varepsilon_{\left\{A \atop B\right\}}(0_{\left\{A \atop B\right\}},z''')\frac{\partial}{\partial z'''}G_{\left\{A \atop B\right\}}(z''',z). \quad (5.10.16)$$

Employing this notation, Eq. (5.10.12) for the A side may be written as

$$A\text{-Side:}\quad G(z'',z') = G_A(z'',z') - \frac{1}{4\pi}G_A(z'',0_A)D(0_A,z') + \frac{1}{4\pi}G(0_A,z')D_A(0_A,z''). \quad (5.10.17)$$

Eqns. (5.10.14) and (5.10.17) determine $G(z'',z')$ everywhere once boundary values are determined for the continuous quantities

$$G(0,z') = G(0_A,z') = G(0_B,z') \quad (5.10.18)$$

and

$$D(0,z') = D(0_A,z') = D(0_B,z'), \quad (5.10.19)$$

which describe continuity of the electrostatic potential (Eq. (5.10.18)) and continuity of the normal component of the electrostatic displacement field (Eq. (5.10.19)) across the interface, respectively. To determine these boundary values, we set $z''\to 0$ in Eqns. (5.10.17) and (5.10.14), obtaining

$$A\text{-Side:}\quad G(0,z') = G_A(0_A,z') - \frac{1}{4\pi}G_A(0,0)D(0,z') + \frac{1}{4\pi}G(0,z')D_A(0,0) \quad (5.10.20)$$

and

$$B\text{-Side:}\quad G(0,z') = \frac{1}{4\pi}G_B(0,0)D(0,z') - \frac{1}{4\pi}G(0,z')D_B(0,0). \tag{5.10.21}$$

These two equations readily yield $G(0,z')$ as

$$G(0,z') = \frac{G_B(0,0)G_A(0,z')}{\left[1-\frac{1}{4\pi}D_A(0,0)\right]G_B(0,0)+\left[1+\frac{1}{4\pi}D_B(0,0)\right]G_A(0,0)}. \tag{5.10.22}$$

Noting that the only property of $G_{\left\{ A \atop B \right\}}$ defined above is that it arises from a unit point charge in the restricted region $\left\{ A \atop B \right\}$, we are free to impose upon it an arbitrary boundary condition at the interface. For convenience, we require the vanishing of $D_{\left\{ A \atop B \right\}}(0,z'') = 0$ at the interface (associated with introducing a subsidiary image in the "other" region $\left\{ B \atop A \right\}$). With this, Eq. (5.10.22) becomes

$$G(0,z') = \frac{G_B(0,0)G_A(0,z')}{G_B(0,0)+G_A(0,0)}, \tag{5.10.23}$$

and Eqns. (5.10.20 and 5.10.21) yield

$$D(0,z') = \frac{4\pi\, G_A(0,z')}{G_B(0,0)+G_A(0,0)}. \tag{5.10.24}$$

Employing these results for $G(0,z')$ and $D(0,z')$ in Eqns. (5.10.14 and 5.10.17) completes the determination of the global electrostatic Coulomb Green's function, G, of the planar heterostructure in all regions of space in terms of the infinitely extended non-local Green's functions of its constituent parts G_A, G_B. The vanishing of the denominator of Eqns. (5.10.23 and 5.10.24) corresponds to a frequency pole representing the occurrence of a new collective electrostatic oscillation associated with the interface—a *surface plasmon*:

$$G_A(0,0) + G_B(0,0) = 0. \tag{5.10.25}$$

In regard to our choice of boundary condition for $G_{\left\{ A \atop B \right\}}$, we can assure the vanishing of $D_{\left\{ A \atop B \right\}}(0,z'')$ at the interface by choosing a subsidiary image of the same unit strength at the "mirror" point in the "other" region across the boundary as

$$G_{\left\{ A \atop B \right\}}(\mathbf{x},\mathbf{x}') = G^{\infty}_{0\left\{ A \atop B \right\}}(\mathbf{x}-\mathbf{x}') + G^{\infty}_{0\left\{ A \atop B \right\}}(\mathbf{x}-\tilde{\mathbf{x}}'), \tag{5.10.26}$$

where $\tilde{\mathbf{x}}' = (\bar{\mathbf{x}}', -z')$ and $\mathbf{x}' = (\bar{\mathbf{x}}', z')$. Here, $G^{\infty}_{0\left\{ A \atop B \right\}}$ is given in 3D $\mathbf{p}$-representation by Eq. (5.10.5) with $\varepsilon \to \varepsilon_{\left\{ A \atop B \right\}}$. Similar remarks apply to D_A and D_B. In lateral wavevector representation ($\bar{\mathbf{x}} - \bar{\mathbf{x}}' \to \bar{\mathbf{p}}$), $G_{\left\{ A \atop B \right\}}$ is given by

$$G_{\left\{ A \atop B \right\}}(z,z';\bar{\mathbf{p}}) = 2\int_{-\infty}^{\infty} dp_z \frac{e^{ip_z(z-z')} + e^{ip_z(z+z')}}{(\bar{p}^2 + p_z^2)\varepsilon_{\left\{ A \atop B \right\}}(\mathbf{p},\omega)}. \tag{5.10.27}$$

Defining an *inverse surface dielectric function* $\varepsilon^{-1}_{\{A \atop B\},\bar{p}}(\omega)$ as follows:

$$G_{\{A \atop B\}}(0,0;\bar{\mathbf{p}}) = 4\int_{-\infty}^{\infty} dp_z \frac{1}{(p_z^2+\bar{p}^2)\varepsilon_{\{A \atop B\}}(\mathbf{p},\omega)} \equiv \frac{4\pi}{\bar{p}}\varepsilon^{-1}_{\{A \atop B\},\bar{p}}(\omega), \tag{5.10.28}$$

the surface plasmon dispersion relation of Eq. (5.10.25) takes the form

$$\int_{-\infty}^{\infty} \frac{dp_z}{(p_z^2+\bar{p}^2)\varepsilon_A(\mathbf{p},\omega)} + \int_{-\infty}^{\infty} \frac{dp_z}{(p_z^2+\bar{p}^2)\varepsilon_B(\mathbf{p},\omega)} = 0, \tag{5.10.29}$$

which includes the role of non-locality in both regions A and B. It was first derived by Fuchs and Kliewer.[13] In the case when region A is a local medium, $\varepsilon_A(\mathbf{p},\omega) \to \varepsilon_A(\omega)$, evaluation of the integral on the right-hand side of Eq. (5.10.27) yields

$$G_A(z,z';\bar{\mathbf{p}}) = \frac{2\pi}{\bar{p}}\frac{(e^{-\bar{p}|z-z'|}+e^{-\bar{p}|z+z'|})}{\epsilon_A(\omega)}, \tag{5.10.30}$$

and, consequently, the dispersion relation of Eq. (5.10.29) for A local and B non-local is given by

$$-\frac{1}{\varepsilon_A(\omega)} = \frac{\bar{p}}{\pi}\int_{-\infty}^{\infty} \frac{dp_z}{(p_z^2+\bar{p}^2)\varepsilon_B(\mathbf{p},\omega)} = \epsilon^{-1}_{B,\bar{p}}(\omega). \tag{5.10.31}$$

In regard to the image potential $V_{Img}(z',z'')$ in region A, it is readily recognized as given by

$$V_{Img}(z',z'') = G(z',z'') - G^{\infty}_{0A}(z',z''),$$

which may be evaluated using Eq. (5.10.17) for $G(z',z'')$ jointly with our choice of the subsidiary image (such that $D_{\{A \atop B\}}(0,z'')=0$), yielding the result (restore $\mathbf{p},\omega$):

$$V_{Img}(z',z'';\bar{p},\omega) = \frac{2\pi}{\bar{p}}\frac{\exp\left[-\bar{p}\left(|z'|+|z''|\right)\right]}{\varepsilon_A(\omega)}\frac{\varepsilon_{B,\bar{p}}(\omega)-\varepsilon_A(\omega)}{\varepsilon_{B,\bar{p}}(\omega)+\varepsilon_A(\omega)} \quad \left[\text{for } z,z'>0 \text{ in region } A\right]. \tag{5.10.32}$$

This result reflects the fact that if region B is an ideal metal (with $|\varepsilon_{B,\bar{p}}(\omega)| \to \infty$), then the image change sited in region B contributes to the electric field in region A with strength (and sign) identical to that of the strength (and sign) of the original charge sited in region A. This is characteristic of perfect reflection of the field in region A at the interface between the regions when B is an ideal metal, with no penetration of the field from A into region B. More generally, Eq. (5.10.32) exhibits the non-local image strength factor,

$$\frac{\varepsilon_{B,\bar{p}}(\omega)-\varepsilon_A(\omega)}{\varepsilon_{B,\bar{p}}(\omega)+\varepsilon_A(\omega)},$$

which applies for region B as an arbitrary non-local medium. (This image strength factor is well known in the local limit, $\varepsilon_{B,\bar{p}}(\omega) \to \varepsilon_B(\omega)$.) On the other hand, if region B is a vacuum and region A is a semiconductor plasma with $\varepsilon_A(\omega) = 1-\omega_p^2/\omega^2$ ($\omega_p = \sqrt{4\pi e^2\rho/m}$ is the bulk plasma frequency with ρ as the bulk equilibrium density and m as the effective mass), then the surface plasmon dispersion relation of Eq. (5.10.31) yields the collective surface plasmon mode as $\omega = \omega_p/\sqrt{2}$, which is also manifested as a frequency pole of the image strength factor above.

5.11 Feynman Path Integrals

In closing this introductory chapter on Green's functions it is appropriate to point out their equivalence with Feynman's path integrals.[14] To exhibit the latter, recall that the time-development operator of Eqns. (3.1.3 and 3.3.1) yields the Green's function as (Eq. (5.1.21)):

$$G_1^{ret}\left(\mathbf{x}, t; \mathbf{x}', t\right) = -i\eta_+(t-t')\left\langle \mathbf{x}|e^{-i(H_0+H')(t-t')}|\mathbf{x}'\right\rangle. \tag{5.11.1}$$

(While this involves $H = H_0 + H'$ independent of time and is stated for retarded Green's functions, the arguments provided below can be generalized to Hamiltonians that do have explicit time dependence, and also apply for multiparticle statistical Green's functions.) Quite generally H_0 and H' do *not* commute, so the exponential on the right of Eq. (5.11.1) *cannot* be factorized. However, if the time interval $t-t'$ is broken into many (N) infinitesimal segments of width $\Delta t = (t-t')/N$ (anticipating that $N \to \infty$ and $\Delta t \to 0$), we have N factors for the time-development operator as

$$e^{-i(H_0+H')t} = e^{-i(H_0+H')\Delta tN} = e^{-i(H_0+H')\Delta t}e^{-i(H_0+H')\Delta t} \ldots e^{-i(H_0+H')\Delta t}, \tag{5.11.2}$$

each of which can be written approximately as

$$e^{-i(H_0+H')\Delta t} \cong 1 - i(H_0 + H')\Delta t \cong (1 - iH_0\delta t)(1 - iH'\Delta t) + \ldots \tag{5.11.3}$$

with the neglect of terms of order $(\Delta t)^2$, which is acceptable as $\Delta t \to 0$. Alternatively, expressed in exponential form,

$$e^{-i(H_0+H')\Delta t} \cong e^{-iH_0\Delta t}e^{-iH'\Delta t}, \tag{5.11.4}$$

wherein higher order terms in Δt are again neglected. Furthermore, anticipating the insertion of $N-1$ complete sets of states, $\int d\mathbf{x}_j|\mathbf{x}_j><\mathbf{x}_j|$, $(i = 1, 2, \ldots, N \to \infty)$ between adjacent exponentials, we have factors such as

$$\left\langle x_j|e^{-i(H_0+H')\Delta t}|\mathbf{x}_{j+1}\right\rangle \cong \left\langle \mathbf{x}_j|e^{-iH_0\Delta t}e^{-iH'\Delta t}|\mathbf{x}_{j+1}\right\rangle, \tag{5.11.5}$$

and considering a *local* potential H' we write

$$\begin{aligned}\left\langle \mathbf{x}_j \middle| e^{-iH_0\Delta t} e^{-iH'\Delta t} \middle| \mathbf{x}_{j+1} \right\rangle &= \int d\mathbf{r} \left\langle \mathbf{x}_j \middle| e^{-iH_0\Delta t} \middle| \mathbf{r} \right\rangle \left\langle \mathbf{r} \middle| e^{-iH'\Delta t} \middle| \mathbf{x}_{j+1} \right\rangle \\ &= \int d\mathbf{r} \left\langle \mathbf{x}_j \middle| e^{-iH_0\Delta t} \middle| \mathbf{r} \right\rangle \delta(\mathbf{r} - \mathbf{x}_{j+1}) e^{-iH'(\mathbf{r})\Delta t}.\end{aligned} \tag{5.11.6}$$

Bearing in mind that $H_0 = H_0(\mathbf{x}, \mathbf{p})$ generally involves momentum as well as position operators, it is appropriate to introduce transforms to momentum representation, yielding[14,15]

$$\begin{aligned}\left\langle x_j \middle| e^{-iH_0\Delta t} e^{-iH'\Delta t} \middle| \mathbf{x}_{j+1} \right\rangle = \int \tfrac{d\mathbf{p}_j}{2\pi} exp\big[&i\mathbf{p}_j(\mathbf{x}_{j+1} - \mathbf{x}_j) \\ &- i\Delta t(H_0(\mathbf{x}_j, \mathbf{p}_j; t_j) - H'(\mathbf{x}_j; t_j))\big].\end{aligned} \tag{5.11.7}$$

Finally, the Feynman path integral takes the form

$$G_1^{ret}\left(\mathbf{x}, t; \mathbf{x},' t'\right) = -i\eta_+(t - t') \sum_{j=1}^{N-1} \int_{-\infty}^{\infty} d\mathbf{x}_j \sum_{j=1}^{N} \int_{-\infty}^{\infty} \frac{d\mathbf{p}_j}{(2\pi)^3} exp\left(\frac{i}{\hbar} S^N\right), \tag{5.11.8}$$

where S^N is defined by

$$S^N = \sum_{j=1}^{N} \left[\mathbf{p}_j(\mathbf{x}_{j+1} - \mathbf{x}_j) - \Delta t H(\mathbf{p}_j; \mathbf{x}_j; t_j)\right], \tag{5.11.9}$$

with the limits $\Delta t \to 0$, $N \to \infty$ understood such that Lim $N\Delta t = t - t'$. Noting that $\mathbf{x}_{j+1} - \mathbf{x}_j$ can be written as $\Delta \mathbf{x} = dt(d\mathbf{x}/dt) = \dot{\mathbf{x}}dt$, the summand of Eq. (5.11.9) is just Δt times the Lagrangian.

There are numerous fine expositions of Feynman's path integrals in addition to his own papers[14] and his book,[15] some of which include many examples of particular interest. Among these are books by Kleinert,[16] Dittrich and Reuter,[17] Swanson,[18] Schulman,[19] Smirnov,[20] Grosche and Steiner,[21] T.D. Lee,[22] Papadopoulos and Devreese,[23] Albeverio, *et al.*,[24] Antoine and Tirapegui,[25] also Popov,[26] Glimm and Jaffe,[27] and others.

5.12 Chapter 5 Problems

P.5.1. Consider an electron confined to motion on a two-dimensional (2D) plane with a Hamiltonian matrix of the form $h^{(1)} = \alpha\boldsymbol{\sigma} \cdot \mathbf{p}$ in the absence of a magnetic field, where $\boldsymbol{\sigma} = (\sigma_1, \sigma_2)$ represents 2D Pauli (pseudo) spin matrices and $\mathbf{p}$ is the 2D momentum operator in position representation (α is a constant). Examine the equation for the retarded 2×2 matrix electron Green's function, G, defined by $(i\frac{\partial}{\partial t} - h^{(1)})G(\mathbf{x}, \mathbf{x}', t - t') = I\delta^{(2)}(\mathbf{x} - \mathbf{x}')\delta(t - t')$ (I is the unit 2×2 matrix and $\mathbf{x} = (x_1, x_2)$, $\hbar \to 1$) in frequency/energy representation ($i\frac{\partial}{\partial t} \to \omega$). *Solve* for all the elements of G.

P.5.2. Consider the application of a strong magnetic field normal to the 2D layer of Problem 5.1 and determine its effect on the Hamiltonian, $h^{(1)}$, ignoring Zeeman splitting (zero magnetic moment). With this, examine the equations for its retarded 2×2 matrix Green's function, G, in frequency/energy representation ($i\frac{\partial}{\partial t} \to \omega$). Determine the equation for G_{11} (the 11 element of the 2×2 matrix G) in which the other elements G_{12}, G_{21}, G_{22} have been *eliminated and explicitly exhibit the role of the angular momentum operator*. (You need *not* solve this for G_{11}.)

P.5.3. For normal materials, consider a three-dimensional (3D) electron in a one-dimensional (1D) asymmetric double quantum well potential of the form $V(z) = -U_1\delta(z - a/2) - U_2\delta(z + a/2)$, where a is the separation between the wells and U_1 and U_2 are real positive constants. Determine the retarded Green's function explicitly and examine its dispersion relation, discussing bonding/antibonding-type state features and the role of the ratio U_1/U_2 in regard to bond-type states.

P.5.4. Consider a normal electron (3D) confined to move between two parallel planes at which its wave function and Green's function are forced to vanish by infinitely high potential walls at $z = \pm a/2$ (as in a quantum well). Determine the corresponding retarded Green's function in terms of the infinite space Green's function (there is no magnetic field involved here). (Hint: Use an image technique to treat the z-direction and transform $\mathbf{x} - \mathbf{x}' \to \mathbf{p}$ in the x-y lateral plane.)

P.5.5. Consider an electron confined to move within a cylinder at which its Green's function and wave function are forced to vanish. Determine the retarded Green's function for this system.

P.5.6. Verify Eqns. (5.1.28–5.1.30) and discuss the density of states, $D(\omega)$, in each case.

P.5.7. Discuss estimates of the energies of bound states arising from attractive 2D and 1D δ-function potentials based on Section 5.3.

P.5.8. Determine the line width, Γ, of Eq. (5.5.13) for strongly Landau-quantized 2D electrons with a Gaussian-scattering potential, $U(\mathbf{r}) = U_0 \exp(\frac{-r^2}{a^2})$, having range a taken to be large, $a \gg r$.

P.5.9. Consider a 3D electron in an infinite periodic crystal lattice of the Krönig–Penney type characterized by the 1D potential (n is an integer and $\delta(x)$ is a 1D Dirac delta function, $U > 0$):

$$V = U \sum_{n=-\infty}^{\infty} \delta(z - na).$$

Determine the retarded Green's function for an electron in this lattice in frequency/energy representation. Elucidate the role of tunneling in the positional dependence of the Green's function. Also, examine the dispersion relation, discussing allowed and forbidden bands and their bounds (with gaps in the energy spectrum); and find the effective mass of the electron in the first allowed band.

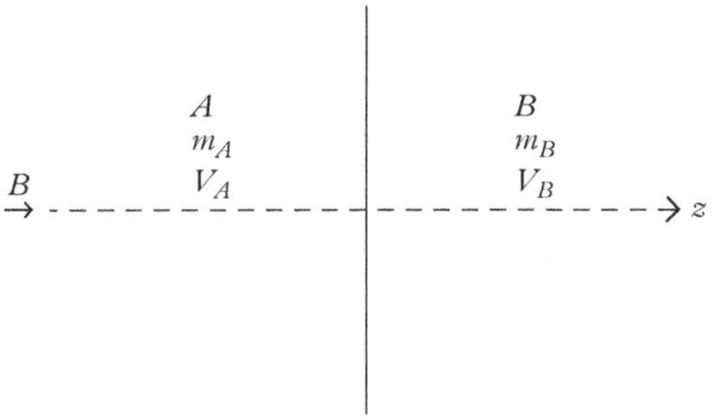

Figure 5.4: *Adjoining regions A, B having different potentials V_A, V_B and masses m_A, m_B.*

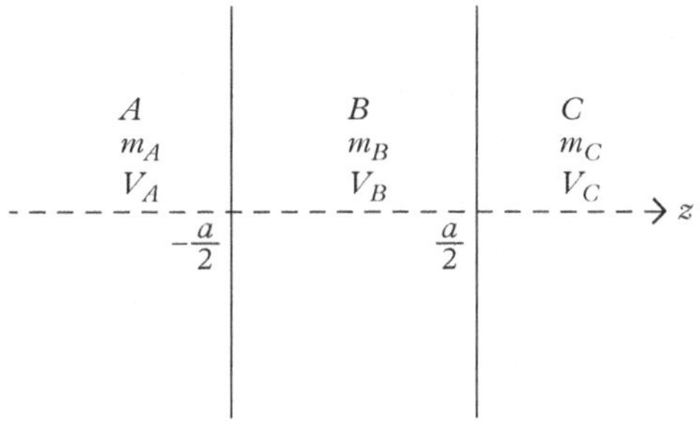

Figure 5.5: *Adjoining regions A, B, C having different potentials V_A, V_B, V_C and masses m_A, m_B, m_C.*

P.5.10. Prove that Eq. (5.10.26) causes $D_A(0, z'')$ to vanish at the interface.

P.5.11. Verify Eqns. (5.10.31) and (5.10.32) and complete the solution for the global non-local electrostatic Green's function.

P.5.12. Determine the surface mode for the single interface problem described by Figure 5.3, Eqns. (5.9.21 and 5.9.22). (No magnetic field.)

P.5.13. Consider the planar heterostructure of Eqns. (5.9.15 and 5.9.16). Determine the retarded Green's function for an electron capable of 3D motion in this heterostructure when a magnetic field **B** is applied normal to the interface of regions A and B (see Figure 5.4). Examine the attendant surface state dispersion relation. (Hint: Use the material of Section 5.6.)

P.5.14. Consider a planar heterostructure as shown in Figure 5.5, in which an electron has different mass and different constant potential in each of the three regions A,B,C (but within a given region, the mass and potential are constant).

Determine the retarded Green's function for an electron capable of 3D motion in this heterostructure in terms of the individual Green's functions of the individual regions as if they were infinitely extended.

P.5.15. Prove that the invariance of the Hamiltonian under a gauge transformation $\lambda(\mathbf{x}, t)$ requires that $\psi(\mathbf{x}, t) \rightarrow e^{\frac{ie\lambda(\mathbf{x},t)}{\hbar c}} \psi(\mathbf{x}, t)$, as stated in the first paragraph of Section 5.6.

P.5.16. Verify Eqns. (5.11.6 and 5.11.7).

5.13 References for Chapter 5

R.5.1. J. Schwinger, "Quantum Kinematics and Dynamics," W.A. Benjamin, Inc. Publ., Section VIII. 6, p.343 (1970).

R.5.2. D.M. Newns, Phys. Rev. *178*, 1123 (1969); P.W. Anderson *124*, 41 (1961).

R.5.3. H. Haug and A.-P. Jauho, "Quantum Kinetics in Transport and Optics of Semiconductors," Springer Publ., 2nd Edition, Chapter 12 (2008).

R.5.4. S.M. Dutra, "Cavity Quantum Electrodynamics," Wiley-Interscience, Appendix F (2005).

R.5.5. T. Ando, A.B. Fowler and F. Stern, "Electronic Properties of Two Dimensional Systems," Rev. Mod. Phys. *54* No.2, 437 Section 6A (1982).

R.5.6. Bateman Manuscript Project, "Higher Transcendental Functions," Vol. II, Editors: A. Erdelyi, *et al.* No.32, p. 195 (1953).

R.5.7. V. Fessatidis, N.J.M. Horing and K. Sabeeh, Philosophical Magazine **B79**, 77 (1999).

R.5.8. N.J.M. Horing, H.L. Cui & G. Fiorenza, Physical Review A *34*, 612 (1986).

R.5.9. M. Ali Omar, "Elementary Solid State Physics," Revised Printing, Addison-Wesley Publ. Chapter 5 (1993).

R.5.10. J. Inglesfield, J. Phys. C: Solid State Phys. 4, L 14 (1971).

R.5.11. F. Garcia-Moliner and V.R. Velasco, "Theory of Single & Multiple Interfaces," World Scientific (1992).

R.5.12. F. Garcia-Moliner and F. Flores, "Introduction to the Theory of Solid Surfaces," Cambridge University Press (1979).

R.5.13. R. Fuchs and K.L. Kliewer, Phys. Rev. *B3*, 2270 (1971).

R.5.14. R.P. Feynman, Rev. Mod. Phys. *20*, 367 (1948). Also, R.P. Feynman, Phys. Rev. *76*, 749 (1949); R.P. Feynman, Phys. Rev. *76*, 769 (1949) and R.P. Feynman, Phys. Rev. *80*; 440 (1950).

R.5.15. R.P. Feynman and A.R. Hibbs, "Quantum Mechanics and Path Integrals," McGraw-Hill Book Co. (1965).

R.5.16. H. Kleinert, "Path Integrals," World Scientific (1995).

R.5.17. W. Dittrich and M. Reuter, "Classical and Quantum Dynamics," Springer (2001).

R.5.18. M.S. Swanson, "Path Integrals and Quantum Processes," Academic Press (1992).

R.5.19. L.S. Shulman, "Techniques and Applications of Path Integration," Wiley-Interscience (1981).

R.5.20. V.A. Smirnov, "Evaluating Feynman Integrals," Springer Tracts in Modern Physics, Vol. 211 (2004).

R.5.21. C. Grosche and F. Steiner, "Handbook of Feynman Path Integrals," Springer (1998).

R.5.22. T.D. Lee, "Particle Physics and Introduction to Field Theory," Vol. 1, Harwood Academic Publishers (1981).

R.5.23. G.J. Papadopoulos and J.T. Devreese, "Path Integrals and Applications" (1977).

R.5.24. S. Albeverio, *et al.*, "Feynman Path Integrals," Lecture Notes in Physics, No. 106, Springer-Verlag (1979).

R.5.25. J.P. Antoine and E. Tirapegui, "Functional Integration," Plenum Press (1980).

R.5.26. V.N. Popov, "Functional Integrals in Quantum Field Theory and Statistical Physics" D. Reidel Publishing Co. (1983).

R.5.27. J. Glimm and A. Jaffe, "Quantum Physics: A Functional Integral Point of View," Springer-Verlag (1981).

6

Quantum Mechanical Ensemble Averages and Statistical Thermodynamics

Abstract

Chapter 6 introduces quantum mechanical ensemble theory by proving the asymptotic equivalence of the quantum mechanical *microcanonical* ensemble average with the quantum *grand canonical* ensemble average for many-particle systems, based on the method of Darwin and Fowler. The procedures involved identify the grand partition function, entropy and other statistical thermodynamic variables, including the grand potential, Helmholtz Free Energy, thermodynamic potential, Gibbs Free Energy, Enthalpy and their relations in accordance with the fundamental laws of thermodynamics. Accompanying saddle-point integrations define temperature (inverse thermal energy) and chemical potential (Fermi energy). The concomitant emergence of quantum statistical mechanics and Bose–Einstein and Fermi–Dirac distribution functions are discussed in detail (including Bose condensation). The magnetic moment is derived from the Helmholtz Free Energy and is expressed in terms of a one-particle retarded Green's function with an imaginary time argument related to inverse thermal energy. This is employed in a discussion of diamagnetism and the de Haas–van Alphen effect.

6.1 Quantum Mechanical Ensemble Averages and Thermodynamics

The average value for an operator $\mathbf{X}$, $< N', E', \xi'|\mathbf{X}|N', E', \xi' >$, in a specific microscopic quantum state of definite macroscopic number N' and energy E', $|N', E', \xi'\rangle$, depends on the microscopic eigenvalues ξ' characterizing that particular state. In fact, the particular microscopic state characterized by ξ' is generally unknown, and considering that the macroscopic system is in contact with an external environment that randomly changes ξ' through interaction, it usually suffices to average over such microscopic states with equal weight. Thus, we seek information about the quantum mechanical *microcanonical* ensemble average, defined as

$$\langle \mathbf{X} \rangle^{N',E'} \equiv \frac{\sum_{\xi'} \langle N', E', \xi' \,|\mathbf{X}|\, N', E', \xi' \rangle}{\sum_{\xi'} \langle N', E', \xi' | N', E', \xi' \rangle}, \tag{6.1.1}$$

Quantum Statistical Field Theory: Schwinger's Variational Method. Norman J. Morgenstern Horing, Oxford University Press (2017). © Norman J. Morgenstern Horing. DOI 10.1093/oso/9780198791942.001.0001

for a macroscopic system of definite number N' and energy E'. For states normalized to unity, the denominator (Tr indicates trace and $Tr_{N',E'}$ indicates trace over states of given N', E' values),

$$\sum_{\xi'} \langle N', E', \xi' | N', E', \xi' \rangle \equiv Tr_{N',E'}\mathbf{I} = \rho(N', E'), \tag{6.1.2}$$

is just the number of microscopic states for given values of N' and E'. This is the *thermodynamic probability* central to the concept of entropy S through the defining equation

$$S = k \ln \rho(N', E'), \tag{6.1.3}$$

where k is the Boltzmann constant and T will denote Kelvin temperature.

Another average, associated with definite temperature and chemical potential, μ, is the quantum mechanical *grand canonical* ensemble average defined by ($\beta = 1/kT; \alpha = -\beta\mu$; Tr denotes the full trace):

$$\langle \mathbf{X} \rangle^{\alpha,\beta} = \frac{Tr\left(e^{-\beta H - \alpha N}\mathbf{X}\right)}{Tr\left(e^{-\beta H - \alpha N}\right)}, \tag{6.1.4}$$

which admits fluctuations of energy and number for systems in contact with heat and particle baths/reservoirs. (The ordinary (*non*-"grand") quantum mechanical canonical ensemble average is defined by the counterpart of Eq. (6.1.4) obtained by setting $\alpha \to 0$. Mathematically, the presence of α merely shifts the origin of the energy scale by $\mu = -\alpha/\beta$.) The normalizing denominator, $e^{W(\alpha,\beta)}$, given by

$$e^{W(\alpha,\beta)} \equiv Tr\left(e^{-\beta H - \alpha N}\right) = \sum_{N',E'} e^{-\alpha N' - \beta E'}\, Tr_{N',E'}\mathbf{I}, \tag{6.1.5}$$

or, alternatively,

$$e^{W(\alpha\,,\,\beta)} = \sum_{N'\,,\,E'} e^{-\alpha N' - \beta E'} \rho(N'\,,\,E') = Z; \quad W = \ln Z \tag{6.1.6}$$

defines the *grand partition function* Z, and $\Omega \equiv -W/\beta$ is called the *grand potential* (also *thermodynamic potential, Gibbs Function*, or *Landau Free Energy*).

For convenience, we view $\langle \mathbf{X} \rangle^{\alpha,\beta}$ as the quotient of boundary values of analytic functions in the lower half complex λ and τ planes for the numerator and denominator of

$$\langle \mathbf{X} \rangle^{i\lambda,i\tau} = \frac{Tr\left(e^{-iH\tau - iN\lambda}\mathbf{X}\right)}{Tr\left(e^{-iH\tau - iN\lambda}\right)}. \tag{6.1.7}$$

In order to relate the various ensemble averages and to verify the interpretation of α and β in terms of temperature and chemical potential, we outline the analysis of microcanonical and grand canonical ensemble averages presented by Martin and Schwinger,[1] based on the method of Darwin and Fowler.[2–6] Defining $e^{W(i\lambda,i\tau)}$ in analogy to Eq. (6.1.6) and in the expectation of a continuum of energy values E', whereas N' can only take integer values, we have

$$e^{W(i\lambda,i\tau)} = \sum_{N'} e^{-iN'\lambda} \int dE' e^{-iE'\tau} \rho(N', E'), \tag{6.1.8}$$

which may be Fourier-inverted as

$$\rho(N', E') = \int_{-\pi}^{\pi} \frac{d\lambda}{2\pi} \int_{-\infty}^{\infty} \frac{d\tau}{2\pi} e^{iN'\lambda + iE'\tau + W(i\lambda,i\tau)}. \tag{6.1.9}$$

Writing

$$e^{W_N(i\tau)} = \int_{-\pi}^{\pi} \frac{d\lambda}{2\pi} e^{iN'\lambda + W(i\lambda,i\tau)}, \tag{6.1.10}$$

the integral

$$\rho(N', E') = \int_{-\infty}^{\infty} \frac{d\tau}{2\pi} e^{iE'\tau + W_N(i\tau)} \tag{6.1.11}$$

may be evaluated approximately for large E' (which gives rise to a rapidly varying exponential) by the method of steepest descents[7] using a contour deformation into the lower half of the complex τ-plane, so that it passes through the sole saddle point at $\tau = -i\beta$, given by

$$\overline{E} = -\frac{\partial W_N(i\tau = \beta)}{\partial \beta} = \frac{\int dE' E' e^{-\beta E'} \rho(N', E')}{\int dE' e^{-\beta E'} \rho(N', E')}. \tag{6.1.12}$$

(The last equality arises from Fourier-inversion of Eq. (6.1.11).) This saddle-point evaluation of Eq. (6.1.11) yields

$$\rho(N', E') = \frac{\exp\left(\beta E' + W_N(\beta)\right)}{\sqrt{2\pi \frac{\partial^2 W_N(\beta)}{\partial \beta^2}}}. \tag{6.1.13}$$

Another approximate steepest descent evaluation, for large N', of $e^{W_N(\beta)}$ (Eq. (6.1.10)) involves contour deformation into the lower half of the complex λ-plane, passing through the sole saddle point at $\lambda = -i\alpha$, given by

$$\overline{N} = -\frac{\partial W(\alpha, \beta)}{\partial \alpha} = \frac{\sum_{N'} N' e^{-\alpha N'} \int dE' e^{-\beta E'} \rho(N', E')}{\sum_{N'} e^{-\alpha N'} \int dE' e^{-\beta E'} \rho(N', E')}. \tag{6.1.14}$$

(The last equality is readily verified using Eq. (6.1.8).) The result is

$$e^{W_N(\beta)} = \frac{\exp\left(\alpha N' + W(\alpha, \beta)\right)}{\sqrt{2\pi \frac{\partial^2 W(\alpha,\beta)}{\partial \alpha^2}}}, \tag{6.1.15}$$

and thus

$$\rho(N',E') = \frac{e^{\beta E'+\alpha N'+W(\alpha,\beta)}}{\sqrt{4\pi^2 \frac{\partial^2 W}{\partial\alpha^2}\frac{\partial^2 W_N}{\partial\beta^2}}}, \tag{6.1.16}$$

which can be rewritten for large N' as (neglecting terms of order $\ln N \ll N$):

$$\rho(N',E') \cong \frac{\exp\left(\alpha N' + \beta E' + W(\alpha,\beta)\right)}{2\pi\sqrt{\frac{\partial^2 W}{\partial\alpha^2}\frac{\partial^2 W}{\partial\beta^2} - \left(\frac{\partial^2 W}{\partial\alpha\partial\beta}\right)^2}}. \tag{6.1.17}$$

Considering $\langle\mathbf{X}\rangle^{i\lambda,i\tau}$ of Eq. (6.1.7), we have

$$\langle\mathbf{X}\rangle^{i\lambda,i\tau}\, e^{W(i\lambda,i\tau)} = \sum_{N',E'} e^{-i\lambda N'-i\tau E'} \sum_{\xi'} \left\langle N',E',\xi'\,|\mathbf{X}|\,N',E',\xi'\right\rangle, \tag{6.1.18}$$

or, using Eqns. (6.1.1) and (6.1.2), which define $\langle\mathbf{X}\rangle^{N',E'}$,

$$\langle\mathbf{X}\rangle^{i\lambda,i\tau}\, e^{W(i\lambda,i\tau)} = \sum_{N'} e^{-i\lambda N'} \int dE' e^{-i\tau E'} \rho(N',E')\,\langle\mathbf{X}\rangle^{N',E'}. \tag{6.1.19}$$

Inverting the Fourier series and integral as above, we have

$$\rho(N',E')\,\langle\mathbf{X}\rangle^{N',E'} = \int_{-\pi}^{\pi}\frac{d\lambda}{2\pi}\int_{-\infty}^{\infty}\frac{d\tau}{2\pi} e^{iN'\lambda+iE'\tau+W(i\lambda,i\tau)}\,\langle\mathbf{X}\rangle^{i\lambda,i\tau}. \tag{6.1.20}$$

The same kind of saddle-point integrations may be employed here as those above, provided that $\langle\mathbf{X}\rangle^{i\lambda,i\tau}$ is slowly varying, such that it suffices to evaluate it at the saddle points $\tau = -i\beta$, $\lambda = -i\alpha$ and extract it from both integrals on the right of Eq. (6.1.20). The remaining integrals just represent $\rho(N',E')$ (Eq. (6.1.9)), which cancels with an identical factor on the left, leaving the approximate result

$$\langle\mathbf{X}\rangle^{N',E'} \approx \langle\mathbf{X}\rangle^{\alpha,\beta} \tag{6.1.21}$$

for large N', E' (provided $\langle\mathbf{X}\rangle^{i\lambda,i\tau}$ varies slowly as a function of λ and τ).

With the neglect of terms of order $\ln N \ll N$ for large N, E, the result of Eq. (6.1.17) for $\rho(N',E')$ may be rewritten approximately as (dropping primes):

$$\ln\rho(N,E) = \alpha N + \beta E + W(\alpha,\beta). \tag{6.1.22}$$

Identifying the *entropy* S according to Eq. (6.1.3), we have

$$\frac{S}{k} = \ln\rho(N,E) = \alpha N + \beta E + W(\alpha,\beta). \tag{6.1.23}$$

Moreover, according to the fundamental laws of thermodynamics,

$$dE = TdS - PdV + \mu dN, \tag{6.1.24}$$

where P is pressure, V is volume, T is Kelvin temperature and μ is the *chemical potential*, defined as the energy needed to add a particle at constant entropy and volume (i.e., $dV = 0$, $dS = 0$),

$$\mu = \left(\frac{\partial E}{\partial N}\right)_{S,V}. \tag{6.1.25}$$

Eq. (6.1.24) also yields

$$P = -\left(\frac{\partial E}{\partial V}\right)_{S,N}. \tag{6.1.26}$$

Considering changes of entropy (Eq. (6.1.23)) through extensive variables,

$$\frac{dS}{k} = \alpha dN + \beta dE + dW, \tag{6.1.27}$$

identity with Eq. (6.1.24) requires that

$$P = kT\left(\frac{\partial W}{\partial V}\right)_{S,N}, \tag{6.1.28}$$

with physical identifications of the saddle points as

$$\beta = \frac{1}{kT} \quad \text{and} \quad \alpha = \frac{-\mu}{kT} = -\mu\beta. \tag{6.1.29}$$

Moreover, we note that Eq. (6.1.23) implies that

$$E = -(\partial W/\partial \beta)_{S,N}. \tag{6.1.30}$$

Since pressure is an intensive variable, integration of Eq. (6.1.28) with respect to volume at constant entropy and number yields

$$W(\alpha, \beta) = V\beta P(\alpha, \beta). \tag{6.1.31}$$

It should be noted that the zero-temperature limit corresponds to the ground state.

Other important thermodynamic variables include the Helmholtz Free Energy, F, defined as[8,9]

$$F = E - TS. \tag{6.1.32}$$

Employing Eq. (6.1.23), we can also write

$$F = -\frac{\alpha}{\beta}N - kTW(\alpha, \beta) \tag{6.1.33}$$

or, using Eq. (6.1.6) and Eq. (6.1.29),

$$F = \mu N - kTlnZ; \quad F - \mu N \equiv \Omega = -kT\ln Z. \tag{6.1.34}$$

In the last relation on the right, we have defined the symbol Ω, termed the *thermodynamic potential* by Landau.

For an infinitesimal reversible process

$$dF = dE - TdS - SdT$$

and using Eq. (6.1.24), we have

$$TdS = dE + PdV - \mu dN, \tag{6.1.35}$$

so that

$$dF = -PdV - SdT + \mu dN. \tag{6.1.36}$$

Thus, for a reversible, isothermal ($dT = 0$) process conserving number ($dN = 0$), we have

$$dF = -PdV, \tag{6.1.37}$$

in which case the change of the Helmholtz Free Energy is given by

$$F_f - F_i = -\int_i^f PdV, \tag{6.1.38}$$

which amounts to the work done on the system during the process.

On the other hand, for a reversible isothermal ($dT = 0$) and isovolumic ($dV = 0$) process,

$$dF = \mu dN, \tag{6.1.39}$$

which indicates that the change of the Helmholtz Free Energy for this process arises solely from the change of number, and if the number is also conserved in this case, then

$$dF = 0 \quad \text{and} \quad F = constant. \tag{6.1.40}$$

In terms of the Helmholtz Free Energy, we may also write the entropy as

$$S = -(\partial F/\partial T)_{V,N} \tag{6.1.41}$$

for constant volume and number, whereas the pressure P is given by

$$P = -(\partial F/\partial V)_{T,N} \tag{6.1.42}$$

for constant temperature and number.

Yet another important thermodynamic variable is the Gibbs Free Energy denoted by the symbol G and defined as

$$G = E - TS + PV = F + PV, \tag{6.1.43}$$

which may be written in terms of the Enthalpy W (also known as the "heat constant") of the body defined as

$$W = E + PV, \tag{6.1.44}$$

so that

$$G = W - TS. \tag{6.1.45}$$

The relations of the various variables defined above in describing thermodynamic states and their variations are discussed in some detail in References 8 and 9. As the discussion of thermodynamics in this section deals extensively with energy and heat transfer, it is obvious that the results provided here can be used to determine specific heats at constant volume and at constant pressure, as well as their relation. This is also well discussed in References 8 and 9. However, the subject of diamagnetic susceptibility is often omitted in modern books on aspects of advanced quantum mechanics, although it is an important physical phenomenon that is strictly of quantum origin. This subject will be treated in Sections 6.3 and 6.4.

6.2 Boson and Fermion Statistics for Non-Interacting Systems: Bose Condensation

Considering non-interacting systems, $H^{(2)} = 0$, we write the Hamiltonian $H^{(1)}$ and number operator N in an energy representation using Eq. (2.2.6) with both a' and b' referring to the spectrum of the one-particle energy eigenstate E_γ, so that $\langle E_{\gamma'} | h^{(1)} | E_\gamma \rangle = E_\gamma \delta_{\gamma,\gamma'}$, whence

$$H^{(1)} = \sum_{E_\gamma} E_\gamma \psi^+(E_\gamma)\psi(E_\gamma) = \sum_{E_\gamma} E_\gamma N(E_\gamma) \tag{6.2.1}$$

and

$$N = \sum_{E_\gamma} \psi^+(E_\gamma)\psi(E_\gamma) = \sum_{E_\gamma} N(E_\gamma), \tag{6.2.2}$$

$N(E_\gamma)$ being the number operator for particles having energy E_γ (Eq. (2.2.33)). The *grand partition function*, Z, is given by Eq. (6.1.6) as ($\alpha = -\mu\beta$):

$$Z = e^{W(\alpha,\beta)} = Tr\left(e^{-\beta(H^{(1)}-\mu N)}\right) = Tr\left(e^{-\beta\sum_{E_\gamma}(E_\gamma-\mu)N(E_\gamma)}\right). \tag{6.2.3}$$

In the absence of interaction, the number subspaces for different E_γ-values are mutually independent, and denoting the sum over all possible occupation numbers $N(E_\gamma) \to n$ for a given total energy by Tr_γ,we have

$$e^{W(\alpha,\beta)} = Tr\left(\prod_{E_\gamma} e^{-\beta(E_\gamma-\mu)N(E_\gamma)}\right) = \prod_{E_\gamma} Tr_\gamma\left(e^{-\beta(E_\gamma-\mu)N(E_\gamma)}\right)$$

$$= \prod_{E_\gamma}\sum_n e^{-\beta(E_\gamma-\mu)n}, \tag{6.2.4}$$

where n is the integer occupation number eigenvalue of the number operator $N(E_\gamma)$ for the particle energy eigenstate E_γ. (The ordinary non-"grand" partition function $\tilde{Z}$ is defined by the counterpart of Eq. (6.2.4) obtained by setting $\mu \to 0$. Similar remarks apply below.)

In the case of Bosons, $n = 0, 1, 2, \cdots \infty$, and the n-sum is just a geometric series, yielding the grand partition function (denoted by Z_B) as

$$Z_B = e^{W(\alpha,\beta)} = \prod_{E_\gamma} \frac{1}{1 - e^{-\beta(E_\gamma-\mu)}}. \tag{6.2.5}$$

For Fermions, $n = 0, 1$ are the only possible values, so the grand partition function Z_F is given by

$$Z_F = e^{W(\alpha,\beta)} = \prod_{E_\gamma}\left(1 + e^{-\beta(E_\gamma-\mu)}\right). \tag{6.2.6}$$

Bearing in mind that the average number $\overline{N}$ may be obtained as $-\frac{\partial W(\alpha,\beta)}{\partial\alpha}$, we determine the grand potential from Eq. (6.2.4) as

$$W = \ln\prod_{E_\gamma} Tr_\gamma\left(e^{-\beta(E_\gamma-\mu)N(E_\gamma)}\right) = \sum_{E_\gamma}\ln Tr_\gamma\left(e^{-\beta(E_\gamma-\mu)N(E_\gamma)}\right) \tag{6.2.7}$$

and differentiate with respect to $\alpha = -\mu\beta$, obtaining

$$\overline{N} = \sum_{E_\gamma}\frac{Tr_\gamma\left(N(E_\gamma)e^{-\beta(E_\gamma-\mu)N(E_\gamma)}\right)}{Tr_\gamma\left(e^{-\beta(E_\gamma-\mu)N(E_\gamma)}\right)} = \sum_{E_\gamma} f(E_\gamma), \tag{6.2.8}$$

where we write the average population of state E_γ as $f(E_\gamma)$. On the other hand, using Eqns. (6.2.5) and (6.2.6), we find

$$\text{Bosons: } \overline{N} = \sum_{E_\gamma}\frac{1}{e^{\beta(E_\gamma-\mu)} - 1}, \tag{6.2.9}$$

$$\text{Fermions: } \overline{N} = \sum_{E_\gamma}\frac{1}{e^{\beta(E_\gamma-\mu)} + 1}. \tag{6.2.10}$$

From comparison with Eq. (6.2.8), we infer the Bose–Einstein and Fermi–Dirac equilibrium statistical distribution functions for Bosons/Fermions in the state E_γ:

$$\text{Bosons: } f(E_\gamma) = \frac{1}{e^{\beta(E_\gamma-\mu)} - 1}, \tag{6.2.11}$$

$$\text{Fermions: } f(E_\gamma) = \frac{1}{e^{\beta(E_\gamma-\mu)} + 1}. \tag{6.2.12}$$

It is readily verified from Eqns. (6.1.30) and (6.2.5 and 6.2.6) that the average energy $E = -(\partial W/\partial\beta)_{S,N}$ is given in both the Boson and Fermion cases by

$$E = \sum_{E_\gamma} E_\gamma f(E_\gamma), \tag{6.2.13}$$

and the specific heat at constant volume, C_v, can be obtained by further differentiation of E with respect to temperature at constant number. Detailed results are available in standard textbooks on quantum statistical mechanics.[2–6] Moreover, the non-degenerate limit is characterized by dominance of the exponential in the denominators of Eq. (6.2.11 and 6.2.12), so that $f(E_\gamma) = e^{-\beta(E_\gamma-\mu)}$ for both Fermions and Bosons in this case.

There is an interesting and very important singular behavior of the Boson statistical distribution at low temperature, due to the fact that its denominator can vanish. Considering an infinite free gas of Bosons having a macroscopic number of particles and single-particle energy eigenvalues $E_\gamma \to q^2/2m$, the lowest energy level is $E_0 = 0$. At zero temperature (in the absence of thermal excitation) the particles populate this lowest state with no restriction regarding number in the case of Bosons. Since there is no energy required at any stage of population for yet another particle to be added to the highly degenerate microscopic energy level $E_0 = 0$, the chemical potential μ is also zero and the denominator of the Boson distribution function of Eq. (6.2.11) vanishes for $E_0 = 0$. The associated divergence reflects the fact that the microscopic level $E_0 = 0$ is accommodating a macroscopic occupation of $N \sim V \to \infty$ particles. This phenomenon is known as a *Bose condensation* and it plays a major role in superfluidity and superconductivity, which will be discussed in Chapter 13. Notwithstanding the continuous nature of the energy spectrum ($E_\gamma \to q^2/2m$) for an infinite gas of free Bosons, the microscopic energy state E_0 requires separate treatment in consequence of its macroscopic population. For finite, low temperature, and/or in the presence of particle–particle interactions, there will be some excitations to populate higher energy levels $E_\gamma \neq 0$, but the Bose-condensed ground state $E_0 = 0$ still retains a macroscopic population and requires separate treatment from the rest of the spectrum. Qualitatively similar remarks apply to single-particle energy spectra for Bosons that are not continuous and may have a nonzero lowest level $E_0 \neq 0$. A Bose condensation can occur at low temperature in this case as well, with the chemical potential $\mu = E_0$ as the energy needed to add one more particle to the macroscopically populated ground state and the concomitant Bose distribution is divergent.

It is important to note that the fluctuation of number in the Bose-condensed state is of the same order as the macroscopic number, n, itself: We have

$$\begin{aligned}\left\langle [N-\langle N\rangle]^2\right\rangle &= \left\langle N^2\right\rangle - \langle N\rangle^2 \\ &= \frac{1}{\beta}\frac{\partial}{\partial\mu}\left(\frac{Tr(N\exp[-\beta(H-\mu N)])}{Tr\exp[-\beta(H-\mu N)]}\right) \\ &= \frac{1}{\beta}\frac{\partial\langle N\rangle}{\partial\mu} \cong \frac{1}{\beta}\frac{\partial n}{\partial\mu},\end{aligned} \tag{6.2.14}$$

since the macroscopic number, n, in the Bose-condensed state is the major contribution to $\langle N\rangle$. Considering the infinite free gas of Bosons with lowest energy level $E_0 = 0$, we need to evaluate $\frac{\partial n}{\partial\mu}$. In order to determine the μ-dependence for this zero-energy state in the appropriate limit, $\mu \to 0$, we approximate the Bose–Einstein distribution as $f(E_0 = 0) = \frac{1}{e^{-\beta\mu}-1} \cong -\frac{1}{\beta\mu}$, so that $\frac{\partial f(E_0=0)}{\partial\mu} = \frac{1}{\beta\mu^2}$ and correspondingly, we obtain

$$\left\langle [N-\langle N\rangle]^2\right\rangle^{1/2} = n, \tag{6.2.15}$$

which confirms that the fluctuation of number in the condensed state is macroscopic, as is its population.

6.3 Helmholtz Free Energy and Magnetic Moment

Considering further the Helmholtz Free Energy given by Eq. (6.1.34) as (on a per-unit-volume basis):

$$F = \mu n - kT\ln Z, \tag{6.3.1}$$

we recall Eq. (6.1.14) and the fact that $W = \ln Z$, from which it is readily seen that

$$\frac{\partial F}{\partial\mu} = 0, \tag{6.3.2}$$

which is useful to determine the relation between average particle number of the system and its chemical potential.

Focusing attention on the case of Fermions ($Z \to Z_F$), Eq. (6.2.6) yields

$$\begin{aligned}F &= \mu n - kT\ln\prod_{E_\gamma}(1+e^{-\beta(E_\gamma-\mu)}) \\ &= \mu n - kT\sum_{E_\gamma}\ln(1+e^{-\beta(E_\gamma-\mu)}).\end{aligned} \tag{6.3.3}$$

(A similar result obtains for Bosons, with $Z \to Z_B$ of Eq. (6.2.5.) In the special case of spin degeneracy, we may write Eq. (6.3.3) alternatively as

$$F = \mu n - 2kT\sum_{E_\gamma}\ln(1+e^{-\beta E_\gamma}), \tag{6.3.4}$$

in which we suppress explicit reference to the uniform shift μ of the non-interacting-single-particle energy origin, corresponding to $Z \to \tilde{Z}$, the ordinary partition function (this does not materially affect the magnetic susceptibility results to be discussed below).

The magnetic moment (per unit volume), $\mathcal{M}$, induced by an applied magnetic field, B_0, is defined as (H is the Hamiltonian density):

$$\mathcal{M} = -\left\langle \frac{\partial H}{\partial B_0} \right\rangle \tag{6.3.5}$$

(not to be confused with $-\partial \langle H \rangle / \partial B_0$). Finally, the magnetic moment per unit volume may be written in terms of the Helmholtz Free Energy as[10,11]

$$\mathcal{M} = -\left(\frac{\partial F}{\partial B_0} \right)_T . \tag{6.3.6}$$

Various differing definitions of magnetic susceptibility appear in the literature as follows:[11]

$$\chi_{iso.} = \left(\frac{\partial \mathcal{M}}{\partial B_0} \right)_T = -\left(\frac{\partial^2 F}{\partial B_0^2} \right)_{T,N} \quad (Isothermal)^{11}, \tag{6.3.7}$$

$$\chi_{adi.} = \left(\frac{\partial \mathcal{M}}{\partial B_0} \right)_{S,N} = -\left\langle \frac{\partial^2 H}{\partial B_0^2} \right\rangle_{S,N} \quad (Adiabatic)^{12} \tag{6.3.8}$$

and A. H. Wilson[13] has employed yet another variant,

$$\chi = \frac{\mathcal{M}}{B_0}. \tag{6.3.9}$$

It is clear that the determination of magnetic moment (and susceptibility) devolves upon evaluation of the Helmholtz Free Energy. This can be approached using Eq. (6.2.3) for Z with Eq. (6.3.1) (or Eq. (6.3.4) for Fermions), which requires knowledge of the entire energy spectrum. However, if that is not immediately available, it is advantageous to formulate an equivalent and useful differential characterization of the quantum partition function, $\tilde{Z}$, for use in Eq. (6.3.1) to determine the Helmholtz Free Energy, which, in turn, yields the magnetization given by Eq. (6.3.6). The differential characterization involved centers about the fact that

$$-\frac{\partial}{\partial \beta} exp(-\beta H) = H exp(-\beta H), \tag{6.3.10}$$

which is intimately related to the quantum partition function through the quantity $\tilde{\Gamma}(\mathbf{x}, \mathbf{x}'; \beta)$ defined by ($| \, 0\rangle$ denotes the ground state):

$$\tilde{\Gamma}(\mathbf{x}, \mathbf{x}'; \beta) = \langle 0 \mid \psi(\mathbf{x}', t) exp(-\beta H) \psi^{\dagger}(\mathbf{x}, t) \mid 0 \rangle, \tag{6.3.11}$$

so that

$$\frac{\partial \tilde{\Gamma}(\mathbf{x},\mathbf{x}';\beta)}{\partial\beta} = -\langle 0 \mid \psi(\mathbf{x}',t) H exp(-\beta H)\psi^{\dagger}(\mathbf{x},t) \mid 0\rangle$$
$$= -\langle 0 \mid \psi(\mathbf{x}',0) H exp(-\beta H)\psi^{\dagger}(\mathbf{x},0) \mid 0\rangle \tag{6.3.12}$$

(which has *no* time dependence for a time-independent Hamiltonian since the two times in the creation/annihilation operators are the same). Installing a complete set of single-particle energy eigenstates ($H \mid E_\gamma\rangle = E_\gamma \mid E_\gamma\rangle$) on the right of Eq. (6.3.12) and recalling Eq. (5.1.18) we have (recall that we have excluded interactions starting in Section 6.2):

$$\frac{\partial \tilde{\Gamma}(\mathbf{x},\mathbf{x}';\beta)}{\partial\beta} = -h_{\mathbf{x}}^{(1)} \sum_{\gamma} \phi_\gamma(\mathbf{x}') exp(-\beta E_\gamma)\phi_\gamma^*(\mathbf{x}), \tag{6.3.13}$$

where $h_{\mathbf{x}}^{(1)}$ and $\phi_\gamma(\mathbf{x})$ represent the single-particle Hamiltonian and its energy eigenfunctions in position representation. Bearing in mind Eq. (6.3.11), we have

$$\frac{\partial \tilde{\Gamma}(\mathbf{x},\mathbf{x}';\beta)}{\partial\beta} = -h_{\mathbf{x}}^{(1)} \tilde{\Gamma}(\mathbf{x},\mathbf{x}';\beta). \tag{6.3.14}$$

Furthermore, in the limit $\beta \to 0$,

$$\tilde{\Gamma}(\mathbf{x},\mathbf{x}';\beta \to 0) = \langle 0 \mid \psi(\mathbf{x}',0)\psi^{\dagger}(\mathbf{x},0) \mid 0\rangle = \delta^{(3)}(\mathbf{x}-\mathbf{x}') \tag{6.3.15}$$

by virtue of the canonical equal-time commutation relations. It is important to note that Eq. (6.3.14) is just the Schrödinger equation with replacement of the time variable by $[t-t' \to -i\beta]$ ($\hbar \to 1$). Moreover, Eq. (6.3.15) means that the differential equation Eq. (6.3.14) is accompanied by the "initial" condition at $\beta = 0$ that is characteristic of retarded Green's functions at initial time. (It should be noted that we are using the term "retarded" here loosely since $\tilde{\Gamma}$ has the internal ψ, ψ^+ structure of a retarded Green's function, but it lacks the time cut-off $\eta_+(t-t')$-function at $t < t'$ due to the absence of time ordering. Correspondingly, it is continued into $t < t'$ as well as $t > t'$ with *no* abrupt discontinuity, which affects its frequency transform.) In view of this we have

$$\tilde{\Gamma}(\mathbf{x},\mathbf{x}';\beta) = iG_{1>}^{ret}(\mathbf{x},\mathbf{x}';[t-t' \to -i\beta]), \tag{6.3.16}$$

which permits us to determine $\tilde{\Gamma}(\mathbf{x},\mathbf{x}';\beta)$ using $G_{1>}^{ret}$ evaluations such as those of Chapter 5. (In this matter, we must also bear in mind that the $G_{1>}^{ret}$-retarded Green's function employed here is averaged in the vacuum state, and does *not* carry a Bose/Fermi distribution factor imposed by the grand canonical ensemble averaging to be introduced in the next chapter, albeit related.) Finally, the result for the quantum partition function may then be obtained from the recognition that with $\mathbf{x} = \mathbf{x}'$, we have

$$\int d^3\mathbf{x}\, \tilde{\Gamma}(\mathbf{x},\mathbf{x};\beta) = \sum_{E_\gamma} \int d^3\mathbf{x} \phi_\gamma(\mathbf{x}) exp(-\beta E_\gamma)\phi_\gamma^*(\mathbf{x})$$
$$= \sum_{E_\gamma} e^{-\beta E_\gamma} \int d^3\mathbf{x} \mid \phi_\gamma(x) \mid^2 = \sum_{E_\gamma} e^{-\beta E_\gamma} = Tr e^{-\beta H} = \tilde{Z}, \tag{6.3.17}$$

since the eigenfunctions are normalized to unity.

The Helmholtz Free Energy in the Fermion case, given by Eq. (6.3.3), involves the function $B(E)$ defined as

$$B(E) = -kT\ln(1 + e^{-\beta E}) = \int_C \frac{ds}{2\pi i} e^{sE} p(s), \tag{6.3.18}$$

which we have written as an inverse Laplace transform on the right-hand side of Eq. (6.3.18) (C denotes the inverse Laplace transform contour, from $-i\infty + \delta$ to $+i\infty + \delta$). Of course, $p(s)$ is the Laplace transform of $B(E)$,

$$p(s) = \int_0^\infty dE e^{-sE} B(E), \tag{6.3.19}$$

and, furthermore, using Eq. (6.3.3) with Eq. (6.3.18), we have

$$F - \mu n = \int_C \frac{ds}{2\pi i}\left(\sum_{E_\gamma} e^{sE_\gamma}\right) p(s) = \int_C \frac{ds}{2\pi i} \tilde{Z}(\beta \to -s) p(s), \tag{6.3.20}$$

where the β-dependence of $\tilde{Z}$ is replaced by $\beta \to -s$. This formulation addresses the inconvenience of the explicit appearance of the ln-function in $B(E)$ by rewriting Eq. (6.3.20) as

$$\begin{aligned} F - \mu n = \int_C \frac{ds}{2\pi i}\left(\frac{\tilde{Z}(\beta \to -s)}{s^2}\right)(s^2 p(s)) &= \int_C \frac{ds}{2\pi i}\frac{\tilde{Z}(\beta \to -s)}{s^2}\int_0^\infty dE\left(\frac{\partial^2 e^{-sE}}{\partial E^2}\right)B(E), \\ &= \int_C \frac{ds'}{2\pi i}\frac{\tilde{Z}(\beta \to s')}{s'^2}\int_0^\infty dE e^{s'E}\frac{\partial^2}{\partial E^2}B(E), \end{aligned} \tag{6.3.21}$$

where we have integrated by parts twice and put $s' = -s$. Since $\frac{\partial B(E)}{\partial E}$ supplants the ln-function of Eq. (6.3.18) by the Fermi distribution function $f_0(E)$, we obtain

$$F - \mu n = \int_C \frac{ds'}{2\pi i}\frac{\tilde{Z}(\beta \to s')}{s'^2}\int_0^\infty dE\, e^{s'E}\frac{\partial f_0(E)}{\partial E}, \tag{6.3.22}$$

and restore the explicit appearance of μ below. This result is particularly convenient at zero temperature where $\frac{\partial f_0(E)}{\partial E} = -\delta(E-\mu)$, but it is also useful at finite temperature, with

$$\frac{\partial f_0(E)}{\partial E} = -\frac{\beta}{4} sech^2\left([E-\mu]\frac{\beta}{2}\right). \tag{6.3.23}$$

Finally, introducing $\tilde{Z}$ from Eqns. (6.3.17) and (6.3.16), we have a fairly general result,

$$F - \mu n = i\int_C \frac{ds'}{2\pi i}\frac{1}{s'^2}\left(\int d^3\mathbf{x} G_{1>}^{ret}(\mathbf{x}, \mathbf{x}, [t-t' \to -is'])\right)\int_0^\infty dE\, e^{s'E}\frac{\partial f_0(E)}{\partial E} \tag{6.3.24}$$

and in the case of spatial translational invariance,

$$F-\mu n=-\frac{i\beta}{4}\int_C \frac{ds'}{2\pi i}\frac{1}{s'^2}G^{ret}_{1>}(0,0;[t-t'\to -is'])\int_0^\infty dEe^{s'E}sech^2\left(\frac{[E-\mu]\beta}{2}\right), \quad (6.3.25)$$

for a non-interacting Fermi system, on a per-unit-volume basis ($Vol.\to 1$).

6.4 Diamagnetic Susceptibility and the de Haas–van Alphen Effect

As indicated at the end of Section 6.1, diamagnetism is strictly of quantum mechanical origin. That there is no *classical* diamagnetism for an electron gas may be surmised from the fact that the Lorentz force cannot do work, and therefore does not participate in the thermal energetics discussed previously. More rigorously, the classical Free Energy per unit volume is given by[14,15]

$$F_{classical}=-nKT\,ln\left\{\int d^3\mathbf{p}\int d^3\mathbf{x}\,exp\left(-\frac{1}{2mkT}\left[\mathbf{p}-\frac{Q}{c}\mathbf{A}\right]^2\right)\right\}, \quad (6.4.1)$$

with $\mathbf{A}$ as the vector potential of the magnetic field and n is the number density. As the classical spectrum is continuous and the $\mathbf{p}$-integration range is $[-\infty<\mathbf{p}<\infty]$, the introduction of the variable

$$\mathbf{P}=\mathbf{p}-\frac{Q}{c}\mathbf{A} \quad (6.4.2)$$

completely eliminates the appearance of the magnetic field from the classical Free Energy and from its derivatives with respect to the magnetic field, so that the magnetic moment and diamagnetic susceptibility vanish. This is the Bohr-van Leeuwen theorem.

The quantization of orbits by a magnetic field reorganizes the energy spectrum, as we have seen in Chapter 5, so the formerly classical continuous spectrum now involves discrete states for motion in the plane perpendicular to the magnetic field and Eqns. (6.4.1 and 6.4.2) are no longer applicable. With this failure of the Bohr-van Leeuwen theorem in a quantizing magnetic field, it is necessary to re-examine the Helmholtz Free Energy on the basis of Eq. (6.3.25). The retarded Green's function involved for a non-interacting spinless ($\bar{\mu}_0\to 0$) electron gas in a constant, uniform quantizing magnetic field was discussed in Chapter 5, Eq. (5.6.25), and adjusting it to have $\mathbf{x}=\mathbf{x}'$, ($\mathbf{R}=0$) and $t-t'\to -is'$, we obtain

$$G^{ret}_{1>}(0,0;[t-t'\to -is'])=\frac{-i}{2}\frac{m^{\frac{3}{2}}\omega_c}{(2\pi)^{\frac{3}{2}}}\frac{1}{s'^{\frac{1}{2}}sinh(\frac{\omega_c s'}{2})}. \quad (6.4.3)$$

Therefore (per unit volume; restore the explicit appearance of μ; multiply by 2 for spin degeneracy):

$$F - \mu n = -\frac{1}{4}\frac{m^{\frac{3}{2}}\omega_c}{(2\pi)^{\frac{3}{2}}}\beta \int_C \frac{ds'}{2\pi i}\frac{1}{s'^{\frac{5}{2}}}\frac{1}{sinh(\frac{\omega_c s'}{2})}\int_0^{\infty} dE e^{s'E} sech^2\left(\frac{[E-\mu]\beta}{2}\right). \tag{6.4.4}$$

In the degenerate limit of zero temperature, we have $\left(\frac{\partial f_0(E)}{\partial E} = -\delta(E-\mu)\right)$:

$$F - \mu n = -\frac{m^{\frac{3}{2}}\omega_c}{(2\pi)^{\frac{3}{2}}}\int_C \frac{ds'}{2\pi i}\frac{e^{\mu s'}}{s'^{\frac{5}{2}} sinh(\frac{\omega_c s'}{2})}. \tag{6.4.5}$$

The inverse Laplace transform on the right of Eq. (6.4.5) can be evaluated in two ways to exhibit the magnetic field dependence of the Helmholtz Free Energy. For low-intermediate magnetic field strengths ($\hbar\omega_c \ll \mu$), we note that the s'-integrand has isolated poles at $s'_n = \pm\frac{2\pi n i}{\hbar\omega_c}$ ($n = 1, 2\ldots$) and a branch point at $s'_0 = 0$, with a branch cut extending to $-\infty$ on the negative real s'-axis. Deforming the s'-contour of integration to surround these singularities, we employ the identity:[16]

$$\frac{1}{sinh\, x} = \sum_{n=-\infty}^{\infty}\frac{(-1)^n}{x - n\pi i}, \tag{6.4.6}$$

which yields oscillatory magnetic field terms as (except at $s' = 0; n \neq 0$):

$$(F-\mu n)_{osc} = -2\left(\frac{m}{2\pi}\right)^{\frac{3}{2}}\sum_{\substack{n=-\infty\\ n\neq 0}}^{\infty}(-1)^n \oint_{C_n}\frac{ds'}{2\pi i}\frac{e^{\mu s'}}{s'^{\frac{5}{2}}}\frac{1}{s' - \frac{2\pi n i}{\omega_c}}$$

$$= -\frac{4m^{\frac{3}{2}}\omega_c^{\frac{5}{2}}}{(2\pi)^4}\sum_{n>0}\frac{(-1)^n}{n^{\frac{5}{2}}}cos\left(\frac{2\pi n\mu}{\omega_c} - \frac{5\pi}{4}\right). \tag{6.4.7}$$

Furthermore, there is also a contribution from the branch point and branch cut that is monotonic in magnetic field strength, given by

$$(F-\mu n)_{mon} = -\frac{m^{\frac{3}{2}}\omega_c}{(2\pi)^{\frac{3}{2}}}\int_{C_0}\frac{ds'}{2\pi i}\frac{e^{\mu s'}}{s'^{\frac{5}{2}} sinh(\frac{\omega_c s'}{2})}, \tag{6.4.8}$$

where C_0 is the Hankel contour, which encircles the entire branch cut on the negative real s'-axis including the branch point at $s' = 0$. This integrand requires careful analysis because its behavior near the s'-origin is singular, but it can be seen to yield a convergent integral by separating the first two terms of the power series expansion of the inverse *sinh*-function, as follows:

$$(F = \mu n)_{mon} = -\frac{m^{\frac{3}{2}}\omega_c}{(2\pi)^{\frac{3}{2}}}\int_{C_0}\frac{ds'}{2\pi i}\frac{e^{\mu s'}}{s'^{\frac{5}{2}}}\left(\frac{2}{\omega_c s'} - \frac{\omega_c s'}{12}\right)$$
$$-\frac{m^{\frac{3}{2}}\omega_c}{(2\pi)^{\frac{3}{2}}}\int_{C_0}\frac{ds'}{2\pi i}\frac{e^{\mu s'}}{s'^{\frac{5}{2}}}\left(\frac{1}{sinh(\frac{\omega_c s'}{2})} - \frac{2}{\omega_c s'} + \frac{\omega_c s'}{12}\right). \qquad (6.4.9)$$

The first term on the right yields gamma functions[17] and the second term is a convergent branch cut integral,

$$(F - \mu n)_{mon} = -\frac{4\sqrt{2}m^{\frac{3}{2}}\mu^{\frac{5}{2}}}{15\pi^2} + \frac{m^{\frac{3}{2}}\mu^{\frac{1}{2}}\omega_c^2}{12\sqrt{2}\pi^2}$$
$$+\left(\frac{m}{2\pi}\right)^{\frac{3}{2}}\frac{\omega_c}{2\pi^2}\int_0^{\infty}d\rho\frac{e^{-\mu\rho}}{\rho^{\frac{5}{2}}}\left(\frac{1}{sinh(\frac{\omega_c\rho}{2})} - \frac{2}{\omega_c\rho} + \frac{\omega_c\rho}{12}\right), \qquad (6.4.10)$$

which is monotonic as a function of ω_c.

A second method for evaluation of the inverse Laplace transform, Eq. (6.4.5), in high magnetic fields ($\hbar\omega_c \sim \mu$) is based on the identity

$$\frac{1}{sinh(x)} = 2\sum_{n=0}^{\infty}e^{-(2n+1)x}, \qquad (6.4.11)$$

which yields

$$\int_C\frac{ds'}{2\pi i}\frac{e^{\mu s'}}{s'^{\frac{5}{2}}\,sinh(\frac{\omega_c s'}{2})} = 2\sum_{n=0}^{\infty}\int_C\frac{ds'}{2\pi i}\frac{e^{s'(\mu-[n+1/2]\omega_c)}}{s'^{\frac{5}{2}}}$$
$$= \frac{2}{\Gamma(5/2)}\sum_{n=0}^{\infty}\eta_+(\mu - [n+1/2]\omega_c)\,[\mu - (n+1/2)\omega_c]^{\frac{3}{2}}. \qquad (6.4.12)$$

Accordingly ($\eta_+(x)$ is the Heaviside unit step function: $\eta_+(x) = 1$ or 0 for $x > 0$ or $x < 0$):

$$F - \mu n = -\sqrt{8}\frac{m^{\frac{3}{2}}\omega_c}{3\pi^2}\sum_{n=0}^{\infty}\eta_+(\mu - [n+1/2]\omega_c)\,[\mu - (n+1/2)\omega_c]^{\frac{3}{2}}, \qquad (6.4.13)$$

which clearly exhibits the effect of higher indexed Landau levels, contributing when $[n+1/2]\omega_c$ is lowered below the Fermi energy as ω_c is reduced.

In summary, the effects of low/intermediate magnetic field quantization of orbits on the Helmholtz Free Energy are given by the sum of $(F-\mu n)_{mon}$ (Eq. 6.4.10 and $(F-\mu n)_{osc}$ (Eq. 6.4.7), which explicitly exhibits oscillatory magnetic field behavior, known as de Haas–van Alphen oscillations in diamagnetic susceptibility measurements.[18] This oscillatory behavior is also in evidence at higher magnetic fields when Landau levels become able to contribute as they flick past the Fermi energy with decreasing magnetic field. On the basis of these results for the Helmholtz Free Energy, the magnetic moment and diamagnetic susceptibility may be readily evaluated using Eqns. 6.3.6–6.3.9. While we have restricted our considerations to the zero-temperature degenerate limit, the same methods can be employed at finite temperatures.[13] Other informative discussions of Landau diamagnetism and de Haas–van Alphen oscillations are provided in Reference 11, Sections 5.6–5.10 and in the book of Jones and March.[15] Similar techniques may be employed to determine the relation between density, n, and chemical potential, μ, in a quantizing magnetic field using Eqns. (6.3.1 and 6.3.2). In this regard, one must distinguish the cases in which the system is isolated (n constant) or in contact with a particle bath/reservoir (μ constant).[19]

6.5 Chapter 6 Problems

P.6.1. Consider particles subject to classical mechanical and classical (non-degenerate, Maxwell–Boltzmann) statistical laws. Prove:

(a) the virial theorem, and

(b) the equipartition theorem.

P.6.2. Consider the grand partition function $e^W = Tr\left(e^{-\beta(H^{(1)}-\mu N)}\right)$ for spinless electrons in a magnetic field subject to classical dynamical and statistical laws. Introducing the magnetic field into the Hamiltonian through its vector potential, with $\mathbf{p} \to \mathbf{p} - \frac{e}{c}\mathbf{A}(\mathbf{x})$ for momentum, and noting the absence of quantization, show that the grand partition function does not depend on the magnetic field. What effect does this have on the magnetic susceptibility?

P.6.3. Verify Eqns. (6.2.5) and (6.2.6) in detail.

P.6.4. Verify Eqns. (6.2.9–6.2.13) in detail.

P.6.5. Prove Eq. (6.2.15).

P.6.6. Considering Eq. (6.3.2), derive an explicit expression for density n in terms of magnetic field strength and chemical potential (Fermi energy) at zero temperature for a 3D non-interacting spinless electron gas in a quantizing magnetic field.

P.6.7. Following the procedures of Section 6.4, determine $F - \mu n$ for a *two*-dimensional non-interacting spinless electron gas in the degenerate limit of zero temperature (*no* magnetic field).

P.6.8. Determine the magnetic moment of the 2D electron gas subject to Landau quantization, but devoid of Zeeman spin-splitting, in two cases: (a) Constant density and (b) constant Fermi energy (held steady by contact with a particle bath/reservoir).

P.6.9. Consider the role of Zeeman spin-splitting: Determine $F - \mu n$ for both the 3D and 2D non-interacting electron gases in a quantizing magnetic field; also density as a function of field and Fermi energy for arbitrary $\overline{\mu}_0 \neq 0$ at zero temperature.

P.6.10. Using the results of Problem 9, derive the magnetic moments of the 3D and 2D non-interacting electron gases for the two cases: (a) Constant density and (b) constant Fermi energy in a particle bath/reservoir.

P.6.11. Show that the two totally integrated terms involved in Eq. (6.3.21) make no contributions in the final evaluation of $F-\mu n$.

P.6.12. In Section 6.1, apply the method of Darwin and Fowler to establish the indicated results.

6.6 References for Chapter 6

R.6.1. P.C. Martin and J. Schwinger, Phys. Rev. *115*, 1342 (1959).

R.6.2. K. Huang, "Statistical Mechanics," 2nd Ed., John Wiley & Sons (1987).

R.6.3. R.K. Pathria, "Statistical Mechanics," Pergamon Press (1972).

R.6.4. L.E. Reichl, "A Modern Course in Statistical Mechanics," Univ. Texas Press, Austin (1980).

R.6.5. R.L. Liboff, "Kinetic Theory," 3rd Ed., Springer (2003).

R.6.6. D. ter Haar, "Elements of Statistical Mechanics," Rinehart, N.Y. (1954).

R.6.7. H. Jeffreys and B. Jeffreys, "Methods of Mathematical Physics," 3rd Ed., Cambridge University Press (1956).

R.6.8. Mark W. Zemansky, "Heat and Thermodynamics," 3rd Ed., McGraw-Hill Book Co., Inc., Sections 11.10–11.12, p.221 (1951). Note that Zemansky denotes the Helmholtz Free Energy by A instead of the more commonly used symbol F which we employ.

R.6.9. L.D. Landau and E.M. Lifshitz, "Statistical Physics," Part I, 3rd Ed., Pergamon Press, Section 15 (1980). (Note that Landau and Lifshitz denote the Gibbs Free Energy by Φ instead of the symbol G employed here.)

R.6.10. Ibid., Eqns. (11.4), (11.3) and (15.11).

R.6.11. H.J. Zeiger and G.W. Pratt, "Magnetic Interactions in Solids," Clarendon Press, Oxford, Eqns. (A.5.17), (A.5.18) and (A.5.19) (1973).

R.6.12. L.E. Reichl, "A Modern Course in Statistical Mechanics," Univ. Texas Press, Austin p.44 (1980); another useful reference is R.K. Pathria, "Statistical Mechanics," Pergamon Press, Section 8.2, p.227 ff. (1972).

R.6.13. A.H. Wilson, "The Theory of Metals," 2nd Ed., Cambridge University Press, Sections 6.6 and 6.7, p.171 (also p.160–175) (1965).

R.6.14. R. Kubo, "Statistical Mechanics," North-Holland, p.171 (1964).

R.6.15. W. Jones and N.H. March, "Theoretical Solid State Physics," Vol. 1, Dover Publ., Section 4.15 (1973).

R.6.16. I.S. Gradshteyn and I.M. Ryzhik, "Tables of Integrals, Series and Products," Academic Press, p.36, #1.422.3 (1965).

R.6.17. Bateman Manuscript Project, "Higher Transcendental Functions," Vol. I, Editor A. Erdelyi, McGraw-Hill, p.13, #1.6.2 (1953).

R.6.18. W.J. de Haas and P.M. van Alphen, Proc. Acad. Sci. Amst. *33*, 1106 (1930).

R.6.19. W. Zawadzki, A. Raymond and M. Kublisa, "Reservoir Model for 2D Electron Gasses in Quantizing Magnetic Fields: A Review," to be published.

7

Thermodynamic Green's Functions and Spectral Structure

Abstract
Multiparticle thermodynamic Green's functions are defined in Chapter 7 in terms of grand canonical ensemble averages of time-ordered products of creation and annihilation operators, and they are interpreted as tracing the amplitude for time-developing correlated interacting particle motions taking place in the background of a *thermal ensemble*. Under equilibrium conditions (absence of time-dependent fields), time-translational invariance prevails, and the one-particle thermal Green's function can be Fourier-transformed in terms of a single frequency variable $t-t' \to \omega$, which leads to a Lehmann spectral representation whose frequency poles describe the energy spectrum. Because of the thermodynamic ensemble averaging (as opposed to vacuum state averaging), this Green's function has finite values for both $t > t'$ and $t < t'$ (in contrast to retarded Green's functions) and the two parts $G_{1>}$ and $G_{1<}$, respectively, obey a simple proportionality relation that facilitates the introduction of a spectral weight function. The latter relation is also interpreted in terms of a periodicity/antiperiodicity property of a modified Green's function in imaginary time capable of a Fourier series representation with imaginary (Matsubara) frequencies. The analytic continuation from imaginary time to real time is fully discussed and the procedure for the solution of the thermodynamic Green's function is delineated in detail, with an important example provided. Related commutator/anticommutator functions are discussed, as well as retarded/advanced Green's functions, and the sum rule for the single-particle spectral weight function is derived. Finally, the statistical thermodynamic information carried in one- and two-particle thermodynamic Green's functions is described, in part.

7.1 Definition and Interpretation of Thermal Green's Functions

Recognizing that microcanonical ensemble averages, $\langle \mathbf{X} \rangle^{N,E}$, are asymptotically equal to grand canonical ensemble averages, $\langle \mathbf{X} \rangle^{\alpha,\beta}$, for large number and energy, one could, in principle, employ either in the description of physical systems. However, the statistical weighting factor $e^{-iH\tau}$ may be interpreted as a time-displacement operator, through imaginary time $\tau = -i\beta$, which greatly simplifies the imposition of side conditions on the

Quantum Statistical Field Theory: Schwinger's Variational Method. Norman J. Morgenstern Horing,
Oxford University Press (2017). © Norman J. Morgenstern Horing. DOI 10.1093/oso/9780198791942.001.0001

partial differential and integral equations characterizing the dynamical development of the system. Correspondingly, an n-particle thermodynamic (thermal) Green's function in position representation is defined by[1] (except where otherwise noted, we employ the Heisenberg picture):

$$G_n^{i\lambda,i\tau}(\mathbf{x}_1, t_1, \ldots, \mathbf{x}_n, t_n; \mathbf{x}_1', t_1', \ldots, \mathbf{x}_n', t_n')$$
$$= (-i)^n \varepsilon \left\langle \left(\psi(\mathbf{x}_1, t_1) \ldots \psi(\mathbf{x}_n, t_n) \psi^+(\mathbf{x}_n', t_n') \ldots \psi^+(\mathbf{x}_1', t_1') \right)_+ \right\rangle^{i\lambda,i\tau}. \tag{7.1.1}$$

As usual, $(\ldots)_+$ denotes time ordering, with the factor having the largest time argument to the far left, and the other factors proceeding to the right in order of decreasing time argument, with the factor having the lowest time argument on the far right. ε is identically $+1$ for Bosons, but it is $+1$ or -1 for Fermions depending on whether the time-ordered factors in Eq. (7.1.1) involve an even or odd permutation of factors to reach the order shown. Examining the operators in Eq. (7.1.1), and bearing in mind that the averaging process takes place in the background of a thermal ensemble, the n creation operators may be viewed as jointly creating n additional particles at $\mathbf{x}_1', t_1', \ldots, \mathbf{x}_n', t_n'$, with $G_n^{i\lambda,i\tau}(1, \ldots n; 1' \ldots n')$ tracing their joint dynamical propagation characteristics to $\mathbf{x}_1, t_1, \ldots, \mathbf{x}_n, t_n$, where they are annihilated by the n annihilation operators; yielding the amplitude for this process with account of their correlated motions due to interparticle interactions. To explore this interpretation in the present multiparticle context, we will consider, in some detail, the one-particle thermodynamic (thermal) Green's function,

$$G_1^{i\lambda,i\tau}(\mathbf{x}_1, t_1; \mathbf{x}_1', t_1') = -i\varepsilon \left\langle \left(\psi(\mathbf{x}_1, t_1) \psi^+(\mathbf{x}_1', t_1') \right)_+ \right\rangle^{i\lambda,i\tau}. \tag{7.1.2}$$

Having discussed the creation/annihilation interpretation of the retarded Green's function description of single-particle dynamics in Chapter 5, we note that it can be extended to multiparticle Green's functions with some care concerning the multiparticle background. In particular, ψ annihilates one particle from an N-particle state yielding a state having $N-1$ particles, not the null state, so the function does *not* vanish for $t_1 < t_1'$ and is not retarded. Moreover, while ψ, ψ^+ always can be understood as annihilation/creation operators changing particle number, the presence of particle–particle interaction imposes limitations on the concept of a single-particle state even in the context of the one-particle thermal Green's function.

7.2 Lehmann Spectral Representation of the One-Particle Thermal Green's Function

Denoting the one-particle thermodynamic Green's function[2,3] by $G_1^{\alpha,\beta} \to G_1^{\mu,\beta}$ ($\mu \equiv -\frac{\alpha}{\beta}$ is the chemical potential), we have

$$G_1^{\mu,\beta}(\mathbf{x}, t; \mathbf{x}', t') = -i\varepsilon e^{-W(\mu,\beta)} Tr\left[e^{-\beta(H-\mu N)} \left(\psi_H(\mathbf{x}, t) \psi_H^+(\mathbf{x}', t') \right)_+ \right]. \tag{7.2.1}$$

Since the field operators obey Heisenberg equations of motion and develop in time under $H = H^{(1)} + H^{(2)}$, we have (in the absence of explicit time dependence; $\hbar \to 1$):

$$\psi_H^+(\mathbf{x}', t') = e^{iH(t'-t)} \psi_H^+(\mathbf{x}', t) e^{-iH(t'-t)}, \tag{7.2.2}$$

and for $t > t'$,

$$G_{1>}^{\mu,\beta}(\mathbf{x},t;\mathbf{x}',t') = -iTr[e^{-\beta\left(H-\mu N+\frac{W}{\beta}\right)} \times\psi(\mathbf{x},t)\,e^{iH(t'-t)}\psi^{+}(\mathbf{x}',t)\,e^{-iH(t'-t)}], \tag{7.2.3}$$

and for $t' > t$,

$$G_{1<}^{\mu,\beta}(\mathbf{x},t;\mathbf{x}',t') = \mp iTr[e^{-\beta\left(H-\mu N+\frac{W}{\beta}\right)} \times e^{iH(t'-t)}\psi^{+}(\mathbf{x}',t)\,e^{-iH(t'-t)}\psi(\mathbf{x},t)]. \tag{7.2.4}$$

The trace "Tr" stands for the sum of all diagonal matrix elements in *exact* eigenstates of total number $N \to N'$ and total energy $H \to E'_{N'}$, so for $G_{1>}^{\mu,\beta}(\mathbf{x},t;\mathbf{x}',t')$ we obtain

$$G_{1>}^{\mu,\beta}(\mathbf{x},t;\mathbf{x}',t') = -i\sum_{N',E'_{N'}}\left\langle N',E'_{N'}\left|\begin{matrix}[e^{-\beta\left(H-\mu N+\frac{W}{\beta}\right)}\psi(\mathbf{x},t)\,e^{iH(t'-t)}\\ \times\psi^{+}(\mathbf{x}',t)\,e^{-iH(t'-t)}]\end{matrix}\right|N',E'_{N'}\right\rangle. \tag{7.2.5}$$

In this, the summation also extends over all other microscopic specifications ξ' for the eigenstates of number and energy as in Chapter 6, and the number-energy eigenstates are taken at time t. Inserting the identity operator $I = \sum_{N'',E''_{N''}}|N'',E''_{N''}\rangle\langle N'',E''_{N''}|$ into the center of the matrix element, and noting that $e^{iH(t'-t)}|N'',E''_{N''}\rangle = e^{iE''_{N''}(t'-t)}|N'',E''_{N''}\rangle$, we have

$$G_{1>}^{\mu,\beta}(\mathbf{x},t;\mathbf{x}',t') = -i\sum_{\substack{N',N'',\\ E'_{N'},E''_{N''}}} e^{-\beta\left(E'_{N'}-\mu N'+\frac{W}{\beta}\right)}e^{i\left(E'_{N'}-E''_{N''}\right)T}\tilde{f}\left(\mathbf{x},\mathbf{x}'\,:';''\right), \tag{7.2.6}$$

which depends only on the time difference $T = (t-t')$ as H has no explicit time dependence, and

$$\tilde{f}\left(\mathbf{x},\mathbf{x}'\,:';''\right) = \langle N',E'_{N'}|\psi(\mathbf{x},t)|N'',E''_{N''}\rangle\langle N'',E''_{N''}|\psi^{+}(\mathbf{x}',t)|N',E'_{N'}\rangle \tag{7.2.7}$$

is independent of t since the number-energy eigenstates and operators are all at the same time t. Furthermore, since ψ^{+},ψ create and annihilate one particle, $\tilde{f}(\mathbf{x},\mathbf{x}'\,:';'')$ vanishes unless $N'' = N' + 1$. Moreover, since particle–particle interactions generally limit the validity of the concept of a single-particle state (since it would have a finite lifetime and decay as energy is transferred to other states by the interaction), the matrix element $\langle N',E'_{N'}|\psi(\mathbf{x},t)|N'',E''_{N''}\rangle$ does not exactly yield a single-particle wave function (which it would do if there were no interactions). However, it is the closest thing to it within the framework of an interacting many-particle system.

A similar analysis of $G_{1<}^{\mu,\beta}(\mathbf{x},t;\mathbf{x}',t')$ yields

$$G_{1<}^{\mu,\beta}(\mathbf{x},t;\mathbf{x}',t') = \mp i\sum_{\substack{N',N'',\\ E'_{N'},E''_{N''}}} e^{-\beta\left(E'_{N'}-\mu N'+\frac{W}{\beta}\right)}e^{-i\left(E'_{N'}-E''_{N''}\right)T}\tilde{f}\left(\mathbf{x},\mathbf{x}'\,:'';'\right), \tag{7.2.8}$$

and interchanging the single prime and double prime sums, we obtain

$$G_{1<}^{\mu,\beta}\left(\mathbf{x},t;\mathbf{x}',t'\right)=\mp i\sum_{\substack{N',N'',\\E'_{N'},E''_{N''}}}e^{-\beta\left(E''_{N''}-\mu N''+\frac{W}{\beta}\right)}e^{i\left(E'_{N'}-E''_{N''}\right)T}\widetilde{f}\left(\mathbf{x},\mathbf{x}'\,:';''\right). \tag{7.2.9}$$

Forming the Fourier transform of G_1 over the entire real-time axis from $-\infty$ to $+\infty$, we have (note that we use the standard Fourier transform notation here, so $G_1^{\mu,\beta}(T)=\int_{-\infty}^{\infty}(d\omega/2\pi)G_1^{\mu,\beta}(\omega)e^{-i\omega T}$ contrary to some practice in which a factor "i" is installed; also recall that the Heaviside unit step function $\eta_+(T)=i\int_{-\infty}^{\infty}(d\omega/2\pi)\,e^{-i\omega T}/\left(\omega+i0^+\right)$ vanishes for $T<0$ and is unity for $T>0$):

$$G_1^{\mu,\beta}\left(\mathbf{x},\mathbf{x}';\omega\right)=\int_{-\infty}^{+\infty}dTe^{i\omega T}\left[\eta_+\left(T\right)G_{1>}^{\mu,\beta}\left(\mathbf{x},\mathbf{x}';T\right)+\eta_+\left(-T\right)G_{1<}^{\mu,\beta}\left(\mathbf{x},\mathbf{x}';T\right)\right], \tag{7.2.10}$$

which results in (denote $\widetilde{f}\left(\mathbf{x},\mathbf{x}'\,:';''\right)\to\widetilde{f}\left(\mathbf{x},\mathbf{x}'\right)$):

$$\begin{aligned}G_1^{\mu,\beta}\left(\mathbf{x},\mathbf{x}';\omega\right)=\mp i&\sum_{\substack{N',N'',\\E'_{N'},E''_{N''}}}e^{-\beta\left(E''_{N''}-\mu N''+\frac{W}{\beta}\right)}\widetilde{f}\left(\mathbf{x},\mathbf{x}'\right)\\&\times\int_{-\infty}^{0}dTe^{i(\omega-i\delta)T}e^{i\left(E'_{N'}-E''_{N''}\right)T}\\&-i\sum_{\substack{N',N'',\\E'_{N'},E''_{N''}}}e^{-\beta\left(E'_{N'}-\mu N'+\frac{W}{\beta}\right)}\widetilde{f}\left(\mathbf{x},\mathbf{x}'\right)\\&\times\int_{0}^{\infty}dTe^{i(\omega+i\delta)T}e^{i\left(E'_{N'}-E''_{N''}\right)T}.\end{aligned} \tag{7.2.11}$$

The exponential factors involving $\omega\to\omega\pm i\delta$ ($\delta\to0^+$) assure convergence at the infinite time limits, and the resulting T-integrals yield frequency-dependent denominators as ($\mathcal{P}$ denotes the principal part integral):

$$\frac{1}{\omega-\Delta E\pm i\delta}=\mathcal{P}\frac{1}{\omega-\Delta E}\mp\pi i\delta\left(\omega-\Delta E\right), \tag{7.2.12}$$

in accordance with the Dirac prescription. Correspondingly, $G_1^{\mu,\beta}\left(\mathbf{x},\mathbf{x}';\omega\right)$ takes the form (set $\omega_{N''N'}\equiv E''_{N''}-E'_{N'}$ and bear in mind that $\hbar\to1$ and $N''-N'\equiv1$ for non-zero contributions):

$$G_1^{\mu,\beta}\left(\mathbf{x},\mathbf{x}';\omega\right) = \sum_{\substack{N',N'',\\ E'_{N'},E''_{N''}}} e^{-\beta\left(E'_{N'}-\mu N'+\frac{W}{\beta}\right)}\widetilde{f}\left(\mathbf{x},\mathbf{x}'\right)\left(1\mp e^{-\beta\left(\omega_{N''N'}-\mu\right)}\right)$$
$$\times\left[\mathcal{P}\frac{1}{\omega-\omega_{N''N'}} - i\pi\,\delta\left(\omega-\omega_{N''N'}\right)\coth^{\pm1}\left(\frac{\beta\left[\omega_{N''N'}-\mu\right]}{2}\right)\right]. \quad (7.2.13)$$

Alternatively expressed, Eq. (7.2.13) may be rewritten as

$$G_1^{\mu,\beta}\left(\mathbf{x},\mathbf{x}';\omega\right) = \int_{-\infty}^{\infty}\frac{d\omega'}{2\pi}A(\mathbf{x},\mathbf{x}';\omega')\left[\mathcal{P}\frac{1}{\omega-\omega'} - i\pi\coth^{\pm1}\left(\frac{[\omega-\mu]\beta}{2}\right)\delta(\omega-\omega')\right], \quad (7.2.14)$$

where the *spectral weight* function, $A\left(\mathbf{x},\mathbf{x}';\omega'\right)$, is defined as

$$A\left(\mathbf{x},\mathbf{x}';\omega'\right) = 2\pi\sum_{\substack{N',N'',\\ E'_{N'},E''_{N''}}} e^{-\beta\left(E'_{N'}-\mu N'+\frac{W}{\beta}\right)}\widetilde{f}\left(\mathbf{x},\mathbf{x}'\,{:}';''\right)\left(1\mp e^{-\beta\left(\omega_{N''N'}-\mu\right)}\right)\delta\left(\omega'-\omega_{N''N'}\right). \quad (7.2.15)$$

Furthermore, for $t > t'$, we have

$$G_{1>}^{\mu,\beta}\left(\mathbf{x},\mathbf{x}';\omega'\right) = \frac{1}{i}\frac{A\left(\mathbf{x},\mathbf{x}';\omega'\right)}{1\mp e^{\beta(\mu-\omega')}} = \frac{1}{i}\left[1\pm f(\omega')\right]A\left(\mathbf{x},\mathbf{x}';\omega'\right), \quad (7.2.16)$$

and for $t < t'$ ($f(\omega)$ is the Bose/Fermi equilibrium distribution function, not to be confused with $\widetilde{f}\left(\mathbf{x},\mathbf{x}'\right)$):

$$G_{1<}^{\mu,\beta}\left(\mathbf{x},\mathbf{x}';\omega'\right) = -\frac{1}{i}\frac{A\left(\mathbf{x},\mathbf{x}';\omega'\right)}{1\mp e^{-\beta(\mu-\omega')}} = \pm\frac{1}{i}f(\omega')A\left(\mathbf{x},\mathbf{x}';\omega'\right). \quad (7.2.17)$$

It should be noted that if $\widetilde{f}\left(\mathbf{x},\mathbf{x}'\right) = \widetilde{f}\left(\mathbf{x}',\mathbf{x}\right)$ then it is real and $A\left(\mathbf{x},\mathbf{x}';\omega'\right)$ is also real. In this case, one can readily verify that $\mathrm{Re}G_1^{\mu,\beta}(\omega)$ and $\mathrm{Im}G_1^{\mu,\beta}(\omega)$ are related by

$$\mathrm{Re}G_1^{\mu,\beta}\left(\mathbf{x},\mathbf{x}';\omega\right) = \mathcal{P}\frac{1}{\pi}\int_{-\infty}^{+\infty}d\omega'\frac{\mathrm{Im}G_1^{\mu,\beta}\left(\mathbf{x},\mathbf{x}';\omega'\right)}{\omega'-\omega}\coth^{\mp1}\left(\frac{(\omega'-\mu)\,\beta}{2}\right) \quad (7.2.18)$$

and

$$A\left(\mathbf{x},\mathbf{x}';\omega'\right) = -2\mathrm{Im}G_1^{\mu,\beta}\left(\mathbf{x},\mathbf{x}';\omega'\right)\coth^{\mp1}\left(\frac{(\omega'-\mu)\,\beta}{2}\right). \quad (7.2.19)$$

Of course, these relations also hold for homogeneous and isotropic media. Quite generally, the spectral weight $A(\omega)$ and $G_1^{\mu,\beta}\left(\mathbf{x},\mathbf{x}';\omega\right)$ are both singular at $\omega = \omega_{N''N'} = E''_{N''} - E'_{N'}$, the frequency locations of the exact multiparticle energy differences of the interacting many-particle system, with $N'' = N' + 1$. Moreover, the residues at the singularities yield

$$\widetilde{f}\left(\mathbf{x},\mathbf{x}'\right) \equiv \widetilde{f}\left(\mathbf{x},\mathbf{x}'\,{:}';''\right) = \left\langle N',E'_{N'}\left|\psi\left(\mathbf{x}\right)\right|N'',E''_{N''}\right\rangle\left\langle N'',E''_{N''}\left|\psi^{+}\left(\mathbf{x}'\right)\right|N',E'_{N'}\right\rangle, \quad (7.2.20)$$

which, crudely speaking, corresponds to the product of a one-particle wave function-like matrix element at **x**, with its complex conjugate at **x**′. Thus, examination of the spectral weight function of the interacting many-particle system yields full dynamical information about the exact energies of excited states through the locations of frequency singularities, and about the wave function-like matrix elements through the residues. The one-particle Green's function also carries statistical information through the grand canonical ensemble average statistical factors in its spectral structure, and this will be discussed below in greater detail.

7.3 Other Spectral Considerations

Considering the fact that $N'' = N'+1$ in order that $\widetilde{f}\,(\mathbf{x}, \mathbf{x}'\ ;';'')$ be non-zero, an inspection of Eq. (7.2.6) for $G_{1>}^{\mu,\beta}\,(\mathbf{x}, t; \mathbf{x}', t')$ and Eq. (7.2.9) for $G_{1<}^{\mu,\beta}\,(\mathbf{x}, t; \mathbf{x}', t')$ shows that the two are proportional with an imaginary time translation through $\tau = -i\beta$,

$$G_{1>}^{\mu,\beta}\left(\mathbf{x}, t+\tau; \mathbf{x}', t'\right) = \pm e^{-\mu\beta}\, G_{1<}^{\mu,\beta}\left(\mathbf{x}, t; \mathbf{x}', t'\right). \tag{7.3.1}$$

Alternatively, one can see that this result also follows at once from the definitions of $G_{1>}^{\mu,\beta}$ and $G_{1<}^{\mu,\beta}$:

$$G_{1>}^{i\lambda,i\tau}\left(\mathbf{x}, t; \mathbf{x}', t'\right) = \frac{1}{i} e^{-W(i\lambda,i\tau)}\, Tr\left[e^{-iN\lambda-iH\tau}\,\psi\left(\mathbf{x}, t\right)\psi^{+}\left(\mathbf{x}', t'\right)\right] \tag{7.3.2}$$

and

$$G_{1<}^{i\lambda,i\tau}\left(\mathbf{x}, t; \mathbf{x}', t'\right) = \pm\frac{1}{i} e^{-W(i\lambda,i\tau)}\, Tr\left[e^{-iN\lambda-iH\tau}\,\psi^{+}\left(\mathbf{x}', t'\right)\psi\left(\mathbf{x}, t\right)\right]. \tag{7.3.3}$$

Using the cyclic invariance of the trace ($TrAB = TrBA$), we have

$$G_{1<}^{i\lambda,i\tau}\left(\mathbf{x}, t; \mathbf{x}', t'\right) = \pm\frac{1}{i} e^{-W(i\lambda,i\tau)}\, Tr\left[\psi\left(\mathbf{x}, t\right) e^{-iN\lambda-iH\tau}\,\psi^{+}\left(\mathbf{x}', t'\right)\right]. \tag{7.3.4}$$

Viewing $e^{-iH\tau}$ as a time-translation operator (through imaginary time $\tau = -i\beta$), and bearing in mind the identity $\psi f(N) = f(N+1)\psi$ (Eq. (2.2.15)), we have

$$\psi\left(\mathbf{x}, t\right) e^{-iN\lambda} e^{-iH\tau} = e^{-i\lambda} e^{-iN\lambda} e^{-iH\tau}\,\psi\left(\mathbf{x}, t+\tau\right), \tag{7.3.5}$$

which immediately yields

$$G_{1<}^{i\lambda,i\tau}\left(\mathbf{x}, t; \mathbf{x}', t'\right) = \pm e^{-i\lambda}\, G_{1>}^{i\lambda,i\tau}\left(\mathbf{x}, t+\tau; \mathbf{x}', t'\right), \tag{7.3.6}$$

as above. In terms of Fourier transforms with respect to the time difference variable $T = t - t'$,

$$G_{1\gtrless}^{i\lambda,i\tau}\left(\mathbf{x},\mathbf{x}';T\right) = \int_{-\infty}^{\infty} \frac{d\omega}{2\pi} e^{-i\omega T} G_{1\gtrless}^{i\lambda,i\tau}\left(\mathbf{x},\mathbf{x}';\omega\right), \tag{7.3.7}$$

we have

$$G_{1<}^{i\lambda,i\tau}\left(\mathbf{x},\mathbf{x}';\omega\right) = \pm e^{-i\lambda-i\omega\tau} G_{1>}^{i\lambda,i\tau}\left(\mathbf{x},\mathbf{x}';\omega\right). \tag{7.3.8}$$

Expressing this in terms of the spectral weight function $A(\omega)$, we recover Eqns. (7.2.16 and 7.2.17),

$$iG_{1\{\gtrless\}}^{\mu,\beta}\left(\mathbf{x},\mathbf{x}';\omega\right) = \begin{Bmatrix} (1 \pm f(\omega)) \\ \pm f(\omega) \end{Bmatrix} A\left(\mathbf{x},\mathbf{x}';\omega\right). \tag{7.3.9}$$

Moreover, again combining $G_{1>}^{\mu,\beta}(T)$ and $G_{1<}^{\mu,\beta}(T)$ with the Heaviside unit step functions $\eta_+(\pm T)$,

$$G_1^{\mu,\beta}\left(\mathbf{x},\mathbf{x}';T\right) = \eta_+(T)G_{1>}^{\mu,\beta}\left(\mathbf{x},\mathbf{x}';T\right) + \eta_+(-T)G_{1<}^{\mu,\beta}\left(\mathbf{x},\mathbf{x}';T\right), \tag{7.3.10}$$

there is yet another equivalent form for $G_1^{\mu,\beta}$ in direct time representation as,[1]

$$G_1^{\mu,\beta}\left(\mathbf{x},\mathbf{x}';T\right) = -\frac{i}{2}\int \frac{d\omega}{2\pi} e^{-i\omega T} A\left(\mathbf{x},\mathbf{x}';\omega\right)\left\{\coth^{\pm 1}\left(\frac{(\omega-\mu)\,\beta}{2}\right) + (\eta_+(T) - \eta_+(-T))\right\}, \tag{7.3.11}$$

in which the ω'-convolution integral of Eq. (7.2.14) has been evaluated explicitly.

7.4 Periodicity/Antiperiodicity in Imaginary Time, Matsubara Frequencies and Analytic Continuation

The results of the preceding section can be understood in terms of a periodicity/antiperiodicity condition on the Green's function in imaginary time. Defining a slightly modified set of Green's functions as $\overline{G}_1$,

$$\overline{G}_1\left(\mathbf{x},t;\mathbf{x}',t'\right) = e^{i\mu(t-t')} G_1^{i\lambda,i\tau}\left(\mathbf{x},t;\mathbf{x}',t'\right) \tag{7.4.1}$$

and $\overline{G}_n$,

$$\begin{aligned}\overline{G}_n&\left(\mathbf{x}_1,t_1,\ldots\mathbf{x}_n,t_n;\mathbf{x}_1',t_1',\ldots\mathbf{x}_n',t_n'\right)\\ &= e^{i\mu(t_1+\cdots+t_n-t_1'-\cdots-t_n')} G_n^{i\lambda,i\tau}\left(\mathbf{x}_1,t_1,\ldots\mathbf{x}_n,t_n;\mathbf{x}_1',t_1',\ldots\mathbf{x}_n',t_n'\right).\end{aligned} \tag{7.4.2}$$

We note that this modification corresponds to shifting the origin of energy per particle by the chemical potential, with the corresponding shift $h^{(1)} \to h^{(1)} - \mu$, or $H \to H - \mu N$. In terms of these modified Green's functions, the relation Eq. (7.3.6) takes the form

$$\overline{G}_{1<}\left(\mathbf{x},\mathbf{x}';T\right) = \pm\overline{G}_{1>}\left(\mathbf{x},\mathbf{x}';T+\tau\right), \tag{7.4.3}$$

or, considering all times extended to the complex domain,

$$\overline{G}_{1}\left(\mathbf{x},\mathbf{x}';T\right) = \pm\overline{G}_{1}\left(\mathbf{x},\mathbf{x}';T+\tau\right). \tag{7.4.4}$$

This boundary condition of periodicity/antiperiodicity in imaginary time uniquely characterizes the thermal averaging involved in the definition of the Green's function. Green's functions defined by different averaging procedures (in terms of matrix elements of time-ordered products of creation and annihilation operators) generally obey the same partial differential equations. They are distinguished from one another only by boundary conditions such as retardation, periodicity/antiperiodicity (as in the case at hand), etc. The boundary condition of periodicity/antiperiodicity on the time variables of the one-particle Green's function is prototypical for the time variables of all n-particle Green's functions,

$$\overline{G}_{n}\left(t_i=\tau\right) = \pm\overline{G}_{n}\left(t_i=0\right), \tag{7.4.5}$$

for each time variable t_i separately, with the other time variables of $\overline{G}_n$ arbitrary. This more general statement of periodicity follows from the recognition that $e^{-\beta H}$ is the translation operator for an imaginary time displacement, as in the case of $G_1^{i\lambda,i\tau}$ (Eq. (7.3.6)) above. A more detailed discussion of the extension to complex times is provided in Reference 1.

The partial differential equations of motion for the Green's functions, which are based on Eq. (3.2.19), may be reformulated into integral equations that incorporate the boundary conditions in their structure. In either formulation, partial differential or integral, the periodicity/antiperiodicity condition may be enforced by employing a Fourier series representation with imaginary times restricted to the fundamental interval $0 \to \tau = -i\beta$, and with frequencies in the Fourier series exponentials, $\exp(-i\pi\nu t/\tau)$ (ν = even or odd integers in the Bose–Einstein or Fermi–Dirac cases, respectively), known as the *Matsubara frequencies*. Thus, with imaginary times restricted to the fundamental imaginary interval $0 \le t_i, t_i' \le \tau = -i\beta$,

$$\begin{aligned}\overline{G}_n\left(\mathbf{x}_1,t_1,\ldots\mathbf{x}_n,t_n;\mathbf{x}_1',t_1',\ldots\mathbf{x}_n',t_n'\right) &= \sum_{\{\nu\}\{\nu'\}}\exp\left[-i\pi\left(\nu_1t_1\cdots+\nu_nt_n-\nu_1't_1'\cdots-\nu_n't_n'\right)/\tau\right]\\ &\quad\times\tau^{-n}\mathcal{G}_n\left(\mathbf{x}_1,\nu_1,\ldots\mathbf{x}_n,\nu_n;\mathbf{x}_1',\nu_1',\ldots\mathbf{x}_n',\nu_n'\right),\end{aligned} \tag{7.4.6}$$

with $2n$ frequency indices $\left\{\nu=\nu_1\ldots,\nu_n;\nu_1'\ldots,\nu_n'\right\}$ conjugate to the $2n$ imaginary time variables. The inverse relation defining the Fourier coefficients is given by

$$\begin{aligned}&\overline{\mathcal{G}}_n\left(\mathbf{x}_1,\nu_1,\ldots\mathbf{x}_n,\nu_n;\mathbf{x}_1',\nu_1',\ldots\mathbf{x}_n',\nu_n'\right)\\ &\quad=\tau^{-n}\int_0^\tau dt_1\cdots dt_n\int_0^\tau dt_1'\cdots dt_n'\exp\left[i\pi\left(\sum_i\nu_it_i-\sum_i\nu_i't_i'\right)\Big/\tau\right]\\ &\quad\quad\times\overline{G}_n\left(\mathbf{x}_1,t_1,\ldots\mathbf{x}_n,t_n;\mathbf{x}_1',t_1',\ldots\mathbf{x}_n',t_n'\right).\end{aligned} \tag{7.4.7}$$

The Fourier coefficients $\mathcal{G}_n(\{\nu,\nu'\})$ vanish for $\{\nu\}\neq\{\nu'\}$ if there is time-translational invariance corresponding to a Hamiltonian that has no explicit time dependence. For example, in the case of the single-particle Green's function with time-translational invariance, the time dependence involves only $t-t'$ and just one set of Matsubara frequencies, ν, not two:

$$\overline{G}_1\left(\mathbf{x},t;\mathbf{x}',t'\right)=\frac{1}{\tau}\sum_{\nu}\exp\left[-i\pi\nu(t-t')/\tau\right]\overline{\mathcal{G}}_1\left(\mathbf{x},\mathbf{x}';\nu\right), \tag{7.4.8}$$

with the Fourier series coefficient ($T=t-t'$),

$$\overline{\mathcal{G}}_1\left(\mathbf{x}_1,\mathbf{x}_2;\nu\right)=\int_0^{\tau}dTe^{i\pi\nu T/\tau}\overline{G}_1\left(\mathbf{x}_1,\mathbf{x}_2,T\right). \tag{7.4.9}$$

Once the equations of motion (with $h^{(1)}\rightarrow h^{(1)}-\mu$) are solved for the Fourier series coefficients, it is necessary to analytically continue the solutions from the imaginary time interval $0\rightarrow\tau=-i\beta$ to the full real-time axis $-\infty\rightarrow\infty$.

The modified one-particle Green's functions under consideration have the form

$$\overline{G}_1\left(\mathbf{x},\mathbf{x}';\omega\right)=\int_{-\infty}^{\infty}\frac{d\omega'}{2\pi}\bar{A}\left(\mathbf{x},\mathbf{x}';\omega'\right)\left[\mathcal{P}\frac{1}{\omega-\omega'}-i\pi\coth^{\pm1}\left(\beta\omega/2\right)\delta\left(\omega-\omega'\right)\right], \tag{7.4.10}$$

with a modified spectral weight given by

$$\overline{A}\left(\mathbf{x},\mathbf{x}';\omega'\right)=A\left(\mathbf{x},\mathbf{x}';\omega'+\mu\right), \tag{7.4.11}$$

as one can readily verify using Eq. (7.2.14). Another equivalent form may be obtained by introducing the manifestly discontinuous auxiliary function (suppress $\mathbf{x},\mathbf{x}'$: $\overline{A}\left(\mathbf{x},\mathbf{x}';\omega'\right)\rightarrow\overline{A}\left(\omega'\right)$, etc.):

$$\overline{g}(z)=\int_{-\infty}^{\infty}\frac{d\omega'}{2\pi}\frac{\overline{A}\left(\omega'\right)}{z-\omega'}, \tag{7.4.12}$$

whose discontinuity across the real z-axis (use $z=\omega\pm i\delta$ and the Dirac prescription) yields

$$\overline{A}\left(\omega\right)=\frac{1}{i}\lim_{\delta\rightarrow0}\left(\overline{g}(\omega-i\delta)-\overline{g}(\omega+i\delta)\right). \tag{7.4.13}$$

Moreover, the definition of $\overline{g}(z)$ (Eq. (7.4.12)) may be used to rewrite Eq. (7.4.10) in direct real-time representation as

$$\overline{G}_1\left(\mathbf{x},\mathbf{x}';T\right)=\int_{-\infty}^{\infty}\frac{d\omega}{2\pi}e^{-i\omega T}\left(\frac{\overline{g}(\omega+i\delta)}{1\mp e^{-i\omega\tau}}+\frac{\overline{g}(\omega-i\delta)}{1\mp e^{i\omega\tau}}\right). \tag{7.4.14}$$

The key to executing the analytic continuation from imaginary times in the integral over the imaginary time contour ($0 \to \tau = -i\beta$) of Eq. (7.4.9) to real times ($-\infty < t < \infty$) is to recognize that the Green's functions, which may be thought of as arising from variational differentiation of a transformation function of the form $\langle\tau|0\rangle = \left\langle 0 \left| e^{-iH\tau} \right| 0\right\rangle$, are automatically "time ordered" along the imaginary time contour $0 \to \tau$ by application of the Schwinger Action Principle (Chapter 3). Such "contour ordering" yields the Fourier series coefficient of Eq. (7.4.9) as

$$\overline{\mathcal{G}}_1\left(\mathbf{x},\mathbf{x}';\nu\right) = \int_0^\tau dT e^{i\pi\nu T/\tau}\overline{G}_{1>}\left(\mathbf{x},\mathbf{x}',T\right), \tag{7.4.15}$$

and substitution of the Fourier transform $\overline{G}_{1>}(\mathbf{x},\mathbf{x}',\omega)$ from Eqns. (7.2.16) and (7.3.7) yields an important link in the analytic continuation:

$$\begin{aligned}\overline{\mathcal{G}}_1\left(\mathbf{x},\mathbf{x}';\nu\right) &= \int_0^\tau dT e^{i\pi\nu T/\tau}\int_{-\infty}^{\infty}\frac{d\omega}{2\pi i}e^{-i\omega T}\frac{\overline{A}(\omega)}{1\mp e^{-i\omega\tau}}\\ &= \int_{-\infty}^{\infty}\frac{d\omega}{2\pi}\frac{\overline{A}(\omega)}{\frac{\pi\nu}{\tau}-\omega} = \overline{g}(z=\pi\nu/\tau).\end{aligned} \tag{7.4.16}$$

This is to say that the determination of $\overline{\mathcal{G}}_1(\nu)$ from the Fourier series representation set in the imaginary time interval $[0,\tau]$ provides us with an evaluation of the auxiliary function $\overline{g}(z=\frac{\pi\nu}{\tau})$ at the discrete set of points $z=\frac{\pi\nu}{\tau}$. However, this identification is not unique since any other function $\overline{g}(e^{i2nz\tau}z)$ has the same values at $z=\frac{\pi\nu}{\tau}$. Baym and Mermin,[4] in their careful analysis of the analyticity properties and the desired continuation, showed that the uniquely correct choice to complete the analytic continuation process is one which assures that $\overline{g}(z)$ is analytic everywhere off the real axis and goes to zero as $z\to\infty$ along any ray in the upper or lower half planes.

In summary, the thermodynamic Green's function may be determined by:

(A) Solving dynamical equations for its Fourier series coefficient $\overline{\mathcal{G}}_1(\nu)$ (any convenient representation, not necessarily position representation);

(B) Obtaining the auxiliary function $\overline{g}(\frac{\pi\nu}{\tau})$ at discrete points $z=\frac{\pi\nu}{\tau}$, using

$$\overline{\mathcal{G}}_1(\nu) = \overline{g}(z=\pi\nu/\tau) = \left[\overline{g}(ze^{i2nz\tau})\right]_{z=\frac{\pi\nu}{\tau}}; \tag{7.4.17}$$

(C) Determining the unique function having the values given in Eq. (7.4.17), which (1) is also analytic everywhere off the real z-axis, and (2) goes to zero as $z\to\infty$ along any ray in the upper or lower half planes.

The auxiliary function $\overline{g}(z)$ thus defined may be employed in Eq. (7.4.13) to determine the spectral weight function $\overline{A}(\omega)$ for use in Eq. (7.4.10) to yield the Fourier transform of the real-time Green's function, or, alternatively, one may use $\overline{g}(z)$ in Eq.(7.4.14) for the real-time Green's function over the infinite real-time axis.

There are a few problems in which the real-time spectral form with weight function $\overline{A}(\omega)$ can be determined directly using the equations of motion. However, more often, this is not feasible and the analytic continuation procedure discussed above is needed.

An example that has been particularly useful is that of an infinite, translationally invariant system having the single-particle energy dispersion, $\varepsilon(p)$, as a function of the magnitude of momentum. In this case, $\overline{G}, \overline{g}, \overline{\mathcal{G}}$ and $\overline{A}$ are all functions of $\mathbf{x}-\mathbf{x}'$ (to the exclusion of $\mathbf{x}+\mathbf{x}'$), and a spatial Fourier transform $\mathbf{x}-\mathbf{x}' \to \mathbf{p}$ yields

$$\overline{A}(\mathbf{p},\omega') = 2\pi\delta(\omega' + \mu - \varepsilon(p)), \tag{7.4.18}$$

so that (Eq. (7.4.12)),

$$\begin{aligned}\overline{g}(\omega \pm i\delta) &= \int_{-\infty}^{\infty} d\omega' \frac{\delta(\omega' + \mu - \varepsilon(p))}{\omega - \omega' \pm i\delta} \\ &= \frac{1}{\omega - (\varepsilon(p) - \mu) \pm i\delta}.\end{aligned} \tag{7.4.19}$$

Eq. (7.4.14) calls for multiplication by factors such as $\frac{1}{1+e^{\mp i\omega\tau}} = \frac{1}{1+e^{\mp\omega\beta}}$ (which we here consider in the Fermi–Dirac case only). It is clear that subsequent integration in a product form with Eq. (7.4.19) will result in the replacement $\omega \to \varepsilon(p)-\mu$, so that the factors may be replaced in the degenerate limit by

$$\frac{1}{1+e^{\mp i\omega\tau}} = \frac{1}{1+e^{\mp(\varepsilon(p)-\mu)\beta}} \to \left\{\begin{matrix}\eta_+(\mu - \varepsilon(p)) \text{ for lower sign} \\ \eta_+(\varepsilon(p) - \mu) \text{ for upper sign}\end{matrix}\right\} \text{ for } \beta \to \infty.$$

In particular, if $\varepsilon(p) = p^2/2m$ and $\mu = p_F^2/2m$ with $\beta \to \infty$ (zero temperature), we have

$$\frac{1}{1+e^{\mp i\omega\tau}} \to \left\{\begin{matrix}\eta_+(p_F - p) \text{ for } + \\ \eta_+(p - p_F) \text{ for } -\end{matrix}\right\}. \tag{7.4.20}$$

Forming $\overline{G}_1(\mathbf{p},\omega)$ for Fermions in the degenerate case using Eq. (7.4.14), we have

$$\begin{aligned}\overline{G}_1(\mathbf{p},\omega) &= \frac{\overline{g}(\omega + i\delta)}{1+e^{-i\omega\tau}} + \frac{\overline{g}(\omega - i\delta)}{1+e^{i\omega\tau}}, \\ &= \frac{\eta_+(p-p_F)}{\omega - \left(\frac{p^2}{2m} - \mu\right) + i\delta} + \frac{\eta_+(p_F - p)}{\omega - \left(\frac{p^2}{2m} - \mu\right) - i\delta}.\end{aligned} \tag{7.4.21}$$

Since the only difference between these two terms is in the sign $\pm i\delta$ in the denominator, this may be rewritten as

$$\overline{G}_1(\mathbf{p},\omega) = \frac{1}{\omega - \left(\frac{p^2}{2m} - \mu\right) + i\delta\, sign(p - p_F)}. \tag{7.4.22}$$

A spatial Fourier transform yields (p_F, v_F are the Fermi momentum, speed, respectively):

$$\overline{G}_1(\mathbf{x}-\mathbf{x}';\omega) = \int \frac{d^3\mathbf{p}}{(2\pi)^3} \frac{e^{i\mathbf{p}\cdot(\mathbf{x}-\mathbf{x}')}}{\omega - \left(\frac{p^2}{2m}-\mu\right) + i\delta sign(p-p_F)}$$

$$= \frac{-i}{(2\pi)^2|\mathbf{x}-\mathbf{x}'|}\int_0^\infty dpp \frac{e^{ip|\mathbf{x}-\mathbf{x}'|} - e^{-ip|\mathbf{x}-\mathbf{x}'|}}{\omega - \left(\frac{p^2}{2m}-\mu\right) + i\delta \mathrm{sign}(|p|-p_F)}$$

$$= \frac{im}{2\pi^2|\mathbf{x}-\mathbf{x}'|}\int_{-\infty}^\infty dp \frac{pe^{ip|\mathbf{x}-\mathbf{x}'|}}{\left(p-\sqrt{2m(\mu+\omega)}\right)\left(p+\sqrt{2m(\mu+\omega)}\right) - i\delta sign(|p|-p_F)}. \tag{7.4.23}$$

The contribution from the pole in the upper half p-plane at $p = \sqrt{2m(\mu+\omega)} + i\delta \rightarrow \sqrt{p_F^2 + 2\omega p_F/v_F} \cong p_F + \omega/v_F$ for excitation energies small compared to the Fermi energy, $\hbar\omega << \mu = p_F^2/2m$, is approximately given for $|p| > p_F$ as

$$G_1(\mathbf{x}-\mathbf{x}';\omega) \cong \frac{-m}{2\pi|\mathbf{x}-\mathbf{x}'|}\exp\left[ip_F|\mathbf{x}-\mathbf{x}'| + i\frac{\omega}{v_F}|\mathbf{x}-\mathbf{x}'|\right]. \tag{7.4.24}$$

This result was used by Gorkov in developing a Green's function theory of superconductivity.[5,6]

7.5 Commutator/Anticommutator Functions; Retarded/Advanced Green's Functions; Spectral Sum Rule

We consider yet another type of propagation function, an ensemble averaged commutator/anticommutator function (suppress positional arguments $\mathbf{x}, \mathbf{x}'$):

$$\mathcal{G}_1(T) = \overline{G}_{1>}(T) - \overline{G}_{1<}(T)$$

$$= e^{i\mu T}\left(G_{1>}^{i\lambda,i\tau}(T) - G_{1<}^{i\lambda,i\tau}(T)\right)$$

$$= -ie^{i\mu T}\left\langle\left(\psi(t)\psi^+(t') \mp \psi^+(t')\psi(t)\right)\right\rangle^{i\lambda,i\tau} \tag{7.5.1}$$

or

$$\mathcal{G}_1(T) = -ie^{i\mu T}\left\langle\{[\psi(t), \psi^+(t')]\}\right\rangle^{i\lambda,i\tau}. \tag{7.5.2}$$

Using Eqns. (7.3.7–7.3.9), we also have

$$\begin{aligned} \mathcal{G}_1(T) &= e^{i\mu T}\int_{-\infty}^{\infty}\frac{d\omega}{2\pi}e^{-i\omega T}\left(G_{1>}^{i\lambda,i\tau}(\omega)-G_{1<}^{i\lambda,i\tau}(\omega)\right) \\ &= e^{i\mu T}\int_{-\infty}^{\infty}\frac{d\omega}{2\pi}e^{-i\omega T}\left(1\mp e^{-i\lambda-i\omega\tau}\right)G_{1>}^{i\lambda,i\tau}(\omega) \end{aligned} \tag{7.5.3}$$

or

$$\mathcal{G}_1(T) = e^{i\mu T}\int_{-\infty}^{\infty}\frac{d\omega}{2\pi i}e^{-i\omega T}A(\omega) = \int_{-\infty}^{\infty}\frac{d\omega}{2\pi i}e^{-i\omega T}\overline{A}(\omega). \tag{7.5.4}$$

Thus, the spectral weight $\overline{A}(\omega)$ is just the Fourier transform of $i\mathcal{G}_1(T) = i(\overline{G}_{1>}(T) - \overline{G}_{1<}(T))$. As such, we note that for null time difference, $T = 0$,

$$\mathcal{G}_1(0) = \int_{-\infty}^{\infty}\frac{d\omega}{2\pi i}\overline{A}(\omega) = -i\left\langle\left\{\left[\psi(t),\psi^+(t)\right]\right\}\right\rangle^{i\lambda,i\tau} \tag{7.5.5}$$

using Eqns. (7.5.2) and (7.5.4). By virtue of the canonical equal-time commutation relations (in position representation), the right-hand side is proportional to $\delta^{(3)}(\mathbf{x}-\mathbf{x}')$, specifically

$$\int_{-\infty}^{\infty}\frac{d\omega}{2\pi}\overline{A}(\mathbf{x},\mathbf{x}';\omega) = \delta^{(3)}\left(\mathbf{x}-\mathbf{x}'\right). \tag{7.5.6}$$

This is a sum rule for the spectral weight function $\overline{A}(\mathbf{x},\mathbf{x}';\omega)$. In the case of spatial translational invariance, $\overline{A}(\mathbf{x},\mathbf{x}';\omega) \to \overline{A}(\mathbf{x}-\mathbf{x}';\omega)$, and Fourier-transforming $\mathbf{x}-\mathbf{x}' \to \mathbf{p}$, we have the corresponding sum rule in momentum representation,

$$\int_{-\infty}^{\infty}\frac{d\omega}{2\pi}\overline{A}(\mathbf{p};\omega) = 1. \tag{7.5.7}$$

There are two other variants of Green's functions associated with $\mathcal{G}_1(T)$:

(A) A retarded Green's function defined as

$$\overline{G}_r(T) = \eta_+(T)\mathcal{G}_1(T) = \eta_+(T)\left(\overline{G}_{1>}(T)-\overline{G}_{1<}(T)\right), \tag{7.5.8}$$

and using Eq. (7.5.4) for $\mathcal{G}_1(T)$, with $\eta_+(T)$ in frequency representation as

$$\eta_\pm(T) = \pm i \int_{-\infty}^{\infty} \frac{d\omega}{2\pi} \frac{e^{-i\omega T}}{\omega \pm i0^+} = 1 \text{ for } T \gtrless 0, \text{ and } \equiv 0 \text{ for } T \lessgtr 0,$$

we have

$$\overline{G}_r(T) = \int \frac{d\omega'}{2\pi} \int \frac{d\omega''}{2\pi} \frac{e^{-i\omega' T} e^{-i\omega'' T}}{\omega' + i\delta} \overline{A}(\omega''). \tag{7.5.9}$$

Fourier-transforming $\overline{G}_r(T) \to \overline{G}_r(\omega)$, we obtain

$$\overline{G}_r(\omega) = \int \frac{d\omega'}{2\pi} \frac{\overline{A}(\omega')}{\omega - \omega' + i\delta} = \overline{g}(\omega + i\delta), \tag{7.5.10}$$

where $\overline{g}(\omega + i\delta)$ is the auxiliary function of Eq. (7.4.12), which is the Matsubara Fourier series coefficient evaluated at $\omega + i\delta$. Correspondingly,

$$\overline{G}_r(T) = \int \frac{d\omega}{2\pi} e^{-i\omega T} \overline{g}(\omega + i\delta). \tag{7.5.11}$$

This relation provides the special connection between periodic functions and their causally retarded counterparts. We also note from Eq. (7.5.10) that $\overline{A}(\omega)$ can be obtained from the retarded Green's function as

$$\overline{A}(\omega) = -2Im\overline{G}_r(\omega) = -2Im\overline{g}(\omega + i\delta). \tag{7.5.12}$$

Alternatively expressed as an *operator* using Eq. (5.1.24), $\overline{A}_{op}(\omega) = 2\pi\delta(\omega - H)$. With Eq. (7.5.12), the spectral weight $\overline{A}(\omega)$ can be obtained from the retarded Green's function, which may be used in Eq. (7.3.8) and (7.3.9) to obtain the thermal Green's function $G_{1\gtrless}$.

(B) An advanced Green's function is defined as

$$\overline{G}_a(T) = -\eta_+(-T)\mathcal{G}_1(T) = \eta_+(-T)\left(\overline{G}_{1<}(T) - \overline{G}_{1>}(T)\right). \tag{7.5.13}$$

Both of these one-particle functions, $\overline{G}_r(T)$ and $\overline{G}_a(T)$, are evaluated in the presence of interparticle interactions that affect the structure of the spectral weight $\overline{A}(\omega)$ (hence the structure of $\overline{G}_{r,a}(T)$).

7.6 Statistical Thermodynamic Information in the One- and Two-Particle Green's Functions

Previously, we indicated that statistical information is also available in the structure of the grand canonical ensemble averaged one-particle Green's function. Considering first the number operator,

$$N = \int d^{(3)}\mathbf{x}\psi^{+}(\mathbf{x},t)\psi(\mathbf{x},t), \tag{7.6.1}$$

the average macroscopic number is given by

$$\langle N\rangle^{\mu,\beta} = \int d^{(3)}\mathbf{x}\left\langle\psi^{+}(\mathbf{x},t)\psi(\mathbf{x},t)\right\rangle^{\mu,\beta} = \pm i\int d^{(3)}\mathbf{x}G_{1<}^{\mu,\beta}(\mathbf{x},t;\mathbf{x},t), \tag{7.6.2}$$

and bearing in mind that (Section 6.1; $\alpha = -\mu\beta$) $\langle N\rangle^{\mu\beta} = \beta^{-1}\partial W/\partial\mu$, we can express W in terms of the single-particle Green's function as

$$W = \pm i\beta\int_{-\infty}^{\mu} d\mu'\int d^3\mathbf{x}G_{1<}^{\mu',\beta}(\mathbf{x},t;\mathbf{x},t) = lnZ, \tag{7.6.3}$$

which also provides the all-important statistical grand partition function Z (Eq. 6.1.6) in terms of the interacting single-particle Green's function alone.

In terms of the spectral weight, we have

$$\langle N\rangle^{\mu,\beta} = \int\frac{d\omega}{2\pi}f(\omega)\int d^{(3)}\mathbf{x}A(\mathbf{x},\mathbf{x};\omega). \tag{7.6.4}$$

The implicit sums over internal variables, such as spin, mean that $\int d^{(3)}\mathbf{x}A(\mathbf{x},\mathbf{x};\omega)$ must be understood to include a spin (etc.) trace as well as the trace integral over position variables. Since $f(\omega)$ is the equilibrium statistical distribution function (Chapter 6), which describes the average number of particles in the energy state $\hbar\omega$, it is clear that $A(\mathbf{x},\mathbf{x};\omega)$ describes the number of microscopic states at position $\mathbf{x}$ per unit frequency/energy interval per unit volume (i.e., the density of states of the interacting quantum system). In momentum representation for a spatially homogeneous system having spatial translational invariance, $\mathbf{x}-\mathbf{x}'\to\mathbf{p}$ under spatial Fourier transformation, and we have

$$\langle N\rangle^{\mu,\beta} = \int\frac{d\omega}{2\pi}f(\omega)\left(\int d^{(3)}\mathbf{x}\right)\int\frac{d^{(3)}\mathbf{p}}{(2\pi)^3}A(\mathbf{p},\omega). \tag{7.6.5}$$

The positional integral simply yields a volume factor, V, so

$$\frac{\langle N\rangle^{\mu,\beta}}{V} = \int\frac{d\omega}{2\pi}f(\omega)\int\frac{d^{(3)}\mathbf{p}}{(2\pi)^3}A(\mathbf{p},\omega), \tag{7.6.6}$$

with $A(\mathbf{p},\omega)$ representing the number of microscopic quantum states having momentum $\mathbf{p}$ per unit frequency/energy interval, per unit volume of momentum space. As a weighing function over the spectrum, $A(\mathbf{p},\omega)$ is normalized to total integrated weight unity by the sum rule, Eq. (7.5.7). Furthermore, (Eq. (6.1.31)):

$$W(\alpha,\beta) = V\beta P(\alpha,\beta)\,, \tag{7.6.7}$$

where $P(\alpha,\beta)$ is the pressure. Hence,

$$\left[\frac{\partial P}{\partial \mu}\right]_{\beta,V} = \frac{1}{V\beta}\frac{\partial W}{\partial \mu} = \frac{\langle N\rangle^{\mu,\beta}}{V} = n(\beta,\mu)\,, \tag{7.6.8}$$

where $n(\beta,\mu)$ is the average particle density as a function of β and μ, given above in terms of the spectral weight (Eqns. (7.6.4 and 7.6.8)). Integrating Eq. (7.6.8) with respect to μ holding β and V fixed, we obtain the pressure in the form

$$P(\beta,\mu) = \int_{-\infty}^{\mu} d\mu' n(\beta,\mu')\,. \tag{7.6.9}$$

This μ'-integral involving the spectral weight (implicitly) will yield an expression for $P(\beta,\mu)$, which constitutes an equation of state for a system with particle–particle interactions to the extent that they are included in the spectral weight. It also provides an explicit expression for the grand potential $W(\alpha,\beta)$ (Eq. (7.6.7)), and, with that, the grand partition function $e^{W(\alpha,\beta)}$.

Considering the average energy of a many-particle system, we have

$$\langle H\rangle^{\mu\beta} = \left\langle H^{(1)}\right\rangle^{\mu\beta} + \left\langle H^{(2)}\right\rangle^{\mu\beta}\,, \tag{7.6.10}$$

where $\left\langle H^{(1)}\right\rangle^{\mu\beta}$ is

$$\left\langle H^{(1)}\right\rangle^{\mu\beta} = \pm i\int d^{(3)}\mathbf{x} Lim_{\substack{t'\to t^+\\ \mathbf{x}'\to\mathbf{x}}} h^{(1)} G_1^{\mu,\beta}\left(\mathbf{x},t;\mathbf{x}',t'\right), \tag{7.6.11}$$

as one can readily verify from the definitions of $H^{(1)}$ and $G_1^{\mu,\beta}$. Similarly, the definitions of $H^{(2)}$ and $G_2^{\mu,\beta}$ yield

$$\left\langle H^{(2)}\right\rangle^{\mu\beta} = \frac{1}{2}\int d^{(3)}\mathbf{x}'\int d^{(3)}\mathbf{x}'' V(\mathbf{x}'-\mathbf{x}'') G_2^{\mu,\beta}\left(\mathbf{x}',t,\mathbf{x}'',t;\mathbf{x}',t^+,\mathbf{x}'',t^+\right). \tag{7.6.12}$$

Using the Green's function equations to be discussed in Chapter 8, it will be seen that $\left\langle H^{(2)}\right\rangle^{\mu\beta}$ can be rewritten in terms of the *single*-particle Green's function $G_1^{\mu,\beta}$ as

$$\left\langle H^{(2)}\right\rangle^{\mu\beta} = \frac{\pm i}{2}\int d^{(3)}\mathbf{x} Lim_{\substack{t'\to t^+\\ \mathbf{x}'\to\mathbf{x}}}\left(i\frac{\partial}{\partial t} - h^{(1)}\right) G_1^{\mu,\beta}\left(\mathbf{x},t;\mathbf{x}',t'\right) \tag{7.6.13}$$

so that the total average energy is given by

$$\langle H\rangle^{\mu\beta} = \frac{\pm i}{2}\int d^{(3)}\mathbf{x}Lim_{\substack{t'\to t^+\\ \mathbf{x}'\to\mathbf{x}}}\left(i\frac{\partial}{\partial t}+h^{(1)}\right)G_1^{\mu,\beta}\left(\mathbf{x},t;\mathbf{x}',t'\right). \tag{7.6.14}$$

This may be expressed in terms of the single-particle spectral weight function as (assume spatial translational invariance; V is volume):

$$\langle H\rangle^{\mu,\beta} = V\int\frac{d\omega}{2\pi}\int\frac{d^{(3)}\mathbf{p}}{(2\pi)^3}f(\omega)A(\mathbf{p},\omega)\frac{\omega+p^2/2m}{2}. \tag{7.6.15}$$

From another standpoint, we can isolate the particle–particle interaction and correlation effects in the two-particle Green's function G_2. Starting with Eq. (4.1.74) and setting $a_1' \equiv a_2'' = a$, and summing over this common index, we obtain a trace on both sides of the equation. Moreover, we consider $t_1 = \tau$ and $t_2 = 0$, so that $\langle a,t_1| = \langle a,t_2|\,e^{-\beta H}$ introduces the equilibrium ensemble under the trace. Shifting the origin of the single-particle energy scale by the chemical potential $\mu = -\alpha/\beta$, we have a description in terms of the grand canonical Gibbs ensemble, and Eq. (4.1.74) takes the form

$$\begin{aligned}\frac{\partial e^W}{\partial\lambda} &= e^W\frac{\partial W}{\partial\lambda}\\ &= -\frac{\beta}{2}\int d^{(3)}\mathbf{x}'\int d^{(3)}\mathbf{x}''\mathcal{V}(\mathbf{x}'-\mathbf{x}'')\,Tr\left(e^{-\beta(H-\mu N)}\psi^+(\mathbf{x}',t)\psi^+(\mathbf{x}'',t)\psi(\mathbf{x}'',t)\psi(\mathbf{x}',t)\right),\end{aligned} \tag{7.6.16}$$

where W is the grand potential (Eq. (6.1.6)). Noting the definition of G_2 (Eq. (7.1.1) with $n = 2$), we have ($\mathcal{V}^{(\lambda)} = \lambda\mathcal{V}$),

$$\frac{\partial W}{\partial\lambda} = -\frac{\beta}{2}\int d^{(3)}\mathbf{x}'\int d^{(3)}\mathbf{x}''\frac{\mathcal{V}^{(\lambda)}(\mathbf{x}'-\mathbf{x}'')}{\lambda}\overline{G}_2^{(\lambda)}(1',1'';1'^+,1''^+)_{t_1'=t_1''}. \tag{7.6.17}$$

Here, we have restored the strength factor to $\mathcal{V}\to\mathcal{V}^{(\lambda)}\to\lambda\mathcal{V}$, and indicate the dependence of $\overline{G}_2\to\overline{G}_2^{(\lambda)}$ on λ. Integration of Eq. (7.6.17) with respect to λ from $\lambda = 0$ to $\lambda = 1$ now separates the part of W that depends on particle–particle interactions from its value, $W(0)$, in the absence of coupling, as

$$W = W(0) - \frac{\beta}{2}\int_0^1\frac{d\lambda}{\lambda}\int d^{(3)}\mathbf{x}'\int d^{(3)}\mathbf{x}''\mathcal{V}^{(\lambda)}(\mathbf{x}'-\mathbf{x}'')\overline{G}_2^{(\lambda)}(1',1'';1'^+,1''^+)_{t_1'=t_1''}. \tag{7.6.18}$$

This representation has been useful in identifying plasmon contributions to W by relating G_2 to the inverse dielectric function. However, another important representation in terms of the *single*-particle Green's function may be obtained by employing Eqns. (7.6.12 and 7.6.13), with the result

$$W = W(0) \mp \frac{i\beta}{2} \int_0^1 \frac{d\lambda}{\lambda} \int d^{(3)}\mathbf{x}\, Lim_{\substack{t' \to t^+ \\ \mathbf{x}' \to \mathbf{x}}} \left(i\frac{\partial}{\partial t} - h^{(1)} \right) G_1^{\mu,\beta}\left(\mathbf{x}, t; \mathbf{x}', t'\right). \qquad (7.6.19)$$

In terms of the *single*-particle spectral weight function, this takes the form (assume spatial translational invariance):

$$W = W(0) - \frac{V\beta}{2} \int_0^1 \frac{d\lambda}{\lambda} \int \frac{d^3\mathbf{p}}{(2\pi)^3} \int \frac{d\omega}{2\pi} (\omega - p^2/2m) f(\omega) A(\mathbf{p}, \omega), \qquad (7.6.20)$$

and all the effects of particle–particle interaction and correlation are embedded in the last term on the right of Eq. (7.6.20). Of course, the first term on the right, $W(0)$, which completely excludes such coupling effects, may be evaluated in terms of the *non*-interacting single-particle retarded Green's function using Eq. (6.3.17), which describes the ordinary partition function as

$$\tilde{Z} = Tre^{-\beta H_0^{(1)}} = i \int d^3\mathbf{x} G_{1>}^{ret}(\mathbf{x}, \mathbf{x}; t - t' \to -i\beta) \qquad (7.6.21)$$

(subscript "0" on $H_0^{(1)}$ means that the field operators in $H^{(1)}$ are devoid of coupling effects). Yet another evaluation of $W(0)$ has already been exhibited in Eq. (6.2.7).

7.7 Chapter 7 Problems

P.7.1. Verify Eq. (7.4.14).

P.7.2. *Graphene* is the name of the two-dimensional (2D) electron system of Chapter 5, Problem 1, with Hamiltonian matrix $h^{(1)} = \alpha\boldsymbol{\sigma} \cdot \mathbf{p}$. Derive the corresponding thermodynamic matrix Green's function for it exhibiting $G_{1>}^{\mu\beta}(\mathbf{x}, \mathbf{x}'; T)$, $G_{1<}^{\mu\beta}(\mathbf{x}, \mathbf{x}'; T)$ and the spectral weight function $\bar{A}(\mathbf{x}, \mathbf{x}'; \omega)$.

P.7.3. For the same Graphene electron system described in Problem 2, derive the relation between density and chemical potential as a function of temperature, and evaluate it explicitly for the zero-temperature degenerate limit. (Hint: Consult Eq. (7.6.5).)

P.7.4. For the same electron system as in Problem 2, derive the counterpart of Eq. (7.6.11) for average energy.

P.7.5. Verify Eqns. (7.2.15–7.2.17).

P.7.6. Considering the retarded Green's function for normal three-dimensional (3D) electrons, $G_1^{ret}(\mathbf{x}, \mathbf{x}'; \omega)$ given in Eq. (5.3.9) subject to a single-point $\delta^{(3)}(\mathbf{x})$-potential centered at the origin ($\mathbf{x}_i \to 0$), derive the corresponding spectral weight function using the "smear" interpretation associated with Eq. (5.3.11).

P.7.7. Determine the relation between the overall total density of states per unit frequency/energy, $D(\omega)$, and the spectral weight function $\bar{A}(\mathbf{p}, \omega)$ in the case of translational invariance.

P.7.8. Problem 2 is concerned with the Green's function for electrons in Graphene and its spectral weight function. Employing the results for the latter, $A(\mathbf{x}_1, \mathbf{x}_2; \omega)$, in 2D position representation construct the "ring diagram," $R(\mathbf{x}_1, \mathbf{x}_2; \omega + i\delta)$, which is at the core of the determination of the polarizability, in accordance with

$$R(\mathbf{x}_1 - \mathbf{x}_2; \omega + i\delta) = i\, Tr \int \frac{d\omega'}{2\pi i} \int \frac{d\omega''}{2\pi i} \frac{f(\omega'') - f(\omega')}{(\omega + \omega'' - \omega' + i\delta)} A(\mathbf{x}_1 - \mathbf{x}_2; \omega') A(\mathbf{x}_2 - \mathbf{x}_1; \omega''),$$

for a spatially homogeneous system. Evaluate this ring diagram, $R(\mathbf{p}; \omega + i\delta)$, in momentum representation using the Graphene spectral weight function. (Tr denotes the 2D matrix (pseudo)spin trace; $f(\omega)$ is the Fermi–Dirac distribution function.)

P.7.9. Considering the retarded magnetic field Green's functions for non-interacting 2D and 3D electron systems, subject to Landau quantization and Zeeman spin-splitting, determine their spectral weight functions and construct their thermodynamic Green's function counterparts at arbitrary temperature.

P.7.10. Determine the appropriate relations for $\langle H \rangle^{\mu\beta}$ and W in terms of the spectral weight function A, which should supplant Eqns. (7.6.15) and (7.6.20) for a system that lacks spatial translational invariance.

7.8 References for Chapter 7

R.7.1. P.C. Martin and J. Schwinger, Phys. Rev. *115*, 1342 (1959).

R.7.2. A.A. Abrikosov, L. Gor'kov and I. Dzyaloshinskii, "Quantum Field Theory Methods in Statistical Physics," 2nd Ed., Pergamon Press (1965).

R.7.3. A.L. Fetter and J.D. Walecka, "Quantum Theory of Many-Particle Systems," McGraw-Hill (1971).

R.7.4. G. Baym and N.D. Mermin, J. Math. Phys. **2**, 232 (1961).

R.7.5. L.P. Gor'kov, Soviet Phys. JETP *7*, 505 (1958); L.P. Gor'kov, Soviet Phys. JETP *9*, 1364 (1959).

R.7.6. A.A. Abrikosov, L. Gor'kov and I. Dzyaloshinskii, "Quantum Field Theory Methods in Statistical Physics," 2nd Ed., Pergamon Press, Chapter VII (1965).

8

Equations of Motion with Particle–Particle Interactions and Approximations

Abstract

Starting with the equation of motion for the field operator $\psi(\mathbf{x}, t)$ of an interacting many-particle system, Chapter 8 develops the single-particle Green's function equation and ultimately the n-particle Green's function equation of motion. In this, the n-particle Green's function is coupled to the $(n + 1)$-particle Green's function through the inter-particle interaction potential (also to the $(n - 1)$-particle Green's function), so an infinite chain of such coupled Green's function equations emerges. Particularly important cases of interest are those of (i) the one-particle Green's function equation with its coupling to the two-particle Green's function; and (ii) the two-particle Green's function equation with its coupling to the three-particle Greens' function. To develop solutions, it is necessary to introduce non-correlation decoupling procedures. To simplify the G_1-equation, an assumed factorization of the involved G_2-function into products of the type $G_1 G_1$ produces the (approximate) closed Hartree and Hartree–Fock equations for the single-particle Green's function. In regard to the G_2-equation, a similar factorization "ansatz" for G_3 into products of the type $G_2 G_1$ yields an approximate closed equation for G_2 that is decoupled from higher order Green's functions. In particular, this non-correlation decoupling procedure in the G_2-equation leads to the *Sum of Ladder Diagrams* integral equation for G_2, and its iterates produce the multiple Born approximation series, carrying finite lifetime collisional implications for single-particle states in G_1. Based on the structure of the G_2 equation, a similar treatment of the G_{11}-equation for the joint propagation of one electron and one hole subject to mutual Coulomb attraction is analyzed and explicitly solved, exhibiting exciton states in which a bound electron–hole system has a discrete hydrogen-like spectrum of energy eigenstates. Its role in single-particle propagation is also discussed in terms of implications for the one-electron self-energy Σ and the T-matrix.

Quantum Statistical Field Theory: Schwinger's Variational Method. Norman J. Morgenstern Horing, Oxford University Press (2017).
  DOI 10.1093/oso/9780198791942.001.0001

8.1 Multiparticle Green's Function Equations of Motion

Recalling the equation of motion, Eq. (3.2.21), for the Heisenberg field operator $\psi_H(\mathbf{x}, t) \to \psi(\mathbf{x}, t)$, we now retain the particle–particle interaction term

$$\begin{aligned} &\left(i\frac{\partial}{\partial t_1} - h^{(1)}\left(\mathbf{x}_1, \tfrac{\hbar}{i}\nabla_1, t_1\right)\right)\psi(\mathbf{x}_1, t_1) \\ &\qquad - \int d^{(3)}\mathbf{x}_2 \mathcal{V}(\mathbf{x}_1 - \mathbf{x}_2)\, \psi^+(\mathbf{x}_2, t_1)\, \psi(\mathbf{x}_2, t_1)\, \psi(\mathbf{x}_1, t_1) = 0. \end{aligned} \tag{8.1.1}$$

Again, multiplying Eq. (8.1.1) by $\psi^+\left(\mathbf{x}_1', t_1'\right)$ from the left or right to form a time-ordered product for $t_1 \neq t_1'$, and taking the ensemble average to obtain a one-particle Green's function equation of motion, which is homogeneous for all times except $t_1 = t_1'$ (we write $h^{(1)}\left(\mathbf{x}_1, \frac{\hbar}{i}\nabla_1, t_1\right) \to h_1^{(1)}$) we have

$$\begin{aligned} &\left(i\frac{\partial}{\partial t_1} - h_1^{(1)}\right) G_1^{i\lambda, i\tau}(\mathbf{x}_1, t_1; \mathbf{x}_1', t_1') \\ &\qquad \mp i \int d^{(3)}\mathbf{x}_2 \mathcal{V}(\mathbf{x}_1 - \mathbf{x}_2)\, G_2^{i\lambda, i\tau}\left(\mathbf{x}_1, t_1, \mathbf{x}_2, t_1; \mathbf{x}_1', t_1', \mathbf{x}_2, t_1^+\right) \\ &\qquad = \delta^{(3)}\left(\mathbf{x}_1 - \mathbf{x}_1'\right) \delta\left(t_1 - t_1'\right). \end{aligned} \tag{8.1.2}$$

The two-particle Green's function, $G_2^{i\lambda, i\tau}$, arises from the particle–particle interaction term in Eq. (8.1.1), which results in the average of a time-ordered product of two ψ's and two ψ^+'s (recall Eq. (7.1.1) for $n = 2$). The Dirac delta function right-hand side vanishes for $t_1 \neq t_1'$, as it must due to the homogeneity of Eq. (8.1.1), and its correctness as an impulsive inhomogeneous driving term for $t_1 = t_1'$ may again be verified by integrating Eq. (8.1.2) with respect to t from $t_1 = t_1' - \delta$ to $t_1 = t_1' + \delta$ ($\delta \to 0^+$), with the result

$$i\left(G_{1>}^{i\lambda, i\tau}\left(\mathbf{x}_1, t_1' + \delta; \mathbf{x}_1', t_1'\right) - G_{1<}^{i\lambda, i\tau}\left(\mathbf{x}_1, t_1' - \delta; \mathbf{x}_1', t_1'\right)\right) = \delta^{(3)}\left(\mathbf{x}_1 - \mathbf{x}_1'\right). \tag{8.1.3}$$

Once more, we have used the fact that $h^{(1)} G_1^{i\lambda, i\tau}$ and $\int d^{(3)}\mathbf{x}_2 \mathcal{V} G_2^{i\lambda, i\tau}$ are finite quantities, which yield negligibly small contributions when integrated over an infinitesimal time interval. Substituting the definitions of $G_{1>}^{i\lambda, i\tau}$ and $G_{1<}^{i\lambda, i\tau}$ into the left-hand side, we obtain

$$\left\langle \left\{ \left[\psi\left(\mathbf{x}_1, t_1'\right), \psi^+\left(\mathbf{x}_1', t_1'\right)\right] \right\} \right\rangle = \delta^{(3)}\left(\mathbf{x}_1 - \mathbf{x}_1'\right), \tag{8.1.4}$$

the correctness of which is assured by the equal-time canonical commutation/anticommutation relations. The equation of motion for $\overline{G}_1 = e^{i\mu(t_1 - t_1')} G_1^{i\lambda, i\tau}$ follows at once from Eq. (8.1.2) as

$$\begin{aligned} &\left(i\frac{\partial}{\partial t_1} + \mu - h_1^{(1)}\right) \overline{G}_1(\mathbf{x}_1, t_1; \mathbf{x}_1', t_1') \\ &\qquad \mp i \int d^{(3)}\mathbf{x}_2 \mathcal{V}(\mathbf{x}_1 - \mathbf{x}_2)\, \overline{G}_2\left(\mathbf{x}_1, t_1, \mathbf{x}_2, t_1; \mathbf{x}_1', t_1', \mathbf{x}_2, t_1^+\right) \\ &\qquad = \delta^{(3)}\left(\mathbf{x}_1 - \mathbf{x}_1'\right) \delta(t_1 - t_1'). \end{aligned} \tag{8.1.5}$$

Clearly, the change $G_1^{i\lambda,i\tau} \to \overline{G}_1$ just shifts the origin of the single-particle energy spectrum by the chemical potential μ. One can derive another equation of motion for this Green's function by starting with the equation of motion for $\frac{\partial\psi^+(\mathbf{x}_1',t_1')}{\partial t_1'}$ (instead of $\frac{\partial\psi(\mathbf{x}_1,t_1)}{\partial t_1}$ as was done above). This would produce the *adjoint* equation for the Green's function, and for G_1 it is given by

$$\begin{aligned}&\left(-i\frac{\partial}{\partial t_1'}+\mu-h_{1'}^{(1)}\right)\overline{G}_1(\mathbf{x}_1,t_1;\mathbf{x}_1',t_1')\\&\quad\mp i\int d^{(3)}\mathbf{x}_2\mathcal{V}\left(\mathbf{x}_1'-\mathbf{x}_2\right)\overline{G}_2(\mathbf{x}_1,t_1,\mathbf{x}_2,t_1';\mathbf{x}_1',t_1',\mathbf{x}_2,t_1'^{+})\\&\quad=\delta^{(3)}(\mathbf{x}_1-\mathbf{x}_1')\delta(t_1-t_1').\end{aligned} \tag{8.1.6}$$

Considering extension into the complex time domain, $\delta\left(t_1-t_1'\right)$ is understood to be defined with respect to integration along the imaginary time interval $0 \to \tau = -i\beta$. Furthermore, returning to Eq. (8.1.1) and multiplying by $(n-1)$ more pairs of $\psi\left(\mathbf{x}_j,t_j\right)\psi^+\left(\mathbf{x}_j',t_j'\right)$ operators ($j = 2\ldots n$) from the left or right appropriate to the time ordering (as well as multiplying by $\psi^+\left(\mathbf{x}_1',t_1'\right)$), we similarly obtain an equation of motion for $\overline{G}_n$ as

$$\begin{aligned}&\left(i\frac{\partial}{\partial t_1}+\mu-h_1^{(1)}\right)\overline{G}_n(\mathbf{x}_1,t_1,\ldots\mathbf{x}_n,t_n;\mathbf{x}_1',t_1',\ldots\mathbf{x}_n',t_n')\\&\quad\mp i\int d^{(3)}\mathbf{x}_{n+1}\mathcal{V}\left(\mathbf{x}_1-\mathbf{x}_{n+1}\right)\overline{G}_{n+1}(\mathbf{x}_1,t_1,\ldots\mathbf{x}_{n+1},t_1;\mathbf{x}_1',t_1',\ldots\mathbf{x}_{n+1},t_1^{+})\\&\quad=\sum_{i=1}^{n}\delta^{(3)}\left(\mathbf{x}_1-\mathbf{x}_i'\right)\delta\left(t_1-t_i'\right)(\pm1)^{i-1}\\&\quad\times\overline{G}_{n-1}(\mathbf{x}_2,t_2,\ldots\mathbf{x}_n,t_n;\mathbf{x}_1',t_1',\ldots\mathbf{x}_{i-1}',t_{i-1}',\mathbf{x}_{i+1}',t_{i+1}',\ldots\mathbf{x}_n',t_n').\end{aligned} \tag{8.1.7}$$

In dealing with multiparticle Green's functions, $\overline{G}_n\left(1,\ldots n;1',\ldots n'\right)$ (here and below we use the notation $1 = \mathbf{x}_1,t_1$, etc.), one can derive other equations focused on $\frac{\partial}{\partial t_j}$ ($j = 2\ldots n$), instead of $\frac{\partial}{\partial t_1}$ as was done above. All such equations and their adjoint equations are correct, albeit asymmetric in their attention to a single time variable. A symmetric partial differential characterization, which treats all time variables on a symmetric footing, is given in the Martin–Schwinger paper.[1]

An important special case of Eq. (8.1.7) is that of the two-particle Green's function, $\overline{G}_2$, which obeys the equation

$$\begin{aligned}&\left(i\frac{\partial}{\partial t_1}+\mu-h_1^{(1)}\right)\overline{G}_2\left(\mathbf{x}_1,t_1,\mathbf{x}_2,t_2;\mathbf{x}_1',t_1',\mathbf{x}_2',t_2'\right)\\&\mp i\int d^{(3)}\mathbf{x}_3\mathcal{V}\left(\mathbf{x}_1-\mathbf{x}_3\right)\overline{G}_3\left(\mathbf{x}_1,t_1,\mathbf{x}_2,t_2,\mathbf{x}_3,t_1;\mathbf{x}_1',t_1',\mathbf{x}_2',t_2',\ldots\mathbf{x}_3,t_1^{+}\right)\\&=\delta^{(3)}(\mathbf{x}_1-\mathbf{x}_1')\delta\left(t_1-t_1'\right)\overline{G}_1\left(\mathbf{x}_2,t_2;\mathbf{x}_2',t_2'\right)\pm\delta^{(3)}\left(\mathbf{x}_1-\mathbf{x}_2'\right)\delta\left(t_1-t_2'\right)\overline{G}_1\left(\mathbf{x}_2,t_2;\mathbf{x}_1',t_1'\right),\end{aligned} \tag{8.1.8}$$

which will be discussed further below. In the *absence* of interaction ($\mathcal{V} \to 0$), this equation may be solved by writing the differential operator acting on $\overline{G}_2$ on the left as $(\overline{G}_1)^{-1}$, and inverting it, with the Hartree–Fock result, $\overline{G}_2(1,2;1',2') = \overline{G}_1(1,1')\,\overline{G}_1(2,2') \pm \overline{G}_1(1,2')\,\overline{G}_1(2,1')$; the first term is the Hartree term and the second (exchange) term is the Fock term.

Contemplating a solution in Fourier series for the imaginary time interval $0 \to \tau = -i\beta$, Martin and Schwinger present the equation for the Fourier series coefficient of the n-particle Green's function (Eq. (8.1.7)) as[1]

$$\begin{aligned}
&[\frac{\pi\nu_1}{\tau} + \mu - h_1^{(1)}]\overline{\mathcal{G}}_n\left(\mathbf{x}_1,\nu_1,\ldots\mathbf{x}_n,\nu_n;\mathbf{x}'_1,\nu'_1,\ldots\mathbf{x}'_n,\nu'_n\right)\\
&\qquad \mp \frac{i}{\tau}\int d^{(3)}\mathbf{x}_{n+1}\mathcal{V}\left(\mathbf{x}_1-\mathbf{x}_{n+1}\right)\sum_{\nu''_1,\nu_{n+1},\nu'_{n+1}}\delta_{\nu_1-\nu''_1,\nu_{n+1}-\nu'_{n+1}}\exp\left(i\nu'_{n+1}0^+\right)\\
&\qquad \times \overline{\mathcal{G}}_{n+1}\left(\mathbf{x}_1,\nu''_1,\ldots\mathbf{x}_{n+1},\nu_{n+1};\mathbf{x}'_1,\nu'_1,\ldots\mathbf{x}'_{n+1},\nu'_{n+1}\right)\\
&\qquad = \sum_i \delta_{\nu_1,\nu'_i}\delta^{(3)}\left(\mathbf{x}_1-\mathbf{x}'_i\right)(\pm 1)^{i-1}\\
&\qquad \times \overline{\mathcal{G}}_{n-1}\left(\mathbf{x}_2,\nu_2,\ldots\mathbf{x}_n,\nu_n;\mathbf{x}'_1,\nu'_1,\ldots\mathbf{x}'_{i-1},\nu'_{i-1},\mathbf{x}'_{i+1},\nu'_{i+1},\ldots\mathbf{x}'_n,\nu'_n\right).
\end{aligned} \tag{8.1.9}$$

In particular, the Fourier series coefficient of the one-particle Green's function obeys the equation

$$\begin{aligned}
&[\frac{\pi\nu_1}{\tau} + \mu - h_1^{(1)}]\overline{\mathcal{G}}_1\left(\mathbf{x}_1,\nu_1;\mathbf{x}'_1,\nu'_1\right) \mp \frac{i}{\tau}\int d^{(3)}\mathbf{x}_2\mathcal{V}\left(\mathbf{x}_1-\mathbf{x}_2\right)\\
&\qquad \times \sum_{\nu''_1,\nu_2,\nu'_2}\delta_{\nu_1-\nu''_1,\nu_2-\nu'_2}\overline{\mathcal{G}}_2\left(\mathbf{x}_1,\nu''_1,\mathbf{x}_2,\nu_2;\mathbf{x}'_1,\nu'_1,\mathbf{x}_2,\nu'_2\right)e^{i\nu'_2 0^+}\\
&\qquad = \delta_{\nu_1,\nu'_1}\delta^{(3)}\left(\mathbf{x}_1-\mathbf{x}'_1\right).
\end{aligned} \tag{8.1.10}$$

Having obtained the Green's function equations, we can complete the verification of the expression for average energy $\langle H\rangle^{\mu,\beta}$ given in Eq. (7.6.10). To start, note that the first term of

$$\langle H\rangle^{\mu,\beta} = \left\langle H^{(1)}\right\rangle^{\mu,\beta} + \left\langle H^{(2)}\right\rangle^{\mu,\beta}, \tag{8.1.11}$$

which is given by

$$\begin{aligned}
\left\langle H^{(1)}\right\rangle^{\mu,\beta} &= \int d^{(3)}\mathbf{x}\left\langle\psi^+(\mathbf{x},t)\,h^{(1)}\psi(\mathbf{x},t)\right\rangle^{\mu,\beta}\\
&= \pm i\int d^{(3)}\mathbf{x}\lim_{\substack{t'\to t^+\\ \mathbf{x}'\to\mathbf{x}}} h^{(1)}G_1^{\mu,\beta}\left(\mathbf{x},t;\mathbf{x}',t'\right),
\end{aligned} \tag{8.1.12}$$

should not be mistaken for a purely single-particle energy contribution because particle–particle interaction effects are embedded in the spectral structure of G_1 through its equation of motion coupling to G_2. Considering next $\left\langle H^{(2)}\right\rangle^{\mu,\beta}$, we have

$$\begin{aligned}\left\langle H^{(2)}\right\rangle^{\mu,\beta} &= \frac{1}{2!}\int d^{(3)}\mathbf{x}'\int d^{(3)}\mathbf{x}''\mathcal{V}\left(\mathbf{x}'-\mathbf{x}''\right)\\ &\times\left\langle\psi^{+}\left(\mathbf{x}',t\right)\psi^{+}\left(\mathbf{x}'',t\right)\psi\left(\mathbf{x}'',t\right)\psi\left(\mathbf{x}',t\right)\right\rangle^{\mu,\beta}\\ &=\frac{1}{2}\int d^{(3)}\mathbf{x}'\int d^{(3)}\mathbf{x}''\mathcal{V}\left(\mathbf{x}'-\mathbf{x}''\right)\\ &\times G_2^{\mu,\beta}\left(\mathbf{x}',t,\mathbf{x}'',t;\mathbf{x}',t^{+},\mathbf{x}'',t^{+}\right).\end{aligned} \tag{8.1.13}$$

However, the right-hand side is intimately related to the $\mathcal{V}G_2$ term of the equation of motion, Eq. (8.1.2), for $G_1^{\mu,\beta}$, which we can use to substitute in place of $\mathcal{V}G_2^{\mu,\beta}$, eliminating its explicit appearance from Eq. (8.1.13):

$$\left\langle H^{(2)}\right\rangle^{\mu,\beta} = \frac{\pm i}{2}\int d^{(3)}\mathbf{x}\lim_{\substack{t'\to t^{+}\\ \mathbf{x}'\to\mathbf{x}}}\left(i\frac{\partial}{\partial t}-h^{(1)}\right)G_1^{\mu,\beta}\left(\mathbf{x},t;\mathbf{x}',t'\right). \tag{8.1.14}$$

Combining Eqns. (8.1.12) and (8.1.13), we obtain the result

$$\langle H\rangle^{\mu,\beta} = \frac{\pm i}{2}\int d^{(3)}\mathbf{x}\lim_{\substack{t'\to t^{+}\\ \mathbf{x}'\to\mathbf{x}}}\left(i\frac{\partial}{\partial t}+h^{(1)}\right)G_1^{\mu,\beta}\left(\mathbf{x},t;\mathbf{x}',t'\right). \tag{8.1.15}$$

8.2 Non-Correlation Decoupling Procedures for G_2: Hartree and Hartree–Fock Approximations

At a glance, it is clear that the Green's function equations of the preceding section are problematic, in that the equation for a given Green's function, say $\overline{G}_n$, is not closed upon that function alone. Rather, it is coupled through the particle–particle interaction potential to the next higher order Green's function, $\overline{G}_{n+1}$ (and to the next lower order Green's function $\overline{G}_{n-1}$). Thus, to obtain sufficient information with which one may hope to solve for $\overline{G}_n$, it is necessary to provide information about $\overline{G}_{n+1}$. But the equation from which one might obtain such information about $\overline{G}_{n+1}$—the equation of motion for $\overline{G}_{n+1}$—requires input information about $\overline{G}_{n+2}$, etc. This chain of coupled equations is infinite, a quantum counterpart of the BBGKY equations of classical statistical mechanics.[2]

To start, note that if there is no interparticle interaction, $\mathcal{V}\to 0$, then the chain of equations is fully decoupled, and the one-particle Green's function equation is determined by the Fourier series coefficient given by Eq. (8.1.10) for $n=1$ as

$$\left[\frac{\pi\nu}{\tau}+\mu-h^{(1)}\right]\overline{\mathcal{G}}_1^0\left(\mathbf{x},\mathbf{x}';\nu\right)=\delta^{(3)}\left(\mathbf{x}-\mathbf{x}'\right), \tag{8.2.1}$$

which is solved symbolically as

$$\overline{\mathcal{G}}_1^0\left(\mathbf{x},\mathbf{x}';\nu\right)=\frac{1}{\frac{\pi\nu}{\tau}+\mu-h^{(1)}}\delta^{(3)}\left(\mathbf{x}-\mathbf{x}'\right), \tag{8.2.2}$$

and the auxiliary function (Eqns. (7.4.16) and (7.4.17)) is given by

$$\overline{g}^0\left(\mathbf{x},\mathbf{x}';z\right) = \frac{1}{z+\mu-h^{(1)}}\delta^{(3)}\left(\mathbf{x}-\mathbf{x}'\right), \tag{8.2.3}$$

yielding the spectral weight $\overrightarrow{A}^0(\omega)$ in position representation as

$$\overrightarrow{A}^0\left(\mathbf{x},\mathbf{x}';\omega\right) = 2\pi\delta\left(\omega+\mu-h^{(1)}\right)\delta^{(3)}\left(\mathbf{x}-\mathbf{x}'\right), \tag{8.2.4}$$

whence the solution for $\overrightarrow{G}_1^0$ is given by (Eq. (7.2.14)):

$$\begin{aligned}\overrightarrow{G}_1^0\left(\mathbf{x},\mathbf{x}';t-t'\right) &= \int_{-\infty}^{\infty}\frac{d\omega}{2\pi}e^{-i\omega(t-t')} \\ &\times\int_{-\infty}^{\infty}\frac{d\omega'}{2\pi}\left[\mathcal{P}\frac{1}{\omega-\omega'}-\pi i\delta\left(\omega-\omega'\right)\tanh^{\mp1}\left(\beta\omega/2\right)\right] \\ &\times 2\pi\delta\left(\omega'+\mu-h^{(1)}\right)\delta^{(3)}\left(\mathbf{x}-\mathbf{x}'\right).\end{aligned} \tag{8.2.5}$$

To put the symbolic solution, Eq. (8.2.4), into tangible terms, we expand $\delta^{(3)}\left(\mathbf{x}-\mathbf{x}'\right) = \left\langle\mathbf{x}|\mathbf{x}'\right\rangle$ in eigenvectors of $h^{(1)}$ using $\mathbf{I} = \sum_{E'}\left|E'\right\rangle\left\langle E'\right|$, recalling that $h^{(1)}$ is in position representation, so that

$$h^{(1)}\left\langle\mathbf{x}|E'\right\rangle = E'\left\langle\mathbf{x}|E'\right\rangle \text{ and } F(h^{(1)})\left\langle\mathbf{x}|E'\right\rangle = F\left(E'\right)\left\langle\mathbf{x}|E'\right\rangle, \tag{8.2.6}$$

for an arbitrary function F. Consequently,

$$\begin{aligned}\overrightarrow{A}^0\left(\mathbf{x},\mathbf{x}';\omega\right) &= 2\pi\delta\left(\omega+\mu-h^{(1)}\right)\sum_{E'}\left\langle\mathbf{x}|E'\right\rangle\left\langle E'|\mathbf{x}'\right\rangle \\ &= 2\pi\sum_{E'}\delta\left(\omega+\mu-E'\right)\phi_{E'}(\mathbf{x})\phi^*_{E'}(\mathbf{x}').\end{aligned} \tag{8.2.7}$$

The concentration of the spectrum at specific single-particle energies, enforced by Dirac delta functions which vanish off those specific energies, is characteristic of non-interacting systems, since interactions generally impart a finite lifetime to the non-interacting energies, represented by a spreading of the delta function into a Lorentzian or more complicated shape (due to transitions induced by the interactions that move particles out of a specific non-interacting energy state into others).

A first procedure to break the infinite chain of $\overline{G}_n$ equations in the presence of interparticle interactions and achieve closure is suggested by the non-interacting wave function discussion of Chapter 1, in which a composite two-particle wave function is composed of the product of two single-particle wave functions, representing the neglect of correlation and interaction of the particles. The corresponding non-correlation statement at the level of the $\overline{G}_1$ equation is to approximate the role of $\overline{G}_2\left(1,2;1',2'\right)$ by a product of the form $\overline{G}_1(1,1')\overline{G}_1(2,2')$. Thus, we have the ordinary Hartree equation for $\overline{G}_1 \to \overline{G}_1^H$ ($1 = \mathbf{x}_1, t_1$, etc.) as

$$\left(i\frac{\partial}{\partial t_1} + \mu - h_1^{(1)}\right)\overline{G}_1^H(1;1') \mp i\int d^{(3)}\mathbf{x}_2 \mathcal{V}(\mathbf{x}_1 - \mathbf{x}_2)\overline{G}_1^H(2;2^+)\overline{G}_1^H(1;1') = \delta(1-1'). \tag{8.2.8}$$

Recalling that the particle density is given by

$$n(\mathbf{x}) = \pm i\overline{G}_1^H(\mathbf{x}, t; \mathbf{x}, t^+), \tag{8.2.9}$$

we have

$$\left(i\frac{\partial}{\partial t_1} + \mu - h_1^{(1)} - \int d^{(3)}\mathbf{x}_2 \mathcal{V}(\mathbf{x}_1 - \mathbf{x}_2)\, n(\mathbf{x}_2)\right)\overline{G}_1^H(1;1') = \delta(1-1'). \tag{8.2.10}$$

It is clear that the interparticle interaction $\mathcal{V}$ contributes an additional time-independent potential to $h_1^{(1)}$ as $\int d^{(3)}\mathbf{x}_2 \mathcal{V}(\mathbf{x}_1 - \mathbf{x}_2)\, n(\mathbf{x}_2)$. This "Hartree" potential is accumulated from all contributing particles through the two-body interaction $\mathcal{V}(\mathbf{x}_1 - \mathbf{x}_2)$, weighted with the local source density $n(\mathbf{x}_2)$. It is an important calculational tool in the theory of atomic structure, where $h_1^{(1)}$ includes the central attractive nuclear Coulomb potential. There is a significant self-consistency problem involved here, since the Hartree potential depends upon $\overline{G}_1^H$ through the density $n(\mathbf{x})$ (Eq. (8.2.9)), but $\overline{G}_1^H$ is itself dependent upon the Hartree potential in the solution of Eq. (8.2.10). In the simple case of spatial translational invariance for a homogenous electron gas, Eq. (8.2.10) is easily solved since $n(\mathbf{x}_2) = n =$ constant, and then the Hartree potential V^H is just a constant,

$$V^H = \int d^{(3)}\mathbf{x}_2 \mathcal{V}(\mathbf{x}_1 - \mathbf{x}_2)\, n(\mathbf{x}_2) = n\int d^{(3)}\mathbf{X}\mathcal{V}(\mathbf{X}) = n\overline{\mathcal{V}}. \tag{8.2.11}$$

This constant ($n\overline{\mathcal{V}}$) just shifts the chemical potential,

$$\mu \to \mu - n\overline{\mathcal{V}}. \tag{8.2.12}$$

With the substitution $\mu \to \mu - n\overline{\mathcal{V}}$ and $h_1^{(1)} \to -\nabla_1^2/2m$, all the non-interacting solutions, Eqns. (8.2.1)–(8.2.5), remain valid for the present case of Hartree potential with $\mathcal{V} \neq 0$. The self-consistency problem is reduced to the determination of the constant shift of the chemical potential here. Thus, for the Hartree approximation in the case of spatial translational invariance, $h_1^{(1)} \to -\nabla_1^2/2m$, the spectral weight is still concentrated at single-particle energies and we have

$$\overline{A}^H(\mathbf{x}, \mathbf{x}'; \omega) = 2\pi\delta\left(\omega + \mu - n\overline{\mathcal{V}} + \frac{\nabla_1^2}{2m}\right)\delta^{(3)}(\mathbf{x} - \mathbf{x}'), \tag{8.2.13}$$

which must be solved self-consistently with

$$n = \pm i\overline{G}_{1<}^H(\mathbf{x}, \mathbf{x}; t - t = 0) = \int \frac{d\omega}{2\pi} f(\omega)\overline{A}^H(\mathbf{x}, \mathbf{x}; \omega). \tag{8.2.14}$$

Fourier-transforming into momentum representation $\mathbf{x}-\mathbf{x}' \to \mathbf{p}$, we have

$$\overline{A}^{H}(\mathbf{p};\omega) = 2\pi\delta\left(\omega + \mu - n\overline{\mathcal{V}} - \frac{p^2}{2m}\right). \tag{8.2.15}$$

A second step in decoupling the chain of $\overline{G}_n$ equations is to improve the Hartree approximation by the addition of a Fock term, corresponding to the symmetrization/antisymmetrization of wave functions using the Slater permanent/determinant of Eq. (1.1.42), to assure that the principle of indistinguishability is satisfied. Thus, we write, approximately,

$$\overline{G}_2\left(1,2;1'2'\right) \approx \overline{G}_1^{HF}\left(1;1'\right)\overline{G}_1^{HF}\left(2;2'\right) \pm \overline{G}_1^{HF}\left(1;2'\right)\overline{G}_1^{HF}\left(2;1'\right), \tag{8.2.16}$$

and, correspondingly, the $\overline{G}_1$-equation takes the form

$$\left(i\frac{\partial}{\partial t_1} + \frac{\nabla_1^2}{2m} + \mu\right)\overline{G}_1^{HF}\left(1;1'\right) - \int d^{(3)}\mathbf{x}_2\mathcal{V}\left(\mathbf{x}_1-\mathbf{x}_2\right)n(\mathbf{x}_2)\overline{G}_1^{HF}\left(1;1'\right)$$

$$-\int d^{(3)}\mathbf{x}_2\mathcal{V}\left(\mathbf{x}_1-\mathbf{x}_2\right)\overline{G}_1^{HF}\left(1;\mathbf{x}_2,t_1^+\right)\overline{G}_1^{HF}\left(\mathbf{x}_2,t_1;1'\right) = \delta\left(1-1'\right), \tag{8.2.17}$$

where we have taken $h_1^{(1)} \to -\nabla_1^2/2m$ (assuming spatial translational invariance) and $n(\mathbf{x}_2) = \pm i\overline{G}_1^{HF}\left(2;2^+\right)$. Again, there is a self-consistency problem since $n(\mathbf{x}_2)$ determines $\overline{G}_1^{HF}$ but is in turn determined by it. Now, however, the time-independent Hartree–Fock potential, V^{HF}, is non-local (unlike the local Hartree potential),

$$V^{HF}\left(\mathbf{x}_1,\mathbf{x}_2\right) = \delta^{(3)}\left(\mathbf{x}_1-\mathbf{x}_2\right)\int d^{(3)}\mathbf{x}_3\mathcal{V}\left(\mathbf{x}_1-\mathbf{x}_3\right)n(\mathbf{x}_3)$$
$$+i\mathcal{V}\left(\mathbf{x}_1-\mathbf{x}_2\right)\overline{G}_{1<}^{HF}\left(1;2\right)_{t_2=t_1}, \tag{8.2.18}$$

entering the $\overline{G}_1^{HF}$ equation in the form

$$\left(i\frac{\partial}{\partial t_1} + \frac{\nabla_1^2}{2m} + \mu\right)\overline{G}_1^{HF}\left(1;1'\right) - \int d^{(3)}\mathbf{x}_2 V^{HF}\left(\mathbf{x}_1,\mathbf{x}_2\right)\overline{G}_1^{HF}\left(\mathbf{x}_2,t_1;1'\right) = \delta\left(1-1'\right). \tag{8.2.19}$$

However, for the case at hand of spatial translational invariance, the Fourier transform to momentum space can be carried out using the convolution theorem, resulting in a *local* equation in momentum space,

$$\left(i\frac{\partial}{\partial t_1} - h\left(\mathbf{p}\right) + \mu\right)\overline{G}_1^{HF}\left(\mathbf{p},t_1-t_1'\right) = \delta\left(t_1-t_1'\right), \tag{8.2.20}$$

where

$$h\left(\mathbf{p}\right) = \frac{p^2}{2m} + n\overline{\mathcal{V}} \pm \int\frac{d^{(3)}\mathbf{p}'}{(2\pi)^3}\mathcal{V}\left(\mathbf{p}-\mathbf{p}'\right)\langle n(\mathbf{p}')\rangle \tag{8.2.21}$$

and

$$\mathcal{V}(\mathbf{p}) = \int d^{(3)}\mathbf{x}\mathcal{V}(\mathbf{x})\, e^{-i\mathbf{p}\cdot\mathbf{x}}, \tag{8.2.22}$$

with the definition

$$\langle n(\mathbf{p})\rangle = \frac{1}{e^{i(h(\mathbf{p})-\mu)\tau} \mp 1}. \tag{8.2.23}$$

The solution yields the spectral weight $A^{HF}(\mathbf{p},\omega)$ in momentum space as

$$A^{HF}(\mathbf{p},\omega) = 2\pi\delta\left(\omega + \mu - h(\mathbf{p})\right), \tag{8.2.24}$$

which again requires a self-consistent solution, since $h(\mathbf{p})$ depends on $\langle n(\mathbf{p})\rangle$, which in turn depends on $h(\mathbf{p})$.

As approximations that do *not* take account of correlated particle motions arising from particle–particle interactions, the Hartree and Hartree–Fock expressions for $\overline{G}_2$ are oversimplified in that they produce solutions for $\overline{G}_1$ having spectral weights $\overline{A}(\omega)$ given by Dirac delta functions of frequency. The time-independent effective potential of the Hartree approximation, which is local in position representation with corresponding well-defined single-particle eigenenergies (within the framework of the self-consistency requirement), has associated eigenfunctions given by

$$\left[h_1^{(1)} - \mu + \int d^{(3)}\mathbf{x}_2 \mathcal{V}(\mathbf{x}_1 - \mathbf{x}_2)\, n(\mathbf{x}_2)\right] \phi^H_{E'_H}(\mathbf{x}_1) = E'_H \phi^H_{E'_H}(\mathbf{x}_1), \tag{8.2.25}$$

where

$$n(\mathbf{x}_2) = \sum_{E'_H} f\left(E'_H\right) \left|\phi^H_{E'_H}(\mathbf{x}_2)\right|^2, \tag{8.2.26}$$

leading to an expression for the Hartree spectral weight $\overline{A}^H(\omega)$ that has the detailed form of Eq. (8.2.7). While the addition of the Fock term renders the time-independent Hartree–Fock potential $V^{HF}(\mathbf{x}_1, \mathbf{x}_2)$ non-local in position space, it is, in fact, local in momentum space. Thus, the effects of the principle of indistinguishability do *not* incorporate correlation effects that limit the lifetime of the self-consistent single-particle energy states. Correspondingly, the spectral weight $\overline{A}^{HF}(\mathbf{p},\omega)$ has frequency delta function form, as in Eq. (8.2.24), for every value of momentum.

8.3 Non-Correlation Decoupling Procedures: Collisional Effects; *Sum of Ladder Diagrams* Integral Equation

In order to determine correlation effects on the joint propagation of two particles due to particle–particle interactions in terms of causing transitions between single-particle energy eigenstates, thus rendering their lifetimes finite, it is necessary to go beyond the Hartree and Hartree–Fock approximations for $\overline{G}_2$. Of course, this entails analysis of the equation of motion for $\overline{G}_2$, which couples to $\overline{G}_3$; and it is at this stage that the next approximation must be made. This equation of motion for $\overline{G}_2$ is given by Eq. (8.1.8), which we write in compact form using the definition $\mathcal{V}(1-3) = \mathcal{V}(\mathbf{x}_1 - \mathbf{x}_3)\,\delta(t_1 - t_3)$ as $(1 = \mathbf{x}_1, t_1;\ d2 = d^{(3)}\mathbf{x}_2 dt_2;\ \delta(1-1') = \delta^{(3)}(\mathbf{x}_1 - \mathbf{x}_1')\,\delta(t_1 - t_1')$, etc.):

$$\left(i\frac{\partial}{\partial t_1} + \mu - h_1^{(1)}\right)\overline{G}_2(1,2;1',2') \mp i\int d3\,\mathcal{V}(1-3)\,\overline{G}_3(1,2,3;1',2',3^+)$$
$$= \delta(1-1')\,\overline{G}_1(2;2') \pm \delta(1-2')\,\overline{G}_1(2;1'). \tag{8.3.1}$$

Considering the $\overline{G}_1$ equation in conjunction with this,

$$\left(i\frac{\partial}{\partial t_1} + \mu - h_1^{(1)}\right)\overline{G}_1(1;1') \mp i\int d3\,\mathcal{V}(1-3)\,\overline{G}_2(1,3;1',3^+) = \delta(1-1'), \tag{8.3.2}$$

we multiply it by $\overline{G}_1(2;2')$ from the right, obtaining

$$\begin{aligned}&\left(i\frac{\partial}{\partial t_1} + \mu - h_1^{(1)}\right)\overline{G}_1(1;1')\,\overline{G}_1(2;2')\\ &\quad \mp i\int d3\,\mathcal{V}(1-3)\,\overline{G}_2(1,3;1',3^+)\,\overline{G}_1(2;2') = \delta(1-1')\,\overline{G}_1(2;2').\end{aligned} \tag{8.3.3}$$

Interchanging the free indices $1' \leftrightarrow 2'$, we also have

$$\begin{aligned}&\left(i\frac{\partial}{\partial t_1} + \mu - h_1^{(1)}\right)\overline{G}_1(1;2')\,\overline{G}_1(2;1')\\ &\quad \mp i\int d3\,\mathcal{V}(1-3)\,\overline{G}_2(1,3;2',3^+)\,\overline{G}_1(2;1') = \delta(1-2')\,\overline{G}_1(2;1').\end{aligned} \tag{8.3.4}$$

Combining Eqns. (8.3.3) and (8.3.4) with a $\pm$ sign for Bosons and Fermions, we have

$$\begin{aligned}&\left(i\frac{\partial}{\partial t_1} + \mu - h_1^{(1)}\right)[\overline{G}_1(1;1')\,\overline{G}_1(2;2') \pm \overline{G}_1(1;2')\,\overline{G}_1(2;1')]\\ &\quad \mp i\int d3\,\mathcal{V}(1-3)\,[\overline{G}_2(1,3;1',3^+)\,\overline{G}_1(2;2') \pm \overline{G}_2(1,3;2',3^+)\,\overline{G}_1(2;1')]\\ &\quad = \delta(1-1')\,\overline{G}_1(2;2') \pm \delta(1-2')\,\overline{G}_1(2;1').\end{aligned} \tag{8.3.5}$$

Subtraction of Eq. (8.3.5) from Eq. (8.3.1) cancels the delta function driving terms on the right-hand sides, yielding a homogeneous equation for the correlated part of $\overline{G}_2$, as

$$\begin{aligned}&\left(i\frac{\partial}{\partial t_1} + \mu - h_1^{(1)}\right)\Big\{\overline{G}_2(1,2;1',2')\\ &\quad -\left[\overline{G}_1(1;1')\,\overline{G}_1(2;2') \pm \overline{G}_1(1;2')\,\overline{G}_1(2;1')\right]\Big\}\\ &\quad \mp i\int d3\,\mathcal{V}(1-3)\,\Big\{\overline{G}_3(1,2,3;1',2',3^+)\\ &\quad -\left[\overline{G}_2(1,3;1',3^+)\,\overline{G}_1(2;2') \pm \overline{G}_2(1,3;2',3^+)\,\overline{G}_1(2;1')\right]\Big\} = 0.\end{aligned} \tag{8.3.6}$$

Eq. (8.3.6) is a fine point of departure to analyze the effects of particle–particle collisions on the correlated propagation of two particles. The problem of approximating $\overline{G}_3$ is akin to the problem of approximating a three-particle wave function in order to learn about the collisional correlation effects in a two-particle wave function. To start, one may approximate a three-particle wave function by a correlated two-particle wave function multiplied by a one-particle wave function, with appropriate symmetrization/antisymmetrization. In this spirit, Martin and Schwinger make the corresponding $\overline{G}_3$ factorization approximation[1]

$$\begin{aligned}\overline{G}_3\left(1,2,3;1',2',3^+\right) \approx\, & \overline{G}_2\left(1,3;1',3^+\right)\overline{G}_1\left(2;2'\right) \\ & \pm \overline{G}_2\left(1,3;2',3^+\right)\overline{G}_1\left(2;1'\right) \\ & \pm \overline{G}_2\left(1,3;1',2'\right)\overline{G}_1\left(2;3\right), \end{aligned} \tag{8.3.7}$$

arguing that this is appropriate for long-ranged slowly varying forces and short-range forces at low densities. It has been successfully applied in the theory of hard sphere interactions and nuclear matter. The resulting equation for $\overline{G}_2$ is the well-known *sum of ladder diagrams* integral equation for $\overline{G}_2$,

$$\begin{aligned}&\left(i\frac{\partial}{\partial t_1}+\mu-h_1^{(1)}\right)\begin{pmatrix}\overline{G}_2\left(1,2;1',2'\right) \\ -\left[\overline{G}_1\left(1;1'\right)\overline{G}_1\left(2;2'\right)\pm\overline{G}_1\left(1;2'\right)\overline{G}_1\left(2;1'\right)\right]\end{pmatrix} \\ &\quad -i\int d3\, \mathcal{V}\left(1-3\right)\overline{G}_2\left(1,3;1',2'\right)\overline{G}_1\left(2;3\right)=0. \end{aligned} \tag{8.3.8}$$

Introducing the uncorrelated Green's function, $\overline{G}_1^0$, which is inverse to the one-particle differential operator

$$\left(i\frac{\partial}{\partial t_1}+\mu-h_1^{(1)}\right)\overline{G}_1^0\left(1;1'\right)=\delta\left(1;1'\right), \tag{8.3.9}$$

Eq. (8.3.8) may be rewritten as

$$\begin{aligned}\overline{G}_2\left(1,2;1',2'\right) =& \left[\overline{G}_1\left(1;1'\right)\overline{G}_1\left(2;2'\right)\pm\overline{G}_1\left(1;2'\right)\overline{G}_1\left(2;1'\right)\right] \\ & +i\int d3\int d4\, \overline{G}_1^0\left(1;4\right)\overline{G}_1\left(2;3\right)\mathcal{V}\left(4-3\right)\overline{G}_2\left(4,3;1',2'\right), \end{aligned} \tag{8.3.10}$$

which takes the form of the *sum of ladder diagrams* integral equation with a further approximation often made by replacing $\overline{G}_1 \to \overline{G}_1^0$ throughout Eq. (8.3.10):

$$\begin{aligned}\overline{G}_2\left(1,2;1',2'\right) =& \left[\overline{G}_1^0\left(1;1'\right)\overline{G}_1^0\left(2;2'\right)\pm\overline{G}_1^0\left(1;2'\right)\overline{G}_1^0\left(2;1'\right)\right] \\ & +i\int d3\int d4\, \overline{G}_1^0\left(1;4\right)\overline{G}_1^0\left(2;3\right)\mathcal{V}\left(4-3\right)\overline{G}_2\left(4,3;1',2'\right). \end{aligned} \tag{8.3.11}$$

(This equation, as other Green's function equations, is understood to be set in the fundamental interval $[0,\tau]$ for each time variable, and the time integration limits are $0 \to \tau$.)

Since $\overline{G}_2 - \overline{G}_1 \overline{G}_1$ describes the correlated joint propagation of two interacting particles, it embodies a detailed description of their scattering by one another. An analysis of it along the lines of conventional scattering theory can be found in references[3–6] in Section 8.6.

One may iterate Eq. (8.3.11) to obtain the sum of ladder diagrams explicitly. The first iterate provides the Born collision approximation given by

$$\begin{aligned}\overline{G}_2(1,2;1',2') &= \overline{G}_1^0(1;1')\,\overline{G}_1^0(2;2') \pm \overline{G}_1^0(1;2')\,\overline{G}_1^0(2;1') \\ &+ i\int_0^\tau d3 \int_0^\tau d4\, \overline{G}_1^0(1;4)\,\overline{G}_1^0(2;3)\,V(4-3)\,\overline{G}_1^0(4;1')\,\overline{G}_1^0(3;2') \\ &\pm i\int_0^\tau d3 \int_0^\tau d4\, \overline{G}_1^0(1;4)\,\overline{G}_1^0(2;3)\,V(4-3)\,\overline{G}_1^0(4;2')\,\overline{G}_1^0(3;1').\end{aligned} \tag{8.3.12}$$

8.4 Electron–Hole Interaction: Exciton States

While electron–electron Coulomb repulsion pervades our considerations in this chapter (and much of this book), it is also of interest to consider the *attractive* interaction between electrons and positively charged *holes* in crystalline solids. Such holes occur as vacant states in valence bands that are almost filled with electrons, and they behave as particles of opposite (positive) charge with respect to charge transport. Moreover, we will treat holes along with conduction band electrons using an effective mass approximation so that their kinetics are those of free particles, but with different effective masses m_e, m_h.

A full, formal derivation of the equations of motion of such a model electron–hole system will be provided in Chapter 11, but it is relatively easy to anticipate the principal results even at this stage, if we restrict our considerations to just the electron–hole Coulomb attraction, neglecting other electron–electron, hole–hole, electron–phonon and hole–phonon interactions. In this regard, the two-particle $H^{(2)}$-Hamiltonian of Eq. (3.2.16) is supplanted by

$$\begin{aligned}H^{(2)} = -\frac{1}{2}\int d^{(3)}\mathbf{x}' \int d^{(3)}\mathbf{x}''\, &\Psi^+(\mathbf{x}',t)\,\Phi^+(\mathbf{x}'',t)\,V(\mathbf{x}'-\mathbf{x}'')\,\Phi(\mathbf{x}'',t) \\ &\Psi(\mathbf{x}',t) - \text{Hermitian conjugate},\end{aligned}$$

or

$$H^{(2)} = -\int d^{(3)}\mathbf{x}' \int d^{(3)}\mathbf{x}''\, \Psi^+(\mathbf{x}',t)\,\Psi(\mathbf{x}',t)\,V(\mathbf{x}'-\mathbf{x}'')\,\Phi^+(\mathbf{x}'',t)\,\Phi(\mathbf{x}'',t), \tag{8.4.1}$$

where $\hat{\Psi}^+$, $\hat{\Psi}$ are creation, annihilation operators of electrons (e), while $\hat{\Phi}^+$, $\hat{\Phi}$ create and annihilate holes (h). The e-operators obey the usual canonical Fermi anticommutation relations among themselves at equal times, as do the h-operators, but both e-operators simply anticommute with each of the h-operators at equal times. Furthermore, the negative sign in $H^{(2)}$ reflects the fact that we are considering Coulomb attraction rather than repulsion. We also shift the single-particle energy origin by the chemical potential μ_e for electrons and μ_h for holes (by which we also account for the energy gap E_g between the valence and conduction bands).

With these few changes, it is apparent that a repetition of our derivations of one-particle and two-particle Green's function equations of motion can be carried out with $H^{(2)}$ given by Eq. (8.4.1) (instead of Eq. (3.2.16)), yielding the following results for the one-electron ($\overline{G}^{e}_{10}$) and one-hole ($\overline{G}^{h}_{01}$) Green's functions:

$$\left(i\frac{\partial}{\partial t_1} + \frac{\nabla_1^2}{2m_e} + \mu_e\right)\overline{G}^{e}_{10}(1,2) = \delta(1,2) + i\int d3'\mathcal{V}(1,3')\overline{G}_{11}(1,3';2,3'^{+}) \qquad (8.4.2)$$

and

$$\left(i\frac{\partial}{\partial t_1'} + \frac{\nabla_{1'}^2}{2m_h} + \mu_h\right)\overline{G}^{h}_{01}(1'2') = \delta(1',2') + i\int d3\mathcal{V}(1',3)\overline{G}_{11}(3,1';3^{+},2'), \qquad (8.4.3)$$

with $\overline{G}_{10}, \overline{G}_{01}$ defined, as usual, by

$$\overline{G}^{e}_{10}(1,2) = -i\varepsilon\left\langle(\hat{\Psi}(1)\hat{\Psi}^{+}(2))_{+}\right\rangle, \qquad (8.4.4)$$

$$\overline{G}^{h}_{01}(1',2') = -i\varepsilon\left\langle(\hat{\Phi}(1')\hat{\Phi}^{+}(2'))_{+}\right\rangle. \qquad (8.4.5)$$

The two-particle (one-electron and one-hole) Green's function $\overline{G}_{11}$ is defined as

$$\overline{G}_{11}(1,2;1',2') = (-i)^2\varepsilon\left\langle(\hat{\Psi}(1)\hat{\Phi}(2)\hat{\Phi}^{+}(2')\hat{\Psi}^{+}(1'))_{+}\right\rangle \qquad (8.4.6)$$

(note that the first two arguments are for annihilation operators of *electrons and holes*, respectively). The Green's function $\overline{G}_{11}$ describes the probability amplitude for the joint propagation of one electron and one hole with attractive Coulomb coupling, with the electron created at space-time point $1'$ and the hole created at space-time point $2'$, propagating to their space-time points of annihilation $1, 2$ for the electron and hole, respectively. Repetition of the derivation of the equations of motion for the two-particle Green's function (with similar approximations) yields the $\overline{G}_{11}$ integral equation in a form analogous to Eq. (8.3.11):

$$\begin{aligned}&\overline{G}_{11}(1,2;1',2')\\ &= \overline{G}^{e}_{10}(1,1')\overline{G}^{h}_{01}(2,2') - i\int d3\int d3'\overline{G}^{e}_{10}(1,3)\overline{G}^{h}_{01}(2,3')\mathcal{V}(3-3')\overline{G}_{11}(3,3';1',2').\end{aligned} \qquad (8.4.7)$$

This is structurally the same as Eqns. (8.3.10 and 8.3.11), except for the absence of the "Fock" term, $\overline{G}^{e}_{10}(1,2')\overline{G}^{h}_{01}(2,1')$, which is to be expected since the electron and hole are, in fact, distinguishable by their different effective masses so that the Pauli principle is not applicable (also, there is a minus sign on the right due to the attractive binding interaction).

The analysis of Eq. (8.4.7) for $\overline{G}_{11}$ may be undertaken along the lines of conventional scattering theory used in the analysis of Eqns. (8.3.11 and 8.3.12), for $\overline{G}_2$. Having referred the reader to References 3 and 4 for the latter, and noting that for the case at hand $\overline{G}_{11}$ differs somewhat because of the difference of electron and hole masses, it is worthwhile to discuss the scattering analysis in some detail here. Introducing a position-space direct product matrix notation, $\langle \mathbf{r}_1^e | \otimes \langle \mathbf{r}_1^h | = \langle \mathbf{r}_1^e, \mathbf{r}_1^h |$, we have $\overline{G}_{11}(1,2;1',2') \equiv \overline{G}_{11}(\mathbf{r}_1^e, t_1, \mathbf{r}_2^h, t_2; \mathbf{r}_{1'}^e, t_{1'}, \mathbf{r}_{2'}^h, t_{2'}) = \left\langle \mathbf{r}_1^e, \mathbf{r}_2^h | \overline{G}_{11}(t_1, t_2; t_{1'}, t_{2'}) | \mathbf{r}_{1'}^e, \mathbf{r}_{2'}^h \right\rangle$, and correspondingly, Eq. (8.4.7) takes the form (we assume time-translational invariance; $\mathcal{V} \to \mathcal{V}_{eh}\delta(t_3 - t_{3'})$ is taken to be instantaneous):

$$\overline{G}_{11}(t_1, t_2; t_{1'}, t_{2'}) = \overline{G}_{10}^e(t_1 - t_{1'})\overline{G}_{01}^h(t_2 - t_{2'})$$
$$-i \int dt_3 \overline{G}_{10}^e(t_1 - t_3)\overline{G}_{01}^h(t_2 - t_3)\mathcal{V}_{eh}\overline{G}_{11}(t_3, t_3; t_{1'}, t_{2'}). \tag{8.4.8}$$

(Positional integrals over the product space are understood in accordance with our direct product-positional matrix notation when products of Green's functions occur.) This equation can be solved for $\overline{G}_{11}(t_1, t_2; t_{1'}, t_{2'})$, with arbitrary time indices on the left, if we determine $\overline{G}_{11}(t_3, t_3; t_{1'}, t_{2'})$ having the first two time indices (those of the electron and hole annihilation operators) equal on the right. The equality of these two time indices is forced by the instantaneous nature of the Coulomb potential, $\mathcal{V}_{eh} \to \delta(t_3 - t_{3'})$. We obtain an integral equation for this quantity by setting $t_1 = t_2$ throughout Eq. (8.4.8), with the result

$$\overline{G}_{11}(t_1, t_1; t_{1'}, t_{2'}) = \overline{G}_{10}^e(t_1 - t_{1'})\overline{G}_{01}^h(t_1 - t_{2'})$$
$$-i \int dt_3 \overline{G}_{10}^e(t_1 - t_3)\overline{G}_{01}^h(t_1 - t_3)\mathcal{V}_{eh}\overline{G}_{11}(t_3, t_3; t_{1'}, t_{2'}). \tag{8.4.9}$$

As an inhomogeneous linear integral equation, Eq. (8.4.8) may be brought into more tractable form by introducing a matrix Ω, which serves as a "Green's function" by replacing the inhomogeneous driving term, $\overline{G}_{10}^e\overline{G}_{01}^h \to \delta(t_1 - t_{1'})\delta(t_2 - t_{2'})$, by the indicated delta functions (Ω is also in the nature of a resolvent), with the defining equation given by

$$\Omega(t_1, t_2; t_{1'}, t_{2'}) = \delta(t_1 - t_{1'})\delta(t_2 - t_{2'})$$
$$-i \int dt_3 \overline{G}_{10}^e(t_1 - t_3)\overline{G}_{01}^h(t_2 - t_3)\mathcal{V}_{eh}\,\Omega(t_3, t_3; t_{1'}, t_{2'}), \tag{8.4.10}$$

so that

$$\overline{G}_{11}(t_1, t_2; t_{1'}, t_{2'}) = \int dt_3 \int dt_{3'} \Omega(t_1, t_2; t_3, t_{3'})\overline{G}_{10}^e(t_3 - t_{1'})\overline{G}_{01}^h(t_{3'} - t_{2'}). \tag{8.4.11}$$

Again, we note that the first two time indices of $\Omega(t_3, t_3; t_{1'}, t_{2'})$ on the right of Eq. (8.4.10) are equal by virtue of the instantaneous nature of the Coulomb potential, $\mathcal{V}_{eh} \to \mathcal{V}_{eh}\delta(t_3 - t_{3'})$. To solve Eq. (8.4.10), we therefore set $t_2 = t_1$ throughout, obtaining

$$\Omega(t_1,t_1;t_{1'},t_{2'}) = \delta(t_1-t_{1'})\delta(t_1-t_{2'}) \\ -i\int dt_3\overline{G}^e_{10}(t_1-t_3)\overline{G}^h_{01}(t_1-t_3)\mathcal{V}_{eh}\,\Omega(t_3,t_3;t_{1'},t_{2'}). \tag{8.4.12}$$

Noting that the driving term of this integral equation (Eq. (8.4.12)) is $\delta(t_1-t_{1'})\delta(t_1-t_{2'}) = \delta(t_1-t_{1'})\delta(t_{1'}-t_{2'})$, it is readily verified (by substitution or iteration) that the solution has the form[1]

$$\Omega(t_1,t_1;t_{1'},t_{2'}) = \Omega(t_1,t_1;t_{1'},t_{1'})\delta(t_{1'}-t_{2'}). \tag{8.4.13}$$

With time-translational invariance, we define

$$\Omega(t_1-t_{1'}) \equiv \Omega(t_1,t_1;t_{1'},t_{1'}), \tag{8.4.14}$$

and find from Eq. (8.4.12) the result

$$\Omega(t_1-t_{1'}) = \delta(t_1-t_{1'}) - i\int dt_3\overline{G}^e_{10}(t_1-t_3)\overline{G}^h_{01}(t_1-t_3)\mathcal{V}_{eh}\,\Omega(t_3-t_{1'}). \tag{8.4.15}$$

In the context of thermal equilibrium, the time arguments and integration range lie in the interval $t_0 \to t_0+\tau = t_0 - i\beta$, and one can proceed using a Matsubara Fourier series analysis (which leads to a Boson spectral structure, since advancing t_1 by τ actually advances *two* time arguments incurring two $\pm$ factors; $(\pm)^2 = +1$), followed by analytic continuation to real times. This is done for electron–electron interaction in the Martin–Schwinger paper.[3–6] To point the way without getting into that much detail, we will instead analyze Eq. (8.4.15) for one electron in attractive Coulomb interaction with one hole in terms of retarded Green's functions using full real-time axis Fourier transforms, $t_1 - t_{1'} \to \omega$, with the result

$$\Omega(\omega) = 1 - i\int_{-\infty}^{\infty}\frac{d\omega_e}{2\pi}\overline{G}^e_{10}(\omega_e)\overline{G}^h_{01}(\omega-\omega_e)\mathcal{V}_{eh}\,\Omega(\omega). \tag{8.4.16}$$

Noting that $\Omega(\omega)$ is free from the ω_e-integral on the right, we have the positional product-space matrix equation,

$$\left[1 + i\left(\int_{-\infty}^{\infty}\frac{d\omega_e}{2\pi}\overline{G}^e_{10}(\omega_e)\overline{G}^h_{01}(\omega-\omega_e)\right)\mathcal{V}_{eh}\right]\Omega(\omega) = 1, \tag{8.4.17}$$

or

$$\Omega(\omega) = \left[1 + i\left(\int_{-\infty}^{\infty}\frac{d\omega_e}{2\pi}\overline{G}^e_{10}(\omega_e)\overline{G}^h_{01}(\omega-\omega_e)\right)\mathcal{V}_{eh}\right]^{-1}. \tag{8.4.18}$$

The ω_e-integral involves single-particle Green's functions G^e_{10} and G^h_{01}, which we take approximately in the absence of Coulomb interaction as ($G^e_{10} \to G^0_{10}$ and $G^h_{01} \to G^0_{01}$; suppress μ_e, μ_h):

$$\overline{G}^0_{10}(\omega) = \frac{1}{\omega - h^{(1)}_e + i\delta}; \quad \overline{G}^0_{01}(\omega) = \frac{1}{\omega - h^{(1)}_h + i\delta}. \tag{8.4.19}$$

Recalling that $h_{e,h}^{(1)}$ is diagonal in position representation, (Eq. (1.2.24)), the ω_e-integral has the form

$$\begin{aligned}
\int_{-\infty}^{\infty} \frac{d\omega_e}{2\pi} \overline{G}_{10}^{0}(\omega_e)\overline{G}_{01}^{0}(\omega-\omega_e) &= \int \frac{d\omega_e}{2\pi} \frac{1}{\omega_e - h_e^{(1)} + i\delta} \frac{1}{\omega - \omega_e - h_h^{(1)} + i\delta'} \\
&= \frac{1}{\omega - h_e^{(1)} - h_h^{(1)} + i\delta} \int \frac{d\omega_e}{2\pi} \left[\frac{1}{\omega_e - h_e^{(1)} + i\delta} - \frac{1}{\omega_e - \omega + h_h^{(1)} - i\delta'} \right] \\
&= \frac{1}{\omega - h_e^{(1)} - h_h^{(1)} + i\delta} \int \frac{d\omega_e}{2\pi} \left[-\pi i\delta(\omega_e - h_e^{(1)}) - (\pi i\delta(\omega_e - \omega + h_h^{(1)})) \right] \\
&= \frac{-i}{\omega - h_e^{(1)} - h_h^{(1)} + i\delta},
\end{aligned} \tag{8.4.20}$$

where the evaluation was carried out in terms of Dirac prescription principal part and δ-function decompositions of the terms on the right. The result for $\Omega(\omega)$ in Eq. (8.4.18) is

$$\begin{aligned}
\Omega(\omega) &= \left[1 + \frac{1}{\omega - h_e^{(1)} - h_h^{(1)} + i\delta} V_{eh} \right]^{-1} \\
&= \left[\omega - h_e^{(1)} - h_h^{(1)} + V_{eh} + i\delta \right]^{-1} \left[\omega - h_e^{(1)} - h_h^{(1)} \right].
\end{aligned} \tag{8.4.21}$$

The frequency poles of $\Omega(\omega)$ define the modes and resonances of $\overline{G}_{11}$, including discrete bound states that occur in consequence of the attractive Coulomb potential V_{eh} between electrons and holes. To put this into concrete terms, we take $h_e^{(1)} = -\nabla_e^2/2m_e + E_g$ and $h_h^{(1)} = -\nabla_h^2/2m_h$ (E_g = gap energy). Defining center of mass and relative coordinates according to

$$\mathbf{R} = \frac{m_e \mathbf{r}_e + m_h \mathbf{r}_h}{m_e + m_h}; \ \mathbf{r} = \mathbf{r}_e - \mathbf{r}_h, \tag{8.4.22}$$

it is readily seen that

$$h_e^{(1)} + h_h^{(1)} = -\frac{\nabla_R^2}{2M} - \frac{\nabla_r^2}{2\mu} + E_g, \tag{8.4.23}$$

where $M = m_e + m_h$ and $\mu^{-1} = m_e^{-1} + m_h^{-1}$ (μ is the reduced mass). The first term, $-\nabla_R^2/2M$, is the kinetic energy of the center-of-mass motion, and the second term, $-\nabla_r^2/2\mu$, is the kinetic energy of the relative electron–hole motion. The latter, combined with the attractive Coulomb potential $V_{eh}(r)$, which depends only on the relative coordinate, yields discrete frequency poles of $\Omega(\omega)$ as given by Eq. (8.4.21) (as well as poles of $\overline{G}_{11}$) that are characteristic of the spectrum of a hydrogenic atom, as in Eq. ((1.1.17)). These discrete frequency poles define *exciton* states as bound states of an electron and hole in the present context, in accordance with the vanishing of the denominator of $\Omega(\omega)$ given by

$$\omega + \frac{\nabla_R^2}{2M} + \left(\frac{\nabla_r^2}{2\mu} + \mathcal{V}_{eh}(r) - E_g\right) = 0. \tag{8.4.24}$$

If, for convenience, we take the center of mass to be stationary and put $E_g \rightarrow 0$, Eq. (8.4.21) may be rewritten as

$$\Omega(\omega) = \left[\omega + \frac{\nabla_r^2}{2\mu} + \mathcal{V}_{eh}(r) + i\delta\right]^{-1} \left[\omega + \frac{\nabla_r^2}{2\mu}\right],$$

or

$$\Omega(\mathbf{r},\mathbf{r}';\omega) = \int d^{(3)}\mathbf{r}''\, G_{hyd}(\mathbf{r},\mathbf{r}'';\omega) \left(G_{\mathcal{V}=0}(\mathbf{r}'',\mathbf{r}';\omega)\right)^{-1}, \tag{8.4.25}$$

where

$$G_{hyd}(\mathbf{r},\mathbf{r}'';\omega) = \left[\omega + \frac{\nabla_r^2}{2\mu} + \mathcal{V}_{eh}(r) + i\delta\right]^{-1} \delta^{(3)}(\mathrm{r}-\mathrm{r}'') \tag{8.4.26}$$

is the retarded one-particle Green's function for a hydrogen atom with the attractive Coulomb force center permanently stationed at the origin, and with the electron mass replaced by the reduced mass. $G_{\mathcal{V}=0}(\mathbf{r},\mathbf{r}'';\omega)$ is its "free" electron counterpart with no attractive Coulomb center. If the associated discrete hydrogen-like energy eigenstates $\phi_n(\mathbf{r})$ have energy eigenvalues E_n, then Eqns. (5.1.20 and 5.1.25) provide G_{hyd} for excitons in position representation as

$$G_{hyd}(\mathbf{r},\mathbf{r}';\omega) = \sum_n \frac{\phi_n(\mathbf{r})\phi_n^*(\mathbf{r}')}{\omega - E_n + i\delta}, \tag{8.4.27}$$

where E_n and ϕ_n are given by Eq. (1.1.17) and Eq. (1.1.16), respectively, with μ in place of m. Such exciton states are observed in the optical spectra of semiconductors. Similar results obtain for substitutional impurity atom levels in semiconductors. An interesting calculation of the influence of an electron–hole plasma on exciton energy levels was carried out by Röpke, *et al.*[7]

The Ω-matrix and its hydrogenic exciton energy level spectrum also play an important role in the self-energy of the *single*-electron (or hole) Green's function equation through its Coulombic coupling to G_{11} in Eq. (8.4.2) (or Eq. (8.4.3)), where we denote the fully coupled one-electron Green's function by $\overline{G}_{10}^e$ to distinguish it from the non-interacting $\overline{G}_{10}^0$ approximation of Eqns. (8.4.19 and 8.4.20). We also use position-space matrix notation while keeping time dependence explicit, rewriting the fully interacting one-electron $\overline{G}_{10}^e$-equation (Eq. (8.4.2), with its Coulomb-coupling to holes) in terms of the Ω matrix using Eq. (8.4.11), with the result

$$\begin{aligned}
&\left(i\frac{\partial}{\partial t_1} - h_e^{(1)} + \mu_e\right)\overline{G}_{10}^e(t_1,t_2) \\
&\quad = \delta(t_1 - t_2) + i\mathcal{V}_{eh}\overline{G}_{11}(t_1,t_1;t_2,t_1^+) \\
&\quad = \delta(t_1 - t_2) + i\mathcal{V}_{eh}\int dt_3 \int dt_{3'}\Omega(t_1,t_1;t_3,t_{3'})\overline{G}_{10}^e(t_3 - t_2)\overline{G}_{01}^h(t_{3'} - t_1^+).
\end{aligned} \tag{8.4.28}$$

Using Eqns. (8.4.13) and (8.4.14), we have

$$\left(i\frac{\partial}{\partial t_1} - h_e^{(1)} + \mu_e\right)\overline{G}_{10}^e(t_1, t_2)$$
$$= \delta(t_1 - t_2) + iV_{eh}\int dt_3 \Omega(t_1 - t_3)\overline{G}_{10}^e(t_3 - t_2)\overline{G}_{01}^h(t_3 - t_1^+). \tag{8.4.29}$$

Furthermore, defining a T-matrix for the problem at hand as

$$T(t_1, t_2) = V_{eh}\,\Omega(t_1, t_2), \tag{8.4.30}$$

we find

$$\left(i\frac{\partial}{\partial t_1} - h_e^{(1)} + \mu_e\right)\overline{G}_{10}^e(t_1 - t_2) = \delta(t_1 - t_2) + i\int dt_3\, T(t_1 - t_3)\overline{G}_{10}^e(t_3 - t_2)\overline{G}_{01}^h(t_3 - t_1^+). \tag{8.4.31}$$

Fourier-transforming to ω-frequency representation this becomes

$$\left(\omega - h_e^{(1)}\right)\overline{G}_{10}^e(\omega) = 1 + i\int \frac{d\omega'}{2\pi} T(\omega')G_{01}^h(\omega' - \omega)\overline{G}_{10}^e(\omega), \tag{8.4.32}$$

and we can identify the one-electron self-energy (Eq. (5.4.25)) as

$$\sum(\omega) = -i\int \frac{d\omega'}{2\pi} T(\omega')\overline{G}_{01}^h(\omega' - \omega)$$
$$= -i\int \frac{d\varpi}{2\pi} T(\omega + \varpi)\overline{G}_{01}^h(\varpi). \tag{8.4.33}$$

The real part of $\sum(\omega)$ shifts the single-electron energy spectrum in consequence of the electron–hole interaction, while the imaginary part of $\sum(\omega)$ imparts a finite lifetime, $\gamma(\omega)$, to electron states due to electron–hole scattering. The generalization to equilibrium thermodynamic Green's functions (for the case of electron–electron interaction, rather than electron–hole) is discussed in detail in References 3–6, which also relate this description of scattering to the optical theorem of conventional scattering theory.

8.5 Chapter 8 Problems

P.8.1. Verify Eqns. (8.1.12–8.1.15).

P.8.2. Verify Eqns. (8.2.17–8.2.26).

P.8.3. Derive the second Born Collision Approximation for $\overline{G}_2$ in terms of $\overline{G}_1^0$ by iterating Eq. (8.3.11) twice. (The first iterate is given by Eq. (8.3.12).)

P.8.4. Repeat the derivation of the equation of motion for the two-particle Green's function (with appropriate approximations) to derive Eq. (8.4.7) for the G_{11} electron–hole Green's function. Why is there no "Fock" term?

P.8.5. Verify all steps in the analysis of time dependence in Eqns. (8.4.8–8.4.33) and explicitly determine the leading terms of the Ω-matrix, the T-matrix and the self-energy.

P.8.6. Considering thermal equilibrium, carry out the analysis of Eq. (8.4.15) in terms of a Matsubara Fourier series and analytically continue the result to the real-time axis. (Hint: See Reference 1, Section 5, and Reference 4, Appendix.)

P.8.7. Discuss the role of exciton implicit in Ω (Eq. (8.4.25)) in T (Eq. (8.4.30)) and in the one-electron self-energy $\sum$ (Eq. (8.4.33)).

P.8.8. Suppose a constant uniform electric field is applied to pull apart the electron and hole of the exciton. How much energy must it supply?

P.8.9. Consider the electron–hole interaction potential $\mathcal{V}$ to be a contact interaction having a $\delta-$ function form, as

$$\mathcal{V}(1-2) = \lambda\delta^{(3)}(\mathbf{r}_1 - \mathbf{r}_2)\delta(t_1 - t_2) = \lambda\delta(1-2).$$

Determine the result for the exciton.

P.8.10. Suppose that the attractive electron–hole interaction $\mathcal{V}$ is that of a simple harmonic oscillator. Determine the resulting G_{11} function and discuss excitons for this system.

P.8.11. In Section 8.4, our discussion addressed a three-dimensional electron–hole system with attractive Coulomb interaction. How would the results change if the e–h system is confined to motion on a two-dimensional planar sheet?

8.6 References for Chapter 8

R.8.1. P.C. Martin and J. Schwinger, Phys. Rev. *115*, 1342, Eq. (6.57) (1959).

R.8.2. K. Huang, "Statistical Mechanics," 2nd Ed., John Wiley & Sons Publ., p. 65 (1987).

R.8.3. P.C. Martin and J. Schwinger, Phys. Rev., *115*, 1342, Section 5 (1959).

R.8.4. L.P. Kadanoff and G. Baym, "Quantum Statistical Mechanics," W.A. Benjamin Inc., Chapter 13 (1962).

R.8.5. N. Ashby, "Applications of Field Theory to Electrical Conductivity," Ph.D. Thesis, Harvard University (1961).

R.8.6. N. Ashby in "Lectures in Theoretical Physics," Boulder, University of Colorado Press (1960).

R.8.7. G. Röpke, T. Seifert, H. Stoltz and R. Zimmermann, Phys. Stat. Sol. *(b) 100*, 215 (1980).

9

Non-Equilibrium Green's Functions: Variational Relations and Approximations for Particle Interactions

Abstract

Chapter 9 opens with the definitions of physical and mathematically convenient nonequilibrium Green's functions and describes their close relationship: In this regard, coupling-correlated (C) single- and multi-particle propagation dynamics are fully addressed. The physical Green's function averages with respect to an initial thermal equilibrium distribution, whereas the mathematically convenient type averages with respect to the actual time-development operator $U(t_0 + \tau, t_0)$, which incorporates an agent (field) driving the system out of equilibrium. The example of linear conductivity is discussed, expressing it in terms of the two-particle equilibrium Green's function (Kubo-type formulation). With the admission of particle sources ($S : \eta, \eta^+$) and non-conservation of number, the non-equilibrium Green's functions are constructed with numbers of creation and annihilation operators that may *differ*. It is shown that these multiparticle Green's functions may be derived as variational derivatives with respect to sources η, η^+ of a generating functional $e^{\mathcal{W}} = Tr\,[U(t_0 + \tau, t_0)]_{CS} \;/\; Tr\,[U(t_0 + \tau, t_0)]_C$, and the *non-interacting* case is seen to yield the n-particle Green's function as a permanent/determinant of single-particle Green's functions, respectively. These variational relations permit the development of a more symmetric formulation of the multiparticle Green's function equations, symmetric in all space-time coordinates. Chapter 9 also introduces *Cumulants* and the *Linked Cluster Theorem*. Furthermore, the Random Phase Approximation (RPA) is derived employing a variational derivative of G_1 with respect to applied potential U to represent the coupling to G_2 in the G_1-equation of motion. A subsequent time-dependent Hartree approximation involving the effective potential V yields the RPA integral equation for the inverse dielectric screening function $K(1,2) = \delta V(1)/\delta U(2)$ (also known as the *sum of ring diagrams*). The density-density correlation function is also discussed. Furthermore, Schwinger's variational differential formulation of perturbation theories for the Green's function, self-energy, vertex operator, and also shielded potential perturbation theory, are reviewed in Chapter 9. Moreover, there is also further discussion of the analytic continuation of integration of products of Green's functions in imaginary time, continued analytically to the real-time

Quantum Statistical Field Theory: Schwinger's Variational Method. Norman J. Morgenstern Horing,
Oxford University Press (2017). © Norman J. Morgenstern Horing. DOI 10.1093/oso/9780198791942.001.0001

axis with time-ordering along the contour of integration in the complex time plane, which leads to the *Langreth Algebra*. Finally, this chapter includes an account of the *Generalized Kadanoff–Baym Ansatz*, which has played an important role in the modern development of quantum transport theory involving highly non-equilibrium dynamics in strong electric fields.

9.1 Non-Equilibrium Green's Functions, Definitions and Variational Relations in Terms of Particle Sources

With the application of an external electromagnetic field, $\mathbf{E}_0(\mathbf{x},t)$, or even just a time-dependent scalar potential, $U(\mathbf{x},t)$ (or a time-dependent magnetic field or other appropriate agent), a many-particle system is driven out of equilibrium. In this situation, the quantized field variables $\psi(1)$, $\psi^+(1')$ obey the usual canonical quantum Hamilton equations discussed in Chapter 4 (Eq. (4.1.64)):

$$\frac{\partial\psi(\mathbf{x},t)}{\partial t} = -i\frac{\delta_l H}{\delta\psi^+(\mathbf{x},t)}, \text{ and } \frac{\partial\psi^+(\mathbf{x},t)}{\partial t} = i\frac{\delta_r H}{\delta\psi(\mathbf{x},t)}, \tag{9.1.1}$$

where the driving fields are embedded in the one-particle part of the Hamiltonian through the replacement $\frac{\hbar}{i}\nabla \to \left[\frac{\hbar}{i}\nabla_1 - \frac{e}{c}\mathbf{A}(1)\right]$ and an additional potential term $U(1)$, so that $H \to H(t) = H_0 + \Delta H(t)$ involves explicit time dependence through $\Delta H(t)$ alone. With these modifications, the resulting equation of motion for $\psi(1)$ has the usual structure of Eq. (3.2.21) with electron–electron interactions, and, consequently, the non-equilibrium single- and multi-particle Green's function equations of motion are similar in form to those discussed in the last few chapters. However, it is important to clarify the appropriate trace-averaging process in the non-equilibrium case. The trace ensemble weighting factor appropriate for the definition of non-equilibrium Green's functions (according to Eqns. (6.1.4) and (7.1.1)), which track the physical time propagation characteristics of a system initially in equilibrium at time t_0, driven out of equilibrium by $\Delta H(t)$ applied starting at t_0, is the initial Gibbs ensemble (we suppress the explicit appearance of μN):

$$e^{-\beta H} \to e^{-\beta H_0} = e^{-iH_0\tau}, \tag{9.1.2}$$

where H_0 is the Hamiltonian *before* $\Delta H(t)$ is turned on. Accordingly, the physical one-particle Green's function $\mathbb{G}_1$ is defined, as before, by

$$\mathbb{G}_1(1,1') = \frac{-i\varepsilon\, Tr\left[e^{-iH_0\tau}\,(\psi(1)\psi^+(1'))_+\right]}{Tr\left[e^{-iH_0\tau}\right]}, \tag{9.1.3}$$

in the full Heisenberg picture (in which $\psi(1)$, $\psi^+(1')$ time-develop under $H_0 + \Delta H(t)$). It is important to note that the ensemble weighting factor $e^{-iH_0\tau}$ used here is *not* the time-development operator for $t = 0 \to \tau$ in the presence of $\Delta H(t)$, so the periodicity/antiperiodicity properties do *not* apply to $\mathbb{G}_1(1,1')$. In due course it will be seen that its variational derivatives are *not* higher order Green's functions of the same type involving time-ordered operator averages, but instead, they are physically retarded commutator counterparts. These facts cause serious problems in the determination of $\mathbb{G}_1(1,1')$ in connection with posing side conditions (no longer periodicity/antiperiodicity) and with the development of perturbation expansions. Therefore, another related Green's function, $G_1(1,1')$, is defined using the actual time-development operator, $U(t_0+\tau,t_0)$, for $H = H_0 + \Delta H(t)$ in place of the Gibbs ensemble weighting factor,

$$G_1(1,1';t_0) = \frac{-i\varepsilon\, Tr\left[U(t_0+\tau,t_0)\,(\psi(1)\psi^+(1'))_+\right]}{Tr\left[U(t_0+\tau,t_0)\right]}, \tag{9.1.4}$$

in the full Heisenberg picture in the presence of time-dependent external fields, which drive the system out of equilibrium. It will be shown below that this G_1 function has variational derivatives that are higher order Green's functions of the same type, facilitating the development of perturbation theories. Moreover, it also has relatively simple periodicity/antiperiodicity properties in time (again suppressing μN):

$$G_1(1,1';t_0)_{t_1=t_0} = \pm G_1(1,1';t_0)_{t_1=t_0+\tau}. \tag{9.1.5}$$

Furthermore, it relates to $\mathbb{G}_1(1,1')$ in the infinitely distant past before $\Delta H(t)$ is turned on, as follows:[1]

$$\lim_{t_0\to-\infty}\left[G_{1\gtrless}(1,1';t_0)\right] = \mathbb{G}_{1\gtrless}(1,1'), \tag{9.1.6}$$

and we will henceforth write it in the limit $t_0\to-\infty$ as $G_{1\gtrless}(1,1')$, or $G_1(1,1')$.

Of course, it is possible to rewrite Eq. (9.1.4) in terms of interaction picture operators $\psi_I(1)$, $\psi_I(1')$ using Eqns. (3.2.5) and (3.2.6), Eq. (3.3.17) and Eqns. (3.3.20) and (3.3.21), and the group properties of the unitary time-development operators. Considering $t_1 > t_1'$, for example,

$$G_{1>}(1,1';t_0) = \frac{-i\varepsilon\, Tr\left[U(t_0+\tau,t_0)U^+(t_1,t_0)\psi_S(\mathbf{x}_1)U(t_1,t_0)U^+(t_1',t_0)\psi_S^+(\mathbf{x}_1')U(t_1',t_0)\right]}{Tr\left[U(t_0+\tau,t_0)\right]} \tag{9.1.7}$$

or

$$G_{1>}(1,1';t_0) = \frac{-i\varepsilon\, Tr\left[e^{-iH_0\tau}V(t_0+\tau,t_1)\psi_I(1)V(t_1,t_1')\psi_I^+(1')V(t_1',t_0)\right]}{Tr\left[U(t_0+\tau,t_0)\right]}. \tag{9.1.8}$$

With $t_1 > t_1'$ confined to the interval $[t_0\to t_0+\tau]$, this expression behaves as if the time-development operator participates in the time ordering of the creation and annihilation operators in the interaction picture (although it does *not* in the Heisenberg picture, Eq. (9.1.4)). In view of this, we write Eq. (9.1.8) in the form

$$G_{1>}(1,1';t_0) = \frac{-i\varepsilon\, Tr\left[e^{-iH_0\tau}(V(t_0+\tau,t_0)\psi_I(1)\psi_I^+(1'))_+\right]}{Tr\left[e^{-iH_0\tau}V(t_0+\tau,t_0)\right]} \tag{9.1.9}$$

in the interaction picture (as employed in Reference 1).[1] Similar considerations apply to $G_{1<}(1,1';t_0)$.

An important example of the difference between the variational derivatives of averages of Heisenberg operators defined with $U(t_0+\tau,t_0)$ in place of the equilibrium ensemble weighting factor, and those defined with the canonical equilibrium ensemble ($e^{-iH_0\tau}$) as weighting factor, is provided by the electrical conductivity. Recalling the current operator of Eq. (2.2.62) (also see Section 3.4):

$$\begin{aligned}\mathbf{J}(\mathbf{x},t) &= \mathbf{J}^{(0)}(\mathbf{x},t)+\mathbf{J}^{(1)}(\mathbf{x},t)\\ &= \frac{\hbar Q}{2mi}\lim_{\mathbf{x}'\to\mathbf{x}}\left[\left(\nabla-\nabla'\right)\psi^{+}(\mathbf{x}',t)\psi(\mathbf{x},t)\right]-\frac{Q^2}{mc}\mathbf{A}(\mathbf{x},t)\,\psi^{+}(\mathbf{x},t)\psi(\mathbf{x},t),\end{aligned} \tag{9.1.10}$$

associated with the electromagnetic interaction part of the Hamiltonian, H',

$$H'(t')=-\frac{1}{c}\int d^{(3)}\mathbf{x}'\mathbf{J}^{(0)}(\mathbf{x}',t')\cdot\mathbf{A}\left(\mathbf{x}',t'\right)-\frac{1}{2c}\int d^{(3)}\mathbf{x}'\mathbf{J}^{(1)}\left(\mathbf{x}',t'\right)\cdot\mathbf{A}\left(\mathbf{x}',t'\right), \tag{9.1.11}$$

we abbreviate the trace average as

$$\frac{Tr\left[U(t_0+\tau,t_0)\mathbf{J}(\mathbf{x},t)\right]}{TrU(t_0+\tau,t_0)}\Rightarrow\frac{\langle t_0+\tau|\mathbf{J}(\mathbf{x},t)|t_0\rangle}{\langle t_0+\tau|t_0\rangle}, \tag{9.1.12}$$

in recognition of the time-development operation of $U(t_0+\tau,t_0)$ from t_0 to $t_0+\tau$. Considering the gauge in which the scalar potential vanishes, we have $\mathbf{E}=\frac{-1}{c}\frac{\partial\mathbf{A}}{\partial t}$, and in frequency representation, $\mathbf{E}=\frac{i\omega}{c}\mathbf{A}$. Correspondingly, we define a tensor that is periodic in imaginary time, $\overline{\mathbf{L}}_p$, which is associated with the conductivity as

$$\overline{\mathbf{L}}_p\left(\mathbf{x},t;\mathbf{x}'t'\right)=\frac{\delta}{\delta\mathbf{A}(\mathbf{x}',t')}\left(\frac{\langle t_0+\tau|\mathbf{J}(\mathbf{x},t)|t_0\rangle}{\langle t_0+\tau|t_0\rangle}\right). \tag{9.1.13}$$

Treating $\mathbf{J}^{(0)}$ first, we find, in accordance with Eq. (4.2.24) (with $Z\to\mathbf{A}$):

$$\begin{aligned}\frac{\delta}{\delta\mathbf{A}(\mathbf{x}',t')}\left(\frac{\left\langle t_0+\tau|\mathbf{J}^{(0)}(\mathbf{x},t)|t_0\right\rangle}{\langle t_0+\tau|t_0\rangle}\right) &= \frac{i}{c}\frac{\left\langle t_0+\tau|\left(\mathbf{J}^{(0)}(\mathbf{x}',t')\mathbf{J}^{(0)}(\mathbf{x},t)\right)_{+}|t_0\right\rangle}{\langle t_0+\tau|t_0\rangle}\\ &\quad-\frac{i}{c}\frac{\left\langle t_0+\tau|\mathbf{J}^{(0)}(\mathbf{x}',t')|t_0\right\rangle}{\langle t_0+\tau|t_0\rangle}\frac{\left\langle t_0+\tau|\mathbf{J}^{(0)}(\mathbf{x},t)|t_0\right\rangle}{\langle t_0+\tau|t_0\rangle}.\end{aligned} \tag{9.1.14}$$

For linear conductivity, the right-hand side is evaluated at zero field, $\mathbf{A}=0$, and the last term vanishes under equilibrium conditions.

Considering next $\mathbf{J}^{(1)}$, we note that $\left\langle t_0+\tau|\mathbf{J}^{(1)}(\mathbf{x},t)|t_0\right\rangle$ may be written with the insertion of unit operators expressed in terms of complete sets as in Eq. (4.2.18), which we denote in terms of a "composition" symbol $\otimes$, as

$$\delta\left\langle t_0+\tau|\mathbf{J}^{(1)}(\mathbf{x},t)|t_0\right\rangle=\delta\left[\langle t_0+\tau|t\rangle\otimes\left\langle t|\mathbf{J}^{(1)}(\mathbf{x},t)|t\right\rangle\otimes\langle t|t_0\rangle\right]. \tag{9.1.15}$$

Noting that the variations $\delta_A \langle t_0 + \tau | t \rangle$ and $\delta_A \langle t | t_0 \rangle$ involve the insertion of $(-\frac{1}{c}\mathbf{J}^{(0)}$ and $\frac{-1}{c}\mathbf{J}^{(1)})$ into the matrix elements, all such terms vanish in the linear conductivity limit because of the presence of $\langle t | \mathbf{J}^{(1)}(\mathbf{x},t) | t \rangle \sim \mathbf{A} \to 0$. The present situation differs from that of Eq. (4.2.24) because of the *non*-vanishing nature of

$$\frac{\delta \langle t | \mathbf{J}^{(1)}(\mathbf{x},t) | t \rangle}{\delta \mathbf{A}(\mathbf{x}',t')} = \left\langle t \left| \frac{\delta \mathbf{J}^{(1)}(\mathbf{x},t)}{\delta \mathbf{A}(\mathbf{x}',t')} \right| t \right\rangle$$
$$= \frac{-Q^2}{mc} \delta^{(3)}(\mathbf{x}-\mathbf{x}')\,\delta(t-t')\,\langle t | \psi^+(\mathbf{x},t)\psi(\mathbf{x},t) | t \rangle, \qquad (9.1.16)$$

due to the explicit linear dependence of $\mathbf{J}^{(1)}$ on $\mathbf{A}$. The linear conductivity result for the periodic tensor $\overline{\mathbf{L}}_p$ (Eqns. (9.1.13–9.1.16)) can be evaluated in terms of equilibrium Green's functions ($\mathbf{A} \to 0$) using the notation $\langle \mathbf{J}(\mathbf{x},t) \rangle \equiv \langle \mathbf{J}(\mathbf{x},t) \rangle_p = \frac{\langle t_0+\tau | \mathbf{J}(\mathbf{x},t) | t_0 \rangle}{\langle t_0 + \tau | t_0 \rangle}$, and it is expressed in terms of the time-ordered periodic current–current correlation function as

$$\overline{\mathbf{L}}_p(\mathbf{x},t;\mathbf{x}',t') = \frac{\delta \langle \mathbf{J}(\mathbf{x},t) \rangle_p}{\delta \mathbf{A}(\mathbf{x}',t')} = \frac{i}{c} \langle (\mathbf{J}^{(0)}(\mathbf{x}',t')\mathbf{J}^{(0)}(\mathbf{x},t))_+ \rangle_p$$
$$-\frac{Q^2}{mc}\delta^{(3)}(\mathbf{x}-\mathbf{x}')\,\delta(t-t')\,\langle \psi^+(\mathbf{x},t)\psi(\mathbf{x},t) \rangle_p . \qquad (9.1.17)$$

(The notation $\langle \cdots \rangle \equiv \langle \cdots \rangle_p$ is used to emphasize averaging with weight $U(t_0+\tau, t_0)$, leading to *periodic* functions.) This periodic result should be compared with the corresponding result of Eq. (3.4.9), in which the current is averaged in an equilibrium state. The variational derivative of Eq. (3.4.9) with respect to $\mathbf{A}(\mathbf{x}',t')$ leads to a causally retarded, *physical* tensor, $\overline{\mathbf{L}}_c$ (counterpart of $\overline{\mathbf{L}}_p$), given by ($\hbar \to 1$),

$$\overline{\mathbf{L}}_c = \frac{i}{c}\eta_+(t-t')\,\langle [\mathbf{J}^{(0)}(\mathbf{x},t),\mathbf{J}^{(0)}(\mathbf{x}',t')] \rangle_c - \frac{Q^2}{mc}\delta^{(3)}(\mathbf{x}-\mathbf{x}')\,\delta(t-t')\,\langle \psi^+(\mathbf{x},t)\psi(\mathbf{x},t) \rangle_c . \qquad (9.1.18)$$

The notation $\langle \cdots \rangle_c$ here denotes averaging the retarded current–current commutator function with the grand canonical ensemble weight factor. This "retarded commutator" response function, $\overline{\mathbf{L}}_c$, provides the linear *physical* conductivity in frequency representation in the gauge in which there is no scalar potential, using the fact that $\mathbf{E} = \frac{i\omega}{c}\mathbf{A}$. Generally, however, it is easier to calculate its periodic counterpart $\overline{\mathbf{L}}_p$ given above in terms of Green's functions, and use it to determine $\overline{\mathbf{L}}_c$, either by examining the individual time orderings of $\overline{\mathbf{L}}_p$ separately and subtracting them to form the commutator function of $\overline{\mathbf{L}}_p$, or by noting the relation of the retarded current–current commutator function to the spectral weight of the periodic function as discussed in Section 7.5 (with the replacements $\psi(\mathbf{x},t) \to \mathbf{J}(\mathbf{x},t)$, $\psi^+(\mathbf{x}',t') \to \mathbf{J}(\mathbf{x}',t')$. Accordingly, we write $\overline{\mathbf{L}}_c(T) = \eta_+(T)\,(L_>(T) - L_<(T))$, and denoting the spectral weight of $\overline{\mathbf{L}}_p$ by $\overline{\mathcal{A}}_L(\omega)$, we have (Eq. (7.5.11)):

$$\overline{\mathbf{L}}_c(T) = \int_{-\infty}^{\infty} \frac{d\omega}{2\pi i} e^{-i\omega T} \int_{-\infty}^{\infty} \frac{d\omega'}{2\pi} \frac{\overline{\mathcal{A}}_L(\omega')}{\omega - \omega' + i\delta} = \int_{-\infty}^{\infty} \frac{d\omega}{2\pi i} e^{-i\omega T} \mathcal{L}(\omega + i\delta), \qquad (9.1.19)$$

where $\mathcal{L}(\omega)$ is the Fourier series (Matsubara) coefficient of the periodic response function $\overline{\mathbf{L}}_p$, continued analytically to all frequencies as the "auxiliary" function evaluated at $\omega + i\delta$. This spectral connection permits the determination of the physical causally retarded linear conductivity from its periodic counterpart, which is more convenient mathematically. (However, the sum rule of the form of Eqns. (7.5.6 and 7.5.7) does not apply here.) It should be noted that the result here, relating linear conductivity to the equilibrium two-particle Green's function representing the current–current correlation function ("Kubo formula"),[2] involves linearization of the current with respect to an *externally applied* electromagnetic field. However, the actual current responds to the effective (screened) field, which, in general, differs from the external field; this difference needs to be taken into account.

Henceforth, we will use the full Heisenberg picture, Eq. (9.1.4), understanding $t_0 \to -\infty$ and suppressing the explicit appearance of t_0, i.e., $G(1,1';t_0) \to G(1,1')$, etc. Multiparticle non-equilibrium Green's functions are also defined with the ensemble weight factor $U(t_0+\tau, t_0)$ in analogy to Eq. (7.1.1),

$$G_n(1,\dots n; 1',\dots n') = (-i)^n \varepsilon \left\langle \left(\psi(1)\dots\psi(n)\psi^+(n')\dots\psi^+(1')\right)_+ \right\rangle, \tag{9.1.20}$$

where ε has the same meaning as in Eq. (7.1.1) (here, $\psi(n) \equiv \psi(\mathbf{x}_n, t_n)$ and $\psi^+(n') = \psi^+(\mathbf{x}'_n, t'_n)$: there are the same number of ψ's as there are of ψ^+'s, but the ψ's have the n-coordinates unprimed, and the ψ^+'s have the n-coordinates primed), and the non-equilibrium average is defined as

$$\left\langle \left(\psi(1)\dots\psi(n)\psi^+(n')\dots\psi^+(1')\right)_+ \right\rangle \equiv \frac{Tr\left[U(t_0+\tau,t_0)(\psi(1)\dots\psi(n)\psi^+(n')\dots\psi^+(1'))_+\right]}{Tr\left[U(t_0+\tau,t_0)\right]}, \tag{9.1.21}$$

in the full Heisenberg picture in the presence of time-dependent external fields that drive the system out of equilibrium. Since $U(t_0+\tau,t_0)$ is the non-equilibrium time-development operator, the defining trace for G_1 is of the form (writing 0 for t_0),

$$G_1(1,1') = \frac{-i\varepsilon \sum_{a'} \left\langle a',\tau \left| (\psi(1)\psi^+(1'))_+ \right| a',0 \right\rangle}{\sum_{a'} \langle a',\tau | a',0\rangle}. \tag{9.1.22}$$

Similar remarks apply to the multiparticle Green's functions G_n,

$$G_n(1,\dots n; 1',\dots n') = (-i)^n \varepsilon \frac{\sum_{a'} \left\langle a',\tau \left| (\psi(1)\dots\psi(n)\psi^+(n')\dots\psi^+(1'))_+ \right| a',0 \right\rangle}{\sum_{a'} \langle a',\tau | a',0\rangle}. \tag{9.1.23}$$

It is clear that the normalizing denominator of Eq. (9.1.23), denoted by

$$e^W \equiv Tr\left[U(t_0+\tau,t_0)\right] = \sum_{a'} \left\langle a',\tau | a',0\right\rangle, \tag{9.1.24}$$

with $U(t_0+\tau,t_0)$ as the time-development operator $t_0 \to t_0+\tau$ (instead of $e^{-\beta(H-\mu N)}$), is in the nature of a non-equilibrium generalization of the grand partition function (Eq. (6.1.6)) and $W \equiv -\beta\Omega$ provides the corresponding non-equilibrium generalization of the grand potential. As in Chapter 4, we distinguish between $e^{W_{CS}}$ in the presence of coupling and sources and e^{W_C} in the presence of coupling without sources, defining

$$e^{\mathcal{W}} = \frac{e^{W_{CS}}}{e^{W_C}} = \frac{Tr\,[U(t_0+\tau,t_0)]_{CS}}{Tr\,[U(t_0+\tau,t_0)]_C} = \frac{\sum_{a'}\langle a',\tau|a',0\rangle_{CS}}{\sum_{a'}\langle a',\tau|a',0\rangle_C}, \tag{9.1.25}$$

with $\mathcal{W} = W_{CS} - W_C$, analogous to Eq. (4.3.5). Since the traces $\sum_{a'}$, which appear in Eq. (9.1.25), are inert under variational differentiation, all of the variational differential results obtained in Section 4.3 for $e^{\mathcal{W}}$ follow here with the time replacements $t_2 \to 0$ and $t_1 \to \tau$ in the case of phonon couplings (Eqns. (4.3.7), (4.3.15) and (4.3.27)) and in the case of particle–particle interactions (Eqns. (4.4.4–4.4.7), (4.4.12), (4.4.20)), and they encompass both cases in (Eq. (4.4.21)). Moreover, it is clear that the constant of variational integration in Eq. (4.4.20), $\langle a_1', t_1|a_2'', t_2\rangle_0 \to Tr\,[U(t_0+\tau,t_0)]_0 = e^{W_0}$, is just the grand partition function in the complete absence of all interactions and all sources, as given by Eqns. (6.2.5) and (6.2.6).

Focusing our considerations on particle sources and particle–particle interactions, we postpone discussion of phonons, phonon interactions and phonon sources to a later chapter. It is convenient to modify the definition of the averaging process of Eq. (9.1.21) (which was defined in the context of fields alone driving the system out of equilibrium) and introduce a useful generalization appropriate to the introduction of particle sources that also drive the system out of equilibrium. Reflecting again on the considerations of Chapter 4, we now define the averaging process for an operator O as

$$\langle O\rangle \equiv \frac{Tr\,[U(t_0+\tau,t_0)O]_{CS}}{Tr\,[U(t_0+\tau,t_0)]_C}, \tag{9.1.26}$$

where the numerator trace is evaluated in the presence of coupling, external fields and particle sources, whereas the denominator trace is evaluated in the presence of coupling and external fields to the exclusion of particle sources. (When we consider phonons later, their particle source will be admitted in both the numerator and denominator traces.) Clearly, in the absence of particle sources, the average of Eq. (9.1.26) reduces to that of Eq. (9.1.21). With this definition, we define a generalized Green's function,

$$G_{\frac{j}{2},\frac{k'}{2}}(1,\ldots m,\ldots j;1',\ldots n',\ldots k') = (-i)^{\frac{j+k'}{2}}\,\varepsilon\left\langle\left(\prod_{m=0}^{j}\psi(\mathbf{x}_m,t_m)\prod_{n=0}^{k'}\psi^+(\mathbf{x}_n',t_n')\right)_+\right\rangle, \tag{9.1.27}$$

where the numbers of ψ's and ψ^+'s may *differ*:

$$\prod_{m=0}^{j}\psi(\mathbf{x}_m,t_m) \equiv \psi(1)\ldots\psi(m)\ldots\psi(j) \tag{9.1.28}$$

and

$$\prod_{n=0}^{k'}\psi^+(\mathbf{x}_n',t_n') \equiv \psi^+(k')\ldots\psi^+(n')\ldots\psi^+(1'). \tag{9.1.29}$$

The space-time arguments of ψ, ψ^+ are indicated by the letter/number index, and ε has the same meaning as above. Thus, $G_{\frac{j}{2},\frac{k'}{2}}$ has j-ψ–operators and k'-ψ^+–operators (moreover, we define $\psi(m = 0) = 1 = \psi^+(n = 0)$). Green's functions with unequal numbers of ψ, ψ^+ operators in the time-ordered product of $\langle(\ldots)_+\rangle$ will vanish if the sources $\eta, \eta^+ \to 0$ vanish, but need *not* vanish otherwise, in particular if particle number is not conserved. Such "anomalous" functions (for example, $G_{\frac{2j}{2},\frac{0}{2}}$ or $G_{\frac{j}{2},\frac{0}{2}}$, etc.) are important in the theory of Boson superfluidity and superconductivity. For $j = k'$, $G_{\frac{j}{2},\frac{j}{2}} \equiv G_j$, so in this case Eq. (9.1.27) is the same as Eq. (9.1.20). (However, note that $G_{\frac{2j}{2},\frac{0}{2}} \neq G_j$, etc.)

Considering particle species subject to number conservation (such that anomalous functions vanish in the limit $\eta, \eta^+ \to 0$), we can determine the dependence on particle sources η, η^+ explicitly by employing Eq. (4.3.12) with particle–particle interactions, excluding phonon interactions at this time. Thus, with the use of Eqns. (4.3.9)–(4.3.12) with $H_S(\mathbf{x}, t) \to H_\eta(\mathbf{x}, t) = \eta^+(\mathbf{x}, t)\psi(\mathbf{x}, t) + \psi^+(\mathbf{x}, t)\eta(\mathbf{x}, t)$ taken as just the particle source part of the Hamiltonian (while neglecting both the phonon coupling and phonon source term as well as the source potential U), we have

$$e^{\mathcal{W}} = \left\langle \left(\exp\left[-i \int_{t_0}^{t_0+\tau} dt' \int d^{(3)}\mathbf{x}' H_\eta(\mathbf{x}', t') \right] \right)_+ \right\rangle_C ,$$

$$= \left\langle \left(\exp\left[-i \int_{t_0}^{t_0+\tau} dt' \int d^{(3)}\mathbf{x}' \left(\eta^+(\mathbf{x}', t')\psi(\mathbf{x}', t') + \psi^+(\mathbf{x}', t')\eta(\mathbf{x}', t') \right) \right] \right)_+ \right\rangle_C . \quad (9.1.30)$$

This may be Taylor-expanded in powers of η, η^+ jointly as (notation: $1 = \mathbf{x}_1, t_1$; $1' = \mathbf{x}'_1, t'_1$; $d1 = d^{(3)}\mathbf{x}_1 dt_1$; $d1' = d^{(3)}\mathbf{x}'_1 dt'_1$, etc.):

$$e^{\mathcal{W}} = \sum_{n=0}^{\infty} \frac{(-i)^n}{(n!)^2} \int d1 \ldots \int dn \int d1' \ldots \int dn' \eta^+(n) \ldots \eta^+(1)$$
$$\varepsilon \langle (\psi(1) \ldots \psi(n)\psi^+(n') \ldots \psi^+(1'))_+ \rangle_C \eta(1') \ldots \eta(n'), \quad (9.1.31)$$

and with Eq. (9.1.20), we have (bear in mind that with number conservation, the number of contributing ψ's and ψ^+'s must be the same, so that n and n' denote the arguments $(\mathbf{x}_n, t_n)$ and $(\mathbf{x}'_n, t'_n)$, respectively, with n and n' referring to the same total number, but one with coordinates unprimed and the other primed):

$$e^{\mathcal{W}} = \sum_{n=0}^{\infty} \int d1 \ldots dn \int d1' \ldots dn' \frac{\eta^+(n) \ldots \eta^+(1)}{n!} \overline{G}_n(1, \ldots n; 1', \ldots n') \frac{\eta(1') \ldots \eta(n')}{n!}, \quad (9.1.32)$$

on the understanding that the $n = 0$ term is unity (i.e., $(e^{\mathcal{W}})_{\eta,\eta^+=0} = 1$) and $\overline{G}_n$ here has $\eta, \eta^+ \to 0$ in Eq. (9.1.32), but $\overline{G}_n$ retains the effects of coupling. Moreover, it is clear that as a coefficient in this Taylor expansion, $\overline{G}_n$ may also be written as

$$\overline{G}_n(1,\dots n;1',\dots n') = \left[\frac{\delta_l}{\delta\eta(n')}\cdots\frac{\delta_l}{\delta\eta(1')}\frac{\delta_l}{\delta\eta^+(1)}\cdots\frac{\delta_l}{\delta\eta^+(n)}e^{\mathcal{W}}\right]_{\eta,\eta^+\to 0}. \tag{9.1.33}$$

With the use of Eq. (9.1.32) and Eq. (9.1.33), the infinite chain of coupled particle–particle Green's function equations given by Eq. (8.1.7) may be succinctly summarized in the form (notation $\zeta \equiv \mu$),

$$\left[\left(i\frac{\partial}{\partial t_1} - h^{(1)} + \zeta\right)\frac{\delta_l}{\delta\eta^+(1)} \mp i\int d2\mathcal{V}(1,2)\frac{\delta_l}{\delta\eta(2^+)}\frac{\delta_l}{\delta\eta^+(2)}\frac{\delta_l}{\delta\eta^+(1)} - \eta(1)\right]e^{\mathcal{W}} = 0, \tag{9.1.34}$$

as one can readily verify, bearing in mind that an inhomogeneous source term η must now be appended in the Green's function equations (Eqns. (8.1.6–8.1.7)), as was done in Eq. (4.2.33).

To gain experience with the application of Eq. (9.1.34), we start with the case of *no* interparticle interactions, $\mathcal{V} \to 0$, so that

$$\left(i\frac{\partial}{\partial t_1} - h^{(1)} + \zeta\right)\frac{\delta_l}{\delta\eta^+(1)}e^{\mathcal{W}} - \eta e^{\mathcal{W}} = 0, \tag{9.1.35}$$

and we employ Eq. (4.3.24),

$$e^{\mathcal{W}} = \exp\left(-i\eta^+\overline{G}_1^0\eta\right), \tag{9.1.36}$$

applying the variational derivative $\delta_l/\delta\eta^+$ of Eq. (9.1.35). (Here, $\overline{G}_1^0$ has been adjusted to originate single-particle energies at ζ in accordance with $h^{(1)} \to h^{(1)} - \zeta$.) With a second variational derivative $\delta_l/\delta\eta$ applied and then taking the limit $\eta, \eta^+ \to 0$, we obtain the expected result,

$$\left(i\frac{\partial}{\partial t_1} - h^{(1)} + \zeta\right)\overline{G}_1^0(1,2) = \delta(1-2). \tag{9.1.37}$$

To determine $\overline{G}_n \to \overline{G}_n^0$ using Eq. (9.1.33), we need to take n variational derivatives of $e^{\mathcal{W}}$ with respect to η and n variational derivatives with respect to η^+, and then set $\eta, \eta^+ \to 0$. This may be conveniently carried out using the Taylor expansion of Eq. (9.1.36),

$$\begin{aligned} e^{\mathcal{W}} &= \sum_{k=0}^{\infty}\frac{\left(-i\eta^+\overline{G}_1^0\eta\right)^k}{k!} \\ &= 1 - \frac{i}{1!}\eta^+\overline{G}_1^0\eta + \frac{(-i)^2}{2!}\eta^+\overline{G}_1^0\eta\eta^+\overline{G}_1^0\eta + \ldots + \frac{(-i)^k}{k!}\eta^+\overline{G}_1^0\eta\eta^+\overline{G}_1^0\eta\ldots\eta^+\overline{G}_1^0\eta + \ldots. \end{aligned} \tag{9.1.38}$$

Clearly, terms having fewer than n $(\eta^+\eta)$-products vanish under the $2n$ variational derivatives, and terms having more than n $(\eta^+\eta)$-products all vanish in the limit $\eta, \eta^+ \to 0$. The only term that contributes to $\overline{G}_n$ has exactly n $(\eta^+\eta)$-products,

$$\eta^+\overline{G}_1^0\eta\eta^+\overline{G}_1^0\eta\ldots\eta^+\overline{G}_1^0\eta$$
$$=\int d\overline{1}\int d\overline{1}'\ldots\int d\overline{n}\int d\overline{n}'\eta^+(\overline{1})\overline{G}_1^0(\overline{1},\overline{1}')\eta(\overline{1}')\eta^+(\overline{2})\overline{G}_1^0(\overline{2},\overline{2}')\eta(\overline{2}')\ldots\eta^+(\overline{n})\overline{G}_1^0(\overline{n},\overline{n}')\eta(\overline{n}'). \tag{9.1.39}$$

Application of n variational derivatives with respect to η^+ and n variational derivatives with respect to η produces $2n$ space-time delta functions, so all $2n$ integrations can be executed, leaving a product of n $\overline{G}_1^0$–factors. To be precise, there are n linear combinations of such products since a single variational derivative produces n additive terms. For example, consider the case of $n = 2$: We have

$$\frac{\delta_l}{\delta\eta(2')}\frac{\delta_l}{\delta\eta(1')}\frac{\delta_l}{\delta\eta^+(1)}\frac{\delta_l}{\delta\eta^+(2)}\int d\overline{1}\int d\overline{1}'\int d\overline{2}\int d\overline{2}'\eta^+(\overline{1})\overline{G}_1^0(\overline{1},\overline{1}')\eta(\overline{1}')\eta^+(\overline{2})\overline{G}_1^0(\overline{2},\overline{2}')\eta(\overline{2}'), \tag{9.1.40}$$

and applying $\delta_l/\delta\eta^+(2)$ with $\delta_l\eta^+(\overline{1})/\delta\eta^+(2) = \delta(\overline{1}-2)$; $\delta_l\eta^+(\overline{2})/\delta\eta^+(2) = \delta(\overline{2}-2)$, we obtain

$$= \frac{\delta_l}{\delta\eta(2')}\frac{\delta_l}{\delta\eta(1')}\frac{\delta_l}{\delta\eta^+(1)}\Big(\int d\overline{1}\int d\overline{1}'\int d\overline{2}\int d\overline{2}'\{\delta(\overline{1}-2)\overline{G}_1^0(\overline{1},\overline{1}')\eta(\overline{1}')\eta^+(\overline{2})\overline{G}_1^0(\overline{2},\overline{2}')\eta(\overline{2}')$$
$$+\eta^+(\overline{1})\overline{G}_1^0(\overline{1},\overline{1}')\eta(\overline{1}')\delta(\overline{2}-2)\overline{G}_1^0(\overline{2},\overline{2}')\eta(\overline{2}')\}\Big). \tag{9.1.41}$$

Applying the next variational derivative, $\delta_l/\delta\eta^+(1)$, a $\pm$ sign arises from the commutation/anticommutation of $\delta\eta^+$ to the far left to form δ_l in such a product. In particular,

$$\frac{\delta_l}{\delta\eta^+(1)}\left[\eta(\bar{1}')\eta^+(\bar{2})\eta(\bar{2}')\right] = \pm\left[\frac{\delta_l\eta^+(\bar{2})}{\delta\eta^+(1)}\right]\eta(\bar{1}')\,\eta(\bar{2}') = \pm\delta(\bar{2}-1)\,\eta(\bar{1}')\,\eta(\bar{2}'). \tag{9.1.42}$$

Continuation of this process in Eq. (9.1.41) yields the expected non-interacting approximation for $\overline{G}_2$ as

$$\overline{G}_2 \to \overline{G}_2^0(1,2;1',2') = \overline{G}_1^0(1,1')\overline{G}_1^0(2,2') \pm \overline{G}_1^0(1,2')\overline{G}_1^0(2,1'), \tag{9.1.43}$$

which is the ordinary Hartree–Fock approximation. The same procedure applied to Eq. (9.1.39) yields $\overline{G}_n \to \overline{G}_n^0$ as

$$\overline{G}_n^0 = \begin{Bmatrix} perm_{(n)} \\ \text{or} \\ det_{(n)} \end{Bmatrix} \overline{G}_1^0, \tag{9.1.44}$$

where $perm_{(n)}$ and $det_{(n)}$ are the permanent and determinant of the $n\times n$ coordinate-indexed matrix $\overline{G}_1^0(i,j)$, respectively. Martin and Schwinger developed improved approximation procedures along these lines[3] for both long-range and short-range particle–particle interaction potentials $\mathcal{V}$ using Eq. (9.1.33).

9.2 Exact Variational Relations in Terms of a Source Potential

The equation of motion for an n-particle Green's function (Eq. (8.1.7) and Eq. (9.1.34)) has been formulated above in terms of a partial time derivative with respect to just *one* of n time variables (for example, $\partial G_n/\partial t_1$), and is thus asymmetric with respect to the other $(n-1)$ time variables, although any of the others could be similarly employed. This situation can be improved by developing exact variational relations in terms of a source potential.

Introducing the source potential, $U(1)$, we employ Eqns. (9.1.24) and (9.1.25) and the associated discussion of e^{W_0} jointly with Eq. (4.4.21) to determine alternative expressions for $e^{\mathcal{W}}$ as (omitting phonon interactions and phonon sources):

$$e^{\mathcal{W}} = e^{-W_C(U)} e^{W_0} \exp\left(-\frac{i}{2}\int_{t_0}^{t_0+\tau} dt' \int d^{(3)}\mathbf{x}' \int d^{(3)}\mathbf{x}'' \frac{\delta_l}{\delta\eta(\mathbf{x}',t'^+)}\frac{\delta_l}{\delta\eta(\mathbf{x}'',t'^+)}\mathcal{V}(\mathbf{x}'-\mathbf{x}'')\right.$$

$$\left.\frac{\delta_l}{\delta\eta^+(\mathbf{x}'',t')}\frac{\delta_l}{\delta\eta^+(\mathbf{x}',t')}\right) \det{}^{\mp 1}(1-\overline{G}_1^0 U)\, e^{-\left(i\eta^+[1-\overline{G}_1^0 U]^{-1}\overline{G}_1^0\eta\right)}, \qquad (9.2.1)$$

where e^{W_0} is the grand partition function in the absence of all sources and couplings (Eq. (6.2.5 and 6.2.6)), and

$$e^{\mathcal{W}} = e^{-W_C(U)} e^{W_0} \exp\left(\frac{i}{2}\int_{t_0}^{t_0+\tau} dt' \int d^{(3)}\mathbf{x}' \int d^{(3)}\mathbf{x}'' \frac{\delta}{\delta U(\mathbf{x}',t')}\mathcal{V}(\mathbf{x}'-\mathbf{x}'')\frac{\delta}{\delta U(\mathbf{x}'',t')}\right)$$

$$\det{}^{\mp 1}(1-\overline{G}_1^0 U)\, e^{-\left(i\eta^+[1-\overline{G}_1^0 U]^{-1}\overline{G}_1^0\eta\right)}, \qquad (9.2.2)$$

where $e^{-W_C(U)} = \left(Tr\,[U(t_0+\tau,t_0)]_C\right)^{-1}$ accounts for coupling in the absence of particle sources in the denominator of Eq. (9.1.25), but it includes the role of the source potential U. It should also be borne in mind that (Eq. (4.4.11)):

$$\overline{G}_U^0 = \left[\left(\overline{G}_1^0\right)^{-1} - U\right]^{-1} = \left[1-\overline{G}_1^0 U\right]^{-1}\overline{G}_1^0. \qquad (9.2.3)$$

Thus, with the use of space-time matrix notation in the exponential arguments, we can rewrite Eq. (9.2.2) compactly as

$$e^{\mathcal{W}} = e^{-W_C(U)} e^{W_0} \exp\left[\frac{i}{2}\frac{\delta}{\delta U}\mathcal{V}\frac{\delta}{\delta U}\right] \det{}^{\mp 1}(1-\overline{G}_1^0 U)\, e^{-i\eta^+\overline{G}_U^0\eta}. \qquad (9.2.4)$$

Again, one may Taylor-expand $e^{\mathcal{W}}$ in powers of η,η^+ jointly and employ Eq. (9.1.32) to identify $\overline{G}_n \to \overline{G}_n(U)$ in the presence of U as

$$\overline{G}_n\left(1\ldots n;1'\ldots n';U\right)=e^{-W_C(U)}e^{W_0}\exp\left[\frac{i}{2}\frac{\delta}{\delta U}\mathcal{V}\frac{\delta}{\delta U}\right]\det{}^{\mp1}(1-\overline{G}_1^0U)\left\{\begin{matrix}perm_{(n)}\\ det_{(n)}\end{matrix}\right\}\overline{G}_U^0\,(i,j)\,,\tag{9.2.5}$$

where we have carried out the evaluation of the last factor on the right using the same procedures involved in obtaining Eq. (9.1.44), but with the inclusion of a source potential U (however, interactions are excluded from $\overline{G}_U^0\,(i,j)$ as their role is determined by the "exp" and "det" factors on the right of Eq. (9.2.5)). The quantity $e^{W_C(U)}$ may be evaluated by recalling that $\left[e^{\mathcal{W}}\right]_{\eta,\eta^+\to 0}=1$ (see Eq. (9.1.32), for example) and applying it in Eq. (9.2.4), whence we obtain

$$e^{W_C(U)}=e^{W_0}\left[\exp\left(\frac{i}{2}\frac{\delta}{\delta U}\mathcal{V}\frac{\delta}{\delta U}\right)\det{}^{\mp1}(1-\overline{G}_1^0U)\right].\tag{9.2.6}$$

Multiplying Eq. (9.2.5) from the left by $e^{W_C(U)}$ and from the right by $e^{-W_C(U)}$, we obtain

$$e^{W_C(U)}\overline{G}_n\left(1\ldots n;1'\ldots n';U\right)e^{-W_C(U)}$$

$$=e^{W_0}\exp\left(\frac{i}{2}\frac{\delta}{\delta U}\mathcal{V}\frac{\delta}{\delta U}\right)\left\{\begin{matrix}perm_{(n)}\\ det_{(n)}\end{matrix}\right\}\overline{G}_U^0\,(i,j')\exp\left(-\frac{i}{2}\frac{\delta}{\delta U}\mathcal{V}\frac{\delta}{\delta U}\right)e^{-W_0}\tag{9.2.7}$$

(note: $\overline{G}_U^0\,(i,j')\equiv\overline{G}_1^0\,(i,j';U)$). Employing the relation

$$\exp\left(\frac{i}{2}\frac{\delta}{\delta U}\mathcal{V}\frac{\delta}{\delta U}\right)f(U)\exp\left(-\frac{i}{2}\frac{\delta}{\delta U}\mathcal{V}\frac{\delta}{\delta U}\right)=f\left(U+i\mathcal{V}\frac{\delta}{\delta U}\right),\tag{9.2.8}$$

we obtain

$$\overline{G}_n\left(1\ldots n;1'\ldots n';U\right)=e^{-W_C(U)}\left\{\begin{matrix}perm_{(n)}\\ det_{(n)}\end{matrix}\right\}\overline{G}_1^0\left(i,j';U+i\mathcal{V}\frac{\delta}{\delta U}\right)e^{W_C(U)}.\tag{9.2.9}$$

Furthermore, applying another relation similar to Eq. (9.2.8),

$$e^{-W_C}f\left(\frac{\delta}{\delta U}\right)e^{W_C}=f\left(\frac{\delta}{\delta U}+\frac{\delta W_C}{\delta U}\right),\tag{9.2.10}$$

we obtain

$$\overline{G}_n\left(1\ldots n;1'\ldots n';U\right)=\left\{\begin{matrix}perm_{(n)}\\ det_{(n)}\end{matrix}\right\}\overline{G}_1^0\left(i,j';U+i\mathcal{V}\frac{\delta}{\delta U}+i\mathcal{V}\frac{\delta W_C}{\delta U}\right).\tag{9.2.11}$$

This equation, which is *symmetric* in all coordinate variable pairs i, j', may be written in differential form noting that the appropriate differential operator for a given pair of variables, say j, j', is that of the *non-interacting* single-particle Green's function but with the single-particle source potential U replaced by $U\longrightarrow U-i\mathcal{V}\frac{\delta}{\delta U}+\frac{\delta W_C}{\delta U}$ (and with no particle sources). Since this is true for all pairs of coordinate variables, we have

$$\prod_{j=1}^{n}\mathcal{F}_j\overline{G}_n(U)=I_n,\tag{9.2.12}$$

with the definitions

$$\mathcal{F} = \left(\left(\overline{G}_1^0 \right)^{-1} - U - i\mathcal{V}\frac{\delta}{\delta U} - i\mathcal{V}\frac{\delta W_C}{\delta U} \right), \tag{9.2.13}$$

and

$$l_n = perm_{(n)} \text{ or } det_{(n)}\, \delta(j, j'). \tag{9.2.14}$$

We have thus arrived at a more symmetric formulation of the equation of motion for $\overline{G}_n$, symmetric in all n-particle space-time coordinate pairs.

9.3 Cumulants and the Linked Cluster Theorem

It is important to recognize the fact that the multiparticle Green's functions $(\overline{G}_n)$ under discussion may be expected to involve parts that contain only products of lower order Green's functions not integrated with the interaction potential, termed *disconnected* parts. This has been seen already in several examples, including Hartree and Hartree–Fock approximations for $\overline{G}_2$. Such disconnected parts may be conveniently eliminated by evaluating the other *connected* parts of the multiparticle Green's functions, C_n, directly as variational derivatives of $\mathcal{W}$ (instead of $e^{\mathcal{W}}$), defining *cumulants* as

$$C_n(1, \dots n; 1', \dots n') = \left[\frac{\delta_l}{\delta\eta(n')} \cdots \frac{\delta_l}{\delta\eta(1')} \frac{\delta_l}{\delta\eta^+(1)} \cdots \frac{\delta_l}{\delta\eta^+(n)} \mathcal{W} \right]_{\eta,\eta^+ \to 0}. \tag{9.3.1}$$

To evaluate this we write Eq. (9.1.32) in the form

$$\mathcal{W} = \ln \left[1 + \sum_{n=1}^{\infty} \int d1 \dots \int dn \int d1' \dots \int dn' \frac{\eta^+(n) \dots \eta^+(1)}{n!} \right.$$
$$\left. \overline{G}_n(1, \dots n; 1', \dots n') \frac{\eta(1') \dots \eta(n')}{n!} \right], \tag{9.3.2}$$

and expand $\mathcal{W}$ in powers of η, η^+ by Taylor-expanding $\ln(1 + x)$ in powers of x:

$$\mathcal{W} \cong \sum_{n=1}^{\infty} \dots - \frac{1}{2} \left(\sum_{n=1}^{\infty} \dots \right)^2 + \dots . \tag{9.3.3}$$

For example, to evaluate C_1 and C_2, anticipating up to four variational derivatives to be followed by setting $\eta, \eta^+ \to 0$, we need only retain terms corresponding to $n = 1, 2$. Thus,

$$
\begin{aligned}
\mathcal{W} \cong & \int d1 \int d1' \eta^+(1)\overline{G}_1(1,1')\,\eta(1') \\
& + \int d1 \int d2 \int d1' \int d2' \frac{\eta^+(2)\eta^+(1)}{2}\overline{G}_2(1,2;1',2')\,\frac{\eta(1')\eta(2')}{2} \\
& - \frac{1}{2}\int d1 \int d1' \eta^+(1)\overline{G}_1(1,1')\,\eta(1') \cdot \int d\overline{1} \int d\overline{1}' \eta^+(\overline{1})\overline{G}_1\left(\overline{1},\overline{1}'\right)\eta(\overline{1}') \\
& + \left(\text{terms of higher order in } \eta, \eta^+\right).
\end{aligned}
\tag{9.3.4}
$$

The determination of $C_1(1,1')$ now follows from Eq. (9.3.1) as

$$
C_1(1,1') = \frac{\delta_l}{\delta\eta(1')}\frac{\delta_l}{\delta\eta^+(1)} \int d\overline{1}\, d\overline{1}' \eta^+(\overline{1})\overline{G}_1\left(\overline{1},\overline{1}'\right)\eta(\overline{1}') = \overline{G}_1(1,1'),
\tag{9.3.5}
$$

since all terms on the right of Eq. (9.3.4) after the first one yield vanishing contributions in the limit $\eta, \eta^+ \to 0$ after variational differentiation. On the other hand, the first term on the right of Eq. (9.3.4) vanishes under the four variational differentiations needed to get C_2, so that

$$
\begin{aligned}
C_2(1,2;1',2') = & \frac{\delta_l}{\delta\eta(2')}\frac{\delta_l}{\delta\eta(1')}\frac{\delta_l}{\delta\eta^+(1)}\frac{\delta_l}{\delta\eta^+(2)} \\
& \left(\frac{1}{4}\int d\overline{1} \int d\overline{1}' \int d\overline{2} \int d\overline{2}' \eta^+(\overline{2})\eta^+(\overline{1})\overline{G}_2\left(\overline{1},\overline{2};\overline{1}',\overline{2}'\right)\eta(\overline{1}')\eta(\overline{2}') \right. \\
& \left. - \frac{1}{2}\int d\overline{1} \int d\overline{1}' \eta^+(\overline{1})\overline{G}_1\left(\overline{1},\overline{1}'\right)\eta(\overline{1}') \int d\overline{2} \int d\overline{2}' \eta^+(\overline{2})\overline{G}_1\left(\overline{2},\overline{2}'\right)\eta(\overline{2}')\right).
\end{aligned}
\tag{9.3.6}
$$

Carrying out variational differentiation in a manner similar to that of Eq. (9.1.42), we have

$$
\frac{\delta_l}{\delta\eta^+(1)}\left(\frac{\delta_l}{\delta\eta^+(2)}\left(\eta^+(\overline{2})\eta^+(\overline{1})\right)\right) = \delta(\overline{2}-2)\delta(\overline{1}-1) \pm \delta(\overline{1}-2)\delta(\overline{2}-1)
\tag{9.3.7}
$$

and

$$
\frac{\delta_l}{\delta\eta(2')}\left(\frac{\delta_l}{\delta\eta(1')}\left(\eta(\overline{1}')\eta(\overline{2}')\right)\right) = \delta(\overline{1}'-1')\delta(\overline{2}'-2') \pm \delta(\overline{1}'-2')\delta(\overline{2}'-1').
\tag{9.3.8}
$$

Thus, Eq. (9.3.6) yields

$$
\begin{aligned}
C_2(1,2;1',2') = & \frac{1}{4}\left\{\overline{G}_2(1,2;1',2') + \overline{G}_2(2,1;2',1') \pm \overline{G}_2(1,2;2',1') \pm \overline{G}_2(2,1;1',2')\right\} \\
& - \frac{1}{2}\left\{\begin{array}{c} \overline{G}_1(1,1')\overline{G}_1(2,2') + \overline{G}_1(2,2')\overline{G}_1(1,1') \\ \pm\overline{G}_1(1,2')\overline{G}_1(2,1') \pm \overline{G}_1(2,1')\overline{G}_1(1,2') \end{array}\right\},
\end{aligned}
\tag{9.3.9}
$$

and since an even or odd number of permutations of the order of indices of a Green's function invokes a factor of ± 1 for Fermions or Bosons, we obtain

$$
C_2(1,2;1',2') = \overline{G}_2(1,2;1',2') - \left(\overline{G}_1(1,1')\overline{G}_1(2,2') \pm \overline{G}_1(1,2')\overline{G}_1(2,1')\right).
\tag{9.3.10}
$$

Thus, the disconnected parts of $\overline{G}_2$, which involve only products of lower order Green's functions $\left(\overline{G}_1\overline{G}_1\right)$, are removed in determining the connected part C_2 according to Eq. (9.3.1). This removal of such disconnected parts in higher order Green's functions, $\overline{G}_n$, is also achieved quite generally in C_n as given by Eq. (9.3.1).

To illustrate the removal of the disconnected parts we examine the grand potential W_C/β in the presence of particle–particle interactions, $\mathcal{V}$, but with no particle or potential sources, $\eta, \eta^+, U \to 0$ (also, no phonon coupling or sources). Considering the dependence of W_C on particle–particle coupling strength $\lambda = \lambda_{ee}$, as in Eq. (7.6.18), we assume analyticity and Taylor-expand it in powers of λ_{ee}:

$$W_C(\lambda_{ee}) = W_0 + \sum_{n=1}^{\infty} \frac{\lambda_{ee}^n}{n!} \left[\frac{\partial^n W_C(\lambda_{ee})}{\partial \lambda_{ee}^n}\right]_{\lambda_{ee}=0}. \tag{9.3.11}$$

The derivatives of $W_C(\lambda_{ee})$ may be evaluated using Eq. (7.6.17) jointly with the Action Principle,

$$\frac{\partial W_C(\lambda_{ee})}{\partial \lambda_{ee}} = \frac{-\beta}{2} \int d^{(3)}\mathbf{x}' \int d^{(3)}\mathbf{x}'' \mathcal{V}(\mathbf{x}'\text{-}\mathbf{x}'') \overline{G}_2^{(\lambda_{ee})}(1', 1'', 1'^+, 1''^+)_{t_1'=t_1''}. \tag{9.3.12}$$

The use of the Action Principle to form successively higher order λ_{ee}-derivatives of $\partial W_C/\partial\lambda_{ee}$ (to form the Taylor series coefficients in Eq. (9.3.11)) necessarily involves the installation of successively higher powers of $\delta\lambda_{ee}H^{(2)}$ into the matrix elements constituting the numerator of $\overline{G}_2^{(\lambda_{ee})}$, yielding higher order Green's functions. However, the λ_{ee}-derivatives of the denominator matrix elements of $\overline{G}_2^{(\lambda_{ee})}$ and of higher order Green's function denominators yield additional subtracted terms constituted of products of lower order Green's functions, thus removing the disconnected parts and leaving only the connected parts, C_n, which cannot be reduced to products of lower order Green's functions. These remarks apply to all higher order Green's functions, G_n, and their connected parts, C_n. In this way, Eq. (9.3.11) yields the *linked cluster theorem*,

$$W_C(\lambda_{ee}) = W_0 + \sum_{n=1}^{\infty} \frac{(i/2)^n}{n!} \int d\,(1) \ldots \int d\,(2n) \mathcal{V}(1,2) \ldots \mathcal{V}(2n-1, 2n) C_{2n}^0\,(1 \ldots 2n, 1 \ldots 2n), \tag{9.3.13}$$

where $C_{2n}^0 \equiv [C_{2n}]_{\lambda_{ee}=0}$ is evaluated at $\lambda_{ee} = 0$ in accordance with Eq. (9.3.11). This *linked cluster* expansion in terms of the *connected* parts, C_{2n}^0, resolves a calculational problem that occurred in the early stages of quantum many-body theory research. W_C and all its λ_{ee}-derivatives are known to be proportional to volume (Eq. (6.1.31)), whereas the disconnected parts, which occur explicitly in a Taylor series in powers of λ_{ee} for e^{W_C}, involve multiple space integrations over products of lower order Green's functions that produce terms exhibiting higher powers of volume. Such terms, when arranged as C_{2n}^0 above, cancel all higher powers of volume beyond the first power, resolving the problem. It is important to note that even with this cancellation, the Taylor expansion of $W_C(\lambda_{ee})$,

Eq. (9.3.13), is not always valid or useful. We will soon discuss a better procedure for evaluating $W_C(\lambda_{ee})$ for a quantum plasma, which brings to light important physical dielectric response phenomenology, such as plasmons and shielding, that are obscured by the Taylor expansion.

9.4 Variational Differential Formulation of the Random Phase Approximation

The equation of motion for $\overline{G}_1(1,1')$, which was derived on the basis of the equation of motion for $\psi(1)$, has the same form with non-equilibrium statistical averaging as it had in the equilibrium case, Eqns. (8.1.1 and 8.1.2), except that we now include the time-dependent source potential $U(1)$ in $h_1^{(1)} \to h_1^{(1)} + U(1)$,

$$\left(i\frac{\partial}{\partial t} - h_1^{(1)} + \zeta - U(1)\right)\overline{G}_1(1,1') \mp i\int d^{(3)}\mathbf{x}_2 \mathcal{V}(\mathbf{x}_1 - \mathbf{x}_2)\overline{G}_2(1,2;1',2^+)_{t_2=t_1} = \delta(1-1'). \tag{9.4.1}$$

The two-particle non-equilibrium Green's function $\overline{G}_2$ may be expressed in terms of a variational derivative of $\overline{G}_1$ using a minor extension of Eqns. (4.2.29 and 4.2.31) with $O_H(t) \to (\psi(1)\psi^+(1'))_+$ and $t_2 = 0 \le t, t' \le \tau = -i\beta$, so that time ordering is along the imaginary time axis. Specifically, Eqns. (4.2.29 and 4.2.31) applied to Eq. (9.1.4) yield

$$\frac{\delta \overline{G}_1(1,1')}{\delta U(2)} = \frac{-i\varepsilon}{Tr\left[U(t_0+\tau,t_0)\right]}\frac{\delta\left(Tr\left[U(t_0+\tau,t_0)\left(\psi(1)\psi^+(1')\right)_+\right]\right)}{\delta U(2)} \tag{9.4.2}$$
$$+\frac{i\varepsilon\, Tr\left[U(t_0+\tau,t_0)\left(\psi(1)\psi^+(1')\right)_+\right]}{TrU(t_0+\tau,t_0)}\frac{\delta TrU(t_0+\tau,t_0)/\delta U(2)}{TrU(t_0+\tau,t_0)},$$

which may be written in terms of $\overline{G}_2$ and $\overline{G}_1$ as

$$\frac{\delta \overline{G}_1(1,1')}{\delta U(2)} = -\overline{G}_2(1,2;1',2^+) + \overline{G}_1(2;2^+)\overline{G}_1(1;1'). \tag{9.4.3}$$

For future reference, we note that if a *non-local* instantaneous single-particle source potential $U(2',2)$ is installed in the Hamiltonian as $\int d^{(3)}\mathbf{x}\int d^{(3)}\mathbf{x}'U(\mathbf{x},\mathbf{x}',t)\,\psi^+(\mathbf{x},t)\,\psi(\mathbf{x}',t)$, then

$$\frac{\delta \overline{G}_1(1,1')}{\delta U(2',2)} = -\overline{G}_2(1,2;1',2'^+) + \overline{G}_1(2;2'^+)\overline{G}_1(1;1') \tag{9.4.4}$$

supplants Eq. (9.4.3).

The $\overline{G}_1\overline{G}_1$ term on the right of Eqns. (9.4.3 and 9.4.4) is due to variational differentiation of the denominator trace of $\delta\overline{G}_1/\delta U$. It is clear that each successive variational derivative of a Green's function with respect to U installs an additional $\psi^+\psi$ product within the time-ordered numerator trace, yielding a Green's function of one higher order, $\frac{\delta^2\overline{G}_1}{\delta U\delta U} \sim \frac{\delta\overline{G}_2}{\delta U} \sim \overline{G}_3$, etc., and the variational derivative of the denominator trace just produces subtracted terms involving only products of lower order Green's functions. Substituting for $\overline{G}_2$ using Eq. (9.4.3) into Eq. (9.4.1), we obtain an equation for $\overline{G}_1$ involving *only* $\overline{G}_1$, albeit with a variational derivative,

$$\left[\left[G_1^0\right]^{-1}(1) - U(1) \mp i\int d2\, V(1-2) G_1(2,2^+) \mp i\int d2\, V(1-2)\frac{\delta}{\delta U(2)}\right] G_1(1,1') \\ = \delta(1-1'), \tag{9.4.5}$$

where $\left[\overline{G}_1^0\right]^{-1}(1) = i\frac{\partial}{\partial t_1} - h_1^{(1)} + \zeta$. It is clear from this that the $\overline{G}_1\overline{G}_1$ term of Eq. (9.4.3) produces a time-dependent generalization of the Hartree approximation since the non-equilibrium density is $\rho(2) = \pm i\overline{G}_1(2,2^+)$, with the effective potential

$$V(1) = U(1) \pm i\int d2\, V(1-2)\overline{G}_1(2,2^+). \tag{9.4.6}$$

This provides a good description of the long-range effects of the electron–electron Coulomb potential for a quantum plasma, even without the variational derivative term, whence we write, approximately,

$$\left[\left[\overline{G}_1^0\right]^{-1}(1) - V(1)\right]\overline{G}_1(1,1') \cong \delta(1-1'), \tag{9.4.7}$$

or, equivalently, in the form of an integral equation,

$$\overline{G}_1(1,1') \cong \overline{G}_1^0(1,1') + \int d2'\overline{G}_1^0(1,2')V(2')\overline{G}_1(2',1'). \tag{9.4.8}$$

Linearizing this time-dependent Hartree approximation, setting $\overline{G}_1 \to \overline{G}_1^0$ on the right, a variational derivative with respect to $V(2)$ yields

$$\frac{\delta\overline{G}_1(1,1')}{\delta V(2)} \cong \overline{G}_1^0(1,2)\overline{G}_1^0(2,1'). \tag{9.4.9}$$

This relation is useful in further consideration of $V(1)$, since we may define a non-local dynamic inverse dielectric function $K(1,2)$ as (Section 4.2, Eq. (9.4.6)):

$$K(1,2) = \frac{\delta V(1)}{\delta U(2)} = \delta(1-2) \pm i\int d2' V(1-2')\frac{\delta\overline{G}_1(2',2'^+)}{\delta U(2)}, \tag{9.4.10}$$

which may be rewritten using the chain rule of variational differentiation as

$$K(1,2) = \delta(1-2) \pm i\int d2'\int d3' V(1-2')\frac{\delta\overline{G}_1(2',2'^+)}{\delta V(3')}K(3',2). \tag{9.4.11}$$

Using Eq. (9.4.9), we obtain the Random Phase Approximation (RPA) integral equation

$$K(1,2) \cong \delta(1-2) \pm i \int d2' \int d3' \mathcal{V}(1-2')\overline{G}_1^0(2',3')\overline{G}_1^0(3',2'^+)K(3',2). \quad (9.4.12)$$

Defining the *ring diagram* R (also called the "polarization operator," Π^0) as

$$R(2',3') = -\frac{\delta\rho(2')}{\delta V(3')} = \mp i\overline{G}_1^0(2',3')\overline{G}_1^0(3',2'^+), \quad (9.4.13)$$

and introducing space-time matrix notation, $K(1,2) = \langle 1\,|K|\,2\rangle$, etc., the RPA integral equation may be written as

$$K = 1 - \mathcal{V}RK, \quad (9.4.14)$$

and infinitely repeated iteration of K on the right-hand side yields

$$K = \sum_{n=0}^{\infty}(-\mathcal{V}R)^n, \quad (9.4.15)$$

which describes the RPA as the *sum of ring diagrams*. Alternatively, we have the formal operator solution of Eq. (9.4.14) as

$$K = (1 + \mathcal{V}R)^{-1}, \quad (9.4.16)$$

from which it is clear that K commutes with $\mathcal{V}R$ so that Eq. (9.4.14) can be written in a form that is sometimes more convenient,

$$K = 1 - K\mathcal{V}R. \quad (9.4.17)$$

The solution of the integral equation, Eq. (9.4.12), will be discussed at length below, but we first consider Eq. (9.4.10) using Eq. (9.4.3) to obtain the exact relation

$$K(1,2) = \delta(1-2) \mp i \int d3\mathcal{V}(1-3)\left[\overline{G}_2(3,2;3^+,2^+) - \overline{G}_1(3,3^+)\overline{G}_1(2,2^+)\right]. \quad (9.4.18)$$

The particular arguments of $\overline{G}_2$ and $\overline{G}_1\overline{G}_1$ on the right permit identification of the equilibrium density-density correlation function $D(3,2)$ as ($\hat{\rho}$ is the density operator here):

$$\begin{aligned} D(3,2) &\equiv -\left[\overline{G}_2(3,2;3^+,2^+) - \overline{G}_1(3,3^+)\overline{G}_1(2,2^+)\right] \\ &= \left\langle\left(\left[\hat{\rho}(3) - \langle\hat{\rho}(3)\rangle^{\mu,\beta}\right]\left[\hat{\rho}(2) - \langle\hat{\rho}(2)\rangle^{\mu,\beta}\right]\right)_+\right\rangle^{\mu,\beta}, \end{aligned} \quad (9.4.19)$$

so that

$$K = 1 \pm i\mathcal{V}D. \quad (9.4.20)$$

An equivalent integral equation is readily derived from Eq. (9.4.14) as

$$D = \pm iR - R\mathcal{V}D, \tag{9.4.21}$$

from which it is clear that $\pm iR$ is the result for D in the absence of interparticle interactions, $\mathcal{V} \to 0$. Moreover, comparing Eq. (9.4.14) and Eq. (9.4.21):

$$D = \pm iRK, \tag{9.4.22}$$

so that D can be determined from K using Eq. (9.4.22) and K can be determined from D with Eq. (9.4.20). Also, when $\mathcal{V}$ is the Coulomb potential (for unit charge),

$$\nabla_1^2 \left[K(1,2) - \delta(1-2)\right] = \mp 4\pi i D(1,2)_{t_1=t_2}. \tag{9.4.23}$$

Thus, $K(1,2)$ and $D(1,2)$ provide equivalent information.

Noting that when t_2 is advanced to $t_2 + \tau$, two arguments in the integrand of Eq. (9.4.18) are advanced through τ, each one invoking a $\pm$ due to periodicity/antiperiodicity, and the combination produces a net $(\pm)^2 = +$ sign, so that $K(1,2)$ is periodic, as well as $D(1,2)$, and thus both have the Boson spectral representations (for *both* Fermion and Boson systems) as described in Chapter 7, Eqns. (7.4.10)–(7.4.16), with a Fourier series coefficient as in Eqns. (7.4.6) and (7.4.7) (but there is a different sum rule for the spectral weight of $K(1,2)$). However, while $K(1,2)$ is very convenient for calculation, it is, in fact, just the Bose-periodic counterpart of the causal physical response function, which is determined by Eqns. (9.4.10 and 9.4.11) with the replacement of the periodic Green's function $G_1(2',2'^+)$ by its physical counterpart $\mathbb{G}_1(2',2'^+)$, whose variational derivative is *not* a $\mathbb{G}_2$-function alone, but rather, it is given by (again suppress the explicit appearance of N):

$$\frac{\delta \mathbb{G}_1(2',2'^+)}{\delta U(2)} = \mp \frac{Tr\left(e^{-iH_0\tau}\left[\rho(2'),\rho(2)\right]\right)}{Tr\left(e^{-iH_0\tau}\right)}\eta_+(t_2'-t_2), \tag{9.4.24}$$

as one can readily verify using Eq. (3.4.3). This is a retarded commutator response function as opposed to a time-ordered one. Again, the physical response corresponding to $K(1,2)$ is determined by the retarded commutator function of Eqns. (7.5.8–7.5.11):

$$K_r(T) = \eta_+(T)\left(K_>(T) - K_<(T)\right), \tag{9.4.25}$$

and denoting the spectral weight of K by $\mathcal{A}_k(\omega)$, we have

$$K_r(T) = \int_{-\infty}^{\infty} \frac{d\omega}{2\pi i} e^{-i\omega T} \int_{-\infty}^{\infty} \frac{d\omega'}{2\pi} \frac{\mathcal{A}_k(\omega')}{\omega - \omega' + i\delta} = \int_{-\infty}^{\infty} \frac{d\omega}{2\pi i} e^{-i\omega T} \mathcal{K}(\omega + i\delta), \tag{9.4.26}$$

where $\mathcal{K}(\omega)$ is the Fourier series coefficient of the periodic response function K, continued analytically to all frequencies as the "auxiliary" function evaluated at $\omega + i\delta$.

This analysis follows that of Section 7.5 as well as that of Section 9.1 for conductivity. (However, the sum rule of the form of Eqns. (7.5.6) and (7.5.7) does not apply here.)

9.5 Variational Differential Formulation of Perturbation Theory

The difficulty of solving interaction problems, such as the particle–particle "self-interaction" $\overline{G}_1$ equation (Eq. (9.4.1) or, equivalently, Eq. (9.4.5)), necessitates the formulation of approximation procedures, or *perturbation theories*, to extract information about the system dynamics. The usual (but not the most useful) starting point is the iteration of Eq. (9.4.5), which we write in integral form as (all time integrals are over the imaginary fundamental time interval $[0, \tau = -i\beta]$ and $d2 \equiv d^4 2 = d^3\mathbf{x}_2\, dt_2$, etc.):

$$\begin{aligned}\overline{G}_1(1,1') = \overline{G}^0_U(1,1') \pm i \int d2 \int d3\, \overline{G}^0_U(1,2)\mathcal{V}(2-3)\overline{G}_1(3,3^+)\overline{G}_1(2,1')\\ \pm i \int d2 \int d3\, \overline{G}^0_U(1,2)\mathcal{V}(2-3)\frac{\delta}{\delta U(3)}\overline{G}_1(2,1'),\end{aligned} \tag{9.5.1}$$

where $\overline{G}^0_U$ is defined by

$$\left[\overline{G}^0_U\right]^{-1}(1) = \left[\overline{G}^0_1\right]^{-1}(1) - U(1). \tag{9.5.2}$$

To generate a solution in powers of $\mathcal{V}$ we start with a first iteration, in which the entire right-hand side of Eq. (9.5.1) is substituted for the last factors $\overline{G}_1$ under the integrals on the right of Eq. (9.5.1), obtaining

$$\begin{aligned}\overline{G}_1(1,1') = \overline{G}^0_U(1,1') \pm i \int d2 \int d3\, \overline{G}^0_U(1,2)\mathcal{V}(2-3)\overline{G}^0_U(3,3^+)\overline{G}^0_U(2,1')\\ \pm i \int d2 \int d3\, \overline{G}^0_U(1,2)\mathcal{V}(2-3)\frac{\delta}{\delta U(3)}\overline{G}^0_U(2,1') + \begin{bmatrix}\text{terms of higher}\\ \text{order in } \mathcal{V}\end{bmatrix}.\end{aligned} \tag{9.5.3}$$

The higher order terms in $\mathcal{V}$ can be generated by repeatedly resubstituting the right-hand side of Eq. (9.5.1) into the last factors $\overline{G}_1$ under the integrals embedded within the terms of higher order in $\mathcal{V}$. When the desired order of $\mathcal{V}$ is reached, the iteration series is truncated, with variational derivatives evaluated using the identity of Eq. (4.5.6) as

$$\frac{\delta \overline{G}^0_U(1,1')}{\delta U(2)} = -\int d3 \int d4\, \overline{G}^0_U(1,3)\left(\frac{\delta\left[\overline{G}^0_U\right]^{-1}(3,4)}{\delta U(2)}\right)\overline{G}^0_U(4,1'). \tag{9.5.4}$$

The variational derivative $\delta\left[\overline{G}^0_U\right]^{-1}(3,4)/\delta U(2)$ may be evaluated from Eq. (9.5.2), and noting that the integral operator version of a differential operator in position representation contains an additional δ-function (Eq. (1.2.24)), we have

$$\frac{\delta\left[\overline{G}^0_U\right]^{-1}(3,4)}{\delta U(2)} = -\delta(3-4)\frac{\delta U(4)}{\delta U(2)} = -\delta(3-4)\delta(4-2), \tag{9.5.5}$$

yielding

$$\frac{\delta\overline{G}^0_U(1,1')}{\delta U(2)} = \overline{G}^0_U(1,2)\overline{G}^0_U(2,1'). \tag{9.5.6}$$

After the variational differentiation is completed, the source potential U is taken to vanish and we obtain

$$\begin{aligned}\overline{G}_1(1,1') = \overline{G}^0_U(1,1') &\pm i\int d2\int d3\,\overline{G}^0_U(1,2)\mathcal{V}(2-3)\overline{G}^0_U(3,3^+)\overline{G}^0_U(2,1')\\ &\pm i\int d2\int d3\,\overline{G}^0_U(1,2)\mathcal{V}(2-3)\overline{G}^0_U(2,3^+)\overline{G}^0_U(3,1') + \left[\begin{matrix}\text{terms of higher}\\ \text{order in }\mathcal{V}\end{matrix}\right]\end{aligned} \tag{9.5.7}$$

expanded to first order in $\mathcal{V}$. Such an expansion in powers of $\mathcal{V}$, where the system has a characteristic time, $\tau \sim \mathcal{V}^{-1}$, between particle–particle scatterings, is in fact a power series in the dimensionless parameter $\frac{t-t'}{\tau}$ associated with $\mathcal{V}(t-t')$, which is useful only for very short times.[4] Similar remarks pertain to dissipative scatterings, which are inadequately described by such a series, leading to the need for a transport equation. In the case at hand, we explore better approximation schemes for particle–particle interactions.

The application of the identity of Eq. (9.5.4) to $\overline{G}_1$ instead of $\overline{G}^0_U$ yields, similarly,

$$\frac{\delta\overline{G}_1(1,1')}{\delta U(2)} = -\int d3\int d4\,\overline{G}_1(1,3)\left(\frac{\delta\left[\overline{G}_1\right]^{-1}(3,4)}{\delta U(2)}\right)\overline{G}_1(4,1'). \tag{9.5.8}$$

This general result may be used to rewrite Eq. (9.5.1) in the form of a *Dyson equation* as

$$\overline{G}_1(1,1') = \overline{G}^0_U(1,1') + \int d2\int d3\,\overline{G}^0_U(1,2)\Sigma(2,3)\overline{G}_1(3,1') \tag{9.5.9}$$

or, using space-time matrix notation,

$$\overline{G}_1 = \overline{G}^0_U + \overline{G}^0_U\Sigma\overline{G}_1 \tag{9.5.10}$$

with the definition of the *self-energy*, $\Sigma(2,3)$, given by

$$\Sigma(2,3) = \pm i\delta(2-3)\int d4V(2-4)\overline{G}_1(4,4^+)$$
$$\mp i\int d4V(2-4)\int d5\overline{G}_1(2,5)\frac{\delta\overline{G}_1^{-1}(5,3)}{\delta U(4)}. \tag{9.5.11}$$

Considering Eq. (9.5.8), in particular the fact that $\overline{G}_1\delta\overline{G}_1^{-1} = -\delta\overline{G}_1\overline{G}_1^{-1}$, it is clear that the self-energy may also be written as

$$\Sigma(2,3) = \pm i\delta(2-3\int d4V(2-4)\overline{G}_1(4,4^+))$$
$$\pm i\int d4V(2-4)\int d5\frac{\delta\overline{G}_1(2,5)}{\delta U(4)}\overline{G}_1^{-1}(5,3). \tag{9.5.12}$$

9.6 Self-Energy and Vertex Operator Perturbation Theory

Multiplying Eq. (9.5.10) from the left by $\left[\overline{G}_U^0\right]^{-1}$ and from the right by $\overline{G}_1^{-1}$, we have

$$\overline{G}_1^{-1} = \left[\overline{G}_U^0\right]^{-1} - \Sigma. \tag{9.6.1}$$

With this, an integro-variational-differential equation for Σ may be obtained using Eq. (9.5.11) as

$$\Sigma(1,2) = \pm i\delta(1-2)\int d3\, V(1-3)\overline{G}_1(3,3^+)$$
$$\mp i\int d3\, V(1-3)\int d4\overline{G}_1(1,4)\frac{\delta\left[\overline{G}_U^0\right]^{-1}(4,2)}{\delta U(3)}$$
$$\pm i\int d3\, V(1-3)\int d4\overline{G}_1(1,4)\frac{\delta\Sigma(4,2)}{\delta U(3)}. \tag{9.6.2}$$

Again, using Eq. (9.5.5), we obtain the equation for Σ as

$$\Sigma(1,2) = \pm i\delta(1-2)\int d3\, V(1-3)\overline{G}_1(3,3^+)$$
$$\pm iV(1-2)\overline{G}_1(1,2)$$
$$\pm i\int d3\, V(1-3)\int d4\,\overline{G}_1(1,4)\frac{\delta\Sigma(4,2)}{\delta U(3)}, \tag{9.6.3}$$

which may be used to develop an iteration series directly for Σ in terms of $\mathcal{V}$ and $\overline{G}_1$ (or, alternatively, in terms of $\mathcal{V}$ and G_1^0)[4] with the leading term (first order in $\mathcal{V}$) given by the first two terms on the right of Eq. (9.6.3) corresponding to the Hartree–Fock approximation.

It is clear from Eq. (9.5.11) that the *vertex operator*, defined as

$$\Gamma(4,2;3) = -\frac{\delta \overline{G}_1^{-1}(4,2)}{\delta U(3)}, \tag{9.6.4}$$

must play a central role in the development of approximations, since it relates to the self-energy Σ by

$$\begin{aligned}\Sigma(1,2) = \pm i\delta(1-2) \int d3\, \mathcal{V}(1-3)\overline{G}_1(3,3^+)\\ \pm i \int d3\, \mathcal{V}(1-3) \int d4\, \overline{G}_1(1,4)\Gamma(4,2;3).\end{aligned} \tag{9.6.5}$$

Considering the vertex operator as a focal point for the development of perturbation theory, we note from Eq. (9.6.1) that

$$\frac{\delta, \overline{G}_1^{-1}(1,2)}{\delta U(3)} = \frac{\delta\left[, \overline{G}_U^0\right]^{-1}(1,2)}{\delta U(3)} - \frac{\delta\Sigma(1,2)}{\delta U(3)}, \tag{9.6.6}$$

or, using Eq. (9.6.4) and Eq. (9.5.5):

$$\Gamma(1,2;3) = \delta(1-2)\delta(2-3) + \frac{\delta\Sigma(1,2)}{\delta U(3)}, \tag{9.6.7}$$

where $\frac{\delta\Sigma(1,2)}{\delta U(3)}$ is, in turn, determined by applying $\frac{\delta}{\delta U(3)}$ to Eq. (9.6.5) (changing the dummy integration variables in Eq. (9.6.5) to avoid 1, 2 and 3). Thus, we obtain an integro-variational-differential equation for Γ as

$$\begin{aligned}\Gamma(1,2;3) = \delta(1-2)\delta(2-3) \pm i\Big\{\delta(1-2) \int d4\, \mathcal{V}(1-4)\frac{\delta\overline{G}_1(4,4^+)}{\delta U(3)}\\ + \int d4\, \mathcal{V}(1-4) \int d5\, \frac{\delta}{\delta U(3)}(\overline{G}_1(1,5)\Gamma(5,2;4))\Big\},\end{aligned} \tag{9.6.8}$$

and bearing in mind that

$$\frac{\delta\overline{G}_1}{\delta U} = -\overline{G}_1\left(\frac{\delta\overline{G}_1^{-1}}{\delta U}\right)\overline{G}_1 = \overline{G}_1\Gamma\overline{G}_1, \tag{9.6.9}$$

we have $(\delta(\overline{G}_1\Gamma) = (\delta\overline{G}_1)\Gamma + \overline{G}_1(\delta\Gamma))$:

$$\Gamma(1,2;3) = \delta(1-2)\delta(2-3) \pm i\left\{\delta(1-2)\int d4\,\mathcal{V}(1-4)\int d5\int d6\,\overline{G}_1(4,5)\Gamma(5,6;3)\right.$$
$$\overline{G}_1(6,4^+) + \int d4\,\mathcal{V}(1-4)\int d5\left[\int d6\int d7\,\overline{G}_1(1,6)\Gamma(6,7;3)\right.$$
$$\left.\left.\overline{G}_1(7,5)\Gamma(5,2;4) + \overline{G}_1(1,5)\frac{\delta\Gamma(5,2;4)}{\delta U(3)}\right]\right\}. \tag{9.6.10}$$

This may be used to generate an iteration series for the vertex operator Γ in powers of $\mathcal{V}$ and $\overline{G}_1$ (alternatively, in terms of $\mathcal{V}$ and $\overline{G}_1^0$), with the leading terms given by

$$\Gamma(1,2;3) = \delta(1-2)\delta(2-3)$$
$$\pm i\left[\mathcal{V}(1,2)\overline{G}_1(1,3)\overline{G}_1(3,2) + \delta(1-2)\int d4\mathcal{V}(1-4)\overline{G}_1(4,3)\overline{G}_1(3,4)\right], \tag{9.6.11}$$

which can be employed to augment the Hartree–Fock approximation embedded in the first two terms on the right of Eq. (9.6.3) for Σ by using Eq. (9.6.7) to obtain $\delta\Sigma/\delta U$ from Γ for use in the third term of Eq. (9.6.3). The use of the vertex operator Γ for generating perturbation theory will be discussed further in the context of shielded potential perturbation theory, below.

9.7 Shielded Potential Perturbation Theory

The introduction of the effective dynamic, non-local (and inhomogeneous) shielded potential $V(1)$ (Eq. (9.4.6)) enables us to rewrite the $\overline{G}_1$ equation in the form

$$\left[\left[\overline{G}_1^0\right]^{-1}(1) - V(1) \mp i\int d2\,\mathcal{V}(1-2)\frac{\delta}{\delta U(2)}\right]\overline{G}_1(1,1') = \delta(1-1'). \tag{9.7.1}$$

This may be expressed in terms of a modified self-energy $M(1,2)$ as

$$\left[\left[\overline{G}_1^0\right]^{-1}(1) - V(1)\right]\overline{G}_1(1,1') - \int d2\,M(1,2)\overline{G}_1(2,1') = \delta(1-1'), \tag{9.7.2}$$

where $M(1,2)$ is defined such that

$$\int d2\,M(1,2)\overline{G}_1(2,1') = \pm i\int d2\,\mathcal{V}(1-2)\frac{\delta\overline{G}_1(1,1')}{\delta U(2)}$$
$$= \pm i\int d2\int d3\,\mathcal{V}(1-2)\frac{\delta V(3)}{\delta U(2)}\frac{\delta\overline{G}_1(1,1')}{\delta V(3)}$$
$$= \pm i\int d3\int d2\,K(3,2)\mathcal{V}(1-2)\frac{\delta\overline{G}_1(1,1')}{\delta V(3)}, \tag{9.7.3}$$

where we have used the chain rule for variational differentiation. Defining the screened particle–particle interaction potential $W(1,3)$ as

$$W(1,3) = \int d2\, K(3,2)\mathcal{V}(2-1) \tag{9.7.4}$$

(noting that $\mathcal{V}(1-2) = \mathcal{V}(2-1)$, generally), we have

$$\int d2\, M(1,2)\overline{G}_1(2,1') = \pm i \int d3\, W(1,3)\frac{\delta \overline{G}_1(1,1')}{\delta V(3)}. \tag{9.7.5}$$

Employing the identity Eq. (9.5.8), we obtain $M(1,2)$ in the form

$$M(1,2) = \mp i \int d3 \int d4\, W(1,3)\overline{G}_1(1,4)\frac{\delta \overline{G}_1^{-1}(4,2)}{\delta V(3)}. \tag{9.7.6}$$

This is similar to the second term of the self-energy of Eq. (9.5.11), with the first "Hartree" term absorbed into the effective potential V of Eq. (9.7.1), which is now the variable of variational differentiation (instead of U in Eq. (9.5.11)). Defining

$$\left[\overline{G}_V^0\right]^{-1} = \left[\overline{G}_1^0\right]^{-1} - V, \tag{9.7.7}$$

we thus have Eq. (9.7.2) as

$$\overline{G}_1^{-1} = \left[\overline{G}_V^0\right]^{-1} - M. \tag{9.7.8}$$

Considering Eq. (9.7.6), it is clear that the appropriate shielded vertex operator is defined by

$$\Gamma_V(4,2;3) = -\frac{\delta \overline{G}_1^{-1}(4,2)}{\delta V(3)}, \tag{9.7.9}$$

so that application of $\frac{\delta}{\delta V}$ to Eq. (9.7.8) yields

$$\Gamma_V(1,2;3) = -\frac{\delta \left[\overline{G}_V^0\right]^{-1}(1,2)}{\delta V(3)} + \frac{\delta M(1,2)}{\delta V(3)}, \tag{9.7.10}$$

or, as above,

$$\Gamma_V(1,2;3) = \delta(1-2)\delta(2-3) + \frac{\delta M(1,2)}{\delta V(3)}. \tag{9.7.11}$$

Taken jointly with Eq. (9.7.6), this yields the integro-variational-differential equation for the shielded vertex operator as

$$\Gamma_V(1,2;3) = \delta(1-2)\delta(2-3) \pm i\frac{\delta}{\delta V(3)}\int d4 \int d5\, W(1,4)\overline{G}_1(1,5)\Gamma_V(5,2;4), \quad (9.7.12)$$

which may be used to generate a shielded potential perturbation series for Γ_V through iteration in complete analogy to the development of Section 9.6.

9.8 Imaginary Time, Contour Time Ordering and the Langreth Algebra

In considering both the equilibrium thermodynamic Green's functions of Chapters 7 and 8, and particularly the non-equilibrium Green's functions of the present chapter, it is important to bear in mind that the periodicity/antiperiodicity condition in imaginary time $t_0 \rightarrow t_0 + \tau = t_0 - i\beta$ defines an imaginary fundamental time interval in which both the differential and integral formulations of the equations of motion are set. For example, the action of the self-energy, Σ, (Eq. (9.5.9)) on a Green's function, $\overline{G}_1$, takes the form

$$\Im = \Sigma\overline{G}_1 \rightarrow \Im\left(1,1'\right) = \int d2\, \Sigma\left(1,2\right)\overline{G}_1\left(2,1'\right), \quad (9.8.1)$$

and with times t_1, t_2 in the interval $t_0 \rightarrow t_0 + \tau$, the t_2-time integration is from t_0 to $t_0 - i\beta$. A procedure for analytic continuation to real time has already been described in detail (Chapter 7) for equilibrium Green's functions having time dependence of the form $\overline{G}_1\left(t_1 - t_2\right)$. Recognizing the importance of integrals such as $\Im = \Sigma\overline{G}$ in Eq. (9.8.1) in both equilibrium and non-equilibrium situations (in the latter case $\overline{G}_1(t_1, t_2)$ depends on t_1 and t_2 separately, not just on their difference), we will describe another procedure for analytic continuation to real time that is universally applicable for such products (integrals), $\Sigma\overline{G}_1$.[5]

Noting that Σ and $\Im$ have symmetry/antisymmetry properties and analytic parts $\Sigma_{\left\{ > \atop < \right\}}, \Im_{\left\{ > \atop < \right\}}$ similar to those of $\overline{G}_1$ and $\overline{G}_{1\left\{ > \atop < \right\}}$, we write the time integral of Eq. (9.8.1) in terms of time interval segments in the case $t_1 < t_1'$ as follows (suppress positional arguments):

$$\begin{aligned}
\Im_<\left(t_1,t_1'\right) &= \int_{t_0}^{t_0-i\beta} dt_2\Sigma\left(t_1,t_2\right)\overline{G}_1\left(t_2,t_1'\right) \\
&= \int_{t_0}^{t_1} dt_2\Sigma_>\left(t_1,t_2\right)\overline{G}_{1<}\left(t_2,t_1'\right) + \int_{t_1}^{t_1'} dt_2\Sigma_<\left(t_1,t_2\right)\overline{G}_{1<}\left(t_2,t_1'\right) \\
&\quad + \int_{t_1'}^{t_0-i\beta} dt_2\Sigma_<\left(t_1,t_2\right)\overline{G}_{1>}\left(t_2,t_1'\right).
\end{aligned} \quad (9.8.2)$$

Taking the limit $t_0 \to -\infty$ so that $\overline{G}_{1\left\{\substack{>\\<}\right\}}$ matches the physical Green's function (Eq. (9.1.6)), and neglecting initial correlations at $t_0 \to -\infty$, we have

$$\mathfrak{I}_<(t_1, t_1') = \int_{-\infty}^{t_1} dt_2 \Sigma_>(t_1, t_2)\, \overline{G}_{1<}(t_2, t_1') - \int_{t_1'}^{t_1} dt_2 \Sigma_<(t_1, t_2)\, \overline{G}_{1<}(t_2, t_1')$$
$$- \int_{-\infty}^{t_1'} dt_2 \Sigma_<(t_1, t_2)\, \overline{G}_{1>}(t_2, t_1').$$

Rewriting the second term on the right as $\int_{t_1'}^{t_1} = \int_{t_1'}^{-\infty} + \int_{-\infty}^{t_1}$, we find that

$$\mathfrak{I}_<(t_1, t_1') = \int_{-\infty}^{t_1} dt_2 \left[\Sigma_>(t_1, t_2) - \Sigma_<(t_1, t_2)\right] \overline{G}_{1<}(t_2, t_1')$$
$$- \int_{-\infty}^{t_1'} dt_2 \Sigma_<(t_1, t_2) \left[\overline{G}_{1>}(t_2, t_1') - \overline{G}_{1<}(t_2, t_1')\right]. \tag{9.8.3}$$

This integration analysis corresponds to deforming the integration contour $[t_0 \to t_0 - i\beta]$, originally in the imaginary direction of the complex t-plane, to a closed time path contour along the real-time axis and back, with time ordering *along the contour* dictating the choices of $G_{\left\{\substack{>\\<}\right\}}$ and $\Sigma_{\left\{\substack{>\\<}\right\}}$ (the analyticity properties of the Green's functions pertinent to the contour deformation from imaginary time to real time are discussed in Reference 3). In this, we also assume that the integration along the final imaginary contour segment $t_0 \to t_0 - i\beta$ can be neglected as $t_0 \to -\infty$, thus implementing the analytic continuation of $\mathfrak{I}$ (Eq. (9.8.1)) from imaginary time to the infinite real-time axis. (The validity of the assumption that the last imaginary contour segment can be neglected is frequently asserted on the basis that a macroscopic system loses memory of its initial state and initial correlations as $t_0 \to -\infty$. Although we take this to be the case, it may not be true generally, as discussed in a critical analysis by V. G. Morozov and G. Röpke.)[6–9] Defining retarded and advanced Green's functions as ($\eta_+(t) = 1, 0$ for $t > 0, t < 0$ is the Heaviside unit step function, Eqns. (7.5.8 and 7.5.12)):

$$\overline{G}_{1r}(1, 1') = \eta_+(t_1 - t_1')\left[\overline{G}_{1>}(1, 1') - \overline{G}_{1<}(1, 1')\right] \text{(retarded)}$$
$$\overline{G}_{1a}(1, 1') = \eta_+(t_1' - t_1)\left[\overline{G}_{1<}(1, 1') - \overline{G}_{1>}(1, 1')\right] \text{(advanced)} \tag{9.8.4}$$

with analogous definitions for Σ_{1r} and Σ_{1a}, we have

$$\mathfrak{I}_{<}\left(t_1,t_1'\right)=\int_{-\infty}^{\infty} dt_2\,\Sigma_r\left(t_1,t_2\right)\overline{G}_{1<}\left(t_2,t_1\right)+\int_{-\infty}^{\infty} dt_2\,\Sigma_{<}\left(t_1,t_2\right)\overline{G}_{1a}\left(t_2,t_1'\right). \tag{9.8.5}$$

Of course, $\overline{G}_{1\left\{\substack{>\\<}\right\}}$ obeys the homogeneous counterpart of the equation of motion for the non-equilibrium Green's function, with no δ-function driving term. However, the $\overline{G}_{1\left\{\substack{r\\a}\right\}}$ equation recovers this δ-function driving term from the time derivative of $\eta_+(t)$, but with initial conditions appropriate to the retarded (r) or advanced (a) character. In terms of a space-time matrix notation, with time integration extended from $-\infty$ to $+\infty$, Eq. (9.8.5) may be written as

$$\mathfrak{I}_{<}=\Sigma_r\overline{G}_{1<}+\Sigma_{<}\overline{G}_{1a}. \tag{9.8.6}$$

Similarly, it is readily shown that

$$\mathfrak{I}_{>}=\Sigma_r\overline{G}_{1>}+\Sigma_{>}\overline{G}_{1a} \tag{9.8.7}$$

and

$$\mathfrak{I}_{r,a}=\Sigma_{r,a}\overline{G}_{1r,a}. \tag{9.8.8}$$

These prescriptions for the analytic continuation from integration in the imaginary time direction ($t_0\rightarrow t_0-i\beta$) to the full infinite real-time axis ($-\infty\rightarrow\infty$) have come to be known as the *Langreth Algebra*. Products of three (or more) matrices of this type, with associated infinite time-axis integration, may be analyzed by successively considering products of two adjacent ones, e.g., $QRS=[(QR)\,S]$ or $QRS=[Q\,(RS)]$.

Products of the type,

$$R\left(t_1,t_2\right)=\overline{G}_1\left(t_1,t_2\right)\overline{G}_1'\left(t_2,t_1\right), \tag{9.8.9}$$

with *no* time integration (for example, the ring diagram of Eq. (9.4.13)) but with t_1,t_2 on the complex time contour, can be analyzed in similar terms, particularly as the two times in the argument of the first Green's function are the same (but in reverse order) as those of the second one. Writing

$$\overline{G}_1\left(t_1,t_2\right)=\eta_+\left(t_1-t_2\right)\overline{G}_{1>}\left(t_1,t_2\right)+\eta_+\left(t_2-t_1\right)\overline{G}_{1<}\left(t_1,t_2\right), \tag{9.8.10}$$

and noting that $\eta_+^2(T)=\eta_+(T)$, while $\eta_+(T)\,\eta_+(-T)\equiv 0$, we have

$$R\left(t_1,t_2\right)=\eta_+\left(t_1-t_2\right)\overline{G}_{1>}\left(t_1,t_2\right)\overline{G}_{1<}'\left(t_2,t_1\right)+\eta_+\left(t_2-t_1\right)\overline{G}_{1<}\left(t_1,t_2\right)\overline{G}_{1>}'\left(t_2,t_1\right), \tag{9.8.11}$$

and

$$R_{>}\left(t_1,t_2\right)=\overline{G}_{1>}\left(t_1,t_2\right)\overline{G}_{1<}'\left(t_2,t_1\right), \tag{9.8.12}$$

$$R_{<}\left(t_1,t_2\right)=\overline{G}_{1<}\left(t_1,t_2\right)\overline{G}_{1>}'\left(t_2,t_1\right). \tag{9.8.13}$$

Forming $R_r(t_1, t_2)$ in accordance with Eq. (9.8.4),

$$R_r(t_1, t_2) = \eta_+(t_1 - t_2)\left[\overline{G}_{1>}(t_1, t_2)\overline{G}'_{1<}(t_2, t_1) - \overline{G}_{1<}(t_1, t_2)\overline{G}'_{1>}(t_2, t_1)\right], \qquad (9.8.14)$$

and adding and subtracting $\eta_+(t_1 - t_2)\overline{G}_{1<}(t_1, t_2)\overline{G}'_{1<}(t_1, t_2)$, we find the result

$$R_r(t_1, t_2) = \overline{G}_{1<}(t_1, t_2)\overline{G}'_{1a}(t_2, t_1) + \overline{G}_{1r}(t_1, t_2)\overline{G}'_{1<}(t_2, t_1). \qquad (9.8.15)$$

Similar results can be obtained for the case in which the time arguments of $\overline{G}_1$ and $\overline{G}'_1$ are identical and in the same order.

9.9 Generalized Kadanoff–Baym Ansatz for Derivation of Quantum Transport Equations

The rigorous derivation of quantum transport equations under non-equilibrium conditions has been an important issue related to non-equilibrium Green's functions, particularly for high electric fields. In this matter, it is desirable to relate the non-equilibrium "lesser" function $\overline{G}_{1<}$ to a Wigner distribution function ρ_w, defined as (set $\tau = t - t'; T = (t + t')/2; \mathbf{R} = (\mathbf{x} + \mathbf{x}')/2; \mathbf{r} = \mathbf{x} - \mathbf{x}'$ and $\mathbf{r} = \mathbf{x} - \mathbf{x}' \to \mathbf{p}$ under Fourier transform):

$$\rho_w(\mathbf{p}, \mathbf{R}, T) = -i\overline{G}_{1<}(\mathbf{p}, \tau = 0, \mathbf{R}, T),$$

in which $\tau = t - t' = 0$, in an effort to facilitate the derivation of the "collision term" of an anticipated quantum kinetic/Boltzmann-like transport equation from the non-equilibrium self-energy $\sum(\mathbf{x}, t; \mathbf{x}'t')$. Considerable study has gone into this matter because the actual Green's function structure of the collision term involves unequal time arguments $t \neq t' (\tau \neq 0)$, whereas $\rho_w(\mathbf{p}, \mathbf{R}, T)$ involves only equal times, $\tau = 0$. However, Lipavsky, Spicka, Velicky and Kalvova developed the "Generalized Kadanoff–Baym Ansatz" approximation[6–9] (GKBA) to deal with it. Their papers explain in detail why it is superior to the earlier Kadanoff–Baym Ansatz[1] (KBA) approximation under strongly non-equilibrium conditions. Here, we offer a very brief account of the GKBA approximation. Writing

$$\overline{G}_{1<}(\mathbf{x}, t; \mathbf{x}', t') = \mp i\langle\psi^+(\mathbf{x}', t')\,\psi(\mathbf{x}, t)\rangle, \qquad (9.9.1)$$

and assuming that, within the context of the fast dynamics of the formation of a collision term, we may neglect both particle–particle and phonon-scattering interactions as an approximation, the time development of the field operator $\psi(\mathbf{x}, t)$ is given by the single-particle equation ($\hbar \to 1$),

$$\left(i\frac{\partial}{\partial t} + \zeta - h^{(1)}\left(\mathbf{x}, \tfrac{\hbar}{i}\nabla, t\right)\right)\psi(\mathbf{x}, t) = 0, \qquad (9.9.2)$$

which may be taken to include strong external fields in $h^{(1)}$ as well as self-consistent field interactions. Considering the associated single-particle retarded Green's function equation in the assumed absence of scattering interactions (beyond self-consistent fields),

$$\left(i\frac{\partial}{\partial t}+\zeta-h^{(1)}\left(\mathbf{x},\tfrac{\hbar}{i}\nabla,t\right)\right)\overline{G}_{1r}^{0}\left(\mathbf{x},t;\mathbf{x}',t'\right)=\delta^{(3)}\left(\mathbf{x}-\mathbf{x}'\right)\delta\left(t-t'\right), \tag{9.9.3}$$

the time development of $\psi\left(\mathbf{x},t\right)$ from $\bar{t}$ to t $(t>\bar{t})$ is given by the *initial value* solution as (see Eqns. (5.1.1–5.1.12) and related discussion):

$$\psi\left(\mathbf{x},t\right)=i\int d^{(3)}\mathbf{x}''\overline{G}_{1r}^{0}\left(\mathbf{x},t;\mathbf{x}'',\bar{t}\right)\psi\left(\mathbf{x}'',\bar{t}\right), \tag{9.9.4}$$

since the action of the differential operator $\left(i\,\partial/\partial t+\zeta-h^{(1)}\right)$ on $\overline{G}_{1r}^{0}\left(\mathbf{x},t;\mathbf{x}'',\bar{t}\right)$ vanishes identically for $t\neq\bar{t}$ in accordance with Eq. (9.9.2), and at $t=\bar{t}^{+}$, $\overline{G}_{1r}^{0}\left(\mathbf{x},t;\mathbf{x}',\bar{t}\right)=\delta^{(3)}\left(\mathbf{x}-\mathbf{x}'\right)/i$. Thus, for $t>\bar{t}$ we can refer $\psi\left(\mathbf{x},t\right)$ back in time to $\psi\left(\mathbf{x}',\bar{t}\right)$ using Eq. (9.9.4). Correspondingly, for $t>t'$,

$$\begin{aligned}\overline{G}_{1<}\left(\mathbf{x},t;\mathbf{x}',t'\right)&=\pm\int d^{3}\mathbf{x}''\overline{G}_{1r}^{0}\left(\mathbf{x},t;\mathbf{x}'',t'\right)\left\langle\psi^{+}\left(\mathbf{x}',t'\right)\psi\left(\mathbf{x}'',t'\right)\right\rangle\\&=i\int d^{3}\mathbf{x}'\overline{G}_{1r}^{0}\left(\mathbf{x},t;\mathbf{x}'',t'\right)\overline{G}_{1<}\left(\mathbf{x}'',t';\mathbf{x}',t'\right);\quad\left(t>t'\right).\end{aligned} \tag{9.9.5}$$

While Eq. (9.9.5) is limited for $t>t'$, it is also necessary to determine the corresponding results for the continued function $\overline{G}_{1<}(\mathbf{x},t;\mathbf{x};t')$ for the case $t<t'$ by examining the time development of the operator $\psi^{+}(\mathbf{x}',t')$. For this purpose we can employ the advanced Green's function $\overline{G}_{1a}^{0}$ counterpart of Eq. (9.9.4) and use arguments similar to those above to obtain the result for both time ranges $t\lessgtr t'$ as

$$\begin{aligned}\overline{G}_{1<}\left(\mathbf{x},t;\mathbf{x}',t'\right)=i\int d^{3}\mathbf{x}''&\left[\overline{G}_{1r}^{0}\left(\mathbf{x},t;\mathbf{x}'',t'\right)\overline{G}_{1<}\left(\mathbf{x}'',t';\mathbf{x}',t'\right)\right.\\&\left.-\overline{G}_{1<}\left(\mathbf{x},t;\mathbf{x}'',t\right)\overline{G}_{1a}^{0}\left(\mathbf{x}'',t;\mathbf{x}',t'\right)\right].\end{aligned} \tag{9.9.6}$$

In an entirely similar way, we also have

$$\begin{aligned}\overline{G}_{1>}\left(\mathbf{x},t;\mathbf{x}',t'\right)=i\int d^{3}\mathbf{x}''&\left[\overline{G}_{1r}^{0}\left(\mathbf{x},t;\mathbf{x}'',t'\right)\overline{G}_{1>}\left(\mathbf{x}'',t';\mathbf{x}',t'\right)\right.\\&\left.-\overline{G}_{1>}\left(\mathbf{x},t;\mathbf{x}'',t\right)\overline{G}_{1a}^{0}\left(\mathbf{x}'',t;\mathbf{x}',t'\right)\right].\end{aligned} \tag{9.9.7}$$

Bearing in mind that

$$\rho_{w}(\mathbf{x},\mathbf{x}';t)=-i\overline{G}_{1<}(\mathbf{x},\mathbf{x}';t,t), \tag{9.9.8}$$

Eq. (9.9.6) may be rewritten as

$$\overline{G}_{1<}(\mathbf{x},t;\mathbf{x}',t') = -\int d^3\mathbf{x}''\left[\overline{G}^0_{1r}(\mathbf{x},t;\mathbf{x}'',t')\rho_w(\mathbf{x}'',\mathbf{x}';t') - \rho_w(\mathbf{x},\mathbf{x}'';t)\overline{G}^0_{1a}(\mathbf{x}'',t;\mathbf{x}',t')\right]. \tag{9.9.9}$$

This GKBA result has facilitated the development of a Generalized Master Equation (GME) that can describe highly non-equilibrium dynamics involving strong external fields (which interfere in scattering processes) and self-consistent fields. Moreover, the GME is the logical progenitor which, upon further approximation, generates a prototypical quantum kinetic/transport Boltzmann-like equation with the attendant collision term under discussion.

References 5–9 present a fully detailed account of the Green's function derivation of highly non-equilibrium quantum transport equations appropriate for strong electric fields using the GKBA approximation. The early development of quantum transport equations using the KBA approximation is discussed in Reference 1. The formulation of *linear* quantum transport coefficients in terms of two-particle equilibrium Green's functions is presented in detail in Reference 3 (Section IV), and was further discussed by N. Ashby.[10] The earliest report of a linear transport formulation in terms of equilibrium two-particle Green's functions was provided by R. Kubo.[2] The formulation of non-equilibrium quantum transport theory undertaken in terms of Wigner distribution functions, often referred to as "phase space distribution functions," must be applied with care because they are not always positive and they are not generally gauge invariant (which is also the case for Green's functions). However, one can construct a gauge invariant counterpart (Ashby thesis, Reference 10, Section II).[10] Density matrix and superoperator techniques have also been employed in the quest for rigorous derivation of quantum transport equations (see Chapter 3 and references therein).

9.10 Chapter 9 Problems

P.9.1. Verify Eqns. (9.1.7–9.1.9).

P.9.2. Examine Eqns. (9.1.30–9.1.34) to verify their validity.

P.9.3. Explain why the sum rule of Eqns. (7.5.6 and 7.5.7) does *not* apply to $\overline{A}_L(\omega)$ in Eq. (9.1.19), and determine what takes its place.

P.9.4. In Section 9.1, we discussed charge current using the gauge in which the scalar potential vanishes, $\mathbf{E} = -\frac{1}{c}\frac{\partial \mathbf{A}}{\partial t}$: Determine the causally retarded conductivity as a function of frequency ω in terms of $\overline{\mathbf{L}}_c$. Also, express Eq. (9.1.17) in terms of $\overline{G}_1$ and $\overline{G}_2$ equilibrium thermodynamic Green's function.

P.9.5. Prove Eq. (9.3.10).

P.9.6. Employing Eq. (9.3.11), carry out the proof of the linked cluster theorem.

P.9.7. Prove the validity of Eqns. (9.4.18–9.4.21).

P.9.8. Verify Eqns. (9.4.24).

P.9.9. Prove Eqns. (9.6.1–9.6.5) for the self-energy.

P.9.10. Prove Eqns. (9.6.10 and 9.6.11) for the vertex operator Γ.

P.9.11. Determine the leading approximation (in powers of $\mathcal{V}$) for the self-energy $\sum$ in ordinary (non-shielded) perturbation theory.

P.9.12. Determine the leading approximation (in powers of $\mathcal{W}$) for the self-energy M in shielded potential perturbation theory. Do the same for Γ_v in Eq. (9.7.12).

P.9.13. Prove Eqns. (9.8.7) and (9.7.8).

9.11 References for Chapter 9

R.9.1. L.P. Kadanoff and G. Baym, "Quantum Statistical Mechanics," W.A. Benjamin, p.96, Eq. (8–21a,b) (1962).

R.9.2. R. Kubo, Can. J.Phys. *31*, 1274 (1956). Also R. Kubo, J. Phys. Soc. Japan *12*, 570 (1957).

R.9.3. P.C. Martin and J. Schwinger, Phys. Rev. *115*, 1342, Section VI (1959).

R.9.4. Reference 1, Eq. (5–21b).

R.9.5. J. Schwinger, J. Math. Phys. *2*, 407 (1961). (Provides the basic background for the considerations in this section).

R.9.6. V. G. Morozov and G. Röpke in "Progress in Nonequilibrium Green's Functions," Editor: M. Bonitz, World Scientific, p.82 (2000).

R.9.7. B. Velicky, A. Kalvova and V. Spicka in "Progress in Nonequilib. Green's Functions III," Editors: M. Bonitz and A. Filinov, J. Phys. Conf. Series **35**, p.1 (2006).

R.9.8. P. Lipavsky, V. Spicka and B. Velicky, Phys. Rev. B., *34*, 6933 (1986). (Including references cited therein.)

R.9.9. V. Spicka, B. Velicky and A. Kalvova, Physica *E29*, p.154, 175, 196 (1999) (Including references cited therein.)

R.9.10. N. Ashby, "Applications of Quantum Field Theory to Electrical Conductivity," Ph.D. thesis, Harvard University, (1961). Also, N. Ashby in "Lectures in Theoretical Physics," Boulder *3*, University of Colorado Press, (1960).

10

Random Phase Approximation Plasma Phenomenology, Semiclassical and Hydrodynamic Models; Electrodynamics

Abstract

Chapter 10 analyzes plasma dielectric response phenomenology in the random phase approximation (RPA), starting with the RPA ring diagram of the polarizability expressed in terms of thermodynamic Green's functions and also in terms of non-interacting single-particle energy eigenfunctions for both homogeneous and inhomogeneous quantum plasmas. Both the direct dielectric function and the dynamic, non-local inverse dielectric screening function are explicitly exhibited by spatial Fourier transformation in cases of translational invariance for three, two and one dimensions (replicating the results of Lindhard (3D) and Stern (2D) in the degenerate limit). This also encompasses the non-local plasmon spectrum and carries the potential to evaluate finite-temperature effects as well as static shielding information (including Friedel oscillations and the Debye–Thomas–Fermi shielding law). The role of a quantizing magnetic field in the screening function is also discussed in detail. Models that are simpler to use analytically, namely the semiclassical and classical limits, and also the hydrodynamic model, are defined and discussed along with surface plasmons. The exchange and correlation energies are formulated in terms of the spectral weight of the RPA screening function (and polarizability) and are discussed. Chapter 10 also presents the formulation of the theory of the van der Waals interaction of two neutral polarizable systems in terms of their two-particle Green's functions (for each of the constituent systems), explaining the emergence of the role of the dynamic, non-local plasma image potential due to screening in the Coulombic interaction energy of the two systems. This is applied to calculation of the van der Waals interaction energy (physisorption) between a neutral atom and a nearby semi-infinite non-local, dynamic metallic plasma. The inverse dielectric screening function K is also shown to play a central role in the power loss of an energetic charged particle passing through a plasma. In consideration of the non-local, dynamic dielectric properties of nanostructures, Chapter 10 also introduces electromagnetic dyadic Green's functions for the Maxwell equations and discusses their relation to the (inverse) dielectric tensor $\overline{\mathbf{K}}$ and effective electromagnetic field $\mathbf{E}$ generated by an incident EM field $\mathbf{E}_0 \equiv \mathbf{D}$; $\mathbf{E} = \overline{\mathbf{K}}\mathbf{D}$. A detailed description of the determination of the RPA dynamic, non-local conductivity

Quantum Statistical Field Theory: Schwinger's Variational Method. Norman J. Morgenstern Horing, Oxford University Press (2017). © Norman J. Morgenstern Horing. DOI 10.1093/oso/9780198791942.001.0001

tensor follows and is applied to a planar quantum well. Kramers–Krönig relations associated with causality, involving conductivity and polarizability, are also discussed. Finally, attention is directed to compound nanostructures, with the development of a method to account for the polarizable electromagnetic response of a multifaceted system having several nanostructured constituent parts (with applications to a bulk semiconductor plasma hosting a planar quantum well, and also to a semiconductor superlattice; in these examples the coupled plasmon spectra and polaritons are discussed).

10.1 Dielectric Response Functions for Inhomogeneous and Homogeneous Systems

The definition $K(1,2) = \frac{\delta V(1)}{\delta U(2)}$, Eq. (9.4.10), may also be stated as

$$V(1) = \int d2K(1,2)U(2), \tag{10.1.1}$$

and the role of K as a non-local, dynamic inverse dielectric (screening) function naturally leads to the interpretation of its inverse, $\varepsilon(1,2)$, in the space-time matrix sense,

$$\int d3\,\varepsilon(1,3)K(3,2) = \delta(1-2); \quad \varepsilon K = K\varepsilon = I, \tag{10.1.2}$$

as a *direct* dielectric function. Considering Eq. (9.4.16), we have

$$\varepsilon = 1 + \mathcal{V}R \tag{10.1.3}$$

or

$$\varepsilon(1,2) = \delta(1-2) + \int d3\mathcal{V}(1-3)R(3,2). \tag{10.1.4}$$

Clearly, the corresponding non-local, dynamic polarizability $\alpha(1,2)$ is given by $(\varepsilon - 1)$ as (Eqns. (9.4.12) and (9.4.13)):

$$\alpha(1,2) = \int d3\mathcal{V}(1-3)R(3,2) = \mp i \int d3\mathcal{V}(1-3)\overline{G}_1^0(3,2)\overline{G}_1^0(2,3^+), \tag{10.1.5}$$

which is the kernel of the RPA integral equation. The solution of the RPA integral equation for the Matsubara Fourier series coefficient of K (as well as that of the polarizability) can be obtained by inserting the Fourier series in imaginary time $[0 \to \tau = -i\beta]$ for each of the two Green's functions of the ring diagram R, and this has been done in detail for a translationally invariant bulk quantum plasma.[1] It is instructive to proceed in a different way here: The time integral involved in α follows at once from the instantaneous nature of the interparticle interaction, $\delta(t_1 - t_3)$, so that

$$\alpha(1,2) = \mp i \int d^3\mathbf{x}_3\mathcal{V}(\mathbf{x}_1 - \mathbf{x}_3)\overline{G}_1^0(\mathbf{x}_3, \mathbf{x}_2; t_1 - t_2)\overline{G}_1^0(\mathbf{x}_2, \mathbf{x}_3; t_2 - t_1^+). \tag{10.1.6}$$

The relevant Fourier series coefficient is that of the ring diagram ($T = t_1 - t_2$):

$$R(\mathbf{x}_3, \mathbf{x}_2; \nu) = \mp i \int_0^{\tau=-i\beta} dT e^{i\frac{\pi\nu T}{\tau}} \overline{G}_1^0(\mathbf{x}_3, \mathbf{x}_2; T)\overline{G}_1^0(\mathbf{x}_2, \mathbf{x}_3; -T) \tag{10.1.7}$$

(where we understand $t_0 \to -\infty$). The Green's functions involved are time-ordered on the integration contour $[T = 0 \to \tau = -i\beta]$, and setting $\frac{\pi\nu}{\tau} \to \omega + i\delta$ for the physical response as indicated above, we deform the contour in the complex time plane through the region of analyticity as indicated in Section 9.8. The time ordering of the Green's functions persists on the deformed contour, and the vertical segment representing the integral from $T = -\infty \to -\infty + \tau$ is neglected (as being so distant in time that it is assumed to be insignificant). This corresponds to the neglect of correlations that are very distant in time (which is not always appropriate, as discussed in Section 9.8). Thus, we have

$$R(\mathbf{x}_3, \mathbf{x}_2; \pi\nu/\tau \to \omega + i\delta) = \mp i \int_0^{\infty} dT e^{i(\omega+i\delta)T} \overline{G}_{1>}^0(\mathbf{x}_3, \mathbf{x}_2; T)\overline{G}_{1<}^0(\mathbf{x}_2, \mathbf{x}_3; -T)$$
$$\mp i \int_{\infty+\tau}^{\tau} dT e^{i(\omega+i\delta)T} \overline{G}_{1>}^0(\mathbf{x}_3, \mathbf{x}_2; T)\overline{G}_{1<}^0(\mathbf{x}_2, \mathbf{x}_3; -T). \tag{10.1.8}$$

In the second integral we translate T through $\tau = -i\beta$ to reach the real-time axis, inducing the interchange $\overline{G}_{1\gtrless}^0(T + \tau) = \pm\overline{G}_{1\lessgtr}^0(T)$ twice, whence

$$R(\mathbf{x}_3, \mathbf{x}_2; \omega + i\delta) = \mp i \int_0^{\infty} dT e^{i(\omega+i\delta)T} \overline{G}_{1>}^0(\mathbf{x}_3, \mathbf{x}_2; T)\overline{G}_{1<}^0(\mathbf{x}_2, \mathbf{x}_3; -T)$$
$$\mp i \int_{\infty}^{0} dT e^{i(\omega+i\delta)T} \overline{G}_{1<}^0(\mathbf{x}_3, \mathbf{x}_2; T)\overline{G}_{1>}^0(\mathbf{x}_2, \mathbf{x}_3; -T). \tag{10.1.9}$$

Using Eqns. (7.3.9) and (7.3.10) we have (denoting the spectral weight of G_1^0 by A):

$$R(\mathbf{x}_3, \mathbf{x}_2; \omega + i\delta) = \int \frac{d\omega'}{2\pi i} \int \frac{d\omega''}{2\pi i} \frac{(1 \pm f(\omega'))f(\omega'')}{\omega + \omega'' - \omega' + i\delta} A(\mathbf{x}_3, \mathbf{x}_2; \omega')A(\mathbf{x}_2, \mathbf{x}_3; \omega'')$$
$$- \int \frac{d\omega'}{2\pi i} \int \frac{d\omega''}{2\pi i} \frac{(1 \pm f(\omega'))f(\omega'')}{\omega + \omega' - \omega'' + i\delta} A(\mathbf{x}_2, \mathbf{x}_3; \omega')A(\mathbf{x}_3, \mathbf{x}_2; \omega''). \tag{10.1.10}$$

The "double-distribution" terms, $f(\omega')f(\omega'')$, are readily seen to cancel by interchanging the dummy integration roles of ω' and ω'' in the last term, whence

$$R(\mathbf{x}_3, \mathbf{x}_2; \omega + i\delta) = \int \frac{d\omega'}{2\pi i} \int \frac{d\omega''}{2\pi i} \frac{f(\omega'') - f(\omega')}{\omega + \omega'' - \omega' + i\delta} A(\mathbf{x}_2, \mathbf{x}_3; \omega'')A(\mathbf{x}_3, \mathbf{x}_2; \omega'). \tag{10.1.11}$$

In the absence of particle–particle interactions we may employ Eq. (8.2.1) for the spectral weight A of G_1^0, obtaining R in terms of single-particle energy eigenstates $\phi_E(\mathbf{x})$ as

$$R(\mathbf{x}_3, \mathbf{x}_2; \omega + i\delta) = \sum_{E'} \sum_{E''} \frac{f(E'' - \mu) - f(E' - \mu)}{\omega + E'' - E' + i\delta} \phi_{E''}(\mathbf{x}_2)\phi^*_{E''}(\mathbf{x}_3)\phi_{E'}(\mathbf{x}_3)\phi^*_{E'}(\mathbf{x}_2). \tag{10.1.12}$$

Below, we employ the notation (note that $f(E)$ and $f(E - \mu)$ are used interchangeably here):

$$R_{\alpha\beta}(\omega) \equiv \frac{f(E_\beta - \mu) - f(E_\alpha - \mu)}{\omega + E_\beta - E_\alpha + i\delta}. \tag{10.1.13}$$

The RPA integral equation (Eq. (9.4.12)) may now be Fourier-time-transformed, $t_1 - t_2 \to \omega$, with the result (put $\phi_{E_\alpha} \to \phi_\alpha$, etc.):

$$\begin{aligned} K(\mathbf{x}_1, \mathbf{x}_2; \omega) &= \delta^{(3)}(\mathbf{x}_1 - \mathbf{x}_2) - \int d^{(3)}\mathbf{x}_3 \int d^{(3)}\mathbf{x}_4 \mathcal{V}(\mathbf{x}_1 - \mathbf{x}_3) R(\mathbf{x}_3, \mathbf{x}_4; \omega + i\delta) K(\mathbf{x}_4, \mathbf{x}_2; \omega) \\ &= \delta^{(3)}(\mathbf{x}_1 - \mathbf{x}_2) - \sum_{\alpha,\beta} R_{\alpha\beta}(\omega) \int d^{(3)}\mathbf{x}_3 \mathcal{V}(\mathbf{x}_1 - \mathbf{x}_3)\phi^*_\beta(\mathbf{x}_3)\phi_\alpha(\mathbf{x}_3) \\ &\quad \times \int d^{(3)}\mathbf{x}_4 \phi^*_\alpha(\mathbf{x}_4)\phi_\beta(\mathbf{x}_4) K(\mathbf{x}_4, \mathbf{x}_2; \omega). \end{aligned} \tag{10.1.14}$$

Applying $\int d^{(3)}\mathbf{x}_1 \phi^*_\mu(\mathbf{x}_1)\phi_\gamma(\mathbf{x}_1)$ to the left-hand side to form an equation for

$$K_{\mu\gamma}(\mathbf{x}_2) = \int d^{(3)}\mathbf{x}_1 \phi^*_\mu(\mathbf{x}_1)\phi_\gamma(\mathbf{x}_1) K(\mathbf{x}_1, \mathbf{x}_2; \omega), \tag{10.1.15}$$

we have the matrix equation

$$K_{\mu\gamma}(\mathbf{x}_2) = \phi^*_\mu(\mathbf{x}_2)\phi_\gamma(\mathbf{x}_2) - \sum_{\alpha,\beta} \mathcal{V}_{\mu\gamma;\alpha\beta} R_{\alpha\beta} K_{\alpha\beta}(\mathbf{x}_2), \tag{10.1.16}$$

where

$$\mathcal{V}_{\mu\gamma;\alpha\beta} = \int d^{(3)}\mathbf{x}_1 \int d^{(3)}\mathbf{x}_3 \phi^*_\mu(\mathbf{x}_1)\phi_\gamma(\mathbf{x}_1) \mathcal{V}(\mathbf{x}_1 - \mathbf{x}_3)\phi^*_\beta(\mathbf{x}_3)\phi_\alpha(\mathbf{x}_3). \tag{10.1.17}$$

The matrix equation, Eq. (10.1.16), may be written in a more symmetric form by defining

$$Q_{\alpha\beta}(\mathbf{x}_2) = R^{1/2}_{\alpha\beta} K_{\alpha\beta}(\mathbf{x}_2); \quad M_{\mu\gamma;\alpha\beta} = R^{1/2}_{\mu\gamma} \mathcal{V}_{\mu\gamma;\alpha\beta} R^{1/2}_{\alpha\beta}, \tag{10.1.18}$$

with the result

$$\sum_{\alpha,\beta} \left(\delta_{\mu\alpha}\delta_{\gamma\beta} + M_{\mu\gamma;\alpha\beta}\right) Q_{\alpha\beta}(\mathbf{x}_2) = R^{1/2}_{\mu\gamma}\phi^*_\mu(\mathbf{x}_2)\phi_\gamma(\mathbf{x}_2). \tag{10.1.19}$$

The only remaining asymmetry is in the inhomogeneous term on the right-hand side, a consequence of the fact that we are analyzing the inverse dielectric function rather than the density correlation function. The formal eigenvector solution for inversion of the matrix $(I+M)$ has been discussed in the literature in connection with dielectric response in the vicinity of a surface.[2] If only a few energy levels are involved, the solution can be obtained numerically, sometimes analytically.

In the case of full three-dimensional (3D) spatial translational invariance, the RPA integral equation, Eq. (9.4.12), can be solved using spatial Fourier transforms for $K(\mathbf{x},\mathbf{x}';\omega) = K(\mathbf{x}-\mathbf{x}';\omega) \rightarrow K(\mathbf{p},\omega)$, and with $h_1^{(1)} \rightarrow -\frac{\nabla_1^2}{2m}$, we have the one-particle spectral weight as $\bar{A}(\mathbf{p},\omega) = 2\pi\delta(\omega+\mu-\frac{p^2}{2m})$. In momentum representation, the RPA integral equation can be solved algebraically to yield

$$K(\mathbf{p},\omega) = \frac{1}{\varepsilon(\mathbf{p},\omega)} = \frac{V(\mathbf{p},\omega)}{U(\mathbf{p},\omega)}, \tag{10.1.20}$$

where ($\mathcal{V}(\mathbf{p}) = \frac{4\pi e^2}{p^2}$ is the 3D Fourier transform of the Coulomb potential):

$$\varepsilon(\mathbf{p},\omega) = 1 + \mathcal{V}(\mathbf{p})R(\mathbf{p},\omega+i\delta), \tag{10.1.21}$$

and (m is the effective mass; spin trace understood),

$$R(\mathbf{p},\omega+i\delta) = -\int \frac{d^{(3)}\mathbf{k}}{(2\pi)^3} f\left(\frac{k^2}{2m}-\mu\right)\left\{\frac{1}{\omega+\frac{k^2}{2m}-\frac{(\mathbf{k}-\mathbf{p})^2}{2m}+i\delta} - \frac{1}{\omega+\frac{(\mathbf{k}-\mathbf{p})^2}{2m}-\frac{k^2}{2m}+i\delta}\right\}. \tag{10.1.22}$$

Separating real and imaginary parts ($\mathcal{P}$ denotes principal part):

$$\mathrm{Re}\,[\varepsilon(\mathbf{p},\omega)] = 1 - \frac{4\pi e^2}{p^2}\int \frac{d^{(3)}\mathbf{k}}{(2\pi)^3} f\left(\frac{k^2}{2m}-\mu\right)\left\{\frac{\mathcal{P}}{\omega+\frac{k^2}{2m}-\frac{(\mathbf{k}-\mathbf{p})^2}{2m}} - \frac{\mathcal{P}}{\omega+\frac{(\mathbf{k}-\mathbf{p})^2}{2m}-\frac{k^2}{2m}}\right\} \tag{10.1.23}$$

and

$$\mathrm{Im}\,[\varepsilon(\mathbf{p},\omega)] = \frac{4\pi e^2}{p^2}\int \frac{d^{(3)}\mathbf{k}}{(2\pi)^3} f\left(\frac{k^2}{2m}-\mu\right)\left\{\delta\left(\omega+\frac{k^2}{2m}-\frac{(\mathbf{k}-\mathbf{p})^2}{2m}\right)\right.$$
$$\left. - \delta\left(\omega+\frac{(\mathbf{k}-\mathbf{p})^2}{2m}-\frac{k^2}{2m}\right)\right\}. \tag{10.1.24}$$

It should be noted that Re ε is an even function of $\mathbf{p}$ and ω, and Im ε is an even function of $\mathbf{p}$, but an odd function of ω. Now,

$$K(\mathbf{p},\omega) = \frac{1}{\mathrm{Re}\,\varepsilon + i\,\mathrm{Im}\,\varepsilon} = \frac{\mathrm{Re}\,\varepsilon}{(\mathrm{Re}\,\varepsilon)^2+(\mathrm{Im}\,\varepsilon)^2} - i\frac{\mathrm{Im}\,\varepsilon}{(\mathrm{Re}\,\varepsilon)^2+(\mathrm{Im}\,\varepsilon)^2}. \tag{10.1.25}$$

Bearing in mind that the spectral weight is given by the discontinuity of the Fourier coefficient (auxiliary function) across the real ω-axis (Eq. (7.4.13)), the spectral weight $\mathcal{A}(\mathbf{p},\omega)$ of the inverse dielectric function is just

$$\mathcal{A}(\mathbf{p},\omega) = \frac{2\ \mathrm{Im}\ \varepsilon}{(\mathrm{Re}\ \varepsilon)^2 + (\mathrm{Im}\ \varepsilon)^2}. \tag{10.1.26}$$

Since Im ε is an odd function of ω, so is $\mathcal{A}(\mathbf{p},\omega)$, and, consequently,

$$\int_{-\infty}^{\infty} d\omega \mathcal{A}(\mathbf{p},\omega) = 0. \tag{10.1.27}$$

From another, more general, point of view, this vanishing is due to the commutation of density operators at different space points at the same time, $[\rho(\mathbf{x}_1,t),\rho(\mathbf{x}_2,t)] = 0$. In the vicinity of a frequency pole of $K(\mathbf{p},\omega)$, or, alternatively, a frequency zero of Re $\varepsilon(\mathbf{p},\omega)$, Re $\varepsilon(\mathbf{p},\omega=\Omega) = 0$, we Taylor-expand

$$\mathrm{Re}\ \varepsilon(\mathbf{p},\omega) \simeq Z^{-1}(\mathbf{p},\omega)(\omega - \Omega), \tag{10.1.28}$$

where

$$Z^{-1}(\mathbf{p},\omega) = \left[\frac{\partial\varepsilon(\mathbf{p},\omega)}{\partial\omega}\right]_{\omega=\Omega}, \tag{10.1.29}$$

and if Im $\varepsilon(\mathbf{p},\omega) \cong$ Im $\varepsilon(\mathbf{p},\Omega)$ is slowly varying in this frequency range we have (define $\gamma(\mathbf{p},\Omega) = 2Z(\mathbf{p},\Omega)\mathrm{Im}\ \varepsilon(\mathbf{p},\Omega)$):

$$\mathcal{A}(\mathbf{p},\omega) = \frac{Z(\mathbf{p},\Omega)\gamma(\mathbf{p},\Omega)}{(\omega-\Omega)^2 + \left(\frac{\gamma(\mathbf{p},\Omega)}{2}\right)^2}. \tag{10.1.30}$$

The almost δ-function resonant character of this Lorentzian line shape as a function of ω indicates the existence of a collective normal mode of the system at frequency Ω, at which the system can oscillate even if the external exciting field becomes vanishingly small (bear in mind that $K(\mathbf{p},\omega) = \frac{V(\mathbf{p},\omega)}{U(\mathbf{p},\omega)}$). The damping of this mode is given by $\gamma(\mathbf{p},\Omega)$, and its excitation amplitude (oscillator strength) is $Z(\mathbf{p},\Omega)$. For the electron plasma, it is called a plasma oscillation, or plasmon, a longitudinal photon in plasma-like media, with the dispersion relation

$$\mathrm{Re}\,[\varepsilon(\mathbf{p},\Omega)] = 1 + \mathcal{V}(\mathbf{p})\mathrm{Re}\,[R(\mathbf{p},\Omega)] = 0. \tag{10.1.31}$$

Moreover, since Re $[\varepsilon(\mathbf{p},\omega)]$ is even in Ω, then $-\Omega$ is also a root, and thus $\mathcal{A}(\mathbf{p},\omega)$ should be written as

$$\mathcal{A}(\mathbf{p},\omega) = \sum_n \left[\frac{Z(\Omega_n)\gamma(\Omega_n)}{(\omega-\Omega_n)^2 + \left(\frac{\gamma(\Omega_n)}{2}\right)^2} - \frac{Z(\Omega_n)\gamma(\Omega_n)}{(\omega+\Omega_n)^2 + \left(\frac{\gamma(\Omega_n)}{2}\right)^2}\right], \tag{10.1.32}$$

where we have also taken account of the fact that $\mathcal{A}(\mathbf{p},\omega)$ is odd in ω, and while Z and γ are also individually odd, their product is even. This formula also takes account of the fact that there may be many roots of the plasmon dispersion relation, indexed by n. In the limit of very small damping, $\gamma(\Omega_n) \to 0$, we have

$$\mathcal{A}(\mathbf{p},\omega) = 2\pi \sum_n Z(\Omega_n)\delta(\omega - \Omega_n) - (\omega \to -\omega), \tag{10.1.33}$$

where $(\omega \to -\omega)$ denotes the preceding quantity with the replacement of ω by $-\omega$. Eq. (10.1.33) describes plasmons having infinite lifetime (zero damping).

10.2 RPA Plasmons and Shielding in 3D, 2D and 1D, Magnetoplasmas, Models and Surface Plasmons

The bulk 3D plasmon dispersion relation may be expressed as (Eqns. (10.1.21)–(10.1.24) and (10.1.31)) (principal part understood; no spin; restore $\hbar$-factors):

$$\mathrm{Re}\,\varepsilon(p,\omega) = 1 - \frac{4\pi e^2}{m}\int \frac{d^{(3)}\mathbf{k}}{(2\pi)^3}\frac{f\left(\frac{k^2}{2m}-\mu\right)}{\hbar^3}\frac{1}{\left(\omega - \frac{\mathbf{k}\cdot\mathbf{p}}{m}\right)^2 - \hbar^2\left(\frac{p^2}{2m}\right)^2} = 0 \tag{10.2.1}$$

(where we have used the 3D Coulomb interaction, $\mathcal{V}(\mathbf{p}) = \frac{4\pi e^2}{p^2}$). The quantum dynamical correction term, $\hbar^2\left(\frac{p^2}{2m}\right)^2$, in the denominator would be absent in a semiclassical model based on a classical collisionless linearized Vlasov–Boltzmann equation coupled to the Poisson equation,[3–5] but with initial statistical averaging over a temperature-dependent Fermi distribution. In the low wave number limit $p \to 0$, the $\mathbf{k}$-integral of Eq. (10.2.1) just yields the 3D equilibrium density ρ/ω^2, with the *local* bulk plasmon result,

$$\omega^2 = 4\pi e^2\rho/m \equiv \omega_p^2, \tag{10.2.2}$$

just as in gas plasma theory. Moreover, the fully non-local $\mathbf{p}$-dependent gas plasma dispersion function is obtained if the quantum dynamical correction is neglected and the temperature-dependent Fermi function $f(\omega)$ is taken in the non-degenerate Maxwell–Boltzmann classical limit. On the other hand, in the low temperature degenerate quantum limit of a Fermi step function, the $\mathbf{k}$-integral of Eq. (10.2.1) yields the Lindhard dielectric function as

$$\begin{aligned}\mathrm{Re}\,\varepsilon(p,\omega) = 1 + \mathcal{V}(p)\frac{mp_F}{2\pi^2\hbar^2}&\left\{1 - \frac{p_F}{2p}\left[1-\left(\frac{m\omega}{\hbar p_F p} - \frac{p}{2p_F}\right)^2\right]\ln\left|\frac{1+\left(\frac{m\omega}{\hbar p_F p} - \frac{p}{2p_F}\right)}{1-\left(\frac{m\omega}{\hbar p_F p} - \frac{p}{2p_F}\right)}\right|\right.\\ &\left.+\frac{p_F}{2p}\left[1-\left(\frac{m\omega}{\hbar p_F p} + \frac{p}{2p_F}\right)^2\right]\ln\left|\frac{1+\left(\frac{m\omega}{\hbar p_F p} + \frac{p}{2p_F}\right)}{1-\left(\frac{m\omega}{\hbar p_F p} + \frac{p}{2p_F}\right)}\right|\right\}\end{aligned} \tag{10.2.3}$$

and the associated plasmon spectrum is of course given by $\mathrm{Re}\,\varepsilon(\mathbf{p},\omega) = 0$. Again, the local plasmon of Eq. (10.2.3) is given by Eq. (10.2.2) and a low wave number power expansion of Eq. (10.2.1) or (10.2.3) yields wave number (p-dependent) quantum corrections with wave number shifting of the local result. In addition, there is a new plasmon resonance solution of Re $\varepsilon(p,\omega) = 0$ given approximately by $\omega = \hbar p p_F/m$, known as "zero sound,"[6] which occurs at zero temperature and differs substantially from ordinary sound.

It is to be expected that in the low wave number limit, $p \to 0$, the local plasma oscillation determined quantum mechanically by the vanishing of Re $\varepsilon(p,\omega) \to 0$ yields $\omega^2 = \omega_p^2$, just as in classical gas plasma theory: This may be understood in terms of the Correspondence Principle, since the quantum mechanical RPA is specifically designed to take account of the long-range aspects of electron–electron Coulomb interaction. To be specific, low wave numbers correspond to long distances, and for electrons interacting over such very long distances the correlation effects are controlled by essentially classical dynamics. This is to say that electron–electron interactions over long distances (i.e., low wave numbers) are in the correspondence limit in which classical features emerge from quantum mechanics.

The imaginary part, Im $[\varepsilon(\mathbf{p},\omega)]$, may be obtained by an explicit integration for all wave numbers and *all* statistical regimes as (no spin):

$$\text{Im}\,[\varepsilon(\mathbf{p},\omega)] = \frac{m^2e^2}{\hbar^4\beta p^3}\ln\left(\frac{1+\exp\left[\mu\beta-\left(\frac{m\omega}{p}-\frac{\hbar p}{2}\right)^2\frac{\beta}{2m}\right]}{1+\exp\left[\mu\beta-\left(\frac{m\omega}{p}+\frac{\hbar p}{2}\right)^2\frac{\beta}{2m}\right]}\right). \tag{10.2.4}$$

In the classical limit, this yields plasmon Landau damping, $\gamma \sim \sqrt{\frac{\pi}{2}}\,(m\beta)^{3/2}\left(\frac{\omega^4}{p^3}\right)e^{-m\omega^2\beta/2p^2}$: In the zero temperature degenerate quantum limit, plasmon decays into electron–hole pairs consistent with energy and momentum conservation is embedded in the result (we approximate $Z \sim \frac{\omega^3}{2\omega_p^2}$ corresponding to the plasmon root $\omega^2 = \omega_p^2$):

$$\begin{aligned}
&\text{for } \infty > \omega > \frac{pp_F}{m} + \frac{\hbar p^2}{2m} \to \gamma \sim 0,\\
&\text{for } \frac{pp_F}{m} + \frac{\hbar p^2}{2m} > \omega > \left|\frac{pp_F}{m} - \frac{\hbar p^2}{2m}\right| \to \gamma \sim \frac{2}{\hbar^4}\frac{e^2m^2\omega^3}{p^3\omega_p^2}\left[\mu - \frac{1}{2m}\left(\frac{m\omega}{p}-\frac{\hbar p}{2}\right)^2\right],\\
&\text{for } \left|\frac{pp_F}{m} - \frac{\hbar p^2}{2m}\right| > \omega > 0 \to \left\{\begin{array}{c}\gamma \sim 0 \text{ if } \frac{\hbar p}{2} > p_F\\ \gamma \sim \frac{2}{\hbar^3}\frac{e^2m^2\omega^4}{p^3\omega_p^2} \text{ if } \frac{\hbar p}{2} < p_F\end{array}\right\},
\end{aligned} \tag{10.2.5}$$

where p_F is the 3D Fermi wave number (momentum), $p_F = \sqrt{2m\mu}/\hbar$.

It is also of interest to examine the shielding of an impurity charge by a quantum plasma. Taking the impurity to be placed at the origin infinitely long ago (to eliminate the transients involved in the establishment of the shielding charge distribution), the effective, shielded potential $V(1)$ is given by

$$V(1) = \int d^4 2K(1-2)U(2), \tag{10.2.6}$$

where $U(2) = Ze/|\mathbf{x}_2|$ for a Coulombic impurity of charge Ze at the origin $\mathbf{x}_1 = 0$. In 3D Fourier representation, we obtain for $T = t_1 - t_2 \to \infty$ the result

$$V(\mathbf{x}_1) = 4\pi Ze \int \frac{d^{(3)}\mathbf{p}}{(2\pi)^3} \frac{e^{i\mathbf{p}\cdot\mathbf{x}_1}}{p^2 (1 + \alpha(\mathbf{p}, \omega = 0))} = 4\pi Ze \int \frac{d^{(3)}\mathbf{p}}{(2\pi)^3} \frac{e^{i\mathbf{p}\cdot\mathbf{x}_1}}{p^2 \varepsilon(\mathbf{p}, \omega = 0)}. \tag{10.2.7}$$

If we employ a semiclassical model of plasma non-locality, neglecting the quantum dynamical correction $\hbar^2 (p^2/2m)^2$ in the polarizability α of Eq. (10.2.1), it may be conveniently rewritten at $\omega = 0$ using Eq. (10.1.22) as (note that $\mathrm{Im}\,[\varepsilon(\mathbf{p}, \omega = 0)] = 0$):

$$\alpha(\mathbf{p}, \omega = 0) = \frac{4\pi e^2}{p^2} \left(\frac{\partial \rho}{\partial \mu}\right)_\beta = \frac{r_{DTF}^{-2}}{p^2}, \tag{10.2.8}$$

whence

$$V(\mathbf{x}_1) = 4\pi Ze \int \frac{d^{(3)}\mathbf{p}}{(2\pi)^3} \frac{e^{i\mathbf{p}\cdot\mathbf{x}_1}}{p^2 + r_{DTF}^{-2}} = \frac{Ze}{|\mathbf{x}_1|} e^{-|\mathbf{x}_1|/r_{DTF}}, \tag{10.2.9}$$

where $r_{DTF} = \left[4\pi e^2 (\partial\rho/\partial\mu)_\beta\right]^{-1/2}$ is the Debye or Thomas–Fermi (DTF) shielding length in the non-degenerate classical or degenerate semiclassical limit.

In the fully quantum mechanical degenerate zero-temperature limit, the static zero-frequency polarizability is

$$\varepsilon(p, \omega = 0) = 1 - V(p) \frac{m p_F}{2\pi^2 \hbar^2} \left\{ \left[-1 + \frac{p^2}{4p_F^2}\right] \ln \left| \frac{1 - \frac{p}{2p_F}}{1 + \frac{p}{2p_F}} \right| \right\}. \tag{10.2.10}$$

In consequence of the log-branch-point singularity at $p = \pm 2p_F$, the wave number integral of Eq. (10.2.7) has a contribution proportional to $\cos(2p_F |\mathbf{x}_1|)/|\mathbf{x}_1|^3$. This *Friedel-oscillatory* contribution dominates over the exponentially decaying DTF shielding law at large distances.[6] Since electron shielding also plays a role in moderating ion interactions, the log-singularity also affects the phonon spectrum, inducing the "Kohn anomaly," a kink in the phonon dispersion curve.[6]

Low-dimensional systems[7] are another important source of quantum effects in solid-state plasmas. A dip in the potential profile at a planar semiconductor–insulator junction (e.g., Si-SiO_2) can give rise to discrete energy levels for motion across the junction with only the lowest such level energetically accessible, while electron motion in that lowest subband can move freely in the two-dimensional (2D) plane of the junction. This creates an "inversion layer" of plasma electrons constrained to 2D motion on a plane. So-called "quantum wells," involving a valley in a planar heterostructure potential profile (e.g., GaAs-AlGaAs), similarly give rise to "size quantization" leaving only the lowest energy levels for motion across the valley energetically accessible, whereas carriers can move freely in the normal plane, thus constituting a 2D plasma.[7] Similar considerations apply to potential profiles that constrain plasma electron motion to a single direction,[8,9] yielding a one-dimensional (1D) "quantum wire" plasma in the single unconstrained direction. Potential profiles that constrain motion in all three directions can support a "zero-dimensional" (0D) quantum dot.[8,9]

For a 2D plasma in the degenerate limit, at zero temperature, the dielectric function is determined by the procedures outlined above as[10] (ρ_{2D} is the 2D equilibrium density; $\bar{r} = (x, y); \bar{p} = (p_x, p_y)$; and the 2D Fourier transform of the Coulomb potential is $\mathcal{V}(p) = 2\pi e^2/p$):

$$\mathrm{Re}\,\varepsilon(\bar{p}, \omega) = 1 + \frac{4\pi m e^2 \rho_{2D}}{\hbar^2 p_F p^2} \left\{ \frac{p}{p_F} - C_- \left[\left(\frac{p}{2p_F} - \frac{m\omega}{\hbar p_F p} \right)^2 - 1 \right]^{1/2} + C_+ \left[\left(\frac{p}{2p_F} + \frac{m\omega}{\hbar p_F p} \right)^2 - 1 \right]^{1/2} \right\}, \tag{10.2.11}$$

and

$$\mathrm{Im}\,\varepsilon(\bar{p}, \omega) = \frac{4\pi m e^2 \rho_{2D}}{\hbar^2 p_F p^2} \left\{ D_- \left[1 - \left(\frac{p}{2p_F} - \frac{m\omega}{\hbar p_F p} \right)^2 \right]^{1/2} - D_+ \left[1 - \left(\frac{p}{2p_F} + \frac{m\omega}{\hbar p_F p} \right)^2 \right]^{1/2} \right\}, \tag{10.2.12}$$

where

$$C_\pm = \text{ sign of } \left[\frac{p}{2p_F} \pm \frac{m\omega}{\hbar p_F p} \right],$$

and

$$D_\pm = 0 \text{ for } \left| \frac{p}{2p_F} \pm \frac{m\omega}{\hbar p_F p} \right| > 1,$$

while

$$C_\pm = 0 \text{ and } D_\pm = 1 \text{ for } \left| \frac{p}{2p_F} \pm \frac{m\omega}{\hbar p_F p} \right| < 1. \tag{10.2.13}$$

Considering low wave numbers, we note that the ratio of the two terms in $C_\pm$ is $m\omega/2\hbar p^2$, which is of order $\hbar\omega/E_F$. For a 2D plasmon, the plasmon energy $\hbar\omega_{2D}$ is expected to become vanishingly small as $p^{1/2}$ even on the basis of classical theory, so the second term of $C_\pm$ is small and $C_\pm \equiv +1$ for both signs. In this case, the square root factors of the $C_\pm$-terms in Eq. (10.2.11) are subtractive and they differ only in the signs within the squared terms ($\pm m\omega/\hbar p p_F \sim \pm\hbar\omega_{2D}/E_F$), which may be taken as small in a Taylor series leading to the local 2D plasmon given by $R_e\varepsilon(p, \omega) = 0$ with the result

$$\omega_{2D}^2 = 2\pi e^2 \rho_{2D} p/m + O(p^2). \tag{10.2.14}$$

Here, the higher wave number corrections ($O(p^2)$) bear quantum effects, while the first term on the right of Eq. (10.2.14) is a classical dynamical result for the 2D plasma (as one should expect in the local limit at low wave number $p << 2p_F$). Again, strong plasmon damping occurs when energy and momentum conservation permit plasmon decay into 2D electron–hole pairs. The latter is described by Im $\varepsilon(p,\omega)$ given in Eq. (10.2.12).

A low wave number analysis of 2D static shielding (ω=0) yields the classical dynamical 2D screening wave number as $p_{DTF}^{(2D)} = 2\pi e^2 \partial\rho_{2D}/\partial\mu$, resulting in a 2D long distance shielded potential as $V(\bar{r},t) \sim (p_{DTF}^2 r^3)^{-1}$ and $p_{DTF} = 2me^2/\hbar^2$ in the degenerate case. However, higher wave numbers involve a branch point and branch cut of $\sqrt{(p/2p_F)^2-1}$, which leads to a long-range 2D Friedel oscillatory[10] contribution as $V(\bar{r},t) \sim \sin(2p_F r)/(2p_F r)^2$.

In the case of a 1D quantum wire plasma, the dielectric function is given in the degenerate limit by ($\hbar \to 1$):[11,12]

$$\varepsilon(p,\omega) = 1 - V(p)\frac{m}{\pi p}\ln\left|\frac{\left(\omega - \frac{p^2}{2m} + \frac{p_F p}{m}\right)\left(\omega + \frac{p^2}{2m} - \frac{p_F p}{m}\right)}{\left(\omega - \frac{p^2}{2m} - \frac{p_F p}{m}\right)\left(\omega + \frac{p^2}{2m} + \frac{p_F p}{m}\right)}\right|$$

$$\cdot -\frac{im}{p}\left[\eta_+\left(1 - \left|\frac{p}{2p_F} - \frac{m\omega}{p_F p}\right|\right) - \eta_+\left(1 - \left|\frac{p}{2p_F} + \frac{m\omega}{p_F p}\right|\right)\right]. \quad (10.2.15)$$

In the present case, the 1D Fourier transform of the Coulomb potential, $V(p)$, is divergent, but can be estimated as $V(p) = e^2 \int_{-\infty}^{\infty} dx e^{ipx}/|x| \sim 2e^2 \ln(1/pa)$, where a is the radius of the quantum wire. The associated low wave number plasmon is given by (ρ_{1D} is the 1D equilibrium density):

$$\omega_{1D}^2 = \frac{2e^2 \rho_{1D} p^2}{m} \ln(1/pa), \quad (10.2.16)$$

and higher order wave number corrections bear quantum effects, etc.

Quantum mechanical effects in plasmas are even more prominent in the presence of a magnetic field inducing circularity of particle orbits in the plane normal to the field. As with other spatial confinement mechanisms, the magnetic field gives rise to quantization of the energy levels (Landau quantization), which produces very interesting structure in the plasmon spectrum, plasmon decay and static shielding. Determining the density perturbation response function, $R(\mathbf{x}_3,\mathbf{x}_2)$, using Eq. (10.1.9) with the retarded 3D magnetic field Green's function of Section 5.6 (employed to construct its thermodynamic counterpart by using the retarded function to derive the associated spectral weight Eq. (7.5.12) and then employing Eq. (7.3.9) for the thermal Green's function), we note that $C(\mathbf{x}_3,\mathbf{x}_2)C(\mathbf{x}_2,\mathbf{x}_3) = 1$, leaving the result translationally invariant so solution may be obtained by a single 3D spatial Fourier transformation, $\mathbf{x}_3 - \mathbf{x}_2 \to \mathbf{p}$. The resulting exact quantum mechanical RPA dielectric function in the presence of an arbitrarily strong uniform, constant ambient magnetic field, $\mathbf{B}$ (z-direction), for a 3D bulk solid-state plasma is given by the process outlined above (with inclusion of the magnetic field in regard to both Landau quantization and Zeeman spin-splitting) as follows:[13]

$$\varepsilon(\mathbf{p},\omega) = 1 - \frac{4\pi e^2}{p^2}\int_0^{\infty} d\omega' \frac{f(\omega')}{\hbar^3}\int_{c-i\infty}^{c+i\infty}\frac{ds}{2\pi i}e^{s\omega'}\frac{\pi^{3/2}}{(2\pi)^3}\sqrt{\frac{2m}{s}}\frac{m\hbar\omega_c}{\tanh(\hbar\omega_c s/2)}\aleph, \quad (10.2.17)$$

where $(\mathbf{p} = \bar{\mathbf{p}}, p_z = p_x, p_y, p_z; \bar{\mathbf{p}} = p_x, p_y;\ \ \omega \to \omega + i\delta)$,

$$\aleph = \frac{2i}{\hbar}\int_0^\infty dT e^{-i\omega T}\left\{\begin{array}{l} \exp\left(\frac{-p_z^2}{8ms}([2T - i\hbar s]^2 + \hbar^2 s^2)\right) \\ \times \exp\left(\frac{\hbar\bar{p}^2}{2m\omega_c}\frac{\cos[(\omega_c/2)(2T - i\hbar s)] - \cosh(\hbar\omega_c s/2)}{\sinh(\hbar\omega_c s/2)}\right) \\ -\exp\left(\frac{-p_z^2}{8ms}([2T + i\hbar s]^2 + \hbar^2 s^2)\right) \\ \times \exp\left(\frac{\hbar\bar{p}^2}{2m\omega_c}\frac{\cos[(\omega_c/2)(2T + i\hbar s)] - \cosh(\hbar\omega_c s/2)}{\sinh(\hbar\omega_c s/2)}\right) \end{array}\right\}, \tag{10.2.18}$$

with ω_c as the cyclotron frequency, m as the effective mass and $\hbar$ is restored. As indicated above, the low wave number limit of this quantum RPA result yields just the plasmon dispersion relation of the linearized Vlasov–Boltzmann equation (which describes self-consistent internal plasma dynamics classically) as $1 = \frac{\omega_p^2 \cos^2\phi}{\omega^2} + \frac{\omega_p^2 \sin^2\phi}{\omega^2 - \omega_c^2}$ (ϕ is the angle of plasmon propagation relative to the magnetic field), with two local principal electrostatic magnetoplasmons $\omega^2 = \frac{1}{2}(\omega_p^2 + \omega_c^2) \pm \frac{1}{2}\sqrt{(\omega_p^2 + \omega_c^2)^2 - 4\omega_p^2\omega_c^2 \sin^2\phi}$. It also yields the same low wave number Debye–Thomas–Fermi static shielding law predicted by the Vlasov–Boltzmann equation, which is devoid of magnetic field corrections as well as quantum corrections at low wave number.[14] This absence of magnetic field corrections at low wave number may also be understood in terms of the Correspondence Principle which mandates that there be no quantum corrections as $p \to 0$. This means that the dynamics are classical and the role of the magnetic field is restricted to the classical Lorentz force, which cannot do work: It is therefore unable to supply energy required to change the charge distribution associated with static shielding. Of course, quantum corrections that occur as higher wave numbers are taken into consideration, and with the reorganization of the energy spectrum by Landau quantization magnetic field effects do appear in static screening at higher wave numbers. It should be noted that the occurrence of a factor $\hbar^3$ under $\int d\omega'$ in Eq. (10.2.17) (as well as some other normalization factors) is actually spurious since $\int d\omega'$ is normalized here in accordance with the expression for the density as a function of the chemical potential in a magnetic field,[13]

$$\rho = 2\int_0^\infty d\omega' \frac{f(\omega')}{\hbar^3}\int_{c-i\infty}^{c+i\infty}\frac{ds}{2\pi i}\, e^{s\omega'}\frac{\pi^{3/2}}{(2\pi)^3}\sqrt{\frac{2m}{s}}\frac{m\hbar\omega_c}{\tanh(\hbar\omega_c s/2)}. \tag{10.2.19}$$

Notwithstanding the essentially classical character of the low wave number behavior of $\varepsilon(\mathbf{p}, \omega)$, one may expect quantum corrections to arise in conjunction with consideration of the higher order wave number terms embodied in the right-hand side of Eqns. (10.2.17) and (10.2.18) for $\varepsilon(\mathbf{p}, \omega)$. The higher order wave number structure bearing quantum corrections affects the plasmon dispersion relation, damping and static shielding as well: New undamped plasmon resonances called Bernstein modes are induced near each multiple of ω_c, both classically and quantum mechanically for $\mathbf{p}$ perpendicular to $\boldsymbol{B}$. The exact nature of these corrections depends on which statistical regime is under consideration and what magnetic field strength is involved.

The most interesting regime is the quantum strong field limit in which all electrons are confined to the lowest Landau level, and $\hbar\omega_c > \mu \sim E_F > k_B T$. Clearly one may expect maximal magnetic field quantum effects in this situation. However, the zero wave number principal plasmon mode dispersion relation again stands unmodified, as one should expect from our earlier discussion. It is the consideration of higher wave number parts of the right-hand sides of Eqns. (10.2.17) and (10.2.18) that reveals important magnetic field quantum effects. An exact evaluation of Eqns. (10.2.17) and (10.2.18) in the quantum strong field limit yields the result[15]

$$\varepsilon(p,\omega) = 1 - \frac{m\omega_p^2}{2\hbar p^2}\sqrt{\frac{m}{2p_z^2\mu}}\exp\left(\frac{-\hbar\bar{p}^2}{2m\omega_c}\right)\sum_{n=0}^{\infty}\frac{1}{n!}\left(\frac{\hbar\bar{p}^2}{2m\omega_c}\right)^n \cdot \left\{\log\left[\frac{\omega - n\omega_c - \hbar p_z^2/2m + \sqrt{2p_z^2\mu/m}}{\omega - n\omega_c - \hbar p_z^2/2m - \sqrt{2p_z^2\mu/m}}\right] + (\omega \to -\omega)\right\}. \qquad (10.2.20)$$

A wave number power expansion of the right-hand side of Eq. (10.2.20) yields wave number shifts of the locations of the two principal plasmon modes that bear magnetic field quantum corrections. It is of even greater interest to maintain the anisotropic log-singularities of Eq. (10.2.20) intact, and investigate their effects on the plasmon spectrum.[16] One then finds that the p_z-dependent log-singularities imply the existence of two plasmon resonances near each value $n\omega_c$, one of which is undamped and the other is mildly damped. These remarks hold for wave vector $\boldsymbol{p}$ not perpendicular to the magnetic field $\boldsymbol{B}$, ($p_z \neq 0$). In the limit of perpendicular propagation ($p_z = 0$) there is just one undamped resonance (Bernstein mode[17]) near each value $n\omega_c$. In regard to the static shielding law, the $\omega \to 0$ limit of the $n = 0$ log-singularity takes the form $\ln|p_z - 2p_F|$. This anisotropic result gives the principal contribution to the quantum strong field counterpart of the Friedel–Kohn "wiggle," and the strong anisotropy destroys the long-range character of the "wiggle."[15]

Needless to say, quantum phenomenology at higher wave numbers also occurs in the plasmon spectrum and static shielding of 2D plasmas in a normal magnetic field.[18–21]

Having indicated, above, the reduction of the RPA to semiclassical/classical limits in which plasma phenomenology is described in terms of the classical collisionless linearized Vlasov–Boltzmann equation coupled to the Poisson equation when averaged with an initial Fermi–Dirac/Maxwell–Boltzmann distribution, it is appropriate to point out that there is yet another, simpler model of dielectric response that describes such dynamic plasmon features and static Thomas–Fermi/Debye plasma shielding. This *hydrodynamic* model[22] is highly tractable analytically, albeit less accurate than the RPA. It often provides considerable insight into complex problems involving electron dielectric response. The hydrodynamic model is based on a classical hydrodynamic equation for the density perturbation $\delta\rho(1)$ and associated velocity $\delta\mathbf{u}(1)$ at a space-time point (1), for which the linearized equation of continuity is

$$\frac{\partial\left(\delta\rho(1)\right)}{\partial t_1} + \rho_0 \nabla_1 \cdot \delta\mathbf{u}(1) = 0. \qquad (10.2.21)$$

In this, the fluid dynamics are governed by Newton's law linearized in the field as (ρ is the perturbed 3D density and ρ_0 is its equilibrium value):

$$\rho_0 \frac{\partial\,(\delta\mathbf{u}(1))}{\partial t_1} = -\rho_0 \frac{e}{m} \nabla_1 \delta V(1) - \frac{1}{m} \nabla_1 \cdot \delta\,\overleftrightarrow{\mathbf{\Pi}}\,(1), \tag{10.2.22}$$

where $\delta\,\overleftrightarrow{\mathbf{\Pi}}\,(1)$ represents stress tensor fluctuations, the divergence of which contributes to force in the form of fluid pressure (δV is related to $\delta\rho$ through the Poisson equation, which is implicitly taken into account in Eq. (10.1.21) by the interparticle Coulomb potential $\mathcal{V}$). The hydrodynamic model under consideration is based on the assumption that

$$\nabla_1 \cdot \delta\,\overleftrightarrow{\mathbf{\Pi}}\,(1) = m\beta^2 \nabla_1 \delta\rho(1), \tag{10.2.23}$$

where β is a parameter having dimensions of velocity (not to be confused with inverse thermal energy $1/\kappa T$). Fourier-transforming with respect to both space and time, we have

$$-\,\omega\delta\rho(\mathbf{p},\omega) + \rho_0 \mathbf{p} \cdot \delta\mathbf{u}(\mathbf{p},\omega) = 0 \tag{10.2.24}$$

and

$$-\,\omega\rho_0\delta\mathbf{u}(\mathbf{p},\omega) = -\frac{e\rho_0}{m}\mathbf{p}\delta V(\mathbf{p},\omega) - \beta^2 \mathbf{p}\delta\rho(\mathbf{p},\omega). \tag{10.2.25}$$

The occurrence of $\mathbf{p} \cdot \delta\mathbf{u}$ in Eq. (10.2.24) may be eliminated by dotting Eq. (10.2.22) by $\mathbf{p}$, and then solving for $\delta\rho/\delta V$: We obtain the hydrodynamic dielectric function using Eq. (10.1.21) jointly with the fact that $R = -\delta\rho/\delta V$, Eq. (9.4.13), with the result

$$\varepsilon(\mathbf{p},\omega) = 1 - \frac{\omega_p^2}{\omega^2 - \beta^2 p^2}, \quad [\omega \to \omega + i\delta]\,. \tag{10.2.26}$$

It is clear that the local plasmon, $\omega = \omega_p$ (for $\mathbf{p} \to 0$), is reproduced by this model, and that Debye–Thomas–Fermi shielding comes about in the long time ($\omega \to 0$) limit, for an appropriate choice of $\beta^2 = r_{DTF}^2\omega_p^2$. There is no Landau damping or plasmon decay into electron–hole pairs in the hydrodynamic model. Furthermore, one cannot correctly represent both non-local corrections to the ω_p-plasmon and r_{DTF} with the same choice of the parameter β. In the degenerate limit of zero temperature, $\beta^2 = v_F^2/3$ (v_F is the Fermi speed) gives the correct Thomas–Fermi shielding length, but $\beta^2 = 3v_F^2/5$ gives the correct non-local correction to the plasmon. Nonetheless, the hydrodynamic model has served well as a tractable guide to solid-state plasma phenomenology in many cases of interest.

As an example, we consider the role of non-locality in the surface plasmon dispersion relation, Eq. (5.10.29), using the hydrodynamic model. To establish the local result first, suppose that $\varepsilon_A(\mathbf{p},\omega) \to \varepsilon_A(\omega) = 1 - \omega_p^2/\omega^2$ is local, as well as ε_B with $\varepsilon_B(\omega) \to 1$. Correspondingly, the surface mode dispersion relation is

$$-\frac{1}{\varepsilon_B} = \frac{1}{\varepsilon_A(\omega)} = \frac{1}{1-\omega_p^2/\omega^2}, \tag{10.2.27}$$

yielding the well-known surface plasmon, $\omega^2 = \omega_p^2/2$. To assess the role of non-locality in the surface plasmon dispersion relation for the case at hand, we again employ Eq. (5.10.29),

$$-1 = \frac{2\bar{p}}{\pi} \int_0^\infty \frac{dp_z}{(p_z^2 + \bar{p}^2)\, \varepsilon_A(\mathbf{p}, \omega)} \tag{10.2.28}$$

using the hydrodynamic model for $\varepsilon_A(\mathbf{p}, \omega)$ as given by Eq. (10.2.26):

$$-1 = \frac{2\bar{p}}{\pi} \int_0^\infty \frac{dp_z}{(p_z^2 + \bar{p}^2)} - \frac{2\bar{p}\omega_p^2}{\pi\beta^2} \int_0^\infty \frac{dp_z}{(p_z^2 + \bar{p}^2)} \frac{1}{(p_z^2 + \bar{p}^2 + (\omega_p^2 - \omega^2))/\beta^2}. \tag{10.2.29}$$

Anticipating that $\omega^2 \sim \omega_p^2/2$, the second integral on the right above is well defined, and carrying it out we obtain

$$1 = \frac{\omega_p^2/2}{\omega_p^2 - (\omega + i\delta)^2} - \frac{\omega_p^2/2}{\omega_p^2 - (\omega + i\delta)^2} \frac{\beta\bar{p}}{\sqrt{\left[\omega_p^2 - (\omega + i\delta)^2 + \beta^2\bar{p}^2\right]^2}}. \tag{10.2.30}$$

The local limit, $\omega = \omega_p/\sqrt{2}$, is recovered if $\bar{p} \to 0$. For non-zero $\bar{p}$, but small in the sense $\bar{p} \ll \omega_p/\beta$, the leading wave number correction to the local surface plasmon is given by

$$\omega^2 \cong \frac{\omega_p^2}{2} + \frac{\beta\bar{p}\omega_p}{\sqrt{2}}. \tag{10.2.31}$$

This linear $\bar{p}$-correction to ω^2 is characteristic of the non-local surface plasmon, and it also appears in more accurate theories such as the RPA. There is no such linear $\bar{p}$-correction to the bulk plasmon. The non-local structure of the surface plasmon has been the focus of intense scrutiny by physicists involved in surface studies.

10.3 Exchange and Correlation Energy

To a large extent, the development of modern quantum many-body theory grew out of the need to overcome problems presented by the occurrence of serious divergencies in the perturbative calculation of correlation energy of an electron plasma having internal Coulomb interactions governed by the Hamiltonian

$$H = \sum_i \frac{\mathbf{p}_i^2}{2m} + \frac{1}{2}\sum_{i \neq j} \frac{e^2}{|\mathbf{r}_i - \mathbf{r}_j|} = H_{el.}^0 + H_{e-e}^{coul.}, \tag{10.3.1}$$

(the first term represents free electrons, and the second term represents the long-range Coulomb interaction among the electrons). The divergencies arose in the straightforward calculation of correlation energy in powers of the Coulomb potential $\mathcal{V}^n \sim 1/p^{2n}$, becoming ever more strongly divergent at $p \to 0$ as n increases. The resolution of this problem lay in the physical recognition that the long range of the Coulomb interaction allows the electrons to behave in a collective manner, and to be specific, self-consistent collective density oscillations (dynamic as well as static screening features) exist in the many-electron plasma, quantum mechanically as well as classically. The significance of these plasma oscillations (plasmons) as elementary excitations of the system was brought into clear focus by the construction of a canonical transformation that put H into an approximately equivalent Hamiltonian of the form[23,24]

$$H \to H^0_{el.} + H^0_{pl.} + H^{shield}_{e\text{–}e} + H_{el.\text{–}pl.}, \tag{10.3.2}$$

which helped to cure the divergence problem encountered in perturbative calculations of the correlation energy. As above, $H^0_{el.}$ represents free electrons. Similarly, $H^0_{pl.}$ represents the energy of free plasmons, the self-consistent collective density oscillations that arise in consequence of the long-range character of the Coulomb electron–electron interaction. The microscopic quantum dynamical mechanism of the plasmon has been discussed in detail above, and it is well known that the local plasma density oscillation (as well as its accompanying electric field oscillation, which may be understood as a longitudinal photon) occurs with the natural frequency given by $\omega_p = (4\pi e^2 \rho/m)^{1/2}$. Associated quantum plasma energies involve multiplication by $\hbar$, so that $H^0_{pl.}$ involves a typical quantum mechanical zero-point plasmon oscillator energy of $\hbar\omega_p/2$ summed over possible plasmon states (local and non-local) contributing to the ground state of the many-electron system. In addition, one has a residual short-range shielded electron–electron interaction term, $H^{shield}_{e\text{–}e}$. Finally, the fact that plasmons are not exact normal modes of the system (that is to say that they are damped and have finite lifetime as elementary excitations) is reflected in the presence of an electron–plasmon interaction term, $H_{el.\text{–}pl.}$, which provides a mechanism for energy exchange between plasmons and electrons. Under equilibrium conditions non-local plasmons yield energy to the electrons in the form of natural damping (plasmon decay into electron–hole pairs quantum mechanically;[23,24] Landau damping classically[4]). For sufficiently high wave number the plasmon is so heavily damped as to be meaningless as an elementary excitation (this delimits possible plasmon states contributing to the $H^0_{pl.}$ ground state mentioned above). However, under appropriate drifting conditions the directionality of energy flow via this mechanism can be reversed, resulting in electrons yielding energy to the plasmons, with concomitant microinstability of the plasmons.[5] All of these plasma features are, in fact, embodied in the structure of the longitudinal dielectric function, ε, or, equivalently, in its inverse, the dynamic, non-local screening function, K, which has been discussed extensively above. This is true both classically and quantum mechanically. Of course, in solids it is generally necessary to describe free-carrier dynamics quantum mechanically.

The correlation energy calculation may be undertaken with Eq. (7.6.18). Separating the part that does not depend on $\mathcal{V}$, the dependence of the grand potential on Coulomb interaction strength $\boldsymbol{\lambda}$ is given by ($\mathcal{V}^{(\lambda)} = \lambda\mathcal{V}$):

$$W - W(0) = -\frac{\beta}{2} \int_0^1 \frac{d\lambda}{\lambda} \int d^{(3)}\mathbf{x}' \int d^{(3)}\mathbf{x}'' \mathcal{V}^{(\lambda)}(\mathbf{x}' - \mathbf{x}'') G_2^{(\lambda)}(1', 1''; 1'^{+}, 1''^{+})_{t_1' = t_1''}. \quad (10.3.3)$$

Recalling Eq. (9.4.18), setting 1 = 2 in it and integrating both sides with respect to $\int d^{(3)}\mathbf{x}_2$ permits the elimination of the explicit appearance of $\int d^{(3)}\mathbf{x}' \int d^{(3)}\mathbf{x}'' \mathcal{V}^{(\lambda)} G_2^{(\lambda)}$ on the right of Eq. (10.3.3) in favor of the inverse dielectric function $K^{(\lambda)}$, yielding the exact relation (we also execute the time integration associated with the instantaneous Coulomb potential $\mathcal{V}(2-3) = \mathcal{V}^{(\lambda)}(\mathbf{x}_2 - \mathbf{x}_3)\delta(t_2 - t_3)$ and relabel indices):

$$W - W(0) = \mp i \frac{\beta}{2} \int_0^1 \frac{d\lambda}{\lambda} \int d^{(3)}\mathbf{x}_2 \left[K^{(\lambda)}(2,2) - \delta(2-2) \right] - (-,-), \quad (10.3.4)$$

where the term $(-,-)$ designates a direct static equilibrium electron–electron interaction term given by

$$(-,-) = \frac{\beta}{2} \int_0^1 \frac{d\lambda}{\lambda} \int d^{(3)}\mathbf{x}_1 \int d^{(3)}\mathbf{x}_2 \mathcal{V}^{(\lambda)}(\mathbf{x}_1 - \mathbf{x}_2) \left[G_1^{(\lambda)}(1, 1^{+}) G_1^{(\lambda)}(2, 2^{+}) \right]_{t_1 = t_2}. \quad (10.3.5)$$

Our further discussion will be limited to the random phase approximation (RPA) and will employ the spectral representation of $\left[K_{RPA}^{(\lambda)} - \delta \right]$ in terms of its associated spectral weight function, $\mathcal{A}_{RPA}^{(\lambda)}(\mathbf{p}, \omega)$, leading to the result (Fourier transform in 3D space and time, $\mathbf{x}' - \mathbf{x}'' \to \mathbf{p}$, $t' - t'' \to \omega$; *vol.* represents volume):

$$W - W(0) = -\frac{\beta(vol.)}{2} \int \frac{d^{(3)}\mathbf{p}}{(2\pi)^3} \int_{-\infty}^{\infty} \frac{d\omega}{2\pi} \int_0^1 \frac{d\lambda}{\lambda} \coth\left(\frac{\hbar\omega\beta}{2} \right) \mathcal{A}_{RPA}^{(\lambda)}(\mathbf{p}, \omega) + (-,-). \quad (10.3.6)$$

Approximating $G_1^{(\lambda)} \sim G_1^0$ (unperturbed), we have $\int_0^1 \frac{d\lambda}{\lambda} \mathcal{V}_{\cdots}^{(\lambda)} \ldots = \mathcal{V} \ldots$, and

$$(-,-) = \frac{\beta}{2} \int d^{(3)}\mathbf{x}' \int d^{(3)}\mathbf{x}'' \mathcal{V}(\mathbf{x}' - \mathbf{x}'') \left[G_1^0(1', 1'^{+}) G_1^0(2'', 2''^{+}) \right]_{t_1 = t_2}. \quad (10.3.7)$$

Apart from the factor β and recalling that $G_1(1, 1^{+}) \sim \langle \rho(1) \rangle$, we see that $(-,-)$ describes the static direct electron–electron equilibrium Coulomb energy, to which one must add the static direct equilibrium Coulomb energy $(+,+)$ of a uniformly smeared positive static background, which ensures overall charge neutrality, and, furthermore, one must add the Coulomb energy $(+,-)$ of the positive background interacting with the electrons. The sum of all such unperturbed equilibrium Coulomb energy terms, $(-,-) + (+,+) + (+,-)$, must vanish for a static system that is uniformly neutral since the net charge at each point is zero. (Note that the product of the two density sums of negatively charged electron and positively charged background densities yields two negative cross terms that exactly cancel the two positive direct terms.)

To identify the *exchange* contribution to W, we replace $G_2^{(\lambda)}$ in Eq. (10.3.3) by its Fock-term, the ring diagram R of Eq. (9.4.13), obtaining the leading contribution to $W_{ex}^{(\lambda)}$ as (again setting $G_1^{(\lambda)} \to G_1^0$):

$$W_{ex} = -\frac{\beta(vol.)}{2}\int \frac{d^{(3)}\mathbf{p}}{(2\pi)^3}\int_{-\infty}^{\infty}\frac{d\omega}{2\pi}\int_0^1 \frac{d\boldsymbol{\lambda}}{\boldsymbol{\lambda}}\coth\left(\frac{\hbar\omega\beta}{2}\right)\mathcal{A}_{HF}^{(\boldsymbol{\lambda})}(\mathbf{p},\omega), \tag{10.3.8}$$

where $\mathcal{A}_{HF}^{(\boldsymbol{\lambda})}$ is the spectral weight of the polarizability $\alpha = \mathcal{V}R$.

The *correlation* part of W, W_{corr}, is given by the right-hand side of Eq. (10.3.4) less W_{ex}, and within the framework of the RPA it is given by

$$W_{corr} = -\frac{\beta(vol.)}{2}\int \frac{d^{(3)}\mathbf{p}}{(2\pi)^3}\int_{-\infty}^{\infty}\frac{d\omega}{2\pi}\int_0^1 \frac{d\boldsymbol{\lambda}}{\boldsymbol{\lambda}}\coth\left(\frac{\hbar\omega\beta}{2}\right)\left[\mathcal{A}_{RPA}^{(\boldsymbol{\lambda})}(\mathbf{p},\omega) - \mathcal{A}_{HF}^{(\boldsymbol{\lambda})}(\mathbf{p},\omega)\right]. \tag{10.3.9}$$

At zero temperature the corresponding correlation energy, $E = -[\partial W/\partial\beta]_{S,N}$, is (note that both $\mathcal{A}^{(\boldsymbol{\lambda})}_{\{RPA \atop -HF\}}(p,\omega)$ and $\coth(\hbar\omega\beta/2)$ are odd functions of ω in Eq. (10.1.26)):

$$E_{corr} = -(vol.)\int \frac{d^{(3)}\mathbf{p}}{(2\pi)^3}\int_0^{\infty}\frac{d\omega}{2\pi}\int_0^1 \frac{d\boldsymbol{\lambda}}{\boldsymbol{\lambda}}\mathcal{A}^{(\boldsymbol{\lambda})}_{\{RPA \atop -HF\}}(\mathbf{p},\omega), \tag{10.3.10}$$

where

$$\mathcal{A}^{(\boldsymbol{\lambda})}_{\{RPA \atop -HF\}}(\mathbf{p},\omega) \equiv \mathcal{A}_{RPA}^{(\boldsymbol{\lambda})}(\mathbf{p},\omega) - \mathcal{A}_{HF}^{(\boldsymbol{\lambda})}(\mathbf{p},\omega). \tag{10.3.11}$$

Recalling Eq. (7.4.13), the spectral weights in Eqns. (10.3.8 and 10.3.9) can be written in terms of the discontinuity of the corresponding auxiliary function $k(\mathbf{p},z)$ across the real frequency z-axis as

$$\mathcal{A}^{(\boldsymbol{\lambda})}_{\{RPA \atop -HF\}}(\mathbf{p},\omega) = \frac{1}{i}\lim_{\delta\to 0}\left[k_{\{RPA \atop -HF\}}(\mathbf{p},\omega - i\delta) - k_{\{RPA \atop -HF\}}(\mathbf{p},\omega + i\delta)\right] \tag{10.3.12}$$

($k(\mathbf{p},\omega + i\delta)$ being the physical, retarded inverse dielectric function). With this, Eq. (10.3.10) can be rewritten as[25] (restore $\hbar$):

$$\frac{E_{corr}}{(vol.)} = -\frac{\hbar}{2\pi i}\int \frac{d^{(3)}\mathbf{p}}{(2\pi)^3}\int_{C\to C'} dz\int_0^1 \frac{d\boldsymbol{\lambda}}{\boldsymbol{\lambda}}k^{(\boldsymbol{\lambda})}_{\{RPA \atop -HF\}}(\mathbf{p},z), \tag{10.3.13}$$

where the contour C encloses the positive real axis, and it may be deformed to the contour C' running along the imaginary $z \to i\omega$ axis, whence we must deal with

$$k^{(\boldsymbol{\lambda})}_{\{RPA \atop -HF\}}(\mathbf{p},i\omega) = \frac{1}{1+\boldsymbol{\lambda}\alpha(\mathbf{p},i\omega)} - (1 - \boldsymbol{\lambda}\alpha(\mathbf{p},i\omega)). \tag{10.3.14}$$

Pine's book[25] clearly explains this manipulation of the frequency (z) integration contour, and addressing next the integration with respect to $\boldsymbol{\lambda}$, Eqns. (10.3.13 and 10.3.14) yield

$$\frac{E_{corr}}{(vol.)} = \frac{\hbar}{4\pi}\int \frac{d^{(3)}\mathbf{p}}{(2\pi)^3}\int_{-\infty}^{\infty} d\omega\left[\ln\left(1+\alpha(\mathbf{p},i\omega)\right) - \alpha(\mathbf{p},i\omega)\right], \tag{10.3.15}$$

and performing the ω-integral by parts, we obtain

$$\frac{E_{corr}}{(vol.)} = \frac{\hbar}{4\pi} \int \frac{d^{(3)}\mathbf{p}}{(2\pi)^3} \int_{-\infty}^{\infty} d\omega \frac{\omega\alpha(\mathbf{p}, i\omega)\,[d\alpha(\mathbf{p}, i\omega)/d\omega]}{1 + \alpha(\mathbf{p}, i\omega)}. \tag{10.3.16}$$

Expansion of the integrand in powers of α leads to a series of terms having increasing powers of p^{-2} mentioned at the beginning of this section, which cause serious divergences of the p-integral as $p \to 0$. Such difficulties are avoided by the closed form of Eq. (10.3.16), which automatically sums the infinite series of divergent terms, eliminating them in favor of a finite result. The evaluation of the correlation energy in the RPA (carrying out the integrations of Eq. (10.3.16)) is very well known and discussed in detail in many books and papers.[23–25] In the high-density limit, it is given by

$$E_{corr}^{RPA} = 0.062 \ln r_s - 0.142 \text{ Rydbergs}, \tag{10.3.17}$$

where $r_s = r_0/a_0$ and r_0 is defined in terms of the electron density as $r_0 = (3/4\pi\rho_0)^{1/3}$ with a_0 as the effective Bohr radius.

10.4 Atom–Surface van der Waals Attraction; Physisorption and the Plasma Image

When a neutral atom is brought into the neighborhood of a metal surface enclosing a semi-infinite sea of electrons, there is an attractive van der Waals (vdW) interaction between them, in some cases leading to adsorption of the atom by the surface, *physisorption*, which is a much weaker atom–surface interaction than the chemisorption discussed in Section 5.2. The van der Waals interaction between an atom and a surface is based on the Coulomb interaction between the atomic electrons and the mobile metallic electrons within the surface, and it *excludes* the possibility of sharing or exchanging electrons by penetration of the atomic electron wave functions into the surface, or vice versa (which is implicitly represented by the off-diagonal Hamiltonian matrix elements of Section 5.2 that are responsible for chemisorption). Moreover, the theory of van der Waals attraction assumes that the atom and its electrons are far enough from the metal electrons within the surface so that exchange effects are negligibly small. These qualitative statements apply to all van der Waals (*vdW*) interactions, including atom–atom, atom–surface and surface–surface. It is also important to remark that the van der Waals interaction has attracted surges of intense scientific interest over many decades because of its fundamental nature. It is of profound importance as an observable manifestation of the uniquely quantum mechanical feature of zero-point photon oscillator energy, including Coulomb interactions representative of longitudinal plasmon-photons in the non-retarded limit of the Casimir interaction.

While all electrons are indistinguishable in principle, the atomic electrons may be thought of as a group distinct from the mobile metallic electrons if their separation is large enough so that exchange effects and wave function penetration (tunneling, sharing) are small. On this basis, we write ψ_a^+, ψ_a to represent creation/annihilation operators for electrons of the atom, and ψ_m^+, ψ_m to represent the corresponding operators for mobile electrons of the metal, with

$$\psi = \psi_a + \psi_m \quad ; \quad \psi^+ = \psi_a^+ + \psi_m^+.$$

Considering them to be mutually independent, we take ψ_a^+, ψ_a to anticommute with ψ_m^+, ψ_m individually. Our non-relativistic treatment for the Coulomb interaction, $\mathcal{V}$, is based on Eq. (3.2.16) ($e \to 1$):

$$\begin{aligned} H^{(2)} = \frac{1}{2!} \int d^{(3)}\mathbf{x}' \int d^{(3)}\mathbf{x}'' \left(\psi_a^+(\mathbf{x}',t) + \psi_m^+(\mathbf{x}',t)\right) \left(\psi_a^+(\mathbf{x}'',t) + \psi_m^+(\mathbf{x}'',t)\right) \\ \mathcal{V}(\mathbf{x}'-\mathbf{x}'') \left(\psi_a(\mathbf{x}'',t) + \psi_m(\mathbf{x}'',t)\right) \left(\psi_a(\mathbf{x}',t) + \psi_m(\mathbf{x}',t)\right). \end{aligned} \tag{10.4.1}$$

Of the sixteen possible products in the integrand of Eq. (10.4.1), one represents the Coulomb interactions (and internal correlations) within the subsystem of atomic (a) electrons, H_{aa}, and, separately, there is a similar one, H_{mm}, for the Coulomb interactions (and internal correlations) within the subsystem of mobile metal (m) electrons. Correspondingly,

$$H^{(2)}_{\left\{ {aa \atop mm} \right\}} = \frac{1}{2!} \int d^{(3)}\mathbf{x}' \int d^{(3)}\mathbf{x}'' \psi^+_{\left\{ {a \atop m} \right\}}(\mathbf{x}',t) \psi^+_{\left\{ {a \atop m} \right\}}(\mathbf{x}'',t) \mathcal{V}(\mathbf{x}'-\mathbf{x}'') \psi_{\left\{ {a \atop m} \right\}}(\mathbf{x}'',t) \psi_{\left\{ {a \atop m} \right\}}(\mathbf{x}',t). \tag{10.4.2}$$

Of the remaining fourteen possible products, eight involve products of three operators of one subsystem and just one of the other (for example, $\psi_m^+\psi_a^+\mathcal{V}\psi_a\psi_a$, etc.) We take these to be negligibly small since such a Hamiltonian term would transfer electrons singly from the atom to the metal or vice versa, which is excluded from our considerations in accordance with the remarks above. The remaining six products involve two operators of the a-electrons and two of the m-electrons. Of these, two are of the type $\psi_a^+\psi_a^+\mathcal{V}\psi_m\psi_m$ or $\psi_m^+\psi_m^+\mathcal{V}\psi_a\psi_a$, which would transfer electrons in pairs from the atom to the metal or vice versa, which is similarly excluded. Of the remaining four products, two are described by

$$\psi^+_{\left\{ {a \atop m} \right\}}(\mathbf{x}',t) \psi^+_{\left\{ {m \atop a} \right\}}(\mathbf{x}'',t) \mathcal{V}(\mathbf{x}'-\mathbf{x}'') \psi_{\left\{ {a \atop m} \right\}}(\mathbf{x}'',t) \psi_{\left\{ {m \atop a} \right\}}(\mathbf{x}',t), \tag{10.4.3}$$

which would annihilate a metal electron at $\mathbf{x}'$ while creating an atomic electron at $\mathbf{x}'$ in its place and simultaneously annihilate an atomic electron at $\mathbf{x}''$ while a metal electron is created in its place at $\mathbf{x}''$, etc. Again, these two terms exchange atomic and metal electrons, processes which are excluded from our present considerations. The final two terms involve just the product of the density operator ρ_a for a-electrons with that of m-electrons, ρ_m, and vice versa, and the two are actually the same since the anticommutation of $\psi_a(\mathbf{x}',t)$ through $\rho_m(\mathbf{x}'',t)$ invokes *two* minus signs; hence they commute, as do $\rho_a(\mathbf{x}',t)$ and $\rho_m(\mathbf{x}'',t)$, etc. Thus the $\rho_a\,\rho_m$ term doubles, canceling the prefactor $\frac{1}{2!}$ in this part of $H^{(2)}$. In summary, van der Waals interaction involves the following Hamiltonian:

$$H^{(2)} = H^{(2)}_{aa} + H^{(2)}_{mm} + H^{(2)}_{am}, \tag{10.4.4}$$

where

$$H_{am}^{(2)} = \int d^{(3)}\mathbf{x}' \int d^{(3)}\mathbf{x}'' \rho_a(\mathbf{x}', t)\mathcal{V}(\mathbf{x}' - \mathbf{x}'')\rho_m(\mathbf{x}'', t). \tag{10.4.5}$$

Here, we will only treat those aspects of atom–metal van der Waals interaction that illustrate the far-reaching utility of the dynamic non-local dielectric screening concept. Consider first the Hamiltonian for the instantaneous Coulomb interaction of the atomic electrons written in the form

$$H_{aa}^{(2)} = \frac{1}{2}\int d^{(3)}\mathbf{x}' \int d^{(3)}\mathbf{x}'' \int dt'' \psi_a^+(\mathbf{x}', t')\psi_a(\mathbf{x}', t')\mathcal{V}(\mathbf{x}'-\mathbf{x}'')\delta(t'-t'')\psi_a^+(\mathbf{x}'', t'')\psi_a(\mathbf{x}'', t''). \tag{10.4.6}$$

In a first approximation, the associated energy (with correlations of the atomic electrons in the presence of the nuclear potential that binds the electrons) may be written as the expectation value of $H_{aa}^{(2)}$ in the ground state of the atom. To develop an approximation directed at incorporating the role of $H_{am}^{(2)}$ (Eq. (10.4.5)) without undertaking a formal perturbation theory (which leads to the same result),[26] we replace $\mathcal{V}(\mathbf{x} - \mathbf{x}')\delta(t - t')$ in Eq. (10.4.6) by the *effective* potential between atomic electrons, $\mathcal{V}_{eff}(\mathbf{x}', \mathbf{x}'', t' - t'')$, part of which is generated by Coulomb interaction between atomic electrons and metal electrons, i.e., the non-local dynamic *image* potential, $\mathcal{V}_{image}(\mathbf{x}', \mathbf{x}'', t' - t'')$ due to the polarization of electrons of the semi-infinite metal, which actually mediates the interaction of the electrons of the atom. Thus, we write their corresponding part of Eq. (10.4.6) as a "van der Waals" term, $H_{vdW}^{(2)}$, given by

$$H_{vdW}^{(2)} = \frac{1}{2}\int d^{(3)}\mathbf{x}' \int d^{(3)}\mathbf{x}'' \int dt'' \psi_a^+(\mathbf{x}', t')\psi_a(\mathbf{x}', t')\mathcal{V}_{image}(\mathbf{x}', \mathbf{x}'', t'-t'')\psi_a^+(\mathbf{x}'', t'')\psi_a(\mathbf{x}'', t''). \tag{10.4.7}$$

Estimating the van der Waals interaction energy contribution from this term as the ground-state expectation value $\left\langle H_{vdW}^{(2)} \right\rangle$, we have the following result valid to second order in the Coulomb interaction between the atomic and metal electrons:

$$E_{vdW}^{(2)} \cong \left\langle H_{vdW}^{(2)} \right\rangle = -\frac{1}{2}\int d^{(3)}\mathbf{x}' \int d^{(3)}\mathbf{x}'' \int dt'' \mathcal{V}_{image}(\mathbf{x}', \mathbf{x}'', t'-t'')G_2^a(\mathbf{x}''t'', \mathbf{x}'t'; \mathbf{x}''t''^+, \mathbf{x}'t'^+). \tag{10.4.8}$$

Here, G_2^a is the two-particle Green's function of the atomic electrons, including the full complement of intra-atom electron–electron interactions, but with no effects from the metal electrons, averaged in the ground state of the atom $|\Phi_0^a\rangle$. The image potential, $\mathcal{V}_{image}$, due to the polarization of the semi-infinite metal electron plasma by the Coulomb field of a point charge outside may be written in terms of the G_2^m-function of the semi-infinite plasma of metal electrons using Eq. (9.4.18) for the inverse dielectric (screening) function K^m of the semi-infinite metal plasma. Discarding G_1^m terms and δ-terms that are not pertinent to this discussion, we have

$$K^m(1,2) \to i\int d3\mathcal{V}(1-3)G_2^m(3,2;3^+,2^+)$$
$$= i\int d^{(3)}\mathbf{x}^{iv}\mathcal{V}(\mathbf{x}'-\mathbf{x}^{iv})G_2^m(\mathbf{x}^{iv}t',\mathbf{x}''t'';\mathbf{x}^{iv}t'^+,\mathbf{x}''t''^+), \tag{10.4.9}$$

where G_2^m includes the effects of mobile metal–metal electron interactions within the surface, and $\mathcal{V}$ is the instantaneous unscreened Coulomb potential. The corresponding image potential, $\mathcal{V}_{image}$, is given by

$$\mathcal{V}_{image}(\mathbf{x}',\mathbf{x}'',t'-t'') = \int d^{(3)}\mathbf{x}'''\left[K^m(\mathbf{x}',\mathbf{x}''';t'-t'')\mathcal{V}(\mathbf{x}'''-\mathbf{x}'')\right] - \mathcal{V}(\mathbf{x}'-\mathbf{x}'')$$
$$= i\int d^{(3)}\mathbf{x}'''\int d^{(3)}\mathbf{x}^{iv}\mathcal{V}(\mathbf{x}'-\mathbf{x}^{iv})G_2^m(\mathbf{x}^{iv}t',\mathbf{x}'''t'';\mathbf{x}^{iv}t'^+,\mathbf{x}'''t''^+)\mathcal{V}(\mathbf{x}'''-\mathbf{x}'')$$
$$-\mathcal{V}(\mathbf{x}'-\mathbf{x}''), \tag{10.4.10}$$

which yields $E_{vdW}^{(2)}$ in an alternative form as

$$E_{vdW}^{(2)} = -\frac{i}{2}\int d^{(3)}\mathbf{x}'\int d^{(3)}\mathbf{x}''\int d^{(3)}\mathbf{x}'''\int d^{(3)}\mathbf{x}^{iv}\int dt''\mathcal{V}(\mathbf{x}'-\mathbf{x}^{iv})$$
$$\times G_2^m(\mathbf{x}^{iv}t',\mathbf{x}'''t'';\mathbf{x}^{iv}t'^+,\mathbf{x}'''t''^+)\mathcal{V}(\mathbf{x}'''-\mathbf{x}'')G_2^a(\mathbf{x}''t'',\mathbf{x}'t';\mathbf{x}''t''^+,\mathbf{x}'t'^+)$$
$$-\text{ direct } \textit{non-image} \text{ counterpart}. \tag{10.4.11}$$

The "non-image" counterpart is embedded in the structure of the atomic electron's internal correlation energy. It does not contribute to van der Waals energy, and, accordingly, is dropped here. Noting that Eq. (10.4.11) is symmetric under the atom↔metal interchange, it is readily seen that another contribution to vdW energy identical to Eq. (10.4.11) comes from an analogous treatment of $H_{mm}^{(2)}$ in place of $H_{aa}^{(2)}$ in Eq. (10.4.6), *eliminating the* 1/2 *prefactor*. This result can also be obtained using straightforward perturbation theory to second order in $H_{am}^{(2)}$ (Eq. (10.4.5)).[26] As in perturbation theory, G_2^m and G_2^a (alternatively, K^m and G_2^a) are to be determined in the absence of atom–metal Coulomb coupling. It is convenient to write the result in the form of Eq. (10.4.8) (but without the eliminated 1/2 prefactor), as

$$E_{vdW}^{(2)} = -\int d^{(3)}\mathbf{x}'\int d^{(3)}\mathbf{x}''\int dt''\mathcal{V}_{image}(\mathbf{x}',\mathbf{x}'',t'-t'')G_2^a(\mathbf{x}''t'',\mathbf{x}'t';\mathbf{x}''t''^+,\mathbf{x}'t'^+), \tag{10.4.12}$$

which suggests interpreting $E_{vdW}^{(2)}$ in terms of a self-energy of the atomic electrons due to screening of their intra-atom electron–electron Coulomb interaction by the metal electrons.

Here, we can fruitfully employ the determination of $\mathcal{V}_{eff} - \mathcal{V} = \mathcal{V}_{image}$ carried out in Section 5.10. In this connection, we denote the effective Coulomb potential in the atom region "A" (in the presence of dynamic, non-local screening by the semi-infinite metal electrons) as $\tilde{\mathcal{G}}(z',z'')$ of Eq. (5.10.17), using script type to avoid confusion with Schrödinger Green's function notation, G, employed for electrons in this section

(Fourier-transformed in the lateral plane and in time), with region B as the metal side ($B \equiv M$). Accordingly, the total effective potential felt by the atomic electrons is given by

$$V_{eff}(z', z'') \equiv \tilde{\mathcal{G}}(z', z'') = \tilde{\mathcal{G}}_A(z', z'') + \frac{1}{4\pi}[\tilde{\mathcal{G}}(0, z'')\tilde{\mathcal{D}}_A(0_A, z') - \tilde{\mathcal{G}}_A(z', 0_A)\tilde{\mathcal{D}}(0, z'')], \tag{10.4.13}$$

where the factors on the right of Eq. (10.4.13) are given by (Eq. (5.10.23)),

$$\tilde{\mathcal{G}}(0, z'') = \frac{\tilde{\mathcal{G}}_B(0,0)\tilde{\mathcal{G}}_A(0, z'')}{\tilde{\mathcal{G}}_A(0,0) + \tilde{\mathcal{G}}_B(0,0)}, \tag{10.4.14}$$

and defining the electric displacement field as (Eq. (5.10.24)):

$$\tilde{\mathcal{D}}_{\left\{ A \atop B \right\}}(z', z'') = -\int_{\left\{ A \atop B \right\}} dz''' \varepsilon_{\left\{ A \atop B \right\}}(z', z''')\frac{\partial}{\partial z'''}\tilde{\mathcal{G}}_{\left\{ A \atop B \right\}}(z''', z''), \tag{10.4.15}$$

we have

$$\tilde{\mathcal{D}}(0, z'') = \frac{4\pi\tilde{\mathcal{G}}_A(0, z'')}{\tilde{\mathcal{G}}_A(0,0) + \tilde{\mathcal{G}}_B(0,0)}. \tag{10.4.16}$$

In these considerations, we have distinguished the infinite space Green's function for region A, $\tilde{\mathcal{G}}_A$, from $\tilde{\mathcal{G}}_A^{\infty}$, as they can differ by a boundary condition. In particular, for convenience, we employed a subsidiary image in this respect to force vanishing of $\tilde{\mathcal{D}}_A(0, z'')$ at the interface. Correspondingly,

$$\tilde{\mathcal{G}}_A(z', z'') = \tilde{\mathcal{G}}_A^{\infty}(z' - z'') + \tilde{\mathcal{G}}_A^{\infty}(z' + z''); \quad \tilde{\mathcal{G}}_A^{\infty}(z' - z'') = \frac{2\pi}{\bar{p}}e^{-\bar{p}|z'-z''|}, \tag{10.4.17}$$

with $\tilde{\mathcal{D}}_A(z', z'') = -\partial\tilde{\mathcal{G}}_A(z', z'')/\partial z'$ since region A is local (vacuum). Furthermore, for the metal surface electron dielectric region B, we have similar relations for $\tilde{\mathcal{G}}_B(z', z'')$ and $\tilde{\mathcal{D}}_B(z', z'')$ involving denominator factors of $\varepsilon_B(\bar{\mathbf{p}}, \omega)$ with a similar subsidiary image. In particular,

$$\tilde{\mathcal{G}}_A(0,0) = \frac{4\pi}{\bar{p}} \quad \text{and} \quad \tilde{\mathcal{G}}_B(0,0) = 4\int_{-\infty}^{\infty} dp_z \frac{1}{(p_z^2 + \bar{p}^2)\varepsilon_B(\mathbf{p}, \omega)} = \frac{4\pi}{\bar{p}}\varepsilon_{\bar{p}}^{-1}(\omega), \tag{10.4.18}$$

since our choice of subsidiary image doubles $\tilde{\mathcal{G}}_{\{A,B\}}(0, z'')$ at the interface $z' = 0$. The last part of Eq. (10.4.18) on the right defines the "surface dielectric function," $\varepsilon_{\bar{p}}(\omega)$, for the semi-infinite non-local region B. Finally, noting that the image potential is given by

$$V_{Img}(z', z'') = V_{eff}(z', z'') - \tilde{\mathcal{G}}_A^{\infty}(z', z'') \equiv \tilde{\mathcal{G}}(z', z'') - \tilde{\mathcal{G}}_A^{\infty}(z', z''), \tag{10.4.19}$$

Eq. (10.4.13) yields

$$V_{Img}(z', z'') = \tilde{\mathcal{G}}_A^{\infty}(z' + z'') + \frac{1}{4\pi}\tilde{\mathcal{G}}(0, z'')\tilde{\mathcal{D}}_A(0, z') - \frac{1}{4\pi}\tilde{\mathcal{G}}_A(z', 0)\tilde{\mathcal{D}}(0, z''), \tag{10.4.20}$$

and the image potential in the atom region A is

$$\mathcal{V}_{Img}(z', z''; \bar{\mathbf{p}}, \omega) = \frac{2\pi}{\bar{p}} \exp\left[-\bar{p}(|z'| + |z''|)\right] \frac{\varepsilon_{\bar{p}}(\omega) - 1}{\varepsilon_{\bar{p}}(\omega) + 1}, \tag{10.4.21}$$

which is well known in the local limit.

The atomic electron G_2^a-function involved in E_{vdW} (Eq. (10.4.12)) has the form

$$G_2^a \to \left\langle \Phi_0^a \left| \rho_a(\mathbf{x}'', t'') \rho_a(\mathbf{x}', t') \right| \Phi_0^a \right\rangle. \tag{10.4.22}$$

Recognizing that the atom has a fixed number of electrons and is electrically neutral (in the absence of ionization), we denote the energy eigenstates of the atomic electrons by $|\Phi_n^a\rangle$, with ground state $|\Phi_0^a\rangle$. These states of the atomic electrons bear the full complement of correlations due to electron–electron interactions of the electrons of the atom. Their completeness is expressed by

$$1 = \sum_n |\Phi_n^a\rangle \langle \Phi_n^a|, \tag{10.4.23}$$

so that

$$G_2^a(\mathbf{x}''t'', \mathbf{x}'t'; \mathbf{x}''t''^+, \mathbf{x}'t'^+) = \sum_n \langle \Phi_0^a | \rho_a(\mathbf{x}'', t'') |\Phi_n^a\rangle \langle \Phi_n^a | \rho_a(\mathbf{x}', t') |\Phi_0^a\rangle. \tag{10.4.24}$$

Employing an atomic electron time-translation operator $e^{-iH_a(t'-t'')}$ to bring the times of the two density operators into coincidence, we have

$$G_2^a(\mathbf{x}''t'', \mathbf{x}'t'; \mathbf{x}''t''^+, \mathbf{x}'t'^+) = \sum_n \langle \Phi_0^a | \rho_a(\mathbf{x}'', t'') |\Phi_n^a\rangle \langle \Phi_n^a | e^{iH_a(t'-t'')} \times \rho_a(\mathbf{x}', t'') e^{-iH_a(t'-t'')} |\Phi_0^a\rangle, \tag{10.4.25}$$

and since

$$H_a |\Phi_n^a\rangle = E_n^a |\Phi_n^a\rangle, \tag{10.4.26}$$

we have

$$G_2^a(\mathbf{x}''t''; \mathbf{x}'t'; \mathbf{x}''t''^+; \mathbf{x}'t'^+) = \sum_n \langle \Phi_0^a | \rho_a(\mathbf{x}'', t'') |\Phi_n^a\rangle \langle \Phi_n^a | \rho_a(\mathbf{x}', t'') |\Phi_0^a\rangle e^{i(E_n^a - E_0^a)(t'-t'')}. \tag{10.4.27}$$

Understanding that both density operators and the energy eigenstates are evaluated at the same time so that the matrix elements are independent of time, we write

$$\langle \Phi_0^a | \rho_a(\mathbf{x}'', t'') |\Phi_n^a\rangle = \langle \rho_a(\mathbf{x}'') \rangle_{0n}$$

and define ($\hbar \to 1$),

$$\omega_{no}^a = E_n^a - E_0^a,$$

whence

$$G_2^a(\mathbf{x}''t'';\mathbf{x}'t';\mathbf{x}''t''^+;\mathbf{x}'t'^+) = \sum_n{}' \langle\rho_a(\mathbf{x}'')\rangle_{0n} \langle\rho_a(\mathbf{x}')\rangle_{n0}\, e^{i\omega_{no}^a(t'-t'')}. \tag{10.4.28}$$

(The prime on the $\Sigma \to \Sigma'$ indicates that the $n = 0$ term is excluded since it has no time dependence and yields a constant, static contribution, which is irrelevant to the van der Waals interaction.) Substitution of this into Eq. (10.4.12) yields $E_{vdW}^{(2)}$ in terms of the Fourier time transform of $V_{image}(t' - t'') \to V_{image}(\omega)$ as:

$$E_{vdW}^{(2)} = \int d^3\mathbf{x}' \int d^3\mathbf{x}'' \int_{-\infty}^{\infty} \frac{d\omega}{2\pi i} V_{image}(\mathbf{x}',\mathbf{x}'';\omega) \sum_n{}' \left(\frac{\langle\rho_a(\mathbf{x}'')\rangle_{0n} \langle\rho_a(\mathbf{x}')\rangle_{n0}}{\omega - \omega_{no}^a} \right), \tag{10.4.29}$$

where we have introduced retardation in performing the time integration. Considering spatial translational invariance in the (lateral) plane $\bar{\mathbf{x}} = (x, y) \to \bar{\mathbf{p}}$ and using Eq. (10.4.21) for $V_{image}(z', z''; \bar{\mathbf{p}}, \omega)$, we obtain

$$\begin{aligned} E_{vdW}^{(2)} = &\int \frac{d^{(2)}\bar{\mathbf{p}}}{(2\pi)^2} \int d^{(3)}\mathbf{x}' \int d^{(3)}\mathbf{x}'' \int_{-\infty}^{\infty} \frac{d\omega}{2\pi i} e^{i\bar{\mathbf{p}}\cdot(\bar{\mathbf{x}}'-\bar{\mathbf{x}}'')} \frac{2\pi}{\bar{p}} e^{-\bar{p}(|z'|+|z''|)} \\ &\times \left(\frac{\varepsilon_{\bar{p}}(\omega) - 1}{\varepsilon_{\bar{p}}(\omega) + 1} \right) \sum_n{}' \left(\frac{\langle\rho_a(\mathbf{x}'')\rangle_{0n} \langle\rho_a(\mathbf{x}')\rangle_{n0}}{\omega - \omega_{no}^a} \right). \end{aligned} \tag{10.4.30}$$

Eq. (10.4.30) may be taken as the starting point of a multipole expansion of $E_{vdW}^{(2)}$ in increasing inverse powers of the distance of the center of the atom, Z, from an origin of coordinates on the metal surface. Denoting the lateral coordinate of the atom center by $\bar{\mathbf{X}}$, we have

$$\begin{aligned} \bar{\mathbf{x}}' &= \bar{\mathbf{X}} + \Delta\bar{\mathbf{x}}' \quad ; \quad \bar{\mathbf{x}}'' = \bar{\mathbf{X}} + \Delta\bar{\mathbf{x}}'' \\ z' &= Z + \Delta z' \quad ; \quad z'' = Z + \Delta z'' \end{aligned}, \tag{10.4.31}$$

and ($Z >> \Delta z$),

$$e^{i\bar{\mathbf{p}}\cdot}\left(\bar{\mathbf{x}}' - \bar{\mathbf{x}}''\right) e^{-\bar{p}(|z'|+|z''|)} = e^{-2\bar{p}|Z|} e^{-\bar{p}(\Delta z'+\Delta z'')} e^{i\bar{\mathbf{p}}\cdot(\Delta\bar{\mathbf{x}}'-\Delta\bar{\mathbf{x}}'')}, \tag{10.4.32}$$

from which a multipole expansion may be undertaken by expanding the exponentials in powers of $\Delta\bar{\mathbf{x}}$ and Δz. We will not present the details of calculation here, but will discuss the results to dipole order. To this order, one must examine matrix elements of the total charge operator for the atom,

$$\widehat{Q} = \int d^3\mathbf{x}' \rho_a(\mathbf{x}') = 0, \tag{10.4.33}$$

which vanishes identically for a neutral atom along with all its matrix elements;

$$\langle\widehat{Q}\rangle_{0n} \equiv 0. \tag{10.4.34}$$

The dipole moment operator for the atom $\vec{\mathcal{D}}$ is defined as

$$\vec{\mathcal{D}} = \int d^3\mathbf{x}'\rho_a(\mathbf{x}')\Delta\mathbf{x}', \tag{10.4.35}$$

and its matrix elements are denoted by

$$\vec{\mathcal{D}}_{0n} = \langle \Phi_0^a | \vec{\mathcal{D}} | \Phi_n^a \rangle = \vec{\mathcal{D}}_{n0}^*. \tag{10.4.36}$$

Similarly, the quadrupole moment operator is written in dyadic form as

$$\hat{\overleftrightarrow{\Theta}} = \int d^{(3)}\mathbf{x}'\rho_a(\mathbf{x}')\left[\Delta\mathbf{x}'\Delta\mathbf{x}' - \frac{1}{3}\mathbf{I}\left|\Delta\mathbf{x}'\right|^2\right], \tag{10.4.37}$$

with matrix elements $\overleftrightarrow{\Theta}_{0n}$, etc. With these definitions, the position-space integrals of Eq. (10.4.30) yield E_{vdW} to dipole–dipole terms as ($\hbar \to 1$):

$$E_{vdW}^{(2)} = \frac{1}{3\pi}\sum_n{}' \left|\vec{\mathcal{D}}_{0n}\right|^2 \int_{-\infty}^{\infty} d\omega \frac{1}{i(\omega - \omega_{no}^a)} \int_0^{\infty} d\bar{p}\,\bar{p}^2 e^{-2\bar{p}|Z|}\left(\frac{1-\varepsilon_{\bar{p}}(\omega)}{1+\varepsilon_{\bar{p}}(\omega)}\right), \tag{10.4.38}$$

where the factor 1/3 arises from averaging over all directions of the spatial orientation of the atom. The $\bar{p}$-integral is of central importance in determining the Z-dependence of E_{vdW}. Defining $\bar{q} = \bar{p}Z$, it takes the form

$$\int_0^{\infty} d\bar{p}\,\bar{p}^2 e^{-2\bar{p}|Z|}\left(\frac{1-\varepsilon_{\bar{p}}(\omega)}{1+\varepsilon_{\bar{p}}(\omega)}\right) = \frac{1}{Z^3}\int_0^{\infty} d\bar{q}\,\bar{q}^2 e^{-2\bar{q}}\left(\frac{1-\varepsilon_{\bar{q}/Z}(\omega)}{1+\varepsilon_{\bar{q}/Z}(\omega)}\right). \tag{10.4.39}$$

It is at once apparent that the inclusion of quadrupole and higher order moments, which involve higher powers of $\Delta\bar{\mathbf{x}}$ and/or Δz, would bring additional powers of $\bar{p}$ with them into the $\bar{p}$-integrand through the expansion of the exponential discussed above. Every such additional power of $\bar{p}$ in the $\bar{p}$-integrand contributes another factor of $1/Z$ to the corresponding integral in $E_{vdW}^{(2)}$, as one readily sees from the substitution $\bar{q} = \bar{p}Z$ above. Thus, it is only the dipole–dipole term of $E_{vdW}^{(2)}$ that varies as $1/Z^3$ and any involvement of quadrupole or higher order multipoles produces terms that vary at least like $1/Z^4$. Even within the framework of the dipole–dipole term, however, the non-locality of $\varepsilon_{\bar{p}}(\omega)$ will also result in terms varying at least like $1/Z^4$. It is only the local limit of $\varepsilon_{\bar{p}}(\omega) \to \varepsilon(\omega)$ that genuinely results in $E_{vdW}^{(2)}$ varying like $1/Z^3$. In this case, Eq. (10.4.36) readily yields the result of Mavroyannis,[27]

$$E_{vdW}^{(2)} = \frac{1}{6\pi Z^3}\sum_n{}' \left|\vec{\mathcal{D}}_{0n}\right|^2 \int_0^{\infty} \frac{du}{u^2 + \left(\omega_{no}^a\right)^2}\left(\frac{1-\varepsilon(iu)}{1+\varepsilon(iu)}\right), \tag{10.4.40}$$

where we have set $\omega = iu$ and deformed the contour of frequency integration. Considering a perfect metal of infinite polarizability, $|\varepsilon(iu) \to \infty|$, the u-integral is elementary, leading to (restore e):

$$E^{(2)}_{vdW} = -\frac{1}{12Z^3}\sum_n{}' \left|\vec{\mathcal{D}}_{0n}\right|^2 = -\frac{e^2}{12Z^3}\left\langle|\Delta\mathbf{x}|^2\right\rangle_0, \qquad (10.4.41)$$

which is the well-known Lennard-Jones[28] result for neutral atom interacting with a perfect metal. It should be borne in mind that a very close approach of the atom to the metal will result in Pauli exclusion exchange effects coming into play, with a concomitant repulsive potential that dominates over the weak van der Waals attraction at very short distances.

10.5 Fast Particle Power Loss in Plasma

A fast charged particle moving through a semiconductor plasma loses energy by exciting plasmons and particle–hole pairs (we ignore losses to phonons). Moving with velocity $\mathbf{v} = v\hat{x}$, the fast particle impresses a Coulomb potential $U(2)$ at a space-time point $2 = (\mathbf{r}_2, t_2)$, which is dynamically and non-locally screened by the plasma, resulting in the effective screened potential $V(1)$ at a space-time point $1 = (\mathbf{r}_1, t_1)$ given by (Eq. (10.1.1)):

$$V(1) = \int d2K(1,2)U(2). \qquad (10.5.1)$$

Here, $K(1,2)$ is the inverse dielectric screening function for the plasma in a 4D position-time matrix representation, such that

$$\int d3K(1,3)\epsilon(3,2) = \delta(1,2), \qquad (10.5.2)$$

with $\epsilon(3,2)$ as the direct dielectric function of the plasma, taken to involve a static background polarizability as well as a mobile plasma. The response dynamics are centered on the perturbed density,

$$\begin{aligned}\rho(1) &= -\int d3R(1,3)V(3)\\ &= -\int d3\int d4R(1,3)K(3,4)U(4),\end{aligned} \qquad (10.5.3)$$

where $R(1,3) = -\delta\rho(1)/\delta V(3)$ is the density-perturbation response function. Introducing the usual random phase approximation (RPA) integral equation

$$K(1,2) = \delta(1-2) - \int d3\,\alpha(1,3)K(3,2), \qquad (10.5.4)$$

the polarizability $\alpha(1,3)$ is written in a form that includes both the free electron plasma response and an additive local, static background contribution $\alpha_0 = \epsilon_0 - 1$ ($\mathcal{V}$ is the interelectron Coulomb interaction of the plasma):

$$\alpha(1,3) = \int d4\mathcal{V}(1-4)R(4,3) + \alpha_0\delta(1-3). \qquad (10.5.5)$$

This results in

$$K(1,2) = \frac{1}{\epsilon_0}\delta(1-2) - \frac{1}{\epsilon_0}\int d3 \int d4\mathcal{V}(1-4)R(4,3)K(3,2), \tag{10.5.6}$$

and applying the Poisson equation, we have

$$\nabla_1^2\left[K(1,2) - \delta(1-2)/\epsilon_0\right] = \frac{4\pi e}{\epsilon_0}\int d3R(1,3)K(3,2), \tag{10.5.7}$$

which yields $\rho(1)$ of Eq. (10.5.3) as

$$\rho(1) = -\frac{\epsilon_0}{4\pi e}\int d4\,\nabla_1^2\left[K(1,4) - \delta(1-4)/\epsilon_0\right]U(4). \tag{10.5.8}$$

Noting that the force (per unit volume) of the passing charged particle on an element of the semiconductor plasma is $-e\rho(1)\nabla_1 V(1)$, the frictional reaction force on the fast particle is therefore given by

$$\mathbf{f} = e\int d^{(3)}1\ \rho(1)\nabla_1 V(1). \tag{10.5.9}$$

Thus, the power loss of the fast particle is $\frac{dW}{dt} = \mathbf{f}\cdot\mathbf{v}$, of course.

As a first example, we focus on a thin 2D quantum well containing a planar sheet of mobile charge that is translationally invariant in the $\bar{\mathbf{r}} = (x, y)$ plane (situated at $z = 0$) and time. Fourier-transforming in the plane of the 2D sheet,

$$K(1,2) = K(\bar{\mathbf{r}}_1 - \bar{\mathbf{r}}_2, z_1, z_2; t_1 - t_2) \to K(\bar{\mathbf{p}}, z_1, z_2; \omega), \tag{10.5.10}$$

with respect to lateral position $\bar{\mathbf{r}}_1 - \bar{\mathbf{r}}_2 \to \bar{\mathbf{p}}$ and time $t_1 - t_2 \to \omega$, the inversion condition (Eq. (10.5.2)) takes the form

$$\int dz_2 K(\bar{\mathbf{p}}, z_1, z_2; \omega)\epsilon(\bar{\mathbf{p}}, z_2, z_3; \omega) = \delta(z_1 - z_3), \tag{10.5.11}$$

with $\epsilon(\bar{\mathbf{p}}, z_1, z_2; \omega)$ given by

$$\epsilon(\bar{\mathbf{p}}, z_1, z_2; \omega) = \delta(z_1 - z_2) + \alpha(\bar{\mathbf{p}}, z_1, z_2; \omega). \tag{10.5.12}$$

For a very thin quantum well with electron motion confined to a single 2D planar sheet, the z-arguments of the density-perturbation response function are localized to the sheet by positional δ-functions as

$$R(\bar{\mathbf{p}}, z_1, z_2; \omega) = \delta(z_1)\delta(z_2)R^{2D}(\bar{\mathbf{p}}, \omega). \tag{10.5.13}$$

In this, $R^{2D}(\overline{\mathbf{p}}, \omega)$ describes the electron density perturbation response properties localized on the 2D sheet, so the 2D electron polarizability on the sheet is $\alpha^{2D}(\overline{\mathbf{p}}, \omega) = 2\pi R^{2D}(\overline{\mathbf{p}}, \omega)/p$. Considering this and Eqns. (10.5.5), (10.5.12) and (10.5.13), we find

$$\epsilon(\overline{\mathbf{p}}, z_1, z_2; \omega) = \epsilon_0 \delta(z_1 - z_2) + \delta(z_2)\alpha^{2D}(\overline{\mathbf{p}}, \omega)\, e^{-p|z_1|} \qquad (10.5.14)$$

($p = |\overline{\mathbf{p}}|$ and $\epsilon_0 = 1 + \alpha_0$ for the background). $\alpha^{2D}(\overline{\mathbf{p}}, \omega)$ is given by the degenerate limit (zero temperature) of the 2D RPA Stern dielectric function[10] of Eqns. (10.2.11 and 10.2.12). To carry out the inversion of the dielectric function, we make the ansatz

$$K(\overline{\mathbf{p}}, z_1, z_2; \omega) = \frac{1}{\epsilon_0}\delta(z_1 - z_2) + \frac{1}{\epsilon_0}\delta(z_2) e^{-p|z_1|}\left[\widetilde{K}^{2D}(\overline{\mathbf{p}}, \omega) - 1\right], \qquad (10.5.15)$$

and determine $\widetilde{K}^{2D}(\overline{\mathbf{p}}, \omega)$ by requiring satisfaction of the inversion condition (Eq. (10.5.11)): Equating coefficients of like positional delta functions leads to

$$\widetilde{K}^{2D}(\overline{\mathbf{p}}, \omega) = \left[1 + \frac{\alpha^{2D}(\overline{\mathbf{p}}, \omega)}{\epsilon_0}\right]^{-1} \equiv \left[\widetilde{\epsilon}^{2D}(\overline{\mathbf{p}}, \omega)\right]^{-1}. \qquad (10.5.16)$$

(The divisor ϵ_0 corresponds to putting $\mathcal{V} \to \widetilde{\mathcal{V}} = \mathcal{V}/\epsilon_0$ or $e^2 \to \widetilde{e}^2 = e^2/\epsilon_0$.)

As an illustration, we employ Eq. (10.5.16) to determine the statically screened Coulomb potential of a point charge at $(0, 0, z_0)$ taking $\epsilon_0 \to 1$:

$$\begin{aligned} V(r, t \to \infty) &= \int \frac{d^{(2)}\overline{\mathbf{p}}}{2\pi} e^{i\overline{\mathbf{p}}\cdot\overline{\mathbf{r}}} \int dz' K(\overline{\mathbf{p}}, z, z'; \omega \to 0) e^{-p|z'-z_0|}/p \\ &= \int \frac{d^{(2)}\overline{\mathbf{p}}}{2\pi} \frac{e^{i\overline{\mathbf{p}}\cdot\overline{\mathbf{r}}}}{p} \left[e^{-p|z-z_0|} - e^{-p(|z|+|z_0|)} \frac{\alpha^{2D}(\overline{\mathbf{p}}, \omega \to 0)}{1 + \alpha^{2D}(\overline{\mathbf{p}}, \omega \to 0)}\right]. \end{aligned} \qquad (10.5.17)$$

In the case of a perfect metal sheet, $\alpha^{2D} \to -\infty$, Eq. (10.5.17) exhibits the potential field of a perfect image of relative strength -1 for points on the same side of the sheet as the original Coulomb site (same sign for z, z_0): It also produces a completely shielded null result at points on the other side of the sheet (different signs for z, z_0), which is to be expected.

A charge Ze moving parallel to the sheet with speed v ($\widehat{\mathbf{i}}$–direction) and at a distance H from it (in the $\widehat{\mathbf{k}}$-direction) impresses the Coulomb potential

$$U(2) = Ze\left|\mathbf{r}_2 - vt_2\widehat{\mathbf{i}} - H\widehat{\mathbf{k}}\right|^{-1} \qquad (10.5.18)$$

on the 2D plasma sheet. Substituting Eqns. (10.5.15) and 10.5.18 into Eqns. (10.5.1) and (10.5.8 and 10.5.9), the power loss per unit time is seen to be

$$\begin{aligned} \frac{dW}{dt} = \mathbf{f}\cdot\mathbf{v} = -Z^2\widetilde{e}^2 v \int \frac{d^2\overline{\mathbf{p}}}{2\pi}\frac{ip_x}{p}\widetilde{K}^{2D}(\overline{\mathbf{p}}, -\overline{\mathbf{p}}\cdot\overline{\mathbf{v}}) \\ \times\left[\widetilde{K}^{2D}(-\overline{\mathbf{p}}, \overline{\mathbf{p}}\cdot\overline{\mathbf{v}}) - 1\right] e^{-2pH}, \end{aligned} \qquad (10.5.19)$$

wherein the frequency argument of $\widetilde{K}^{2D}$ is replaced by $\omega \to \pm\overline{\mathbf{p}}\cdot\overline{\mathbf{v}}$. Since both the real and imaginary parts of $\widetilde{K}^{2D}(\overline{\mathbf{p}},\omega)$ are even functions of $\overline{\mathbf{p}}$, while the real part is even in ω but the imaginary part is odd in ω, the $\widetilde{K}^{2D}\widetilde{K}^{2D}-$ term of Eq. (10.5.19) has an odd p_x-integrand and vanishes. Similarly, the remaining $\widetilde{K}^{2D}-$ term has its real part vanish, and therefore the result involves only the imaginary part of $\widetilde{K}^{2D}$:

$$\frac{dW}{dt} = Z^2\tilde{e}^2 v \int \frac{d^2\overline{\mathbf{p}}}{2\pi} e^{-2pH} \frac{p_x}{p} Im\, \widetilde{K}^{2D}(\overline{\mathbf{p}}, -\overline{\mathbf{p}}\cdot\overline{\mathbf{v}}) \tag{10.5.20}$$

or, equivalently,

$$\frac{dW}{dt} = \frac{Z^2\tilde{e}^2}{2\pi v} \int_{-\infty}^{\infty} dp_y \int_{-\infty}^{\infty} d\omega \frac{\omega}{p} e^{-2pH} Im\widetilde{K}^{2D}(\bar{p}, -\omega), \tag{10.5.21}$$

where

$$p \to \left[p_y^2 + \left(\frac{\omega}{v}\right)^2\right]^{1/2} \tag{10.5.22}$$

and

$$\mathrm{Im}\widetilde{K}^{2D}(\overline{\mathbf{p}}, -\omega) = -\tilde{\epsilon}_2^{2D}(\overline{\mathbf{p}},\omega)\left[\tilde{\epsilon}_1^{2D}(\overline{\mathbf{p}},\omega)^2 + \tilde{\epsilon}_2^{2D}(\overline{\mathbf{p}},\omega)^2\right]^{-1}. \tag{10.5.23}$$

Here,

$$\left(\widetilde{K}^{2D}\right)^{-1} = \tilde{\epsilon}^{2D} = \tilde{\epsilon}_1^{2D} + i\tilde{\epsilon}_2^{2D} \tag{10.5.24}$$

is given by the real and imaginary parts of the Stern 2D RPA dielectric function in the degenerate limit (zero temperature; *sign* $x = 1, 0, -1$ for $x > 0$, $x = 0$, $x < 0$, respectively):

$$\tilde{\epsilon}_1^{2D}(\overline{\mathbf{p}},\omega) = 1 + \frac{2\tilde{e}^2 m^2}{\hbar^3 p^3}\left\{\frac{\hbar p^2}{m} - \sum_{\pm} sign\left[\frac{\hbar p^2}{2m} \pm \omega\right]\right.$$
$$\left.\eta_+\left[\left[\frac{\hbar p^2}{2m} \pm \omega\right]^2 - \frac{2p^2\zeta}{m}\right]\left[\left[\frac{\hbar p^2}{2m} \pm \omega\right]^2 - \frac{2p^2\zeta}{m}\right]^{1/2}\right\} \tag{10.5.25}$$

and

$$\tilde{\epsilon}_2^{2D}(\overline{\mathbf{p}},\omega) = \frac{2\tilde{e}^2 m^2}{\hbar^3 p^3}\sum_{\pm}(\pm)\,\eta_+\left[\frac{2p^2\zeta}{m} - \left[\frac{\hbar p^2}{2m} \mp \omega\right]^2\right]\left[\frac{2p^2\zeta}{m} - \left[\frac{\hbar p^2}{2m} \mp \omega\right]^2\right]^{1/2}. \tag{10.5.26}$$

In this, ζ is the Fermi energy, and $m, \tilde{e} = e/\sqrt{\epsilon_0}$ are the electron effective mass and charge, respectively, with $\eta_+(x)$ as the Heaviside unit step function. Henceforth, explicit reference to the overhead tilde is suppressed and we understand $e^2 \Rightarrow e^2/\epsilon_0$.

In the low-velocity limit we expand

$$K^{2D}(\overline{\mathbf{p}}, -\overline{\mathbf{p}}\cdot\overline{\mathbf{v}}) \cong K^{2D}(\overline{\mathbf{p}}, 0) - p_x v \frac{\partial K^{2D}}{\partial \omega}(\overline{\mathbf{p}}, \omega = 0). \quad (10.5.27)$$

The first term does not contribute to $dW/dt = \mathbf{f}\cdot\mathbf{v}$ because the p_x integrand is odd. The second term yields

$$\frac{dW}{dt} \cong -Z^2 e^2 v^2 \int \frac{d^2\overline{\mathbf{p}}}{2\pi} e^{-2pH} \frac{p_x^2}{p} Im\left[\frac{\partial K^{2D}}{\partial \omega}(\overline{\mathbf{p}}, \omega = 0)\right]$$
$$\cong \frac{Z^2 e^2 v^2}{2} \int_0^\infty dp\, e^{-2pH} p^2 \frac{\partial \epsilon_2^{2D}(\overline{\mathbf{p}}, \omega = 0)}{\partial \omega}\left[\epsilon_1^{2D}(\overline{\mathbf{p}}, 0)\right]^{-2}, \quad (10.5.28)$$

due to the fact that the contributions of $\epsilon_2^{2D}(\overline{\mathbf{p}}, \omega = 0)$ and $(\partial/\partial\omega)\,\epsilon_1^{2D}(\overline{\mathbf{p}}, \omega = 0)$ vanish as odd functions of ω. Further calculation based on Eqns. (10.5.25) and (10.5.26) produces the result (p_F is the Fermi wave number):

$$\frac{dW}{dt} = \frac{2Z^2 e^4 m^2 v^2}{\hbar^3} \int_0^{2p_F} dp \frac{e^{-2pH}}{\left[(2p_F)^2 - p^2\right]^{1/2}} \left[1 + \frac{2e^2 m}{\hbar^2 p}\right]^{-2}. \quad (10.5.29)$$

Taking the Bohr radius $a_0 \cong \hbar^2/me^2$ to be the smallest characteristic length of the system $\left(a_0 \ll H, p_F^{-1}\right)$, the resulting *stopping power*, $dW/dl = v^{-1} dW/dt$, is given by

$$\frac{dW}{dl} = \frac{Z^2 \hbar v}{8} \frac{\partial^2}{\partial H^2} \int_0^{2p_F} dp \frac{e^{-2pH}}{\left[(2p_F)^2 - p^2\right]^{1/2}}. \quad (10.5.30)$$

The final integration of Eq. (10.5.30) is expressed in terms of modified Bessel functions $I_n(x)$ of the first kind and modified Struve functions $L_n(x)$:

$$\frac{dW}{dl} = \pi Z^2 \hbar p_F^2 v \{[I_2(4p_F H) - L_2(4p_F H)]$$
$$+ [I_1(4p_F H) - L_1(4p_F H)]/4p_F H - (8p_F H/3\pi)\}. \quad (10.5.31)$$

In the high-velocity limit, the frequency argument $\omega = -p_x v$ of $K^{2D}(\overline{\mathbf{p}}, -\overline{\mathbf{p}}\cdot\overline{v})$ in Eq. (10.5.21) is sufficiently large that we may use the local "2D plasmon" limit for dielectric response (0^+ is a positive infinitesimal):

$$ImK^{2D}(\overline{\mathbf{p}},-\overline{\mathbf{p}}\cdot\overline{v}) \to -ImK^{2D}(0,\overline{\mathbf{p}}\cdot\overline{v})$$
$$= -Im\left[1-\frac{2\pi e^2\rho^{2D}p}{m(\omega+i0^+)^2}\right]^{-1}_{\omega=p_x v}, \tag{10.5.32}$$

or, alternatively,

$$ImK^{2D}(\overline{\mathbf{p}},-\overline{\mathbf{p}}\cdot\overline{v}) = \pi\delta\left(1-2\pi e^2\rho^{2D}p/m\omega^2\right)_{\omega=p_x v} sign(p_x). \tag{10.5.33}$$

Therefore, in 2D polar coordinates, $\frac{dW}{dt}$ is given by

$$\frac{dW}{dt} = -2Z^2e^2\int_0^\infty dp\, pe^{-2pH}\int_0^{\pi/2} d\theta\cos\theta\;\delta\left(1-2\pi e^2\rho^{2D}/mv^2p\cos^2\theta\right) \tag{10.5.34}$$

and the Dirac δ-function on the right-hand side of Eq. (10.5.34) may be written as

$$\delta(\ldots) = \frac{1}{2}\eta_+\left(p-2\pi e^2\rho^{2D}/mv^2\right)\left[\frac{\cos^2\theta_0}{1-\cos^2\theta_0}\right]^{1/2}\delta(\theta-\theta_0), \tag{10.5.35}$$

where $\cos^2\theta_0 = 2\pi e^2\rho^{2D}/mv^2p$ for $0\le\theta_0\le\pi/2$. Thus the θ integral yields

$$\frac{dW}{dt} = -\frac{\pi Z^2e^4\rho^{2D}}{mv}\frac{\partial}{\partial H}\int_{p_L}^\infty dp\,e^{-2pH}\left[p\left[p-p_L\right]\right]^{-1/2}, \tag{10.5.36}$$

with the lower limit $p_L = 2\pi e^2\rho^{2D}/mv^2$; and the p-integral may be evaluated in terms of modified Bessel functions of the third kind, $K_n(x)$, with the result

$$\frac{dW}{dt} = \frac{2\pi^2Z^2e^6\left(\rho^{2D}\right)^2}{m^2v^3}\exp\left(\frac{2\pi e^2\rho^{2D}H}{mv^2}\right)\left[K_0\left[\frac{2\pi e^2\rho^{2D}H}{mv^2}\right]+K_1\left[\frac{2\pi e^2\rho^{2D}H}{mv^2}\right]\right], \tag{10.5.37}$$

which diverges as $H\to 0$. In the high-velocity limit, we have

$$\left|\frac{dW}{dt}\right| = \frac{\pi Z^2e^4\rho^{2D}}{mvH}. \tag{10.5.38}$$

The divergence as $H\to 0$ may be understood as resulting from the fact that the 2D plasma sheet has been taken to have "zero width." The inclusion of a finite width b for the sheet yields a convergent result as $H\to 0$, but for distances $H >> b$, the present "zero-width" result is valid and meaningful. The case of arbitrary velocity requires a numerical evaluation of $\frac{dW}{dt}$, which has been carried out in the literature.

Considering next the case of charged particle power loss to a uniform bulk 3D semiconductor plasma, we Fourier-transform $K(1,2)$ with respect to *all* positional variables,

$\mathbf{r}_1 - \mathbf{r}_2 \to \mathbf{p}$, as well as time, $t_1 - t_2 \to \omega$, noting that the inversion is algebraic in this representation:

$$K(\mathbf{p}, \omega) = 1/\varepsilon(\mathbf{p}, \omega)\,. \tag{10.5.39}$$

The Coulomb potential impressed on the plasma by the passing charged particle (Eq. (10.5.18)) is given in 3D **p**-representation by

$$U(\mathbf{p}, \omega) = \frac{8\pi^2 Ze}{|\mathbf{p}|^2}\delta(\omega - \mathbf{p}\cdot\mathbf{v})\,. \tag{10.5.40}$$

Proceeding as above, the resistive force is given by

$$\mathbf{f} = -\frac{Z^2e^2}{2\pi^2}\int d^{(3)}\mathbf{p} i\mathbf{p}\,|\mathbf{p}|^{-2} K(\mathbf{p}, -\mathbf{p}\cdot\mathbf{v})\,[K(-\mathbf{p}, \mathbf{p}\cdot\mathbf{v}) - 1]\,, \tag{10.5.41}$$

and employing the even/odd properties of $K = K_1 + iK_2$, we have

$$\mathbf{f} = -\frac{Z^2e^2}{2\pi^2}\int d^{(3)}\mathbf{p}\mathbf{p}\,|\mathbf{p}|^{-2} K_2(\mathbf{p}, -\mathbf{p}\cdot\mathbf{v})\,, \tag{10.5.42}$$

so the power loss of the passing particle is

$$dW/dt = \mathbf{f}\cdot\mathbf{v} = -\frac{Z^2e^2}{2\pi^2}\int d^{(3)}\mathbf{p}\mathbf{p}\cdot\mathbf{v}K_2(\mathbf{p}, -\mathbf{p}\cdot\mathbf{v})\,/\,|\mathbf{p}|^2\,. \tag{10.5.43}$$

Considering the high-velocity limit,

$$K(\mathbf{p}, \omega) = \frac{\omega^2}{(\omega + i0^+)^2 - \omega_p^2}\;;\; K_2 = -\pi\omega_p^2\delta\left(\omega^2 - \omega_p^2\right)\,\text{sign}\,(\omega), \tag{10.5.44}$$

Eq. (10.5.43) yields the result

$$dW/dt = -Z^2e^2\omega_p^2\int_0^{p_F} dp\int_0^1 d\mu\;\mu\;\delta\left(\mu - \frac{\omega_p}{pv}\right) = -\frac{Z^2e^2\omega_p^2}{v}\int_0^{p_F}\frac{dp}{p}\eta_+\left(1 - \frac{\omega_p}{pv}\right). \tag{10.5.45}$$

In this, we have cut off the p-integral at p_F since the plasma–pole approximation of Eq. (10.5.44) is not valid for higher p-values. The p-integration yields the power loss as

$$dW/dt = \frac{-Z^2e^2\omega_p^2}{v}\,ln\left(\frac{p_F v}{\omega_p}\right), \tag{10.5.46}$$

which corresponds to the *Bethe stopping power* formula when expressed in terms of energy loss per unit length.

Although we have focused on local plasma–pole approximations, it is clear that the formulation of dW/dt in terms of $K(1,2)$ can accommodate spatial non-locality as well as spatial inhomogeneity and dynamic frequency dependence.

10.6 Electrodynamic Dielectric Response Functions for Quantum Plasmas: Dyadic Green's Functions

Thus far, our considerations of dielectric response have been limited to the case of non-retarded, instantaneous longitudinal potentials, ignoring the finite velocity of light on dynamic electromagnetic plasma phenomena in semiconductors and nanostructures. To address the electrodynamic case we will introduce an electromagnetic (EM) *dyadic* (tensor) photon Green's function-propagator and also present an alternative description in terms of a dynamic, non-local, inverse *dielectric tensor* (which may also be spatially inhomogeneous) related to the EM dyadic Green's function. Starting with the Maxwell equations (here, $\mathbf{J} \equiv \langle \mathbf{J}_{op} \rangle$ is the c-number current obtained by averaging the current operator, $\mathbf{J}_{op}$ (which was written as $\mathbf{J}$ in Chapter 3) and ρ is the charge density):

$$\begin{aligned} \nabla \cdot \mathbf{E} &= 4\pi\rho, \quad & \nabla \times \mathbf{E} + \frac{1}{c}\frac{\partial \mathbf{B}}{\partial t} &= 0, \\ \nabla \cdot \mathbf{B} &= 0, \quad & \nabla \times \mathbf{B} &= \frac{4\pi}{c}\mathbf{J} + \frac{1}{c}\frac{\partial \mathbf{E}}{\partial t}, \end{aligned} \tag{10.6.1}$$

and forming the curl $(\nabla\times)$ of Faraday's law for $\nabla \times \mathbf{E}$ and substituting for the ensuing appearance of $\nabla \times \mathbf{B}$ using Ampere's law (note the identity $\nabla \times (\nabla \times \mathbf{E}) = \nabla(\nabla \cdot \mathbf{E}) - \nabla^2\mathbf{E})$) leads to the vector Helmholtz wave equation for $\mathbf{E}$ as

$$\left[\bar{\mathbf{I}}\left(\nabla^2 - \frac{1}{c^2}\frac{\partial^2}{\partial t^2}\right) - \nabla\nabla\right] \cdot \mathbf{E} = \frac{4\pi}{c^2}\frac{\partial \mathbf{J}}{\partial t}. \tag{10.6.2}$$

Here $\bar{\mathbf{I}} = \widehat{x}\widehat{x} + \widehat{y}\widehat{y} + \widehat{z}\widehat{z}$ is the unit dyad (3 x 3 unit matrix; $\widehat{x}, \widehat{y}, \widehat{z}$ are the Cartesian coordinate unit vectors). Eq. (10.6.2) remains unchanged in the presence of constant, uniform electric ($\mathbf{E}_0$) and magnetic ($\mathbf{B}_0$) fields since all space and time derivatives of such fields vanish and their sources ($\rho_0, \mathbf{J}_0$) are understood to be infinitely distant, playing no direct role in the dynamics of the electromagnetic wave fields (although they do affect particle dynamics through the Schrödinger or classical dynamical equations involved in the polarization of induced media charges and currents). In frequency representation, $t \to \omega$, we have

$$\left[\bar{\mathbf{I}}\left(\nabla^2 + \frac{\omega^2}{c^2}\right) - \nabla\nabla\right] \cdot \mathbf{E}(\mathbf{r},\omega) = -\frac{4\pi i\omega}{c^2}\mathbf{J}(\mathbf{r},\omega). \tag{10.6.3}$$

(Note that we denote the position variable by $\mathbf{r} = x\widehat{x} + y\widehat{y} + z\widehat{z}$ here (instead of $\mathbf{x}$). Generally, we suppress the ω-dependence of the electric field and the EM Green's functions, writing $\mathbf{E}(\mathbf{r},\omega) \to \mathbf{E}(\mathbf{r})$, etc.)

The EM dyadic Green's function, $\overline{\mathfrak{G}}_0^0$, in vacuum is defined by the equation ($1 = \mathbf{r}_1, t_1$):

$$\left[\bar{\mathbf{I}}\left(\nabla^2 - \frac{1}{c^2}\frac{\partial^2}{\partial t^2}\right) - \boldsymbol{\nabla\nabla}\right] \cdot \overline{\mathfrak{G}}_0^0(1,1') = -\bar{\mathbf{I}}\delta(1-1'), \tag{10.6.4}$$

and in frequency representation ($t - t' \to \omega$), we have

$$\left[\bar{\mathbf{I}}\left(\nabla^2 + \frac{\omega^2}{c^2}\right) - \boldsymbol{\nabla\nabla}\right] \cdot \overline{\mathfrak{G}}_0^0(\mathbf{r},\mathbf{r}';\ \omega) = -\bar{\mathbf{I}}\delta^{(3)}(\mathbf{r}-\mathbf{r}'). \tag{10.6.5}$$

(Note that our dyadic Green's functions are defined here with a *negative* sign on the right-hand side, so they differ by that sign from those of some other works, including our own.)[29–31] Eq. (10.6.5) may be solved by applying the divergence ($\boldsymbol{\nabla}\cdot$), and noting that $\boldsymbol{\nabla}\cdot\bar{\mathbf{I}} = \boldsymbol{\nabla}$ and $\boldsymbol{\nabla}\cdot\boldsymbol{\nabla\nabla} = \nabla^2\boldsymbol{\nabla}$, we find that

$$\boldsymbol{\nabla}\cdot\overline{\mathfrak{G}}_0^0(\mathbf{r},\mathbf{r}';\omega) = -\frac{c^2}{\omega^2}\boldsymbol{\nabla}\delta^{(3)}(\mathbf{r}-\mathbf{r}'). \tag{10.6.6}$$

Substitution of this into the $\boldsymbol{\nabla\nabla}\cdot\overline{\mathfrak{G}}_0^0$ term of Eq. (10.6.5) yields

$$\bar{\mathbf{I}}\left(\nabla^2 + \frac{\omega^2}{c^2}\right)\overline{\mathfrak{G}}_0^0(\mathbf{r},\mathbf{r}';\omega) = -\left(\bar{\mathbf{I}} + \frac{c^2}{\omega^2}\boldsymbol{\nabla\nabla}\right)\delta^{(3)}(\mathbf{r}-\mathbf{r}'). \tag{10.6.7}$$

Introducing the *scalar* Helmholtz Green's function $\mathfrak{g}_0^0$ defined by

$$\left(\nabla^2 + \frac{\omega^2}{c^2}\right)\mathfrak{g}_0^0(\mathbf{r},\mathbf{r}';\omega) = -\delta^{(3)}(\mathbf{r}-\mathbf{r}'), \tag{10.6.8}$$

Eq. (10.6.7) may be rewritten in the form

$$\left(\nabla^2 + \frac{\omega^2}{c^2}\right)\left[\overline{\mathfrak{G}}_0^0(\mathbf{r},\mathbf{r}';\omega) - \left(\bar{\mathbf{I}} + \frac{c^2}{\omega^2}\boldsymbol{\nabla\nabla}\right)\mathfrak{g}_0^0(\mathbf{r},\mathbf{r}';\omega)\right] = 0, \tag{10.6.9}$$

which has the Levine–Schwinger solution,[30, 31]

$$\overline{\mathfrak{G}}_0^0(\mathbf{r},\mathbf{r}';\omega) = \left(\bar{\mathbf{I}} + \frac{c^2}{\omega^2}\boldsymbol{\nabla\nabla}\right)\mathfrak{g}_0^0(\mathbf{r},\mathbf{r}';\omega), \tag{10.6.10}$$

where, for infinite space, we have (with outgoing wave boundary conditions) $\mathfrak{g}_0^0(\mathbf{r},\mathbf{r}';\omega) = \exp\left[i\frac{\omega}{c}\left|\mathbf{r}-\mathbf{r}'\right|\right]/\left(4\pi\left|\mathbf{r}-\mathbf{r}'\right|\right)$ as the free-space 3D scalar Green's function of the Helmholtz equation.

For a linear medium, the current density $\mathbf{J}(1)$ is composed of an induced current $\mathbf{J}_{ind}(1)$ (linearly related to $\mathbf{E}$ by a generally dynamic, non-local and inhomogeneous conductivity tensor $\overline{\boldsymbol{\sigma}}$, $\mathbf{J}_{ind}(1) = \int d2\, \overline{\boldsymbol{\sigma}}(1,2) \cdot \mathbf{E}(2)$), as well as an external current density $\mathbf{J}_{ext}(1)$,

$$\mathbf{J}(1) = \int d2\, \overline{\boldsymbol{\sigma}}(1,2) \cdot \mathbf{E}(2) + \mathbf{J}_{ext}(1). \tag{10.6.11}$$

Moreover, the conductivity tensor may be written as

$$\overline{\boldsymbol{\sigma}}(1,2) = \frac{\delta \mathbf{J}(1)}{\delta \mathbf{E}(2)}. \tag{10.6.12}$$

Considering a system composed of several distinct charged components, for example a bulk (3D) component (such as an ionic diatomic lattice and/or a 3D quantum plasma) and a nanostructure (nano) component (such as a 2D or 1D quantum plasma and/or low-dimensional phonons), we will examine the case in which the charge and current distributions of the various components are approximately mutually independent of each other. This approximation neglects change and distortion of the charge/current distributions of the various components due to each other's presence resulting from mutual interaction and correlations between the components, and neglects charge transfer between the components (as occurs in quantum tunneling). Moreover, frequency ranges corresponding to quantum transitions from the energy states of one component to those of the other component are also excluded from consideration in this approximation. Within this framework, the independent charge and current densities are additive,

$$\begin{aligned} \rho &= \rho_{3D} + \rho_{nano}, \\ \mathbf{J} &= \mathbf{J}_{3D} + \mathbf{J}_{nano}. \end{aligned} \tag{10.6.13}$$

While we undertake this discussion in the context of a composite 3D and nano-system, the considerations also apply to systems composed of two or more 3D components, or two or more nano-components provided only that the conditions of validity described above are well satisfied. Pursuant to Eq. (10.6.13) we have the longitudinal polarizability as

$$\alpha = \mathcal{V}R = -\mathcal{V}\frac{\delta \rho}{\delta V} = \alpha_{3D} + \alpha_{nano}, \tag{10.6.14}$$

and the conductivity tensor is given by

$$\overline{\boldsymbol{\sigma}} = \frac{\delta \mathbf{J}}{\delta \mathbf{E}} = \overline{\boldsymbol{\sigma}}_{3D} + \overline{\boldsymbol{\sigma}}_{nano}, \tag{10.6.15}$$

where the independent "3D" and "nano" polarizabilities and conductivities are each determined in the absence of the other member of the composite system. (Although we focus here on the electrodynamic response, similar considerations obviously apply to longitudinal electrostatic response.) Combining Eqns. (10.6.2, 11 and 15), we have

$$\left[\bar{\mathbf{I}}\left(\nabla^2 - \frac{1}{c^2}\frac{\partial^2}{\partial t^2}\right) - \nabla\nabla\right]\cdot \mathbf{E}(1) - \frac{4\pi}{c^2}\frac{\partial}{\partial t}\int d1'\left[\bar{\boldsymbol{\sigma}}_{3D}(1,1') + \bar{\boldsymbol{\sigma}}_{nano}(1,1')\right]\cdot \mathbf{E}(1') = \frac{4\pi}{c^2}\frac{\partial}{\partial t}\mathbf{J}_{ext}(1). \tag{10.6.16}$$

The associated *dyadic Green's function* $\bar{\mathfrak{G}}$ for the composite system is defined to satisfy the same integro-differential equation, but with $-\bar{I}\delta(1-1') = -\bar{I}\delta^{(3)}(\mathbf{r}-\mathbf{r}')\delta(t-t')$ replacing $(4\pi/c^2)\,\partial\mathbf{J}_{ext}(1)/\partial t$ on the right-hand side:

$$\left[\bar{\mathbf{I}}\left(\nabla_1^2 - \frac{1}{c^2}\frac{\partial^2}{\partial t_1^2}\right) - \nabla_1\nabla_1\right]\cdot \bar{\mathfrak{G}}(1,1') - \frac{4\pi}{c^2}\frac{\partial}{\partial t_1}\int d1''\left[\bar{\boldsymbol{\sigma}}_{3D}(1,1'') + \bar{\boldsymbol{\sigma}}_{nano}(1,1'')\right]\cdot \bar{\mathfrak{G}}(1'',1') = -\bar{\mathbf{I}}\delta(1-1'). \tag{10.6.17}$$

In the case of time-translational invariance for $\bar{\boldsymbol{\sigma}}_{3D}$ and $\bar{\boldsymbol{\sigma}}_{nano}$, we employ the Fourier time transform $(t-t' \to \omega)$, with the resulting integro-differential equation:

$$\left[\bar{\mathbf{I}}\left(\nabla_{\mathbf{r}}^2 + \frac{\omega^2}{c^2}\right) - \nabla_{\mathbf{r}}\nabla_{\mathbf{r}}\right]\cdot \bar{\mathfrak{G}}\left(\mathbf{r},\mathbf{r}';\omega\right) + \frac{4\pi i\omega}{c^2}\int d^{(3)}\mathbf{r}''\left[\bar{\boldsymbol{\sigma}}_{3D}\left(\mathbf{r},\mathbf{r}'';\omega\right) + \bar{\boldsymbol{\sigma}}_{nano}\left(\mathbf{r},\mathbf{r}'';\omega\right)\right]\cdot \bar{\mathfrak{G}}\left(\mathbf{r}'',\mathbf{r}';\omega\right) = -\bar{\mathbf{I}}\delta^{(3)}\left(\mathbf{r}-\mathbf{r}'\right). \tag{10.6.18}$$

Considering linear systems, $\bar{\boldsymbol{\sigma}}_{3D}$, $\bar{\boldsymbol{\sigma}}_{nano}$ do not depend on $\mathbf{E}$, and in accordance with principle of superposition, $\mathbf{E}(1)$ of Eq. (10.6.16) may be written in terms of $\bar{\mathfrak{G}}$ as

$$\mathbf{E}(1) = -\frac{4\pi}{c^2}\int d2\,\bar{\mathfrak{G}}(1,2)\cdot \frac{\partial}{\partial t_2}\mathbf{J}_{ext}(2) + \mathbf{E}_0(1). \tag{10.6.19}$$

Here, $\mathbf{E}_0(1)$ is an external field that satisfies the homogeneous counterpart of Eq. (10.6.16), namely $\left(\bar{\mathfrak{G}}\right)^{-1}\cdot \mathbf{E}_0 = 0$ (corresponding to sources of $\mathbf{E}_0$ that are infinitely distant). The solution for $\mathbf{E}$ in $\mathbf{r}$-ω representation has the form

$$\mathbf{E}(\mathbf{r}_1,\omega) = \frac{4\pi i\omega}{c^2}\int d^{(3)}\mathbf{r}_2\bar{\mathfrak{G}}(\mathbf{r}_1,\mathbf{r}_2;\omega)\cdot \mathbf{J}_{ext}(\mathbf{r}_2,\omega) + \mathbf{E}_0(\mathbf{r}_1,\omega). \tag{10.6.20}$$

Since $\bar{\mathfrak{G}}$ describes the responding electric field due to a unit-impulsive source element (Eqns. (10.6.17 and 10.6.18)), it may be considered as the *photon propagator* of the composite system composed of a 3D medium jointly with a nanostructure element embedded in it.

To develop further insight into the EM dyadic Green's function $\overline{\mathfrak{G}}$ and rewrite its integro-differential equation (Eq. (10.6.18)) in a more convenient integral equation form, we represent it symbolically (in position-frequency matrix notation) as follows: Starting with Eq. (10.6.5), we write

$$\left(\overline{\mathfrak{G}}_0^0\right)^{-1} = -\left[\bar{\mathbf{I}}\left(\nabla^2 + \frac{\omega^2}{c^2}\right) - \nabla\nabla\right] \tag{10.6.21}$$

as the inverse of the 3D EM dyadic Green's function in vacuum. Defining $\overline{\mathfrak{G}}_{3D}^{-1}$ by

$$\overline{\mathfrak{G}}_{3D}^{-1} = \left(\overline{\mathfrak{G}}_0^0\right)^{-1} - \frac{4\pi i\omega}{c^2}\overline{\boldsymbol{\sigma}}_{3D}, \tag{10.6.22}$$

as the inverse of the EM dyadic Green's function for a 3D plasma-like medium, the role of $\overline{\mathfrak{G}}_{3D}$ as a bulk photon propagator may be seen by multiplying Eq. (10.6.22) by $\overline{\mathfrak{G}}_{3D}$ from the right, leading to

$$\left[\left(\overline{\mathfrak{G}}_0^0\right)^{-1} - \frac{4\pi i\omega}{c^2}\overline{\boldsymbol{\sigma}}_{3D}\right] \cdot \overline{\mathfrak{G}}_{3D} = \bar{\mathbf{I}}, \tag{10.6.23}$$

which, in position-time representation, is the integro-differential equation

$$\left[\bar{\mathbf{I}}\left(\nabla_1^2 - \frac{1}{c^2}\frac{\partial^2}{\partial t_1^2}\right) - \nabla_1\nabla_1\right]\cdot\overline{\mathfrak{G}}_{3D}(1,1') - \frac{4\pi}{c^2}\frac{\partial}{\partial t_1}\int d\,1''\overline{\boldsymbol{\sigma}}_{3D}(1,1'')\cdot\overline{\mathfrak{G}}_{3D}(1'',1') = -\bar{\mathbf{I}}\delta(1-1'). \tag{10.6.24}$$

With Eq. (10.6.21) for $\left(\overline{\mathfrak{G}}_0^0\right)^{-1}$, Eq. (10.6.18) for $\overline{\mathfrak{G}}$ takes the form

$$\left(\overline{\mathfrak{G}}_0^0\right)^{-1} \cdot \overline{\mathfrak{G}} = \bar{\mathbf{I}} + \frac{4\pi i\omega}{c^2}\left[\overline{\boldsymbol{\sigma}}_{3D} + \overline{\boldsymbol{\sigma}}_{nano}\right] \cdot \overline{\mathfrak{G}}, \tag{10.6.25}$$

and multiplication by $\overline{\mathfrak{G}}_0^0$ from the left leads to the dyadic Green's function integral equation as (position-frequency matrix notation):

$$\overline{\mathfrak{G}} = \overline{\mathfrak{G}}_0^0 + \frac{4\pi i\omega}{c^2}\overline{\mathfrak{G}}_0^0 \cdot \left[\overline{\boldsymbol{\sigma}}_{3D} + \overline{\boldsymbol{\sigma}}_{nano}\right] \cdot \overline{\mathfrak{G}}. \tag{10.6.26}$$

It is also possible to express $\overline{\mathfrak{G}}$ in terms of $\overline{\mathfrak{G}}_{3D}$ as follows: Considering Eq. (10.6.18) jointly with Eq. (10.6.22) yields

$$\overline{\mathfrak{G}}_{3D}^{-1} \cdot \overline{\mathfrak{G}} = \bar{\mathbf{I}} + \frac{4\pi i\omega}{c^2}\overline{\boldsymbol{\sigma}}_{nano} \cdot \overline{\mathfrak{G}}, \tag{10.6.27}$$

and multiplication by $\overline{\mathfrak{G}}_{3D}$ from the left yields

$$\overline{\mathfrak{G}} = \overline{\mathfrak{G}}_{3D} + \frac{4\pi i\omega}{c^2}\overline{\mathfrak{G}}_{3D} \cdot \overline{\boldsymbol{\sigma}}_{nano} \cdot \overline{\mathfrak{G}}. \tag{10.6.28}$$

Here, $(4\pi i\omega/c^2)\overline{\mathfrak{G}}_{3D} \cdot \overline{\sigma}_{nano}$ is the kernel of this integral equation for $\overline{\mathfrak{G}}$. Written in position-frequency representation, this equation reads as

$$\overline{\mathfrak{G}}(\mathbf{r}_1, \mathbf{r}_2; \omega) = \overline{\mathfrak{G}}_{3D}(\mathbf{r}_1, \mathbf{r}_2; \omega) + \frac{4\pi i\omega}{c^2} \int d^{(3)}\mathbf{r}_1' \int d^{(3)}\mathbf{r}_2' \, \overline{\mathfrak{G}}_{3D}(\mathbf{r}_1, \mathbf{r}_1'; \omega) \cdot \overline{\sigma}_{nano}(\mathbf{r}_1', \mathbf{r}_2'; \omega) \cdot \overline{\mathfrak{G}}(\mathbf{r}_2', \mathbf{r}_2; \omega), \tag{10.6.29}$$

constituting the final form of the integral equation for the EM dyadic Green's function for the composite system.

To explore the EM response to an incident field $\mathbf{E}_0$, it is useful to introduce a dynamic, non-local dyadic function $\overline{\mathbf{K}}(1, 1')$ through the relation:[29]

$$\overline{\mathfrak{G}}(1, 1') = \int d2 \, \overline{\mathbf{K}}(1, 2) \cdot \overline{\mathfrak{G}}_0^0(2, 1'), \quad (\overline{\mathfrak{G}} = \overline{\mathbf{K}} \cdot \overline{\mathfrak{G}}_0^0). \tag{10.6.30}$$

Substituting Eq. (10.6.30) into Eq. (10.6.26) and multiplying by $\left(\overline{\mathfrak{G}}_0^0\right)^{-1}$ from the right yields an integral equation for $\overline{\mathbf{K}}$ (in position-frequency matrix notation) as

$$\overline{\mathbf{K}} = \overline{\mathbf{I}} + \frac{4\pi i\omega}{c^2} \overline{\mathfrak{G}}_0^0 \cdot [\overline{\sigma}_{3D} + \overline{\sigma}_{nano}] \cdot \overline{\mathbf{K}}. \tag{10.6.31}$$

To better understand the nature of $\overline{\mathbf{K}}$, we consider Eq. (10.6.16) in the form (again, in position-frequency matrix notation):

$$\left(\overline{\mathfrak{G}}_0^0\right)^{-1} \cdot \mathbf{E} = \frac{4\pi i\omega}{c^2} \left([\overline{\sigma}_{3D} + \overline{\sigma}_{nano}] \cdot \mathbf{E} + \mathbf{J}_{ext}\right), \tag{10.6.32}$$

and multiplication by $\overline{\mathfrak{G}}_0^0$ from the left yields

$$\mathbf{E} = \frac{4\pi i\omega}{c^2} \overline{\mathfrak{G}}_0^0 \cdot \left([\overline{\sigma}_{3D} + \overline{\sigma}_{nano}] \cdot \mathbf{E} + \mathbf{J}_{ext}\right) + \mathbf{E}_0. \tag{10.6.33}$$

In explicit position-time representation, this integral equation is given by

$$\mathbf{E}(1) = -\frac{4\pi}{c^2} \int d1' \int d2' \, \overline{\mathfrak{G}}_0^0(1, 1') \cdot \frac{\partial}{\partial t_1'} \left[\overline{\sigma}_{3D}(1', 2') + \overline{\sigma}_{nano}(1', 2')\right] \cdot \mathbf{E}(2') - \frac{4\pi}{c^2} \int d1' \, \overline{\mathfrak{G}}_0^0(1, 1') \cdot \frac{\partial}{\partial t_1'} \mathbf{J}_{ext}(1') + \mathbf{E}_0(1), \tag{10.6.34}$$

where $\mathbf{E}_0$ is an external, incident field, satisfying the homogeneous equation, $\left(\overline{\mathfrak{G}}\right)^{-1} \cdot \mathbf{E}_0 = 0$. Taking the variational derivative of the total effective electric field $\mathbf{E}(1)$ of the composite system, with respect to the external, incident field $\mathbf{E}_0(2)$,

$$\frac{\delta \mathbf{E}(1)}{\delta \mathbf{E}_0(2)} \Longleftrightarrow \frac{\delta \mathbf{E}_i(1)}{\delta \mathbf{E}_{0j}(2)}, \quad i, j = 1, 2, 3,$$

and bearing in mind that $[\delta\mathbf{E}_0(1)/\delta\mathbf{E}_0(2)] = \bar{\mathbf{I}}\delta(1-2)$ and $\int d1'\,\overline{\mathfrak{G}}_0^0(1,1')\cdot(\partial/\partial t_1')\,\mathbf{J}_{ext}(1')$ has no dependence on $\mathbf{E}_0(2)$ (yielding a vanishing differential contribution), the variational differentiation of Eq. (10.6.34) yields

$$\frac{\delta\mathbf{E}(1)}{\delta\mathbf{E}_0(2)} = \bar{\mathbf{I}}\delta(1-2) - \frac{4\pi}{c^2}\int d1'\int d2'\,\overline{\mathfrak{G}}_0^0(1,1')\frac{\partial}{\partial t_1'}\left[\bar{\boldsymbol{\sigma}}_{3D}(1',2') + \bar{\boldsymbol{\sigma}}_{nano}(1',2')\right]\frac{\delta\mathbf{E}(2')}{\delta\mathbf{E}_0(2)}. \tag{10.6.35}$$

In position-frequency matrix notation, this integral equation,

$$\frac{\delta\mathbf{E}}{\delta\mathbf{E}_0} = \bar{\mathbf{I}} + \frac{4\pi i\omega}{c^2}\overline{\mathfrak{G}}_0^0\cdot\left[\bar{\boldsymbol{\sigma}}_{3D} + \bar{\boldsymbol{\sigma}}_{nano}\right]\cdot\frac{\delta\mathbf{E}}{\delta\mathbf{E}_0}, \tag{10.6.36}$$

is identical to that of Eq. (10.6.31) with the recognition that

$$\overline{\mathbf{K}} \equiv \frac{\delta\mathbf{E}}{\delta\mathbf{E}_0} \Longleftrightarrow \mathbf{K}_{ij} \equiv \frac{\delta\mathbf{E}_i}{\delta\mathbf{E}_{0j}};\quad i,j = 1,2,3. \tag{10.6.37}$$

In integral form, Eq. (10.6.37) reads

$$\mathbf{E}(1) = \int d2\,\overline{\mathbf{K}}(1,2)\cdot\mathbf{E}_0(2). \tag{10.6.38}$$

Physically, $\overline{\mathbf{K}}(1,2)$ describes the evolution of an arbitrary applied field $\mathbf{E}_0(2)$ into the effective field $\mathbf{E}(1)$, and it clearly has the meaning of a dynamic, non-local (*inverse*) dielectric tensor. It may also be expressed in terms of the actual photon propagator using Eq. (10.6.30),

$$\overline{\mathbf{K}}(1,1') = \int d2\,\overline{\mathfrak{G}}(1,2)\cdot\left[\overline{\mathfrak{G}}_0^0(2,1')\right]^{-1}. \tag{10.6.39}$$

As $\overline{\mathbf{K}}$ traces the dynamical development of an impressed field $\mathbf{E}_0(2)$ into the effective field $\mathbf{E}(1)$ through Eq. (10.6.38) or $\overline{\mathbf{K}}(1,2) = \delta\mathbf{E}(1)/\delta\mathbf{E}_0(2)$, it is quite appropriate to describe it as an *inverse* dielectric tensor, to be consistent with the terminology of the inverse dielectric (screening) function $K(1,2) = \delta V(1)/\delta U_0(2)$ in the longitudinal case. In both cases, $\overline{\mathbf{K}}(1,2)$ and $K(1,2)$ describe the progressive "dressing" of their respective impressed fields as they propagate into the actual, effective fields.

The determination of either $\overline{\mathfrak{G}}$ or $\overline{\mathbf{K}}$ from its associated integral equation (i.e., Eq. (10.6.26 and 10.6.29) for the photon Green's function or Eq. (10.6.31) for the *inverse* dielectric tensor, respectively) can be used to determine the other by employing Eqns. (10.6.39) or (10.6.30). Since these dyadic-tensor propagators describe the dynamic, non-local screening phenomena of the retarded electromagnetic field, their frequency poles identify the coupled-mode dispersion relations for transverse and longitudinal plasmon-polariton collective electromagnetic normal modes of the composite system. The residues at these poles describe the corresponding relative excitation amplitudes (oscillator strengths) of the retarded collective normal modes, as in the case of longitudinal electrostatic plasmon modes.

In summary, the EM field $\mathbf{E}$ generated by an incident wave field $\mathbf{E}_0$ is given by

$$\mathbf{E} = \overline{\mathbf{K}}\mathbf{E}_0 = \overline{\mathfrak{G}}\left[\overline{\mathfrak{G}}_0^0\right]^{-1}\mathbf{E}_0 = \overline{\mathfrak{G}}\left(\overline{\mathfrak{G}}^{-1} + \frac{4\pi i\omega}{c^2}\left[\overline{\boldsymbol{\sigma}}_{3D} + \overline{\boldsymbol{\sigma}}_{nano}\right]\right)\mathbf{E}_0$$

$$= \mathbf{E}_0 + \frac{4\pi i\omega}{c^2}\overline{\mathfrak{G}}\left[\overline{\boldsymbol{\sigma}}_{3D} + \overline{\boldsymbol{\sigma}}_{nano}\right]\mathbf{E}_0,$$

where we have used Eq. (10.6.26). If there is an external driving current $\mathbf{J}_{ext}$ in addition to $\mathbf{E}_0$, one must add a term proportional to $\frac{4\pi i\omega}{c^2}\overline{\mathfrak{G}}_0^0\mathbf{J}_{ext}$ on the right-hand side of this equation.

Considering the EM dyadic Green's function propagator above, we examine Eq. (10.6.23) in the case of a homogeneous and translationally invariant, but non-local and dynamic, infinite 3D medium. Employing a 3D wave vector representation $\mathbf{r}-\mathbf{r}' = \mathbf{R} \rightarrow \mathbf{q}$,

$$\overline{\mathfrak{G}}_{3D}(\mathbf{q};\omega) = \int d\mathbf{R}\,\overline{\mathfrak{G}}_{3D}(\mathbf{R};\omega)e^{-i\mathbf{q}\cdot\mathbf{R}}, \tag{10.6.40}$$

Eq. (10.6.24) takes the form

$$\left[\overline{\mathbf{I}}\left(-q^2 + \frac{\omega^2}{c^2}\right) + \mathbf{q}\mathbf{q} + \frac{4\pi i\omega}{c^2}\overline{\boldsymbol{\sigma}}_{3D}(\mathbf{q},\omega)\right]\cdot\overline{\mathfrak{G}}_{3D}(\mathbf{q},\omega) = -\overline{\mathbf{I}}. \tag{10.6.41}$$

In three dimensions the bulk conductivity tensor $\overline{\boldsymbol{\sigma}}_{3D}(\mathbf{q},\omega)$ is related to the non-local wave vector- and frequency-dependent dielectric tensor $\overline{\boldsymbol{\varepsilon}}_{3D}(\mathbf{q},\omega)$ of the 3D plasma-like medium as follows (this relation is not applicable as written to confined lower dimensional systems):[32]

$$\overline{\boldsymbol{\varepsilon}}_{3D}(\mathbf{q},\omega) = \overline{\mathbf{I}} + \frac{4\pi i}{\omega}\overline{\boldsymbol{\sigma}}_{3D}(\mathbf{q},\omega). \tag{10.6.42}$$

In the absence of a magnetic field or other anisotropy,

$$\overline{\boldsymbol{\varepsilon}}_{3D}(\mathbf{q},\omega) = \left(\overline{\mathbf{I}} - \frac{\mathbf{q}\mathbf{q}}{q^2}\right)\varepsilon_T(\mathbf{q},\omega) + \frac{\mathbf{q}\mathbf{q}}{q^2}\varepsilon_L(\mathbf{q},\omega), \tag{10.6.43}$$

where $\varepsilon_T(\mathbf{q},\omega)$ and $\varepsilon_L(\mathbf{q},\omega)$ are the transverse and longitudinal dielectric functions, respectively. Eq. (10.6.41) may be solved for $\overline{\mathfrak{G}}_{3D}$ algebraically as

$$\overline{\mathfrak{G}}_{3D}(\mathbf{q},\omega) = \frac{\left(\overline{\mathbf{I}} - \mathbf{q}\mathbf{q}/q^2\right)}{\frac{\omega^2}{c^2}\varepsilon_T(\mathbf{q},\omega) - q^2} + \frac{\mathbf{q}\mathbf{q}/q^2}{\frac{\omega^2}{c^2}\varepsilon_L(\mathbf{q},\omega)}. \tag{10.6.44}$$

The denominators on the right of this equation show that longitudinal collective modes (such as plasmons and/or LO phonons, etc.) are described by $\varepsilon_L(\mathbf{q},\omega) = 0$, corresponding to frequency poles of the second term of $\overline{\mathfrak{G}}_{3D}(\mathbf{q},\omega)$ in Eq. (10.6.44). The transverse collective modes, given by the frequency poles of the first term of $\overline{\mathfrak{G}}_{3D}(\mathbf{q},\omega)$,

$$\omega^2 \varepsilon_T(\mathbf{q},\omega) = c^2 q^2, \tag{10.6.45}$$

are *polaritons*. Physically, these modes are photons ($\omega = cq$) coupled—*hybridized*—with the polarization modes of the medium. (In this context, we employ the term "polariton" in an unusually general sense, to include medium polarization modes that may not have any wave number dependence, and we also include longitudinal as well as transverse EM-matter modes.)

In vacuum, $\varepsilon_T = \varepsilon_L = 1$, $\left[\overline{\boldsymbol{\varepsilon}}_{3D}(\mathbf{q},\omega) = \overline{\mathbf{I}}\right]$, and the 3D photon Green's function takes the form

$$\overline{\mathfrak{G}}^0_0(\mathbf{q},\omega) = \left(\overline{\mathbf{I}} - \frac{\mathbf{q}\mathbf{q}}{\left(\omega^2/c^2\right)}\right)\left(\omega^2/c^2 - q^2\right)^{-1}. \tag{10.6.46}$$

Recognizing $\left(\omega^2/c^2 - q^2\right)^{-1}$ as the (negative of the) 3D Green's function for the *scalar* Helmholtz equation in vacuum, in wave number and frequency representation, $\mathfrak{g}^0_0(\mathbf{q},\omega)$, we may rewrite Eq. (10.6.46) as

$$\overline{\mathfrak{G}}^0_0(\mathbf{q},\omega) = -\left(\overline{\mathbf{I}} - \frac{\mathbf{q}\mathbf{q}}{\left(\omega^2/c^2\right)}\right)\mathfrak{g}^0_0(\mathbf{q},\omega). \tag{10.6.47}$$

This is just the Levine–Schwinger relation, Eq. (10.6.10) in wave number-frequency representation. Of course, the pole at $\omega = cq$ represents a photon in vacuum.

To exhibit an example of coupling of the photon with a polarization mode, we describe the medium by the local plasma dielectric function

$$\varepsilon_T(\mathbf{q}\to 0,\omega) = \varepsilon_L(\mathbf{q}\to 0,\omega) = 1 - \frac{\omega_p^2}{\omega^2} \to \varepsilon_\infty - \frac{\omega_p^2}{\omega^2}, \tag{10.6.48}$$

where we have made allowance for a time-independent high-frequency background dielectric constant (ε_∞) in solids. Then the transverse collective mode dispersion relation (Eq. (10.6.45)) takes the form

$$\omega^2 = \frac{\omega_p^2}{\varepsilon_\infty} + \frac{c^2 q^2}{\varepsilon_\infty}, \tag{10.6.49}$$

which represents the coupling of the photon in medium (having frequency $cq/\sqrt{\varepsilon_\infty}$) hybridized with the plasmon in medium (having frequency $\omega_p/\sqrt{\varepsilon_\infty}$), with moderation (shielding) of electric charge strength by the medium's background dielectric constant, ε_∞.

10.7 Dynamic, Non-Local Conductivity and Inverse Dielectric Tensors

The determination of the dynamic, non-local and possibly inhomogeneous conductivity tensor $\overline{\boldsymbol{\sigma}}(1,2)$ involves the non-equilibrium averages of the charge and current density operators, with the change density as

$$\rho(1) = e\langle\psi^+(1)\psi(1)\rangle, \tag{10.7.1}$$

and the current density is given by (see Eq. (9.1.10)ff. and Section 3.4; $\mathbf{A}$ is the vector potential; $\hbar \to 1$):

$$\begin{aligned} \mathbf{J}(1) &= \frac{e}{2im}\langle[\psi^+(1)\,(\nabla_1\psi(1)) - (\nabla_1\psi^+(1))\,\psi(1)]\rangle - \frac{e^2}{mc}\mathbf{A}(1)\,\langle\psi^+(1)\psi(1)\rangle \\ &= \frac{e}{2im}\lim_{1'\to 1}\left(\nabla_1 - \nabla_1' - \frac{i2e}{c}\mathbf{A}(1)\right)\langle\psi^+(1')\psi(1)\rangle. \end{aligned} \tag{10.7.2}$$

Expressed in terms of the non-equilibrium one-electron Green's function G_1, which includes the effects of the impressed electric field, we have

$$\rho(1) = -ie\lim_{1'\to 1} G_1(1,1'^+) \tag{10.7.3}$$

and

$$\mathbf{J}(1) = -\frac{e}{2m}\lim_{1'\to 1}\left(\nabla_1 - \nabla_1' - \frac{i2e}{c}\mathbf{A}(1)\right) G_1(1,1'^+). \tag{10.7.4}$$

The linear conductivity $\overline{\boldsymbol{\sigma}}$ may be identified in terms of the relation

$$\mathbf{J}(1) = \int d2\,\overline{\boldsymbol{\sigma}}(1,2)\,\mathbf{E}(2) \quad \text{or} \quad \overline{\boldsymbol{\sigma}}(1,2) = \delta\mathbf{J}(1)/\delta\mathbf{E}(2), \tag{10.7.5}$$

and employing the gauge in which $\mathbf{E} = -(1/c)\ \partial\mathbf{A}/\partial t \to (i\omega/c)\mathbf{A}$, we have

$$\overline{\boldsymbol{\sigma}}(\mathbf{r}_1,\mathbf{r}_2;\omega) = \left(\frac{c}{i\omega}\right)\frac{\delta\mathbf{J}(\mathbf{r}_1,\omega)}{\delta\mathbf{A}(\mathbf{r}_2,\omega)}. \tag{10.7.6}$$

Noting the explicit appearance of $\mathbf{A}$ in Eq. (10.7.4) and in the non-equilibrium equation for G_1, which has the field $(\mathbf{A},\phi)$-dependent replacement $\nabla \to \nabla - i\frac{e}{c}\mathbf{A}$ in $h^{(1)}$,

$$\left\{i\frac{\partial}{\partial t_1} + \zeta - \frac{1}{2m}\left(-i\nabla_1 - \frac{e}{c}\mathbf{A}(1)\right)^2 - e\phi(1)\right\} G_1(1,1') = \delta(1-1'), \tag{10.7.7}$$

it is convenient to work with the tensor $\overline{\mathbf{L}}_p$,

$$\overline{\mathbf{L}}_p(1,2) = \left.\frac{\delta \mathbf{J}(1)}{\delta \mathbf{A}(2)}\right]_{\mathbf{A}(2)\to 0}.$$

Since $\phi(1)$ does not generally involve $\mathbf{A}$, it does not directly contribute to the formal structure of $\overline{\mathbf{L}}_p$: However, the scattering interaction potentials embedded in $\phi(1)$ impact directly on $G_1(1,1'^+)$ and hence on $\delta G_1(1,1'^+)/\delta\mathbf{A}(2)$, playing a central role in limiting conductivity. However, this issue will not be pursued here as our focus is on collective modes.

For linear media, the conductivity and $\overline{\mathbf{L}}_p$ do not depend on $\mathbf{E}$, and after the variational differentiation is carried out they may be described in terms of equilibrium Green's functions. Thus,

$$\overline{\mathbf{L}}_p(1,2) = -\frac{e^2}{mc}\overline{\mathbf{I}}n_0(1)\,\delta(1-2) - \frac{e}{2m}\lim_{1'\to 1}\left(\nabla_1 - \nabla_1'\right)\left.\frac{\delta G_1(1,1'^+)}{\delta\mathbf{A}(2)}\right]_{\mathbf{A}(2)=0}, \qquad (10.7.8)$$

where $n_0(1)$ is the equilibrium density, $\overline{\mathbf{I}}$ is the unit 3x3 dyadic and $[\delta G_1/\delta\mathbf{A}]_{\mathbf{A}=0}$ can be evaluated exactly in terms of an equilibrium two-particle Green's function (Section 8.1), in particular, the current–current correlation function. However, in the absence of collision mechanisms, and for the purposes of electromagnetic plasma wave analysis, we will employ an approximation in the nature of the RPA. Introducing the non-interacting Green's function $G_1^0(1,1')$ in the absence of external fields and interactions,

$$\left(i\frac{\partial}{\partial t_1} + \zeta + \frac{\nabla_1^2}{2m}\right)G_1^0(1,1') = \delta(1-1'), \qquad (10.7.9)$$

we have

$$\left[G_1^0(1,1')\right]^{-1} = \left(i\frac{\partial}{\partial t_1} + \zeta + \frac{\nabla_1^2}{2m}\right)\delta(1-1') \qquad (10.7.10)$$

as the inverse of the non-interacting Green's function G_1^0. With this, Eq. (10.7.7) may be rewritten in position-time matrix notation as

$$\left\{\left[G_1^0\right]^{-1} - \frac{e^2}{2mc^2}A^2 + \frac{e}{2imc}(\nabla\cdot\mathbf{A} + \mathbf{A}\cdot\nabla) - e\phi\right\}G_1 = I. \qquad (10.7.11)$$

Multiplication by G_1^0 from the left yields G_1 as

$$G_1 = G_1^0 - G_1^0\left\{\frac{e}{2imc}(\nabla\cdot\mathbf{A} + \mathbf{A}\cdot\nabla) - \frac{e^2}{2mc^2}A^2 - e\phi\right\}G_1, \qquad (10.7.12)$$

or, explicitly, we have

$$G_1(1,1') = G_1^0(1,1') - \int d2' G_1^0(1,2')$$
$$\left\{\frac{e}{2imc}(\nabla_2' \cdot \mathbf{A}(2') + \mathbf{A}(2') \cdot \nabla_2') - \frac{e^2}{2mc^2}\left[\mathbf{A}(2')\right]^2 - e\phi(2')\right\} G_1(2',1'). \tag{10.7.13}$$

Taking the variational derivative of G_1 in Eq. (10.7.13), the linear approximation is given by

$$\left.\frac{\delta G_1(1,1'^{+})}{\delta \mathbf{A}(2)}\right]_{\mathbf{A}(2)=0;\phi=0} = -\frac{e}{2imc} \lim_{2'\to 2} \left(\nabla_2 - \nabla_2'\right) G_1^0(1,2')G_1^0(2,1'), \tag{10.7.14}$$

in which we have replaced G_1 by G_1^0 on the right as the linear approximation and integrated the first ∇_2' on the right of Eq. (10.7.13) by parts. Substituting this into Eq. (10.7.8), we thus obtain the response tensor of the system as

$$\overline{\mathbf{L}}_p(1,2) = -\frac{e^2}{mc}\left\{\overline{\mathbf{I}}n_0(1)\delta(1-2) - \frac{1}{4m}\lim_{\substack{1'\to 1\\ 2'\to 2}}(\nabla_1 - \nabla_1')(\nabla_2 - \nabla_2')\mathcal{R}(1,2;1',2')\right\}, \tag{10.7.15}$$

which is periodic in imaginary time with a Boson spectral structure: Here, we have defined

$$\mathcal{R}(1,2;1',2') = -iG_1^0(1,2')G_1^0(2,1'), \tag{10.7.16}$$

in analogy to the non-retarded electrostatic ring diagram density perturbation response function $R(1,2) = -iG_1^0(1,2)G_1^0(2,1)$, to which the specific relation is given by

$$\lim_{1'\to 1}\lim_{2'\to 2}\mathcal{R}(1,2;1',2') \equiv R(1,2).$$

It should be noted that the presence of an ambient, time-independent, magnetic field $\mathbf{B} = \boldsymbol{\nabla} \times \mathbf{A}_M$ contributes an additional magnetic vector potential term $\mathbf{A}_M = \mathbf{A}_{total} - \mathbf{A}$, which is involved in the Hamiltonian H and associated velocity/momentum operators as $\mathbf{v} = \frac{1}{m}\left(\frac{1}{i}\boldsymbol{\nabla} - \frac{e\mathbf{A}_M}{c}\right)$. In determining the linear current response to $\mathbf{A}$ (which may also be undertaken using $\mathbf{J}_{op}(1) = -\delta H/\delta(\mathbf{A}(1)/c)$ with subsequent evaluation of $\overline{\mathbf{L}}_p(1,2) = [\delta\mathbf{J}(1)/\delta\mathbf{A}(2)]_{\mathbf{A}(2)\to 0}$; recall $\mathbf{J} = \langle\mathbf{J}_{op}\rangle$), it is advantageous to retain the magnetic vector potential $\mathbf{A}_M$ within the kinetic part of H. The result, which incorporates the role of the ambient magnetic field in linear response, is given in terms of matrix elements as

$$[\overline{\mathbf{L}}_p(1,2)]_{ij} = -\frac{e^2}{mc}n_0(1)\delta_{ij}\delta(1-2)$$
$$-\frac{e^2}{c}\lim_{1'\to 1}\lim_{2'\to 2}\left(\frac{v_i^*(1') + v_i(1)}{2}\frac{v_j^*(2') + v_j(2)}{2}\right)R(1,2;1',2').$$

A thorough discussion of this and its implications for the magnetoconductivity tensor and the associated coupled photon-magnetoplasmon spectrum subject to Landau quantization (in both 3D and 2D) is available in the literature.[18–21,33]

Considering the periodic response tensor of Eq. (10.7.15), its imaginary time Matsubara Fourier series coefficients take the form

$$\overline{\mathbf{L}}_p(\mathbf{r}_1,\mathbf{r}_2;\pi\nu/\tau) = -\frac{e^2}{mc}\left\{\overline{\mathbf{I}}n_0(\mathbf{r}_1)\delta^{(3)}(\mathbf{r}_1-\mathbf{r}_2) - \frac{1}{4m}\lim_{\substack{1'\to 1\\ 2'\to 2}}\left(\nabla_1-\nabla_1'\right)\left(\nabla_2-\nabla_2'\right)\mathcal{R}\left(\mathbf{r}_1,\mathbf{r}_2;\mathbf{r}_1',\mathbf{r}_2';\pi\nu/\tau\right)\right\}, \quad (10.7.17)$$

where $\mathcal{R}\left(\mathbf{r}_1,\mathbf{r}_2;\mathbf{r}_1',\mathbf{r}_2';\frac{\pi\nu}{\tau}\right)$ may be written as ($T = t_1 - t_2$; this periodic response function has Bosonic spectral structure):

$$\mathcal{R}\left(\mathbf{r}_1,\mathbf{r}_2;\mathbf{r}_1',\mathbf{r}_2';\pi\nu/\tau\right) = -i\int_0^\tau dT e^{iT(2\pi\nu/\tau)}G_1^0(\mathbf{r}_1,\mathbf{r}_2';T)G_1^0(\mathbf{r}_2,\mathbf{r}_1';-T). \quad (10.7.18)$$

Note that $\lim_{1'\to 1}\lim_{2'\to 2}\mathcal{R}\left(\mathbf{r}_1,\mathbf{r}_2;\mathbf{r}_1',\mathbf{r}_2';\frac{\pi\nu}{\tau}\right) \equiv R(\mathbf{r}_1,\mathbf{r}_2;\frac{\pi\nu}{\tau})$ is the Fourier series coefficient of the non-interacting density perturbation response function. Changing the integration variable $T' \to \tau - T$ and employing the antiperiodicity properties of the one-electron Green's functions, $G_1^0(\tau - T) = -G_1^0(-T)$ and $G_1^0(T-\tau) = -G_1^0(T)$, $\mathcal{R}\left(\mathbf{r}_1,\mathbf{r}_2;\mathbf{r}_1',\mathbf{r}_2';\frac{\pi\nu}{\tau}\right)$ may be rewritten as

$$\mathcal{R}\left(\mathbf{r}_1,\mathbf{r}_2;\mathbf{r}_1',\mathbf{r}_2';\pi\nu/\tau\right) = -i\int_0^\tau dT e^{-iT(2\pi\nu/\tau)}G_{1<}^0(\mathbf{r}_1,\mathbf{r}_2';-T)G_{1>}^0(\mathbf{r}_2,\mathbf{r}_1';T), \quad (10.7.19)$$

with the "greater" and "lesser" Green's functions defined in Chapter 7:

$$G_{1\gtrless}^0(T) = G_1^0(T) \text{ for } T \gtrless 0. \quad (10.7.20)$$

Bearing in mind the discussions of Eq. (7.5.10) and Eq. (9.1.19), the causal response function $\overline{\mathbf{L}}_c(\mathbf{r}_1,\mathbf{r}_2;\omega)$ may be obtained from the periodic response function $\overline{\mathbf{L}}_p(\mathbf{r}_1,\mathbf{r}_2;\frac{\pi\nu}{\tau})$ by substituting $(2\pi\nu/\tau) \to (\omega + i0^+)$ into Eq. (10.7.17). Consequently, the causal response function is given by

$$\mathcal{R}\left(\mathbf{r}_1,\mathbf{r}_2;\mathbf{r}_1',\mathbf{r}_2';\pi\nu/\tau \to \omega + i0^+\right) = -i\int_0^\tau dT e^{-i(\omega+i0^+)T}G_{1<}^0(\mathbf{r}_1,\mathbf{r}_2';-T)G_{1>}^0(\mathbf{r}_2,\mathbf{r}_1';T). \quad (10.7.21)$$

The time integration path may be written in terms of the convergent semi-infinite time-axis integrals as

$$\int_0^\tau dT = \int_0^{-\infty} dT + \int_{-\infty}^{-\infty+\tau} dT + \int_{-\infty+\tau}^{\tau} dT, \quad (10.7.22)$$

where the limit $-\infty$ assures convergence due to the factor $i0^+$ in the exponential and we ignore the second term on the right-hand side, since $e^{-i(\omega+i0^+)T} \to 0$ as $T \to -\infty$. Moreover, with the change of variable $T = T' + \tau$ (also $T'' = -T'$) in the third term and $T' = -T$ in the first term, $\mathcal{R}\left(\mathbf{r}_1, \mathbf{r}_2; \mathbf{r}'_1, \mathbf{r}'_2; \frac{\pi\nu}{\tau}\right)$ may be rewritten as

$$\mathcal{R}_c\left(\mathbf{r}_1, \mathbf{r}_2; \mathbf{r}'_1, \mathbf{r}'_2; \omega\right) = \mathcal{R}_{(-)}\left(\mathbf{r}_1, \mathbf{r}_2; \mathbf{r}'_1, \mathbf{r}'_2; \omega\right) - \mathcal{R}_{(+)}\left(\mathbf{r}_1, \mathbf{r}_2; \mathbf{r}'_1, \mathbf{r}'_2; \omega\right), \qquad (10.7.23)$$

where $\mathcal{R}_{(\pm)}\left(\mathbf{r}_1, \mathbf{r}_2; \mathbf{r}'_1, \mathbf{r}'_2; \omega\right)$ is given by

$$\mathcal{R}_{(\pm)}\left(\mathbf{r}_1, \mathbf{r}_2; \mathbf{r}'_1, \mathbf{r}'_2; \omega\right) = -i \int_0^\infty dT\, e^{i(\omega+i0^+)T}\, G^0_{1\lessgtr}(\mathbf{r}_1, \mathbf{r}'_2; T)\, G^0_{1\gtrless}(\mathbf{r}_2, \mathbf{r}'_1; -T). \qquad (10.7.24)$$

Thus, the causal response function $\overline{\mathbf{L}}_c(\mathbf{r}_1, \mathbf{r}_2; \omega)$ is given by

$$\overline{\mathbf{L}}_c(\mathbf{r}_1, \mathbf{r}_2; \omega) = \\ -\frac{e^2}{mc}\left\{\overline{\mathbf{I}}n_0(\mathbf{r}_1)\delta^{(3)}(\mathbf{r}_1 - \mathbf{r}_2) - \frac{1}{4m}\lim_{\substack{1'\to 1\\ 2'\to 2}}(\nabla_1 - \nabla'_1)(\nabla_2 - \nabla'_2)\mathcal{R}_c\left(\mathbf{r}_1, \mathbf{r}_2; \mathbf{r}'_1, \mathbf{r}'_2; \omega\right)\right\}, \qquad (10.7.25)$$

in which $\mathcal{R}_c\left(\mathbf{r}_1, \mathbf{r}_2; \mathbf{r}'_1, \mathbf{r}'_2; \omega\right)$ is defined by Eqns. (10.7.23 and 10.7.24). Finally, the causal linear response tensor $\overline{\mathbf{L}}_c(\mathbf{r}_1, \mathbf{r}_2; \omega)$ is related to the causal linear conductivity tensor $\overline{\boldsymbol{\sigma}}_c(\mathbf{r}_1, \mathbf{r}_2; \omega)$ by $\overline{\boldsymbol{\sigma}}_c(\mathbf{r}_1, \mathbf{r}_2; \omega) = (c/i\omega)\overline{\mathbf{L}}_c(\mathbf{r}_1, \mathbf{r}_2; \omega)$ (for our choice of gauge in which the scalar potential vanishes, $\phi = 0$, with $\mathbf{E} = -(1/c)(\partial \mathbf{A}/\partial t) \to (i\omega/c)\mathbf{A}$). Therefore, the linear conductivity tensor $\overline{\boldsymbol{\sigma}}$ is

$$\overline{\boldsymbol{\sigma}}_c(\mathbf{r}_1, \mathbf{r}_2; \omega) = \frac{ie^2}{m\omega}\left\{\overline{\mathbf{I}}n_0(\mathbf{r}_1)\delta(\mathbf{r}_1 - \mathbf{r}_2) - \frac{1}{4m}\lim_{\substack{1'\to 1\\ 2'\to 2}}(\nabla_1 - \nabla'_1)(\nabla_2 - \nabla'_2)\mathcal{R}_c\left(\mathbf{r}_1, \mathbf{r}_2; \mathbf{r}'_1, \mathbf{r}'_2; \omega\right)\right\}. \qquad (10.7.26)$$

Considering that the study of nanostructures involves the analysis of $\overline{\boldsymbol{\sigma}}$ for spatially confined systems, we provide an illustration by evaluating $\overline{\boldsymbol{\sigma}}$ for a single planar quantum well involving real subband wave functions $\xi_\alpha(z)$ as z-dependent energy eigenstates for motion across the well, having subband energies E_α and Green's function of the form

$$G^0_{1\gtrless}(\mathbf{r}_1, \mathbf{r}_2; t_1, t_2) = \sum_\alpha \xi_\alpha(z_1)\xi_\alpha(z_2)g^\alpha_\gtrless(\overline{\mathbf{r}}_1 - \overline{\mathbf{r}}_2; T). \qquad (10.7.27)$$

Here, α denotes the subband index of an electron state confined in the z-direction and $g^\alpha_\gtrless(\overline{\mathbf{r}}_1 - \overline{\mathbf{r}}_2; T)$ is the associated 2D part of the thermodynamic Schrödinger Green's function for motion along the translationally invariant x-y plane of the quantum well (the chemical potential measured relative to the α lateral subband energy is $\zeta_\alpha = \zeta - E_\alpha$ and $T = (t_1 - t_2)$):

$$g^{\alpha}_{\{\gtrless\}}(\bar{\mathbf{r}}_1;\bar{\mathbf{r}}_2;T) = -i\int_{-\infty}^{\infty}\frac{d^{(2)}\bar{\mathbf{q}}}{(2\pi)^2}e^{i\bar{\mathbf{q}}\cdot(\bar{\mathbf{r}}_1-\bar{\mathbf{r}}_2)}e^{-i(\varepsilon_{\bar{\mathbf{q}}}+E_\alpha)T}\begin{Bmatrix}1-f_0(\epsilon_{\bar{\mathbf{q}}}+E_\alpha)\\ -f_0(\epsilon_{\bar{\mathbf{q}}}+E_\alpha)\end{Bmatrix}; \qquad (10.7.28)$$

and $f_0(\varepsilon_{\bar{\mathbf{q}}}+E_\alpha) = \left\{\exp\left[(\varepsilon_{\bar{\mathbf{q}}}+E_\alpha-\zeta)/kT\right]+1\right\}^{-1}$ is the Fermi distribution function with chemical potential ζ, and $\varepsilon_{\bar{\mathbf{q}}} = \hbar^2q^2/2m$ represents the part of single-electron kinetic energy along the quantum well plane. In momentum representation for the translationally invariant x-y plane, $\bar{\mathbf{r}}_1-\bar{\mathbf{r}}_2 \to \bar{\mathbf{q}}$, we obtain

$$G^0_{1\gtrless}(z_1,z_2;\bar{\mathbf{q}};T) = -i\sum_\alpha \xi_\alpha(z_1)\xi_\alpha(z_2)e^{-i(\varepsilon_{\bar{\mathbf{q}}}+E_\alpha)T}\begin{Bmatrix}1-f_0(\epsilon_{\bar{\mathbf{q}}}+E_\alpha)\\ -f_0(\epsilon_{\bar{\mathbf{q}}}+E_\alpha)\end{Bmatrix}. \qquad (10.7.29)$$

In the case of 3D translational invariance, we have

$$\xi_\alpha(z) \to \frac{e^{iq_zz}}{\sqrt{2\pi}} \text{ and } E_\alpha \to q_z^2/2m. \qquad (10.7.30)$$

Eq. (10.7.26) will be employed jointly with Eq. (10.7.29) in the determination of the conductivity tensors of nanostructures having planar quantum wells. Fourier-transforming along the translationally invariant x-y plane, the associated conductivity tensor, $\bar{\boldsymbol{\sigma}}_{QW}$, is given by ($n_0(z_1)$ denotes the equilibrium density of the quantum well at z_1):

$$\bar{\boldsymbol{\sigma}}_{QW}(z_1,z_2;\bar{\mathbf{q}},\omega) = \frac{ie^2}{m\omega}\left\{\bar{\mathbf{I}}n_0(z_1)\delta(z_1-z_2)+\frac{1}{4m}\lim_{z_1'\to z_1}\lim_{z_2'\to z_2}\int_{-\infty}^{\infty}\frac{d^{(2)}\bar{\mathbf{q}}'}{(2\pi)^2}\bar{\mathbf{F}}\,\mathcal{R}\left(z_1,z_2;z_1',z_2',\bar{\mathbf{q}};\omega\right)\right\}, \qquad (10.7.31)$$

and the dyadic-tensor operator $\bar{\mathbf{F}}$ and $\mathcal{R}\left(z_1,z_2;z_1',z_2',\bar{\mathbf{q}};\omega\right)$ may be written as ($\widehat{z}$ is a unit vector in the z-direction):

$$\bar{\mathbf{F}} = \left[i\left(2\bar{\mathbf{q}}'-\bar{\mathbf{q}}\right)+\widehat{z}\left(\frac{\partial}{\partial z_1}-\frac{\partial}{\partial z_1'}\right)\right]\left[i\left(2\bar{\mathbf{q}}'-\bar{\mathbf{q}}\right)+\widehat{z}\left(\frac{\partial}{\partial z_2}-\frac{\partial}{\partial z_2'}\right)\right], \qquad (10.7.32)$$

with

$$\mathcal{R}\left(z_1,z_2;z_1',z_2',\bar{\mathbf{q}};\omega\right) = \mathcal{R}_{(-)}\left(z_1,z_2;z_1',z_2',\bar{\mathbf{q}};\omega\right)-\mathcal{R}_{(+)}\left(z_1,z_2;z_1',z_2',\bar{\mathbf{q}};\omega\right) \qquad (10.7.33)$$

and

$$\mathcal{R}_{(\pm)}\left(z_1,z_2;z_1',z_2',\bar{\mathbf{q}},\omega\right) = -i\int_0^{\infty}dT\,e^{i(\omega+i0^+)T}\,G^0_{1\lessgtr}(z_1,z_2';\bar{\mathbf{q}}',T)G^0_{1\gtrless}(z_2,z_1';\bar{\mathbf{q}}'-\bar{\mathbf{q}},-T). \qquad (10.7.34)$$

It should be noted that the $\overline{\mathbf{q}}'$-dependence of $\mathcal{R}\left(z_1, z_2; z_1', z_2', \overline{\mathbf{q}}; \omega\right)$ is not explicitly exhibited since it will eventually be integrated out in $\overline{\boldsymbol{\sigma}}_{QW}$ (see Eq. (10.7.31)). Furthermore, $n_0(z_1)$ has the form

$$n_0(z_1) = -iTr\left[G^0_{1<}(1,1)\right] = 2\sum_{\alpha} |\xi_\alpha(z_1)|^2 \int_{-\infty}^{\infty} \frac{d^{(2)}\overline{\mathbf{q}}'}{(2\pi)^2} f_0\left(\epsilon_{\overline{\mathbf{q}}'} + E_\alpha\right), \tag{10.7.35}$$

and employing Eq. (10.7.29) in Eqns. (10.7.34 and 10.7.33) yields

$$\mathcal{R} \equiv \mathcal{R}\left(z_1, z_2; z_1', z_2', \overline{\mathbf{q}}; \omega\right) = \sum_{\alpha,\beta} R_{\alpha\beta}\, \xi_\alpha(z_1)\xi_\beta(z_1')\xi_\alpha(z_2')\xi_\beta(z_2), \tag{10.7.36}$$

where the matrix element $R_{\alpha\beta}$ of the density perturbation response function of the quantum well is given by

$$R_{\alpha\beta} \equiv R_{\alpha\beta}(\overline{\mathbf{q}}', \overline{\mathbf{q}}; \omega) = 2\frac{f_0\left(\epsilon_{\overline{\mathbf{q}}'} + E_\alpha\right) - f_0\left(\epsilon_{\overline{\mathbf{q}}'-\overline{\mathbf{q}}} + E_\beta\right)}{\hbar\omega - \epsilon_{\overline{\mathbf{q}}'} + \epsilon_{\overline{\mathbf{q}}'-\overline{\mathbf{q}}} + E_\beta - E_\alpha + i0^+}. \tag{10.7.37}$$

The factor 2 in Eqns. (10.7.35) and (10.7.37) arises from the spin sum over degenerate electron states. Substitution of Eq. (10.7.36) into Eq. (10.7.31) yields

$$\overline{\boldsymbol{\sigma}}_{QW}(z_1, z_2; \overline{\mathbf{q}}, \omega) = \frac{ie^2}{m\omega}\Bigg[\overline{\mathbf{I}} n_0(z_1)\delta(z_1 - z_2) + \frac{1}{4m} \lim_{z_1' \to z_1} \lim_{z_2' \to z_2} \sum_{\alpha,\beta} \int_{-\infty}^{\infty} \frac{d^{(2)}\overline{\mathbf{q}}'}{(2\pi)^2} R_{\alpha\beta}(\overline{\mathbf{q}}', \overline{\mathbf{q}}; \omega) \times \overline{\mathbf{F}}\left[\xi_\alpha(z_1)\xi_\beta(z_1')\xi_\alpha(z_2')\xi_\beta(z_2)\right]\Bigg]. \tag{10.7.38}$$

Considering the case in which the intersubband energy denominators $\hbar\omega + E_\beta - E_\alpha$ are large for $\alpha \neq \beta$, we may neglect intersubband transitions in the wells, whence $R_{\alpha\beta} \to \delta_{\alpha\beta} R_{\alpha\alpha}$, and the conductivity tensor of the quantum well nanostructure reduces to

$$\overline{\boldsymbol{\sigma}}_{QW}(z_1, z_2; \overline{\mathbf{q}}, \omega) = \frac{ie^2}{m\omega}\Bigg[\overline{\mathbf{I}} n_0(z_1)\delta(z_1 - z_2) + \frac{1}{4m} \lim_{z_1' \to z_1} \lim_{z_2' \to z_2} \sum_{\alpha} \int_{-\infty}^{\infty} \frac{d^{(2)}\overline{\mathbf{q}}'}{(2\pi)^2} R_{\alpha\alpha}(\overline{\mathbf{q}}', \overline{\mathbf{q}}; \omega) \times \overline{\mathbf{F}}\left[\xi_\alpha(z_1)\xi_\alpha(z_1')\xi_\alpha(z_2')\xi_\alpha(z_2)\right]\Bigg], \tag{10.7.39}$$

with $R_{\alpha\alpha}(\overline{\mathbf{q}}', \overline{\mathbf{q}}; \omega)$ now devoid of the large-denominator, small intersubband terms:

$$R_{\alpha\alpha}(\overline{\mathbf{q}}', \overline{\mathbf{q}}; \omega) = 2\frac{f_0\left(\epsilon_{\overline{\mathbf{q}}'} + E_\alpha\right) - f_0\left(\epsilon_{\overline{\mathbf{q}}'-\overline{\mathbf{q}}} + E_\alpha\right)}{\hbar\omega - \epsilon_{\overline{\mathbf{q}}'} + \epsilon_{\overline{\mathbf{q}}'-\overline{\mathbf{q}}} + i0^+}. \tag{10.7.40}$$

Considering the action of $\overline{\mathbf{F}}$ on subband wave functions (ignoring intersubband transitions) with the limits indicated in Eq. (10.7.39), we obtain

$$\lim_{z_1'\to z_1}\lim_{z_2'\to z_2}\overline{\mathbf{F}}\left[\xi_\alpha(z_1)\xi_\alpha(z_1')\xi_\alpha(z_2')\xi_\alpha(z_2)\right] = -(2\overline{\mathbf{q}}'-\overline{\mathbf{q}})(2\overline{\mathbf{q}}'-\overline{\mathbf{q}})\,|\xi_\alpha(z_1)|^2\,|\xi_\alpha(z_2)|^2\,,$$

because the action of the derivative terms in $\overline{\mathbf{F}}$ on the subband wave functions yields vanishing results, e.g.,

$$\begin{aligned}&i\hat{z}(2\overline{\mathbf{q}}'-\overline{\mathbf{q}})\lim_{z_1'\to z_1}\lim_{z_2'\to z_2}\left(\frac{\partial}{\partial z_1}-\frac{\partial}{\partial z_1'}\right)\left[\xi_\alpha(z_1)\xi_\alpha(z_1')\xi_\alpha(z_2')\xi_\alpha(z_2)\right]\\ &= i\hat{z}(2\overline{\mathbf{q}}'-\overline{\mathbf{q}})\left(\frac{\partial\xi_\alpha(z_1)}{\partial z_1}\xi_\alpha(z_1')-\xi_\alpha(z_1)\frac{\partial\xi_\alpha(z_1')}{\partial z_1'}\right)_{z_1'=z_1}|\xi_\alpha(z_2)|^2 = 0.\end{aligned}$$

The resulting conductivity tensor for a quantum well with multiple subbands is given by

$$\begin{aligned}\overline{\boldsymbol{\sigma}}_{QW}(z_1,z_2;\overline{\mathbf{q}},\omega) = \frac{ie^2}{m\omega}\Bigg\{&\overline{\mathbf{I}}n_0(z_1)\delta(z_1-z_2)-\frac{1}{m}\sum_\alpha|\xi_\alpha(z_1)|^2\,|\xi_\alpha(z_2)|^2\\ &\times\int_{-\infty}^{\infty}\frac{d^{(2)}\overline{\mathbf{q}}'}{(2\pi)^2}(\overline{\mathbf{q}}'-\overline{\mathbf{q}}/2)(\overline{\mathbf{q}}'-\overline{\mathbf{q}}/2)R_{\alpha\alpha}(\overline{\mathbf{q}}',\overline{\mathbf{q}};\omega)\Bigg\}, \qquad (10.7.41)\end{aligned}$$

provided the intersubband transitions are negligible. For the case in which only the lowest quantum well subband is populated in the degenerate limit ($T = 0K$), $\alpha = 1$ and we suppress reference to this index, so that

$$\begin{aligned}\overline{\boldsymbol{\sigma}}_{QW}(z_1,z_2;\overline{\mathbf{q}},\omega) = \frac{ie^2}{m\omega}\Bigg\{&\overline{\mathbf{I}}n_0(z_1)\delta(z_1-z_2)-\frac{1}{m}|\xi(z_1)|^2\,|\xi(z_2)|^2\\ &\times\int_{-\infty}^{\infty}\frac{d^{(2)}\overline{\mathbf{q}}'}{(2\pi)^2}(\overline{\mathbf{q}}'-\overline{\mathbf{q}}/2)(\overline{\mathbf{q}}'-\overline{\mathbf{q}}/2)R(\overline{\mathbf{q}}',\overline{\mathbf{q}};\omega)\Bigg\}. \qquad (10.7.42)\end{aligned}$$

Assuming the quantum well to be thin ($|q_\perp|\,b \ll 1$, where $q_\perp = \left[(\omega^2/c^2)\varepsilon(\omega)-\overline{\mathbf{q}}^2\right]^{1/2}$ and b is the well width), we expect that $|\xi(z_1)|^2 \cong \delta(z_1)$, and the equilibrium density $n_0(z)$ of the quantum well (Eq. (10.7.35)) may be written as

$$n_0(z_1) = \delta(z_1)n_0 \quad ; \quad n_0 = 2\int_{-\infty}^{\infty}\frac{d^{(2)}\overline{\mathbf{q}}}{(2\pi)^2}f_0(\epsilon_{\overline{\mathbf{q}}}+E_0) \qquad (10.7.43)$$

(E_0 is the lowest subband energy in the z-direction). Correspondingly, we obtain the conductivity tensor of a thin quantum well as

$$\overline{\boldsymbol{\sigma}}_{QW}(z_1,z_2;\overline{\mathbf{q}},\omega) = \delta(z_1)\delta(z_2)\overline{\boldsymbol{\sigma}}_{2D}(\overline{\mathbf{q}},\omega) \qquad (10.7.44)$$

(note that $n_0(z_1)\delta(z_1-z_2) = n_0\delta(z_1)\delta(z_1-z_2) = n_0\delta(z_1)\delta(z_2)$) with $\overline{\boldsymbol{\sigma}}_{2D}(\mathbf{q},\omega)$ as the 2D conductivity tensor of the well (in lateral wave number and frequency representation), given by

$$\overline{\boldsymbol{\sigma}}_{2D}(\overline{\mathbf{q}},\omega) = \frac{ie^2}{m\omega}\left\{\overline{\mathbf{I}}n_0 - \frac{1}{m}\int_{-\infty}^{\infty}\frac{d^{(2)}\overline{\mathbf{q}}'}{(2\pi)^2}(\overline{\mathbf{q}}'-\overline{\mathbf{q}}/2)(\overline{\mathbf{q}}'-\overline{\mathbf{q}}/2)R(\overline{\mathbf{q}}',\overline{\mathbf{q}};\omega)\right\}. \quad (10.7.45)$$

However, note that $\overline{\mathbf{I}}$ is the 3D unit dyadic.

Of course, it is also of interest to examine the 3D bulk conductivity tensor, $\overline{\boldsymbol{\sigma}}_{3D}(\overline{\mathbf{q}},\omega)$. Returning to Eq. (10.7.38) and using the 3D substitutions of Eq. (10.7.30), we obtain (n_{3D} is the 3D density):

$$\begin{aligned}\overline{\boldsymbol{\sigma}}_{3D}(\mathbf{q},\omega) = \frac{ie^2}{m\omega}\Bigg[\overline{\mathbf{I}}n_{3D} - \frac{1}{2m}\int\frac{d^{(3)}\mathbf{p}}{(2\pi)^3}f_0(\varepsilon_{\mathbf{p}})\\ \times\left\{\frac{(2\mathbf{p}+\mathbf{q})(2\mathbf{p}+\mathbf{q})}{\frac{q^2}{2m}+\frac{\mathbf{p}\cdot\mathbf{q}}{m}-\omega-i0^+}+\frac{(2\mathbf{p}-\mathbf{q})(2\mathbf{p}-\mathbf{q})}{\frac{q^2}{2m}-\frac{\mathbf{p}\cdot\mathbf{q}}{m}+\omega+i0^+}\right\}\Bigg].\end{aligned} \quad (10.7.46)$$

The corresponding dynamic, non-local 3D bulk dielectric tensor is given by Eq. (10.6.42) as

$$\begin{aligned}\overline{\boldsymbol{\varepsilon}}_{3D}(\mathbf{q},\omega) = \overline{\mathbf{I}} - \frac{4\pi e^2}{m\omega^2}\Bigg[\overline{\mathbf{I}}n_{3D} - \frac{1}{2m}\int\frac{d^{(3)}\mathbf{p}}{(2\pi)^3}f_0(\varepsilon_{\mathbf{p}})\\ \times\left\{\frac{(2\mathbf{p}+\mathbf{q})(2\mathbf{p}+\mathbf{q})}{\frac{q^2}{2m}+\frac{\mathbf{p}\cdot\mathbf{q}}{m}-\omega-i0^+}+\frac{(2\mathbf{p}-\mathbf{q})(2\mathbf{p}-\mathbf{q})}{\frac{q^2}{2m}-\frac{\mathbf{p}\cdot\mathbf{q}}{m}+\omega+i0^+}\right\}\Bigg].\end{aligned} \quad (10.7.47)$$

The statistical integrals in Eq. (10.7.47) are elementary in the degenerate limit ($T = 0K$) and evaluation of the real parts yields logarithmic functions, with integration of the imaginary parts rendered immediate by the Dirac delta functions associated with $\pm i0^+$ in the denominators. A semiclassical model of the 3D-bulk plasma dielectric tensor based on a linearized, collisionless Boltzmann equation coupled to the Maxwell equations, but averaged over an initial Fermi–Dirac distribution, $f_0(\varepsilon_{\mathbf{p}})$ to enforce the Pauli exclusion principle, may be obtained from Eq. (10.7.47) by dropping the $\frac{q^2}{2m}$-terms in the denominators of the integrands. The classical limit can then be obtained by putting $f_0(\varepsilon_{\mathbf{p}})$ into the Maxwellian limit.

Yet another interesting description of non-local 3D plasma electromagnetics is based on the hydrodynamic model of plasma non-locality, Eqns. (10.2.21–10.2.26), written in the form

$$-\omega\delta\rho(\mathbf{q},\omega) + \mathbf{q}\cdot\mathbf{J}(\mathbf{q},\omega) = 0, \quad (10.7.48)$$

$$\omega\mathbf{J}(\mathbf{q},\omega) = i\frac{e^2\rho_0}{m}\mathbf{E}(\mathbf{q},\omega) + \beta^2\mathbf{q}\delta\rho(\mathbf{q},\omega). \quad (10.7.49)$$

Using Eq. (10.7.48) to eliminate $\delta\rho$ from Eq. (10.7.49), we obtain

$$\left(\bar{\mathbf{I}} - \frac{\beta^2 \mathbf{q}\mathbf{q}}{\omega^2}\right) \cdot \mathbf{J}(\mathbf{q},\omega) = \frac{ie^2\rho_0}{m\omega}\mathbf{E}(\mathbf{q},\omega), \tag{10.7.50}$$

which yields the 3D hydrodynamic conductivity tensor as

$$\bar{\sigma}_{3D}(\mathbf{q},\omega) = \frac{ie^2\rho_0}{m\omega}\bar{\mathbf{I}} + \frac{i\beta^2 e^2\rho_0}{m\omega}\frac{\mathbf{q}\mathbf{q}}{\omega^2 - \beta^2 q^2}, \tag{10.7.51}$$

and, correspondingly, the bulk hydrodynamic dielectric tensor is given by

$$\bar{\varepsilon}_{3D}(\mathbf{q},\omega) = \left(1 - \frac{\omega_p^2}{\omega^2}\right)\bar{\mathbf{I}} - \frac{\beta^2\omega_p^2}{\omega^2}\frac{\mathbf{q}\mathbf{q}}{\omega^2 - \beta^2 q^2}, \tag{10.7.52}$$

or, with $\bar{\mathbf{I}} = \bar{\mathbf{I}} - \widehat{qq} + \widehat{qq}$ ($\widehat{q} = \mathbf{q}/q$), we have

$$\bar{\varepsilon}_{3D}(\mathbf{q},\omega) = \left(1 - \frac{\omega_p^2}{\omega^2}\right)\left(\bar{\mathbf{I}} - \widehat{qq}\right) + \widehat{qq}\cdot\left(1 - \frac{\omega_p^2}{\omega^2 - \beta^2 q^2}\right). \tag{10.7.53}$$

Noting that the longitudinal bulk dielectric function of Eq. (10.6.43) is given by

$$\varepsilon_L(\mathbf{q},\omega) = \widehat{q}\cdot\bar{\varepsilon}_{3D}(\mathbf{q},\omega)\cdot\widehat{q} = 1 - \frac{\omega_p^2}{\omega^2 - \beta^2 q^2}, \tag{10.7.54}$$

we similarly find the transverse bulk dielectric function as

$$\varepsilon_T(\mathbf{q},\omega) = 1 - \frac{\omega_p^2}{\omega^2}, \tag{10.7.55}$$

for the hydrodynamic model. Thus we obtain $\bar{\mathfrak{G}}_{3D}(\mathbf{q},\omega)$ of Eq. (10.6.44) for this model as

$$\bar{\mathfrak{G}}_{3D}(\mathbf{q},\omega) = \left(\bar{\mathbf{I}} - \widehat{qq}\right)\frac{c^2}{\left[\omega^2 - \left(\omega_p^2 + c^2q^2\right)\right]} + \left(\widehat{qq}\right)\frac{c^2\left(1 - \beta^2q^2/\omega^2\right)}{\omega^2 - \left(\omega_p^2 + \beta^2q^2\right)}. \tag{10.7.56}$$

The collective electromagnetic-plasma modes arising from the polarization of the electron gas in the hydrodynamic model are given by the frequency poles of this result; namely a coupled transverse EM-plasma "polariton" wave which hybridizes light with a plasmon, as given by

$$\omega^2 = \omega_p^2 + c^2q^2, \tag{10.7.57}$$

and a longitudinal plasma mode that is non-local,

$$\omega^2 = \omega_p^2 + \beta^2 q^2, \tag{10.7.58}$$

which is implicit in Eq. (10.2.26). While the local limit has already been discussed in Eq. (10.6.49), the present analysis shows that hydrodynamic non-locality enters only into the longitudinal mode, and not at all into the transverse mode, at least within the framework of our oversimplified description of the hydrodynamic model. Of course, use of the RPA in Eq. (10.7.1) would produce much more interesting detail. Notwithstanding the fact that these modes could have been discussed on an even simpler basis, our explicit determination of the EM Green's function also supplies the residues at the frequency poles, which provide their excitation amplitudes.

Finally, it is worthwhile to comment that the enforcement of causality associated with the installation of a Heaviside cutoff function in time,

$$\tilde{\overline{\sigma}}(\mathbf{r}-\mathbf{r}',\ t-t') = 2\eta_+(t-t')\overline{\sigma}(\mathbf{r}-\mathbf{r}',\ t-t'), \tag{10.7.59}$$

implies the addition of a small imaginary part in frequency representation, $\omega \to \omega + i\epsilon$. Kramers–Krönig relations then arise using the Dirac prescription as[34]

$$\tilde{\overline{\sigma}}(\mathbf{q},\omega) = \frac{i}{\pi}\int_{\infty}^{\infty} d\omega' \frac{\overline{\sigma}(\mathbf{q},\omega')}{\omega-\omega'+i\varepsilon} = \overline{\sigma}(\mathbf{q},\omega) - i\omega\overline{\alpha}(\mathbf{q},\omega), \tag{10.7.60}$$

where the conductivity $\overline{\sigma}$ and polarizability $\overline{\alpha}(\mathbf{q},\omega)$ are related as

$$\overline{\alpha}(\mathbf{q},\omega) = \frac{2}{\pi}\mathcal{P}\int_0^{\infty} \frac{d\omega\overline{\sigma}(\mathbf{q},\omega')}{\omega'^2-\omega^2}. \tag{10.7.61}$$

The underlying assumptions of analyticity, etc., involved in such Kramers–Krönig relations for susceptibilities,

$$Re\tilde{\overline{\sigma}}(\mathbf{q},\omega) = \mathcal{P}\int_{-\infty}^{\infty} \frac{d\omega'}{\pi}\frac{Im\tilde{\overline{\sigma}}(\mathbf{q},\omega')}{\omega'-\omega} \quad \text{and} \quad Im\tilde{\overline{\sigma}}(\mathbf{q},\omega) = -\mathcal{P}\int_{-\infty}^{\infty} \frac{d\omega'}{\pi}\frac{Re\tilde{\overline{\sigma}}(\mathbf{q},\omega')}{\omega'-\omega}, \tag{10.7.62}$$

and their limitations, have been carefully analyzed by Martin,[35] and are also discussed in some detail by Landau and Lifshitz.[36]

10.8 Polaritons in Compound Nanostructures

To illustrate the analysis of polaritons in compound nanostructures along the lines described in the text between Eqns. (10.6.12) and (10.6.29), we consider the example of a compound nanostructure composed of a single 2D quantum well embedded in a 3D host semiconductor plasma. In this, we employ the integral equation, Eq. (10.6.29), for the EM dyadic Green's function $\overline{\mathfrak{G}}(\mathbf{r}_1,\mathbf{r}_2;\omega)$ using $\overline{\mathfrak{G}}_{3D}(\mathbf{r}_1,\mathbf{r}_2;\omega)$ given by Eq. (10.6.44) in the special case of a dynamic 3D bulk medium that is local, with $\varepsilon_T(\omega) = \varepsilon_L(\omega) \equiv \varepsilon(\omega)$, as well as $\overline{\boldsymbol{\sigma}}_{nano} = \overline{\boldsymbol{\sigma}}_{QW}$ for a thin quantum well. Since we have spatial translational invariance in the $\overline{\mathbf{r}} = x$-$y$ plane, we Fourier-transform $\overline{\mathbf{r}}_1 - \overline{\mathbf{r}}_2 \to \overline{\mathbf{q}}$, and write the integral equation in the form (suppress $\overline{\mathbf{q}}, \omega$):[29]

$$\overline{\mathfrak{G}}(z_1, z_2) = \overline{\mathfrak{G}}_{3D}(z_1 - z_2) + \frac{4\pi i\omega}{c^2} \int dz_1' \int dz_2' \, \overline{\mathfrak{G}}_{3D}(z_1 - z_1') \cdot \overline{\boldsymbol{\sigma}}_{QW}(z_1', z_2') \cdot \overline{\mathfrak{G}}(z_2', z_2). \tag{10.8.1}$$

Noting the $\delta(z_1')\,\delta(z_2')$-form of $\overline{\boldsymbol{\sigma}}_{QW}(z_1', z_2')$ in Eq. (10.7.44) we have a great simplification of the integral equation,

$$\overline{\mathfrak{G}}(z_1, z_2) = \overline{\mathfrak{G}}_{3D}(z_1 - z_2) + \frac{4\pi i\omega}{c^2} \overline{\mathfrak{G}}_{3D}(z_1, 0)\, \overline{\boldsymbol{\sigma}}_{2D}(\overline{\mathbf{q}}, \omega)\, \overline{\mathfrak{G}}(0, z_2). \tag{10.8.2}$$

To solve, we set $z_1 = 0$, obtaining

$$\overline{\mathfrak{G}}(0, z_2) = \left[\overline{\mathbf{I}} - \frac{4\pi i\omega}{c^2} \overline{\mathfrak{G}}_{3D}(0, 0)\, \overline{\boldsymbol{\sigma}}_{2D}(\overline{\mathbf{q}}, \omega) \right]^{-1} \overline{\mathfrak{G}}_{3D}(-z_2), \tag{10.8.3}$$

with the result

$$\begin{aligned} \overline{\mathfrak{G}}(z_1, z_2) &= \overline{\mathfrak{G}}_{3D}(z_1 - z_2) \\ &+ \frac{4\pi i\omega}{c^2} \overline{\mathfrak{G}}_{3D}(z_1, 0)\, \overline{\boldsymbol{\sigma}}_{2D}(\overline{\mathbf{q}}, \omega) \left[\overline{\mathbf{I}} - \frac{4\pi i\omega}{c^2} \overline{\mathfrak{G}}_{3D}(0, 0)\, \overline{\boldsymbol{\sigma}}_{2D}(\overline{\mathbf{q}}, \omega) \right]^{-1} \overline{\mathfrak{G}}_{3D}(-z_2). \end{aligned} \tag{10.8.4}$$

For the spatially local dynamic bulk medium of our example ($\varepsilon(\omega) = \varepsilon_T(\omega) = \varepsilon_L(\omega)$), the 3D electromagnetic dyadic Green's function of Eq. (10.6.44) may be written as

$$\overline{\mathfrak{G}}_{3D}(\mathbf{q}, \omega) = \frac{\left(\omega^2 \varepsilon(\omega)/c^2\right) \overline{\mathbf{I}} - \mathbf{q}\mathbf{q}}{\left(\omega^2 \varepsilon(\omega)/c^2\right) \left[\left(\omega^2 \varepsilon(\omega)/c^2\right) - q^2 \right]}, \tag{10.8.5}$$

with the only change from the free space case given by the replacement $c^2 \to c^2/\varepsilon(\omega)$. As Eq. (10.8.4) requires $\overline{\mathfrak{G}}_{3D}$ in z-representation,

$$\overline{\mathfrak{G}}_{3D}(z_1, z_2; \overline{\mathbf{q}}, \omega) = \frac{1}{2\pi} \int_{-\infty}^{\infty} dq_z e^{iq_z(z_1 - z_2)}\, \overline{\mathfrak{G}}_{3D}(\mathbf{q}, \omega), \tag{10.8.6}$$

we employ the Levine–Schwinger solution of Eq. (10.6.10) with $c^2 \to c^2/\varepsilon(\omega)$, which takes the mixed $(z, \overline{\mathbf{q}})$ representation form ($z = z_1 - z_2$):

$$\overline{\mathfrak{G}}_{3D}(z; \overline{\mathbf{q}}, \omega) = \left\{ \overline{\mathbf{I}} - \left(\frac{c^2}{\omega^2 \varepsilon(\omega)} \right) \left[\overline{\mathbf{q}\mathbf{q}} + (\overline{\mathbf{q}\hat{\mathbf{z}}} + \overline{\hat{\mathbf{z}}\mathbf{q}}) \left(\frac{1}{i} \frac{\partial}{\partial z} \right) + \overline{\hat{\mathbf{z}}\hat{\mathbf{z}}} \left(\frac{1}{i} \frac{\partial}{\partial z} \right)^2 \right] \right\} \mathfrak{g}(z), \tag{10.8.7}$$

with $\mathfrak{g}(z) \equiv \mathfrak{g}(z;\overline{\mathbf{q}},\omega)$ as the scalar Helmholtz Green's function in the mixed $(z,\overline{\mathbf{q}})$ representation given by $\left[q_\perp^2 = (\omega^2/c^2)\,\varepsilon(\omega) - \overline{q}^2 \text{ and } \omega \to \omega + i0^+\right]$:

$$\mathfrak{g}(z) = -\frac{1}{2\pi}\int_{-\infty}^{\infty} dq_z \frac{e^{iq_z z}}{q_\perp^2 - q_z^2} = -\frac{1}{2iq_\perp} e^{iq_\perp |z|}. \tag{10.8.8}$$

Evaluating the derivatives in Eq. (10.8.7), we have

$$\left[\mathfrak{g}, \frac{1}{i}\mathfrak{g}', \frac{1}{i^2}\mathfrak{g}''\right] = -\frac{1}{2iq_\perp}\left[1, q_\perp sign(z), q_\perp^2 - 2iq_\perp \delta(z)\right] e^{iq_\perp |z|}, \tag{10.8.9}$$

(where a prime on the left-hand side indicates differentiation with respect to z) and obtain $\overline{\mathfrak{G}}_{3D}(z;\overline{\mathbf{q}},\omega)$ as[29]

$$\begin{aligned}\overline{\mathfrak{G}}_{3D}(z;\overline{\mathbf{q}},\omega) = &-\left(\frac{1}{2iq_\perp}\right)\left(\frac{c^2}{\omega^2\varepsilon(\omega)}\right)\\ &\times\left\{\frac{\omega^2\varepsilon(\omega)}{c^2}\overline{\mathbf{I}} - \overline{\mathbf{q}}\overline{\mathbf{q}} - q_\perp sign(z)(\overline{\mathbf{q}}\widehat{z} + \widehat{z}\overline{\mathbf{q}}) - \widehat{z}\widehat{z}\left[q_\perp^2 - 2iq_\perp\delta(z)\right]\right\} e^{iq_\perp |z|},\end{aligned} \tag{10.8.10}$$

with $sign(z) = 1, 0, -1$ for $z > 0$, $z = 0$, $z < 0$. Evaluating $\overline{\mathfrak{G}}_{3D}$ at $z = 0$ (noting that $\overline{\mathbf{q}}\widehat{z}$-terms vanish since $sign(0) = 0$) and choosing a coordinate system in which $q_x \equiv 0$, we may write

$$\begin{aligned}&\overline{\mathfrak{G}}_{3D}(z = 0;\overline{\mathbf{q}},\omega)\\ &= -\frac{1}{2iq_\perp}\left\{\overline{\mathbf{I}}_{2D} - \frac{q_y^2 c^2}{\omega^2\varepsilon(\omega)}\widehat{y}\widehat{y} + \widehat{z}\widehat{z}\left(1 - \frac{c^2}{\omega^2\varepsilon(\omega)}\left[q_\perp^2 - 2iq_\perp\delta(0)\right]\right)\right\},\end{aligned} \tag{10.8.11}$$

where $\overline{\mathbf{I}}_{2D} = \widehat{x}\widehat{x} + \widehat{y}\widehat{y}$ is the 2D unit dyadic of the x–y plane. Alternatively expressed (using $q_\perp^2 = \left[(\omega^2\varepsilon(\omega)/c^2) - \overline{\mathbf{q}}^2\right]$), we have

$$\overline{\mathfrak{G}}_{3D}(z = 0;\overline{\mathbf{q}},\omega) = -\frac{1}{2iq_\perp}\left\{\widehat{x}\widehat{x} + \frac{q_\perp^2 c^2}{\omega^2\varepsilon(\omega)}\widehat{y}\widehat{y} + \frac{q_y^2 c^2}{\omega^2\varepsilon(\omega)}\widehat{z}\widehat{z}\right\} - \frac{c^2}{\omega^2\varepsilon(\omega)}\delta(0)\widehat{z}\widehat{z}. \tag{10.8.12}$$

The EM wave dispersion relation is determined from the frequency poles of Eq. (10.8.5) as well as Eq. (10.8.12), with the "light pole" arising from $\overline{\mathfrak{G}}_{3D}$ as $q_\perp = 0 \Longrightarrow \omega^2 = c^2 q_y^2/\varepsilon(\omega)$. This couples light with bulk plasmons of the host semiconductor for $\varepsilon(\omega) = 1 - \omega_p^2/\omega^2$ as $\omega^2 = c^2 q_y^2 + \omega_p^2$. Other EM modes arise from the poles of the inverse dyadic $\left[\overline{\mathbf{I}} - \frac{4\pi i\omega}{c^2}\overline{\mathfrak{G}}_{3D}(0,0)\,\overline{\boldsymbol{\sigma}}_{2D}(\overline{\mathbf{q}},\omega)\right]^{-1}$ (Eq. (10.8.4)). Noting that $\overline{\boldsymbol{\sigma}}_{2D}(\overline{\mathbf{q}},\omega)$, given by Eq. (10.7.45), has the form

$$\overline{\boldsymbol{\sigma}}_{2D}(\overline{\mathbf{q}},\omega) = \frac{ie^2}{m\omega}\left\{n_0\widehat{z}\widehat{z} + n_0\overline{\mathbf{I}}_{2D} + \overline{\boldsymbol{\sigma}}'_{2D}(\overline{\mathbf{q}},\omega)\right\}, \tag{10.8.13}$$

where $\overline{\boldsymbol{\sigma}}'_{2D}(\overline{\mathbf{q}},\omega)$ is a 2D dyadic in the x-y plane given by

$$\overline{\boldsymbol{\sigma}}'_{2D}(\overline{\mathbf{q}},\omega) = -\frac{1}{m}\int_{-\infty}^{\infty}\frac{d^{(2)}\overline{\mathbf{q}}'}{(2\pi)^2}\left(\overline{\mathbf{q}}'-\frac{\overline{\mathbf{q}}}{2}\right)\left(\overline{\mathbf{q}}'-\frac{\overline{\mathbf{q}}}{2}\right)R\left(\overline{\mathbf{q}}',\overline{\mathbf{q}};\omega\right), \tag{10.8.14}$$

it is readily verified that $\overline{\boldsymbol{\sigma}}'_{2D}(\overline{\mathbf{q}},\omega)$ vanishes in the local limit (see Eqns. (10.7.36 and 10.7.37)). In consequence of Eqns. (10.8.12 and 10.8.13) we have

$$\begin{aligned}
&\overline{\mathfrak{G}}_{3D}(z=0,\overline{\mathbf{q}};\omega)\,\overline{\boldsymbol{\sigma}}_{2D}(\overline{\mathbf{q}},\omega)\\
&= -\frac{e^2}{2m\omega q_\perp}\left\{n_0\widehat{x}\widehat{x} + \frac{n_0c^2}{\omega^2\varepsilon(\omega)}q_\perp^2\widehat{y}\widehat{y} + \frac{n_0c^2}{\omega^2\varepsilon(\omega)}q_y^2\widehat{z}\widehat{z} + \left(\widehat{x}\widehat{x} + \frac{c^2}{\omega^2\varepsilon(\omega)}q_\perp^2\widehat{y}\widehat{y}\right)\overline{\boldsymbol{\sigma}}'_{2D}(\overline{\mathbf{q}},\omega)\right\}\\
&-\frac{ie^2n_0c^2}{m\omega^3\varepsilon}\delta(0)\widehat{z}\widehat{z}.
\end{aligned} \tag{10.8.15}$$

It should be noted that the $\widehat{z}\widehat{z}$ element at the end of Eq. (10.8.15) diverges as $\delta(0)$. Objectionable as this may seem, it simplifies the inversion process since the inverse of a 3×3 matrix A,

$$A = \begin{pmatrix} a_{xx} & a_{xy} & a_{xz}\\ a_{yx} & a_{yy} & a_{yz}\\ a_{zx} & a_{zy} & a_{zz}\end{pmatrix}, \tag{10.8.16}$$

having $a_{zz}\to\infty$ involves a determinant $\widetilde{\Delta}$ whose dominant term is

$$\lim_{a_{zz}\to\infty}\widetilde{\Delta} = a_{zz}(a_{xx}a_{yy} - a_{xy}a_{yx}). \tag{10.8.17}$$

The effect of this denominator on the minors and parts of minors of A that do *not* contain $a_{zz}\to\infty$ is to force the vanishing of their contributions to the inverse, while those minors and parts that *do* involve a_{zz} produce a finite contribution independent of $a_{zz}\to\infty$. It is straightforward to verify that the result of inversion is

$$\lim_{a_{zz}\to\infty} A^{-1} = \frac{1}{\Delta}\begin{pmatrix} a_{yy} & -a_{xy} & 0\\ -a_{yx} & a_{xx} & 0\\ 0 & 0 & 0\end{pmatrix} = \begin{bmatrix} \left(A_{2D}^{-1}\right) & 0\\ & 0\\ 0\ \ 0 & 0\end{bmatrix}, \tag{10.8.18}$$

where Δ is the *reduced determinant* of matrix A,

$$\Delta \equiv (a_{xx}a_{yy} - a_{xy}a_{yx}), \tag{10.8.19}$$

and A_{2D}^{-1} is the inverse of the finite 2×2 block of A. Thus, we may write this as a 2×2 block of the inverse matrix involved in Eq. (10.8.4, right-hand side), as

$$\left[\bar{\mathbf{I}} - \frac{4\pi i\omega}{c^2}\bar{\mathfrak{G}}_{3D}(z_1 = 0)\bar{\boldsymbol{\sigma}}_{2D}(\bar{\mathbf{q}}, \omega)\right]^{-1}$$
$$= \left[\bar{\mathbf{I}}_{2D} + \frac{i\omega_{2D}^2}{c^2 q_\perp q_y}\left\{\widehat{xx} + \frac{c^2}{\omega^2 \varepsilon(\omega)} q_\perp^2 \widehat{yy} + \frac{1}{n_0}\left(\widehat{xx} + \frac{c^2 q_\perp^2 \widehat{yy}}{\omega^2 \varepsilon(\omega)}\right)\bar{\boldsymbol{\sigma}}'_{2D}(\bar{\mathbf{q}}, \omega)\right\}\right]^{-1}, \tag{10.8.20}$$

(for the 2×2 block), with all other elements (third row and third column) of the 3×3 matrix vanishing. Here, $\omega_{2D}^2 = 2\pi e^2 n_0 q_y / m$ is the local 2D plasma frequency of the thin quantum well sheet. As indicated above, in the local limit $\bar{\boldsymbol{\sigma}}'_{QW}(\bar{\mathbf{q}}, \omega)$ vanishes, and, correspondingly, the EM wave dispersion relation associated with the vanishing reduced determinant of Eq. (10.8.19) factors into two parts:

$$(A)\ 1 + \frac{i\omega_{2D}^2}{c^2 q_\perp q_y} = 0 \tag{10.8.21}$$

and

$$(B)\ 1 + \frac{i\omega_{2D}^2 q_\perp}{\omega^2 q_y \varepsilon(\omega)} = 0. \tag{10.8.22}$$

If the host semiconductor has a local 3D bulk plasma ($\varepsilon(\omega) = 1 - \omega_p^2/\omega^2$) in addition to the embedded quantum well plasma, the solution of dispersion relation (A) couples light with both the bulk plasmon and the 2D-plasmon jointly, as (provided $\omega^2 > 0$):

$$(A)\ \omega^2 = c^2 q_y^2 + \omega_p^2 - \omega_{2D}^4 / c^2 q_y^2. \tag{10.8.23}$$

Furthermore, for the same host plasma, dispersion relation (B) also couples light with both the 3D-bulk and 2D-plasmons as (again, $\omega^2 > 0$):

$$(B)\ \omega^2 = \omega_p^2 - \frac{\omega_{2D}^4}{2q_y^2 c^2} \pm \omega_{2D}^2 \sqrt{1 + \omega_{2D}^4 / 4 q_y^4 c^4}. \tag{10.8.24}$$

In the electrostatic limit, this confirms the known coupling of 2D and 3D plasmons:

$$\omega^2 \to \omega_p^2 + \omega_{2D}^2. \tag{10.8.25}$$

Furthermore, if there is no 3D-bulk host plasma, $\omega_p \to 0$, Eq. (10.8.24) reduces to Stern's retarded result[37] (with $\pm \to +$ for stability).

In summary, with the neglect of $\bar{\boldsymbol{\sigma}}'_{2D}$ and the $\hat{z}\hat{z}$-element in the inversion of Eq. (10.8.20), the dyadic Green's function of Eq. (10.8.4) in $[\bar{\mathbf{q}}, z_1, z_2, \omega]$-representation is given by (suppress $\bar{\mathbf{q}}, \omega$):

$$\overline{\mathfrak{G}}(z_1, z_2) = \overline{\mathfrak{G}}_{3D}(z_1 - z_2) + \frac{4\pi i\omega}{c^2}\overline{\mathfrak{G}}_{3D}(z_1, 0)\overline{\boldsymbol{\sigma}}_{2D}(\mathbf{q}, \omega)\left[\frac{\hat{x}\hat{x}}{1 + \frac{i\omega_{2D}^2}{c^2 q_\perp q_y}} + \frac{\hat{y}\hat{y}}{1 + \frac{i\omega_{2D}^2 q_\perp}{\omega^2 \varepsilon(\omega) q_y}}\right]\overline{\mathfrak{G}}_{3D}(-z_2), \tag{10.8.26}$$

where we have executed the matrix inversion of Eq. (10.8.20). However, it is worth noting that the divergence of the z-z element of $\overline{\mathfrak{G}}_{3D}$ in Eqns. (10.8.12 and 10.8.15) is an artifact of our unphysical assumption of an infinitely thin quantum well, yielding a z-z term with $\delta(0) \to \infty$. For a thin quantum well, this deficiency can be remedied by estimating $\delta(0) \to \frac{1}{L_z}$, with L_z as the small but finite width of the well. Eq. (10.8.11) then yields the z-z element of $\overline{\mathfrak{G}}_{3D}$ as

$$\overline{\mathfrak{G}}_{3D}^{zz}(z = 0; \overline{q}; \omega) = -\hat{z}\hat{z}\left(\frac{1}{2iq_\perp}\left[1 - \frac{c^2}{\omega^2 \varepsilon(\omega)}\left(q_\perp^2 - 2i\frac{q_\perp}{L_z}\right)\right]\right). \tag{10.8.27}$$

Taken jointly with the first term on the right of Eq. (10.8.13), $\overline{\sigma}_{2D}^{zz}(\overline{q}, \omega) = \hat{z}\hat{z}ie^2 n_0/m\omega$, this yields the dispersion relation of yet another coupled mode (but $\overline{\boldsymbol{\sigma}}'_{2D}$ is neglected).

Another example of interest is the Type I superlattice, which we model as a periodic array of identical thin 2D sheets of mobile electrons that can move on a given sheet, but not between sheets. The conductivity of this system may be represented by summing analogous contributions of the type of Eq. (10.7.44) for each of the superlattice sheets situated at $z = na$ (a is the superlattice period, $n = -\infty \ldots -1, 0, 1 \ldots \infty$), assuming that there is negligible tunneling between superlattice sheets and that their separation a is much larger than their width. Thus, in accordance with the discussion following Eq. (10.6.12), we write the superlattice conductivity, $\overline{\boldsymbol{\sigma}}_{SL}$, as

$$\overline{\boldsymbol{\sigma}}_{SL}(z_1, z_2; \overline{\mathbf{q}}, \omega) = \sum_{n=-\infty}^{\infty} \delta(z_1 - na)\delta(z_2 - na)\overline{\boldsymbol{\sigma}}_{2D}(\overline{\mathbf{q}}, \omega), \tag{10.8.28}$$

where $\overline{\boldsymbol{\sigma}}_{2D}(\overline{\mathbf{q}}, \omega)$ is given by Eq. (10.7.45), and its local limit is $\overline{\boldsymbol{\sigma}}_{2d}(\overline{\mathbf{q}}, \omega) = \frac{ie^2 n_0}{m\omega}\overline{\mathbf{I}}$, where n_0 is the common 2D equilibrium density of each of the identical superlattice sheets (see Eqns. (10.8.13 and 10.8.14, ff.)). Again employing Eq. (10.6.29) with $\overline{\boldsymbol{\sigma}}_{nano} \to \overline{\boldsymbol{\sigma}}_{SL}$, we obtain the superlattice generalization of Eq. (10.8.2) as

$$\overline{\mathfrak{G}}(z_1, z_2) = \overline{\mathfrak{G}}_{3D}(z_1 - z_2) + \frac{4\pi i\omega}{c^2}\sum_{n=-\infty}^{\infty}\overline{\mathfrak{G}}_{3D}(z_1 - na)\cdot\overline{\boldsymbol{\sigma}}_{2D}\cdot\overline{\mathfrak{G}}(na, z_2), \tag{10.8.29}$$

and setting $z_1 = ma$, we have a matrix equation,

$$\overline{\mathfrak{G}}(ma, z_2) = \overline{\mathfrak{G}}_{3D}(ma - z_2) + \frac{4\pi i\omega}{c^2}\sum_{n=-\infty}^{\infty}\overline{\mathfrak{G}}_{3D}(ma - na)\cdot\overline{\boldsymbol{\sigma}}_{2D}\cdot\overline{\mathfrak{G}}(na, z_2). \tag{10.8.30}$$

Considering lattice periodicity, we examine a Fourier series having $\overline{\mathfrak{G}}(na, z_2)$ as coefficients: Denoting this Fourier series involving $\overline{\mathfrak{G}}(ma, z_2)$ by $\overline{\mathfrak{G}}(q, z_2)$ and that of $\overline{\mathfrak{G}}_{3D}(ma - na)$ by $\overline{\mathfrak{G}}_{3D}(q')$ in Eq. (10.8.30), i.e.:

$$\overline{\mathfrak{G}}(q, z_2) = \sum_{m=-\infty}^{\infty} e^{iqam}\overline{\mathfrak{G}}(ma, z_2) \quad ; \quad \overline{\mathfrak{G}}(ma, z_2) = \frac{a}{2\pi}\int_{-\pi/a}^{\pi/a} dq e^{-iqam}\overline{\mathfrak{G}}(q, z_2),$$

and regarding z_2 as a fixed parameter, we have

$$\overline{\mathfrak{G}}(ma, z_2) = \overline{\mathfrak{G}}_{3D}(ma - z_2) + \left(\frac{4\pi i\omega}{c^2}\right)\left(\frac{a}{2\pi}\right)^2 \times \int_{-\pi/a}^{\pi/a} dq' \int_{-\pi/a}^{\pi/a} dq\, e^{-imaq'} \sum_{n=-\infty}^{\infty} e^{ina(q'-q)}\overline{\mathfrak{G}}_{3D}(q') \cdot \overline{\boldsymbol{\sigma}}_{2D} \cdot \overline{\mathfrak{G}}(q, z_2). \qquad (10.8.31)$$

Employing the Poisson summation formula, $\left[\frac{1}{a}\sum_{n=-\infty}^{\infty} \exp\left(2\pi\, inz/a\right) = \sum_{n=-\infty}^{\infty} \delta(z - na)\right]$, the sum over n is

$$\sum_{n=-\infty}^{\infty} e^{in(q'-q)a} = \frac{2\pi}{a}\sum_{n=-\infty}^{\infty} \delta(q' - q + 2\pi\, n/a) = \frac{2\pi}{a}\delta(q' - q), \qquad (10.8.32)$$

where we have used the fact that q, q' are restricted to the interval $(-\pi/a, \pi/a)$, so only the $n = 0$ Dirac delta function term need be considered. Carrying out the q'-integration of Eq. (10.8.29), we have

$$\overline{\mathfrak{G}}(ma, z_2) = \overline{\mathfrak{G}}_{3D}(ma - z_2) + \frac{4\pi i\omega}{c^2}\frac{a}{2\pi}\int_{-\pi/a}^{\pi/a} dq\, e^{-imaq}\overline{\mathfrak{G}}_{3D}(q)\overline{\boldsymbol{\sigma}}_{2D}\overline{\mathfrak{G}}(q, z_2). \qquad (10.8.33)$$

Furthermore, in terms of $\overline{\mathfrak{G}}(q, z_2)$, this result reads as

$$\overline{\mathfrak{G}}(q, z_2) = \overline{\mathfrak{G}}_{3D}(q, z_2) + \frac{4\pi i\omega}{c^2}\overline{\mathfrak{G}}_{3D}(q)\overline{\boldsymbol{\sigma}}_{2D}\overline{\mathfrak{G}}(q, z_2), \qquad (10.8.34)$$

which may be solved algebraically as

$$\overline{\mathfrak{G}}(q, z_2) = \overline{\Delta}^{-1}\overline{\mathfrak{G}}_{3D}(q, z_2), \qquad (10.8.35)$$

where

$$\overline{\Delta}(q) = \overline{\mathbf{I}} - \frac{4\pi i\omega}{c^2}\overline{\mathfrak{G}}_{3D}(q) \cdot \overline{\boldsymbol{\sigma}}_{2D}. \qquad (10.8.36)$$

Inverting $\overline{\mathfrak{G}}(q, z_2)$ back into $\overline{\mathfrak{G}}(na, z_2)$ and substituting it into Eq. (10.8.29), we obtain the dyadic photon propagator for an infinite superlattice of 2D quantum wells embedded in a 3D bulk medium as

$$\overline{\mathfrak{G}}(z_1, z_2) = \overline{\mathfrak{G}}_{3D}(z_1 - z_2) + \frac{4\pi i\omega}{c^2} \sum_{n=-\infty}^{\infty} \overline{\mathfrak{G}}_{3D}(z_1 - na) \cdot \overline{\boldsymbol{\sigma}}_{2D}$$
$$\times \sum_{m=-\infty}^{\infty} \left(\frac{a}{2\pi}\right) \left(\int_{-\pi/a}^{\pi/a} dq e^{iqa(m-n)} \overline{\Delta}^{-1}(q)\right) \cdot \overline{\mathfrak{G}}_{3D}(ma - z_2). \qquad (10.8.37)$$

To determine the normal modes of the superlattice, it is necessary to examine the frequency poles of $\overline{\Delta}^{-1}$, which may be identified from the zeros of $\det\overline{\Delta}$,

$$\det \overline{\Delta} = \det \left(\bar{\mathbf{I}} - \frac{4\pi i\omega}{c^2} \overline{\mathfrak{G}}_{3D}(q), \overline{\boldsymbol{\sigma}}_{2D})\right) = \det \left(\bar{\mathbf{I}} - \frac{4\pi i\omega}{c^2} \sum_{m=-\infty}^{\infty} e^{iqam} \overline{\mathfrak{G}}_{3D}(ma, z_2) \overline{\boldsymbol{\sigma}}_{2D}\right), \qquad (10.8.38)$$

where $\overline{\mathfrak{G}}_{3D}(z_1, z_2)$ is given in Section 10.6.

Considering the z-Fourier series coefficient $\overline{\mathfrak{G}}_{3D}(q) \equiv \overline{\mathfrak{G}}_{3D}(q, \overline{\mathbf{q}}) = \sum_{m=-\infty}^{\infty} e^{iqam} \overline{\mathfrak{G}}_{3D}(ma, \overline{\mathbf{q}})$ with $\overline{\mathfrak{G}}_{3D}(z = ma, \overline{\mathbf{q}})$ given by Eq. (10.8.10), and employing the following identities to carry out the indicated summations:

$$S_1 \equiv \sum_{m=-\infty}^{\infty} e^{iq_\perp |m| a} e^{iqma} = -i \frac{\sin(q_\perp a)}{\cos(q_\perp a) - \cos(qa)}; \ \text{Re}(iq_\perp) < 0, \qquad (10.8.39)$$

$$S_2 \equiv \sum_{m=-\infty}^{\infty} sign(ma) e^{iq_\perp |m| a} e^{iqma} = i \frac{\sin(qa)}{\cos(q_\perp a) - \cos(qa)}; \quad \text{Re}(iq_\perp) < 0, \qquad (10.8.40)$$

$$S_3 \equiv \sum_{m=-\infty}^{\infty} \delta(ma) e^{iq_\perp |m| a} e^{iqma} = \infty, \quad [\delta(0) \to \infty], \qquad (10.8.41)$$

the matrix $\overline{\mathfrak{G}}_{3D}(q)$ is given by

$$\overline{\mathfrak{G}}_{3D}(q) = -\left(\frac{1}{2iq_\perp}\right)\left(\frac{c^2}{\omega^2 \varepsilon}\right) \begin{pmatrix} \frac{\omega^2}{c^2}\varepsilon S_1 & 0 & 0 \\ 0 & q_\perp^2 S_1 & -q_\perp q_y S_2 \\ 0 & -q_\perp q_y S_2 & \infty \end{pmatrix} \qquad (10.8.42)$$

(it should be noted that the $\widehat{zz}$ component is infinite because $S_3 \to \infty$). Employing Eq. (10.8.42) into Eq. (10.8.36) yields

$$\overline{\Delta}(q) = \begin{pmatrix} \Delta_{xx} & 0 & 0 \\ 0 & \Delta_{yy} & \Delta_{yz} \\ 0 & \Delta_{zy} & \Delta_{zz} \end{pmatrix}$$
$$= \begin{pmatrix} 1 + \alpha \frac{\omega^2}{c^2} \varepsilon S_1 \sigma_{2D}^{xx} & 0 & 0 \\ 0 & 1 + \alpha q_\perp^2 S_1 \sigma_{2D}^{yy} & -\alpha q_\perp q_y S_2 \sigma_{2D}^{zz} \\ 0 & -\alpha q_\perp q_y S_2 \sigma_{2D}^{yy} & \infty \end{pmatrix}, \qquad (10.8.43)$$

with $\alpha = (2\pi/q_\perp\omega\varepsilon)$. Inverting the dyadic $\overline{\Delta}(q)$, we have

$$\overline{\Delta}^{-1}(q) = \frac{1}{\Delta}\begin{pmatrix} \Delta_{yy} & 0 & 0 \\ 0 & \Delta_{xx} & 0 \\ 0 & 0 & 0 \end{pmatrix}, \tag{10.8.44}$$

with Δ as the reduced determinant of $\overline{\Delta}(q)$, namely

$$\Delta = \Delta_{xx}\Delta_{yy}. \tag{10.8.45}$$

The EM superlattice spectrum consists of bulk modes from the frequency poles of $\overline{\mathfrak{G}}_{3D}$ in Eq. (10.8.37) and also quasi-2D modes associated with the frequency roots of Δ_{xx} and Δ_{yy}. The vanishing of Δ_{xx} yields

$$\Delta_{xx} = 0 = 1 - i\frac{2\pi\omega}{q_\perp c^2\varepsilon(\omega)}\frac{\sin(q_\perp a)}{\cos(q_\perp a) - \cos qa}\sigma_{QW}^{xx},$$

and in the local limit we have

$$\Delta_{xx} = 0 = 1 + \frac{\omega_{2D}^2\mathcal{S}}{c^2 q_\perp q\varepsilon(\omega)} \quad \text{where} \quad \mathcal{S} \equiv \frac{\sin(q_\perp a)}{\cos(q_\perp a) - \cos qa}. \tag{10.8.46}$$

In the limit $qa \to \infty$, $q_\perp \to iq$, $\mathcal{S} \to i$ and we recover Eq. (10.8.21) for a single 2D planar quantum well. The vanishing of Δ_{yy} is given by

$$\Delta_{yy} = 0 = 1 + \frac{2\pi q_\perp}{\omega\varepsilon(\omega)}S_1\sigma_{QW}^{yy}, \tag{10.8.47}$$

and in the local limit we have

$$\Delta_{yy} = 0 = 1 + \frac{\omega_{2D}^2}{\omega^2\varepsilon(\omega)}\frac{q_\perp}{q}\mathcal{S}. \tag{10.8.48}$$

In the limit $qa \to \infty$, we recover Eq. (10.8.22). More generally, there is a (relatively) low frequency ($\omega < cq$) root of Eq. (10.8.43) given by (we assume that it is coupled to a background bulk 3D plasmon of the host semiconductor):

$$\omega^2 = \omega_p^2 + \omega_{2D}^2\frac{\sinh(qa)}{\cosh(qa) - \cos(qa)}. \tag{10.8.49}$$

This dispersion relation represents a band of mode frequencies since there are solutions for every value of q in the range $\frac{-\pi}{a} \le q \le \frac{\pi}{a}$. In the absence of a bulk 3D plasma this reduces to the well-known results of Fetter[38] and Olego, *et al.*[39].

While we have limited our discussion to local, planar, thin quantum wells, it is clear that the formulation presented here can incorporate the roles of non-locality and finite width of the wells in the polariton modes. Such hybridizations of photons with polarization modes of the semiconductor medium may include bulk and surface plasmons and phonons, superlattice, 2D and 1D quantum wire plasmons, quantum dot collective density oscillations, as well as any combination of these features in a nanostructured device. One needs only to construct the pertinent dyadic EM Green's function *without* the given nanostructure feature and use it in place of $\overline{\mathfrak{G}}_{3D}$ in Eq. (10.6.29), jointly with the conductivity associated with *that feature alone* put in place of $\overline{\boldsymbol{\sigma}}_{nano}$ in the kernel of the integral equation (Eq. (10.6.29)).

10.9 Chapter 10 Problems

P.10.1. Show that Eq. (10.1.27) can be ascribed to the commutation of density operators at different space points at the same time.

P.10.2. (a) Prove Eq. (10.2.1) for the RPA plasmon dispersion relation of a 3D plasma of spinless electrons at arbitrary temperature. (b) Verify Eq. (10.2.3) at zero temperature and determine the leading non-local shift of the plasma frequency in a low wave number power expansion.

P.10.3. Consider the role of a normal magnetic field on a plasma of electrons confined to move on a 2D plane at $z = 0$: Derive the associated RPA dielectric function $\varepsilon(\mathbf{p}, \omega)$ at arbitrary temperature.

P.10.4. Further to Problem 3 above, describe (qualitatively) the plasmon spectrum and the associated plasmon damping.

P.10.5. (a) Derive the 2D static shielded potential $V(\mathbf{x}_1)$ analogous to the 3D result of Eq. (10.2.7). Show that there is a Friedel oscillatory long-range contribution of the form $V(\mathbf{r}, t) \sim \sin(2p_F r)/(2p_F r)^2$ at low temperature. (No magnetic field.) (b) Verify Eq. (10.5.17) and the remarks immediately following it.

P.10.6. Derive the *hydrodynamic* dielectric function $\varepsilon(\mathbf{p}, \omega)$ in the presence of a classical magnetic field (no Landau quantization) analogous to Eq. (10.2.26).

P.10.7. Determine the leading terms of the exchange and correlation energies discussed in Section 10.3 at zero temperature (no magnetic field).

P.10.8. Derive the Levine–Schwinger EM dyadic Green's function of Eq. (10.6.10) and verify the associated free-space 3D scalar Helmholtz Green's function.

P.10.9. Verify the equation for $\mathbf{E} = \bar{\mathbf{K}}\mathbf{E}_0$ (Eq. (10.6.38)) and the associated remarks following Eq. (10.6.39) (including the summary paragraph below it).

P.10.10. Prove Eq. (10.6.44).

P.10.11. Verify Eqns. (10.7.27 and 10.7.28).

P.10.12. Prove Eqns. (10.7.46 and 10.7.47).

P.10.13. The discussion of polaritons in a compound nanostructure was focused on the dispersion relations (A) and (B) of Eqns. (10.8.23 and 10.8.24), which couple light with both bulk and 2D plasmons jointly. However, it is noted below that for a thin quantum well of finite width L_z, there is yet another coupled mode: Determine the dispersion relation of this mode.

P.10.14. Carry out the multipole expansion following Eq. (10.4.31) in detail to obtain Eqns. (10.4, 10.39–10.41) for $E_{vdw}^{(2)}$ to dipole–dipole terms.

P.10.15. The paragraph between Eqns. (10.7.16) and (10.7.17) suggests an alternative way of deriving $\overline{\mathrm{L}}_p(1,2)$, including an ambient quantizing magnetic field. Carry out this derivation and verify the associated result.

10.10 References for Chapter 10

R.10.1. L.P. Kadanoff and G. Baym, "Quantum Statistical Mechanics," W.A. Benjamin, Appendix (1962).
R.10.2. P.J. Feibelman, C.B. Duke and A. Bagchi, Phys. Rev. *B5*, 2436 (1972).
R.10.3. A.I. Akhiezer, "Collective Oscillations in a Plasma," Pergamon Press (1967).
R.10.4. L.D. Landau, "Collected Papers," Pergamon Press, p.445 (1965).
R.10.5. J.D. Jackson, J. Nucl. Energy, Part C *1*, 171 (1960).
R.10.6. A.L. Fetter and J.D. Walecka, "Quantum Theory of Many-Particle Systems," McGraw Hill (1971). (See p. 183–7 (zero sound); pp. 175–9 (Friedel oscillation); p. 411 (Kohn anomaly)
R.10.7. Paul Harrison, "Quantum Wells, Wires & Dots," 2nd Ed., John Wiley & Sons (2005).
R.10.8. T. Chakraborty, "Quantum Dots," Elsevier Science B.V. (1999).
R.10.9. S.M. Reimann and M. Manninen, Rev. Mod. Phys. *74*, 1283 (2002).
R.10.10. F. Stern, Phys. Rev. Lett. *18*, 546 (1967). Also, see T. Ando, A.B. Fowler and F. Stern, Rev. Mod. Phys. *54*, 437 (1982) (see pp. 449–452.)
R.10.11. P.F. Williams and A.N. Bloch, Phys. Rev. *B10*, 1097 (1974).
R.10.12. M.M. Mohan and A.Griffin, Phys. Rev. *B32*, 2030 (1985).
R.10.13. N.J.M. Horing, Annals of Physics (NY) *31*, 1 (1965); Also, N.J.M Horing, Phys. Rev. *136*, A494 (1964).
R.10.14. N.J.M. Horing, J. Phys. Soc. Japan *21*, Supp., 704 (1966); Also, N.J.M. Horing, Annals of Physics (NY) *68*, 337 (1971), ibid: *54*, 405 (1969).
R.10.15. N.J.M. Horing, Phys. Rev *186*, 434 (1969).
R.10.16. N.J.M. Horing, "Plasmon Resonances in the Quantum Strong Field Limit," in "The Many Body Problem," Editors: L.M. Garrido, A. Cruz and T.W. Priest, Plenum Press, p.307 (1969).
R.10.17. I.B. Bernstein, Phys. Rev. *109*, 10 (1958).
R.10.18. N.J.M. Horing and M. Yildiz, Phys. Lett. *A44*, 386 (1973).
R.10.19. N.J.M. Horing, M. Orman and M. Yildiz, Phys. Lett. *A48*, 7 (1974).
R.10.20. N.J.M. Horing and M. Yildiz, Annals of Physics (NY) *97*, 216 (1976).
R.10.21. N.J.M. Horing, M. Orman, E. Kamen and M.L. Glasser, Phys. Lett. *85A*, 378 (1981).
R.10.22. A.L. Fetter, Annals of Phys. (NY) *81*, 367 (1973).
R.10.23. D. Pines "Elementary Excitations in Solids," W.A. Benjamin, Inc. (1964).
R.10.24. D. Pines "The Many-Body Problem," W.A. Benjamin, Inc. (1961). (Including original papers reproduced therein.)
R.10.25. Ref.R.10.23, Appendix C.
R.10.26. N.J.M. Horing and L.Y. Chen, Phys. Rev. A *66*, 042905 (2002).

R.10.27. C. Mavroyannis, Mol. Phys. *6*, 593 (1963).
R.10.28. J.E. Lennard-Jones, Trans. Faraday Soc. *28*, 333 (1932).
R.10.29. N.J.M. Horing, T.Y Bagaeva and V.V. Popov, J. Opt. Soc. Amer. *24*, 2428 (2007).
R.10.30. H. Levine and J. Schwinger, Phys. Rev. *74*, 958 (1948); H. Levine and J. Schwinger, "On the Theory of Diffraction by an Aperture in an Infinite Plane Screen. II.," Phys. Rev. *75*, 1423 (1949).
R.10.31. H. Levine and J. Schwinger, Commun. Pure & Appl. Math. *3*, 355–391 (1950).
R.10.32. J.D. Jackson, "Classical Electrodynamics," 2nd Edition, Wiley, p.287 (1975).
R.10.33. M. Orman, "Electromagnetic Response of a Quantized Magnetoplasma in Two and Three Dimensions," Stevens Institute of Technology, Ph.D. Thesis (1974).
R.10.34. P.C. Martin and J. Schwinger, Phys. Rev. **115**, 1342, Eqns. (4.56 and 57) (1959).
R.10.35. P.C. Martin, Phys. Rev. *161*, 143 (1967).
R.10.36. L.D. Landau and E.M. Lifshitz, "Statistical Physics, Part I," Section 123, Pergamon Press (1980).
R.10.37. F. Stern, Phys. Rev. Lett. *18*, 546 (1967). Also, T. Ando, A.B. Fowler and F. Stern, Rev. Mod. Phys. *54*, 437 (1982).
R.10.38. A.L. Fetter, Ann. Phys, (N.Y.) *88*, 1 (1974).
R.10.39. D. Olego, A. Pinczuk, A.C. Gossard, Phys. Rev. B *26*, 7867 (1982).

11

Interacting Electron–Hole–Phonon System

Abstract
Chapter 11 is focused on the use of variational differential techniques and the Schwinger Action Principle to derive coupled-field Green's function equations governing the quantum dynamics of a multi-component system having multiple interactions, illustrating some features of an interacting electron–hole–phonon system in a crystalline solid. The coupled Green's function equations involve five interactions (electron–electron, hole–hole, electron–hole, electron–phonon and hole–phonon), which, taken jointly, would be challenging to represent with Feynman diagrammatics. The chapter opens with a detailed description of a model Hamiltonian, which features electron–phonon and hole–phonon interactions in a small-oscillation approximation for the ion-core lattice (in addition to the Coulombic repulsions among electrons and among holes and Coulombic attraction between electrons and holes). Starting with quantum Hamilton equations of motion for the various creation and annihilation operators and their non-equilibrium average/expectation values, variational differentiation with respect to particle sources leads to a chain of coupled Green's function equations involving differing species of Green's functions. For example, the one-electron Green's function equation is coupled to the two-electron Green's function (as seen in earlier chapters) and also to the one-electron/one-hole Green's function, and it is also coupled to the Green's function for one-electron propagation influenced by a non-trivial phonon field. Similar remarks apply to the one-hole Green's function equation. To obtain requisite information for the involved higher order Green's functions, their equations of motion are derived by applying further variational differentiation with respect to appropriate sources, resulting, for example, in a two-electron Green's function equation coupled to a three-electron Green's function and a two-electron/one-hole Green's function, as well as to the Green's function for two-electron propagation influenced by a non-trivial phonon field (ditto for the two-hole Green's function equation). Similar "chain" couplings occur for the one-electron/one-hole Green's function equation. Moreover, the Green's function equation for single-electron propagation influenced by a non-trivial phonon field is derived by variational differentiation of the one-electron Green's function equation with respect to the phonon source, etc. Chapter 11 also introduces the one-phonon Green's function using the quantum Hamilton equations of motion for the ionic coordinate displacements and conjugate momenta with non-equilibrium averaging. In this model calculation holes are ignored, and the role of electron coupling in phonon propagation

Quantum Statistical Field Theory: Schwinger's Variational Method. Norman J. Morgenstern Horing, Oxford University Press (2017). DOI 10.1093/oso/9780198791942.001.0001

is emphasized and seen to result in dynamic, non-local electron screening of the phonon spectrum, with attendant hybridization of the ion and electron plasma modes, a Bohm–Staver phonon mode and the Kohn anomaly. (There is also a brief review of non-interacting phonon Green's functions for the uninitiated.) Another important feature of the single-electron Green's function with phonon coupling shown in Chapter 11 is that its equation of motion can be rewritten in the usual form as coupled to the two-electron Green's function—*even in the absence of both interelectron Coulomb interactions and electron–hole interactions*—with an effective time-dependent electron–electron interaction potential mediated by the one-phonon Green's function, whence the *polaron* arises as an electron propagating jointly with its induced lattice polarization. An alternative formulation of the coupled Green's function equations of motion for the electron–hole–phonon model system with five interactions is also discussed in the interest of facilitating useful approximation procedures. It is applied in a further analysis of electron–hole interaction dynamics and the development of a generalized shielded potential approximation with an electron–hole interaction operator. Finally, the inverse dielectric screening response function for the fully coupled electron–hole–phonon system is discussed in detail. The associated polarizability is identified with its various contributions arising from Coulombic electron–electron and hole–hole interactions and others arising from phonon-mediated electron–electron and hole–hole interactions and with the role of the electron–hole interaction duly noted; there is yet another polarizability contribution arising from the effect of the self-consistent field on ionic displacements. The associated collective mode dispersion relation hybridizes all the modes that would be associated with these polarizability contributions individually. Lastly, there is a brief discussion of the (theoretical) origin of the exciton–plasmon interaction.

11.1 Electron–Hole–Phonon Interaction Hamiltonian

Crystalline solids[1–4]—metals, insulators and semiconductors—generally occur with their atoms positioned in a regular periodic cellular lattice array in equilibrium. The electron energy eigenstate spectrum forms into *bands* due to the periodic lattice potential of the atoms, with the bands of allowed energy eigenvalues separated by intervening *gap* regions in which no energy eigenvalues can occur. Mobile electrons in a partially empty *conduction band* are free to move through the crystal carrying current, leaving behind the charged ion cores of their parent atoms. Moreover, in the case of a *valence band*, which is almost filled with electrons, a vacant state behaves as a particle of opposite (positive) charge, a *hole*, with respect to charge transport. In metals it is relatively easy to excite electrons between neighboring conduction band states with small amounts of energy, whereas in insulators the large energy required to excite electrons across a large gap between a fully populated valence band and an empty conduction band is prohibitive. Semiconductors, with their remarkable and exceptionally useful conduction and optical properties, stand between metals and insulators. In their quiescent state, their valence band is fully occupied and their conduction band is completely vacant. However, modest excitation due to thermal activity can excite electrons across the gap into the conduction band leaving vacant state holes in the valence band. Such mobile electrons and holes can also occur by ionization or trapping of an electron by impurity atom dopants embedded in the structure of the semiconductor. The model we eventually employ for electrons and holes in actual calculations is restricted to the effective mass approximation, which severely limits its validity in representing solid-state phenomenology in general. Nevertheless, the effective mass regime is itself one of considerable interest and it provides a fine vehicle for illustration of the coupled-field Green's function equations governing the quantum dynamics of multiple interacting quantum systems, while addressing several important features of solid-state physics.

The ion cores of the atoms, positioned in a periodic lattice array in equilibrium, interact by means of their charge both with each other and with the mobile electron and hole charge carriers. In regard to the charge interaction of the ion cores among themselves, they can execute oscillatory motions about their equilibrium positions. The interaction u of the kth ion of the lth unit cell (equilibrium position $\mathbf{R}_{lk}$) having displacement operator (i^{th} component) ${}^{i}Q_{lk}(t)$, with the k'^{th} ion of the l'^{th} unit cell (equilibrium position $\mathbf{R}_{l'k'}$) having displacement operator (i'^{th} component) ${}^{i'}Q_{l'k'}(t)$, is described by a Hamiltonian term involving their mutual potential energy as[1–4]

$$H_{ii} = \frac{1}{2}\sum_{l,k,l',k'} u\left(\mathbf{R}_{lk} + \mathbf{Q}_{lk} - \mathbf{R}_{l'k'} - \mathbf{Q}_{l'k'}\right), \quad (k \neq k' \,;\, l \neq l') \tag{11.1.1}$$

(ion charge number, magnitude and sign will be installed as needed below). Anticipating the description of small oscillations of the ion-core lattice, a Taylor expansion of H_{ii} in powers of $\mathbf{Q}_{lk}$ and $\mathbf{Q}_{l'k'}$ jointly yields a constant zero-order term that can be ignored, also first-order terms that vanish by definition of the equilibrium positions of the ions and second-order terms that are of major importance in defining the possible small lattice oscillations of the system, whose quanta are termed *phonons*. The second-order terms are given by

$$H_{ii} \cong \frac{1}{2}\sum_{\substack{i,l,k\\ i',l',k'}} {}^{i}Q_{lk}\,{}^{j'}Q_{l'k'}\,(\nabla_i\nabla_{i'}u) = \frac{1}{2}\sum_{\substack{i,l,k\\ i',l',k'}} \Xi_{ii'}(l,k;l',k')\,{}^{i}Q_{lk}\,{}^{i'}Q_{l'k'}\,, \tag{11.1.2}$$

where the real, symmetric coefficient, $\nabla_i\nabla_{i'}\,u \equiv \Xi_{ii'} \equiv \Xi_{i'i}$, is evaluated at the equilibrium positions of the ion cores. Higher order terms, which we will not discuss, are responsible for phonon–phonon interactions, which become active as the small oscillation approximation breaks down. Of course, there is also a kinetic energy term in the Hamiltonian for ion motion (M_k is the ion mass):

$$H_i = \frac{1}{2}\sum_{i,l,k} M_k\,{}^{i}\dot{Q}_{lk}\,{}^{i}\dot{Q}_{lk} = \frac{1}{2}\sum_{i,l,k}\frac{{}^{i}P_{lk}\,{}^{i}P_{lk}}{M_k}\,, \tag{11.1.3}$$

where ${}^{i}\dot{Q}_{lk} \equiv d\,{}^{i}Q_{lk}/dt$ and ${}^{i}P_{lk} \equiv M_k\,{}^{i}\dot{Q}_{lk_2}$. The free phonon Hamiltonian H_p encompasses both the ion kinetic energy H_i and H_{ii} as

$$H_p = H_i + H_{ii}\,, \tag{11.1.4}$$

which obviously describes coupled oscillator motion, with the normal modes being free phonons whose eigenenergies are just the characteristic normal mode frequencies (modulo $\hbar$). In the "first quantized" description we employ for ionic motion (many solid-state physics textbooks provide the second quantized description for phonons), we have the canonical equal-time commutation relations for ${}^iQ_{lk}$ and its conjugate momentum ${}^iP_{lk} = M_k\, {}^i\dot{Q}_{lk}$ as

$$\left[{}^iQ_{lk}(t),\, {}^{i'}P_{l'k'}(t)\right] = i\,\delta_{ii'}\,\delta_{kk'}\,\delta_{ll'} \tag{11.1.5}$$

and

$$\left[{}^iQ_{lk}(t),\, {}^{i'}Q_{l'k'}(t)\right] = 0 = \left[{}^iP_{lk}(t),\, {}^{i'}P_{l'k'}(t)\right]. \tag{11.1.6}$$

All components of ${}^iQ_{lk}$ and ${}^iP_{lk}$ commute with the electron and hole creation and annihilation operators at the same time.

The charge coupling of the ion cores with mobile electrons (e) and holes (h) may be described by an interaction Hamiltonian written in terms of the potential projected by the ion cores on an element of electron or hole density as ($\hat{\Psi}^+, \hat{\Psi}$ and $\hat{\Phi}^+, \hat{\Phi}$ are the creation and annihilation operators for electrons and holes, respectively, as in Section 8.4.):

$$H_{\left\{\substack{e-i \\ h-i}\right\}} = \sum_{l,k} \int d^{(3)}\mathbf{x}_1 \left\{\begin{matrix}\hat{\Psi}^+(1)\hat{\Psi}(1) \\ \hat{\Phi}^+(1)\hat{\Phi}(1)\end{matrix}\right\} u\left(\mathbf{x}_1 - \mathbf{R}_{lk} - \mathbf{Q}_{lk}\right). \tag{11.1.7}$$

For small ion oscillations, the potential u may be expanded to linear order in ${}^i\mathbf{Q}_{lk}$ as (neglecting the irrelevant zero-order term)

$$H_{\left\{\substack{e-i \\ h-i}\right\}} = \sum_i \sum_{l,k} \int d^{(3)}\mathbf{x}_1 \left\{\begin{matrix}\hat{\Psi}^+(1)\hat{\Psi}(1) \\ \hat{\Phi}^+(1)\hat{\Phi}(1)\end{matrix}\right\} {}^i\gamma_{lk}^{\left\{\substack{e \\ h}\right\}}(\mathbf{x}_1)\, {}^iQ_{lk}(t_1), \tag{11.1.8}$$

where (note that we allow that the effective electron–ion and hole–ion interactions ${}^i\gamma_{lk}^{\left\{\substack{e \\ h}\right\}}(\mathbf{x}_1)$ may differ):

$${}^i\gamma_{lk}^{\left\{\substack{e \\ h}\right\}}(\mathbf{x}_1) = \nabla_{1i}\, u(\mathbf{x}_1 - \mathbf{R}_{lk}). \tag{11.1.9}$$

Larger ion oscillations would involve the expansion of $H_{\left\{\substack{e-i \\ h-i}\right\}}$ to second order in ${}^iQ_{lk}(t_1)$ by expanding $u(\mathbf{r}_1 - \mathbf{R}_{lk} - \mathbf{Q}_{lk})$ to second order. However, that is not pertinent to our present considerations. (As indicated above, we now employ Ψ^+, Ψ (Φ^+, Φ) as mobile electron (hole) creation, annihilation operators rather than ψ^+, ψ employed previously, since the latter would apply to *all* band energy eigenstates in the present context.) Including electron–electron ($e - e$) and hole–hole ($h - h$) Coulomb interactions together with the present considerations, we write the Hamiltonian H of the fully interacting electron–hole–phonon system as ($\mathcal{V}$ is the usual particle–particle interaction potential):

$$
\begin{aligned}
H(t_1) = &\int d^{(3)}\mathbf{x}_1\, \hat{\Psi}^+(1)\, h_{100}\, \hat{\Psi}(1) + \int d^{(3)}\mathbf{x}_1 \hat{\Phi}^+(1) h_{001} \hat{\Phi}(1) + H_p(t_1) \\
&+ \frac{1}{2}\int d^{(3)}\mathbf{x}_1 \int d^{(3)}\mathbf{x}_2\, \hat{\Psi}^+(1)\, \hat{\Psi}^+(\mathbf{x}_2, t_1) \mathcal{V}(\mathbf{x}_1 - \mathbf{x}_2)\, \hat{\Psi}(\mathbf{x}_2, t_1)\, \hat{\Psi}(1) \\
&+ \frac{1}{2}\int d^{(3)}\mathbf{x}_1 \int d^{(3)}\mathbf{x}_2\, \hat{\Phi}^+(1)\, \hat{\Phi}^+(\mathbf{x}_2, t_1) \mathcal{V}(\mathbf{x}_1 - \mathbf{x}_2)\, \hat{\Phi}(\mathbf{x}_2, t_1)\, \hat{\Phi}(1) \\
&- \int d^{(3)}\mathbf{x}_1 \int d^{(3)}\mathbf{x}_2 \hat{\Psi}^+(1)\, \hat{\Psi}(1)\, \mathcal{V}(\mathbf{x}_1 - \mathbf{x}_2)\, \hat{\Phi}^+(\mathbf{x}_2, t_1)\, \hat{\Phi}(1) \\
&- \sum_i \sum_{l,k} \int d^{(3)}\mathbf{x}_1\, {}^i\gamma^e_{lk}(\mathbf{x}_1)^i\, Q_{lk}(t_1)\, \hat{\Psi}^+(1)\, \hat{\Psi}(1) \\
&- \sum_i \sum_{l,k} \int d^{(3)}\mathbf{x}_1\, {}^i\gamma^h_{lk}(\mathbf{x}_1)^i\, Q_{lk}(t_1)\, \hat{\Phi}^+(1)\, \hat{\Phi}(1).
\end{aligned}
\tag{11.1.10}
$$

Here, h_{100}, h_{001} (first and second terms on the right of Eq. (11.1.10)) represent the single electron, hole band Hamiltonians in position representation, including kinetic energy, possibly time-dependent external electric and magnetic fields, and the periodic lattice potential, which we will represent in terms of effective masses[1–4] for the electrons and holes for illustrative purposes. As indicated above, the third term, H_p, is the free phonon Hamiltonian, Eq. (11.1.4). Furthermore, the fourth and fifth terms are the Hamiltonian terms describing repulsive *e-e* and *h-h* Coulomb interactions, respectively. The electron–hole (*e-h*) Coulomb attraction, which gives rise to important effects in solids, including *e-h* drag and the *exciton e-h* bound states discussed in Section 8.4, is represented in the sixth term of the Hamiltonian on the right of Eq. (11.1.10). Moreover, the seventh and eighth terms are the parts of the Hamiltonian that describe electron–phonon and hole–phonon interactions, respectively. The derivation of this interaction Hamiltonian, which is briefly described qualitatively above, is carried out in full in Reference 3. In addition, we introduce a *source* Hamiltonian as $H'(t_1) = H'_B(t_1) + H'_F(t_1)$, where $H'_B(t_1)$ is a *Boson source* Hamiltonian term, which includes the source potential U associated with the electron/hole densities (excluding the ion-core densities) and the phonon particle source ${}^i\mathcal{J}_{lk}$ as

$$
H'_B(t_1) = \int d^{(3)}\mathbf{x}_1 \left[\hat{\Psi}^+(1)\ \hat{\Psi}(1) - \hat{\Phi}^+(1)\, \hat{\Phi}(1)\right]\ U(1) + \sum_{ilk} {}^i\mathcal{J}_{lk}(t_1)\, {}^iQ_{lk}(t_1), \tag{11.1.11}
$$

and the Fermi particle source Hamiltonian term is given by

$$
\begin{aligned}
H'_F(t_1) = &\int d^{(3)}\mathbf{x}_1 \left[\hat{\Psi}^+(1)\, \xi(1) + \xi^+(1)\hat{\Psi}(1)\right] \\
&+ \int d^{(3)}\mathbf{x}_1 \left[\hat{\Phi}^+(1)\, \zeta(1) + \zeta^+(1)\hat{\Phi}(1)\right].
\end{aligned}
\tag{11.1.12}
$$

The source potential $eU \to U$, the phonon source ${}^i\mathcal{J}_{lk}$ and the Fermion particle sources

$$
\eta, \eta^+ \to \begin{Bmatrix} \xi, \xi^+ \\ \zeta, \zeta^+ \end{Bmatrix} \begin{matrix} \leftarrow \text{(electrons)} \\ \leftarrow \text{(holes)} \end{matrix} \tag{11.1.13}
$$

have all been described in Chapter 4. The complete Hamiltonian, H_c, is given by

$$H_c(t_1) = H(t_1) + H'(t_1) = H(t_1) + H'_B(t_1) + H'_F(t_1). \tag{11.1.14}$$

11.2 Coupled Electron–Hole–Phonon Green's Functions and Equations

The averaging process which we will use to define non-equilibrium Green's functions in the presence of Fermi particle (F), Bose phonon and potential (B) sources is denoted in a manner similar to that of the non-equilibrium Green's function average of Eq. (9.1.26), except that we now include Bose phonon sources in both the numerator and denominator traces, so that

$$\langle O \rangle \equiv \frac{Tr\,[U(t_0+\tau,t_0)O]_{C,B,F}}{Tr\,[U(t_0+\tau,t_0)]_{C,B}}. \tag{11.2.1}$$

Here, the numerator trace is evaluated in the presence of all five couplings (subscript C: e-h, e-p, h-p, e-e, h-h) and it also includes the Bose phonon and external field (potential) sources ($H'_B(t)$) as well as the Fermi (e, h) particle sources ($H'_F(t)$). The denominator trace, however, is evaluated in the presence of the five couplings and Bose phonon and external field sources, while *excluding* all Fermi particle sources. (The inclusion of the phonon and external field sources in the denominator trace is somewhat unusual, but it proves to be useful later. In any case, the end results will be the same with or without them in the denominator when *all* sources are finally taken to vanish, after all variational derivatives and other manipulations involved in developing equations of motion are completed. This final vanishing of sources is mandated by the requirement that $< 1 >= 1$.)

We define a generalized electron–hole–phonon non-equilibrium Green's function with the average of Eq. (11.2.1) as (the letter argument of a creation or annihilation operator here denotes its space-time argument, for example, $\hat{\Psi}(m) \equiv \hat{\Psi}(\mathbf{x}_m, t_m)$, etc.; furthermore, we introduce the convention $\hat{\Psi}(0) = \hat{\Psi}^+(0) = \hat{\Phi}(0) = \hat{\Phi}^+(0) = \overleftrightarrow{Q}(\bar{t}_0) \equiv 1$):

$$\begin{aligned}
&\overleftrightarrow{G}_{\frac{i}{2},\frac{i'}{2};j;\frac{k''}{2},\frac{k'''}{2}}(1\ldots m\ldots i;1'\ldots m'\ldots i'//\bar{t}_1\ldots\bar{t}_q\ldots\bar{t}_{2j}//1''\ldots n''\ldots k'';1'''\ldots n'''\ldots k''')\\
&= \varepsilon_{i,i',k'',k'''}(1\ldots m\ldots i;1''\ldots n''\ldots k'';1'''\ldots n'''\ldots k''';1'\ldots m'\ldots i').(-i)^{\frac{(i+i'+k''+k'''+2j)}{2}}\\
&\quad\times\left\langle\left(\Pi_{m=0}^{i}\hat{\Psi}(m)\ \Pi_{n''=0}^{k''}\hat{\Phi}(n'')|:\Pi_{q=0}^{2j}\overleftrightarrow{Q}(\bar{t}_q)\ \Pi_{n'''=0}^{k'''}\hat{\Phi}^+(n''')\ \Pi_{m'=0}^{i'}\hat{\Psi}^+(m')\right)_+\right\rangle,
\end{aligned} \tag{11.2.2}$$

where i, i' are the total numbers of $\hat{\Psi}$, $\hat{\Psi}^+$operators in the average $\langle\ldots\rangle$, respectively; k'', k''' are the total numbers of $\hat{\Phi}$, $\hat{\Phi}^+$ operators, respectively; $2j$ is the total number of $\overleftrightarrow{Q}$-phonon operators (a total number 0 signals the absence of such operators from the structure and their replacement by unity); and recall that $\overleftrightarrow{Q}$ is the ion displacement vector that is described by the third-rank tensor elements ${}^{i}Q_{lk}$ and $\varepsilon_{i,i';k'',k'''} = (-1)^{P}$, where P is the number of permutations that reduces the actual time-ordered sequence of Fermion creation and annihilation operators to the sequence described by the argument of $\varepsilon_{i,i';k'',k'''}$ ignoring the presence of ${}^{i}Q_{lk}$-operators. Note that this Green's function need not have equal numbers of $\Psi(\Phi)$ and $\Psi^{+}(\Phi^{+})$ operators to be non-vanishing, because it is dependent on particle sources, and, correspondingly, particle number is not necessarily conserved. For the case in which there are equal numbers of $\Psi(\Phi)$ and $\Psi^{+}(\Phi^{+})$ operators, $i' = i$ and $k'' = k''' = k$, we will use the shorter notation,

$$\overleftrightarrow{G}_{\frac{i}{2},\frac{i}{2};j;\frac{k''}{2},\frac{k''}{2}} \rightarrow \overleftrightarrow{G}_{i,j,k}. \tag{11.2.3}$$

(It should be noted that we are using i here in three different contexts: i is the number of $\hat{\Psi}$-operators, and it also denotes the ith component of $\overleftrightarrow{Q} \rightarrow {}^{i}Q_{lk}$ as well as $(-1)^{1/2}$. The three contexts differ greatly, so we trust that this will not cause confusion to the reader.) In Eq. (11.2.2), the double slash punctuation (//) appearing in the argument of $\overleftrightarrow{G}_{\frac{i}{2},\frac{i'}{2};j;\frac{k''}{2},\frac{k'''}{2}}$ and the overhead double arrow tensor symbol only serve the purpose of separating the arguments of the state variables of different particle species and denoting the tensorial character of $\overleftrightarrow{Q}$, respectively. They will be omitted when clearly unnecessary in some cases, such as $G_{i,0,0} = G_{\frac{i}{2},\frac{i}{2};0;\frac{0}{2},\frac{0}{2}}$. Furthermore, it should be noted that the Green's function $\overline{G}_{11}$ of Eq. (8.4.6), which was defined in the absence of all sources as well as of phonons, is defined with a different order of arguments from the definition used in this chapter, specifically,

$$\overline{G}_{11}(1,2;1',2') = (-i)^{2}\,\varepsilon_{e}\,\varepsilon_{h}\,\big\langle(\Psi(1)\Phi(2)\Phi^{+}(2')\Psi^{+}(1')_{+}\big\rangle \tag{11.2.4}$$

$$= G_{\frac{1}{2},\frac{1}{2};0;\frac{1}{2},\frac{1}{2}} = G_{1,0,1}(1,1';2,2'). \tag{11.2.5}$$

Apart from this notational difference, the one-particle/one-hole Green's functions appear the same in the absence of sources and phonons. Finally, for convenience, we also define

$$\overleftrightarrow{v}^{e}_{p}(1,2) = \{{}^{i}\gamma^{e}_{lk}(\mathbf{x}_{1})\}\,\delta\,(1,2), \tag{11.2.6}$$

and $$\overleftrightarrow{v}^{h}_{p}(1,2) = \{{}^{i}\gamma^{h}_{lk}(\mathbf{x}_{1})\}\,\delta\,(1,2), \tag{11.2.7}$$

where $\overleftrightarrow{v}^{e,h}_{p}$ is a tensor of rank three, having indices i, l, k, etc.

Considering the complexity of the problem at hand, which has five distinct interactions (*e-h*, *e-p*, *h-p*, *e-e*, *h-h*), the inclusion of the Fermi particle sources only in the numerator trace of the average, as defined in Eq. (11.2.1) (while excluding them from the denominator trace), facilitates the generation of higher order electron and hole Green's function equations of motion by variational differentiation of equations of motion for lower order electron/hole Green's functions. After obtaining the requisite order of Fermi Green's function equations of motion, we will set the Fermi particle sources to zero, and then use the remaining phonon source to generate the needed phonon Green's function equations of motion, and determine effective potentials for the system. The resulting coupled Green's function equations are the same as those that would be generated if the Fermi particle sources were included in the denominator trace (as well as in the numerator trace) when *all* sources are set to zero, whereupon particle number conservation is restored causing averages of products of odd numbers of Fermi operators to vanish.

The Hamilton equations of motion for the electron field variables may be written as (see Eq. (4.1.64), for example):

$$\frac{\partial \hat{\Psi}}{\partial t} = -i\frac{\delta_l H_c}{\delta \hat{\Psi}^+}; \quad \frac{\partial \hat{\Psi}^+}{\partial t} = i\frac{\delta_r H_c}{\delta \hat{\Psi}}, \tag{11.2.8}$$

and for holes we similarly obtain

$$\frac{\partial \hat{\Phi}}{\partial t} = -i\frac{\delta_l H_c}{\delta \hat{\Phi}^+}; \quad \frac{\partial \hat{\Phi}^+}{\partial t} = i\frac{\delta_r H_c}{\delta \hat{\Phi}}. \tag{11.2.9}$$

Averaging Eqns. (11.2.8) and (11.2.9) in accordance with Eq. (11.2.1), we have

$$\frac{\partial \left\langle \hat{\Psi} \right\rangle}{\partial t} = -i\left\langle \frac{\delta_l H_c}{\delta \hat{\Psi}^+} \right\rangle; \quad \frac{\partial \left\langle \hat{\Psi}^+ \right\rangle}{\partial t} = i\left\langle \frac{\delta_r H_c}{\delta \hat{\Psi}} \right\rangle, \tag{11.2.10}$$

and

$$\frac{\partial \left\langle \hat{\Phi} \right\rangle}{\partial t} = -i\left\langle \frac{\delta_l H_c}{\delta \hat{\Phi}^+} \right\rangle; \quad \frac{\partial \left\langle \hat{\Phi}^+ \right\rangle}{\partial t} = i\left\langle \frac{\delta_r H_c}{\delta \hat{\Phi}} \right\rangle. \tag{11.2.11}$$

The fact that $\left\langle \hat{\Psi} \right\rangle \neq 0$ and $\left\langle \hat{\Phi} \right\rangle \neq 0$ do not vanish is a manifestation that carrier number is not conserved in the presence of particle sources. The construction of an equation of motion for an electron (or hole) Green's function may now proceed using variational differentiation with respect to particle sources, $\xi(\mathbf{x}', t')$, as in Eq. (4.2.36) for t in the interval $t_0 \to t_0 + \tau$ (bearing in mind that ξ-particle sources occur only in the numerator of the average),

$$\begin{aligned}
\frac{\delta_l}{\delta \xi(\mathbf{x}', t')}\left(\frac{\partial \left\langle \hat{\Psi}(\mathbf{x}, t) \right\rangle}{\partial t}\right) &= \frac{\partial}{\partial t}\frac{\delta_l \left\langle \hat{\Psi}(\mathbf{x}, t) \right\rangle}{\delta \xi(\mathbf{x}', t')} \\
&= -i\frac{\partial}{\partial t}\left[\varepsilon\left\langle (\hat{\Psi}(\mathbf{x}, t)\, \hat{\Psi}^+(\mathbf{x}', t'))_+ \right\rangle\right] \\
&= \frac{\partial}{\partial t}\left[G_{\frac{1}{2},\frac{1}{2};0;\frac{0}{2},\frac{0}{2}}(\mathbf{x}, t; \mathbf{x}', t')\right] \\
&= -i\frac{\delta_l}{\delta \xi(\mathbf{x}', t')}\left\langle \frac{\delta_l H_c}{\delta \hat{\Psi}^+} \right\rangle.
\end{aligned} \tag{11.2.12}$$

Noting that (Eq. (11.2.3)),

$$G_{1,0,0}(\mathbf{x},t;\mathbf{x}',t') = G_{\frac{1}{2},\frac{1}{2};0;\frac{0}{2},\frac{0}{2}}(\mathbf{x},t;\mathbf{x}',t') = -i\varepsilon\left\langle(\hat{\Psi}(\mathbf{x},t)\,\hat{\Psi}^{+}(\mathbf{x}',t'))_{+}\right\rangle, \qquad (11.2.13)$$

we have the one-electron equation of motion with finite Fermi particle sources as

$$i\frac{\partial}{\partial t}G_{1,0,0}(\mathbf{x},t;\mathbf{x}',t') = \left[\frac{\delta_l}{\delta\xi(\mathbf{x}',t')}\left\langle\frac{\delta_l H_c}{\delta\hat{\Psi}^{+}(\mathbf{x},t)}\right\rangle\right]. \qquad (11.2.14)$$

We may evaluate $\left\langle\delta_l H_c/\delta\hat{\Psi}^{+}(\mathbf{x},t)\right\rangle$ by reading off coefficients of $\delta\hat{\Psi}^{+}$ from $\delta H_c = \delta H + \delta H'$ using Eqns. (11.1.10–11.1.14) for H_c considered subject to a variation of $\hat{\Psi}^{+}\to\hat{\Psi}^{+}+\delta\hat{\Psi}^{+}$ (with $\delta\hat{\Psi}^{+}$ anticommuted to the far left, for the left variational derivative). For example, the first term of H in Eqns. (11.1.10–11.1.12) yields a contribution $\left\langle\delta_l H_c/\delta\hat{\Psi}^{+}(\mathbf{x},t)\right\rangle \to h_{100}\hat{\Psi}(\mathbf{x},t)$, with the integration facilitated by the mutual independence of $\hat{\Psi}^{+}(\mathbf{x},t)$ and $\hat{\Psi}^{+}(\mathbf{x}',t)$ expressed as $\boldsymbol{\delta}\hat{\Psi}^{+}(1')/\boldsymbol{\delta}\hat{\Psi}^{+}(1) = \delta^{(3)}(\mathbf{x}_1'-\mathbf{x}_1)$ (for $t_1 = t_1'$) as in Eqns. (4.1.63 and 4.1.66). Furthermore, the *e-e* interaction Hamiltonian yields two first-order terms proportional to a single-space integral of $\hat{\Psi}^{+}\mathcal{V}\hat{\Psi}\hat{\Psi}$, with the two terms jointly canceling the 1/2 prefactor (terms of second order in $\delta\hat{\Psi}^{+}$ are negligible) as in Eqns. (4.1.69 and 4.1.70). The electron–hole interaction term yields a single-space integral of $\hat{\Psi}\mathcal{V}\hat{\Phi}^{+}\hat{\Phi}$, and the electron–phonon interaction term yields a term of the form $\sum {}^{i}\gamma_{lk}^{e}\,{}^{i}Q_{lk}\,\hat{\Psi}$. Further contributions from H' to $\frac{\delta_l H_c}{\delta\hat{\Psi}^{+}(\mathbf{x},t)}$ arise from the source potential U, which yields a term of the form $\hat{\Psi}U$ (which simply combines with the first term), and, finally, the particle source term $\xi(\mathbf{x},t)$. Collecting these terms and averaging as in Eq. (11.2.1) we have

$$\begin{aligned}\left\langle\frac{\delta_l H_c}{\delta\hat{\Psi}^{+}(\mathbf{x},t)}\right\rangle &= [h_{100}+U(\mathbf{x},t)]\left\langle\hat{\Psi}(\mathbf{x},t)\right\rangle + \int d^{(3)}\mathbf{x}_2\left\langle\hat{\Psi}^{+}(\mathbf{x}_2,t)\,\mathcal{V}(\mathbf{x}-\mathbf{x}_2)\,\hat{\Psi}(\mathbf{x}_2,t)\,\hat{\Psi}(\mathbf{x},t)\right\rangle\\ &\quad-\int d^{(3)}\mathbf{x}_2\left\langle\hat{\Psi}(\mathbf{x},t)\,\mathcal{V}(\mathbf{x}-\mathbf{x}_2)\,\hat{\Phi}^{+}(\mathbf{x}_2,t)\,\hat{\Phi}(\mathbf{x}_2,t)\right\rangle\\ &\quad-\sum_i\sum_{l,k}{}^{i}\gamma_{lk}^{e}(\mathbf{x})\left\langle{}^{i}Q_{lk}(t)\,\hat{\Psi}(\mathbf{x},t)\right\rangle + \xi(\mathbf{x},t). \qquad (11.2.15)\end{aligned}$$

Considering the variational derivative of Eq. (11.2.15) with respect to the electron particle source $\xi(\mathbf{x}',t')$ and bearing in mind that only the numerator of the average of Eq. (11.2.1) involves Fermi particle sources, we employ the variational differential procedures of Chapters 4 and 9. Following Section 4.3, we note that the action of $\delta_r/\delta\xi = \delta_l/\delta\xi$ on the average of an odd number of Fermi variables just installs an additional $\hat{\Psi}^{+}$ factor in the average (along with a factor $(-i\varepsilon)$) subject to time ordering. Accordingly, we obtain the result below for finite Fermi particle sources:

$$\frac{\delta_l}{\delta\xi(\mathbf{x}',t')}\left\langle\frac{\delta_l H_c}{\delta\hat{\Psi}^+(\mathbf{x},t)}\right\rangle = [h_{100}+U(\mathbf{x},t)]\left\{-i\varepsilon\left\langle\left(\hat{\Psi}(\mathbf{x},t)\hat{\Psi}^+(\mathbf{x}',t')\right)_+\right\rangle\right\}$$
$$+\int d^{(3)}\mathbf{x}_2\left\{-i\varepsilon\left\langle\left(\hat{\Psi}^+(\mathbf{x}',t')\hat{\Psi}^+(\mathbf{x}_2,t^+)\mathcal{V}(\mathbf{x}-\mathbf{x}_2)\hat{\Psi}(\mathbf{x}_2,t)\hat{\Psi}(\mathbf{x},t)\right)_+\right\rangle\right\}$$
$$-\int d^{(3)}\mathbf{x}_2\left\{-i\varepsilon\left\langle\left(\hat{\Psi}^+(\mathbf{x}',t')\hat{\Psi}(\mathbf{x},t^+)\mathcal{V}(\mathbf{x}-\mathbf{x}_2)\hat{\Phi}^+(\mathbf{x}_2,t^+)\hat{\Phi}(\mathbf{x}_2,t)\right)_+\right\rangle\right\}$$
$$-\sum_i\sum_{l,k}{}^i\gamma^e_{lk}(\mathbf{x})\left\{-i\varepsilon\left\langle\left(\hat{\Psi}^+(\mathbf{x}',t')\,{}^iQ_{lk}(t)\,\hat{\Psi}(\mathbf{x},t)\right)_+\right\rangle\right\}+\delta(\mathbf{x}-\mathbf{x}')\delta(t-t'), \tag{11.2.16}$$

with the last term arising from $\delta_l\xi(\mathbf{x},t)/\delta\xi(\mathbf{x}',t') = \delta^{(3)}(\mathbf{x}-\mathbf{x}')\delta(t-t')$. In terms of the generalized Green's functions defined in Eq. (11.2.2), this yields the equation of motion for the one-electron Green's function $G_{1,0,0}(1,2)$ (in 4D notation; $1 = \mathbf{x}_1, t_1$) with finite Fermion particle sources as

$$\left[i\frac{\partial}{\partial t_1}-h_{100}(1)-U(1)\right]G_{1,0,0}(1,2)-i^{1/2}\int \overleftrightarrow{v}^e_p(1,3)\,\overleftrightarrow{G}_{1,\frac{1}{2},0}(3,2//t_3)$$
$$+\,i\int\mathcal{V}(1,3)\left[G_{2,0,0}(1,3;2,3^+)-G_{1,0,1}(1,2;3,3^+)\right]=\delta(1,2), \tag{11.2.17}$$

where $\overleftrightarrow{v}^e_p(1,2)\equiv\left[{}^i\gamma^e_{lk}(\mathbf{x}_1)\right]\delta(1,2)$ and repeated indices indicate integration over associated space-time coordinates (all space and $t_0\to t_0+\tau$ for the time integral; of course, the i,l,k indices implicit in $\overleftrightarrow{v}^e_p(1,2)$ are also to be summed jointly with the corresponding repeated indices of Q_{lk} within the structure of $\overleftrightarrow{G}_{1,\frac{1}{2},0}$). In the presence of finite Fermion particle sources, $G_{2,0,0}(1,2;1',2')$ is the non-equilibrium counterpart of the two-electron Green's function discussed in Chapters 8 and 9, which describes the joint probability amplitude for two interacting electrons to propagate from their creation points $1'$, $2'$ to their annihilation points $1,2$. Moreover, $G_{1,0,1}(1,2;1',2')$ is the one-electron/one-hole Green's function, which describes the joint probability amplitude for one-electron interacting with one-hole, as they both propagate from their creation points $2,2'$ to their annihilation points $1,1'$. Furthermore, $\overleftrightarrow{G}_{1,\frac{1}{2},0}(1,2//t_3)$ is the Green's function describing the probability amplitude for one-electron propagation as influenced by a non-trivial, finite phonon field. Moreover, all the Green's functions here are in the presence of finite Fermi sources as well as finite Bose sources. Clearly, the equation of motion for the one-hole Green's function G_{001} can be developed in a completely analogous manner to that of one-electron, with the result

$$\left[i\frac{\partial}{\partial t_1}-h_{001}(1)+U(1)\right]G_{0,0,1}(1,2)-i^{\frac{1}{2}}\int\overleftrightarrow{v}^h_p(1,3)\,\overleftrightarrow{G}_{0,\frac{1}{2},1}(t_3//3\,,2)$$
$$+\,i\int\mathcal{V}(1,3)\left[G_{0,0,2}(1,3;2,3^+)-G_{1,0,1}(3,3^+;1,2)\right]=\delta(1,2) \tag{11.2.18}$$

(again, in the presence of finite Fermion and Boson particle sources), where $\overleftrightarrow{v}^h_p(1,2)\equiv i\left[{}^i\gamma^h_{lk}(\mathbf{x}_1)\right]\delta(1,2)$, with corresponding interpretations of the two-hole Green's function

$G_{0,0,2}$ and the one-hole Green's function $\overleftrightarrow{G}_{0,\frac{1}{2},1}$ as influenced by a non-trivial, finite phonon field.

The development of the two-electron Green's function ($G_{2,0,0}$) equation of motion may proceed by subjecting Eq. (11.2.17) to $\frac{\delta_l}{\delta\xi^+(1')}\frac{\delta_l}{\delta\xi(2')}$ variations, since this will place a second $\hat{\Psi}\hat{\Psi}^+$ pair into the numerator trace of $G_{1,0,0}$, yielding a $G_{2,0,0}$ as

$$\frac{\delta_l}{\delta\xi^+(1')}\frac{\delta_l}{\delta\xi(2')}G_{1,0,0}(1,2) \sim G_{2,0,0}(1,1';2,2'). \tag{11.2.19}$$

However, one must also note that with two $\hat{\Psi}^+$'s and two $\hat{\Psi}$'s, there are now two possible ways that an equal-time $\hat{\Psi}^+\hat{\Psi}$- product can occur in the numerator trace. Each of them is in a time-ordered structure, and can form a "greater minus lesser" combination upon infinitesimal time integration across the equal time point (as in Eqns. (5.1.3 and 5.1.4), which, in turn, results in a spatial $\delta^{(3)}$-function in accordance with the canonical commutation/anticommutation relations. Of course, the differential time characterization also involves a time δ-function as in Eq. ((5.1.3)). Bearing in mind that the residual numerator traces have one $\hat{\Psi}^+\hat{\Psi}$ pair, the four-dimensional δ-functions multiply G_1-functions. The two such δG_1 product terms for equal time combine with the G_2 of Eq. (11.2.19) yielding the result as (Section 4.3):

$$\left[i\frac{\partial}{\partial t_1} - h_{100}(1) - U(1)\right] G_{2,0,0}(1,1';2,2') - i^{\frac{1}{2}}\int \overleftrightarrow{v}^{e}_{p}(1,3)\,\overleftrightarrow{G}_{2,\frac{1}{2},0}(1,1';2,2'//t_3)$$
$$+i\int V(1,3)\left[G_{3,0,0}(1,1',3;2,2',3^+) - G_{2,0,1}(1,1';2,2';3,3^+)\right]$$
$$= \delta(1,2)\,G_{1,0,0}(1',2') - \delta(1,2')\,G_{1,0,0}(1',2), \tag{11.2.20}$$

where the occurrence of the three-electron Green's function, $G_{3,0,0}$, resulting from the action of the two variational derivatives on $G_{2,0,0}$ on the right of Eq. (11.2.17) is to be expected from Chapters 8 and 9. $G_{2,0,1}$ represents the joint probability amplitude for two electrons and one hole, all propagating with mutual interaction from their points of creation to their points of annihilation. Furthermore, $G_{2,\frac{1}{2},0}$ represents the joint probability amplitude for two electrons propagating and interacting as influenced by a non-trivial, finite phonon field.

Completely analogous considerations apply to the two-hole Green's function, $G_{0,0,2}$, starting with Eq. (11.2.18) for $G_{0,0,1}$,

$$\frac{\delta_l}{\delta\zeta^+(1')}\frac{\delta_l}{\delta\zeta(2')}G_{0,0,1}(1,2) \sim G_{0,0,2}(1,1';2,2'), \tag{11.2.21}$$

with the resulting equation of motion as

$$\left[i\frac{\partial}{\partial t_1} - h_{001}(1) + U(1)\right] G_{0,0,2}(1,1';2,2') - i^{\frac{1}{2}}\int \overleftrightarrow{v}^{h}_{p}(1,3)\,\overleftrightarrow{G}_{0,\frac{1}{2},2}(t_3//1,1';2,2')$$
$$+i\int V(1,3)\left[G_{0,0,3}(1,1',3;2,2',3^+) - G_{1,0,2}(3,3';1,1';2,2')\right]$$
$$= \delta(1,2)\,G_{0,0,1}(1',2') - \delta(1,2')\,G_{0,0,1}(1',2), \tag{11.2.22}$$

and with corresponding interpretations of the three-hole Green's function $G_{0,0,3}$ and the two-hole/one-electron Green's function $G_{1,0,2}$ as well as the two-hole Green's function influenced by a non-trivial, finite phonon field, $G_{0,\frac{1}{2},2}$.

The equation of motion for the joint electron–hole propagation function, $G_{1,0,1}$, may be obtained by applying $\frac{\delta_l}{\delta\zeta^+(1')}\frac{\delta_l}{\delta\zeta(2')}$ to Eq. (11.2.17), since

$$\frac{\delta_l}{\delta\zeta^+(1')}\frac{\delta_l}{\delta\zeta(2')}G_{1,0,0}(1,2) \sim G_{1,0,1}(1,2;1',2'), \tag{11.2.23}$$

with the result

$$\left[i\frac{\partial}{\partial t_1} - h_{100}(1) - U(1)\right] G_{1,0,1}(1,2;1'2') - i^{\frac{1}{2}}\int \overleftrightarrow{v}^{e}_{p}(1,3)\ \overleftrightarrow{G}_{1,\frac{1}{2},1}(1,2//t_3//1',2')$$
$$+i\int \mathcal{V}(1,3)\left[G_{2,0,1}(1,3;2,3^+;1',2') - G_{1,0,2}(1,2;1',3;2',3^+)\right]$$
$$= \delta(1,2)\,G_{001}(1',2'), \tag{11.2.24}$$

where $\overleftrightarrow{G}_{1,\frac{1}{2},1}$ represents the joint probability propagation amplitude for one-electron interacting with one-hole in the presence of a non-trivial, finite phonon field, etc.

As the one-electron Green's function equation (Eq. (11.2.17)) involves $\overleftrightarrow{G}_{1,\frac{1}{2},0}$, we develop an equation for it by applying $\delta/\delta\,{}^{i}\mathcal{J}_{lk}(t)$ to Eq. (11.2.17), since this installs a factor ${}^{i}Q_{lk}(t)$ within the time-ordered matrix elements of $G_{1,0,0}$ needed to construct $\overleftrightarrow{G}_{1,\frac{1}{2},0}$, i.e.:

$$\frac{\delta G_{1,0,0}(1,2)}{\delta\,{}^{i}\mathcal{J}_{lk}(t)} \sim \overleftrightarrow{G}_{1,\frac{1}{2},0}(1,2//t), \tag{11.2.25}$$

with the result

$$\left[i\frac{\partial}{\partial t_1} - h_{100}(1) - U(1)\right] \overleftrightarrow{G}_{1,\frac{1}{2},0}(1,2//t) - i^{\frac{1}{2}}\int \overleftrightarrow{v}^{e}_{p}(1,3)\ \overleftrightarrow{G}_{1,1,0}(3,2//t_3//t)$$
$$+i\int \mathcal{V}(1,3)\left[\overleftrightarrow{G}_{2,\frac{1}{2},0}(1,3;2,3^+//t) - \overleftrightarrow{G}_{1,\frac{1}{2},1}(1,2//t//3,3^+)\right] = \delta(1,2)\left\langle \overleftrightarrow{Q}(t)\right\rangle, \tag{11.2.26}$$

where $\overleftrightarrow{G}_{1,1,0}$ describes the joint probability amplitude for propagation of one-electron and one-phonon in interaction, and $\overleftrightarrow{G}_{2,\frac{1}{2},0}$ describes the propagation of two electrons in the presence of a non-trivial, finite phonon field. The right-hand side of Eq. (11.2.26) arises from the recognition that *all* the various Green's functions on the left-hand side of Eq. (11.2.17) involve the *same* denominator trace $\left(D_B \equiv Tr\left[U(t_0+\tau,t_0)\right]_{C,B}\right)$ in their definitions as averages (Eq. (11.2.1)). Therefore, including differential operations (such as $\partial/\partial t_1$, h_{100}, etc.) in the total numerator trace (N_B) on the left, Eq. (11.2.17) may be written symbolically as $N_B D_B^{-1} = \delta(1,2)$. Applying $\delta/\delta\, {}^i\mathcal{J}_{lk}$ to this, the right-hand side vanishes $\left[\delta(\delta(1,2))/\delta\, {}^i\mathcal{J}_{lk} \equiv 0\right]$, and we thus obtain the identity

$$\left(\frac{\delta N_B}{\delta\, {}^i\mathcal{J}_{lk}}\right) D_B^{-1} = N_B D_B^{-2}\left(\frac{\delta D_B}{\delta\, {}^i\mathcal{J}_{lk}}\right) = \left(\frac{N_B}{D_B}\right)\frac{1}{D_B}\frac{\delta D_B}{\delta\, {}^i\mathcal{J}_{lk}}. \tag{11.2.27}$$

But the left-hand side of Eq. (11.2.27) is just the left-hand side of Eq. (11.2.26), and $N_B D_B^{-1} = \delta\,(1,2)$ and $(1/D_B)\left(\delta D_B/\delta\, {}^i\mathcal{J}_{lk}\right) = \left\langle \overleftrightarrow{Q}\,(k)\right\rangle$ (with a general cancellation of factors $\left[(-i)^{1/2}\varepsilon\right]$ arising from $\left[\delta/\delta\, {}^i\mathcal{J}_{lk}\right]$ across the entire equation). In this, we see that the right-hand side of Eqns. (11.2.26 and 11.2.27) arises directly from the dependence of the denominator trace D_B on the Boson source ${}^i\mathcal{J}_{lk}$.

11.3 Electron–Phonon Interaction Phenomenology

While the electron–electron interaction has been a principal focus of our considerations in this book and the hole–hole interaction follows suit (with some physical aspects of the electron–hole interaction discussed in Section 8.4), very little attention has been given to the phonon field and its interactions with electrons and holes. These phonon interactions, treated only on a formal basis thus far, are at the heart of solid-state physics and are responsible for many interesting and important physical phenomena. As the electron–phonon and hole–phonon interactions have similar structure, we will explore some physical aspects of just one of them, the electron–phonon interaction.

The one-phonon Green's function is defined in terms of a special case of Eq. (11.2.2), with the symbol $\overleftrightarrow{D}$ (not to be confused with D_B of the preceding section) as

$$\overleftrightarrow{D}\,(t_1,t_2) = \overleftrightarrow{G}_{0,0;1;0,0}\,(t_1,t_2) = -i\left\langle\left(\overleftrightarrow{Q}\,(t_1)\,\overleftrightarrow{Q}\,(t_2)\right)_+\right\rangle. \tag{11.3.1}$$

Since the $\overleftrightarrow{Q}$'s obey canonical commutation (rather than anticommutation) relations Eqns. (11.1.5 and 11.1.6), they are Boson field variables and $\overleftrightarrow{D}\,(t_1,t_2)$ has Boson spectral structure under equilibrium conditions. The quantum Hamilton equations of motion for phonons may be obtained using Eqns. (4.1.35) and (11.1.2–11.1.11) as

$${}^i\dot{Q}_{lk} = \frac{\partial\left(H + H'(t_1)\right)}{\partial\, {}^iP_{lk}} = \frac{{}^iP_{lk}}{M_k} \tag{11.3.2}$$

and (omitting phonon–hole coupling),

$$\begin{aligned} {}^i\dot{P}_{lk} = -\frac{\partial\left(H + H'(t_1)\right)}{\partial\, {}^iQ_{lk}} &= -\sum\nolimits_{i',l',k'} \Xi_{ii'}\left(l,k;l',k'\right)\, {}^{i'}Q_{l'k'}\,(t_1) \\ &+ \int d^{(3)}\mathbf{x}_1\, {}^i\gamma^e_{lk}\,(\mathbf{x}_1)\,\hat{\Psi}^+(1)\hat{\Psi}(1) - {}^i\mathcal{J}_{lk}\,(t_1)\,. \end{aligned} \tag{11.3.3}$$

Combining Eqns. (11.3.2) and (11.3.3), we have

$$M_k\,{}^i\ddot{Q}_{lk}(t_1)+\sum\nolimits_{i'',l'',k''}\Xi_{ii''}\left(l,k;l'',k''\right)\,{}^{i''}Q_{l''k''}(t_1)$$
$$=\int d^{(3)}\mathbf{x}_1\,{}^i\gamma^e_{lk}(\mathbf{x}_1)\,\hat{\Psi}^+(1)\hat{\Psi}(1)-{}^i\mathcal{J}_{lk}(t_1)\,. \qquad (11.3.4)$$

Averaging Eq. (11.3.4) in accordance with Eq. (11.2.1) (wherein we now eliminate the Bose phonon sources from the normalizing denominator, ($D_B = Tr\,[U(t_0+\tau,t_0)]_{C,B} \to Tr\,[U(t_0+\tau,t_0)]_C$), for the purposes of this section) and taking the variational derivative with respect to ${}^{i'}\mathcal{J}_{l'k'}(t_2)$ to install another factor ${}^{i'}Q_{l'k'}$ inside the time-ordered average, we obtain the equation of motion for the one-phonon Green's function as

$$M_k\frac{d^2D_{ii'}}{dt_1^2}(l,k,t_1;l',k',t_2)+\sum\nolimits_{i'',l'',k''}\Xi_{ii''}\left(l,k;l'',k''\right)D_{i''i'}(l'',k'',t_1;l',k',t_2)$$
$$=-i\int d^3\mathbf{x}_1\,{}^i\gamma^e_{lk}(\mathbf{x}_1)\left\langle\hat{\Psi}^+(1)\,{}^{i'}Q_{l'k'}(t_2)\,\hat{\Psi}(1)\right\rangle-\delta_{ii'}\delta_{ll'}\delta_{kk'}\delta\left(t_1-t_2\right). \qquad (11.3.5)$$

In the absence of phonon interactions with electrons (holes) $\gamma^{e(h)} \to 0$, the phonon Green's function, $D \to D^0$, becomes that of Eqns. (4.3.26 and 4.3.27) and (4.4.21), satisfying the equation

$$M_k\frac{d^2}{dt^2}D^0_{ii'}\left(l,k,t_1;l',k',t_2\right)+\sum\nolimits_{i'',l'',k''}\Xi_{ii''}\left(l,k;l'',k''\right)D^0_{i''i'}\left(l'',k'',t_1;l',k',t_2\right)$$
$$=-\delta_{ii'}\delta_{ll'}\delta_{kk'}\delta\left(t_1-t_2\right). \qquad (11.3.6)$$

Eq. (11.3.6) is a matrix equation for the Green's function of a system of coupled harmonic oscillators, and under conditions of time-translational invariance, it has the spectral structure of a Boson Green's function with the auxiliary function $D^0_{ii'}(t_1-t_2) \to \mathcal{D}^0_{ii'}(z)$ of Eq. (7.4.17). Moreover, further simplification may be achieved by invoking the periodicity of the crystal lattice under translations through a lattice vector, so that[1–3]

$$\mathcal{D}^0_{ii'}(l,k;l',k';z)=\mathcal{D}^0_{ii'}(k,k';\mathbf{R}_l-\mathbf{R}_{l'};z)$$
$$=\Omega\int'_{Bz}\frac{d^{(3)}\mathbf{q}}{(2\pi)^3}e^{i\mathbf{q}\cdot(\mathbf{R}_l-\mathbf{R}_{l'})}\mathcal{D}^0_{ii'}(k,k';\mathbf{q};z), \qquad (11.3.7)$$

which can accommodate a crystal lattice having several atoms per unit cell, and the inverse of Eq. (11.3.7) is the Fourier series

$$\mathcal{D}^0_{ii'}(k,k';\mathbf{q};z)=\sum_{\mathbf{R}_l-\mathbf{R}_{l'}}e^{-i\mathbf{q}\cdot(\mathbf{R}_l-\mathbf{R}_{l'})}\mathcal{D}^0_{ii'}(k,k';\mathbf{R}_l-\mathbf{R}_{l'};z). \qquad (11.3.8)$$

In this, the three non-coplanar real-space basis vectors of the direct lattice, $\mathbf{a}_1, \mathbf{a}_2, \mathbf{a}_3$ (which are generally *not* orthogonal), form the edges of the unit cell of the direct lattice having volume $\Omega_a = \mathbf{a}_1 \cdot \mathbf{a}_2 \times \mathbf{a}_3$, and $\Omega = L^3\Omega_a$ is the total volume of the crystal, which has $L^3 = N$ unit cells. Furthermore,[5–7] $\int'_{Bz} \frac{d^{(3)}\mathbf{q}}{(2\pi)^3}$ denotes integration over the first Brillouin zone (Bz) in reciprocal space (spanned by basis vectors $\mathbf{b}_1$, $\mathbf{b}_2$, $\mathbf{b}_3$ so that $\mathbf{a}_i \cdot \mathbf{b}_j = 2\pi\delta_{ij}$; i.e., $\mathbf{b}_1 = \frac{2\pi}{\Omega_a}[\mathbf{a}_2 \times \mathbf{a}_3]$ and cyclic permutations). Of course, Ξ has similar representations due to lattice periodicity. The resulting matrix equation for $\mathcal{D}^0_{ii'}(k, k'; \mathbf{q}; z)$ has sharply reduced dimensions (namely $3A \times 3A$ where A is the number of atoms in the unit cell), and using the Poisson sum formula (extended to the lattice under consideration), Eq. (11.3.6) becomes

$$M_k\, z^2 \mathcal{D}^0_{ii'}(k, k'; \mathbf{q}; z) - \sum_{i''k''} \Xi_{ii''}(k, k''; \mathbf{q})\, \mathcal{D}^0_{i''i'}(k'', k'; \mathbf{q}; z) = \delta_{ii'}\delta_{kk'}, \tag{11.3.9}$$

where

$$\Xi(k, k''; \mathbf{q}) = \sum_{\mathbf{R}_l - \mathbf{R}_{l''}} \Xi_{ii''}(l, k; l'', k'')e^{-i\mathbf{q}\cdot(\mathbf{R}_l - \mathbf{R}_{l''})} \tag{11.3.10}$$

is summed over interacting ion pair relative equilibrium positions, $(\mathbf{R}_l - \mathbf{R}_{l''})$. Introducing the *dynamical matrix* Θ as

$$\Theta_{ii''}(k, k''; \mathbf{q}) \equiv M_k^{-\frac{1}{2}}\, \Xi(k, k''; \mathbf{q}) M_{k''}^{-\frac{1}{2}}, \tag{11.3.11}$$

and also defining

$$\Delta_{ii'}(k, k'; \mathbf{q}; z) \equiv M_k^{\frac{1}{2}} \mathcal{D}^0_{ii'}(k, k'; \mathbf{q}; z) M_{k'}^{\frac{1}{2}}, \tag{11.3.12}$$

we may rewrite Eq. (11.3.9) as

$$z^2 \Delta_{ii'}(k, k'; \mathbf{q}; z) - \sum_{i''k''} \Theta_{ii''}(k, k''; \mathbf{q})\, \Delta_{i''i'}(k'', k'; \mathbf{q}; z) = \delta_{ii'}\delta_{kk'}. \tag{11.3.13}$$

In $3A$-dimensional matrix notation this takes the form

$$z^2 \Delta(\mathbf{q}; z) - \Theta(\mathbf{q})\, \Delta(\mathbf{q}; z) = I. \tag{11.3.14}$$

Since Θ is a real, symmetric (Hermitian) matrix, it has eigenvectors ($3A$-dimensional) $\hat{e}_j$ as

$$\Theta(\mathbf{q})\, \hat{e}_j(\mathbf{q}) = \omega_j^2(\mathbf{q})\hat{e}_j(\mathbf{q}), \tag{11.3.15}$$

with real eigenvalues, $\omega_j^2(\mathbf{q})$ (which are positive for stable phonon modes). The associated completeness relation is

$$I = \sum_j \hat{e}_j(\mathbf{q})\hat{e}_j^{+}(\mathbf{q}), \tag{11.3.16}$$

and the orthogonality relation is

$$\hat{e}_i^{+} \cdot \hat{e}_j = \delta_{ij}. \tag{11.3.17}$$

It readily follows that the eigenvector expansion of Δ is given by

$$\Delta(\mathbf{q}, z) = \sum_j \frac{\hat{e}_j(\mathbf{q})\hat{e}_j^+(\mathbf{q})}{z^2 - \omega_j^2(\mathbf{q})}. \tag{11.3.18}$$

Taken together with Eqns. (11.3.7) and (11.3.12), this result for Δ provides a general solution for $D^0_{ii'}(l, k, t_1; l', k', t_2)$.

Considering a one-dimensional ionic diatomic lattice having atoms with masses M_+, M_- occurring alternately on a line with equal and opposite charges, and equal spacing, a, we will treat longitudinal oscillations polarized in the $\mathbf{q}$-direction and omit the polarization index ($\Delta_{ii'} \rightarrow \Delta$, $\Theta_{ii'} \rightarrow \Theta$), so that Eq. (11.3.13) becomes a two-dimensional matrix equation in k, k':

$$z^2 \Delta(k, k'; \mathbf{q}; z) - \sum\nolimits_{k''} \Theta(k, k''; \mathbf{q})\, \Delta(k'', k'; \mathbf{q}; z) = \delta_{kk'};$$
$$\left(k; k'; k'' = \left\{ \begin{matrix} + \ for\ M_+ \\ - \ for\ M_- \end{matrix} \right\}\right). \tag{11.3.19}$$

This represents crystal dynamics of particular importance for polar semiconductors, for which we assume only nearest-neighbor interactions with a uniform spring constant α, such that (Eqns. (11.3.10 and 11.3.11)):

$$\Theta(k, k''; \mathbf{q}) \Longrightarrow \begin{pmatrix} 2\alpha M_+^{-1} & -\alpha(M_+M_-)^{-\frac{1}{2}}(1 + e^{-iqa}) \\ -\alpha(M_+M_-)^{-\frac{1}{2}}(1 + e^{iqa}) & 2\alpha M_-^{-1} \end{pmatrix}. \tag{11.3.20}$$

While this model is only relevant to short-range interactions and is inadequate for Coulomb interactions among the ions, it is popular for its simplicity and we explore it accordingly. The corresponding equation (Eq. (11.3.14)) for the matrix Δ is given by

$$\begin{pmatrix} z^2 - 2\alpha M_+^{-1} & \alpha(M_+M_-)^{-\frac{1}{2}}(1 + e^{-iqa}) \\ \alpha(M_+M_-)^{-\frac{1}{2}}(1 + e^{iqa}) & z^2 - 2\alpha M_-^{-1} \end{pmatrix} \begin{pmatrix} \Delta_{++} & \Delta_{+-} \\ \Delta_{-+} & \Delta_{--} \end{pmatrix} = \begin{pmatrix} 1 & 0 \\ 0 & 1 \end{pmatrix}. \tag{11.3.21}$$

Inverting the first matrix on the left, we obtain Δ as

$$\begin{pmatrix} \Delta_{++} & \Delta_{+-} \\ \Delta_{-+} & \Delta_{--} \end{pmatrix} = \frac{1}{S} \begin{pmatrix} z^2 - 2\alpha M_-^{-1} & -\alpha(M_+M_-)^{-\frac{1}{2}}(1 + e^{-iqa}) \\ -\alpha(M_+M_-)^{-\frac{1}{2}}(1 + e^{iqa}) & z^2 - 2\alpha M_+^{-1} \end{pmatrix}, \tag{11.3.22}$$

where S is the determinant,

$$S = \left(z^2 - 2\alpha M_+^{-1}\right)\left(z^2 - 2\alpha M_-^{-1}\right) - \alpha^2 M_+^{-1} M_-^{-1}(1 + e^{iqa})(1 + e^{-iqa}). \tag{11.3.23}$$

The collective phonon modes given by $S = 0$ are $\left(\mu^{-1} \equiv M_+^{-1} + M_-^{-1}\right)$:

$$z^2 \rightarrow \omega^2_{phonon} = \alpha\mu^{-1}\left(1 \pm \sqrt{1 - 4\mu^2 M_+^{-1} M_-^{-1} \sin^2(qa/2)}\right). \tag{11.3.24}$$

The lower root ($\pm \to -$) is the longitudinal acoustic (LA) phonon, which has the value for $qa/2 < 1$ given by

$$\omega_{LA}^2 \to \frac{\alpha a^2 q^2}{2(M_1 + M_2)} \qquad \text{(zone center)}, \tag{11.3.25}$$

whereas the upper root ($\pm \to +$) is the longitudinal optical (LO) phonon given for $qa \to 0$ as

$$\omega_{LO}^2 \to 2\alpha/\mu \qquad \text{(zone center)}. \tag{11.3.26}$$

The corresponding LA and LO phonon frequencies are given at the zone edge ($qa = \pi$) by (take $M_+ > M_-$):

$$\omega_{LA}^2 \cong 2\alpha/M_+ \qquad \text{(zone edge)}, \tag{11.3.27}$$

$$\omega_{LO}^2 \cong 2\alpha/M_- \qquad \text{(zone edge)}. \tag{11.3.28}$$

One of the most important features of the ionic lattice is its electric polarizability. Denoting the alternating ion charges by $Z_k \to Z, -Z$, the equation of motion for the ensemble-averaged ionic displacements $\langle Q_k \rangle$ (neglecting electron–phonon interaction) is given by Eq. (11.3.6) with the replacement $D^0 \to \langle Q_k \rangle$ on the left and the replacement $-\delta \to Z_k E$ on the right, where E is the applied electric field. Therefore, D^0 may be employed to solve this inhomogeneous equation (in $\mathbf{q}, z$ representation for the 1D lattice under consideration, with $\mathbf{q} \to 0$ for a spatially uniform driving field E of frequency $z \to \omega$). The pertinent 2×2 matrix for $\mathcal{D}^0$ may be constructed from the matrix Δ of Eq. (11.3.22) using Eq. (11.3.12), with the $q \to 0$ result,

$$\begin{aligned} \mathcal{D}^0(k, k'; q \to 0; \omega) &= (M_k M_{k'})^{-\frac{1}{2}} \Delta(k, k'; q \to 0; z) \\ &= \frac{1}{S} \begin{pmatrix} \dfrac{z^2 - 2\alpha M_-^{-1}}{M_+} & \dfrac{-2\alpha}{M_+ M_-} \\ \dfrac{-2\alpha}{M_+ M_-} & \dfrac{z^2 - 2\alpha M_+^{-1}}{M_-} \end{pmatrix}, \end{aligned} \tag{11.3.29}$$

where

$$S \to z^2(z^2 - 2\alpha/\mu). \tag{11.3.30}$$

Correspondingly, we obtain the phonon Green's function solution for $\langle Q \rangle$ as

$$\begin{aligned} \binom{\langle Q_+ \rangle}{\langle Q_- \rangle} &= -\mathcal{D}^0(k, k'; q \to 0; \omega) \binom{ZE}{-ZE} \\ &= \frac{-ZE}{\omega^2(\omega^2 - 2\alpha/\mu)} \begin{pmatrix} \dfrac{\omega^2 - 2\alpha M_-^{-1}}{M_+} & \dfrac{-2\alpha}{M_+ M_-} \\ \dfrac{-2\alpha}{M_+ M_-} & \dfrac{\omega^2 - 2\alpha M_+^{-1}}{M_-} \end{pmatrix} \binom{1}{-1}, \end{aligned} \tag{11.3.31}$$

where we have suppressed the common time-oscillatory factor of $\langle Q_K \rangle$ and of **E**. The relative displacement of $+Z$ and $-Z$ is thus given by

$$\langle Q_+ \rangle - \langle Q_- \rangle = \frac{ZE}{\mu(2\alpha/\mu - \omega^2)}. \qquad (11.3.32)$$

This yields the ionic (phonon) polarizability as $\frac{Z^2}{\Omega_a \mu(2\alpha/\mu - \omega^2)}$, where Ω_a is the volume of the unit cell of the diatomic lattice. Obviously, the phonon Green's function for this system, D^0, can be constructed for arbitrary wave number q using Eqns. (11.3.22) and (11.3.12).

The coupling of phonon propagation to the electron field occurs on the right of Eq. (11.3.5), embedded in the integrand factor:

$$\left\langle \hat{\Psi}^+(1)\, {}^{i'}Q_{l'k'}(t_2) \hat{\Psi}(1) \right\rangle = i \frac{\delta \left\langle \hat{\Psi}^+(1)\, \hat{\Psi}(1) \right\rangle}{\delta^{i'} \mathcal{J}_{l'k'}(t_2)} = i \frac{\delta \langle \rho_e(1) \rangle}{\delta^{i'} \mathcal{J}_{l'k'}(t_2)}, \qquad (11.3.33)$$

where $\langle \rho_e(1) \rangle$ is the electron density as it is perturbed in the presence of the phonon field. To gain an appreciation for the physical mechanism of the interaction, we rewrite Eq. (11.3.5) using the Poisson equation,

$$\langle \rho_e(1) \rangle = -\frac{1}{4\pi} \nabla_1^2 \left(V(1) - U(1) \right), \qquad (11.3.34)$$

where $V(1)$ is the effective potential generated by the phonon potential field $U(1)$ perturbing the electron density. Considering that the electron–electron Coulomb interaction is active, it provides dynamic non-local screening through the electron inverse dielectric function $K_e(1,2)$ in accordance with the relation

$$V(1) = \int d2\, K_e(1,2)\, U(2), \qquad (11.3.35)$$

where we now treat $U(2)$ as the potential impressed on the electron system by the ions (phonons), which is given to first order in the non-equilibrium average $\langle {}^iQ_{lk} \rangle$ by Eq. (11.1.8),

$$U(2) \Longrightarrow \sum_{i,l,k} {}^i\gamma_{lk}^e(\mathbf{x}_2) \left\langle {}^iQ_{lk}(t_2) \right\rangle. \qquad (11.3.36)$$

Combining Eqns. (11.3.33–11.3.36), we have

$$\langle \hat{\Psi}^+(1)^{i'} Q_{l'k'}(t_2)\, \hat{\Psi}(1) \rangle$$

$$= \frac{-i}{4\pi} \nabla_1^2 \int d2'' \left(K_e(1,2'') - \delta(1-2'') \right) \sum\nolimits_{i'',l'',k''} {}^{i''}\gamma_{l''k''}^e(\mathbf{x}_2'') \frac{\delta \left\langle {}^{i''}Q_{l''k''}(t_2'') \right\rangle}{\delta^{i'} \mathcal{J}_{l'k'}(t_2)}$$

$$= \frac{-i}{4\pi} \nabla_1^2 \int d2'' \left(K_e(1,2'') - \delta(1-2'') \right) \sum\nolimits_{i'',l'',k''} {}^{i''}\gamma_{l''k''}^e(\mathbf{x}_2'')\, D_{i''i'}(l'',k'',t_2'';l',k',t_2). \qquad (11.3.37)$$

Correspondingly, Eq. (11.3.5) takes the form

$$M_k \frac{d^2}{dt_1^2} D_{ii'}(l,k,t_1;l',k',t_2) + \sum\nolimits_{i'',l'',k''} \Xi_{ii''}\left(l,k;l'',k''\right) D_{i''i'}\left(l'',k'',t_1;l',k',t_2\right)$$

$$= \frac{-1}{4\pi} \int d^{(3)}\mathbf{x}_1\, {}^{i}\gamma^{e}_{lk}(\mathbf{x}_1)\, \nabla_1^2 \int d2'' \left(K_e\left(1,2''\right) - \delta\left(1-2''\right)\right)$$

$$\times \sum\nolimits_{i'',l'',k''} {}^{i''}\gamma^{e}_{l''k''}\left(\mathbf{x}_2''\right) D_{i''i'}\left(l'',k'',t_2'';l',k',t_2\right) - \delta_{ii'}\delta_{ll'}\delta_{kk'}\delta\left(t_1-t_2\right). \tag{11.3.38}$$

The local terms on the right arising from $\delta\left(1-2''\right)$ involve

$$\nabla_1^2\, {}^{i''}\gamma^{e}_{l''k''}\left(\mathbf{x}_2'' \to \mathbf{x}_1\right) = \nabla_1^2 \nabla_{1,i''} u\left(\mathbf{x}_1 - \mathbf{R}_{l''k''}\right) = \nabla_{1,i''}\nabla_1^2 u\left(\mathbf{x}_1 - \mathbf{R}_{l''k''}\right) : \tag{11.3.39}$$

For the bare Coulomb ion potential, $u\left(\mathbf{x}_1 - \mathbf{R}_{l''k''}\right) \simeq \left(|\mathbf{x}_1 - \mathbf{R}_{l''k''}|\right)^{-1}$, we have $\nabla_1^2 u \simeq -4\pi\delta^{(3)}\left(\mathbf{x}_1 - \mathbf{R}_{l''k''}\right)$, which facilitates the $\mathbf{x}_1$-integration. Furthermore, $\nabla_{1,i''}$ involved in Eq. (11.3.39) can be integrated by parts on the right-hand side of Eq. (11.3.38) to act on ${}^{i}\gamma^{e}_{lk}(\mathbf{x}_1)$, reproducing $\Xi_{ii''}$, which may be absorbed into the terms of $\Xi_{ii''}(l,k;l,k)$ on the left of Eq. (11.3.38), with a *full cancellation* of such terms, leaving only the screened terms on the right of Eq. (11.3.38). Disposing of the local $\delta(1-2'')$ terms in this way and considering the case in which there is just *one* species of ion in the crystal lattice, the k-index is superfluous and may be dropped, whence we have

$$M\frac{d^2}{dt_1^2} D_{ii'}\left(l,t_1;l',t_2\right) = -\delta_{ii'}\delta_{ll'}\delta\left(t_1-t_2\right)$$

$$-\frac{1}{4\pi}\int d^{(3)}\mathbf{x}_1\, {}^{i}\gamma^{e}_{l}(\mathbf{x}_1)\,\nabla_1^2 \int d2'' K_e\left(1,2''\right) \sum\nolimits_{i'',l''} {}^{i''}\gamma^{e}_{l''}\left(\mathbf{x}_2''\right) D_{i''i'}\left(l'',t_2'';l',t_2\right). \tag{11.3.40}$$

In the case of equilibrium conditions and time-translational invariance, Eq. (11.3.40) may be Fourier-analyzed as a Boson Green's function to obtain the auxiliary function $D_{ii'}(t-t') \rightarrow \mathcal{D}_{ii'}(z)$ of Eq. (7.4.17). Furthermore, since the system under consideration is a crystal lattice, which is periodic under translation through a lattice vector, $\mathcal{D}_{ii'}(l,l';z)$ is a function only of $(\mathbf{R}_l - \mathbf{R}_{l'})$ and can be represented by a spatial Fourier series transform as $\mathcal{D}_{ii'}(\mathbf{q},z) = \sum_{R_l-R_{l'}} e^{-i\mathbf{q}\cdot(\mathbf{R}_l-\mathbf{R}_{l'})}\mathcal{D}_{ii'}(\mathbf{R}_l-\mathbf{R}_{l'},z)$ and $\mathcal{D}_{ii'}(\mathbf{R}_l-\mathbf{R}_{l'},z) = \Omega\int_{Bz}' \frac{d^{(3)}\mathbf{q}}{(2\pi)^3} e^{i\mathbf{q}\cdot(\mathbf{R}_l-\mathbf{R}_{l'})}\mathcal{D}_{ii'}(\mathbf{q},z)$ (as in Eqns. (11.3.7 and 11.3.8)), which permits the explicit algebraic solution of Eq. (11.3.40) as (recall Eq. (11.1.9) defining ${}^{i}\gamma^{e}_{lk}(\mathbf{x}_1) = \nabla_{1i}u(\mathbf{x}_1 - \mathbf{R}_{lk})$ and $\Xi_{ii'} = \nabla_i\nabla_{i'}u$):

$$\mathcal{D}^{-1}_{ii'}(\mathbf{q},z) = Mz^2\delta_{ii'} - \frac{1}{4\pi\Omega_a}\sum_{\mathbf{p}}\Big[(\mathbf{p}+\mathbf{q})_i(\mathbf{p}+\mathbf{q})_{i'}(\mathbf{p}+\mathbf{q})^2 u^2(\mathbf{p}+\mathbf{q})\,K_e(\mathbf{p}+\mathbf{q},z)$$

$$-p^2 p_i p_{i'} u^2(\mathbf{p})\,K_e(\mathbf{p},0)\Big], \tag{11.3.41}$$

where the ionic potential is screened, the **p**'s are reciprocal lattice vectors and a variant of the convolution theorem has been employed jointly with the Poisson sum formula,

$$\sum_{\mathbf{R}_l} e^{-i\mathbf{q}\cdot\mathbf{R}_l} = \frac{(2\pi)^3}{\Omega_a} \sum_{\mathbf{p}} \delta(\mathbf{q}-\mathbf{p}). \tag{11.3.42}$$

The last term of Eq. (11.3.41) involving $K_e(\mathbf{p}, 0)$ appears in consequence of the fact that a zero frequency, long wavelength ($\mathbf{q} \to 0$) motion of the ions is equivalent to a uniform spatial translation of the crystal lattice as a whole.[8,9] The terms of Eq. (11.3.41) involving non-zero reciprocal lattice vectors ($\mathbf{p} \neq 0$), known as *Umklapp* terms, are generally small compared to the $\mathbf{p} = 0$ term because u and K_e generally diminish as $\mathbf{p}$ increases. The role of electron screening (K_e) in phonon propagation can therefore be assessed by retaining only the $\mathbf{p} = 0$ term,

$$\mathcal{D}^{-1}_{ii'} = Mz^2\delta_{ii'} - \frac{1}{4\pi\Omega_a} q_i q_{i'} \mathbf{q}^2 u^2(\mathbf{q})\, K_e(\mathbf{q}, z). \tag{11.3.43}$$

The phonon collective mode frequencies, as modified by electron screening, may be identified in terms of the frequency poles $z \to \omega$, of $\mathcal{D}_{ii'}$ or, equivalently, obtained from the vanishing of the determinant of the right-hand side of Eq. (11.3.43). Alternatively, the phonon frequencies may be obtained from the vanishing eigenvalues of the $\mathcal{D}_{ii'}$ matrix on the right of Eq. (11.3.43) written in dyadic form as ($\widehat{\mathbf{q}} \equiv \mathbf{q}/q$),

$$\overleftrightarrow{\mathcal{D}}^{-1} = M\omega^2 \overleftrightarrow{I} - \frac{4\pi(\mathcal{Z}e)^2}{\Omega_a} K_e(q, \omega)\, \widehat{\mathbf{q}}\widehat{\mathbf{q}}, \tag{11.3.44}$$

where we have taken $u(q) = 4\pi(\mathcal{Z}e)^2/q^2$ as the unscreened ion–ion Coulomb interaction ($\mathcal{Z}$ is the charge number of the ion) and we assume isotropy, $K_e(\mathbf{q}, \omega) = K_e(q, \omega)$, of the electron screening. It is readily verified that $\overleftrightarrow{\mathcal{D}}^{-1}$ (Eq. (11.3.44)) has a longitudinal eigenvector $\widehat{\mathbf{q}} \equiv \mathbf{q}/q$ with eigenvalue described by

$$\overleftrightarrow{\mathcal{D}}^{-1}\widehat{\mathbf{q}} = \left(M\omega^2 - \frac{4\pi(\mathcal{Z}e)^2}{\Omega_a} K_e(q, \omega)\right)\widehat{\mathbf{q}}. \tag{11.3.45}$$

The longitudinal phonon dispersion relation is therefore given by

$$M\omega^2 - \frac{4\pi(\mathcal{Z}e)^2}{\Omega_a} K_e(q, \omega) = 0. \tag{11.3.46}$$

In the high-frequency regime, $K_e(q,\omega) \to \frac{\omega^2}{\omega^2-\omega_p^2}$, where ω_p^2 is the electron plasma frequency. This yields the phonon frequency (coupled to the electron plasma) as

$$\omega^2 = \omega_p^2 + \frac{4\pi(\mathcal{Z}e)^2}{M\Omega_a}, \tag{11.3.47}$$

representing a hybridization of the electron plasma mode with the ion plasma mode (corresponding to ion density Ω_a^{-1} for the single ion per unit cell under consideration). It should be noted that our high-frequency, low-wave number approximation $K_e(q,\omega) \to \omega^2/(\omega^2-\omega_p^2)$ ignores $Im\, K_e(q,\omega)$, whose non-vanishing value at higher wave numbers will provide damping of this hybrid mode. On the other hand, at low frequencies $K_e(q,0)$ is real and there is no damping. There are two interesting electron shielding features in the phonon spectrum in this limit, which are embedded in the analytic structure of $K_e(q,0)$ for $q_F \gg q_{DTF}$. These arise from the neighborhoods of $q \sim 2q_F$ and $q \sim q_{DTF}$. The log-singularity of $K_e(q,0)$ near $q \sim 2q_F$ produces a "kink" in the phonon dispersion relation of Eq. (11.3.46), which is known as the *Kohn anomaly*. On the other hand, the low-wave number, low-frequency *Bohm–Staver* phonon dispersion relation emerges from Eq. (11.3.46) for q in the vicinity of the Debye–Thomas–Fermi wave number (q_{DTF}) as

$$\omega^2 = \frac{4\pi\,(\mathcal{Z}e)^2}{M\Omega_a}\frac{q^2}{q_{DTF}^2}, \tag{11.3.48}$$

which is in the nature of a longitudinal sound mode. Having provided an introduction to the physical content of the phonon Green's function with electron screening, we refer the reader to the literature[5–11] for further discussion of such phenomenology, in particular the determination of elastic constants of crystal and the role of electrical conductivity in the screening of the phonon spectrum.

Another important physical feature of the electron–phonon interaction is its role in the electron Green's function, which leads to the *polaron*, an electron propagating jointly with its induced lattice polarization. Again ignoring holes and electron–electron interaction, the electron propagation characteristics are described by $G_{100} \to G_1$ of Eq. (11.2.17) as $[h_{100} \to h_1, U \to 0]$:

$$\left(i\frac{\partial}{\partial t_1} - h_1\right) G_1(1,2) - i^{\frac{1}{2}} \int d3\, \overleftrightarrow{v}^{e}_{p}(1,3)\, \overleftrightarrow{G}_{1,\frac{1}{2},0}(3,2//t_3) = \delta(1,2). \tag{11.3.49}$$

While we already have one form of the equation of motion for $\overleftrightarrow{G}_{1,\frac{1}{2},0}(1,2//t_3)$ given by Eq. (11.2.26), it is advantageous to derive another form focused on the phonon time variable t_3 of the pertinent average $\left\langle \hat{\Psi}^+(2)\,{}^i Q_{lk}(t_3)\hat{\Psi}(1)\right\rangle$. Rewriting Eq. (11.3.4) in (i,l,k)-tensor form using the uncoupled phonon Green's function $\overleftrightarrow{D^0}(t_1,t_2)$ of Eq. (11.3.6) (with ${}^i\mathcal{J}_{lk} \to 0$), we have

$$\int dt' \left[\overleftrightarrow{D^0}(t_1 - t')\right]^{-1} \overleftrightarrow{Q}(t') = -\int d^{(3)}\mathbf{x}'\, \overleftrightarrow{\gamma^e}(\mathbf{x}')\, \hat{\Psi}^+(1')\, \hat{\Psi}(1'). \tag{11.3.50}$$

Multiplying by $\hat{\Psi}^+(2)$ from the left and by $\hat{\Psi}(1)$ from the right and averaging, etc., we may write

$$\begin{aligned}&\int dt' \left[\overleftrightarrow{D^0}(t_3 - t')\right]^{-1} \left\langle \hat{\Psi}^+(2)\overleftrightarrow{Q}(t')\hat{\Psi}(1)\right\rangle \\ &= -\left\langle \hat{\Psi}^+(2)\int d^{(3)}\,\mathbf{x}'\,\overleftrightarrow{\gamma^e}(\mathbf{x}')\hat{\Psi}^+(\mathbf{x}',t_3)\hat{\Psi}(\mathbf{x}',t_3)\hat{\Psi}(1)\right\rangle,\end{aligned} \tag{11.3.51}$$

and inverting the inverse phonon Green's function $\left[\overleftrightarrow{D^0}\right]^{-1}$, we obtain

$$\left\langle \hat{\Psi}^+(2)\overleftrightarrow{Q}(t_3)\hat{\Psi}(1)\right\rangle = -\int dt'\,\overleftrightarrow{D^0}(t_3-t')\int d^{(3)}\mathbf{x}'\,\overleftrightarrow{\gamma^e}(\mathbf{x}')\left\langle \hat{\Psi}^+(2)\hat{\Psi}^+(\mathbf{x}',t')\hat{\Psi}(\mathbf{x}',t')\hat{\Psi}(1)\right\rangle. \tag{11.3.52}$$

Substitution of Eq. (11.3.52) into Eq. (11.3.49) yields the result

$$\left(i\frac{\partial}{\partial t_1} - h_1^{(1)}\right) G_1(1,2) + i\int d3\widetilde{\mathcal{V}}(1,3)\, G_2(1,3;2,3^+) = \delta(1,2), \tag{11.3.53}$$

where the entire effect of coupling to the phonon field on the one-electron Green's function is embedded in an effective time-dependent electron–electron interaction potential $\widetilde{\mathcal{V}}$ given by

$$\begin{aligned}\widetilde{\mathcal{V}}(1,3) &= \overleftrightarrow{\gamma^e}(\mathbf{x}_1)\,\overleftrightarrow{D^0}(t_1-t_3)\,\overleftrightarrow{\gamma^e}(\mathbf{x}_3)\\ &= \sum_{\substack{i,l,k\\ i',l',k'}} {}^{i}\gamma^e_{lk}(\mathbf{x}_1) D^0_{ii'}(l,k;t_1-t_3;l',k')\,{}^{i'}\gamma^e_{l'k'}(\mathbf{x}_3).\end{aligned} \tag{11.3.54}$$

With these results, the determination of the electron Green's function, G_1, in the presence of phonon coupling may be carried out using the analytical/approximation methods described in earlier chapters for two-body interactions, but the non-locality in time of the effective electron–electron interaction $\widetilde{\mathcal{V}}$ must now be taken into account. (This result differs in appearance from that of second-quantized formulations employing non-Hermitian phonon creation/annihilation operators because we employ Hermitian ion-displacement operators $\overleftrightarrow{Q}(t)$.)[12]

11.4 Alternative Formulation of the Equations of Motion

In the interest of developing a more symmetrical formulation of the equations of motion for the interacting electron–hole–phonon system,[13,14] and facilitating useful approximation procedures, it is desirable to examine an alternative description of the coupled field equations in which higher order and mixed Green's functions are expressed in terms of variational derivatives of G_{100} with respect to the phonon source $\overleftrightarrow{\mathcal{J}}$ and source potential U. In this context, we set the Fermion particle sources $\xi, \xi^+, \zeta, \zeta^+$ to zero, but retain functional dependence on U and ${}^{i}\mathcal{J}_{lk}$ in all the Green's functions, in both their numerator and denominator traces,

$$\overleftrightarrow{G}_{i,j,k} = \overleftrightarrow{G}_{i,j,k}\left(U, \overleftrightarrow{\mathcal{J}}\right). \tag{11.4.1}$$

Recalling the discussion associated with Eqns. (9.1.24 and 9.1.25), the denominator trace, $Tr\,[U(t_0+\tau, t_0)]$, may be understood as a generalization of the grand partition function to a non-equilibrium state generated by the Boson sources $\overleftrightarrow{\mathcal{J}}$ and U, and we define the corresponding non-equilibrium generalization of the grand potential as $\Omega_{C,B} = -W_{C,B}/\beta$ with

$$W_{C,B} = \ln\, Tr\, [U(t_0 + \tau,\, t_0)]_{C,B}\,, \tag{11.4.2}$$

Reflecting on the derivation of Eq. (4.3.7) jointly with Eq. (4.3.8) (or Eq. (4.4.2)), it is clear that completely analogous results apply here. Moreover, we have (note Eq. (11.1.11), first term on the right):

$$\begin{aligned}\frac{\delta W_{C,B}}{\delta U(1)} &= \left(Tr\, [U(t_0 + \tau,\, t_0)]_{C,B}\right)^{-1} \delta \left(Tr\, [U(t_0 + \tau,\, t_0)]_{C,B}\right) / \delta U(1) \\ &= \frac{-i\, Tr\left[U(t_0 + \tau, t_0)(\hat{\Psi}^+(1)\,\hat{\Psi}(1) - \hat{\Phi}^+(1)\,\hat{\Phi}(1))\right]_{C,B}}{Tr\, [U(t_0 + \tau,\, t_0)]_{C,B}},\end{aligned} \tag{11.4.3}$$

by the Action Principle. Consequently, we obtain

$$\frac{\delta W_{C,B}}{\delta U(1)} = -i\left\langle\left(\hat{\Psi}^+(1)\,\hat{\Psi}(1) - \hat{\Phi}^+(1)\,\hat{\Phi}(1)\right)\right\rangle_{C,B} = -\left[G_{1,0,0}(1,1^+)_{C,B} - G_{0,0,1}(1,1^+)\right]_{C,B}. \tag{11.4.4}$$

Bearing in mind that both the numerator and denominator traces of the Green's functions depend in U and ${}^{i}\mathcal{J}_{lk}$, we employ Eq. (4.2.24) to apply $\delta/\delta U(3)$ to $G_{1,0,0}(1,2)$ with the result (we suppress the explicit subscript notation C,B, understanding it where appropriate, and consider only Fermions here):

$$\begin{aligned}\frac{\delta G_{1,0,0}(1,2)}{\delta U(3)} &= -\left[G_{2,0,0}(1,3;2,3^+) - G_{1,0,1}(1,2;3;3^+)\right] \\ &\quad + \left[\left(G_{1,0,0}(1,2)\left[G_{1,0,0}(3,3^+) - G_{0,0,1}(3,3^+)\right]\right)\right].\end{aligned} \tag{11.4.5}$$

With these results, the $G_{1,0,0}$-equation of motion (Eq. (11.2.17)) may be rewritten as

$$\begin{aligned}&\left[i\frac{\partial}{\partial t_1} - h_{100}(1) - V_{100}(1)\right] G_{1,0,0}(1,2) - i^{\frac{1}{2}} \int \overleftrightarrow{v}^{e}_{p}(1,3)\, \overleftrightarrow{G}_{1;\frac{1}{2};0}(3,2//t_3) \\ &\quad - i\int \mathcal{V}(1,3)\frac{\delta G_{1,0,0}(1,2)}{\delta U(3)} = \delta(1,2),\end{aligned} \tag{11.4.6}$$

where the time-dependent one-electron Hartree potential is

$$V_{100}(1) = U(1) - i\int \mathcal{V}(1,3)\left[G_{1,0,0}(3,3^+) - G_{0,0,1}(3,3^+)\right] \tag{11.4.7}$$

and recalling Eq. (11.4.4), we have

$$V_{100}(1) = U(1) + i\int \mathcal{V}(1,3)\frac{\delta W_{C,B}}{\delta U(3)}. \tag{11.4.8}$$

Moreover, further variational differentiation with respect to ${}^{i}\mathcal{J}_{lk}$ yields

$$\frac{\delta G_{1,0,0}(1,2)}{\delta\, {}^{i}\mathcal{J}_{lk}(t_3)} = (-i)^{\frac{1}{2}}\left[\overleftrightarrow{G}_{1;\frac{1}{2};0}(1,2//t_3) - G_{1,0,0}(1,2)\,\overleftrightarrow{G}_{0;\frac{1}{2};0}(//t_3//)\right] \tag{11.4.9}$$

$$= (-i)^{\frac{1}{2}}\overleftrightarrow{G}_{1;\frac{1}{2};0}(1,2//t_3) - G_{1,0,0}(1,2)\frac{\delta W_{C,B}}{\delta\, {}^{i}\mathcal{J}_{lk}(t_3)}, \tag{11.4.10}$$

where we have used

$$\frac{\delta W_{C,B}}{\delta\, {}^{i}\mathcal{J}_{lk}(t_3)} = -i\left\langle {}^{i}Q_{lk}(t_3)\right\rangle = (-i)^{\frac{1}{2}}G_{0;\frac{1}{2};0}. \tag{11.4.11}$$

Thus, we reformulate the one-electron G_{100}-equation of motion as

$$\left[i\frac{\partial}{\partial t_1} - h_{100}(1) - V'_{100}(1)\right]G_{1,0,0}(1,2) - i\int \overleftrightarrow{v}^{e}_{p}(1,3)\frac{\delta G_{1,0,0}(1,2)}{\delta \overleftrightarrow{\mathcal{J}}(t_3)}$$
$$- i\int \mathcal{V}(1,3)\frac{\delta}{\delta U(3)}G_{1,0,0}(1,2) = \delta(1,2), \tag{11.4.12}$$

with the one-electron Hartree potential given as

$$V'_{100}(1) = U(1) + i\int \mathcal{V}(1,3)\frac{\delta W_{C,B}}{\delta U(3)} + i\int \overleftrightarrow{v}^{e}_{p}(1,3)\frac{\delta W_{C,B}}{\delta \overleftrightarrow{\mathcal{J}}(t_3)}. \tag{11.4.13}$$

In a similar manner, the one-hole $G_{0,0,1}$-equation of motion is reformulated as

$$\left[i\frac{\partial}{\partial t_1} - h_{001}(1) - V'_{001}(1)\right]G_{0,0,1}(1,2) - i\int \overleftrightarrow{v}^{h}_{p}(1,3)\frac{\delta G_{0,0,1}(1,2)}{\delta \overleftrightarrow{\mathcal{J}}(t_3)}$$
$$- i\int \mathcal{V}(1,3)\frac{\delta}{\delta U(3)}G_{0,0,1}(1,2) = \delta(1,2), \tag{11.4.14}$$

with the one-hole time-dependent Hartree potential given as

$$V'_{001}(1) = -U(1) + i\int \mathcal{V}(1,3)\frac{\delta W_{C,B}}{\delta U(3)} + i\int \overleftrightarrow{v}^{h}_{p}(1,3)\frac{\delta W_{C,B}}{\delta \overleftrightarrow{\mathcal{J}}(t_3)}. \tag{11.4.15}$$

It is convenient to define row vectors,

$$\mathbf{v}_{100}(1,3) = \left[\mathcal{V}(1,3),\ \overleftrightarrow{v}^{e}_{p}(1,3)\right], \tag{11.4.16}$$

$$\mathbf{v}_{001}(1,3) = \left[\mathcal{V}(1,3),\ \overleftrightarrow{v}^{h}_{p}(1,3)\right], \tag{11.4.17}$$

and also define column vectors,

$$\mathbf{Y}_{100}(3) = \begin{bmatrix} U(3) \\ \overleftrightarrow{\mathcal{J}}(t_3) \end{bmatrix}, \tag{11.4.18}$$

$$\mathbf{Y}_{001}(3) = \begin{bmatrix} -U(3) \\ \overleftrightarrow{\mathcal{J}}(t_3) \end{bmatrix}, \tag{11.4.19}$$

and, in addition, define operators $\hat{F}_{100}$ and $\hat{F}_{001}$ as

$$\hat{F}_{100}(1,3) = \left[i\frac{\partial}{\partial t_1} - h_{100}(1) - V'_{100}(1) - i\mathbf{v}_{100}(1,a)\frac{\delta}{\delta \mathbf{Y}_{100}(a)} \right] \delta(1,3) \tag{11.4.20}$$

and

$$\hat{F}_{001}(1,3) = \left[i\frac{\partial}{\partial t_1} - h_{001}(1) - V'_{001}(1) - i\mathbf{v}_{001}(1,a)\frac{\delta}{\delta \mathbf{Y}_{001}(a)} \right] \delta(1,3), \tag{11.4.21}$$

where the repeated index a is summed to represent the space-time matrix product of $\mathbf{v}_{100}(1,3)$ with $\frac{\delta}{\delta \mathbf{Y}_{100}(3)}$ *and* the product of the row vectors of Eqns. (11.4.16 and 11.4.17) multiplying the column vectors of Eqns. (11.4.18 and 11.4.19) as defined above. Furthermore, if we view the operators $\hat{F}_{100}$, $\hat{F}_{001}$ and $G_{1,0,0}$, $G_{0,0,1}$ as matrices in their space-time indices, then the Eqns. (11.4.12) and (11.4.14) for $G_{1,0,0}$ and $G_{0,0,1}$ may be written in the compact matrix forms

$$\hat{F}_{100} G_{1,0,0} = I_{10} \tag{11.4.22}$$

and

$$\hat{F}_{001} G_{0,0,1} = I_{01}, \tag{11.4.23}$$

where I_{10} is the 4D δ-function in position-time representation of the electron coordinates, $\langle 1e \,|\, I_{10} \,|\, 2e \rangle = \delta(1,2)$, and a similar hole definition applies to I_{01}, $\langle 1h \,|\, I_{01} \,|\, 2h \rangle = \delta(1,2)$. The two-electron/hole Green's function equations of motion for $G_{2,0,0}$ and $G_{0,0,2}$ (Eqns. (11.2.20) and (11.2.22)) can also be written in such a compact form with the introduction of the symbol $A \otimes B$ to denote the direct product of two space-time matrices A and B. In this context, we write $det\big[A(1,2)B(1',2')\big] = A(1,2)B(1',2') - A(1,2')B(1',2)$ and the compact forms for the two-electron/two-hole Green's function equations are given by

$$\hat{F}_{100} G_{2,0,0} = \det(I_{10} \otimes G_{1,0,0}) \tag{11.4.24}$$

and

$$\hat{F}_{001} G_{0,0,2} = \det(G_{0,0,1} \otimes I_{01}), \tag{11.4.25}$$

where we understand the direct product representation of $G_{2,0,0}$ to be defined as $\big\langle 1,2 \,|\, G_{2,0,0} \,|\, 1',2' \big\rangle = \big\langle 1 \,|\big\langle 2 \,|\, G_{2,0,0} \,|\, 1' \big\rangle \,|\, 2' \big\rangle \equiv G_{2,0,0}(1,2;1',2')$, with a similar definition for $G_{0,0,2}$. Similarly, we obtain the equation for $G_{1;\frac{1}{2};0}$ from Eq. (11.2.26) as

$$\hat{F}_{100}\overleftrightarrow{G}_{1;\frac{1}{2};0} = I_{10} \otimes \left\langle \overleftrightarrow{Q} \right\rangle, \tag{11.4.26}$$

and Eq. (11.2.24) for the joint electron/hole propagator may be written in compact form as

$$\hat{F}_{100}\overleftrightarrow{G}_{1,0,1} = I_{10} \otimes G_{0,0,1}. \tag{11.4.27}$$

These equations, Eqns. (11.3.22)–(11.3.27), are exact for the fully interacting electron–hole–phonon system.

11.5 Analysis of Electron–Hole Interaction Dynamics

It is at once apparent that Eq. (11.4.27) for the joint electron/hole propagator, $G_{1,0,1}$, is unnecessarily asymmetric with respect to electrons and holes, as $\hat{F}_{100}$ refers only to electrons and not at all to holes. This is easily remedied by operating from the left with $\hat{F}_{001}$, which refers only to holes, with the more symmetric result (we define $I_{11} = I_{10} \otimes I_{01}$):

$$\hat{F}_{001}\hat{F}_{100}G_{1,0,1} = I_{10} \otimes I_{01} = I_{11}, \tag{11.5.1}$$

since $\hat{F}_{001}G_{001} = I_{01}$ by Eq. (11.4.23).

Considering the variational differential operator $\hat{F}_{100}$ (Eq. (11.4.20)) as a space-time matrix,

$$\hat{F}_{100} = i\frac{\partial}{\partial t} - h_{100} - V'_{100} - i\mathbf{v}_{100}\frac{\delta}{\delta \mathbf{Y}_{100}}, \tag{11.5.2}$$

we examine its action on any quantity X which is a functional of $\mathbf{Y}_{100}$, applying the usual rule for differentiation of a product to variational derivatives, obtaining

$$\begin{aligned}
\hat{F}_{100}X &= \hat{F}_{100}\left(G_{1,0,0}G^{-1}_{1,0,0}X\right) \\
&= \left(i\frac{\partial}{\partial t} - h_{100} - V'_{100}\right) G_{1,0,0}G^{-1}_{1,0,0}X \\
&\quad - i\mathbf{v}_{100}\frac{\delta G_{1,0,0}}{\delta \mathbf{Y}_{100}}G^{-1}_{1,0,0}X - i\mathbf{v}_{100}G_{1,0,0}\frac{\delta}{\delta \mathbf{Y}_{100}}\left(G^{-1}_{1,0,0}X\right).
\end{aligned} \tag{11.5.3}$$

Since $\hat{F}_{100}G_{1,0,0} = I_{10}$ is the net coefficient of $G^{-1}_{1,0,0}X$ in the first two terms on the right of Eq. (11.5.3), we may rewrite it as

$$\hat{F}_{100}X = \left(I_{10} - i\mathbf{v}_{100}G_{1,0,0}\frac{\delta}{\delta \mathbf{Y}_{100}}\right)\left(G^{-1}_{1,0,0}X\right). \tag{11.5.4}$$

With a similar result for $\hat{F}_{001}$, we may write Eq. (11.5.1) in the form

$$\left\{\left(I_{01} - i\mathbf{v}_{001}G_{0,0,1}\frac{\delta}{\delta \mathbf{Y}_{001}}\right)G^{-1}_{0,0,1}\right\}\left\{\left(I_{10} - i\mathbf{v}_{100}G_{1,0,0}\frac{\delta}{\delta \mathbf{Y}_{100}}\right)G^{-1}_{1,0,0}\right\}G_{101} = I_{11}. \tag{11.5.5}$$

Multiplying both sides by $\left(I_{01} - i\mathbf{v}_{001}\, G_{0,0,1} \frac{\delta}{\delta \mathbf{Y}_{001}}\right)^{-1}$ and noting that the action of this operator on I_{11} on the right just returns I_{11} since $\delta I_{11}/\delta \mathbf{Y}_{100} \equiv 0$, we have

$$G_{0,0,1}^{-1} \left\{ \left(I_{10} - i\mathbf{v}_{100}\, G_{1,0,0} \frac{\delta}{\delta \mathbf{Y}_{100}} \right) G_{1,0,0}^{-1} \right\} G_{1,0,1} = I_{11}\,, \tag{11.5.6}$$

and using the rule for variational differentiation of a product, we find

$$\left(\left[I_{10} - i\mathbf{v}_{100}\, G_{1,0,0} \frac{\delta}{\delta \mathbf{Y}_{100}} \right] G_{0,0,1}^{-1} G_{1,0,0}^{-1} G_{1,0,1} \right) + i\mathbf{v}_{100} G_{1,0,0} G_{1,0,0}^{-1} \left[\frac{\delta G_{0,0,1}^{-1}}{\delta \mathbf{Y}_{100}} \right] G_{1,0,1} = I_{11}. \tag{11.5.7}$$

We now multiply both sides of Eq. (11.5.7) by $\left(I_{10} - i\mathbf{v}_{100}\, G_{1,0,0} \frac{\delta}{\delta \mathbf{Y}_{100}}\right)^{-1}$, obtaining

$$\left\{ G_{0,0,1}^{-1} G_{1,0,0}^{-1} + \left(I_{10} - i\mathbf{v}_{100}\, G_{1,0,0} \frac{\delta}{\delta \mathbf{Y}_{100}} \right)^{-1} i\mathbf{v}_{100} G_{1,0,0}\, G_{1,0,0}^{-1}\, {}^{100}\mathbf{\Gamma}_{001} \right\} G_{1,0,1}$$

$$= \left[G_{0,0,1}^{-1} G_{1,0,0}^{-1} - \widehat{\mathfrak{I}}_{11} \right] G_{1,0,1} = \left(I_{10} - i\mathbf{v}_{100}\, G_{1,0,0} \frac{\delta}{\delta \mathbf{Y}_{100}} \right)^{-1} I_{11} = I_{11} \tag{11.5.8}$$

since I_{11} has no dependence on $\mathbf{Y}_{100}$. Here, we have defined the variational differential electron–hole interaction operator $\widehat{\mathfrak{I}}_{11}$ as

$$\widehat{\mathfrak{I}}_{11} \equiv - \left\{ \left(I_{10} - i\mathbf{v}_{100}\, G_{1,0,0} \frac{\delta}{\delta \mathbf{Y}_{100}} \right)^{-1} i\mathbf{v}_{100} G_{1,0,0} \right\} G_{1,0,0}^{-1}\, {}^{100}\mathbf{\Gamma}_{001}, \tag{11.5.9}$$

and ${}^{100}\mathbf{\Gamma}_{001}(1,2;3)$ is defined by

$${}^{100}\mathbf{\Gamma}_{001}(1,2;3) = -\frac{\delta G_{0,0,1}^{-1}(1,2)}{\delta \mathbf{Y}_{100}(3)}. \tag{11.5.10}$$

(${}^{100}\mathbf{\Gamma}_{001}(1,2;3)$ has the sense of a *screened cross-species* vertex function.)

To further explore the highly complicated character of the electron–hole interaction operator $\widehat{\mathfrak{I}}_{11}$, we invert the action of $\hat{F}_{100}$ in Eq. (11.5.4) and set $X = V'_{100}$, obtaining (bear in mind that $(AB)^{-1} \equiv B^{-1}A^{-1}$ for operators):

$$\hat{F}_{100}^{-1} V'_{100} = G_{1,0,0} \left[I_{10} - i\mathbf{v}_{100}\, G_{1,0,0} \frac{\delta}{\delta \mathbf{Y}_{100}} \right]^{-1} V'_{100}. \tag{11.5.11}$$

Recalling the definition of $\hat{F}_{100}$ (Eq. (11.5.2)), we also have

$$\hat{F}_{100}^{-1} V'_{100} = \left[\left(G_{1,0,0}^{0} \right)^{-1} - V'_{100} - i\mathbf{v}_{100} \frac{\delta}{\delta \mathbf{Y}_{100}} \right]^{-1} V'_{100}. \tag{11.5.12}$$

Anticipating an approximation scheme that neglects the further application of variational derivatives (relatively weak correlation), the far right-hand side of Eq. (11.5.12) may be augmented by adding $i\mathbf{v}_{100}\delta/\delta\mathbf{Y}_{100}$ to V'_{100}. In this context the added term has negligible effects, and, correspondingly, we may rewrite Eq. (11.5.12) as[13,14]

$$\hat{F}_{100}^{-1}V'_{100} = \left[\left(G^0_{1,0,0}\right)^{-1} - V'_{100} - i\mathbf{v}_{100}\frac{\delta}{\delta\mathbf{Y}_{100}}\right]^{-1}\left(V'_{100} + i\mathbf{v}_{100}\frac{\delta}{\delta\mathbf{Y}_{100}}\right). \tag{11.5.13}$$

It is clear that the two operator factors on the right of Eq. (11.5.13) now commute, so that

$$\begin{aligned}\hat{F}_{100}^{-1}V'_{100} &= \left(V'_{100} + i\mathbf{v}_{100}\frac{\delta}{\delta\mathbf{Y}_{100}}\right)\left[\left(G^0_{1,0,0}\right)^{-1} - V'_{100} - i\mathbf{v}_{100}\frac{\delta}{\delta\mathbf{Y}_{100}}\right]^{-1}\\ &= \left(V'_{100} + i\mathbf{v}_{100}\frac{\delta}{\delta\mathbf{Y}_{100}}\right)G_{1,0,0},\end{aligned} \tag{11.5.14}$$

where we have invoked the reciprocal character of $\hat{F}_{100}$ and $G_{1,0,0}$ (Eq. (11.4.22)).

An instructive alternative derivation of Eq. (11.5.14) can be based on writing the $G_{1,0,0}$ equation in the form

$$G_{1,0,0} = G^0_{1,0,0} + G^0_{1,0,0}\left[V'_{100} + i\mathbf{v}\frac{\delta}{\delta\mathbf{Y}_{100}}\right]G_{1,0,0}, \tag{11.5.15}$$

where $G^0_{1,0,0}$ is defined by

$$\left(i\frac{\partial}{\partial t} - h_{100}\right)G^0_{1,0,0} = I_{10}. \tag{11.5.16}$$

Assuming the existence of an iteration series solution, infinitely repeated iteration yields $G_{1,0,0}$ as

$$G_{1,0,0} = \sum_{n=0}^{\infty}\Pi^n_{m=0}\,G^0_{1,0,0}\left(\left[V'_{100} + i\mathbf{v}_{100}\frac{\delta}{\delta\mathbf{Y}_{100}}\right]G^0_{1,0,0}\right)^m, \tag{11.5.17}$$

from which the commutability of $\left[V'_{100} + i\mathbf{v}_{100}\frac{\delta}{\delta\mathbf{Y}_{100}}\right]$ with $G_{1,0,0}$ is clear since $G^0_{1,0,0}$ does not depend on either V'_{100} or $\mathbf{Y}_{100}$. Considering the reciprocal character of $G_{1,0,0}$ and $\hat{F}_{100}$, it is clear that $\left[V'_{100} + i\mathbf{v}_{100}\frac{\delta}{\delta\mathbf{Y}_{100}}\right]$ also commutes with $\hat{F}_{100}^{-1}$. Thus,

$$\hat{F}_{100}^{-1}\left[V'_{100} + i\mathbf{v}_{100}\frac{\delta}{\delta\mathbf{Y}_{100}}\right] = \left[V'_{100} + i\mathbf{v}_{100}\frac{\delta}{\delta\mathbf{Y}_{100}}\right]G_{1,0,0}\,, \tag{11.5.18}$$

and again invoking the argument leading to Eq. (11.5.13), we drop the last $\delta/\delta\mathbf{Y}_{100}$ term on the left-hand side, obtaining Eq. (11.5.14).

Comparing Eqns. (11.5.11) and (11.5.14), we have

$$\left(V'_{100} + i\mathbf{v}_{100}\frac{\delta}{\delta\mathbf{Y}_{100}}\right) G_{1,0,0} = G_{1,0,0}\left[I_{10} - i\mathbf{v}_{100}G_{1,0,0}\frac{\delta}{\delta\mathbf{Y}_{100}}\right]^{-1} V'_{100}. \tag{11.5.19}$$

Subtracting $V'_{100}\,G_{1,0,0}$ from both sides yields

$$i\mathbf{v}_{100}\frac{\delta G_{1,0,0}}{\delta\mathbf{Y}_{100}} = G_{1,0,0}\left[I_{10} - i\mathbf{v}_{100}G_{1,0,0}\frac{\delta}{\delta\mathbf{Y}_{100}}\right]^{-1} i\mathbf{v}_{100}G_{1,0,0}\frac{\delta V'_{100}}{\delta\mathbf{Y}_{100}}. \tag{11.5.20}$$

The left-hand side may be re-expressed as (Eq. (4.5.6)),

$$i\mathbf{v}_{100}\frac{\delta G_{1,0,0}}{\delta\mathbf{Y}_{100}} = -i\mathbf{v}_{100}G_{1,0,0}\frac{\delta G^{-1}_{1,0,0}}{\delta\mathbf{Y}_{100}}G_{1,0,0}, \tag{11.5.21}$$

and comparison with Eq. (11.5.20) yields

$$\left[I_{10} - i\mathbf{v}_{100}G_{1,0,0}\frac{\delta}{\delta\mathbf{Y}_{100}}\right]^{-1} i\mathbf{v}_{100}G_{1,0,0}\frac{\delta V'_{100}}{\delta\mathbf{Y}_{100}} = -i\mathbf{v}_{100}\frac{\delta G^{-1}_{1,0,0}}{\delta\mathbf{Y}_{100}}G_{1,0,0}. \tag{11.5.22}$$

This relation will be seen to provide a more useful expression for the electron–hole interaction operator $\widehat{\mathfrak{I}}_{11}$ (in place of Eq. (11.5.9)) in the next section.

11.6 Generalized Shielded Potential Approximation: Electron–Hole Interaction Operator

In a generalization of the shielded potential approximation to the fully interacting electron–hole–phonon system under conditions of high density, we assume that $G_{1,0,0}(G_{0,0,1})$ can be treated as a functional of the effective potential $V'_{100}(V'_{001})$ alone,

$$G_{1,0,0} = G_{1,0,0}(V'_{100}) \tag{11.6.1}$$

and

$$G_{0,0,1} = G_{0,0,1}(V'_{001}). \tag{11.6.2}$$

Thus,

$$\frac{\delta G^{-1}_{1,0,0}}{\delta\mathbf{Y}_{100}} = \frac{\delta G^{-1}_{1,0,0}}{\delta V'_{100}}\frac{\delta\, V'_{100}}{\delta\mathbf{Y}_{100}}, \tag{11.6.3}$$

and defining

$$\mathbf{v}_{100}\frac{\delta V'_{100}}{\delta \mathbf{Y}_{100}} = \widetilde{v}_{100}, \tag{11.6.4}$$

we obtain from Eq. (11.5.22),

$$\left[I_{10} - i\mathbf{v}_{100}G_{1,0,0}\frac{\delta}{\delta \mathbf{Y}_{100}}\right]^{-1} i\widetilde{v}_{100}G_{1,0,0} = -i\,\widetilde{v}_{100}\frac{\delta G^{-1}_{1,0,0}}{\delta V'_{100}}G_{1,0,0}, \tag{11.6.5}$$

where we have used the chain rule for variational differentiation on the right-hand side. To explore the significance of the shielded potential approximation in the context of $G_{1,0,1}$ we recall the action of $\widehat{\mathfrak{I}}_{11}$ on $G_{1,0,1}$ from Eq. (11.5.9) as

$$\widehat{\mathfrak{I}}_{11}\, G_{1,0,1} = \left\{\left(I_{10} - i\mathbf{v}_{100}\, G_{1,0,0}\frac{\delta}{\delta \mathbf{Y}_{100}}\right)^{-1} i\,\widetilde{v}_{100}\, G_{1,0,0}\right\} G^{-1}_{1,0,0}\, {}^{100}\Gamma'_{001}\, G_{1,0,1}, \tag{11.6.6}$$

where we have employed Eqns. (11.5.9 and 11.5.10) to write

$${}^{100}\mathbf{\Gamma}_{001} = -\frac{\delta G^{-1}_{0,0,1}}{\delta \mathbf{Y}_{100}} = -\frac{\delta V'_{100}}{\delta \mathbf{Y}_{100}}\, {}^{100}\Gamma'_{001}, \tag{11.6.7}$$

with the definitions

$${}^{100}\Gamma'_{001} = -\frac{\delta G^{-1}_{0,0,1}}{\delta V'_{100}} \quad \text{and} \quad {}^{100}\Gamma'_{100} = -\frac{\delta G^{-1}_{1,0,0}}{\delta V'_{100}}, \tag{11.6.8}$$

and we have also used Eq. (11.6.4) in Eq. (11.6.6).

In principle, the variational derivative on the right-hand side of Eq. (11.6.6) acts on $G_{1,0,1}$ to produce higher order coupled Green's functions. However, estimates indicate that such terms are negligibly small under high-density conditions in the generalized shielded potential approximation. To start, we note that the factors to the right of the curly brackets in Eq. (11.6.6) may be estimated using $G_{1,0,1} \sim G_{1,0,0}G_{0,0,1}$ as

$$\begin{aligned} G^{-1}_{1,0,0}\, {}^{100}\Gamma'_{001}\, G_{1,0,1} &\sim G^{-1}_{1,0,0}\, {}^{100}\Gamma'_{001}\, G_{1,0,0}\, G_{0,0,1} \\ &\sim {}^{100}\Gamma'_{001}\, G_{0,0,1} \sim \frac{\delta G^{-1}_{0,0,1}}{\delta V'_{100}} G_{0,0,1}. \end{aligned} \tag{11.6.9}$$

Recalling that V'_{100} and V'_{001} differ mainly by the sign of U, such that $\delta V'_{001}/\delta V'_{100} \sim 0(1)$, we have

$$G^{-1}_{1,0,0}\, {}^{100}\Gamma'_{001}\, G_{1,0,1} \sim \frac{\delta G^{-1}_{0,0,1}}{\delta V'_{001}} G_{0,0,1}. \tag{11.6.10}$$

However, considering the modified self-energy M_{001} of holes, defined by

$$G_{0,0,1}^{-1} = \left[G_{0,0,1}^{0}\right]^{-1} - V'_{001} - M_{001}, \tag{11.6.11}$$

an analysis for holes similar to that of Eq. (9.7.6) for electrons may be undertaken using Eqns. (11.4.21–11.4.23), from which M_{001} may be identified as

$$M_{001} \sim -i\widetilde{v}_{001}\, G_{0,0,1}\frac{\delta G_{0,0,1}^{-1}}{\delta V'_{001}}. \tag{11.6.12}$$

As in Eq. (9.7.11) for electrons, the principal feature of the leading vertex approximation for holes is that the variation of M_{001} with respect to V'_{001} is negligibly small at high densities,

$$\frac{\delta M_{001}}{\delta V'_{001}} \sim \frac{\delta M_{001}}{\delta V'_{100}} \sim \frac{\delta M_{001}}{\delta \mathbf{Y}_{100}} \ll 1. \tag{11.6.13}$$

However, it is clear from Eqns. (11.6.9) and (11.6.12) that (except for the factor $\widetilde{v}_{001}$):

$$G_{1,0,0}^{-1}\, {}^{100}\Gamma'_{001}\, G_{1,0,1} \sim M_{001}, \tag{11.6.14}$$

so that

$$\frac{\delta}{\delta \mathbf{Y}_{100}}\left(G_{1,0,0}^{-1}\, {}^{100}\Gamma'_{001}\, G_{1,0,1}\right) \ll 1, \tag{11.6.15}$$

and thus variational differentiation to the right of the curly brackets of Eq. (11.6.6) is negligible. We therefore indicate that, within the framework of the shielded potential approximation, such variational differentiation can be stopped past the curly brackets, introducing the "stop" symbol ‡, in the equation for the electron–hole interaction operator, $\widehat{\mathfrak{I}}_{11}$,

$$\widehat{\mathfrak{I}}_{11}\, G_{1,0,1} \cong \left\{\left(I_{10} - i\mathbf{v}_{100}\, G_{1,0,0}\frac{\delta}{\delta \mathbf{Y}_{100}}\right)^{-1} i\widetilde{v}_{100}\, G_{1,0,0}\right\} \ddagger\, G_{1,0,0}^{-1}\, {}^{100}\Gamma'_{001}\, G_{1,0,1}. \tag{11.6.16}$$

Alternatively expressed,

$$\widehat{\mathfrak{I}}_{11}\, G_{1,0,1} \equiv \mathfrak{I}_{11}\, G_{1,0,1} \tag{11.6.17}$$

with $\mathfrak{I}_{11}$ defined as the curly bracket itself, with variational differentiation stopped at its end, multiplying $G_{1,0,0}^{-1}\, {}^{100}\Gamma'_{001}$,

$$\mathfrak{I}_{11} = \left\{\left[I_{10} - i\mathbf{v}_{100}\, G_{1,0,0}\frac{\delta}{\delta \mathbf{Y}_{100}}\right]^{-1} i\widetilde{v}_{100}\, G_{1,0,0}\right\} \ddagger\, G_{1,0,0}^{-1}\, {}^{100}\Gamma'_{100}. \tag{11.6.18}$$

Employing Eq. (11.5.22), $\mathfrak{I}_{11}$ may be written as

$$\mathfrak{I}_{11} = -i\mathbf{v}_{100} \frac{\delta G^{-1}_{1,0,0}}{\delta \mathbf{Y}_{100}} G_{1,0,0} G^{-1}_{1,0,0}\, {}^{100}\Gamma'_{100} \tag{11.6.19}$$

or, finally, the effective 4-point electron–hole interaction is given by

$$\mathfrak{I}_{11} = i\widetilde{v}_{100} \frac{\delta G^{-1}_{1,0,0}}{\delta V'_{100}}\, {}^{100}\Gamma'_{001} = -i\widetilde{v}_{100}\, {}^{100}\Gamma'_{100}\, {}^{100}\Gamma'_{001}, \tag{11.6.20}$$

with $G_{1,0,1}$ to be determined from Eq. (11.5.8) in the form

$$G_{1,0,1} = G_{1,0,0} G_{0,0,1} + i\, G_{1,0,0} G_{0,0,1} \mathfrak{I}_{11}\, G_{1,0,1}, \tag{11.6.21}$$

within the framework of the generalized shielded potential approximation.

A similar analysis of the two-electron Green's function, $G_{2,0,0}$, equation of motion (Eq. (11.4.24)), which provided the basis for the foregoing electron–hole considerations, was carried out in Reference 13 (Section 6), with analogous results.

11.7 Dielectric Response of the Electron–Hole–Phonon System

The effective potential $V(1)$ of an electron–hole–phonon system due to an impressed potential $U(1)$ is given by Eqns. (11.1.8) and (11.3.36) as (recall that ${}^{i}\gamma_{lk}(\mathbf{x}_1) = \nabla_{1i}\, u(\mathbf{x}_1 - \mathbf{R}_{lk})$ and we denote its matrix as $\overleftrightarrow{\gamma_p}(\mathbf{x}_1)$):

$$V(1) = U(1) + \int d3\, \mathcal{V}(1-3)\, [\langle \rho_e(3)\rangle - \langle \rho_h(3)\rangle] + \overleftrightarrow{\gamma_p}(\mathbf{x}_1) \left\langle \overleftrightarrow{Q}(t_1) \right\rangle, \tag{11.7.1}$$

since the electrons and holes are oppositely charged and the potential projected by the ions is represented by $\overleftrightarrow{\gamma_p}(\mathbf{x}_1) \left\langle \overleftrightarrow{Q}(t_1) \right\rangle$. To determine the inverse dielectric response function $K(1,2)$, it is necessary to take the variational derivative $\delta/\delta U(2)$ of Eq. (11.7.1). Bearing in mind that the source potential U couples to both the electron and hole densities in H'_B (Eq. (11.1.11)), this variational differentiation yields the exact result

$$K(1,2) = \frac{\delta V(1)}{\delta U(2)} = \delta(1-2) + \int d3\, \mathcal{V}(1-3) \left[\frac{\delta \langle \rho_e(3)\rangle}{\delta U(2)} - \frac{\delta \langle \rho_h(3)\rangle}{\delta U(2)} \right] + \overleftrightarrow{\gamma_p}(\mathbf{x}_1) \frac{\delta \left\langle \overleftrightarrow{Q}(t_1) \right\rangle}{\delta U(2)}, \tag{11.7.2}$$

where

$$\frac{\delta\left\langle \rho_e(3)\right\rangle}{\delta U(2)} = -i\begin{Bmatrix} -G_{2,0,0}(3,2;3^+,2^+) + G_{1,0,0}(2,2^+)G_{1,0,0}(3,3^+) \\ +G_{1,0,1}(3,2;3^+,2^+) - G_{0,0,1}(2,2^+)G_{1,0,0}(3,3^+) \end{Bmatrix} \tag{11.7.3}$$

and

$$\frac{\delta\left\langle \rho_h(3)\right\rangle}{\delta U(2)} = -i\begin{Bmatrix} +G_{0,0,2}(3,2;3^+,2^+) - G_{0,0,1}(2,2^+)G_{0,0,1}(3,3^+) \\ -G_{1,0,1}(3,2;3^+,2^+) + G_{1,0,0}(2,2^+)G_{0,0,1}(3,3^+) \end{Bmatrix}, \tag{11.7.4}$$

which include the terms arising from $\delta/\delta U$ acting on the normalizing denominator traces of the averages $\langle \varrho_e\rangle$, $\langle \varrho_h\rangle$.

In regard to the last term on the right of Eq. (11.7.2) involving $\delta\left\langle \overleftrightarrow{Q}(t_1)\right\rangle/\delta U(2)$, we consider the average of Eq. (11.3.4) with the recognition that the force ${}^i\mathcal{F}_{lk}(t)$ is to be identified with the self-consistent electric field $-[\partial V_k/\partial x_i]_{lk}$ evaluated $\mathbf{R}_{lk}$, with the result for an ionic crystal ($V_k \to Z_k V$ here),

$$M_k\frac{d^2}{dt_1^2}\left\langle {}^iQ_{lk}(t_1)\right\rangle + \sum\nolimits_{i''l''k''}\Xi_{ii''}(l,k;l'',k'')\left\langle {}^{i''}Q_{l''k''}(t_1)\right\rangle$$
$$= \int d^{(3)}\mathbf{x}_1'\left({}^i\gamma_{lk}^{(e)}(\mathbf{x}_1')\left\langle\hat{\Psi}^+(1')\hat{\Psi}(1')\right\rangle + {}^i\gamma_{lk}^{(h)}(\mathbf{x}_1')\left\langle\hat{\Phi}^+(1')\hat{\Phi}(1')\right\rangle\right) - \left[\frac{\partial V_k}{\partial x_i}\right]_{lk}, \tag{11.7.5}$$

where we have restored the role of holes in Eq. (11.3.4) on an equal footing with that of electrons. Employing the free-phonon Green's function D^0 of Eq. (11.3.6) to convert Eq. (11.7.5) to integral form (and noting that D^0 is driven by a negative δ-function), we obtain

$$\left\langle {}^iQ_{lk}(t_1)\right\rangle = -\sum\nolimits_{i''l''k''}\int dt' D^0_{ii''}(l,k,t_1;l'',k'',t')\int d^{(3)}\mathbf{x}_1'\left[{}^i\gamma_p^{(e)}(\mathbf{x}_1')\left\langle \varrho_e(1')\right\rangle\right.$$
$$\left. + {}^i\gamma_p^{(h)}(\mathbf{x}_1')\left\langle \varrho_h(1')\right\rangle\right] + \left\langle {}^iQ_{lk}(t_1)\right\rangle_{III}, \tag{11.7.6}$$

with the last term on the right defined as

$$\left\langle {}^iQ_{lk}(t_1)\right\rangle_{III} \equiv \sum\nolimits_{i''l''k''}\int dt' D^0_{ii''}(l,k,t_1;l'',k'',t')\int d^{(3)}\mathbf{x}'\,\delta^{(3)}(\mathbf{x}'-\mathbf{R}_{l''k''})\frac{\partial V_k(\mathbf{x}',t')}{\partial x_i'}. \tag{11.7.7}$$

This last term on the right represents the direct effect of the effective field on the ion displacement, which was considered above for a 1D ionic diatomic lattice in Eqns. (11.3.19–11.3.32). Proceeding in position-time representation we have

$$\overleftrightarrow{\gamma}_p(\mathbf{x}_1)\left\langle\overleftrightarrow{Q}(t_1)\right\rangle = -\overleftrightarrow{\gamma}_p(\mathbf{x}_1)\int d1'\,\overleftrightarrow{D}^0(t_1-t_1')\left\{\overleftrightarrow{\gamma}_p^{(e)}(\mathbf{x}_1')\langle\varrho_e(1')\rangle + \overleftrightarrow{\gamma}_p^{(h)}(\mathbf{x}_1')\langle\varrho_h(1')\rangle\right\}$$
$$+\overleftrightarrow{\gamma}_p(\mathbf{x}_1)\left\langle\overleftrightarrow{Q}(t_1)\right\rangle_{III}. \qquad (11.7.8)$$

Considering next the action of $\delta/\delta U(2)$, we have

$$\overleftrightarrow{\gamma}_p(\mathbf{x}_1)\frac{\delta\left\langle\overleftrightarrow{Q}(t_1)\right\rangle}{\delta U(2)} = -\overleftrightarrow{\gamma}_p(\mathbf{x}_1)\int d1'\,\overleftrightarrow{D}^0(t_1-t_1')\left\{\overleftrightarrow{\gamma}_p^{(e)}(\mathbf{x}_1')\frac{\delta\langle\varrho_e(1')\rangle}{\delta U(2)}\right.$$
$$\left.+\overleftrightarrow{\gamma}_p^{(h)}(\mathbf{x}_1')\frac{\delta\langle\varrho_h(1')\rangle}{\delta U(2)}\right\} + \overleftrightarrow{\gamma}_p(\mathbf{x}_1)\frac{\delta\left\langle\overleftrightarrow{Q}(t_1)\right\rangle_{III}}{\delta U(2)}, \qquad (11.7.9)$$

and invoking the chain rule of variational differentiation Eq. (11.7.2) yields ($\delta V(1')/\delta U(2) = K(1',2)$):

$$K(1,2) = \delta(1-2) + \int d3d4\left[\mathcal{V}(1-3)\left(\frac{\delta\langle\varrho_e(3)\rangle}{\delta V(4)} - \frac{\delta\langle\varrho_h(3)\rangle}{\delta V(4)}\right)\right.$$
$$-\overleftrightarrow{\gamma}_p(\mathbf{x}_1)\int d3d4\,\overleftrightarrow{D}^0(t_1-t_3)\left\{\overleftrightarrow{\gamma}_p^{(e)}(\mathbf{x}_3)\frac{\delta\langle\varrho_e(3)\rangle}{\delta V(4)}\right.$$
$$\left.\left.+\overleftrightarrow{\gamma}_p^{(h)}(\mathbf{x}_3)\frac{\delta\langle\varrho_h(3)\rangle}{\delta V(4)}\right\}\right]K(4,2) + \quad \overleftrightarrow{\gamma}_p(\mathbf{x}_1)\int d4\frac{\delta\left\langle\overleftrightarrow{Q}(t_1)\right\rangle_{III}}{\delta V(4)}K(4,2).$$
$$(11.7.10)$$

The polarizability of the system may be identified from Eq. (11.7.10) as

$$\alpha(1,2) = -\int d3\mathcal{V}(1-3)\left(\frac{\delta\langle\varrho_e(3)\rangle}{\delta V(2)} - \frac{\delta\langle\varrho_h(3)\rangle}{\delta V(2)}\right)$$
$$+\overleftrightarrow{\gamma}_p(\mathbf{x}_1)\int d3\,\overleftrightarrow{D}^0(t_1-t_3)\left\{\overleftrightarrow{\gamma}_p^{(e)}(\mathbf{x}_3)\frac{\delta\langle\varrho_e(3)\rangle}{\delta V(2)} + \overleftrightarrow{\gamma}_p^{(h)}(\mathbf{x}_3)\frac{\delta\langle\varrho_h(3)\rangle}{\delta V(2)}\right\}$$
$$-\overleftrightarrow{\gamma}_p(\mathbf{x}_1)\frac{\delta\left\langle\overleftrightarrow{Q}(t_1)\right\rangle_{III}}{\delta V(2)}. \qquad (11.7.11)$$

The first terms on the right of Eq. (11.7.11) involving $\mathcal{V}(1-3)$ represent the usual contributions to the polarizability through *e-e* and *h-h* Coulombic interactions (which have been thoroughly discussed in the RPA in earlier chapters), except to say that hole plasmons involving the hole effective mass and density will now couple to the electron plasmons through the collective mode dispersion relation $\varepsilon = 1 + \alpha = 0$. Furthermore, the second set of terms on the right of Eq. (11.7.11) represent polarizability contributions induced by *e-e* and *h-h* phonon-mediated interactions, which are not instantaneous, but nevertheless contribute plasma-like collective modes that are also coupled to the

Coulomb-induced modes just discussed. In this context, the perturbed electron/hole densities $\left(\delta\left\langle \varrho_{e,h}\right\rangle / \delta V\right)$ in turn perturb the ionic displacements (emitting a phonon), which modify the ions' projected field, thus impacting the electron/hole densities yet again (with the absorption of the phonon).

It is interesting to note that while $G_{1,0,1}$ does not explicitly appear in $\langle \varrho_e \rangle$ (or $\langle \varrho_h \rangle$) in Eq. (11.7.11) for α, it nevertheless does influence the perturbed densities (and hence the polarizabilities) through their defining single-particle Green's functions, $\varrho \sim G_1$. For example, in the case of the electron density, Eq. (11.4.6) may be written in integral form as

$$\begin{aligned}\langle \varrho_e \rangle \sim G_{1,0,0}(1,2) = G^0_{1,0,0}(1,2) &+ \int G^0_{1,0,0}(1,1')V_{100}(1')G_{1,0,0}(1',2) \\ &+ i\int G^0_{1,0,0}(1,1')\mathcal{V}(1',3)\frac{\delta G_{1,0,0}(1',2)}{\delta U(3)} \\ &+ i^{\frac{1}{2}}\int G^0_{1,0,0}(1,1')\overleftrightarrow{v}^{\,e}_{p}(1',3)\overleftrightarrow{G}_{1,\frac{1}{2},0}(1',2//t_3),\end{aligned} \tag{11.7.12}$$

but the involvement of $G_{1,0,1}$ becomes apparent through Eq. (11.4.5),

$$\begin{aligned}\frac{\delta G_{1,0,0}(1,2)}{\delta U(3)} &= -\left[G_{2,0,0}\left(1,3;2,3^+\right) - G_{1,0,1}\left(1,2;3;3^+\right)\right] \\ &\quad + \left[\left(G_{1,0,0}\left(1,2\right)\left[G_{1,0,0}\left(3;3^+\right) - G_{0,0,1}\left(3,3^+\right)\right]\right)\right].\end{aligned} \tag{11.7.13}$$

We note in passing that the incorporation of $G_{1,0,1}$ in the self-consistent development of dielectric response would provide for hybridization of exciton modes with plasmons in the collective mode spectrum.

Finally, the last contribution to α in Eq. (11.7.11) occurs even in the absence of both electrons and holes, and is associated with the direct effect of the self-consistent field on ionic displacements: For example, the realization of this ionic polarizability was presented above in Section 11.3 (following Eq. (11.3.32)) for the special case of a 1D ionic diatomic lattice in the local limit as

$$\alpha_{\substack{ionic\\ phonon}}(\mathbf{q},\omega) = \frac{Z^2}{\Omega_a\,\mu\,(2\alpha/\mu - \omega^2)}. \tag{11.7.14}$$

Of course, the longitudinal dispersion relation, $\varepsilon = 1 + \alpha = 0$, hybridizes all the collective oscillations due to both Coulombic- and phonon-induced *e-e* and *h-h* interactions, along with the purely ionic modes (such as those associated with the ion polarizability of Eq. (11.7.11)), yielding a very rich spectrum. A great deal of further detail about the methods discussed here, and the resulting coupled quantum-dynamic Green's function equations may be found in References 13 and 14.

11.8 Exciton-Plasmon Interaction

In regard to our note above concerning the role of excitons in the response of an electron–hole system to external field excitation, we now drop the phonon couplings and employ the notation of Chapters 8 and 9, which states that the inverse dielectric function $K(1,2)$ (which determines the effective potential $V(1)$ at $1 = \mathbf{r}_1,\ t_1$ due to an impressed potential $U(2)$ at $2 = \mathbf{r}_2,\ t_2$; $K(1,2) = \delta V(1)/\delta U(2)$ or $V(1) = \int d2K(1,2)U(2)$) is given by

$$K(1,2) = \delta(1-2) - i\int d3\mathcal{V}(1-3)D(3,2), \tag{11.8.1}$$

where $D(3,2)$ is related to the charge-density $(\hat{\rho}\hat{\rho})$ autocorrelation function as

$$\begin{aligned} D(3,2) &= -\left[\overline{G}_2(3,2:3^+,2^+) - \overline{G}_1(3,3^+)\overline{G}_1(2,2^+)\right] \\ &= \left\langle\left(\left[\hat{\rho}(3) - \langle\hat{\rho}(3)\rangle\right]\left[\hat{\rho}(2) - \langle\hat{\rho}(2)\rangle\right]\right)_+\right\rangle \end{aligned} \tag{11.8.2}$$

(all averages $\langle\ldots\rangle$ are thermal averages and $K(1,2)$ as written here is the periodic function, which is related to its causal counterpart by spectral considerations (see Section 7.5)). Writing the charge density operator $\hat{\rho}$ in terms of its electron and hole components ($e \to 1$)

$$\hat{\rho} = \hat{\rho}_e - \hat{\rho}_h; \quad \hat{\rho}_e = \Psi^+\Psi; \quad \hat{\rho}_h = \Phi^+\Phi \tag{11.8.3}$$

we may rewrite Eq. (11.8.2) in terms of one- and two-electron and -hole Green's functions as

$$\begin{aligned} D(3,2) = {} & 2\overline{G}_{11}(3,2:3^+,2^+) - \overline{G}_{01}(2,2^+)\overline{G}_{10}(3,3^+) - \overline{G}_{10}(2,2^+)\overline{G}_{01}(3,3^+) \\ & - \overline{G}_{20}(3,2:3^+,2^+) + \overline{G}_{10}(2,2^+)\overline{G}_{10}(3,3^+) \\ & - \overline{G}_{02}(3,2:3^+,2^+) + \overline{G}_{01}(2,2^+)\overline{G}_{01}(3,3^+). \end{aligned} \tag{11.8.4}$$

The leading term on the right of Eq. (11.8.4),

$$D(3,2) \simeq 2\overline{G}_{11}(3,2:3^+,2^+), \tag{11.8.5}$$

was seen in Chapter 8 to involve frequency poles describing hydrogen-like excitons, which will clearly participate in response to excitation by external fields through Eq. (11.8.1). (These result from the ladder-sum integral equation we have employed for $\overline{G}_{11}$ involving the electron–hole interaction to all orders.) However, $\overline{G}_{20}$ and $\overline{G}_{02}$ in the second and third lines on the right of Eq. (11.8.4) introduce electron and hole plasmons, respectively, into the dielectric response of Eq. (11.8.1), which then hybridize with the excitons of $\overline{G}_{11}$ to produce coupled collective modes constituted of excitons interacting with electron/hole plasmons. In the frequency regions where the exciton and plasmon energies (frequencies) are close, there is considerable shifting (and consequent separation) of the coupled modes, with the appearance of a gap in the frequency spectrum.

11.9 Chapter 11 Problems

P.11.1. Derive Eq. (11.2.26) taking account of the variation of the normalizing denominator trace D_B of $\overleftrightarrow{G}_{1,0,0}$ with respect to ${}^{i}\mathcal{J}_{lk}$.

P.11.2. Consider the ensemble average of Eq. (11.3.4) for the displacement $< {}^{i}Q^{\circ}_{lk}(t_1) >$ in the absence of the electron–phonon interaction term, but driven by an external field ${}^{i}\mathcal{J}_{lk}(t_1) = {}^{i}F_{lk}(t_1)$. Determine the solution for the displacement in this driving field using the Green's function $D^{\circ}_{ii'}(l, k, t_1; l', k', t'_2)$ of Eq. (11.3.6). Use this to justify the remarks following Eq. (11.3.28) through the result of Eqns. (11.3.31 and 11.3.32), including the suppressed time-oscillatory factor.

P.11.3. Consider the one-dimensional longitudinal free phonon Green's function (devoid of electron interactions) $D^0(q;\omega)$ for a linear diatomic lattice, which was discussed in Section 11.3 for $q \to 0$: Construct $D^0(q;\omega)$ for finite q explicitly.

P.11.4. Use the result of Problem 2 above to determine the q-dependence of the associated polarizability of the linear diatomic lattice.

P.11.5. Verify Eq. (11.3.38) using our formulation in Section 11.3 in which the normalizing denominator $D_B = Tr\,[U(t_0 + \tau, t_0)]_C$ is devoid of Boson/phonon sources (discussion under Eq. (11.3.4)). What is the corresponding result if D_B does contain Boson/phonon sources and how does it relate to Eq. (11.3.38) when ${}^{i}\mathcal{J}_{lk} \to 0$?

P.11.6. Verify Eqns. (11.4.24–11.4.27).

P.11.7. Prove Eqns. (11.7.3,4).

P.11.8. The ionic polarizability $\alpha(\mathbf{q}, \omega)$ cited in Eq. (11.7.14) is actually valid only for $q = 0$. Determine it for arbitrary $q \neq 0$.

11.10 References for Chapter 11

R.11.1. M. Ali Omar, "Elementary Solid State Physics," Addison-Wesley (1975).

R.11.2. C. Kittel, "Introduction to Solid State Physics," 7th Ed. Wiley (1996).

R.11.3. J. Callaway, "Quantum Theory of the Solid State," 2nd Ed. Academic Press (1991).

R.11.4. R. Enderlein and N.J.M. Horing, "Fundamentals of Semiconductor Physics and Devices," World Scientific (1997).

R.11.5. A.A. Maradudin, E.W. Montroll and G.H. Weiss, "Theory of Lattice Dynamics in the Harmonic Approximation," Solid State Physics Series Suppl. 3, Eds: F. Seitz and D. Turnbull, Academic Press (1963).

R.11.6. J. Ziman, "Principles of the Theory of Solids," Cambridge University Press (1964).

R.11.7. M. Born and K. Huang, "Dynamical Theory of Crystal Lattices," Oxford University Press (1954).

R.11.8. G. Baym, Ph.D. Thesis, Harvard University, p.41–43 (1960). (This reference provides a fuller account of some of the topics discussed in this chapter.)

R.11.9. S.K. Joshi and A.K. Rajagopal, "Lattice Dynamics of Metals," in "Solid State Physics, Advances in Research and Applications," Eds: F. Seitz, D. Turnbull and H. Ehrenreich, Vol. *22*, Academic Press (1968).

R.11.10. T. Staver, Ph.D. Thesis, Princeton University (1952).

R.11.11. D. Pines in "Solid State Physics," Vol. *1*, Eds: F. Seitz and D. Turnbull, Academic Press, p.455 (1955).

R.11.12. R. Puff and G.D. Whitfield, "Application of Green's Function Techniques to the Polaron Problem," in "Polarons and Excitons," Eds: C.G. Kuper and G.D. Whitfield, Plenum Press, p.171 (1962).

R.11.13. P.C. Martin and J. Schwinger, Phys. Rev. *115*, 1342, Section VI (1959).

R.11.14. H.C. Tso and N.J.M. Horing, Phys. Rev. B *44*, 8886 (1991).

12

Graphene

Abstract
Chapter 12 introduces Graphene, which is a two-dimensional "Dirac-like" material in the sense that its energy spectrum resembles that of a relativistic electron/positron (hole) described by the Dirac equation (having zero mass in this case). Its device-friendly properties of high electron mobility and excellent sensitivity as a sensor have attracted a huge world-wide research effort since its discovery about ten years ago. In this chapter, the associated retarded Graphene Green's function is treated and the dynamic, non-local dielectric function is discussed in the degenerate limit of zero temperature. The effects of a quantizing magnetic field on the Green's function of a Graphene sheet and on its energy spectrum are derived in detail: Also the magnetic field Green's function and energy spectrum of a Graphene sheet with a quantum dot (modeled by a 2D Dirac delta-function potential) are thoroughly examined. Furthermore, Chapter 12 similarly addresses the problem of a Graphene antidot lattice in a magnetic field, discussing the Green's function for propagation along the lattice axis, with a formulation of the associated eigen-energy dispersion relation. Finally, magnetic Landau quantization effects on the statistical thermodynamics of Graphene, including its Free Energy and magnetic moment, are also treated in Chapter 12 and are seen to exhibit magnetic oscillatory features.

12.1 Introduction

Graphene is a single-atom-thick two-dimensional planar layer of carbon atoms in a hexagonal honey-comb lattice composed of two superposed triangular sub-lattices. Since the first report in 2004 of its unusual device-friendly material properties[1] it has continuously received enormous attention, both experimental and theoretical, because of its potential to bring about another technological revolution in nanoelectronics and computers. A. K. Geim and K. S. Novoselov were awarded the Nobel Prize in 2010 for its discovery.[1] The relevant properties of Graphene include:

- High mobility at elevated temperature,[2] reaching 200,000 cm^2/Vs, over two orders of magnitude higher than that of silicon-based materials, over 20 times that of GaAs and over twice that of InSb;
- High electron density, about 10^{13} cm^{-2} in a single subband;
- Long carrier mean-free-path, l $\sim$400 nm, at room temperature, opening the possibility of Graphene-based ballistic devices;

Quantum Statistical Field Theory: Schwinger's Variational Method. Norman J. Morgenstern Horing,
Oxford University Press (2017). © Norman J. Morgenstern Horing. DOI 10.1093/oso/9780198791942.001.0001

- Stable to high temperatures, ~3000K;
- Quantum Hall effect occurs at room temperature in Graphene;
- The planar form of Graphene makes it likely to be amenable to commercial fabrication techniques in the style of the highly developed top-down CMOS compatible process flows, a substantial advantage over carbon nanotubes that are difficult to integrate into electronic devices and are difficult to produce in consistent sizes and electronic properties.

All of these properties make Graphene an exceptionally promising material for future nanoelectronic devices. Such applications of Graphene are already in progress. These include:

- **Sensors**—F. Schedin, K.S. Novoselov, *et al.*[3] reported that Graphene-based chemical sensors are capable of detecting minute concentrations (1 part per billion) of various active gases and allow us to discern *individual* events when a molecule attaches to the sensor's surface. Such high sensitivity comes about because the high 2D surface/volume ratio maximizes the role of adsorbed molecules as donors/acceptors, coupled with high conductivity and low noise;
- **Spin valve**—A simple spin valve structure has already been fabricated[4] employing Graphene to provide the spin transport medium between ferromagnetic electrodes (using advantageous properties of Graphene in regard to long spin lifetime, low spin orbit coupling and high conductivity);
- **Electromechanical resonator**—J.S. Bunch, *et al.*[5] demonstrated that Graphene in contact with a gold electrode can be used to electrostatically actuate an electromechanical resonator (employing a 2D Graphene sheet suspended over a trench in a SiO_2 substrate). The motion can be activated by an rf gate voltage superposed on a dc-voltage applied to the Graphene sheet, or by optical actuation using a laser focused on the sheet;
- **FET**—Using Graphene, "proof-of-principle" FET transistors, loop devices and circuitry have already been produced by W. de Heer's group;[6]
- **Quantum interference device**—A quantum interference device using a ring-shaped Graphene structure was built to manipulate electron wave interference effects.

The obvious promise of Graphene for new and improved electronic devices has resulted in a flood of research papers on its transport and optical properties, and on its technological prospects.

12.2 Graphene Hamiltonian, Green's Function and Dielectric Function

On a fundamental level, the extraordinary physical features of Graphene arise from its unusual band structure, in which the conduction and valence bands touch at two nodal zero-gap ("Dirac") points (K, K') in the first Brillouin zone,[7,8] with the resulting low-energy electron/hole energy dispersion relation proportional to momentum (rather than its square). This single-particle energy dispersion, linear in momentum, is analogous to the energy dispersion of relativistic electrons having no mass, and thus likens Graphene electron and hole carriers to massless relativistic "Dirac" Fermions.[9] The associated Hamiltonian is written using a "pseudospin" notation that distinguishes the electron part of the spectrum from the hole part in terms of a spin-like variable. Using the usual 2D Pauli spin matrices ($\boldsymbol{\sigma} = \sigma_x, \sigma_y$) and momentum $\mathbf{p} = p_x, p_y$ on the Graphene plane, the Graphene Hamiltonian in pseudospin representation is given by

$$\check{h}_1 = \gamma \mathbf{p} \cdot \boldsymbol{\sigma} = \gamma \begin{pmatrix} 0 & p_x - sgn(\nu) i p_y \\ p_x + sgn(\nu) i p_y & 0 \end{pmatrix}, \tag{12.2.1}$$

where

$$sgn(\nu) = \begin{Bmatrix} 1 & \text{for } K \\ -1 & \text{for } K' \end{Bmatrix}, \tag{12.2.2}$$

which correspond to the two zero-gap "Dirac" points, K and K', and γ is derived from Graphene band structure parameters as $\gamma = 3\alpha a/2$ (α is the hopping parameter in the tight-binding approximation and a is the lattice spacing): γ is in the nature of a constant Fermi velocity and is independent of density. Comparable to the case of massless relativistic neutrinos, pseudo*helicity*,[9] the component of pseudospin in the momentum direction, commutes with $\check{h}_1$: its eigenvectors form a basis in which $\hat{h}_1$ is diagonal. Employing the transformation from pseudospin basis to pseudohelicity basis,

$$U_p = \frac{1}{\gamma p} \begin{pmatrix} \frac{p_x - sgn(\nu) i p_y}{\gamma p} & \frac{p_x + sgn(\nu) i p_y}{\gamma p} \end{pmatrix}, \tag{12.2.3}$$

$\check{h}_1 \to \hat{h}_1$ is diagonalized as

$$\hat{h}_1 = \left[U_p\right]^+ \check{h}_1 U_p = diag[\varepsilon_1(p), \varepsilon_2(p)], \tag{12.2.4}$$

where ($\mu = 1, 2$)

$$\varepsilon_\mu = (-1)^{\mu+1} \gamma p. \tag{12.2.5}$$

Since the Hamiltonian $\check{h}_1$ in pseudospin representation is a 2×2 matrix, so is the Graphene Green's function. The retarded Green's function matrix satisfies the matrix counterpart of Eq. (5.1.3) ($\overleftrightarrow{I}$ is the unit 2×2 pseudospin matrix; $\hbar \to 1$):

$$\left(i\overleftrightarrow{I} \frac{\partial}{\partial t} - \check{h}_1 \right) \overleftrightarrow{G_1}^{ret}(\mathbf{p}, t) = \overleftrightarrow{I} \delta(t - t'), \tag{12.2.6}$$

and in position-frequency representation, it takes the form ($p_x \to X = x - x'$; $p_y \to Y = y - y'$; $T = t - t' \to \omega$; define $\gamma_\nu \equiv \gamma sgn(\nu) \equiv 1_\nu \gamma$ with $1_\nu = +1$ for K and $1_\nu = -1$ for K')

$$\left[\overleftrightarrow{I}\omega - \gamma\sigma_x \frac{1}{i}\frac{\partial}{\partial X} - \gamma_\nu \sigma_y \frac{1}{i}\frac{\partial}{\partial Y}\right] \overleftrightarrow{G_1}^{ret}(\mathbf{R}, \omega) = \overleftrightarrow{I}\delta(X)\delta(Y). \tag{12.2.7}$$

In terms of individual matrix elements, we have

$$\omega G^{ret}_{1xx} - \left[\gamma \frac{1}{i}\frac{\partial}{\partial X} - \gamma_\nu \frac{\partial}{\partial Y}\right] G^{ret}_{1yx} = \delta(X)\delta(Y) \tag{12.2.8}$$

and

$$\omega G^{ret}_{1yx} = \left[\gamma \frac{1}{i}\frac{\partial}{\partial X} + \gamma_\nu \frac{\partial}{\partial Y}\right] G^{ret}_{1xx}, \tag{12.2.9}$$

and there are similar equations for G^{ret}_{1yy} and G^{ret}_{1xy}. The resulting Green's function elements in pseudospin representation are

$$G^{ret}_{1xx}(\mathbf{p}, \omega) = G^{ret}_{1yy}(\mathbf{p}, \omega) = \omega/(\omega^2 - \gamma^2 p^2); \tag{12.2.10}$$

$$G^{ret}_{1xy}(\mathbf{p}, \omega) = G^{ret*}_{1yx}(\mathbf{p}, \omega) = \gamma(p_x - ip_y)/(\omega^2 - \gamma^2 p^2); \tag{12.2.11}$$

and the 2×2 spectral weight matrix, $\overleftrightarrow{A}(\mathbf{p}, \omega)$, may be determined using the results of Chapter 7 as

$$\overleftrightarrow{A}(\mathbf{p}, \omega) = -2Im[\overleftrightarrow{G_1}^{ret}(\mathbf{p}, \omega)]. \tag{12.2.12}$$

Correspondingly, the equilibrium thermodynamic matrix Green's function for Graphene is given by

$$\overleftrightarrow{G_1}^{\{\lessgtr\}}(\mathbf{p}, \omega) = i \begin{Bmatrix} f(\omega) \\ -1 + f(\omega) \end{Bmatrix} \overleftrightarrow{A}(\boldsymbol{p}, \omega). \tag{12.2.13}$$

On the two-dimensional Graphene plane, the polarizability α is given in wave number-frequency representation by

$$\alpha(\mathbf{p}, \omega + i0^+) = -\mathcal{V}(\mathbf{p}) R(\mathbf{p}, \omega), \tag{12.2.14}$$

where the 2D Coulomb potential $\mathcal{V}(\mathbf{p})$ is given by

$$\mathcal{V}(\mathbf{p}) = \frac{2\pi e^2}{p}. \tag{12.2.15}$$

In the random phase approximation, the density perturbation response function, $R = \delta\rho/\delta V_{eff}$, takes the form

$$R(\mathbf{p}, \omega + i0^+) = \Im_< - \Im^*_>, \tag{12.2.16}$$

with

$$\Im_{>} = \int_0^{\infty} dt e^{-i(\omega - i0^+)t} \int \frac{d^2q}{(2\pi)^2} Tr[\overset{\leftrightarrow}{G}_{1<}(\mathbf{q}; -t)\overset{\leftrightarrow}{G}_{1>}(\mathbf{q}-\mathbf{p}; t)], \tag{12.2.17}$$

and (Tr represents the pseudospin trace),

$$\Im_{<} = \int_{-\infty}^{0} dt e^{-i(\omega + i0^+)t} \int \frac{d^2q}{(2\pi)^2} Tr[\overset{\leftrightarrow}{G}_{1>}(\bar{\mathbf{q}}; -t)\overset{\leftrightarrow}{G}_{1<}(\mathbf{q}-\mathbf{p}; t)]. \tag{12.2.18}$$

Alternatively, one can employ Eq. (10.1.12) for $R(\mathbf{r}_3 - \mathbf{r}_2; \omega + i0^+)$ in position-frequency representation and Fourier-transform to wave number representation using the 2D Graphene eigenfunctions and eigenvalues of $\breve{h}_1$ given by Eqns. (12.2.1) and (12.2.2). The resulting RPA density perturbation response function, $\delta\rho/\delta V_{eff} = R(\mathbf{p}, \omega) \equiv D_0\widetilde{R}(x, v)$, for Graphene in the $T = 0$ degenerate limit is given by[10–13] (we use the notation and results of Reference 13 with dimensionless frequency and wave number variables defined by $\nu = \omega/E_F = \omega/\mu$ and $x = p/p_F$, respectively, and $D_0 = \gamma^{-1}\sqrt{g_s g_v \rho_{2D}/\pi}$; g_s and g_v are spin and valley degeneracies, $g_s = g_v = 2$; $\hbar \to 1$):

$$\widetilde{R}(x, \nu) = \widetilde{R}^+(x, \nu) + \widetilde{R}^-(x, \nu), \quad \text{with} \quad \widetilde{R}^+(x, \nu) = \widetilde{R}_1^+(x, \nu)\theta(\nu - x) + \widetilde{R}_2^+(x, \nu)\theta(x - \nu), \tag{12.2.19}$$

and ($\theta(z) \equiv \eta_+(z) =$ Heaviside unit step function) where (define $\widetilde{\Pi} \equiv -\widetilde{R}$),

$$\begin{aligned} -\text{Re}\,\widetilde{R}_1^+(x, \nu) = \text{Re}\,\widetilde{\Pi}_1^+(x, \nu) = 1 - \frac{1}{8\sqrt{\nu^2 - x^2}} \Big\{ & f_1(x, \nu)\,\theta\big(|2 + \nu| - x\big) \\ & + \text{sgn}(\nu - 2 + x)\, f_1(x, -\nu)\,\theta\big(|2 - \nu| - x\big) \\ & + f_2(x, \nu)\,[\theta(x + 2 - \nu) + \theta(2 - x - \nu)] \Big\}, \end{aligned} \tag{12.2.20}$$

$$\begin{aligned} -\text{Re}\,\widetilde{R}_2^+(x, \nu) = \text{Re}\,\widetilde{\Pi}_2^+(x, \nu) = 1 - \frac{1}{8\sqrt{x^2 - \nu^2}} \Big\{ & f_3(x, \nu)\,\theta\big(x - |\nu + 2|\big) \\ & + f_3(x, -\nu)\,\theta\big(x - |\nu - 2|\big) \\ & + \frac{\pi x^2}{2}\,\big[\theta\big(|\nu + 2| - x\big) + \theta\big(|\nu - 2| - x\big)\big] \Big\}, \end{aligned} \tag{12.2.21}$$

$$\begin{aligned} -\text{Im}\,\widetilde{R}_1^+(x, \nu) = \text{Im}\,\widetilde{\Pi}_1^+(x, \nu) = \frac{-1}{8\sqrt{\nu^2 - x^2}} \Big\{ & f_3(x, -\nu)\,\theta\big(x - |\nu - 2|\big) \\ & + \frac{\pi x^2}{2}\,[\theta(x + 2 - \nu) + \theta(2 - x - \nu)] \Big\}, \end{aligned} \tag{12.2.22}$$

$$-\text{Im}\,\widetilde{R}_2^+(x, \nu) = \text{Im}\,\widetilde{\Pi}_2^+(x, \nu) = \frac{\theta(\nu - x + 2)}{8\sqrt{x^2 - \nu^2}}\,[f_4(x, \nu) - f_4(x, -\nu)\,\theta(2 - x - \nu)], \tag{12.2.23}$$

and

$$-\widetilde{R}^-(x, \nu) = \frac{\pi x^2 \theta(x - \nu)}{8\sqrt{x^2 - \nu^2}} + i\frac{\pi x^2 \theta(\nu - x)}{8\sqrt{\nu^2 - x^2}}. \tag{12.2.24}$$

The quantities $f_1(x,\nu)$, $f_2(x,\nu)$, $f_3(x,\nu)$, $f_4(x,\nu)$ are defined as

$$f_1(x,\nu) = (2+\nu)\sqrt{(2+\nu)^2 - x^2} - x^2 \ln \frac{\sqrt{(2+\nu)^2 - x^2} + (2+\nu)}{\left|\sqrt{\nu^2 - x^2} + \nu\right|}, \quad (12.2.25)$$

$$f_2(x,\nu) = x^2 \ln \frac{\nu - \sqrt{\nu^2 - x^2}}{x}, \quad (12.2.26)$$

$$f_3(x,\nu) = (2+\nu)\sqrt{x^2 - (2+\nu)^2} + x^2 \sin^{-1} \frac{2+\nu}{x}, \quad (12.2.27)$$

$$f_4(x,\nu) = (2+\nu)\sqrt{(2+\nu)^2 - x^2} - x^2 \ln \frac{\sqrt{(2+\nu)^2 - x^2} + (2+\nu)}{x}. \quad (12.2.28)$$

Considering the above results in the zero-frequency static limit, the RPA shielding dielectric function for Graphene is given by (zero temperature):

$$\varepsilon(p,\omega=0) = 1 + \frac{p_{TF}}{p}\left\{\begin{array}{ll} 1 & \text{if } p \le 2p_F \\ 1 + \frac{\pi p}{8p_F} - \frac{1}{2}\sqrt{1 - \left(\frac{2p_F}{p}\right)^2} - \frac{p}{4p_F}\sin^{-1}\left(\frac{2p_F}{p}\right) & \text{if } p > 2p_F \end{array}\right\}, \quad (12.2.29)$$

with the 2D Thomas–Fermi shielding wave number for Graphene as (κ is the static background dielectric constant):

$$p_{TF} = 4e^2 p_F/(\hbar\kappa\gamma). \quad (12.2.30)$$

This determines the shielded Coulombic impurity potential for Graphene in wave number representation in the degenerate limit ($T = 0K$) as

$$V_{shield}(p,\omega=0) = \frac{2\pi e^2}{p\varepsilon(p,\omega=0)}, \quad (12.2.31)$$

which has been employed in transport calculations.[14]

Of course, the vanishing of $\varepsilon(p,\omega)$, or

$$-\alpha(p,\omega+i0^+) = V(p)R(p,\omega) = 1, \quad (12.2.32)$$

with $R(p,\omega)$ given above, identifies Graphene plasmons , and in the local limit ($p \to 0$) the fundamental plasmon mode is given by

$$\omega = bp^{1/2}, \quad (12.2.33)$$

where

$$b = (g_s g_v e^2 E_F/2\kappa)^{1/2}. \quad (12.2.34)$$

Having the fundamental plasmon frequency (Eq. (12.2.33)) proportional to the square root of wave number is typical of local 2D plasmons. However, the coefficient b is proportional to $E_F^{1/2}$, which is proportional to the fourth root of density, $\rho_{2D}^{1/4}$, for Graphene, which is quite different from the $\rho_{2D}^{1/2}$-dependence of the fundamental plasma frequency of a normal 2D plasma. Moreover, the leading wave number correction, given by

$$\omega = bp^{1/2}[1 - p_{TF}p/8p_F^2], \tag{12.2.35}$$

lowers the plasma frequency in Graphene, whereas the corresponding wave number correction would increase the fundamental plasma frequency of a normal 2D plasma.

12.3 Graphene Green's Function in a Perpendicular Magnetic Field: Landau Quantization

To introduce the role of a constant, uniform magnetic field of arbitrary strength normal to the 2D Graphene sheet, we again employ Eq. (12.2.6) and Eqns. (12.2.1) and (12.2.2) with the replacement $\mathbf{p} \to \mathbf{p} - e\mathbf{A}$ and $\mathbf{A} = \frac{1}{2}\mathbf{B} \times \mathbf{r}$ ($\hbar \to 1, c \to 1$). Once again, the requirement of gauge invariance (Eqns. (5.6.1–5.6.11)) yields

$$\overleftrightarrow{G_1}^{ret}(\mathbf{r}, \mathbf{r}'; t, t') = C(\mathbf{r}, \mathbf{r}')\overleftrightarrow{G_1'}(\mathbf{r} - \mathbf{r}'; t - t'), \tag{12.3.1}$$

with the matrix $\overleftrightarrow{G_1'}(\mathbf{r} - \mathbf{r}'; t - t')$ spatially translationally invariant and gauge invariant. $G_1'(\mathbf{r} - \mathbf{r}'; t - t')$ satisfies the equation ($\mathbf{R} = \mathbf{r} - \mathbf{r}'$, $T = t - t'$; suppress the overhead double arrow indicating a 2×2 matrix and the subscript 1 on $\overleftrightarrow{G_1'} \to G'$, and on the unit matrix $\overleftrightarrow{I} \to I$):

$$\left(i\frac{\partial}{\partial T} - \gamma\boldsymbol{\sigma} \cdot \left[\frac{1}{i}\frac{\partial}{\partial \mathbf{R}} - \frac{e}{2}\mathbf{B} \times \mathbf{R}\right]\right) G'(\mathbf{R}, T) = I\delta^{(2)}(\mathbf{R})\delta(T), \tag{12.3.2}$$

and the factor $C(\mathbf{r}, \mathbf{r}')$ incorporates all non-spatially translationally invariant structure (and the arbitrary gauge function $\phi(r)$) as

$$C(\mathbf{r}, \mathbf{r}') = \exp\left[\frac{ie}{2\hbar c}\mathbf{r} \cdot \mathbf{B} \times \mathbf{r}' - \phi(\mathbf{r}) + \phi(\mathbf{r}')\right]. \tag{12.3.3}$$

Landau quantization in Graphene electron dynamics is incorporated in the solution of Eq. (12.3.2) for $G'(\mathbf{R}, T)$, which is written in ω-frequency representation as (define $\gamma_\nu = \gamma\, sign(\nu)$):

$$\left[\omega - \gamma\sigma_x\Pi_{XY} - \gamma_\nu\sigma_y\Pi_{YX}\right] G'(\mathbf{R}, \omega) = I\delta(X)\delta(Y), \tag{12.3.4}$$

where

$$\Pi_{XY} \equiv \frac{1}{i}\frac{\partial}{\partial X} + \frac{eB}{2}Y \text{ and } \Pi_{YX} \equiv \frac{1}{i}\frac{\partial}{\partial Y} - \frac{eB}{2}X. \tag{12.3.5}$$

The matrix elements of Eq. (12.3.4) are given by

$$\omega G'_{11} - [\gamma \Pi_{XY} - i\gamma_\upsilon \Pi_{YX}]\, G'_{21} = \delta(X)\delta(Y), \tag{12.3.6}$$

$$\omega G'_{21} = [\gamma \Pi_{XY} + i\gamma_\upsilon \Pi_{YX}]\, G'_{11}, \tag{12.3.7}$$

and

$$\omega G'_{22} - [\gamma \Pi_{XY} + i\gamma_\upsilon \Pi_{YX}]\, G'_{12} = \delta(X)\delta(Y), \tag{12.3.8}$$

$$\omega G'_{12} = [\gamma \Pi_{XY} - i\gamma_\upsilon \Pi_{YX}]\, G'_{22}. \tag{12.3.9}$$

Combining Eqns. (12.3.6) and (12.3.7), we have

$$\left[\omega - \frac{1}{\omega}\left\{\gamma^2\left(\Pi_{XY}^2 + \Pi_{YX}^2\right) + i\gamma\gamma_\upsilon\left(\Pi_{XY}\Pi_{YX} - \Pi_{YX}\Pi_{XY}\right)\right\}\right] G'_{11}(X, Y; \omega)$$
$$= \delta(X)\delta(Y). \tag{12.3.10}$$

Noting the identity

$$\Pi_{XY}\Pi_{YX} - \Pi_{YX}\Pi_{XY} \equiv -\frac{1}{i}eB, \tag{12.3.11}$$

and, using Eq. (12.3.5) to evaluate Π_{XY}^2 and Π_{YX}^2, the equation for G'_{11} takes the form

$$\left[\omega + \frac{\gamma\gamma_\upsilon}{\omega}(eB)\right] G'_{11}(X, Y; \omega)$$
$$+\frac{\gamma^2}{\omega}\left\{\frac{\partial^2}{\partial X^2} + \frac{\partial^2}{\partial Y^2} - \left(\frac{eB}{2}\right)^2 [X^2 + Y^2] + \frac{eB}{i}\left(X\frac{\partial}{\partial Y} - Y\frac{\partial}{\partial X}\right)\right\} G'_{11}(X, Y; \omega)$$
$$= \delta(X)\delta(Y). \tag{12.3.12}$$

Introducing the definition

$$L_Z = \frac{1}{i}\left(X\frac{\partial}{\partial Y} - Y\frac{\partial}{\partial X}\right) = L_z + L_{z'} \tag{12.3.13}$$

(L_z is the angular momentum operator) there is a considerable simplification by noting that $L_Z G'(\mathbf{R}, T) = 0$, as in Eqns. (5.6.13–5.6.15). To recapitulate the proof, it is shown most easily by choosing the gauge $\phi = 0$, for then $L_Z G'(\mathbf{R}, T) = C^{-1}(\mathbf{x}, \mathbf{x}')L_Z \overleftrightarrow{G_1}^{ret}(\mathbf{x}, t; \mathbf{x}', t')$ because $L_Z C(\mathbf{x}, \mathbf{x}') = 0$. The vanishing of

$$L_Z \overleftrightarrow{G_1}^{ret}(\mathbf{x}, t; \mathbf{x}', t') = (L_z + L_{z'})\overleftrightarrow{G_1}^{ret}(\mathbf{x}, t; \mathbf{x}', t')$$

it is evident from the bilinear dependence of $\overleftrightarrow{G_1}^{ret}(\mathbf{x},t;\mathbf{x}',t')$ on the field operators, $\psi(\mathbf{x},t)$, $\psi^+(\mathbf{x}',t')$, and the concomitant interpretation of $\overleftrightarrow{G_1}^{ret}(\mathbf{x},t;\mathbf{x}',t')$ in terms of the creation of a particle at $(\mathbf{x},t)$, which is subsequently annihilated at $(\mathbf{x}',t')$. The action of L_z on $\overleftrightarrow{G_1}^{ret}(\mathbf{x},t;\mathbf{x}',t')$ may be regarded as measuring the component of orbital angular momentum in the direction of the magnetic field associated with the particle at the point of creation, while the action of $L_{z'}$ must be regarded as measuring the negative of it at the point of annihilation because $L_{z'}$ is acting on the adjoint operator. Therefore, $L_Z = L_z + L_{z'}$ measures the loss of orbital angular momentum in the direction of the magnetic field suffered by the particle during its propagation from $(\mathbf{x},t)$ to $(\mathbf{x}',t')$, and this must vanish because we are dealing with a conserved quantity. Thus, in frequency representation, we obtain

$$L_Z G'_{11}(X,Y;\omega) = 0. \tag{12.3.14}$$

An alternative proof is developed by employing an eigenfunction expansion of the Green's function given by Eq. (5.1.20):

$$\overleftrightarrow{G_1}^{ret}(\mathbf{x},t;\mathbf{x}',t') = -i\eta_+(t-t')\sum_{E,l_z}\Psi_{E,l_z}(\mathbf{x})\Psi^*_{E,l_z}(\mathbf{x}')e^{-iE(t-t')}.$$

In this, the fact that the angular momentum component in the direction of the magnetic field (L_z) is conserved permits us to employ the simultaneous eigenfunctions $\Psi_{E,l_z}(\mathbf{x})$ of the Hamiltonian and L_z, summing over the entire energy spectrum (E) jointly with the angular momentum eigenvalues (l_z). The latter are given by Eq. (1.1.14) with L_z, expressed in terms of azimuthal angle $\tilde{\phi}$ in the x-y plane as

$$L_z = \frac{\hbar}{i}\frac{\partial}{\partial\tilde{\phi}}; \quad L_{z'} = \frac{\hbar}{i}\frac{\partial}{\partial\tilde{\phi}'},$$

so we have its eigenvalue equation and solution as ($\hbar \to 1$):

$$\frac{1}{i}\frac{\partial\Psi_{E,l_z}}{\partial\tilde{\phi}} = l_z\Psi_{E,l_z} \to \Psi_{E,l_z} \sim \overline{\Psi}_{E,l_z}e^{il_z\tilde{\phi}},$$

and $l_z = m$ are appropriate integers with $\overline{\Psi}_{E,l_z} \to \overline{\Psi}_{E,m}$ independent of $\tilde{\phi}$. Consequently, the eigenfunction expansion of $\overleftrightarrow{G_1}^{ret}$ takes the form (ω-representation):

$$\overleftrightarrow{G_1}^{ret}(\mathbf{x},\mathbf{x}';\omega) = \sum_{E,m}\frac{\overline{\Psi}_{E,m}(\mathbf{x})\overline{\Psi}^*_{E,m}(\mathbf{x}')}{\omega+i\epsilon-E}e^{im\phi}e^{-im\phi'}.$$

Since $\overline{\Psi}$, $\overline{\Psi}^*$ are devoid of azimuthal angle dependence, it is immediately clear that $L_z\overleftrightarrow{G_1}^{ret} \to im\overleftrightarrow{G_1}^{ret}$ and $L'_z\overleftrightarrow{G_1}^{ret} \to -im\overleftrightarrow{G_1}^{ret}$ within the sum over the simultaneous eigenvalue spectrum, leading to cancellation and confirming the result

$$(L_z + L_{z'})\overleftrightarrow{G_1}^{ret}(\mathbf{x}_1,\mathbf{x}'_1;\omega) = L_Z\overleftrightarrow{G_1}^{ret}(\mathbf{x}_1,\mathbf{x}'_1;\omega) = 0.$$

In view of this, Eq. (12.3.12) is given by

$$\left[\omega + \frac{\gamma\gamma_\upsilon}{\omega}eB\right] G'_{11}(X,Y;\omega) + \frac{\gamma^2}{\omega}\left\{\frac{\partial^2}{\partial X^2} + \frac{\partial^2}{\partial Y^2} - \left(\frac{eB}{2}\right)^2 [X^2+Y^2]\right\} G'_{11}(X,Y;\omega)$$
$$= \delta(X)\delta(Y) \qquad (12.3.15)$$

Introducing the definitions

$$\Omega = \omega + \frac{\gamma\gamma_\upsilon}{\omega}eB; \qquad \mathcal{M} = \frac{\omega}{2\gamma^2}; \qquad \Omega_c = \frac{2\gamma^2}{\omega}eB, \qquad (12.3.16)$$

we have

$$\Omega G'_{11}(\mathbf{R},\Omega) + \left[\frac{1}{2\mathcal{M}}\nabla^2_{\mathbf{R}} - \mathcal{M}\frac{\Omega_c^2}{8}\mathbf{R}^2\right] G'_{11}(\mathbf{R},\Omega) = \delta^{(2)}(\mathbf{R}), \qquad (12.3.17)$$

which is just the Green's function equation for an isotropic two-dimensional harmonic oscillator (with the impulsive Dirac δ-function driving term having its source point at the origin) in position-frequency representation. The retarded solution is given in position-time (τ) representation by[15] ($\eta_+(\tau)$ is the Heaviside unit step function):

$$G'_{11}(\mathbf{R};\tau) = -\eta_+(\tau)\frac{\mathcal{M}\Omega_c}{4\pi\,\sin(\Omega_c\tau/2)}\exp\left\{\frac{i\mathcal{M}\Omega_c[X^2+Y^2]}{4\tan(\Omega_c\tau/2)}\right\}. \qquad (12.3.18)$$

Transforming to Ω-representation, we have

$$G'_{11}(\mathbf{R};\Omega) = \frac{-\mathcal{M}\Omega_c}{4\pi}\int_0^\infty d\tau\frac{e^{i\Omega\tau}}{\sin(\Omega_c\tau/2)}\exp\left\{\frac{i\mathcal{M}\Omega_c[X^2+Y^2]}{4\tan(\Omega_c\tau/2)}\right\}. \qquad (12.3.19)$$

or

$$G'_{11}(\mathbf{R};\omega) = -\frac{eB}{4\pi}\int_0^\infty d\tau\frac{\exp[i(\omega+\gamma\gamma_\upsilon eB/\omega)\tau]}{\sin(\gamma^2 eB\tau/\omega)}\exp\left\{\frac{ieB[X^2+Y^2]}{4\tan(\gamma^2 eB\tau/\omega)}\right\}. \qquad (12.3.20)$$

Putting $\tilde{T} \equiv \tau/\omega$,

$$\int_0^\infty d\tau \ldots = \omega\int_0^\infty d\tilde{T}\exp\left[i\tilde{T}\left(\omega^2 + \gamma\gamma_\upsilon eB\right)\right]\frac{1}{\sin(\gamma^2 eB\tilde{T})}\exp\left\{\frac{ieB[X^2+Y^2]}{4\tan(\gamma^2 eB\tilde{T})}\right\},$$
$$(12.3.21)$$

expansion of the integrand as a generator[16] of Laguerre polynomials, L_n, yields

$$\int_0^\infty d\tau \ldots = 2i\omega \exp\left(-\frac{eB}{4}\left[X^2+Y^2\right]\right)\int_0^\infty d\tilde{T} \exp\left[i\tilde{T}\left(\omega^2+\gamma\gamma_\nu eB\right)\right]$$
$$\times \sum_{n=0}^{\infty} L_n\left(\frac{eB}{2}\left[X^2+Y^2\right]\right) e^{-i\left(n+\frac{1}{2}\right)2\gamma^2 eB\tilde{T}}. \qquad (12.3.22)$$

The Dirac points are distinguished by $\gamma_\nu = \gamma sgn(\nu) \equiv \gamma 1_\nu$ with $1_\nu = +1$ for K and $1_\nu = -1$ for K'. The result for $G'_{11}(\vec{R}, \omega)$ is given by

$$G'_{11}(\mathbf{R};\omega) = \frac{eB}{2\pi}\omega \exp\left(-\frac{eB}{4}\left[X^2+Y^2\right]\right)\sum_{n=0}^{\infty}\frac{L_n\left(\frac{eB}{2}\left[X^2+Y^2\right]\right)}{\omega^2-2\gamma^2 eB(n+[1-1_\nu]/2)}. \qquad (12.3.23)$$

Similar treatment of Eqns. (12.3.8) and (12.3.9) yields the equation for $G'_{22}(X, Y; \omega)$ as

$$\left[\omega - \frac{1}{\omega}\left\{\gamma^2\left(\Pi_{XY}^2+\Pi_{YX}^2\right)+i\gamma\gamma_\nu\left(\Pi_{YX}\Pi_{XY}-\Pi_{XY}\Pi_{YX}\right)\right\}\right]G'_{22}(X,Y;\omega)$$
$$= \delta(X)\delta(Y), \qquad (12.3.24)$$

and because of the sign reversal in the last term in brackets on the left, we identify $\Omega = \omega - \frac{\gamma\gamma_\nu}{\omega}eB$ in Eq. (12.3.18), obtaining

$$\left[\omega - \frac{\gamma\gamma_\nu}{\omega}eB\right]G'_{22}(X,Y;\omega) + \frac{\gamma^2}{\omega}\left\{\frac{\partial^2}{\partial X^2}+\frac{\partial^2}{\partial Y^2}-\left(\frac{eB}{2}\right)^2\left[X^2+Y^2\right]\right\}G'_{22}(X,Y;\omega)$$
$$= \delta(X)\delta(Y). \qquad (12.3.25)$$

The result for $G'_{22}(\mathbf{R}, \omega)$ is given by

$$G'_{22}(\mathbf{R};\omega) = \frac{eB}{2\pi}\omega \exp\left(-\frac{eB}{4}\left[X^2+Y^2\right]\right)\sum_{n=0}^{\infty}\frac{L_n\left(\frac{eB}{2}\left[X^2+Y^2\right]\right)}{\omega^2-2\gamma^2 eB(n+[1+1_\nu]/2)}. \qquad (12.3.26)$$

The eigen-energy spectrum of the Graphene sheet is obtained from the frequency poles of Eq. (12.3.23) and (12.3.26). It is given by

$$\omega' = \pm\sqrt{2n\gamma^2 eB} \text{ and } \omega' = \pm\sqrt{2(n+1)\gamma^2 eB}, \qquad (12.3.27)$$

as was first found by Ando.[17] The residues at the poles describe the Landau quantized Graphene eigenfunctions.[18]

The off-diagonal elements G'_{21} and G'_{12} are readily obtained from G'_{11} and G'_{22} by Eqns. (12.3.7) and (12.3.9). In summary, we have

$$G'_{\substack{11\\22}}(\mathbf{R};\omega) = \frac{eB}{2\pi}\omega e^{-\frac{eB}{4}R^2}\sum_{n=0}^{\infty}\frac{L_n\left(\frac{eB}{2}R^2\right)}{\omega^2 - 2\gamma^2 eB(n+[1\mp 1_\nu]/2)} \tag{12.3.28}$$

($\mp 1_\nu$ is -1 for G'_{11} and $\mp 1_\nu$ is $+1$ for G'_{22}), and using the identity[16]

$$\frac{\partial}{\partial z}L_n(z) = -L^1_{n-1}(z), \tag{12.3.29}$$

where $L^m_n(z)$ are generalized Laguerre polynomials, we obtain

$$\begin{aligned}\frac{\partial G'_{\substack{11\\22}}(\mathbf{R};\omega)}{\partial\{XorY\}} &= -\frac{(eB)^2}{2\pi}\omega\{XorY\}e^{-\frac{eB}{4}R^2}\\ &\times\sum_{n=0}^{\infty}\frac{\frac{1}{2}L_n\left(\frac{eB}{2}R^2\right)+L^1_{n-1}\left(\frac{eB}{2}R^2\right)}{\omega^2-2\gamma^2 eB(n+[1\mp 1_\nu]/2)}.\end{aligned} \tag{12.3.30}$$

Eqns. (12.3.7 and 12.3.9) therefore yield

$$G'_{21}(\mathbf{R},\omega) = \gamma eB[iX - Y]\mathcal{Q}(\mathbf{R},\omega), \tag{12.3.31}$$

where

$$\mathcal{Q}(\mathbf{R},\omega) = \frac{eB}{2\pi}e^{-\frac{eB}{4}R^2}\sum_{n=0}^{\infty}\frac{L^1_{n-1}\left(\frac{eB}{2}R^2\right)}{D_n(\omega)} \tag{12.3.32}$$

and

$$\begin{aligned}D_n(\omega) &= \omega^2 - 2\gamma^2 eB\,(n+[1\mp 1_\nu]/2)\\ &= \left(\omega-\sqrt{2\gamma^2 eB\,(n+[1\mp 1_\nu]/2)}\right)\left(\omega+\sqrt{2\gamma^2 eB\,(n+[1\mp 1_\nu]/2)}\right).\end{aligned} \tag{12.3.33}$$

Another interesting representation of the Graphene Green's function in a magnetic field may be obtained by writing Eq. (12.3.17) for G'_{11} in circular coordinates as (similar procedure applies for G'_{22})

$$\begin{aligned}\left\{\frac{1}{R}\frac{\partial}{\partial R}\left[R\frac{\partial}{\partial R}\right]-\frac{\mathcal{M}^2\Omega_c^2R^2}{4}+2\mathcal{M}\Omega+\frac{1}{R^2}\frac{\partial^2}{\partial\theta^2}\right\}G'_{11}(\mathbf{R};\Omega)&\\ = 2\mathcal{M}\left(\frac{2\delta(R)}{R}\right)\delta(\theta),&\end{aligned} \tag{12.3.34}$$

where we have inserted a factor of 2 with $\delta(R)$ in parentheses to compensate for the semi-infinite range of $R = [0 \to \infty]$. Noting that $G'_{11}(\mathbf{R};\omega)$ is shown above to depend only on the magnitude of radius R and not on angle, $\partial^2 G'_{11}/\partial\theta^2 \equiv 0$, and averaging over θ ($\frac{1}{2\pi}\int_0^{2\pi} d\theta$) yields the "Bessel Wave Equation" for $G'_{11}(R,\Omega)$ as[19] (with a $\delta(R)$ driving term)

$$\left[\frac{\partial^2}{\partial R^2} + \frac{1}{R}\frac{\partial}{\partial R} + \alpha^2 R^2 + q^2\right] G_{11}(R;\Omega) = \frac{2\mathcal{M}}{\pi}\frac{\delta(R)}{R}, \tag{12.3.35}$$

with $\alpha^2 = -\mathcal{M}^2\Omega_c^2/4$ and $q^2 = 2\mathcal{M}\Omega$. The The two Frobenius *Bessel Wave Function* solutions (order 0) of the homogeneous counterpart of Eq. (12.3.35) for $R > 0$ are exhibited in full detail in Reference 19. They are denoted here by $Z(R)$, having small R behavior as $Z_1(R) \simeq 1$, and the second solution has the form $Z_2(R) \cong Z_1(R) lnR$. Thus, the solution of the homogeneous radial equation may be written as a linear combination of Z_1 and Z_2:

$$G_{11}(R,\Omega) = AZ_1(R) + BZ_2(R), \tag{12.3.36}$$

subject to the "half-axis" Dirac $\delta(R)$-function condition of Eq. (12.3.35) at small $R \to \varepsilon \to 0^+$. Integration of Eq. (12.3.35) with respect to R from $R = 0$ to $R = \varepsilon$ (after multiplication by R) yields

$$\frac{\partial}{\partial\varepsilon} G_{11}(\varepsilon;\Omega) \cong \frac{\mathcal{M}}{\pi\varepsilon} \quad or \quad G_{11}(\varepsilon;\Omega) \cong \frac{\mathcal{M}}{\pi} ln\varepsilon. \tag{12.3.37}$$

This $ln\, R \to ln\varepsilon$ behavior for $R \to \varepsilon \to 0^+$ is indicative of the second Frobenius solution of the Bessel Wave Equation, specifically,

$$G'_{11}(R;\Omega) = \frac{\mathcal{M}}{\pi} Z_2(R) \tag{12.3.38}$$

This determination of $G'_{11}(\mathbf{R},\Omega)$ is also applicable to G'_{22}; and G'_{12} and G'_{21} can be determined as indicated above.

12.4 Graphene Quantum Dot: Magnetic Field Green's Function

Interest has arisen in newly conceived electronic devices based on charge transport through quantum dot systems, in which each dot would support just one single energetically accessible subband energy level in the absence of a magnetic field. In connection with this, the role of a magnetic field has often been represented merely in terms of a Peierls phase factor, such as $C(\mathbf{r},\mathbf{r}')$ given by Eq. (5.6.33). However, this neglects Landau quantization. To address this issue for Graphene, we consider a two-dimensional delta-function potential to represent the potential profile of a quantum dot at the origin as ($\hbar \to 1$):

$$U(\mathbf{r}) = \alpha\delta^{(2)}(\mathbf{r}), \ (\alpha < 0), \tag{12.4.1}$$

which would support just one level (lightly damped) that is energetically accessible in the absence of a magnetic field. Here, $\alpha = \int d^3\mathbf{r}U(\mathbf{r}) < 0$ is essentially the product of the confining potential well depth and the area of the dot. Such Dirac delta-function model potentials have been used in the quantum theory of scattering (and in Chapter 5), yielding useful information. The present model $\delta^{(2)}(\mathbf{r})$-potential is employed to describe the case in which the dot radius is the smallest length of the system. Of course, this model does not account for the actual geometrical structure of the dot but it correctly predicts the proliferation of Landau-quantized states into which the single zero-field level splinters when the magnetic field is applied. The Green's function for the Graphene sheet with the inclusion of the $\delta^{(2)}(\mathbf{r})$ dot potential well, G_{dot}, is determined by the integral equation (frequency representation):

$$G_{dot}(\mathbf{r}_1, \mathbf{r}_2; \omega) = G(\mathbf{r}_1, \mathbf{r}_2; \omega) + \alpha \int d^2\mathbf{r}_3 G(\mathbf{r}_1, \mathbf{r}_3; \omega)\delta^{(2)}(\mathbf{r}_3) G_{dot}(\mathbf{r}_3, \mathbf{r}_2; \omega), \tag{12.4.2}$$

where G is the Landau-quantized electron Green's function for the 2D Graphene sheet *in the absence of the quantum well.* Carrying out the $\mathbf{r}_3$-integration, Eq. (12.4.2) becomes

$$G_{dot}(\mathbf{r}_1, \mathbf{r}_2; \omega) = G(\mathbf{r}_1, \mathbf{r}_2; \omega) + \alpha G(\mathbf{r}_1, 0; \omega) G_{dot}(0, \mathbf{r}_2; \omega). \tag{12.4.3}$$

Putting $\mathbf{r}_1 \to 0$ in Eq. (12.4.3), algebraic solution for $G_{dot}(0, \mathbf{r}_2; \omega)$ yields the full solution for G_{dot} in the form

$$G_{dot}(\mathbf{r}_1, \mathbf{r}_2; \omega) = G(\mathbf{r}_1, \mathbf{r}_2; \omega) + \alpha G(\mathbf{r}_1, 0; \omega) \left[I - \alpha G(0, 0; \omega)\right]^{-1} G(0, \mathbf{r}_2; \omega) \tag{12.4.4}$$

(again, I is the 2×2 unit matrix). The first Green's function on the right, $G(\mathbf{r}_1, \mathbf{r}_2; \omega)$, describes Landau-quantized Graphene carrier dynamics on the infinite 2D host sheet with no quantum dot, and the second term introduces the effects of the quantum well "dot." The inversion of $[I - \alpha G(0, 0; \omega)]$ yields the 2×2 matrix

$$\left[I - \alpha G(0, 0; \omega)\right]^{-1} = \Delta^{-1} \begin{pmatrix} 1 - \alpha G'_{22} & \alpha G'_{12} \\ \alpha G'_{21} & 1 - \alpha G'_{11} \end{pmatrix}, \tag{12.4.5}$$

where

$$\Delta = \left[\left(1 - \alpha G'_{11}\right)\left(1 - \alpha G'_{22}\right) - \alpha^2 G'_{12} G'_{21}\right]. \tag{12.4.6}$$

In this, we have noted that $G(0, 0; \omega) = G'(0, 0; \omega)$ involves $C(\mathbf{r}_1, \mathbf{r}_2) = C(0, 0) \equiv 1$; furthermore, the Peierls phase factor $C(\mathbf{r}_1, \mathbf{r}_2)$ remains active in $G(\mathbf{r}_1, \mathbf{r}_2; \omega) = C(\mathbf{r}_1, \mathbf{r}_2) G'(\mathbf{r}_1, \mathbf{r}_2; \omega)$ in Eq. (12.4.4).

In Eqns. (12.4.5) and (12.4.6) all of the positional arguments of the elements of $G'(0,0;\omega)$ vanish. It is clear from Eq. (12.3.21) that for $X = Y = 0$, this leads to divergence at the lower limit of the $\tilde{T}$-integral. The divergence stems from the artificiality of representing particle confinement by the $\delta^{(2)}(\mathbf{r})$-potential well. Actually, the well has a small but finite radius a and the integral equation, Eq. (12.4.2), should be solved more carefully with $U(\mathbf{r})$ occupying a small but finite region. In this context, we consider the formal solution Eq. (12.4.4) as being "smeared" over the radius a with the replacement

$$G'(0,0;\omega) \Rightarrow G'(a;\omega), \tag{12.4.7}$$

thus eliminating the divergence problem. In the present context, with the radius a as the smallest length parameter, we note that $G'_{12} \sim G'_{21} \sim (\gamma eBa/\omega) G'_{11} \sim (\gamma eBa/\omega) G'_{22}$, so that G'_{12} and G'_{21} can be neglected in comparison with G'_{11} and G'_{22} because of the smallness of a. (Taking $B = 500$ Gauss and $a \sim 10^{-9}$m with $\gamma = 10^6$m/s, the dimensionless ratio is $\gamma eBa/\hbar\bar{\omega} \sim .8 \times 10^{-2}$, where $\bar{\omega} \equiv (2\gamma^2 eB/\hbar)^{1/2} \sim 10^{13}$/s.)

Employing the results of Section 12.3, the dispersion relation for the energy levels of a Landau-quantized Graphene dot in a magnetic field,

$$\Delta = 0, \tag{12.4.8}$$

takes the form

$$\Delta \cong [1 - \alpha G'_{11}(a;\omega)][1 - \alpha G'_{22}(a,\omega)] = 0, \tag{12.4.9}$$

where we have neglected G'_{12} and G'_{21} as explained above. Correspondingly,

$$1 = \alpha \frac{eB}{2\pi} \omega e^{-\frac{eBa^2}{4}} \sum_{n=0}^{\infty} \frac{L_n\left(\frac{eBa^2}{2}\right)}{\omega^2 - 2\gamma^2 eB(n + [1 \mp 1_\nu]/2)}. \tag{12.4.10}$$

This series converges slowly since[16] $L_n(x) \to \frac{e^{\frac{x}{2}}}{\sqrt{\pi}} \frac{\cos(2\sqrt{nx} - \frac{\pi}{4})}{(nx)^{\frac{1}{4}}}$ as $n \to \infty$, with the denominator on the right of Eq. (12.4.10) contributing a factor $\frac{1}{n}$. It is instructive to install typical Graphene parameters (above) in Eq. (12.4.10), with $\frac{eBa^2}{2\hbar} = .36 \times 10^{-4}$; $\exp(-\frac{eBa^2}{4\hbar}) \cong 1$; and $\alpha eB/2\pi\bar{\omega}\hbar^2 \sim .7 \times 10^{-3}$; yielding

$$\left(\frac{\alpha eB\omega}{2\pi\hbar^2\bar{\omega}^2}\right)^{-1} \sim \left(\frac{\alpha eB}{2\pi\hbar^2\bar{\omega}}\right)^{-1} = 1.4 \times 10^3 = \sum_{n=0}^{\infty} \frac{L_n(.36 \times 10^{-4})}{(\omega/\bar{\omega})^2 - (n + [1 \mp 1_\nu]/2)}. \tag{12.4.11}$$

The left-hand side, 1.4×10^3, is large, while the numerators of the terms on the right are relatively small $\left(L_n(.36 \times 10^{-4}) \sim 1\right)$ and diminish as n increases. Correspondingly, as n increases, the roots involve $\omega^2/\bar{\omega}^2$ in the denominators on the right-hand side having to closely approach the poles at $\pm(n + [1 \mp 1_\nu]/2))^{1/2}$ in order to establish equality with the large left-hand side; in which case the single pole for that particular n-value *alone* determines the approximate position of the energy level root. In this estimate, as n increases, the Graphene dot energy levels are close to the frequency poles of the individual terms of the series. Accordingly, we write Eq. (12.4.10) approximately as $(\omega \to \omega_n)$:

$$\omega_n^2 - \omega_n \frac{\alpha eB}{2\pi} e^{-eBa^2/4} L_n\left(\tfrac{eBa^2}{2}\right) - 2\gamma^2 eB(n + [1 \mp 1_\nu]/2) = 0, \quad (12.4.12)$$

with the solutions $(n = 0, 1, 2 \ldots \infty)$

$$\omega_{n\pm} = \frac{\alpha eB}{4\pi} e^{-eBa^2/4} L_n\left(\tfrac{eBa^2}{2}\right) \pm \frac{1}{2}\sqrt{\left(\tfrac{\alpha eB}{2\pi}\right)^2 e^{-eBa^2/2}\left(L_n\left(\tfrac{eBa^2}{2}\right)\right)^2 + 8\gamma^2 eB(n + [1 \mp 1_\nu]/2)}. \quad (12.4.13)$$

Careful numerical calculations of the Landau-quantized Graphene quantum dot energy levels have been carried out on the basis of Eqns. (12.4.6–12.4.8)[23].

The splintering of the electron and hole spectra of Graphene quantum dots into a plethora of distinct states by Landau quantization is embodied in the discrete integer values of n. This is reflected in the many roots of the dispersion relation for Graphene dots exhibited here in Section 12.4, as well as for the infinite Graphene sheet discussed in Section 12.3. It should be borne in mind that the energy levels of the dots differ from those of the sheet by terms involving the parameter α, which characterizes the dot and distinguishes it from the sheet.

12.5 Graphene Antidot Lattice in a Magnetic Field

The 2D delta-function representation of a quantum dot potential well may also be employed to represent a quantum *antidot* potential barrier by reversing the sign of α in Eq. (12.4.1):

$$U(\mathbf{r}) = \alpha\delta^2(\mathbf{r}), (\alpha > 0), \quad (12.5.1)$$

with α representing the product of the potential barrier height times its cross-sectional area, a^2. To discuss the effects of a quantizing magnetic field $\mathbf{B} = B\hat{z}$ on a Krönig–Penney-type model[24] (generalized to two dimensions to accommodate the magnetic field) of a one-dimensional lattice array of antidot potential barriers uniformly spaced at $x_n = nd$ with $y_n \equiv 0$ (n = all integers; d is the period of the antidot array), we construct the appropriate 2×2 Graphene matrix Green's function G_{KP}, which is determined by the 2D integral equation ($\mathbf{r} = x, y\,; \mathbf{r}' = x', y'$):

$$G_{\mathrm{KP}}(\mathbf{r}_1,\mathbf{r}_2;\omega) = G(\mathbf{r}_1,\mathbf{r}_2;\omega) + \int d^2\mathbf{r}'\, G(\mathbf{r}_1,\mathbf{r}';\omega)U(\mathbf{r}')G_{\mathrm{KP}}(\mathbf{r}',\mathbf{r}_2;\omega). \quad (12.5.2)$$

Here, $G(\mathbf{r}_1,\mathbf{r}_2;\omega)$ is the Graphene matrix Green's function for 2D electrons in a perpendicular magnetic field in the absence of any potential barriers as discussed in Section 12.3, and, in accordance with the Krönig–Penney model for an antidot lattice, we introduce an infinite periodic lattice array of identical potential barriers on the x-axis ($y = 0$) as

$$U(\mathbf{r}) = \alpha \sum_{n=-\infty}^{\infty} \delta(x - nd)\delta(y). \tag{12.5.3}$$

As above, $\alpha>0$ is the product of a typical antidot potential barrier height (U_0) times its area (a^2) (all antidots are taken to have the same α value). With these Dirac delta-functions, the x'-integral (as well as the y'-integral) of Eq. (12.5.2) is immediate and we obtain

$$G_{KP}(x_1, x_2; y_1, y_2; \omega) = G(x_1, x_2; y_1, y_2; \omega) + \alpha \sum_{n=-\infty}^{\infty} G(x_1, nd; y_1, 0; \omega) \times G_{KP}(nd; x_2; 0, y_2; \omega). \tag{12.5.4}$$

Suppressing ω henceforth, we note that the solution for $G_{KP}(x_1, x_2; y_1, y_2)$ devolves upon the determination of $G_{KP}(nd, x_2; 0, y_2)$ at a discrete set of values. Therefore, setting $x_1 = md$, and $y_1 = 0$, we have

$$G_{KP}(md, x_2; 0, y_2) = G(md, x_2; 0, y_2) + \alpha \sum_{n=-\infty}^{\infty} G(md; nd; 0, 0) G_{KP}(nd, x_2; 0, y_2). \tag{12.5.5}$$

While $C(md, x_2; 0, y_2)$ is an essential part of $G(md, x_2; 0, y_2)$ since $G(\mathbf{r}_1, \mathbf{r}_2) = C(\mathbf{r}_1, \mathbf{r}_2) G'(\mathbf{r}_1 - \mathbf{r}_2)$, the factor $G(md, nd; 0, 0)$ involves only $C(md, nd; 0, 0) \equiv 1$ (the two position vector arguments are parallel), so we have

$$G(md; nd; 0, 0) = \dot{G}([m - n]d), \tag{12.5.6}$$

which is a translationally invariant 2×2 Graphene matrix Green's function in position representation, as indicated by an overhead dot. Suppressing x_2 and y_2 for the moment, these equations may be solved using the periodicity of the lattice in a Fourier series defined by (r = integers here):

$$\tilde{G}_{KP}(p) = \sum_{r=-\infty}^{\infty} e^{ipdr} G_{KP}(rd) \tag{12.5.7}$$

and

$$G_{KP}(md) = \frac{d}{2\pi} \int_{-\pi/d}^{\pi/d} dp\, e^{-ipdm} \tilde{G}_{KP}(p), \tag{12.5.8}$$

with

$$\dot{G}([m-n]d) = \frac{d}{2\pi}\int_{-\pi/d}^{\pi/d} dp\, e^{-ipd[m-n]}\dot{\tilde{G}}(p). \tag{12.5.9}$$

Correspondingly, Eq. (12.5.5) becomes

$$\begin{aligned}\tilde{G}_{\text{KP}}(p) &= \tilde{G}(p) + \alpha\dot{\tilde{G}}(p)\sum_{n=-\infty}^{\infty} e^{ipdn} G_{\text{KP}}(nd) \\ &= \tilde{G}(p) + \frac{\alpha d}{2\pi}\dot{\tilde{G}}(p)\int_{-\pi/d}^{\pi/d} dq\left(\sum_{n=-\infty}^{\infty} e^{in[p-q]d}\right)\tilde{G}_{\text{KP}}(q).\end{aligned} \tag{12.5.10}$$

Employing the Poisson Sum Formula as

$$\sum_{n=-\infty}^{\infty} e^{in[p-q]d} = \frac{2\pi}{d}\sum_{m=-\infty}^{\infty}\delta(p-q-2\pi m/d),$$

we have

$$\tilde{G}_{\text{KP}}(p) = \tilde{G}(p) + \alpha\sum_{m=-\infty}^{\infty}\int_{-\pi/d}^{\pi/d} dq\,\delta(p-q-2\pi m/d)\dot{\tilde{G}}(p)\tilde{G}_{\text{KP}}(q). \tag{12.5.11}$$

Since the q-integral is extended only over the first Brillouin zone,

$$\tilde{G}_{\text{KP}}(p) = \tilde{G}(p) + \alpha\dot{\tilde{G}}(p)\tilde{G}_{\text{KP}}(p), \tag{12.5.12}$$

we have the solution given by (restore ω,x_2,y_2):

$$\tilde{G}_{\text{KP}}(p;x_2;0,y_2;\omega) = \left[I-\alpha\dot{\tilde{G}}(p;0,0;\omega)\right]^{-1}\tilde{G}(p;x_2;0,y_2;\omega). \tag{12.5.13}$$

Taken jointly with Eq. (12.5.8) and Eq. (12.5.4), the result of Eq. (12.5.13) provides the basis for the determination of the Green's function for the 2D Krönig–Penney-like model for a 1D antidot Graphene lattice in a normal magnetic field (and, of course, for zero field). In position representation, it is given by

$$\begin{aligned}G_{\text{KP}}(x_1,x_2;y_1,y_2;\omega) &= G(x_1,x_2;y_1,y_2;\omega) + \alpha\sum_{n=-\infty}^{\infty} G(x_1,nd;y_1,0;\omega) \\ &\times\frac{d}{2\pi}\int_{-\frac{\pi}{d}}^{\frac{\pi}{d}} dp e^{-ipdn}\left[I-\alpha\dot{\tilde{G}}(p;0,0;\omega)\right]^{-1}\tilde{G}(p;x_2;0,y_2;\omega),\end{aligned} \tag{12.5.14}$$

with G as defined in Section 12.3 (and $\dot{\ddot{G}}$ determined from it as indicated above). Of course, the eigen-energy dispersion relation embodying the effects of Landau quantization is given by

$$det[I - \alpha\dot{\ddot{G}}(p; 0, 0; \omega)] = 0, \tag{12.5.15}$$

in which the Peierls phase factor is unity (with the pseudospin trace understood). Although the dispersion relation is effectively translationally invariant, the full final Green's function (involving $C(\mathbf{r}_1, \mathbf{r}_2)$-factors with y_1 and y_2 non-zero) is not.

12.6 Graphene Statistical Thermodynamics and Magnetic Moment

To construct the equilibrium thermodynamic Green's function for Graphene we note that its spectral weight, $A(\omega)$, may be understood in terms of a *matrix* in accordance with the relations following from Eq. (7.5.12),

$$A(\omega) = -2ImG_r(\omega). \tag{12.6.1}$$

The retarded Graphene matrix Green's function, $G_r(\omega) \equiv \overleftrightarrow{G}{}^{ret}_1$, is given in position-frequency representation as

$$G_r(\mathbf{r}, \mathbf{r}'; \omega) = C(\mathbf{r}, \mathbf{r}')G'(\mathbf{R}; \omega), \tag{12.6.2}$$

where $\mathbf{R} = \mathbf{r} - \mathbf{r}'$ and $C(\mathbf{r}, \mathbf{r}')$ is defined in Eq. (12.3.3). The diagonal elements of $G'(\mathbf{R}, \omega)$ are given by Eq. (12.3.28), which may be conveniently written as

$$G'_{\substack{11\\22}}(\mathbf{R}; \omega) = \frac{eB}{4\pi} e^{\frac{-eBR^2}{4}} \sum_{n=0}^{\infty} L_n\left(\tfrac{eBR^2}{2}\right)\left\{\frac{1}{\omega - \omega_{n\mp}} + \frac{1}{\omega + \omega_{n\mp}}\right\}, \tag{12.6.3}$$

with the eigenfrequencies

$$\omega \to \omega_{n\mp} = \sqrt{2\gamma^2 eB(n + [1 \mp 1_\nu]/2)} \quad \text{and} \quad \omega \to -\omega_{n\mp} = -\sqrt{2\gamma^2 eB(n + [1 \mp 1_\nu]/2)}. \tag{12.6.4}$$

Consequently,

$$ImG'_{\substack{11\\22}}(\mathbf{R}; \omega) = -\frac{eB}{4} e^{\frac{-eBR^2}{4}} \sum_{n=0}^{\infty} L_n\left(\tfrac{eBR^2}{2}\right)\{\delta(\omega - \omega_{n\mp}) + \delta(\omega + \omega_{n\mp})\}, \tag{12.6.5}$$

and bearing in mind that the density of states (Eq. (5.1.27)) is given by

$$D(\omega) = -\frac{1}{\pi} Tr\, ImG_r(\omega) = -\frac{1}{\pi} tr\, Im \int d\mathbf{r} < \mathbf{r}|G_r(\omega)|\mathbf{r} >, \tag{12.6.6}$$

where tr is the pseudospin trace, we have (noting that $C(\mathbf{r},\mathbf{r}) = 1$):

$$D(\omega) = -\frac{1}{\pi} tr\, ImG_r(\mathbf{r},\mathbf{r};\omega) = -\frac{1}{\pi} tr\, ImG'_r(0,0;\omega) = -\frac{1}{\pi} Im(G'_{11}(0,0;\omega) + G'_{22}(0,0;\omega)), \tag{12.6.7}$$

on a per-unit-area basis (the r-integration yields an area factor). Thus, the density of states for Graphene in a magnetic field may be expressed as ($\mathbf{R} = \mathbf{r} - \mathbf{r} = 0; L_n(0) \equiv 1$):

$$D(\omega) = \frac{eB}{4\pi} \sum_{n=0}^{\infty} \sum_{\pm',\mp} \delta(\omega \pm' \omega_{n\mp}). \tag{12.6.8}$$

The sum over $\pm'$ accounts for the two terms in curly brackets in Eq. (12.6.5), the sum over $\mp$ accounts for the 2 terms of the pseudospin trace.

Considering the 2×2 matrix for the spectral weight A given by

$$A(\omega) = -2ImG_r(\omega), \tag{12.6.9}$$

we obtain its diagonal elements from Eq. (12.6.5) as

$$A_{\substack{11\\22}}(\mathbf{r},\mathbf{r}';\omega) = \frac{eB}{2} C(\mathbf{r},\mathbf{r}')e^{-\frac{eB}{4}R^2} \sum_{n=0}^{\infty} L_n\left(\tfrac{eB}{2}R^2\right) \{\delta(\omega - \omega_{n\mp}) + \delta(\omega + \omega_{n\mp})\}. \tag{12.6.10}$$

The corresponding diagonal elements of the thermodynamic Greens function, $G^{\mu,\beta}_{\substack{11\\22}\{\gtrless\}}$, may now be obtained as

$$iG^{\mu,\beta}_{\substack{11\\22}\{\gtrless\}}(\mathbf{r},\mathbf{r}';\omega) = \begin{Bmatrix} (1-f(\omega)) \\ -f(\omega) \end{Bmatrix} \frac{eB}{2} C(\mathbf{r},\mathbf{r}')e^{-\frac{eB}{4}R^2} \sum_{n=0}^{\infty} \sum_{\pm'} L_n\left(\tfrac{eB}{2}R^2\right) \delta(\omega \pm' \omega_{n\mp}), \tag{12.6.11}$$

with $f(\omega) = (1 + e^{\beta(\omega-\mu)})^{-1}$.

Since the Helmholtz Free Energy shown in Eq. (6.3.25) involves summing over *all* states, the 2×2 matrix retarded Green's function presently under consideration must be understood as contributing to F in terms of its trace (which is invariant with respect to change of basis/representation), i.e.:

$$F - \mu n = -\frac{1}{4} \int_C \frac{ds'}{2\pi i} \frac{1}{s'^2} i\, tr\, G^{ret}_{1>}\left(0,0;\left[t-t' \to -is'\right]\right) \int_0^{\infty} dE\, e^{s'E} \beta sech^2\left(\frac{[E-\mu]\beta}{2}\right),$$

(per unit area) where $\beta sech^2\left(\frac{[E-\mu]\beta}{2}\right) = -4\frac{\partial f(E)}{\partial E} \to 4\delta(E-\mu)$ in the degenerate limit of zero temperature, and we have noted that the "retarded" Green's function $G_r \equiv G^{ret}_{1>}$ of Eqns.(6.3.16) → (6.3.25) lacks the abrupt time cut-off $\eta_+(t-t')$ due to the absence of time ordering, so it is actually just the $G^{ret}_{1>}$-function extended continuously into $t < t'$ as well as $t > t'$, and its frequency transform involves only energy $\delta(\omega - E_n)$-functions (to the exclusion of "principal part"-functions).

In this limit,

$$F - \mu n = -\int \frac{ds'}{2\pi i}\frac{e^{\mu s'}}{s'^2} i \, tr \, G^{ret}_{1>}(0,0; T \to -is'). \tag{12.6.12}$$

Employing the result for the retarded Graphene Green's function (Eq.(12.6.3)) in position/frequency representation and transforming to time representation, we maintain the pseudospin trace as above, obtaining

$$tr \, G'^{ret}_{1>}(\mathbf{R}, T) = \frac{eB}{4\pi} e^{-\frac{eB}{4}R^2} \sum_{n=0}^{\infty} L_n\left(\tfrac{eB}{2}R^2\right) \sum_{\pm',\mp} \int_{-\infty}^{\infty} \frac{d\omega}{2\pi} \frac{e^{-i\omega T}}{\omega \pm' \omega_{n\mp}}. \tag{12.6.13}$$

Setting $z = (\omega \pm' \omega_{n\mp})\, T$, the ω-integral is given by[25]

$$\int_{-\infty}^{\infty} \frac{d\omega}{2\pi} \frac{e^{-i\omega T}}{\omega \pm' \omega_{n\mp}} = -ie^{\pm' i T \omega_{n\mp}}. \tag{12.6.14}$$

and $tr \, G'^{ret}_{1>}(\mathbf{R}, T) = -i\frac{eB}{2\pi} e^{-\frac{eB}{4}R^2} \sum_{n=0}^{\infty} L_n\left(\frac{eB}{2}R^2\right) \sum_{\mp} \cos(\omega_{n\mp} T)$. Therefore, with $R \to 0$ and $T \to -is'$, Eq. (12.6.12) yields the Free Energy for Graphene in a magnetic field at zero temperature as $(C(\mathbf{r}, \mathbf{r}) = 0)$:

$$F - \mu n = -\frac{eB}{4\pi} \sum_{n=0}^{\infty} \sum_{\pm',\mp} \int_C \frac{ds'}{2\pi i} \frac{e^{s'(\mu \pm' \omega_{n\mp})}}{s'^2}. \tag{12.6.15}$$

Evaluation of this inverse Laplace transform contour integral by residues yields

$$F - \mu n = -\frac{eB}{4\pi} \sum_{n=0}^{\infty} \sum_{\pm',\mp} \eta_+\left(\mu \pm' \omega_{n\mp}\right)\left(\mu \pm' \omega_{n\mp}\right), \tag{12.6.16}$$

with $\eta_+(x)$ as the Heaviside unit step function.

Defining ΔF as the right-hand side of Eq. (12.6.16), we have its contribution to the magnetic moment of Graphene at zero temperature as (per unit area; Eq. (6.3.6); note that $x \, d(\eta_+(x))/dx = x\delta(x) = 0$; $\hbar \to 1$):

$$\Delta M = -\frac{\partial(\Delta F)}{\partial B}\bigg|_{T=0} = \frac{e}{4\pi}\sum_{n=0}^{\infty}\sum_{\pm',\mp}\eta_+\left(\mu \pm' \omega_{n\mp}\right)\left[\left(\mu \pm' \omega_{n\mp}\right) \pm' B\frac{\partial\omega_{n\mp}}{\partial B}\right]. \quad (12.6.17)$$

Noting Eq. (12.6.4), we finally obtain ΔM as

$$\Delta M = \frac{e}{4\pi}\sum_{n=0}^{\infty}\sum_{\pm',\mp}\eta_+\left(\mu \pm' \omega_{n\mp}\right)\left(\mu \pm' \tfrac{3}{2}\omega_{n\mp}\right), \quad (12.6.18)$$

which explicitly exhibits oscillatory features in Graphene as Landau-quantized levels pass through the Fermi level, μ, as the magnetic field strength varies. In conformance with the precepts of the grand canonical ensemble, with which we are working, we consider the system to be in contact with a particle bath so that μ is a constant, and, correspondingly, the density n on the left of Eq. (12.6.16) also contributes oscillatory behavior as a function of B. Finally, the magnetic susceptibility, arising from $\partial(\Delta M)/\partial B$ by differentiating Eq. (12.6.18), exhibits rather strong low-field behavior as $B^{-1/2}$. This has been carefully examined in the literature, and mechanisms that limit increase as $B \to 0$ (including disorder or relatively small thermal wavelength) have been analyzed in detail.[26–28]

12.7 Further Remarks and Developments

In recognition of the immensity of the commercial impact that Graphene (and similar materials) could have as a successor to silicon in electronics and computers, physicists, materials scientists, engineers, chemists and technologists from every part of the world have concentrated research efforts on it. In just ten years a vast literature has resulted, which is far beyond the scope of this book to describe. However, to help the interested reader get access to recent developments, we provide references to some key reviews, articles and books.

By way of introduction to the subject, there is an interesting paper of F. Wilczek's introductory remarks at the Nobel Symposium 148 on "Graphene and Quantum Matter" in Sweden in June 2010,[29] and another *Physics Today* article about the Nobel laureates, Geim and Novelselov[30] (whose own article in *Nature* should be consulted[31]). The reader will also find informative introductory articles on the general features of Graphene by Castro Neto,[32] Geim and MacDonald[33] and Katsnelson.[34] There are particularly interesting articles on new directions by Castro Neto and Novoselov,[35] and on Graphene electronics by Yung, *et al.*[36]

There are major review articles on the electronic properties of Graphene in general by Castro Neto, *et al.*[37], also one by Abergel, *et al.*[38] focused on theory and another by Andrei, *et al.*[39] with the perspective of scanning tunneling microscopy and magneto-transport. The electronic properties of bilayer Graphene are reviewed by McCann and Koshino,[40] and also by Killi, *et al.* with an emphasis on superlattices.[41] Furthermore, there are major review articles on electron–electron interactions, plasmons and screening in Graphene by Kotov, *et al.*,[42] and another by Levitov, *et al.*[43] The extension of such studies to gapped monolayer and bilayer Graphene is reported by Pyatkovskiy[44] and Triola and Rossi[45] (respectively). Yet another paper by Bludov, *et al.* focuses on surface plasmon-polaritons in Graphene.[46] Plasmonics in Dirac systems is discussed by Stauber[47] and also by Grigorenko, *et al.*[48], and the optical nano-imaging of gate-turnable Graphene plasmons is addressed by Chen, *et al.*[49]

The basic magnetic properties of Graphene carriers and the relativistic quantum Hall effect are reviewed by Goerbig,[50] with extension to magnetoplasmons and the particle–hole excitation spectrum in Graphene by Roldan, *et al.*[51] Furthermore, Chakraborty and Apalkov review these magnetic features for bilayer as well as monolayer Graphene.[52] The magnetoplasmon spectrum in Graphene has been discussed in detail by Berman, *et al.*,[53] and in two papers by Roldan, *et al.*[54,55]

As one of Graphene's features of central interest, its transport properties have been the focus of review articles by Peres[56,57] and Das Sarma, *et al.*[58], and its anomalies have been discussed by Landsteiner, *et al.*[59] The subject of AC conductivity is addressed by Gusynin, *et al.*[60] Charge transport in disordered Graphene is treated by Cresti, *et al.*[61] and by Mucciolo and Lewenkopf[62], also by Ping, *et al.*[63], with a random resistor network model of Graphene minimal conductivity discussed by Cheianov, *et al.*[64] Recent progress in ballistic transport and electron optics in Graphene with an inhomogeneous magnetic field is reviewed by Agrawal, *et al.*[65]

The subject of Graphene photonics and optoelectronics is reviewed by Bonaccorso, *et al.*[66] Furthermore, the thermal, thermoelectric and thermo-electromechanical properties of Graphene are discussed by Xu, *et al.*,[67] Pop, *et al.*,[68] and Prabhakar, *et al.*[69]

Needless to say, the fact that Graphene electrons and holes obey the Dirac equation makes it an easily accessible "laboratory" for exploring the phenomenology of quantum electrodynamics and quantum field theory, both theoretically and experimentally. In this matter, the electrons and holes of Graphene are fully analogous to the electrons and positrons of relativistic quantum theory, with electrons having positive energy states and holes (positrons) having negative energy states, but with null energy gap. The interested reader can consult Sakurai's book[9] for background. A short non-technical introduction is provided by Fialkovsky and Vasilevich.[70] One particularly interesting phenomenon is the Klein paradox, in which tunneling through a potential barrier involves a transition from a positive-energy (electron) state to negative-energy (hole) state, with particle propagation in a classically forbidden region (where it would otherwise decay). This was predicted to occur in Graphene by Katsnelson, *et al.*[71] and was further discussed by DeLeo and Rotelli.[72] Another particularly interesting phenomenon is Zitterbewegung, the trembling motion of a wave packet involving interference of positive- and negative-energy components in the wave packet. Valuable discussions of this in Graphene are provided by Katsnelson[73] and Rusin and Zawadski.[74] Graphene van der Waals interactions with an adatom and with a double layer were addressed by Horing[75] and also by Sharma, *et al.*[76] The generalization to retardation, to accommodate Casimir interactions between Graphene layers, was discussed by Sernelius.[77] Lamb-shift-like phenomena in bilayer Graphene in a magnetic field are treated by Shizuya.[78] The bandgap induced in Graphene by vacuum fluctuations is discussed by Kibis,[79] and Chamon, *et al.*[80] review 36 possible gap-opening instabilities in Graphene. There are many other quantum field theoretic papers involving Graphene, including one on gauge fields by Vozmediano, *et al.*[81] and another on effective field theory by Barnes, *et al.*,[82] and yet another on topological invariants by Monaco and Panati,[83] but it is beyond the scope of this book to offer further discussion of this subject.

In addition to the review articles and papers cited above, the reader should be aware that there are books on Graphene, some of which are composed of collections of review articles and some whole journal issues composed of review articles. The latter include the "Graphene Review" issue of *Solid State Communications*[84] containing 18 review articles and the "Electronic and Photonic Properties of Graphene Layers and Carbon Nanoribbons" issue of *Philosophical Transactions of the Royal Society A*[85] with 11 review articles. The book by Aoki and Dresselhaus[86] has ten review articles. The reader should also consult the recent fine introductory book by Foa Torres, *et al.*[87] on Graphene-based nanomaterials. In connection with nanomaterials, there is an interesting review of Graphene quantum dots and spin qubits by Recher and Trauzettel.[88] Furthermore, there are recent books on the fundamentals of Graphene by Katsnelson[89] and by Warner, *et al.*[90], and another by Wolf.[91]

Graphene production was reviewed by Soldano, *et al.*[92] and its synthesis is discussed by Zhu, *et al.*[93] Moreover, Graphene synthesis is the focus of books by Rao and Sood,[94] by Choi, *et al.*[95] and also by Koratkar.[96] Other books focused on Graphene synthesis include one on Graphene optoelectronics by Rashid and Yusoff[97] and another on Graphene nanoelectronics by Raza.[98]

Of course, there is an enormous concentration of interest and work on Graphene electronics and devices. In addition to Reference 33, there is a review article on Graphene transistors by Lemme[99] and a book on Graphene nanoelectronics by Murali,[100] with another book on device applications by Shafraniuk[101] to come soon. Moreover, Graphene and carbon nanotube field effect transistors/devices are discussed in the books by Caine[102], and Wong and Akinwande.[103] Other aspects of Graphene-based materials and their features relating to chemistry may be found in books by Alwarappan and Kumar[104] and by Enoki and Ando.[105] In this regard, the functionalization of Graphene is treated in the book by Georgakialis[106] and its applications in energy storage and conversion is the subject of the book by Liu and Zhou.[107]

All of this merely scratches the surface of the vast literature on Graphene. Furthermore, a six-volume "Handbook of Graphene Science"[108] is in preparation by CRC Press for publication in the near future. The six volumes will treat fabrication methods, nanostructures and atomic arrangement, electrical and optical properties, mechanical and chemical properties, size-dependent properties, and applications and industrialization. It is likely that the flood of literature on Graphene and similar materials (Silicene, etc.) will continue unabated for some time to come.

12.8 Chapter 12 Problems

P.12.1. Determine the eigenvectors and eigenvalues of $\check{h}_1$ (Eq. (12.2.1)) in the pseudospin representation.

P.12.2. Construct the transformation from pseudospin basis to pseudovelocity basis and carry out the diagonalization to $\hat{h}_1$.

P.12.3. Explains the presence of the trace: (Tr) in the RPA density perturbation response function.

P.12.4. Derive the Eqns. (12.3.1–12.3.4) for the Green's function in a normal Graphene magnetic field **B**.

P.12.5. Use the retarded Graphene Green's function in a magnetic field discussed in Section 12.3 to construct the corresponding thermodynamic Green's function at arbitrary temperature.

P.12.6. There has been recent consideration of potentials proportional to the derivative of a Dirac delta-function, $U(x) = \alpha\delta'(x) \equiv \alpha d(\delta(x))/dx$ (α is a constant). Considering Graphene in the absence of a magnetic field, determine the Graphene Green's function in the presence of such a one-dimensional potential, and examine its effects on the transmission of an electron wave across the point $x = 0$.

P.12.7. Further to Problem 6, do the same for an ordinary one-dimensional Dirac delta-function potential $U(x) = \alpha\delta(x)$ in regard to the associated Graphene Green's function and the transmission of an electron wave across the point $x = 0$.

P.12.8. Determine the Helmholtz Free Energy F for Graphene in a magnetic field at arbitrary temperature.

P.12.9. Pursuant to Problem 8, determine the magnetic moment of Graphene as a function of temperature.

P.12.10. Determine the eigen-energy spectrum for a Graphene antidot lattice by solving Eq. (12.5.15) for null magnetic field.

P.12.11. Same as Problem 10, but for the case of a very high magnetic field in which only the lowest Landau level is energetically accessible, and all other Landau levels are very far removed in energy and can be neglected.

P.12.12. Show that Eq. (12.5.6) remains valid even if $y_1 = y_2 = constant \neq 0$.

P.12.13. The studies of linear current response and collective modes in Graphene (with or without a magnetic field) ultimately devolve upon the determination of a current–current correlation function, which involves an equilibrium two-particle Green's function. A leading approximation to this (without scattering) was discussed in Section 10.7. However, the result in Eq. (10.7.17) is not directly applicable to Graphene since its "relativistic" Hamiltonian is linear in the momentum operator (whereas Section 10.7 involved the quadratic momentum dependence of the non-relativistic Hamiltonian). Derive the appropriate counterpart of Eq. (10.7.17) for Graphene using its Hamiltonian to calculate $\mathbf{J}_{op} = -\delta H/\delta(\mathbf{A}(1)/c)$ with subsequent evaluation of $\overline{\mathbf{L}}_p(1,2) = [\delta\mathbf{J}(1)/\delta\mathbf{A}(2)]_{\mathbf{A}(2)\to 0}$. (Bear in mind that $\mathbf{J} = \langle\mathbf{J}_{op}\rangle$.)

12.9 References for Chapter 12

R.12.1. K.S. Novoselov, A.K. Geim, S.V. Morozov, D. Jiang, Y. Zhang, S.V. Dubonos, I.V. Crigorieva and A.A. Firsov, Science **306**, 666 arXiv:cond-mat/0410550 [cond-mat.mtrl-sci] (2004); also see A.H. Castro Neto, F. Guinea, N.M.R. Peres, K.S. Novoselov, A.K. Geim, arXiv:0709.1163v1[cond-mat. other] (2007). A particularly instructive presentation may be found in "Graphene: Carbon in Two Dimensions," M.I. Katsnelson, Cambridge University Press (2012).

R.12.2. S.V. Morozov, K.S. Novoselov, M.I. Katsnelson, F. Shedin, D.C. Elias, J.A. Jaszczak and A.K. Geim, Phys. Rev. Lett. **100**, 016602 (2008).

R.12.3. F. Schedin, A.K. Geim, S.V. Morozov, D. Jiang, E.H. Hill, P. Blake and K.S. Novoselov, arXiv.cond-mat/0610809 (2007).

R.12.4. E.W. Hill, A.K. Geim, K.S. Novoselov, F. Schedin and P. Blake, IEEE Trans. Magnetics **42**, 2694 (2006).

R.12.5. J.S. Bunch, A.M. van der Zande, S.S. Verbridge and I.W. Frank, Science **315**, 490 (2007).

R.12.6. Georgia Tech Research News - http://gtresearchnews.gatech.edu/newsrelease/graphene.html; C. Berger, Z. Song, X. Li, X. Wu, N. Brown, C. Naud, D. Mayou, T. Li, J. Hass, A.N. Machenkov, E.H. Conrad, P.N. First and W.A. de Heer, Science **312**, 1191 (2006); also, M.C.Lemme, IEEE Electron Device Letters **28**, 282 (2007).

R.12.7. P.R. Wallace, Phys. Rev. **71**, 622 (1947).

R.12.8. G.W. Semenoff, Phys. Rev. Lett. **53**, 2449 (1984); F.D.M. Haldane, *ibid* **61**, 2015 (1988).

R.12.9. J.J. Sakurai, "Advanced Quantum Mechanics," Addison-Wesley (1967).

R.12.10. K.W.-K. Shung, Phys. Rev. B**34**, 979 (1986).

R.12.11. T. Ando, J. Phys. Soc. Japan, **75**, 074716 (2006).

R.12.12. B. Wunsch, T. Stauber, F. Sols & F. Guinea, New J. Phys. **8**, 318 (2006).

R.12.13. E.H. Hwang and S. Das Sarma, Phys. Rev. B**75**, 205418 (2007).

R.12.14. S.Y. Liu, X.L. Lei and N.J.M. Horing, J. Appl. Phys. **104**, 043705 (2008) and references cited therein.

R.12.15. N.J.M. Horing, Annals of Physics (NY) **31**, 1 (1965) for the 3D magnetized quantum plasma; Also, N.J.M. Horing, M.M. Yildiz, Annals of Physics (NY) **97**, 216 (1976) for the 2D case; [use Eq. (10) with no Fermi averaging, and execute the $\mathbf{p} \to \mathbf{R}$ Fourier transform]

R.12.16. Bateman Manuscript Project, "Higher Transcendental Functions," Vol. 2, Editors: A. Erdelyi, W. Magnus, F. Oberhettinger and F.G. Tricomi, McGraw-Hill, p.189 (1953).

R.12.17. T. Ando, J. Phys. Soc. Japan **74**, p.777 (2005).

R.12.18. P.M. Morse and H. Feshbach, "Method of Theoretical Physics I," McGraw-Hill, Section 7.2 (1953).

R.12.19. P. Moon and E. Spencer, "Field Theory Handbook," 2nd Ed., Springer-Verlag, pp.171ff, also p.154 bottom (1971).

R.12.20. Bateman Manuscript Project, "Higher Transcendental Functions," Vol. 2, Editors: A. Erdelyi, W. Magnus, F. Oberhettinger and F.G. Tricomi, McGraw-Hill, p.13 (1953).

R.12.21. Bateman Manuscript Project, "Higher Transcendental Functions," Vol. 2, Editors: A. Erdelyi, W. Magnus, F. Oberhettinger and F.G. Tricomi, McGraw-Hill, Chapter VII (1953).

R.12.22. Bateman Manuscript Project, "Higher Transcendental Functions," Vol. 2, Editors: A. Erdelyi, W. Magnus, F. Oberhettinger and F.G. Tricomi, McGraw-Hill, p.24, Eq. 4 (1953).
R.12.23. N.J.M Horing and S.Y. Liu, unpublished.
R.12.24. C. Kittel, "Introduction to Solid State Physics" 7th Ed., Wiley (1991).
R.12.25. M. Boas, "Mathematical Methods in the Physical Sciences," 2nd Ed., Wiley (1983).
R.12.26. M. Nakamura, Phys. Rev. B **76**, 113301 (2007).
R.12.27. A. Ghosal, P. Goswami and S. Chakravarty, Phys. Rev. B **75**, 115123 (2007).
R.12.28. M. Koshino and T. Ando, Phys. Rev. B 75, 235333 (2007).
R.12.29. F. Wilczek, arXiv: 1109.1523v2 [cond-mat.mes-hall] (2011).
R.12.30. M. Wilson, Physics Today, *63*, p.14 (2010).
R.12.31. A.K. Geim and K.S. Novoselov, Nature Materials, 183 (2007).
R.12.32. A.H. Castro Neto, Materials Today **13**, 1 (2010).
R.12.33. A.K. Geim and A.H. MacDonald, Physics Today, Aug p.35 (2007).
R.12.34. M.I. Katsnelson, arXiv:0612534v1 [cond-mat.mes-hall] (2006).
R.12.35. A.H. Castro Neto and K. Novoselov, Rep. Prog. Phys. **74**, 082501 (2011).
R.12.36. K.C. Yung, *et al.*, arXiv:1309.0205v1 [cond-mat.mtrl-sci] (2013).
R.12.37. A.H. Castro-Neto, F. Guinea, N.M.R. Peres, K.S. Novoselov and A.K. Geim, Rev. Mod. Phys. *81*, 109 (2009).
R.12.38. D.S.L. Abergel, V. Apalkov, J. Berashevich, K. Ziegler and T. Chakraborty, Adv. Phys. *59*, p.261-482 (2010).
R.12.39. E. Y. Andrei, G. Li and X. Du, Rep. Prog. Phys. **75**, 056501 (2012).
R.12.40. E. McCann and M. Koshino, arXiv:1205.6953v1 [cond-mat.mes-hall] (2012).
R.12.41. M. Killi, S. Wu and A. Paramekanti, arXiv:1205.6813v1 [cond-mat.mes-hall] (2012).
R.12.42. V.N. Kotov, B. Uchoa, V.M. Pereira, F. Guinea and A.H. Castro Neto. Rev. Mod. Phys. *84*, p.1067 (2012).
R.12.43. L.S. Levitov, A.V. Shtyk and M.V. Feigelman, arXiv:1302.5036v1 [cond-mat.mes-hall] (2013).
R.12.44. P.K Pyatkovskiy, J. Phys. Conds. Matter **21**, 025506 (2009).
R.12.45. C. Triola and E. Rossi, arXiv:1205.5272v1 [cond-mat.mes-hall] (2012).
R.12.46. Yu. V. Bludov, A. Ferreira, N.M.R. Peres and M.I. Vasilevskiy, arXiv:1302.2317v1 [cond-mat.mes-hall] (2013).
R.12.47. T. Stauber, arXiv:1310.4296v1 [cond-mat.mes-hall] (2013).
R.12.48. A.N. Grigorenko, M. Polini and K.S. Novoselov, Nature Photon, 6, 749 (2012).
R.12.49. J. Chen, M. Badioli, P. Alonso-González, S. Thongrattanasiri, F. Huth, J. Osmond, *et al.*, arXiv:1202.4996v2 [cond-mat.mtrl-sci] (2012).
R.12.50. M.O. Goerbig, Rev. Mod. Phys. 83, p.1193, (2011).

R.12.51. R. Roldan, M.O. Goerbig and J.N. Fuchs, arXiv:0909.2825v2 [cond-mat.mes-hall] (2009). Also, Semicond. Sci. Technol. **25**, 034005 (2010).

R.12.52. T. Chakraborty and V.M. Apalkov, arXiv:1302.2940v1 [cond-mat.mes-hall] (2013).

R.12.53. O.L. Berman, G. Gumbs and Y. Lozovik, arXiv:0808.3128v1 [cond-mat.mes-hall] (2008).

R.12.54. R. Roldan, J.N. Fuchs and M.O. Goerbig. Phys. Rev. B **80**, 085408 (2009).

R.12.55. R. Roldan, M.O. Goerbig and J.N. Fuchs, arXiv:1012.0779v2 [cond-mat.mes-hall] (2011).

R.12.56. N.M.R. Peres, J. Phys. Condens. Matter **21**, 323201 (2009).

R.12.57. N.M.R. Peres, Rev. Mod. Phys. **82**, 2673 (2010).

R.12.58. S. Das Sarma, S. Adam, E.H. Hwang and E. Rossi, Rev. Mod. Phys. *83*, p.407 (2011).

R.12.59. K. Landsteiner, E. Megías and F. Peña-Benitez, arXiv:1207.5808v2 [hep-th] (2012).

R.12.60. V.P. Gusynin, S.V. Shaparov and J.P. Carbotte, arXiv:0706.3016v2 [cond-mat.mes-hall] (2007).

R.12.61. A. Cresti, N. Nemec, B. Biel, G. Niebler, F. Triozon, G. Cuniberi and S. Roche, NanoRes **1**, 361 (2008).

R.12.62. E.R. Mucciolo and C.H. LewenKopf, arXiv:1006.0255v2 [cond-mat.mes-hall] (2010).

R.12.63. J. Ping, I. Yudhistira, N. Ramakrishnan, S. Cho, S. Adam and M. S. Fuhrer, arXiv:1401.5094v1 [cond-mat.mes-hall] (2014).

R.12.64. V.V. Cheianov, V.L. Falco, B.I. Altshuler and I.L Aleiner, Phys. Rev, Lett **99**, 176801 (2007).

R.12.65. N. Agrawal, S. Ghosh and M. Sharma, arXiv:1301.7707v1 [cond-mat.mes-hall] (2013).

R.12.66. F. Bonaccorso, Z. Sun, T. Hasan and A.C. Ferrari, Nature Photonics **4**, 611 (2010).

R.12.67. Y. Xu, Z. Li and W. Duan, arXiv:1401.3415v1 [cond-mat.mtrl-sci] (2014).

R.12.68. E. Pop, V. Varshney and A.K. Roy, MRS Bull. *37*, p.1273–1281, (2012).

R.12.69. S. Prabhakar, R. Melnik, L.L. Bonilla and S. Badu, arXiv:1401.4608v1 [cond-mat.mes-hall] (2014).

R.12.70. I.V. Fialkovsky and D.V. Vasilevich, arXiv:1111.3017v1 [hep-th] (2011).

R.12.71. M.I. Katsnelson, K.S. Novselov and A. K. Geim, Nature **2**, 620 (2006).

R.12.72. S. De Leo and P. Rotelli, J. Phys. A: Math. Theor. **44**, 475305 (2011).

R.12.73. M.I. Katsnelson, Eur. Phys. J. B **51**, 157 (2006).

R.12.74. T.M. Rusin and W. Zawadski, arXiv:1308.0671v1 [cond-met.mes-hall] (2013).

R.12.75. N.J.M. Horing in “Quantum Field Theory Under the Influence of External Conditions,” Editor: K. Milton, World Scientific, 317 (2010).

R.12.76. A. Sharma, P. Harnish, A. Sylvester, V.N. Kotov, A.H. Castro Neto, arXiv:1402.3369v1 [cond-mat.intrl-sci] (2014).

R.12.77. B.E. Sernelius, Phys. Rev. *B85*, 195427 (2012).
R.12.78. K. Shizuya, arXiv:1205.6330v2 [cond-mat.mes-hall] (2012).
R.12.79. O.V. Kibis, O. Kyriienko and I.A. Shelykh, arXiv:1108.5478v1 [cond-mat.mes-hall] (2011).
R.12.80. C. Chamon, C.Y. Hou, C. Mudry, S. Ryu and L. Santas, arXiv:1206.0807v1 [cond-mat.mes-hall] (2012).
R.12.81. M.A.H. Vozmediano, M.I. Katsnelson and F. Guinea, arXiv:1003.5179v1 [cond-mat.mes-hall] (2010).
R.12.82. E. Barnes, E.H. Wang, R.E. Throkmorton and S. Das Sarma, arXiv:1401.7011v2 [cond-mat.mes-hall] (2014).
R.12.83. D. Monaco and G. Panati, arXiv: 1306.3904v3 [math-ph] (2014).
R.12.84. S. Das Sarma, P. Kin, A.K. Geim and A.H. MacDonald, Editors, "GrapheneReview," Solid State Comm., Vol. *143*, pp.1–125 (2007).
R.12.85. G. Gumbs, D. Huang and O. Roslyak, Editors, "Electronic and Photonic Propoerties of Graphene Layers and Carbon Nanoribbons," Philosophical Transactions of the Royal Society A, Vol.*368*, pp.5351–5556 (2010).
R.12.86. H. Aoki and M.S. Dresselhaus, Editors, "Physics of Graphene," Springer (2014).
R.12.87. L.E.F. Foa Torres, S. Roche and J.-S. Charlier, "Introduction to Graphene-based Nanomaterials: From Electronic Structure to Quantum Transport," Cambridge University Press (2013).
R.12.88. P. Recher and B. Trausettel. Nanotechnology **21**, 1 (2010).
R.12.89. M.I. Katsnelson, "Graphene: Carbon in Two Dimensions," Cambridge University Press (2012).
R.12.90. J.H. Warner, F. Schaffel, M.H. Rumelli and A. Bachmatiuk, "Graphene: Fundamentals and Emergent Applications," Elsevier (2013).
R.12.91. E.L. Wolf, "Graphene: A New Paradigm in Condensed Matter and Device Physics," Oxford University Press (2014).
R.12.92. C. Soldano, A. Mahmood, and E. Dujardin, Carbon **48**, 2127 (2010).
R.12.93. Y. Zhu, S. Murali, W. Cai, X. Li, J.W. Suk, J.R. Potts, and R.S. Ruoff, Advanced Materials **22** (Wiley-VCH) 3906 (2010).
R.12.94. C.N.R. Rao and A.L. Sood, Editors, "Graphene: Synthesis, Properties & Phenomena," Wiley-VCH (2013).
R.12.95. W. Choi and J.W. Lee, Editors, "Graphene Synthesis and Applications," CRC Press (2012).
R.12.96. N. Koratker, "Graphene in Composite Materials: Synthesis, Characterization & Applications," DEStech Publ. (2013).
R.12.97. A. Rashid and M. Yusoff, Editors, "Graphene Optoelectronics: Synthesis, Characterization, Properties & Applications," Wiley-VCH (2014).
R.12.98. H. Raza, Editor, "Graphene Nanoelectronics: Metrology, Synthesis, Properties & Applications," Springer (2012).
R.12.99. M.C. Lemme, Solid State Phenomena **156–158**, 499 (2010).

R.12.100. R. Murali, Editor, "Graphene Nanoelectronics: From Materials to Circuits," Springer (2012).

R.12.101. S. Shafraniuk, "Graphene: Fundamentals, Device & Applications," Pan Stanford Publ. (2015).

R.12.102. T.H. Caine, Editor, "Graphene and Carbon Nanotube Field Effect Transistors," NOVA Publ. (2011).

R.12.103. H.P. Wong and D. Akinwande, "Carbon Nanotube & Device Physics," Cambridge University Press (2011).

R.12.104. S. Alwarappan and A. Kumar, "Graphene-Based Materials: Science & Technology," CRC Press (2013).

R.12.105. T. Enoki and T. Ando, "Physics & Chemistry of Graphene: Graphene to Nanographene," CRC Press (2013).

R.12.106. V. Georgakilas, "Functionalization of Graphene," Wiley-VCH (2014).

R.12.107. Z. Liu and X. Zhou, "Graphene: Energy Storage and Conversion Applications," CRC Press (2014).

R.12.108. "Handbook of Graphene Science," CRC Press, in preparation for publication.

13

Superfluidity and Superconductivity

Abstract
Chapter 13 addresses the theory of superfluids and superconductors, involving the phenomenology of Bose condensation with macroscopic occupation of the lowest microscopic energy level. Bogoliubov and Penrose asserted that this entails the introduction of a part of the field operator that behaves as a c-number in the macroscopic limit ($<\psi(\mathbf{x},t)>\neq 0$), in apparent violation of number conservation, but consonant with the fact that the fluctuation of number in the Bose-condensed state is of the same order as number itself. In a first approximation for the dynamics of the (decoupled) condensed state, $<\psi>=\psi-\tilde{\psi}$, the non-linear Gross–Pitaevskii equation is derived for this *condensate wave function* $<\psi>$. It is seen to be in the nature of a self-consistent Hartree equation, which facilitates the identification of the *coherence length* and the core region of *vortex* motion. The *non*-condensate Green's function $\tilde{G}_1(1,1')=-i<(\tilde{\psi}(1)\tilde{\psi}^+(1'))_+>$ and the non-vanishing anomalous correlation function $\tilde{F}^*(2,1')=-i<(\tilde{\psi}^+(2)\tilde{\psi}^+(1'))_+>$ are also discussed, and their coupled equations of motion (describing the dynamics and elementary excitations of the non-condensate states) are derived in conjunction with a discussion of Landau's criterion for viscosity. The interplay of the macroscopic population of the Bose-condensed state with these non-condensate elementary excitations is discussed as well. Associated concepts of *off-diagonal long-range order* and the interpretation of $<\psi>$ as a *superfluid order parameter* are also introduced. Anderson's proposal of the Bose-condensed state as a phase-coherent wave packet superposition of number states subject to a coarse-grain average is introduced as a resolution of issues relating to number conservation.

In regard to superconductivity, it is bound *Cooper pairs* of electrons that are capable of Bose condensation and superfluid behavior. Correspondingly, the two-particle Green's function has a term involving a product of two anomalous bound-Cooper-pair condensate wave functions of the type $F(1,2)=-i<(\psi(1)\psi(2)_+>\neq 0$, such that $G_2(1,2;1',2')=F(1,2)F^+(1',2')+\tilde{G}_2(1,2;1',2')$. Here, $\tilde{G}_2$ describes the dynamics of the *non*-superfluid-condensate states and their excitations, while non-vanishing F,F^+ again concretely express off-diagonal long-range order and can also be interpreted in terms of a phase-coherent wave packet superposition of Cooper-pair states. The introduction of this form of G_2 into the G_1-equation couples the condensed state with the excitations of the non-condensate states. Taken jointly with the derivation of the dynamical equation for $F(1,2)$, the analysis of superconductivity in this chapter starts with the derivation of the celebrated Gorkov equations, which encompass many of its important features. These include the Bardeen–Cooper–Schrieffer (BCS) energy

Quantum Statistical Field Theory: Schwinger's Variational Method. Norman J. Morgenstern Horing,
Oxford University Press (2017). DOI 10.1093/oso/9780198791942.001.0001

gap and the critical temperature for superconductivity, as well as the Bogoliubov-de Gennes energy eigenfunction equations for Bogoliubon excitations in superconductors. The thermodynamics of superconductors is also discussed, along with the critical magnetic field. The approximate incorporation of a weak magnetic field into the Gorkov equations leads to Ginzburg–Landau theory and a non-linear Schrödinger-like equation for the pair wave function and the associated supercurrent, along with identification of the Cooper-pair density. Furthermore, Chapter 13 addresses the issue of the lack of gauge invariance of the London theory, notwithstanding its correct prediction of the Meissner effect (exclusion of magnetic flux from a superconductor). In an elegant analysis of the role of electromagnetic potentials and gauge transformations in the Gorkov equations facilitated by variational differentiation (due to Ambegaokar and Kadanoff), it is shown that the magnitude of the gap parameter $\overline{\Delta}$ does not change to linear order in perturbing potentials and the associated phase change of $\overline{\Delta}$ can be removed by *re-gauging* the potentials. The resulting expression for the (super)current is a manifestly gauge invariant generalization of the London equation, reducing to it exactly for a transverse vector potential. It is also shown that consistency with the equation of continuity implies the existence of an acoustic normal mode oscillation predicted by Anderson, even in the absence of external fields; and this is supplanted by the plasmon collective mode when the long-range Coulomb interaction is taken into account. Chapter 13 also discusses the penetration of quantized magnetic flux lines into Type II superconductors and the classification of superconductors based on the ratio of magnetic penetration length l to the coherence length ξ: $\kappa = l/\xi$ ($\kappa < 1/\sqrt{2}$ for Type I; $\kappa > 1/\sqrt{2}$ for Type II) in the Ginzburg–Landau equations. The interaction of parallel flux lines is also discussed. Finally, Chapter 13 addresses the subject of Josephson tunneling between superconductors in terms of a transfer Hamiltonian, which is seen to depend on the phase difference of the pair wave functions of the two superconductors on opposite sides of a barrier. This tunneling supercurrent is an oscillatory function of the phase difference even in the absence of external fields, and is further modified by the application of such fields.

13.1 Superfluid Field Operators and the Gross–Pitaevskii Equations

The phenomenon of superfluidity will be addressed here under the assumption of weak particle–particle interactions. While this is not accurate for liquid He^4, particularly in regard to its ground state properties, it is nevertheless useful in the analysis of the excitation spectrum. Also, it provides a fine vehicle to explore the phenomenology of Bose condensation, which is characterized by the macroscopic occupation of the lowest microscopic energy level, denoted by $\varepsilon_0 = 0$ (Section 6.2), at low temperature. The concomitant necessity to distinguish this macroscopically occupied level from the others can be accommodated by separating the corresponding part of the annihilation operator $\psi(\mathbf{x}, t)$, for example in Eq. (3.2.25), isolating the term $\zeta_0(\mathbf{x})$ having $E_\gamma \to \varepsilon_0 = 0$, as follows:

$$\psi(\mathbf{x}, t) = \zeta_0(\mathbf{x}) + \widetilde{\psi}(\mathbf{x}, t) \tag{13.1.1}$$

(there is no time dependence in $\zeta_0(\mathbf{x})$ since we take the lowest energy level as $\varepsilon_0 \to 0$). Here, $\widetilde{\psi}(\mathbf{x}, t)$ involves the annihilation operators for all states other than the Bose-condensed one to which $\zeta_0(\mathbf{x})$ relates. Denoting the creation and annihilation operators for occupation of the Bose-condensed state by a_0^+ and a_0, respectively, we have its macroscopic occupation number n given by

$$\langle n|a_0^+ a_0|n\rangle = n \sim V \to \infty \tag{13.1.2}$$

(V is volume). However, in position representation, with a uniform macroscopic density lodged principally in the Bose-condensed state, $\varepsilon_0 = 0$, we have

$$\langle n|\psi^+(\mathbf{x}, t)\psi(\mathbf{x}, t)|n\rangle \to \langle n|\zeta_0^+(\mathbf{x})\zeta_0(\mathbf{x})|n\rangle = \frac{n}{V}, \tag{13.1.3}$$

so it is clear that $\zeta_0(\mathbf{x})$ and $\zeta_0^+(\mathbf{x})$ each carry a factor $V^{-1/2}$,

$$\zeta_0(\mathbf{x}) = \frac{a_0}{\sqrt{V}}; \quad \zeta_0^+(\mathbf{x}) = \frac{a_0^+}{\sqrt{V}}. \tag{13.1.4}$$

Correspondingly, the canonical Boson commutation relations are given by

$$\left[a_0, a_0^+\right] = 1 \quad \text{and/or} \quad \left[\zeta_0(\mathbf{x}), \zeta_0^+(\mathbf{x})\right] = \frac{1}{V}, \tag{13.1.5}$$

from which we find that $\zeta_0(\mathbf{x})$ and $\zeta_0^+(\mathbf{x})$ essentially *commute* in the macroscopic limit, $V \to \infty$, and they can therefore be treated as c-numbers, at least in a first approximation for the Bose-condensed state. This bold assertion, attributed to N. N. Bogoliubov[1a] and O. Penrose,[1b] means that number is not conserved. For example, understanding $\zeta_0(\mathbf{x})$ as a c-number, we have $\langle\zeta_0(\mathbf{x})\rangle = \zeta_0(\mathbf{x}) \neq 0$, whereas it would be expected to vanish under strict number conservation. This may be understood physically in terms of the recognition that the fluctuation of number in the Bose-condensed state is of the same order as the number, n, itself. Recalling Eq. (6.2.15) we have

$$\left\langle [N - \langle N\rangle]^2\right\rangle^{1/2} = n, \tag{13.1.6}$$

confirming a significant departure from number conservation.

The implications of the c-number nature of ζ_0 for superfluid Green's functions are immediate. To start, we note that since $\zeta_0(\mathbf{x})$ is the only part of $\psi(\mathbf{x}, t)$ that behaves as a c-number, we have

$$\langle\psi(\mathbf{x}, t)\rangle = \zeta_0(\mathbf{x}) \neq 0 \text{ and } \langle\widetilde{\psi}(\mathbf{x}, t)\rangle = 0. \tag{13.1.7}$$

$\langle\psi(\mathbf{x}, t)\rangle$ is called the *condensate wave function* and also the *superfluid order parameter*, and $\widetilde{\psi}(\mathbf{x}, t)$ is the deviation from it that annihilates non-condensate particles. Thus, we have the existence of a non-trivial "anomalous" Green's function of the type, $G_{\frac{1}{2},\frac{0}{2}}(\mathbf{x}, t) \sim \langle\psi(\mathbf{x}, t)\rangle$ as defined in Eq. (11.2.2), and, in summary,

$$\psi(\mathbf{x}, t) = \langle\psi(\mathbf{x}, t)\rangle + \widetilde{\psi}(\mathbf{x}, t). \tag{13.1.8}$$

Bearing in mind Eq. (13.1.3), we have the superfluid density as $\frac{n(\mathbf{x})}{V} \equiv \rho_s(\mathbf{x}) = |\langle\psi(\mathbf{x},t)\rangle|^2$, and for an inhomogeneous system $\langle\psi(\mathbf{x},t)\rangle = \sqrt{\rho_s(\mathbf{x},t)}e^{i\phi(\mathbf{x},t)}$, where $\phi(\mathbf{x})$ is real, as is $\rho_s(\mathbf{x})$.

Considering interactions, the equation of motion for $\psi(\mathbf{x}_1,t_1) = \psi(1)$ is given by (Eq. (3.2.21); $\mathcal{V}(1-1') = \mathcal{V}(\mathbf{x}_1-\mathbf{x}_1')\delta(t_1-t_1')$; $1 = (\mathbf{x}_1,t_1)$),

$$\left(i\frac{\partial}{\partial t_1} - h_1^{(1)}\right)\psi(1) - \int d1'\,\mathcal{V}(1-1')\psi^+(1')\psi(1')\psi(1) = 0, \tag{13.1.9}$$

which may be averaged using the fact that $\langle\widetilde{\psi}(1)\rangle = 0$, with the result

$$\begin{aligned}\left(i\frac{\partial}{\partial t_1} - h_1^{(1)}\right)\langle\psi(1)\rangle &= \int d1'\mathcal{V}(1-1') \\ &\quad \left\langle\left(\langle\psi(1')\rangle^* + \widetilde{\psi}^+(1')\right)\left(\langle\psi(1')\rangle + \widetilde{\psi}(1')\right)\left(\langle\psi(1)\rangle + \widetilde{\psi}(1)\right)\right\rangle \\ &= \int d1'\mathcal{V}(1-1') \\ &\quad \left\{\begin{matrix}|\langle\psi(1')\rangle|^2\langle\psi(1)\rangle + \langle\widetilde{\psi}^+(1')\widetilde{\psi}(1')\widetilde{\psi}(1)\rangle + \langle\psi(1')\rangle^*\langle\widetilde{\psi}(1')\widetilde{\psi}(1)\rangle \\ + \langle\psi(1')\rangle\langle\widetilde{\psi}^+(1')\widetilde{\psi}(1)\rangle + \langle\psi(1)\rangle\langle\widetilde{\psi}^+(1')\widetilde{\psi}(1)\rangle\end{matrix}\right\}.\end{aligned} \tag{13.1.10}$$

Of the five terms in curly brackets on the right of Eq. 13.1.10, the last four involve $\widetilde{\psi}, \widetilde{\psi}^+$-operators that act on the populations of non-Bose-condensed states, which we take to be small compared to the macroscopic population of the Bose-condensed state itself at low temperature with weak interactions. Therefore, the first term on the right is dominant, and we have

$$\left(i\frac{\partial}{\partial t_1} - h_1^{(1)}\right)\langle\psi(1)\rangle = \int d2\;\mathcal{V}(1-2)\;|\langle\psi(2)\rangle|^2\langle\psi(1)\rangle, \tag{13.1.11}$$

which, together with its adjoint equation (complex conjugate), constitute the Gross–Pitaevskii equations describing the dynamics of the condensed state in a first approximation. (Although it is insightful, this analysis is not always realistic, as in the case of HeII.) Eq. (13.1.11) is in the nature of a self-consistent Hartree equation for the condensate wave function $\langle\psi(1)\rangle$. Allowing that the energy of the lowest state, ε_0, may not be zero, then the condition for macroscopic population ($f(\varepsilon_0) \to \infty$) requires that $\mu = \varepsilon_0$, so the time dependence of this singular Bose-condensed state is given by

$$\langle\psi(1)\rangle = \langle\psi(\mathbf{x})\rangle\,e^{-i\mu t_1} = \langle\psi(\mathbf{x})\rangle\,e^{-i\varepsilon_0 t_1}. \tag{13.1.12}$$

Accordingly, Eq. (13.1.11) takes the form

$$\left(\varepsilon_0 - h_1^{(1)}\right)\langle\psi(\mathbf{x}_1)\rangle = \int d^{(3)}\mathbf{x}_2\mathcal{V}(\mathbf{x}_1-\mathbf{x}_2)|\langle\psi(\mathbf{x}_2)\rangle|^2\langle\psi(\mathbf{x}_1)\rangle, \tag{13.1.13}$$

and for a model short-range delta-function potential $\mathcal{V}(\mathbf{x}_1-\mathbf{x}_2) = \lambda\delta^{(3)}(\mathbf{x}_1-\mathbf{x}_2)$, we have

$$\left(\varepsilon_0 - h_1^{(1)}\right)\langle\psi(\mathbf{x}_1)\rangle = \lambda|\langle\psi(\mathbf{x}_1)\rangle|^2\langle\psi(\mathbf{x}_1)\rangle. \tag{13.1.14}$$

From this we can estimate that, for the case of a homogeneous system, $\varepsilon_0 = \mu = \lambda\varrho_s$, which yields the superfluid density as just $\varrho_s = \frac{\mu}{\lambda}$. It is clear from this that the particle–particle interaction strength (measured here by λ) plays an important role in the determination of the superfluid density as a (finite) fraction of the total density. For an *inhomogeneous* superfluid bounded by infinite walls that force $\langle\psi(\mathbf{x})\rangle$ to vanish at $x = 0, L$, while $h_1^{(1)}(\mathbf{x}) = \frac{-\nabla^2}{2m}$ between the walls, the behavior of $\langle\psi(\mathbf{x})\rangle$ near the walls is determined by

$$\left(\varepsilon_0 + \frac{\nabla^2}{2m}\right)\langle\psi(\mathbf{x})\rangle = 0, \tag{13.1.15}$$

so that $\langle\psi(\mathbf{x})\rangle$ rises from zero over a distance $l_0 = (2m\varepsilon_0)^{-\frac{1}{2}} = (2m\mu)^{-\frac{1}{2}}$ ($\sim 10^{-8}$ cm for He^4), the *coherence length*. Deep in the central region for $L \gg l_0$, $\langle\psi(\mathbf{x})\rangle$ becomes constant and we again obtain $\varepsilon_0 = \mu = \lambda\varrho_s$, as above. In the absence of interaction ($\lambda \to 0$) there would be no such central region of uniform superfluid, presenting a conceptual problem.

To address superfluid motion, we consider the relation

$$\langle\psi(\mathbf{x},t)\rangle = \sqrt{\varrho_s(\mathbf{x},t)}e^{i\phi(\mathbf{x},t)} = \sqrt{\varrho_s(\mathbf{x},t)}e^{i[\phi(\mathbf{x})-\varepsilon_0 t]}, \tag{13.1.16}$$

and recall that particle current is given by Eq. (1.1.31) as (no charge, $\mathbf{A} \to 0$; also $\hbar \to 1$),

$$\begin{aligned}\mathbf{J} &= \frac{1}{2m}\left[\langle\psi(\mathbf{x},t)\rangle^*\left(\frac{1}{i}\nabla\langle\psi(\mathbf{x},t)\rangle\right) - \left(\frac{1}{i}\nabla\langle\psi(\mathbf{x},t)\rangle^*\right)\langle\psi(\mathbf{x},t)\rangle\right]\\ &= \frac{\varrho_s(\mathbf{x},t)}{m}\nabla\phi(\mathbf{x},t) = \varrho_s(\mathbf{x},t)\mathbf{v}_s(\mathbf{x},t),\end{aligned} \tag{13.1.17}$$

where we identify the superfluid velocity as

$$\mathbf{v}_s(\mathbf{x},t) = \frac{1}{m}\nabla\phi(\mathbf{x},t). \tag{13.1.18}$$

In the case of uniform motion $\phi(\mathbf{x}) = m\mathbf{v}_s \cdot \mathbf{x}$, and Eq. (13.1.14) takes the form $\left(\varepsilon_0 - \frac{1}{2}m\mathbf{v}_s^2\right) = \lambda\varrho_s$. More generally, $\mathbf{v}_s(\mathbf{x},t)$ is irrotational, $\nabla\times\mathbf{v}_s(\mathbf{x},t) = \frac{1}{m}\nabla\times\nabla\phi(\mathbf{x}) = 0$, and using a reference origin $\mathbf{x}_0$, we have

$$\phi(\mathbf{x}) = m\int_{\mathbf{x}_0}^{\mathbf{x}} d\mathbf{x}'\cdot\mathbf{v}_s(\mathbf{x}'). \tag{13.1.19}$$

The circulation line integral about a closed loop vanishes except for enclosed singular points, in particular *vortices* and multiply connected regions, which are characterized by quantized circulation,

$$m\oint d\mathbf{x}'\cdot\mathbf{v}_s(\mathbf{x}') = 2\pi j, (j = \text{integer}), \tag{13.1.20}$$

to accommodate the single-valuedness of $\langle\psi(\mathbf{x},t)\rangle$. The integer j measures the strength of the vortex. To examine vortex motion in greater detail, particularly for rotating bucket experiments, we employ cylindrical coordinates (r,θ) with the z-axis on the line of the vortex center. Taking the vortex **A** to have strength j, and assuming circular symmetry, we set $\phi(\mathbf{x}) = j\theta$ and Eq. (13.1.14) has the form

$$\left(\varepsilon_0 - \frac{1}{2}m(\mathbf{v}_s(r))^2 + \frac{1}{r}\frac{d}{dr}\left(r\frac{d}{dr}\right)\right)\sqrt{\varrho_s(r)} = \lambda\left(\varrho_s(r)\right)^{\frac{3}{2}}. \tag{13.1.21}$$

With $\phi(\mathbf{x}) = j\theta$, we have $\mathbf{v}_s = \left(\frac{j}{rm}\right)\widehat{\theta}$ in the azimuthal direction (unit vector $\widehat{\theta}$) and $\mathbf{J} = \left(\frac{j\varrho_s(r)}{rm}\right)\widehat{\theta}$. $\varrho_s(r)$ is determined by Eq. (13.1.21), vanishing as $r\to 0$ to avoid a singularity. There is a core region of the vortex over which $\varrho_s(r)$ rises from zero (again with a characteristic length $l_0 = 1/\sqrt{2m\varepsilon_0}$), and outside the core $\varrho_s(r)$ is readily determined from Eq. (13.1.21) as

$$\varrho_s(r) = \frac{1}{\lambda}\left[\varepsilon_0 - \frac{1}{2}m(\mathbf{v}_s(r))^2\right], \quad \text{for } r \gg l_0. \tag{13.1.22}$$

The physical necessity for finite interactions ($\lambda \neq 0$) is clear from the fact that $\varrho_s(r)$ above (Eq. (13.1.22)) would diverge if $\lambda\to 0$. This is further discussed in Reference 2, along with other aspects of the theory of superfluids and superconductors.[2]

13.2 Superfluid Green's Functions and Elementary Excitations

To analyze the interplay of the macroscopic population of the Bose-condensed state with the elementary excitations[3] of the lightly populated non-condensate states, we need an equation for $\widetilde{\psi}$, which may be obtained by subtracting Eq. (13.1.11) for $\langle\psi(1)\rangle$ from Eq. (13.1.9) for $\psi(1) = \langle\psi(1)\rangle + \widetilde{\psi}$. Of the many resulting terms we need retain only those of the lowest (linear) order in $\widetilde{\psi}^+$and $\widetilde{\psi}$, since their effects in single acts of populating or depopulating the relatively lightly populated non-condensate states are larger than the effects of neglected terms, which involve multiple acts of populating/depopulating the non-condensate states, although both are small compared to the macroscopic population of the condensate at low temperature and weak particle–particle interaction. This linearization yields the *Bogoliubov approximation* as

$$\left(i\frac{\partial}{\partial t_1} - h_1^{(1)}\right)\widetilde{\psi}(1) - \int d1''\mathcal{V}(1-1'')\left\{\begin{matrix}|\langle\psi(1'')\rangle|^2\,\widetilde{\psi}(1) + \langle\psi(1'')\rangle^*\langle\psi(1)\rangle\,\widetilde{\psi}(1'') \\ +\langle\psi(1'')\rangle\langle\psi(1)\rangle\,\widetilde{\psi}^+(1'')\end{matrix}\right\} = 0, \tag{13.2.1}$$

with its adjoint equation given by

$$\left(-i\frac{\partial}{\partial t_1'} - h_{1'}^{(1)}\right)\tilde{\psi}^+(1') - \int d1''\mathcal{V}(1'-1'')\left\{\begin{array}{c}|\langle\psi(1'')\rangle|^2\,\tilde{\psi}^+(1') + \langle\psi(1'')\rangle\langle\psi(1')\rangle^*\,\tilde{\psi}^+(1'') \\ +\,\langle\psi(1')\rangle^*\langle\psi(1'')\rangle^*\,\tilde{\psi}(1'')\end{array}\right\} = 0. \tag{13.2.2}$$

Superfluid Green's functions may be constructed in the usual way following Eq. (11.2.2) as averages of time-ordered products of the type

$$\begin{aligned}\langle(\psi(1)\psi^+(1'))_+\rangle &= \langle([\langle\psi(1)\rangle + \tilde{\psi}(1)][\langle\psi(1')\rangle^* + \tilde{\psi}^+(1')])_+\rangle \\ &= \langle\psi(1)\rangle\langle\psi(1')\rangle^* + \langle(\tilde{\psi}(1)\tilde{\psi}^+(1'))_+\rangle,\end{aligned} \tag{13.2.3}$$

where we have noted that $\langle\tilde{\psi}\rangle = 0$. Accordingly, superfluid Green's functions have the form

$$G_1(1,1') = -i\langle\psi(1)\rangle\langle\psi(1')\rangle^* + \tilde{G}_1(1,1'), \tag{13.2.4}$$

where the non-condensate Green's function, $\tilde{G}_1(1,1')$, is defined in the usual way, but with the replacements $\psi \to \tilde{\psi}$ and $\psi^+ \to \tilde{\psi}^+$. Therefore, the equation for $\tilde{G}_1$ may be obtained by multiplying Eq. (13.2.1) by $\tilde{\psi}^+(1')$, with time ordering and averaging, and the result for $\tilde{G}_1$ is given by the equation of motion,

$$\begin{aligned}\left(i\frac{\partial}{\partial t_1} - h_1^{(1)}\right)\tilde{G}_1(1,1') - \int d2\mathcal{V}(1-2)&\left\{\begin{array}{c}|\langle\psi(2)\rangle|^2\tilde{G}_1(1,1') + \langle\psi(2)\rangle^*\langle\psi(1)\rangle\,\tilde{G}_1(2,1') \\ +\,\langle\psi(2)\rangle\langle\psi(1)\rangle\,\tilde{F}^*(2,1')\end{array}\right\} \\ &= \delta(1,1').\end{aligned} \tag{13.2.5}$$

Here, we have introduced another anomalous correlation function as $\tilde{F}^*(2,1') = -i\langle(\tilde{\psi}^+(2)\tilde{\psi}^+(1'))_+\rangle$ (the symbol ">" or "<" applies when $\psi^+(1')$ is multiplied here on the right or left, respectively); its equation of motion may be obtained by multiplying Eq. (13.2.2) by $\tilde{\psi}^+(2)$ and averaging, with the result

$$\begin{aligned}\left(-i\frac{\partial}{\partial t_1} - h_1^{(1)}\right)\tilde{F}^*(1,2) - \int d1'\mathcal{V}(1-1')&\left\{\begin{array}{c}|\langle\psi(1')\rangle|^2\,\tilde{F}^*(1,2) + \langle\psi(1')\rangle\langle\psi(1)\rangle^*\,\tilde{F}^*(1',2) \\ +\,\langle\psi(1)\rangle^*\langle\psi(1')\rangle^*\,\tilde{G}_1(1',2)\end{array}\right\} \\ &= 0.\end{aligned} \tag{13.2.6}$$

Considering time-translational invariance for a uniform superfluid, the non-condensate Green's function has the form $\tilde{G}_1(\mathbf{x}_1,\mathbf{x}_1';t_1-t_1')$, and since $\mathcal{V}(1-2) \sim \delta(t_1-t_2)$ is instantaneous the factor $\langle\psi(2)\rangle\langle\psi(1)\rangle$ in the last term on the right of Eq. (13.2.5) has time dependence given by (bearing in mind Eq. (13.1.12) and $|\langle\psi(1)\rangle| \sim \varrho_s^{1/2}$):

$$\langle\psi(2)\rangle\langle\psi(1)\rangle = \varrho_s e^{-i\mu(t_2+t_1)} \to \varrho_s e^{-i2\mu t_2}. \tag{13.2.7}$$

Therefore, time-translational invariance of $G_1(\mathbf{x}_1,\mathbf{x}_1';t_1-t_1')$ carries the implication, through Eq. (13.2.5), of time-translational invariance for the function defined by

$$\tilde{\mathcal{F}}^*(\mathbf{x}_2,\mathbf{x}_1';t_2-t_1') \equiv e^{-i2\mu t_2}\tilde{F}^*(2,1')_{t_2=t_1}. \tag{13.2.8}$$

Thus, we can Fourier-analyze Eq. (13.2.5) with $t_1 - t_1' \to \omega$, and since we address the uniform case with $h_1^{(1)} = \frac{-\nabla_1^2}{2m}$ we also invoke a spatial Fourier transform, $\mathbf{x}_1 - \mathbf{x}_1' \to \mathbf{p}$, with the result

$$\left(\omega - \frac{p^2}{2m} - \varrho_s\left[\mathcal{V}(0) + \mathcal{V}(\mathbf{p})\right]\right)\widetilde{G}_1(\mathbf{p},\omega) - \varrho_s\mathcal{V}(\mathbf{p})\widetilde{\mathcal{F}}^*(\mathbf{p},\omega) = 1. \qquad (13.2.9)$$

To treat Eq. (13.2.6) on a corresponding basis, it is necessary to employ Eq. (13.2.8) for $\widetilde{F}^*$, which leads to

$$\left(2\mu - \omega - \frac{p^2}{2m} - \varrho_s\left[\mathcal{V}(0) + \mathcal{V}(\mathbf{p})\right]\right)\widetilde{\mathcal{F}}^*_{\gtrless}(\mathbf{p},\omega) - \varrho_s\mathcal{V}(\mathbf{p})\widetilde{G}_{1\gtrless}(\mathbf{p}_1,\omega) = 0. \quad (13.2.10)$$

This system of linear equations, Eqns. (13.2.9 and 13.2.10), can be employed to determine $\widetilde{G}_1$ and $\widetilde{\mathcal{F}}^*$. Furthermore, the elementary excitations of the non-condensed system are readily determined by the condition that the homogeneous counterparts of Eqns. (13.2.9 and 13.2.10) have a non-trivial solution. Writing the excitation energy relative to μ as $\varepsilon(p) = \omega - \mu$, and recalling that $\mu = \varrho_s\lambda \cong \varrho_s\mathcal{V}(0)$, we obtain

$$\varepsilon(p)^2 = (\omega - \mu)^2 = \left[\frac{p^2}{2m} + \varrho_s\mathcal{V}(p)\right]^2 - (\varrho_s\mathcal{V}(p))^2. \qquad (13.2.11)$$

For low wave numbers, $\varepsilon(p) \cong p\sqrt{\varrho_s\mathcal{V}(0)/m}$, which is sound-like, whereas for large wave numbers, $\varepsilon(p) \cong p^2/2m$, which is particle-like.

The specific detail of this $\varepsilon(p)$ spectrum is of key importance in regard to the non-viscous flow property of superfluids. As the uniform motion of a superfluid with velocity $\mathbf{v}_s$ involves a macroscopic mass, M_s, and macroscopic momentum, $\mathbf{P}_s = M_s\mathbf{v}_s$ (because of the macroscopic number of particles in the condensed state), it is very difficult to slow it. Moreover, purely internal condensate particle–particle interactions will not, by themselves, provide a resistive mechanism to center-of-mass motion. However, the elementary excitations arising from particle–particle interactions between condensate and non-condensate states can transfer, throughout the superfluid, resistive/viscous effects arising elsewhere (in particular, from motion relative to the walls containing the superfluid) by the energy and momentum that is transferred from the superfluid in the creation of such excitations. Landau[4,5] and Tisza[6] pointed out that when conservation laws prohibit the creation of such elementary excitations, there is no mechanism for resistance and the superfluid moves with no viscosity. In particular, the change of energy of a superfluid, $E = P_s^2/2M_s$, is related to its change of momentum $\mathbf{P}_s = M_s\mathbf{v}_s$, by $\delta E = \mathbf{P}_s \cdot \delta\mathbf{P}_s/M_s = \mathbf{v}_s \cdot \delta\mathbf{P}_s$, and the creation of the excitation involves energy and momentum conservation in the form $\delta E = -\varepsilon(p)$, $\delta\mathbf{P}_s = -\mathbf{p}$, so that

$$\varepsilon(p) = \mathbf{v}_s \cdot \mathbf{p} \le v_s p. \qquad (13.2.12)$$

Thus, the creation of such an excitation is only possible if $v_s \ge \varepsilon(p)/p$, so we have a critical speed, $\varepsilon(p)/p$, below which there are no excitations, no viscosity and superfluidity occurs and is unimpeded. Above the critical speed, the elementary excitations do exist and transmit viscous effects that destroy superfluidity. Eq. (13.2.12) is known as the *Landau criterion* for inviscid superfluid flow. For a non-interacting Boson gas with $\varepsilon(p) = \frac{p^2}{2m}$, we have $\frac{\varepsilon(p)}{p} = \frac{p}{2m}$, which has a critical speed of zero, so viscous excitations can occur at all finite speeds eliminating superfluid behavior. Further confirmation that the excitations responsible for viscosity *cannot* occur in the absence of condensate–non-condensate interactions may be had from the Gross–Pitaevskii equation (Eq. 13.1.11), which describes the condensate in the absence of such interactions. Its energetics can be examined employing the condensate Green's function sans interactions, $G_1(1,1') = -i\langle\psi(1)\rangle\langle\psi(1')\rangle^*$ (first term on the right of Eq. (13.2.4)) in the general energy expression of Eq. (7.6.14), with the result ($\mathcal{V}(1-1') = U_0\delta(1-1')$):

$$\langle H\rangle = \int d^{(3)}\mathbf{x}\left\{|\vec{\nabla}\langle\psi(1)\rangle|^2/2m + (U_0/2)|\langle\psi(1)\rangle|^4\right\}.$$

For the uniform, spatially translationally invariant case at hand, the plane wave solution $\langle\psi(\mathbf{x},t)\rangle = \sqrt{\rho_s}\, exp\,[i(\mathbf{p}\cdot\mathbf{x}-\omega t)]$ yields the energy per particle as $\varepsilon(\mathbf{p}) = p^2/2m + U_0\rho_s/2$. As the last term, $U_0\rho_s/2$, is a constant, it is tempting to dismiss it as a part of the chemical potential relative to which excitation energies are measured. However, this term is significant because its absence would leave $\varepsilon(p) \to p^2/2m$, which was shown above to be incapable of superfluid behavior because of its vanishing critical speed, whereas we clearly understand the Gross–Pitaevskii model to represent a perfect superfluid. The resolution of this matter lies in the recognition that the term $U_0\rho_s/2$ is an essential part of $\varepsilon(p)$ for the Gross–Pitaevskii superfluid model because it represents the internal condensate particle–particle interaction energy at the heart of this superfluid model. In fact, this point is driven home by noting that the associated critical speed for $\varepsilon(p) = p^2/2m + U_0\rho_s/2$ is infinity (rather than zero) confirming that there are no viscous excitations at any finite speed that could degrade the superfluidity of the Gross–Pitaevskii model, due to its complete lack of condensate–non-condensate particle–particle interactions.

A more realistic $\varepsilon(p)$ spectrum with interaction is depicted in Figure 13.1. As $p \to 0$, we have $\varepsilon(p) \cong p\sqrt{\varrho_s \mathcal{V}(0)/m}$, and $\varepsilon(p)/p$ is essentially the speed of the sound-like excitation. There is also a minimum, ε_0, of $\varepsilon(p)$ in the vicinity of a point p_0, where the dispersion law is (approximately) a displaced parabola. The excitations in this region are called rotons, and have critical speed $\varepsilon(p)/p \cong \varepsilon_0/p_0$. (There are also quantized vortex ring excitations.) For speeds below the lowest of the critical speeds of the various excitations, inviscid superfluidity exists.

Superfluid Green's functions are discussed in much greater detail by Fetter and Walecka[7a] from the point of view of an effective Hamiltonian K_{eff} embodying the

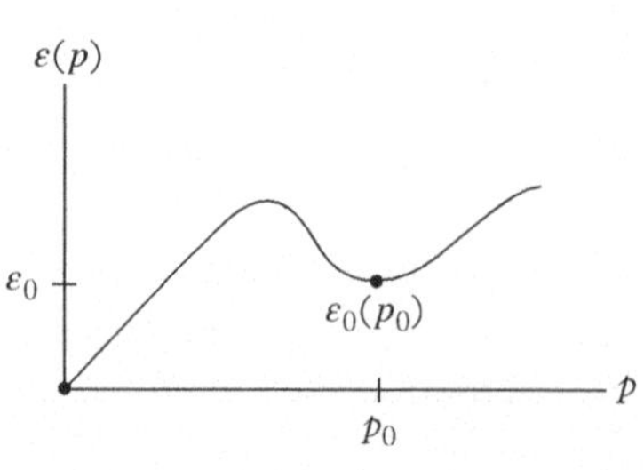

Figure 13.1: *Schematic illustration of* $\varepsilon(p)$ *as a function of* p *for liquid Helium.*

Bogoliubov assertion (in the form of Eq. (13.1.11) as a side condition for the condensate wave function), and the failure of K_{eff} to commute with the number operator reflects the breakdown of number conservation. The authoritative presentation of Abrikosov, Gorkov and Dzyaloshinskii[7b] is also highly informative and among the best. Moreover, a very general analysis of superfluid Green's functions is provided in Reference 3 (Hohenberg and Martin) using an external Boson source term (as in Eqns. (4.2.33 and 4.2.34)) in the Hamiltonian, which results in an inhomogeneous driving term in the equations of motion and also encompasses number *non*-conservation.

The Bose condensation of a macroscopic number of particles into the lowest state, ε_0, carries the implication of special, singular spectral structure for the single-particle Green's function $G_1(1,1')$ of Eq. (7.2.17). Specifically, the corresponding spectral weight in position representation $A\left(\mathbf{x}_1,\mathbf{x}_1',\omega\right)$ is given by

$$\frac{A\left(\mathbf{x}_1,\mathbf{x}_1',\omega\right)}{e^{\alpha+\beta\omega}-1} = 2\pi\,\delta(\omega-\varepsilon_0)\,\langle\psi(\mathbf{x}_1)\rangle\,\langle\psi(\mathbf{x}_1')\rangle^* + \frac{\widetilde{A}\left(\mathbf{x}_1,\mathbf{x}_1',\omega\right)}{e^{(\alpha+\beta\omega)}-1} \qquad (13.2.13)$$

(in the notation of Chapter 6 for α, β), where the first term on the right, proportional to $\delta(\omega-\varepsilon_0)$, represents the role of the Bose-condensed state and $\widetilde{A}\left(\mathbf{x}_1,\mathbf{x}_1',\omega\right)$ is the spectral weight function for the rest of the spectrum.

The special prominence of the Bose-condensed state can also be represented from another point of view. Considering expectation values $\langle\xi,n|\ldots|\xi,n\rangle$ in the lowest eigenstate of $H-\mu N$, we write

$$\begin{aligned} G_{1>}(1,1') &= -i\left\langle\xi,n|\psi(1)\psi^+(1')|\xi,n\right\rangle, \\ G_{1<}(1,1') &= -i\left\langle\xi,n|\psi^+(1')\psi(1)|\xi,n\right\rangle \end{aligned} \qquad (13.2.14)$$

(suppressing the appearance of temperature-dependent statistical factors). Inserting a complete set of states for $(n+1)$ particles, $\sum_{\xi'}|\xi',n+1><\xi',n+1|$, between $\psi(1)$ and $\psi^+(1')$ in $G_{1>}(1,1')$, we have

$$iG_{1>}(1,1') = \sum_{\xi'}\left\langle\xi,n|\psi(1)|\xi',n+1\right\rangle\left\langle\xi',n+1|\psi^+(1')|\xi,n\right\rangle, \qquad (13.2.15)$$

since strict observance of the creation, annihilation properties of ψ^+, ψ and orthogonality of states having different particle numbers assures us that intermediate states of number other than $(n+1)$ cannot contribute. One of the intermediate states, $|\xi,n+1\rangle$, is essentially identical to the condensed state ($\xi' \sim \xi$) except for the presence of an additional particle in the condensate. Separating its contribution,

$$\begin{aligned} iG_{1>}(1,1') &= \langle\xi,n|\psi(1)|\xi,n+1\rangle\left\langle\xi,n+1|\psi^+(1')|\xi,n\right\rangle \\ &\quad + \sum_{\xi'(\neq\xi)}\left\langle\xi,n|\psi(1)|\xi',n+1\right\rangle\left\langle\xi',n+1|\psi^+(1')|\xi,n\right\rangle, \end{aligned} \qquad (13.2.16)$$

we may think of the condensate wave function as

$$\langle\psi(1)\rangle = \langle\xi,n|\psi(1)|\xi,n+1\rangle \qquad (13.2.17)$$

or

$$\langle \psi(1')\rangle^* = \langle \xi, n+1|\psi^+(1')|\xi, n\rangle, \tag{13.2.18}$$

and $\widetilde{G}_{1>}(1,1')$ of Eq. (13.2.4) is given by

$$i\widetilde{G}_{1>}(1,1') = \sum_{\xi'(\neq\xi)} \langle \xi, n|\psi(1)|\xi', n+1\rangle\langle \xi', n+1|\psi^+(1')|\xi, n\rangle, \tag{13.2.19}$$

representing the one-particle Green's function encompassing the roles of all states other than that of the superfluid condensate. In Eqns. (13.2.17 and 13.2.18), the condensate wave function $\langle\psi(1)\rangle$ is interpreted as an off-diagonal number-space matrix element of $\psi(1)$, and since $\langle\psi(\mathbf{x},t)\rangle \sim \sqrt{\rho_s(\mathbf{x},t)}$ it is clear that

$$G_{1>}(1,1') = -i\,\langle\psi(\mathbf{x},t)\rangle\,\langle\psi(\mathbf{x}',t')\rangle^* + \widetilde{G}_{1>}(1,1') \tag{13.2.20}$$

does *not vanish* for large spatial separations of $\mathbf{x}$ and $\mathbf{x}'$, as do normal Green's functions and $\widetilde{G}_{1>}(1,1')$. Correspondingly, this description of superfluid coherence in terms of $\langle\psi(\mathbf{x},t)\rangle\,\langle\psi(\mathbf{x}',t')\rangle^*$ is said to represent *off-diagonal long-range order*,[8] with $\langle\psi(\mathbf{x},t)\rangle$ as the *order parameter*. Similar results obtain for $G_{1<}$ and $\widetilde{G}_{1<}$. Of course, the superfluid Green's function equations are the same as the ones above.

At this point, we have introduced two seemingly disparate points of view concerning the condensate wave function $\langle\psi(\mathbf{x},t)\rangle$. The non-vanishing of $\langle\psi(\mathbf{x},t)\rangle$ in Eqns. (13.1.8 and 13.1.10) was interpreted as a clear indication that number is not conserved. On the other hand, the interpretation of Eq. (13.2.17), $\langle\psi(\mathbf{x},t)\rangle = \langle\xi,n|\psi(1)|\xi,n+1\rangle$, is associated with strict observance of the creation/annihilation properties and orthogonality of states having different particle numbers, which is an aspect of a number-conserving theory. The key to reconciling these two points of view lies in the fact that the Bose-condensed state has huge number fluctuations (Eq. (13.1.6)), and consequently one really needs to view the condensed state in terms of wave packets that are superpositions of states having different number eigenvalues, so that they sample its strongly fluctuating number in describing the average $\langle\psi(\mathbf{x},t)\rangle$. Considering a stationary Bose-condensed ideal gas, we denote the lowest (zero-momentum and energy) state having macroscopic n-particle occupation as $|n\rangle \sim (a_0^+)^n|0\rangle$ ($|0\rangle$ is the vacuum state). Anticipating a coarse-grain average over a microscopically large but macroscopically small volume ΔV, we write (note that $\int_{\Delta V} d^{(3)}\mathbf{x} \sim \Delta V$):

$$a_0 = (\Delta V)^{-\frac{1}{2}} \int_{\Delta V} d^{(3)}\mathbf{x}\,\psi(\mathbf{x})$$

(bearing in mind from Eq. (13.1.4) that $\psi(\mathbf{x}) \sim \zeta_0(\mathbf{x}) \sim (\Delta V)^{-1/2} a_0$ for the condensed state; so that $a_0 \sim (\Delta V)^0 = 1$); moreover, following the normalization discussion at the end of Section 2.2, we have $a_0|n\rangle = n^{\frac{1}{2}}|n-1\rangle$. P. W. Anderson proposed that the Bose-condensed (BC) state may be written as a wave packet superposition of the number states $|n\rangle$, with coefficients, c_n, as

$$|BC\rangle = \sum_n c_n|n\rangle \tag{13.2.21}$$

(with the assumption that ΔV may be made large enough so that c_n and $|n\rangle$ do not depend significantly on the variables of the rest of the system, thus obtaining a local description). The coarse-grain average of $\langle\psi(\mathbf{x})\rangle = \langle BC|\psi(x)|BC\rangle$ in the BC state yields the average condensate wave function as

$$\langle\psi(\mathbf{x})\rangle \rightarrow \frac{1}{\Delta V}\int_{\Delta V} d^{(3)}\mathbf{x}\,\langle\psi(\mathbf{x})\rangle = \frac{1}{\Delta V}\sum_{n,n'} c^*_{n'}c_n \int_{\Delta V} d^{(3)}\mathbf{x}\,\langle n'|\psi(\mathbf{x})|n\rangle, \tag{13.2.22}$$

and since $\psi(\mathbf{x})$ annihilates just one particle, only those terms having $n' = n-1$ contribute, so that

$$\langle\psi(\mathbf{x})\rangle \Rightarrow \sum_n c^*_{n-1}c_n\frac{1}{\Delta V}\int d^{(3)}\mathbf{x}\,\langle n-1|\psi(\mathbf{x})|n\rangle$$

$$\Rightarrow \sum_n c^*_{n-1}c_n(\Delta V)^{-1/2}\,\langle n-1|a_0|n\rangle = \sum_n c^*_{n-1}c_n\left(\frac{n}{\Delta V}\right)^{\frac{1}{2}}. \tag{13.2.23}$$

The phases of the number states $|n\rangle$ are chosen to assure reality of the matrix element. A particular choice of c_n, which alone embodies phase, was made by Anderson to preserve phase coherence as

$$c_n = (2\pi\Delta n)^{-\frac{1}{2}}\exp\left[-\frac{1}{2}\frac{(n-\overline{n})^2}{(\Delta n)^2}\right]e^{i\phi n}, \tag{13.2.24}$$

where $\overline{n}$ is the central number value and Δn measures the spread of number values in the wave packet. This choice of c_n yields the same phase ϕ for every term of the wave packet, since $c^*_{n-1}c_n \sim e^{i\phi}$, independent of n. Consequently, the entire wave packet has the same phase,

$$\langle\psi(\mathbf{x})\rangle = |\langle\psi(\mathbf{x})\rangle|e^{i\phi}, \tag{13.2.25}$$

and the constructive interference associated with Anderson's choice of c_n, Eq. (13.2.24), builds a large wave packet (as opposed to its diminution that would occur for randomly chosen phases with concomitant destructive interference). In summary, this phase-coherent wave packet describes a prominent, non-vanishing value of $\langle\psi(\mathbf{x})\rangle$ while strictly observing the creation/annihilation properties of the field operators by means of the superposition of states of differing number values (in a spread of width Δn about a central number value n), with each contribution having the same phase, ϕ. Correspondingly, the condensate wave function given by this packet description is a state of definite phase, albeit not one of definite number.

The representation of $\langle\psi\rangle$ in terms of a phase coherent superposition of a packet of number states is discussed in detail in "Basic Notions of Condensed Matter Physics" by Anderson,[9a] and in his review article,[9b] where the ramifications of the breakdown of number conservation and the canonically conjugate roles of number and phase are explored in depth. Such spontaneous symmetry-breaking has elicited a great deal of attention in the world of physics over the past half-century and many important advances have been made, but we will not address that subject here.

13.3 Superconductivity: BCS Theory; Anomalous Green's Functions and Gorkov Equations

Superconductivity[7,10–23] involves electric charge conduction by electrons, which, as Fermions, cannot experience a Bose condensation. Nevertheless, the exceptionally long life of undriven *supercurrents* at low temperature (which has been measured in *years* in metallic rings) suggests a frictionless flow characteristic of a superfluid. For example, the resistance of mercury essentially vanishes as temperature is lowered below about 4.25°K, and an ambient magnetic field is expelled from the metal below that critical temperature: the *Meissner effect*. The key to a physical understanding of these phenomena came with the development of the Bardeen–Cooper–Schrieffer (BCS) theory of superconductivity. Cooper found that metallic electrons of opposite spin and opposite momenta experience an attractive interaction mediated by the exchange of a phonon in the vicinity of the Fermi energy, and it is stronger than Coulomb repulsion. This attraction results in a bound state of the electrons, termed a *Cooper pair*, which has net spin zero and is therefore a Boson capable of Bose condensation and superfluid behavior. Correspondingly, we are led to consider a two-particle Green's function, G_2, in a manner similar to that of Eq. (13.2.16) for G_1 (but with the Green's functions as matrices in spin-space, having subscripts (+,–) denoting the z-component of spin (up, down), respectively), and using the same number state notation we are involved with pertinent matrix elements of the type

$$G_2^<(1,2;1'2') \Rightarrow \langle \xi, n|\psi_+^+(2')\psi_-^+(1')\psi_-(1)\psi_+(2)|\xi, n\rangle . \tag{13.3.1}$$

As in the case of superfluidity discussed above, the condensed state is here understood to be viewed in terms of a phase-coherent wave packet of (Cooper-pair) number states. With due attention to the creation, annihilation properties of ψ^+, ψ and the orthogonality of states of different particle numbers, the insertion of a complete set of $n-2$ particle states in the middle of the matrix element on the right of Eq. (13.3.1) yields

$$\begin{aligned} G_2^<(1,2;1'2') &= \langle \xi, n|\psi_+^+(2')\psi_-^+(1')|\xi, n-2\rangle \langle \xi, n-2|\psi_-(1)\psi_+(2)|\xi, n\rangle \\ &\quad + \sum_{\xi'(\neq \xi)} \langle \xi, n|\psi_+^+(2')\psi_-^+(1')|\xi', n-2\rangle \langle \xi', n-2|\psi_-(1)\psi_+(2)|\xi, n\rangle . \end{aligned} \tag{13.3.2}$$

The matrix element $\langle \xi, n|\psi_+^+(1')\psi_-^+(2')|\xi, n-2\rangle$ between essentially identical states ($\xi = \xi'$) that only differ by two particles represents the amplitude for creating two particles of equal and opposite spin in the condensate of Cooper pairs. For equal space-time arguments it is proportional to the wave function of the bound pair, the "pair wave function," with double charge $e^* = 2e$ for two electrons. The last term on the right of Eq. (13.3.2) describes the two-particle Green's function $\widetilde{G}_2^<$ encompassing the roles of all states other than the Cooper-pair condensed superconducting state; for example, $\widetilde{G}_2$ includes Hartree and Hartree–Fock terms, and more. In accordance with this discussion, we introduce "anomalous" Green's functions of the type $G_{\frac{2}{2},\frac{0}{2};0;\frac{0}{2},\frac{0}{2}}$ and $G_{\frac{0}{2},\frac{2}{2};0;\frac{0}{2},\frac{0}{2}}$ (Eq. (11.2.2)) as

$$F(1,2) = -i\langle(\psi_-(1)\psi_+(2))_+\rangle \tag{13.3.3}$$

and

$$F^+(1',2') = i\langle(\psi_+^+(2')\psi_-^+(1'))_+\rangle, \tag{13.3.4}$$

whose non-vanishing values are associated with the breakdown of number conservation, and as in the case of superfluids, they may be understood in terms of a phase-coherent wave packet superposition of Cooper-pair states having a peaked spread of number values. Furthermore, we again have off-diagonal long-range order associated with the condensed superconducting/superfluid state. With the definitions of Eqns. (13.3.3 and 13.3.4), we may write Eq. (13.3.2) as

$$G_2(1,2;1',2') = F(1,2)F^+(1',2') + \widetilde{G}_2(1,2;1',2'). \tag{13.3.5}$$

Bearing in mind the antiperiodicity property of Fermion Green's functions, (Chapter 7), we recall that the advance of the time argument of ψ through $\tau = -i\beta$ (in the context of a Green's function) yields a factor $-e^{-\beta\mu}$, whereas the corresponding advance of the time argument of ψ^+ yields a factor $-e^{\beta\mu}$. This also applies to anomalous Green's functions, so that

$$F(1,2)|_{t_1=\tau} = -e^{-\beta\mu}F(1,2)|_{t_1=0} \tag{13.3.6}$$

and

$$F^+(1',2')|_{t_1'=\tau} = -e^{\beta\mu}F^+(1',2')|_{t_1'=0}. \tag{13.3.7}$$

However,

$$F(1,1)|_{t_1=\tau} = e^{-2\beta\mu}F(1,1)|_{t_1=0} \tag{13.3.8}$$

and

$$F^+(1',1')|_{t_1'=\tau} = e^{2\beta\mu}F^+(1',1')|_{t_1'=0}, \tag{13.3.9}$$

since *two* time arguments are advanced through $\tau = -i\beta$ together, involving two ψ^+-operators in $F^+(1',1')$.

Recalling the G_1-equation of motion as

$$\left(i\frac{\partial}{\partial t_1} - h_1^{(1)}\right) G_1\left(\mathbf{x}_1,t_1;\mathbf{x}_1',t_1'\right) + i\int d^{(3)}\mathbf{x}_2 V\left(\mathbf{x}_1-\mathbf{x}_2\right) G_2\left(\mathbf{x}_1t_1,\mathbf{x}_2t_1;\mathbf{x}_1't_1',\mathbf{x}_2t_1^+\right)$$
$$= \delta^{(3)}\left(\mathbf{x}_1-\mathbf{x}_1'\right)\delta(t_1-t_1'), \tag{13.3.10}$$

it is conventional to model the attractive phonon-mediated electron–electron interaction in terms of a short-range δ-function hard-core potential, as

$$\mathcal{V}(\mathbf{x}_1 - \mathbf{x}_2) = -\lambda\delta^{(3)}(\mathbf{x}_1 - \mathbf{x}_2), \tag{13.3.11}$$

where λ is a positive constant measuring the electron–electron interaction strength, with the implicit understanding that it vanishes for momenta $\mathbf{p}$ outside the range $|\,p^2 - p_F^2|/2m < \hbar\omega_D$ (ω_D is the Debye frequency), so that it acts as a projection mechanism, accepting only momenta near the Fermi surface while eliminating those outside that range. This yields the G_1-equation as

$$\left(i\frac{\partial}{\partial t_1} - h_1^{(1)}\right) G_1\left(\mathbf{x}_1, t_1; \mathbf{x}_1', t_1'\right) - i\lambda G_2\left(\mathbf{x}_1, t_1, \mathbf{x}_1 t_1; \mathbf{x}_1', t_1', \mathbf{x}_1, t_1\right)$$
$$= \delta^{(3)}\left(\mathbf{x}_1 - \mathbf{x}_1'\right)\delta(t_1 - t_1'), \tag{13.3.12}$$

and with Eq. (13.3.5), this becomes

$$\left(i\frac{\partial}{\partial t_1} - h_1^{(1)}\right) G_1\left(\mathbf{x}_1, t_1; \mathbf{x}_1', t_1'\right) - i\lambda F\left(\mathbf{x}_1, t_1; \mathbf{x}_1, t_1\right) F^+\left(\mathbf{x}_1, t_1; \mathbf{x}_1', t_1'\right)$$
$$-i\lambda\widetilde{G}_2\left(\mathbf{x}_1 t_1, \mathbf{x}_1 t_1; \mathbf{x}_1' t_1', \mathbf{x}_1 t_1\right) = \delta^{(3)}\left(\mathbf{x}_1 - \mathbf{x}_1'\right)\delta(t_1 - t_1'). \tag{13.3.13}$$

Considering time-translational invariance with $G_1\left(\mathbf{x}_1 t_1; \mathbf{x}_1' t_1'\right) = G_1\left(\mathbf{x}_1, \mathbf{x}_1'; t_1 - t_1'\right)$ as a function of $(t_1 - t_1')$ alone, to the exclusion of $(t_1 + t_1')$, and noting that the time translation of $F\left(\mathbf{x}_1 t_1; \mathbf{x}_1 t_1\right)$ through τ yields a factor $e^{-2i\tau\mu}$ (Eq. (13.3.8)), so that its time dependence has the form

$$F\left(\mathbf{x}_1 t_1; \mathbf{x}_1 t_1\right) = F\left(\mathbf{x}_1, \mathbf{x}_1\right) e^{-2i\mu t_1} \quad and \quad F^+(\mathbf{x}_1 t_1; \mathbf{x}_1 t_1) = F^+(\mathbf{x}_1, \mathbf{x}_1) e^{2i\mu t_1}, \tag{13.3.14}$$

we conclude from Eq. (13.3.12) that $\left[e^{-2i\mu t_1} F^+\left(\mathbf{x}_1 t_1; \mathbf{x}_1' t_1'\right)\right]$ is a function of $(t_1 - t_1')$ alone to the exclusion of $(t_1 + t_1')$, and may correspondingly be analyzed in terms of a Matsubara Fourier series in imaginary time with Fourier coefficient $\mathcal{F}^+(\mathbf{x}', \mathbf{x}; \nu)$ under equilibrium conditions, as

$$e^{-2i\mu t_1} F^+\left(\mathbf{x}_1, t_1; \mathbf{x}_1', t_1'\right) = \frac{1}{\tau} \sum_{\{\nu\ odd\}} \mathcal{F}^+(\mathbf{x}_1, \mathbf{x}_1'; \nu) e^{-i\pi\nu(t_1 - t_1')/\tau}, \tag{13.3.15}$$

and for $t_1' = t_1$ and $\mathbf{x}_1' = \mathbf{x}_1$, a similar relation for F yields

$$F\left(\mathbf{x}_1 t_1; \mathbf{x}_1 t_1\right) = \frac{1}{\tau} e^{-2i\mu t_1} \sum_{\{\nu\ odd\}} \mathcal{F}(\mathbf{x}_1, \mathbf{x}_1; \nu). \tag{13.3.16}$$

Defining

$$\Delta(1) = -i\lambda F(1,1); \quad \Delta^*(1) = i\lambda F^+(1,1), \tag{13.3.17}$$

we have $\Delta^*(1)|_{t_1} = \Delta^*(\mathbf{x}_1) e^{i2\mu t_1}$ and $\Delta(1)|_{t_1} = \Delta(\mathbf{x}_1) e^{-i2\mu t_1}$; and the time dependence of $\Delta^*(1) \sim e^{-i2\mu t_1}$ will be absorbed to form the time-translationally invariant function $\mathcal{F}^+$ on the right of Eq. (13.3.15) in the context of Eq. (13.3.13) (and Eq. (13.3.18)).

Anticipating that $\widetilde{G}_2$ is relatively small at low temperature with weak interactions among thinly populated *non*-condensate states, we write Eq. (13.3.13) neglecting it, in the form

$$\left(i\frac{\partial}{\partial t_1} - h_1^{(1)}\right) G_1\left(\mathbf{x}_1, t_1; \mathbf{x}_1', t_1'\right) - i\lambda F\left(\mathbf{x}_1, t_1; \mathbf{x}_1, t_1\right) F^+\left(\mathbf{x}_1, t_1; \mathbf{x}_1', t_1'\right)$$
$$= \delta^{(3)}\left(\mathbf{x}_1 - \mathbf{x}_1'\right)\delta(t_1 - t_1'), \qquad (13.3.18)$$

which can be generalized to accommodate non-equilibrium (as well as equilibrium) in the presence of time-dependent fields (potentials) in $h_1^{(1)}$. The corresponding equation of motion for F^+ may be derived from the field equation for ψ^+ (multiplying it by ψ^+, time ordering and averaging), in a manner similar to that involved in Eq. (13.2.10) (but in position-time representation for F^+ coupled to G_1, not $\widetilde{G}_1$), again *without* a Dirac delta-function driving term, with the result

$$\left(i\frac{\partial}{\partial t_1} + h_1^{(1)+}\right) F_1^+\left(\mathbf{x}_1, t_1; \mathbf{x}_1', t_1'\right) + i\lambda F^+\left(\mathbf{x}_1, t_1; \mathbf{x}_1, t_1\right) G_1\left(\mathbf{x}_1, t_1; \mathbf{x}_1', t_1'\right) = 0 \quad (13.3.19)$$

(the significance of $h_1^{(1)+}$ is discussed below, Eq. (13.3.23)). Eqns. (13.3.18 and 13.3.19) embody Gorkov's formulation of the BCS theory of superconductivity.[7,11,12] In the case of equilibrium we may also expand G_1 in a Matsubara Fourier series, $G_1(\mathbf{x}_1, \mathbf{x}_1'; t_1 - t_1') = \frac{1}{\tau}\Sigma_{\{\nu\ odd\}} e^{-i\pi\nu(t_1 - t_1')/\tau}\, \mathcal{G}_1(\mathbf{x}_1, \mathbf{x}_1'; \nu)$, and the resulting equations for the Fourier coefficients $\mathcal{F}(\mathbf{x}_1, \mathbf{x}_2; \nu)$ and $\mathcal{G}_1(\mathbf{x}_1, \mathbf{x}_2; \nu)$ are readily derived from Eqns. (13.3.16–13.3.19) as

$$\left(\frac{\pi\nu}{\tau} - h_1^{(1)}\right)\mathcal{G}_1(\mathbf{x}_1, \mathbf{x}_1'; \nu) + \Delta(\mathbf{x}_1)\mathcal{F}^+(\mathbf{x}_1, \mathbf{x}_1'; \nu) = \delta^{(3)}(\mathbf{x}_1 - \mathbf{x}_1') \qquad (13.3.20)$$

and

$$\left(\frac{\pi\nu}{\tau} + h_1^{(1)+} - 2\mu\right)\mathcal{F}^+(\mathbf{x}_1, \mathbf{x}_1'; \nu) + \Delta^*(\mathbf{x}_1)\mathcal{G}_1(\mathbf{x}_1, \mathbf{x}_1'; \nu) = 0. \qquad (13.3.21)$$

If single-particle energies are measured relative to μ, the Gorkov equations (Eqns. (13.3.18 and 13.3.19)) for the associated functions (indicated by an overhead bar) may be written as ($i\partial/\partial t \to i\partial/\partial t + \mu$; note $\overline{\Delta}(1) \equiv \Delta(\mathbf{x}_1)$ here):

$$\left(i\frac{\partial}{\partial t_1} - h_1^{(1)} + \mu\right)\overline{G}_1(1, 1') + \overline{\Delta}(1)\overline{F}^+(1, 1') = \delta(1 - 1') \qquad (13.3.22)$$

and

$$\left(i\frac{\partial}{\partial t_1} + h_1^{(1)+} - \mu\right)\overline{F}^+(1, 1') + \overline{\Delta}^*(1)\overline{G}_1(1, 1') = 0. \qquad (13.3.23)$$

which accommodate external electromagnetic fields and non-equilibrium conditions with the replacements $\nabla \to \nabla - ie\mathbf{A}/c$ in $h_1^{(1)}$ and $\nabla \to \nabla + ie\mathbf{A}/c$ in $h_1^{(1)+}$ ($\mathbf{A}$ is the vector potential) and $i\frac{\partial}{\partial t_1} \to i\frac{\partial}{\partial t_1} + eU$ in Eq. (13.3.22) and $i\frac{\partial}{\partial t_1} \to i\frac{\partial}{\partial t_1} - eU$ in Eq. (13.3.23). In terms of the equilibrium Matsubara Fourier series coefficients, we have

$$\left(\frac{\pi\nu}{\tau} - h_1^{(1)} + \mu\right)\overline{\mathcal{G}}_1(\mathbf{x}_1, \mathbf{x}_1'; \nu) + \overline{\Delta}(\mathbf{x}_1)\overline{\mathcal{F}}^+(\mathbf{x}_1, \mathbf{x}_1'; \nu) = \delta^{(3)}(\mathbf{x}_1 - \mathbf{x}_1') \quad (13.3.24)$$

and

$$\left(\frac{\pi\nu}{\tau} + h_1^{(1)+} - \mu\right)\overline{\mathcal{F}}^+(\mathbf{x}_1, \mathbf{x}_1'; \nu) + \overline{\Delta}^*(\mathbf{x}_1)\overline{\mathcal{G}}_1(\mathbf{x}_1, \mathbf{x}_1'; \nu) = 0. \quad (13.3.25)$$

Nambu[13,14] succinctly rewrote the Gorkov equations in a matrix form, defining the differential operator matrix Λ,

$$\Lambda \equiv \begin{pmatrix} i\dfrac{\partial}{\partial t_1} - h_1^{(1)} + \mu & \overline{\Delta}(1) \\ \overline{\Delta}^*(1) & i\dfrac{\partial}{\partial t_1} + h_1^{(1)+} - \mu \end{pmatrix}, \quad (13.3.26)$$

and also defining the matrix $\widehat{\mathfrak{G}}$ of normal and anomalous Green's functions for superconductors,

$$\widehat{\mathfrak{G}}(1, 1') = \begin{pmatrix} \overline{G}_1(1, 1') & \overline{F}(1, 1') \\ \overline{F}^+(1, 1') & -\overline{G}_1(1', 1) \end{pmatrix}, \quad (13.3.27)$$

with the result ($\widehat{I}$ is the unit 2×2 matrix),

$$\Lambda\widehat{\mathfrak{G}}(1, 1') = \widehat{I}\delta(1 - 1'). \quad (13.3.28)$$

An alternative derivation of the Gorkov equations involving an effective Hamiltonian, in which the $H^{(2)}$-interaction term is factorized into an anomalous Green's function and a pair of field operators, is presented in Reference 7. Number non-conservation is evident in the failure of this effective Hamiltonian to commute with the number operator and it is again appropriate to consider the Bose-condensed superconducting Cooper-pair states associated with the non-vanishing anomalous Green's functions as phase-coherent wave packet superpositions of (pair) number states.

13.4 Excitation Spectrum, Gap; Isotope Effect

Considering an infinite uniform superconductor in equilibrium[7] (momentum representation, $h_1^{(1)} \to p^2/2m$), Eqns. (13.3.24 and 13.3.25) become

$$\left(\frac{\pi\nu}{\tau} - \frac{p^2}{2m} + \mu\right)\overline{\mathcal{G}}_1(\mathbf{p}; \nu) + \overline{\Delta}\overline{\mathcal{F}}^+(\mathbf{p}; \nu) = 1, \quad (13.4.1)$$

$$\left(\frac{\pi\nu}{\tau} + \frac{p^2}{2m} - \mu\right)\overline{\mathcal{F}}^+(\mathbf{p}; \nu) + \overline{\Delta}^*\overline{\mathcal{G}}_1(\mathbf{p}; \nu) = 0, \quad (13.4.2)$$

where $\overline{\boldsymbol{\Delta}} \equiv \overline{\boldsymbol{\Delta}}(\mathbf{x}_1)$ is independent of $\mathbf{x}_1$, and according to Eqns. (13.3.17 and 13.3.15), we have

$$\overline{\mathbf{\Delta}} = \overline{\mathbf{\Delta}}(\mathbf{x}_1) = -\frac{i\lambda}{\tau}\sum_{\{\nu\ odd\}}\overline{\mathcal{F}}(\mathbf{x}_1 - \mathbf{x}_1 = 0;\nu). \tag{13.4.3}$$

A bit of simple algebra yields the solutions of Eqns. (13.4.1 and 13.4.2) as

$$\overline{\mathcal{G}}_1(\mathbf{p};\nu) = \frac{\frac{\pi\nu}{\tau} + \frac{p^2}{2m} - \mu}{\left[\frac{\pi\nu}{\tau}\right]^2 - \left[\frac{p^2}{2m} - \mu\right]^2 - \left[\overline{\Delta}\right]^2} \tag{13.4.4}$$

and

$$\overline{\mathcal{F}}^+(\mathbf{p};\nu) = -\frac{\overline{\Delta}^*}{\left[\frac{\pi\nu}{\tau}\right]^2 - \left[\frac{p^2}{2m} - \mu\right]^2 - \left[\overline{\Delta}\right]^2}. \tag{13.4.5}$$

The retarded counterpart of these functions may be obtained with the replacement $\frac{\pi\nu}{\tau} \to \omega + i\delta$ (Eq. 7.5.11), so that

$$\overline{\mathcal{G}}_1(\mathbf{p};\omega + i\delta) = \frac{\omega + \frac{p^2}{2m} - \mu}{[\omega]^2 - \left[\frac{p^2}{2m} - \mu\right]^2 - \left[\overline{\Delta}\right]^2 + i\delta} \tag{13.4.6}$$

and

$$\overline{\mathcal{F}}^+(\mathbf{p};\omega + i\delta) = -\frac{\overline{\Delta}^*}{[\omega]^2 - \left[\frac{p^2}{2m} - \mu\right]^2 - \left[\overline{\Delta}\right]^2 + i\delta}, \tag{13.4.7}$$

from which it is apparent that the superconductor excitation spectrum is given by

$$\omega = \pm\sqrt{\left[\frac{p^2}{2m} - \mu\right]^2 + |\overline{\Delta}|^2}. \tag{13.4.8}$$

These elementary excitations above the ground state of the superconductor are known as *Bogoliubons* and the fact that there is a finite gap $|\overline{\Delta}|$ renders them difficult to access unless sufficient energy is supplied. It is correspondingly difficult (energetically) to disturb the stability of the Cooper-pair Bose-condensed superconducting ground state.

The determination of the gap $\overline{\Delta}$ in the uniform equilibrium case is readily carried out using Eqns. (13.4.3 and 13.4.5), with the result

$$\overline{\Delta} = -\overline{\Delta}\frac{\lambda}{\beta}\sum_{\{\nu\ odd\}}\int\frac{d^{(3)}\mathbf{p}}{(2\pi)^3}\frac{1}{\left[\frac{\pi\nu}{\tau}\right]^2 - \left[\frac{p^2}{2m} - \mu\right]^2 - \left[\overline{\Delta}\right]^2}. \tag{13.4.9}$$

The sum $\sum_\nu$ can be carried out using the Mittag-Leffler expansion of $\tan\left(\frac{\pi x}{2}\right) = \frac{4x}{\pi}\sum_{\{\nu\ odd\}}\frac{1}{\nu^2-x^2}$, and introducing the definition $\varepsilon \equiv p^2/2m - \mu$, we have

$$1 = \lambda \int_0^{\omega_D} d\varepsilon N(\varepsilon)\frac{\tanh\left(\frac{\beta}{2}\sqrt{\varepsilon^2 + |\bar{\Delta}|^2}\right)}{\sqrt{\varepsilon^2 + |\bar{\Delta}|^2}}, \tag{13.4.10}$$

where $N(\varepsilon)$ is the density of states for electrons of one spin species of the degenerate spin states, $\int \frac{d^{(3)}\mathrm{p}}{(2\pi)^3} \to \int d\varepsilon N(\varepsilon)$, with $N(\mu) = \frac{mp_F}{2\pi^2}$. Since we understand the phonon-mediated attractive electron–electron interaction responsible for superconductivity to be effective in a neighborhood of the Fermi energy of width $|\varepsilon| \sim \hbar\omega_D \to \omega_D$, we approximate $N(\varepsilon) \to N(\mu)$ and determine the gap at $T = 0$ using

$$1 = \lambda N(\mu)\int_0^{\omega_D} d\varepsilon\frac{\tanh\left(\frac{\beta}{2}\sqrt{\varepsilon^2 + |\bar{\Delta}|^2}\right)}{\sqrt{\varepsilon^2 + |\bar{\Delta}|^2}} \to \lambda N(\mu)\int_0^{\omega_D} d\varepsilon\frac{1}{\sqrt{\varepsilon^2 + |\bar{\Delta}|^2}}, \tag{13.4.11}$$

which yields $|\bar{\Delta}|$ in the case of weak interaction $\lambda N(\mu) \ll 1$ as

$$1 = \lambda N(\mu)\,\mathrm{arcsinh}\,\frac{\omega_D}{|\bar{\Delta}|} \Longrightarrow |\bar{\Delta}| \cong 2\omega_D \exp(-1/\lambda N(\mu)). \tag{13.4.12}$$

This is the celebrated gap equation of the Bardeen–Cooper–Schrieffer (BCS) theory of superconductivity. (The temperature dependence of the gap $|\bar{\Delta}|$ can be obtained from Eq. (13.4.11) by retaining β on the left.)[15] On the other hand, we can also obtain the critical temperature T_c at which $|\bar{\Delta}|$ vanishes from Eq. (13.4.11) as

$$1 = \lambda N(\mu)\int_0^{\omega_D} d\varepsilon\frac{\tanh\left(\frac{\beta_c\varepsilon}{2}\right)}{\varepsilon} \Longrightarrow 1 = \lambda N(\mu)\int_0^{\frac{\beta_c\omega_D}{2}} dx\frac{\tanh x}{x}, \tag{13.4.13}$$

where we have set $x \equiv \beta_c\varepsilon/2$ and $\beta_c \equiv 1/kT_c$. The last integral has been evaluated[7] with the result ($\bar{\Delta}$ is the gap at $T = 0$):

$$T_c = 1.14\frac{\omega_D}{k_B}\exp\left(-\frac{1}{\lambda N(\mu)}\right) = 0.57\frac{|\bar{\Delta}|}{k_B}. \tag{13.4.14}$$

This result explains the *isotope effect*, since the Debye frequency, ω_D, is proportional to the inverse square root of the mass of the ions of the periodic lattice, and so is $T_c \sim \omega_D$ according to Eq. (13.4.14), in agreement with experimental data.

It is, of course, also of interest to determine the stationary time-independent energy eigenfunctions associated with the homogeneous Gorkov equations, Eqns. (13.3.22, 13.3.23, 13.3.26 and 13.3.27), with $i\frac{\partial}{\partial t_1} \to E$. Introducing the eigenvector $\binom{u}{v}$ in Nambu's 2D space, we have

$$\begin{pmatrix} -h_1^{(1)} + \mu & \overline{\Delta}(1) \\ \overline{\Delta}^*(1) & h_1^{(1)+} - \mu \end{pmatrix} \begin{pmatrix} u(\mathbf{x}_1) \\ v(\mathbf{x}_1) \end{pmatrix} = -E \begin{pmatrix} u(\mathbf{x}_1) \\ v(\mathbf{x}_1) \end{pmatrix}, \tag{13.4.15}$$

or

$$\begin{aligned} \left(h_1^{(1)} - \mu\right) u(\mathbf{x}_1) - \overline{\Delta}(\mathbf{x}_1) v(\mathbf{x}) &= Eu(\mathbf{x}_1) \\ \left(-h_1^{(1)+} + \mu\right) v(\mathbf{x}_1) - \overline{\Delta}^*(\mathbf{x}_1) u(\mathbf{x}_1) &= Ev(\mathbf{x}_1). \end{aligned} \tag{13.4.16}$$

These are the *Bogoliubov-de Gennes equations* (except for the sign of $\overline{\Delta}$, which is determined by the consistency condition, Eq. (13.4.10), in which the sign is irrelevant). These equations can also be derived directly from the equations of motion for the field operators using the effective Hamiltonian approach with the $H^{(2)}$-interaction factorized into an anomalous Green's function and a pair of field operators. As both $h_1^{(1)}$ and $h_1^{(1)+}$ are Hermitian, the 2×2 matrix on the left of Eq. (13.4.15) is Hermitian and the eigenvectors are orthogonal. Moreover, with $\binom{u}{v}$ an eigenvector for eigenvalue E, we also have $\binom{v*}{u*}$ an eigenvector for eigenvalue $-E$. These eigenvectors may be used to form the matrix of a unitary transformation (the *Bogoliubov transformation*) that diagonalizes the effective Hamiltonian. For a translationally invariant system in the absence of a magnetic field, the eigenenergies are the Bogoliubons of Eq. (13.4.8) and the eigenstates $\binom{u}{v}$ follow from Eqns. (13.4.15 and 13.4.16).

13.5 Thermodynamics of Superconductors

The dependence of the grand potential W on the attractive phonon-mediated electron–electron interaction that induces Cooper-pair formation, condensation and superconductivity is conveniently analyzed using Eq. (7.6.18),

$$W - W(0) = -\frac{\beta}{2} \int_0^1 \frac{d\lambda'}{\lambda'} \int d^{(3)}\mathbf{x}' \int d^{(3)}\mathbf{x}'' \mathcal{V}^{(\lambda')} \left(\mathbf{x}' - \mathbf{x}''\right) \overline{G}_2^{(\lambda')} (1', 1''; 1'^+, 1''^+)_{t_1' = t_1''}. \tag{13.5.1}$$

Using Eq. (13.3.5) with the neglect of $\widetilde{G}_2$, we obtain

$$\begin{aligned} W - W(0) &= -\frac{\beta}{2} \int_0^1 \frac{d\lambda'}{\lambda'} \int d^{(3)}\mathbf{x}' \int d^{(3)}\mathbf{x}'' \mathcal{V}^{(\lambda')} \left(\mathbf{x}' - \mathbf{x}''\right) F(1', 1'') F^+(1', 1'') \\ &= \frac{\beta}{2} \int_0^1 d\lambda' \int d^{(3)}\mathbf{x}' F(1', 1') F^+(1', 1'), \end{aligned} \tag{13.5.2}$$

where we have used $\mathcal{V}^{(\lambda')}(\mathbf{x}' - \mathbf{x}'') = -\lambda' \delta^{(3)} (\mathbf{x}' - \mathbf{x}'')$. Furthermore, for a translationally invariant system the final $\mathbf{x}'$-integral yields a volume factor, and with Eq. (13.3.17), we have

$$\frac{W - W(0)}{Vol} = \frac{\beta}{2} \int_0^1 \frac{d\lambda'}{\lambda'^2} |\overline{\Delta}|^2 = -\frac{\beta}{2} \int_0^1 d\left(\frac{1}{\lambda'}\right) |\overline{\Delta}|^2, \tag{13.5.3}$$

which converges since $\overline{\Delta} = \overline{\Delta}(\lambda')$ vanishes rapidly as $\lambda' \to 0$. Considering Eq. (13.4.11) at $T = 0$,

$$\frac{1}{\lambda'} = N(\mu) \int_0^{\omega_D} d\varepsilon \frac{1}{\sqrt{\varepsilon^2 + |\overline{\Delta}|^2}}$$

and

$$d\left(\frac{1}{\lambda'}\right) = N(\mu) \int_0^{\omega_D} d\varepsilon \; d(|\overline{\Delta}|^2) \frac{d\left(\varepsilon^2 + |\overline{\Delta}|^2\right)^{-1/2}}{d\left(|\overline{\Delta}|^2\right)}, \tag{13.5.4}$$

where the range of energy integration $0 \to \omega_D$ is chosen in accordance with the fact that the phonon-mediated attractive electron–electron interaction is effective only within a range $\hbar\omega_D$ of the Fermi surface. Setting $z = |\overline{\Delta}|^2$, Eqns. (13.5.3 and 13.5.4) yield

$$\begin{aligned} \frac{W - W(0)}{Vol} &= -\frac{\beta}{2} N(\mu) \int_0^{\omega_D} d\varepsilon \int_0^{|\overline{\Delta}|^2} dz\, z \frac{d\left(\varepsilon^2 + z\right)^{-1/2}}{dz} \\ &= -\frac{\beta}{2} N(\mu) \int_0^{\omega_D} d\varepsilon \left\{ \frac{|\overline{\Delta}|^2}{\sqrt{\varepsilon^2 + |\overline{\Delta}|^2}} - 2\left(\sqrt{\varepsilon^2 + |\overline{\Delta}|^2} - \varepsilon\right) \right\}, \end{aligned} \tag{13.5.5}$$

where the z-integral was done by a parts integration. Recognizing that $|\overline{\Delta}| \ll \varepsilon$ Eq. (13.4.12), the final ε-integration is approximately given by[7,15]

$$\frac{W - W(0)}{Vol} \cong \frac{\beta}{2} N(\mu) \overline{\Delta}^2, \tag{13.5.6}$$

where $\overline{\Delta}$ is real and evaluated at $T = 0$. This yields the BCS condensate energy (relative to the normal ground state; Eq. (6.1.30)) as

$$\frac{E - E(0)}{Vol} \cong -\frac{1}{2} N(\mu) \overline{\Delta}^2. \tag{13.5.7}$$

Since this represents the energy density of the superconducting state at $T = 0$, the application of a magnetic field H_c may be expected to destroy superconductivity by supplying a matching energy density,

$$\frac{H_c^2}{8\pi} \cong \frac{1}{2} N(\mu) \overline{\Delta}^2, \tag{13.5.8}$$

so the critical magnetic field to restore the normal state at $T = 0$ is taken as

$$H_c \cong 2\sqrt{\pi N(\mu)}\, \overline{\Delta}. \tag{13.5.9}$$

A more detailed analysis using Eq. (13.5.3) and the β-dependence of Eq. (13.4.11) yields the temperature dependence of the critical field, $H_c(T)$.[7,15]

13.6 Ginzburg–Landau Theory

Notwithstanding the fact that the Ginzburg–Landau theory[7b,15,16] was originally introduced on a phenomenological basis with a well-reasoned ansatz for the Free Energy functional of the superconducting phase in thermodynamic equilibrium in the presence of a magnetic field, we address this subject here by examining the more general equilibrium Gorkov equations for the Matsubara Fourier series coefficients, Eqns. (13.3.24 and 13.3.25), in the magnetic field with vector potential[16] $\mathbf{A}(\mathbf{x})$ as

$$\left[\frac{\pi\nu}{\tau}+\left(\frac{\left[\nabla_1-\frac{ie}{c}\mathbf{A}(\mathbf{x}_1)\right]^2}{2m}\right)+\mu\right]\overline{\mathcal{G}}_1(\mathbf{x}_1,\mathbf{x}_1';\nu)+\overline{\Delta}(\mathbf{x}_1)\overline{\mathcal{F}}^+(\mathbf{x}_1,\mathbf{x}_1';\nu)=\delta^{(3)}(\mathbf{x}_1-\mathbf{x}_1'),\tag{13.6.1}$$

and

$$\left[-\frac{\pi\nu}{\tau}+\left(\frac{\left[\nabla_1+\frac{ie}{c}\mathbf{A}(\mathbf{x}_1)\right]^2}{2m}\right)+\mu\right]\overline{\mathcal{F}}^+(\mathbf{x}_1,\mathbf{x}_1';\nu)-\overline{\Delta}^*(\mathbf{x}_1)\overline{\mathcal{G}}_1(\mathbf{x}_1,\mathbf{x}_1';\nu)=0,\tag{13.6.2}$$

where

$$\overline{\Delta}^*(\mathbf{x}_1)=\frac{i\lambda}{\tau}\sum_{\{\nu\ odd\}}\overline{\mathcal{F}}^+(\mathbf{x}_1,\mathbf{x}_1;\nu).\tag{13.6.3}$$

Defining a decoupled Green's function, $\overline{\mathcal{G}}_1^0$, that includes the magnetic field (but excludes particle–particle coupling) by

$$\left[\frac{\pi\nu}{\tau}+\left(\frac{\left[\nabla_1-\frac{ie}{c}\mathbf{A}(\mathbf{x}_1)\right]^2}{2m}\right)+\mu\right]\overline{\mathcal{G}}_1^0(\mathbf{x}_1,\mathbf{x}_1';\nu)=\delta^{(3)}(\mathbf{x}_1-\mathbf{x}_1'),\tag{13.6.4}$$

and defining a "τ-reversed" counterpart, $\widehat{\overline{\mathcal{G}}_1^0}$, by

$$\left[-\frac{\pi\nu}{\tau}+\left(\frac{\left[\nabla_1+\frac{ie}{c}\mathbf{A}(\mathbf{x}_1)\right]^2}{2m}\right)+\mu\right]\widehat{\overline{\mathcal{G}}_1^0}(\mathbf{x}_1,\mathbf{x}_1';\nu)=\delta^{(3)}(\mathbf{x}_1-\mathbf{x}_1')\tag{13.6.5}$$

(note that $\widehat{\overline{\mathcal{G}}_1^0}$ differs from $\overline{\mathcal{G}}_1^0$ only by reversal of the signs of the frequency variable and the vector potential, alternatively, they are complex conjugates ($\tau=-i\beta$)), we may rewrite Eqns. (13.6.1 and 13.6.2) as

$$\overline{\mathcal{G}}_1(\mathbf{x}_1,\mathbf{x}_1';\nu)=\overline{\mathcal{G}}_1^0(\mathbf{x}_1,\mathbf{x}_1';\nu)-\int d^{(3)}\mathbf{x}_2\overline{\mathcal{G}}_1^0(\mathbf{x}_1,\mathbf{x}_2;\nu)\overline{\Delta}(\mathbf{x}_2)\overline{\mathcal{F}}^+(\mathbf{x}_2,\mathbf{x}_1';\nu)\tag{13.6.6}$$

and

$$\overline{\mathcal{F}}^{+}(\mathbf{x}_1,\mathbf{x}_1';\nu) = \int d^{(3)}\mathbf{x}_2\, \widehat{\overline{\mathcal{G}}_1^0}(\mathbf{x}_1,\mathbf{x}_2;\nu)\overline{\Delta}^*(\mathbf{x}_2)\,\overline{\mathcal{G}}_1(\mathbf{x}_2,\mathbf{x}_1';\nu). \tag{13.6.7}$$

Substitution of Eq. (13.6.7) into Eq. (13.6.6) yields

$$\begin{aligned}\overline{\mathcal{G}}_1(\mathbf{x}_1,\mathbf{x}_1';\nu) &= \overline{\mathcal{G}}_1^0(\mathbf{x}_1,\mathbf{x}_1';\nu)\\ &-\int d^{(3)}\mathbf{x}_2 \int d^{(3)}\mathbf{x}_2'\, \overline{\mathcal{G}}_1^0(\mathbf{x}_1,\mathbf{x}_2;\nu)\overline{\Delta}(\mathbf{x}_2)\,\widehat{\overline{\mathcal{G}}_1^0}(\mathbf{x}_2,\mathbf{x}_2';\nu)\overline{\Delta}^*(\mathbf{x}_2')\,\overline{\mathcal{G}}_1(\mathbf{x}_2',\mathbf{x}_1';\nu),\end{aligned} \tag{13.6.8}$$

and substitution of this into Eq. (13.6.7) yields

$$\begin{aligned}\overline{\mathcal{F}}^{+}(\mathbf{x}_1,\mathbf{x}_1';\nu) &= \int d^{(3)}\mathbf{x}_2\, \widehat{\overline{\mathcal{G}}_1^0}(\mathbf{x}_1,\mathbf{x}_2;\nu)\overline{\Delta}^*(\mathbf{x}_2)\,\overline{\mathcal{G}}_1^0(\mathbf{x}_2,\mathbf{x}_1';\nu) - \int d^{(3)}\mathbf{x}_2 \int d^{(3)}\mathbf{x}_2' \int d^{(3)}\mathbf{x}_2''\\ &\times \widehat{\overline{\mathcal{G}}_1^0}(\mathbf{x}_1,\mathbf{x}_2;\nu)\overline{\Delta}^*(\mathbf{x}_2)\,\overline{\mathcal{G}}_1^0(\mathbf{x}_2,\mathbf{x}_2';\nu)\overline{\Delta}(\mathbf{x}_2')\,\widehat{\overline{\mathcal{G}}_1^0}(\mathbf{x}_2',\mathbf{x}_2'';\nu)\overline{\Delta}^*(\mathbf{x}_2'')\,\overline{\mathcal{G}}_1(\mathbf{x}_2'',\mathbf{x}_1';\nu).\end{aligned} \tag{13.6.9}$$

Eqns. (13.6.8 and 13.6.9) are exact coupled integral equations within the Gorkov framework. On the assumption that the phonon-mediated electron–electron interaction is weak, so that $\overline{\Delta}(\mathbf{x}_1)$, $\overline{\Delta}^*(\mathbf{x}_1)$ are small, we approximate $\overline{\mathcal{G}}_1 \to \overline{\mathcal{G}}_1^0$ on the far right of both Eqns. (13.6.8) and (13.6.9), reducing the solutions to integration of these iterations, except for the consistent determination of $\overline{\Delta}(\mathbf{x})$ and $\overline{\Delta}^*(\mathbf{x})$, which will be assumed to be slowly varying functions of $\mathbf{x}$. To proceed, we employ Eq. (13.6.3) for $\overline{\Delta}^*(\mathbf{x})$ jointly with $\overline{\mathcal{F}}^{+}$ to obtain a non-linear integral equation for $\overline{\Delta}^*(\mathbf{x})$:

$$\begin{aligned}\overline{\Delta}^*(\mathbf{x}_1) &= \frac{i\lambda}{\tau}\sum_{\{odd\ \nu\}}\Bigg[\int d^{(3)}\mathbf{x}_2\, \widehat{\overline{\mathcal{G}}_1^0}(\mathbf{x}_1,\mathbf{x}_2;\nu)\overline{\Delta}^*(\mathbf{x}_2)\,\overline{\mathcal{G}}_1^0(\mathbf{x}_2,\mathbf{x}_1;\nu)\\ &-\int d^{(3)}\mathbf{x}_2 \int d^{(3)}\mathbf{x}_2' \int d^{(3)}\mathbf{x}_2''\\ &\times \widehat{\overline{\mathcal{G}}_1^0}(\mathbf{x}_1,\mathbf{x}_2;\nu)\overline{\Delta}^*(\mathbf{x}_2)\,\overline{\mathcal{G}}_1^0(\mathbf{x}_2,\mathbf{x}_2';\nu)\overline{\Delta}(\mathbf{x}_2')\,\widehat{\overline{\mathcal{G}}_1^0}(\mathbf{x}_2',\mathbf{x}_2'';\nu)\overline{\Delta}^*(\mathbf{x}_2'')\,\overline{\mathcal{G}}_1^0(\mathbf{x}_2'',\mathbf{x}_1;\nu)\Bigg].\end{aligned} \tag{13.6.10}$$

Gorkov's analysis of Eq. (13.6.10) proceeds with the recognition that all of the Green's functions involve a weak, slowly varying magnetic field with a phase factor $C(\mathbf{x}_1,\mathbf{x}_2)$ (Eq. (5.6.33)),

$$\overline{\mathcal{G}}_1^0(\mathbf{x}_1,\mathbf{x}_2;\nu) = C(\mathbf{x}_1,\mathbf{x}_2)\overline{\mathcal{G}}_1'(\mathbf{x}_1-\mathbf{x}_2;\nu);\ C(\mathbf{x}_1,\mathbf{x}_2) = \exp\left(\frac{ie}{\hbar c}\int_{\mathbf{x}_2}^{\mathbf{x}_1} d\mathbf{x}\cdot\mathbf{A}(\mathbf{x})\right), \tag{13.6.11}$$

and under the condition of slow variation, $\overline{\mathcal{G}}_1'$ is approximately spatially translationally invariant. Furthermore, with the condition of weakness, $\hbar\omega_c \ll \mu$, Landau quantization effects are negligible and $\overline{\mathcal{G}}_1'(\mathbf{x}_1 - \mathbf{x}_2; \nu)$ can be taken in the field free limit, which we denote by $\overline{\mathcal{G}}_0^0(\mathbf{x}_1 - \mathbf{x}_2; \nu)$, so that

$$\overline{\mathcal{G}}_1^0(\mathbf{x}_1, \mathbf{x}_2; \nu) = \left[\exp\left(\frac{ie}{\hbar c}\int_{\mathbf{x}_2}^{\mathbf{x}_1} d\mathbf{x}\cdot\mathbf{A}(\mathbf{x})\right)\right]\overline{\mathcal{G}}_0^0(\mathbf{x}_1 - \mathbf{x}_2; \nu). \tag{13.6.12}$$

As $\overline{\mathcal{G}}_0^0$ is assumed to be devoid of quantum magnetic field effects and the phase factor $\exp\left[\frac{ie}{\hbar c}\int_{\mathbf{x}_2}^{\mathbf{x}_1} d\mathbf{x}\cdot\mathbf{A}(\mathbf{x})\right]$ describes the lack of spatial translational invariance involved in the changing momentum direction of circular electron motion in a magnetic field, Eq. (13.6.12) is considered a classical approximation. Since the Meissner effect entails the presence of a magnetic field in a superconductor being limited to a shallow depth of penetration δ, this weak-field approximation scheme also means that the radius of curvature of the circular magnetic orbit is large compared to the penetration depth.

Recalling that $\widehat{\overline{\mathcal{G}}_1^0}$ differs from $\overline{\mathcal{G}}_1^0$ only by reversal of the signs of the frequency variable and the vector potential, the first integrand on the right of Eq. (13.6.10) has the form

$$\begin{aligned}&\widehat{\overline{\mathcal{G}}_1^0}(\mathbf{x}_1, \mathbf{x}_2; \nu)\overline{\Delta}^*(\mathbf{x}_2)\overline{\mathcal{G}}_1^0(\mathbf{x}_2, \mathbf{x}_1; \nu)\\ &\quad = \exp\left[\frac{(2e)i}{\hbar c}\int_{\mathbf{x}_1}^{\mathbf{x}_2} d\mathbf{x}\cdot\mathbf{A}(\mathbf{x})\right]\widehat{\overline{\mathcal{G}}_0^0}(\mathbf{x}_1, \mathbf{x}_2; \nu)\overline{\Delta}^*(\mathbf{x}_2)\overline{\mathcal{G}}_0^0(\mathbf{x}_2, \mathbf{x}_1; \nu).\end{aligned} \tag{13.6.13}$$

The explicit appearance of "2*e*" in the exponent (which arises from the two Green's functions appearing jointly) is a clear analytic manifestation that the effective charge of a Cooper pair is $2e$. It should also be noted that the approximation of slow variation means that $\mathbf{A}(\mathbf{x})$ is also slowly varying, so that $\int_{\mathbf{x}_1}^{\mathbf{x}2} d\mathbf{x}\cdot\mathbf{A}(\mathbf{x}) \cong \mathbf{A}(\mathbf{x}_1)\cdot(\mathbf{x}_2 - \mathbf{x}_1)$, whence

$$\begin{aligned}&\widehat{\overline{\mathcal{G}}_1^0}(\mathbf{x}_1, \mathbf{x}_2; \nu)\overline{\Delta}^*(\mathbf{x}_2)\overline{\mathcal{G}}_1^0(\mathbf{x}_2, \mathbf{x}_1; \nu)\\ &\quad = \exp\left[\frac{-(2e)i}{\hbar c}\mathbf{A}(\mathbf{x}_1)\cdot(\mathbf{x}_1 - \mathbf{x}_2)\right]\widehat{\overline{\mathcal{G}}_0^0}(\mathbf{x}_1 - \mathbf{x}_2; \nu)\overline{\Delta}^*(\mathbf{x}_2)\overline{\mathcal{G}}_0^0(\mathbf{x}_2 - \mathbf{x}_1; \nu) \qquad (13.6.14)\\ &\quad = \widehat{\overline{\mathcal{G}}_0^0}(\mathbf{x}_1 - \mathbf{x}_2; \nu)\overline{\mathcal{G}}_0^0(\mathbf{x}_2 - \mathbf{x}_1; \nu)\exp\left[(\mathbf{x}_2 - \mathbf{x}_1)\cdot\left(\boldsymbol{\nabla}_1 + \frac{2ei}{\hbar c}\mathbf{A}(\mathbf{x}_1)\right)\right]\overline{\Delta}^*(\mathbf{x}_1),\end{aligned} \tag{13.6.15}$$

where we have used the space-translation operator $\overline{\Delta}^*(\mathbf{x}_2) = e^{(\mathbf{x}_2 - \mathbf{x}_1)\cdot\boldsymbol{\nabla}_1}\overline{\Delta}^*(\mathbf{x}_1)$ (Chapter 4), and noted that $\exp\left[\frac{-(2e)i}{\hbar c}\mathbf{A}(\mathbf{x}_1)\cdot(\mathbf{x}_1 - \mathbf{x}_2)\right]\exp\left[(\mathbf{x}_2 - \mathbf{x}_1)\cdot\boldsymbol{\nabla}_1\right] = \exp\left[(\mathbf{x}_2 - \mathbf{x}_1)\cdot\left(\boldsymbol{\nabla}_1 + \frac{2ei}{\hbar c}\mathbf{A}(\mathbf{x}_1)\right)\right]$ since $\mathbf{A}(\mathbf{x}_1)$ is slowly varying and essentially commutes with $\boldsymbol{\nabla}_1$ (see Reference 15 for a fuller, more general derivation). Thus, the first term on the right of Eq. (13.6.10) may be written as (set $\mathbf{R} = \mathbf{x}_2 - \mathbf{x}_1$):

$$\left\{\begin{matrix} \textit{First Term} \\ \textit{of Eq. } (13.6.10) \end{matrix}\right\} =$$
$$-\frac{\lambda}{\beta}\sum_{\{odd\,\nu\}}\int d^{(3)}\mathbf{R}\,\widehat{\overline{\mathcal{G}}_0^0}(-\mathbf{R};\nu)\overline{\mathcal{G}}_0^0(\mathbf{R};\nu)\exp\left[\mathbf{R}\cdot\left(\nabla_1+\frac{2ei}{\hbar c}\mathbf{A}(\mathbf{x}_1)\right)\right]\overline{\Delta}^*(\mathbf{x}_1). \tag{13.6.16}$$

Employing Eq. (7.4.24) for the zero-field normal Green's function with $\omega \to \frac{\pi\nu}{\tau} \to \frac{i\pi\nu}{\beta}$, and recalling that $\widehat{\overline{\mathcal{G}}_0^0}$ is the complex conjugate of $\overline{\mathcal{G}}_0^0$, the right-hand side of Eq. (13.6.16) involves the geometric series ($\nu_{odd} \to 2n+1$),

$$\begin{aligned}\sum_{\{\nu\ odd\}}\widehat{\overline{\mathcal{G}}_0^0}(-\mathbf{R};\nu)\overline{\mathcal{G}}_0^0(\mathbf{R};\nu) &= \frac{m^2}{(2\pi R)^2}\sum_{\{\nu\ odd\}}e^{-\frac{2\pi R}{v_F\beta}\nu} = \frac{m^2}{(2\pi R)^2}e^{-\frac{2\pi R}{v_F\beta}}\sum_{n=0}^{\infty}e^{-\frac{4\pi R}{v_F\beta}n} \\ &= \frac{m^2}{8\pi^2R^2}\frac{1}{\sinh\left(\dfrac{2\pi R}{v_F\beta}\right)}.\end{aligned} \tag{13.6.17}$$

Thus, Eq. (13.6.16) becomes

$$\left\{\begin{matrix} \textit{First Term} \\ \textit{of Eq. } (13.6.10) \end{matrix}\right\} = -\frac{\lambda m^2}{8\pi^2\beta}\int d^{(3)}\mathbf{R}\frac{\exp\left[\mathbf{R}\cdot\left(\nabla_1+\frac{2ei}{\hbar c}\mathbf{A}(\mathbf{x}_1)\right)\right]}{R^2\sinh\left(\dfrac{2\pi R}{v_F\beta}\right)}\overline{\Delta}^*(\mathbf{x}_1). \tag{13.6.18}$$

In the approximation under consideration, $\mathbf{A}(\mathbf{x}_1)$ is slowly varying and small as well as $\overline{\Delta}^*(\mathbf{x})$ so we expand the exponential to second order, noting that the first-order integrand is an odd function of $\mathbf{R}$ yielding null contribution. The resulting zero-order and second-order terms are given by

$$\left\{\begin{matrix} \textit{First Term} \\ \textit{of Eq. } (13.6.10) \\ \textit{-zero order} \end{matrix}\right\} = -\frac{\lambda m^2}{8\pi^2\beta}\int d^{(3)}\mathbf{R}\frac{1}{R^2\sinh\left(\dfrac{2\pi R}{v_F\beta}\right)}\overline{\Delta}^*(\mathbf{x}_1) \tag{13.6.19}$$

and

$$\left\{\begin{matrix} \textit{First Term} \\ \textit{of Eq. } (13.6.10) \\ \textit{-second order} \end{matrix}\right\} = -\frac{\lambda m^2}{16\pi^2\beta}\int d^{(3)}\mathbf{R}\frac{1}{R^2\sinh\left(\dfrac{2\pi R}{v_F\beta}\right)}\left[\mathbf{R}\cdot\left(\nabla_1+\frac{2ei}{\hbar c}\mathbf{A}(\mathbf{x}_1)\right)\right]^2\overline{\Delta}^*(\mathbf{x}_1). \tag{13.6.20}$$

The zero-order term of Eq. (13.6.19) is divergent as $R \to 0$ because no provision has yet been made to take account of the restriction that our assumed δ-function interaction potential is to vanish outside an energy region of width $\hbar\omega_D$ about the Fermi energy, μ. This problem is readily resolved by recognizing that this zero-order coefficient of $\overline{\Delta}^*(\mathbf{x}_1)$

is just the coefficient of $-\frac{\lambda}{\beta}\sum_{\nu\, odd}\overline{\mathcal{F}}^{+}(\mathbf{x}_1,\mathbf{x}_1;\nu)$ (Eq. (13.6.3)) expanded to first order in $\overline{\Delta}^{*}(\mathbf{x}_1)$, evaluated in the absence of a gap: It is the coefficient of $\overline{\Delta}^{*}=\overline{\Delta}$ calculated on the right-hand sides of Eqns. (13.4.9 and 13.4.10) with $\overline{\Delta}\to 0$ in the ε-integrand (for arbitrary temperature T) as

$$\lambda\int_0^{\omega_D} d\varepsilon N(\varepsilon)\frac{\tanh\left(\frac{\beta}{2}\sqrt{\varepsilon^2+|\overline{\Delta}|^2}\right)}{\sqrt{\varepsilon^2+|\overline{\Delta}|^2}}\longrightarrow \lambda N(\mu)\int_0^{\omega_D}\frac{d\varepsilon}{\varepsilon}\tanh\left(\frac{\beta\varepsilon}{2}\right)=\lambda N(\mu)\ln\left(\frac{1.14\hbar\omega_D}{k_BT}\right)$$
$$=\lambda N(\mu)\ln\left(\frac{1.14\hbar\omega_D}{k_BT_c}\right)-\lambda N(\mu)\ln\left(1+\frac{T-T_c}{T_c}\right)\cong 1+\lambda N(\mu)\frac{T_c-T}{T_c},\quad (13.6.21)$$

where we have also used the defining relation for T_c, Eqns. (13.4.13 and 13.4.14), and have expanded the result to first order in $[(T_c-T)/T_c]$ considering that the approximation of slow variation is only reasonable in the vicinity of T_c. Thus, the zero-order term is

$$\left\{\begin{array}{c}\textit{First Term}\\ \textit{of Eq. (13.6.10)}\\ \textit{-zero order}\end{array}\right\}\cong\left(1+\lambda N(\mu)\frac{T_c-T}{T_c}\right)\overline{\Delta}^{*}(\mathbf{x}_1).\qquad (13.6.22)$$

The "1" term on the right of Eq. (13.6.22) just cancels $\overline{\Delta}^{*}(\mathbf{x}_1)$ on the left of Eq. (13.6.10). The second-order term (Eq. (13.6.20)) is convergent and is given in terms of a closed-form integration leading to the Riemann Zeta function, $\zeta(3)$, as

$$\left\{\begin{array}{c}\textit{First Term}\\ \textit{of Eq. (13.6.10)}\\ \textit{-second order}\end{array}\right\}=\frac{\lambda m^2 v_F^3\beta^2}{3(2\pi)^4}\int_0^\infty dx\frac{x^2}{\sinh x}\left(\nabla_1+\frac{2ei}{\hbar c}\mathbf{A}(\mathbf{x}_1)\right)^2\overline{\Delta}^{*}(\mathbf{x}_1),$$
$$=\frac{7}{6}\frac{\lambda m^2 v_F^3\beta^2\zeta(3)}{(2\pi)^4}\left(\nabla_1+\frac{2ei}{\hbar c}\mathbf{A}(\mathbf{x}_1)\right)^2\overline{\Delta}^{*}(\mathbf{x}_1).\qquad (13.6.23)$$

Since the second (and last) term on the right of Eq. (13.6.10) is already third order in $\overline{\Delta},\overline{\Delta}^{*}$, it suffices to neglect further positional variation in $\overline{\Delta},\overline{\Delta}^{*}$ and approximate the Green's functions by their weak (zero) field limit, $\overline{\mathcal{G}}_1^0\to\overline{\mathcal{G}}_0^0$, with the result

$$\left\{\begin{array}{c}\textit{Second Term}\\ \textit{of Eq. (13.6.10)}\end{array}\right\}\cong\frac{\lambda}{\beta}\overline{\Delta}^{*}(\mathbf{x}_1)|\overline{\Delta}^{*}(\mathbf{x}_1)|^2\sum_{\{odd\,\nu\}}\int d^{(3)}\mathbf{x}_2\int d^{(3)}\mathbf{x}_2'\int d^{(3)}\mathbf{x}_2''$$
$$\left\{\widehat{\mathcal{G}}_0^0(\mathbf{x}_1-\mathbf{x}_2;\nu)\,\overline{\mathcal{G}}_0^0(\mathbf{x}_2-\mathbf{x}_2';\nu)\,\widehat{\mathcal{G}}_0^0(\mathbf{x}_2'-\mathbf{x}_2'';\nu)\,\overline{\mathcal{G}}_0^0(\mathbf{x}_2''-\mathbf{x}_1;\nu)\right\}.$$
$$(13.6.24)$$

As the sum and integrals on the right of Eq. (13.6.24) all involve only zero-field Green's functions exclusively, we can again evaluate this quantity by appealing to Eq. (13.6.3) for $-\frac{\lambda}{\beta}\sum_{\{odd\ \nu\}}\overline{\mathcal{F}}^{+}(\mathbf{x}_1,\mathbf{x}_1;\nu)$ as it is evaluated for zero field on the right of Eqns. (13.4.9–13.4.11), but this time picking off the coefficient of the cubic term $\sim \overline{\Delta}|\Delta|^2$, with the result

$$\left\{\begin{array}{c}\textit{Second Term}\\ \textit{of Eq. (13.6.10)}\end{array}\right\} \cong \overline{\Delta}^*(\mathbf{x}_1)|\overline{\Delta}(\mathbf{x}_1)|^2\lambda N(\mu)$$

$$\times\left[\text{coefficient of } |\overline{\Delta}|^2 \text{ in } \int_0^{\omega_D} d\varepsilon\left(\frac{\tanh\left(\frac{\beta\varepsilon}{2}\sqrt{1+\left(\frac{|\overline{\Delta}|}{\varepsilon}\right)^2}\right)}{\varepsilon\sqrt{1+\left(\frac{|\overline{\Delta}|}{\varepsilon}\right)^2}}-\frac{\tanh\left(\frac{\beta\varepsilon}{2}\right)}{\varepsilon}\right)\right],$$

$$\cong \overline{\Delta}^*(\mathbf{x}_1)|\overline{\Delta}(\mathbf{x}_1)|^2\cdot\lambda N(\mu)\frac{\beta}{4}\int_0^{\omega_D}\frac{d\varepsilon}{2\varepsilon^2}\left(\operatorname{sech}^2\left(\frac{\beta\varepsilon}{2}\right)-\frac{\tanh\left(\frac{\beta\varepsilon}{2}\right)}{\beta\varepsilon/2}\right), \quad (13.6.25)$$

where we have subtracted off the zero-order term evaluated above and expanded the integrand to exhibit it in a form that is readily seen to be convergent at the lower limit. The integrated result is

$$\left\{\begin{array}{c}\textit{Second Term}\\ \textit{of Eq. (13.6.10)}\end{array}\right\} \cong -\overline{\Delta}^*(\mathbf{x}_1)|\overline{\Delta}(\mathbf{x}_1)|^2\lambda N(\mu)\beta_c^2\frac{7\zeta(3)}{8\pi^2}. \quad (13.6.26)$$

Combining Eqns. (13.6.10, 13.6.22, 13.6.23 and 13.6.26), we finally obtain the result (recall $N(\mu) = mp_F/2\pi^2$):

$$\left(1-\frac{T}{T_c}\right)\overline{\Delta}^*(\mathbf{x}_1)+\frac{7\zeta(3)v_F^2\beta_c^2}{48\pi^2}\left(\nabla_1+\frac{2ei}{\hbar c}\mathbf{A}(\mathbf{x}_1)\right)^2\overline{\Delta}^*(\mathbf{x}_1)-\frac{7\zeta(3)\beta_c^2}{8\pi^2}\overline{\Delta}^*(\mathbf{x}_1)|\overline{\Delta}(\mathbf{x}_1)|^2=0, \quad (13.6.27)$$

and since this is valid just below T_c ($T_c - T \ll T_c$), we have set $\beta \to \beta_c = 1/kT_c$ in the second and third terms on the left (but not the first). Bearing in mind that $\overline{\Delta}^*(\mathbf{x}_1) = i\lambda F^+(1,1)$ is proportional to the amplitude for creating a Cooper pair at $\mathbf{x}_1$, Eq. (13.6.27) can be interpreted as a non-linear Schrödinger-like equation for the associated pair wave function $\Psi^*(\mathbf{x}_1)$ normalized as

$$\Psi^*(\mathbf{x}_1) = \sqrt{\frac{7\zeta(3)\beta_c^2 n}{8\pi^2}}\,\overline{\Delta}^*(\mathbf{x}_1), \quad (13.6.28)$$

where n is the total number density of electrons. Thus, Eq. (13.6.27) is rewritten in the form

$$\frac{1}{4m}\left(\nabla_1+i\frac{2e}{\hbar c}\mathbf{A}(\mathbf{x}_1)\right)^2\Psi^*(\mathbf{x}_1)+\frac{1}{R}\left[\frac{T_c-T}{T_c}-\frac{1}{n}|\Psi(\mathbf{x}_1)|^2\right]\Psi^*(\mathbf{x}_1)=0, \quad (13.6.29)$$

where the derived Cooper-pair mass "$2m$" appears appropriately, and its charge "$2e$" also appears as mandated by the preceding equations, and

$$R = \frac{7\zeta(3)\beta_c^2 \mu}{6\pi^2}. \tag{13.6.30}$$

A straightforward calculation of the supercurrent density $\mathbf{J}_s$ using Eq. (10.7.4) with the Green's function of Eq. (13.6.6) yields the result that could be anticipated in terms of the pair wave function (Eq. (1.1.31)):

$$\mathbf{J}_s(\mathbf{x}_1) = -\frac{i2e}{4m}\left(\Psi^*(\mathbf{x}_1)\nabla_1\Psi(\mathbf{x}_1) - \Psi(\mathbf{x}_1)\nabla_1\Psi^*(\mathbf{x}_1)\right) - \frac{(2e)^2}{2mc}\mathbf{A}(\mathbf{x}_1)|\Psi(\mathbf{x}_1)|^2. \tag{13.6.31}$$

Since $|\Psi(\mathbf{x}_1)|^2$, in the coefficient of $\mathbf{A}(\mathbf{x}_1)$ on the right of Eq. 13.6.31, plays the role of an effective superconducting Cooper-pair density, n_s, we can identify it using Eq. (13.6.28) as

$$n_s = \frac{7\zeta(3)|\Delta|^2\beta_c^2}{8\pi^2}n. \tag{13.6.32}$$

(n is the equilibrium density for electrons for a single spin species.) Although Landau quantization has been eliminated in our approximation scheme, linearization of the Ginzburg–Landau equations ((Eq. 13.6.29) dropping the $|\Psi^2|$– term) results in an eigenvalue equation completely analogous to that of Landau quantization.[17]

The Ginzburg–Landau equations clearly describe the dynamics of the superconducting bound Cooper pairs as those of single quantum particles of mass $2m$ and charge $2e$. They were originally proposed on a phenomenological basis with parameters that are accurately determined microscopically by the Gorkov theory. While the derivation presented here is restricted by several approximations, further work has been done to lift some of the restrictions and extend the theory, particularly in regard to broadening the temperature range of applicability and to include the effects of random impurity scatterings (Reference 15, p.314), "dirty"[17,18] superconductors (Reference 17, Section 8-1.1; Reference 18, Section 8-3), paramagnetic [15,17,18] impurities (Reference 15, p.357; Reference 17, Section 8-2; Reference 18, Chapter 8) and gapless superconductivity[19,20] (Reference 19, Chapter 18; Reference 20, Chapter 2). Note that Eq. (13.6.29) *cannot* be extended to time-dependent fields by simply adding a Schrödinger-type term $i\partial\Psi(\mathbf{x}_1, t)/\partial t$ on the left-hand side.[15,16,20] (Reference 15, p.363; Reference 20 p.36 and references cited therein). However, the proper development of time-dependent Ginzburg–Landau theory is discussed in the books by Tinkham (Reference 17, Section 8-3) and Gulian and Zharkov (Reference 20, Chapter 7), and in the article by Werthamer in Parks' book (Reference 12b, Chapter 6, Section IV).

13.7 London Theory; Meissner Effect

Having initiated the discussion of superconductivity on a microscopic basis as involving superfluid motion of the condensate of Cooper pairs and the superconducting energy gap, it is now appropriate to point out its connection to the earlier phenomenological theory advanced by F. and H. London to explain the Meissner effect and "perfect" conductivity. The London theory devolves upon the assumption that the pair wave function

$\Psi(\mathbf{x}_1)$ is "rigid" under perturbations by a transverse vector potential $\mathbf{A}(\mathbf{x}_1)$, such that $\nabla_1 \cdot \mathbf{A}(\mathbf{x}_1) = 0$. In this context, "rigidity" means that $\Psi(\mathbf{x}_1)$ retains the value it had in the absence of a magnetic field even when a (weak) magnetic field is applied. In this case, the first term on the right of Eq. (13.6.31) vanishes, leading to

$$\mathbf{J}_s(\mathbf{x}_1) = -\frac{(2e)^2 n_s}{2mc}\mathbf{A}(\mathbf{x}_1) = -\frac{2e^2 n_s}{mc}\mathbf{A}(\mathbf{x}_1). \tag{13.7.1}$$

(This differs from London's original proposed equation only in the recognition that the Cooper pairs—which were unknown when London's theory was advanced—have charge $2e$ and mass $2m$.) It is immediately apparent that the London theory is tied to a specific gauge, which violates the requirement that $\mathbf{J}_s$ be gauge invariant and defies charge/current conservation. Notwithstanding this unusual commitment to a specific gauge, the London theory had a major success in predicting the Meissner effect: that magnetic flux is expelled from superconductors. Considering the magnetostatic equation $\nabla \times \mathbf{B} = \frac{4\pi}{c}\mathbf{J}_s$, we have

$$\nabla \times (\nabla \times \mathbf{B}) = \nabla(\nabla \cdot \mathbf{B}) - \nabla^2\mathbf{B} = \frac{4\pi}{c}\nabla \times \mathbf{J}_s = -\frac{4\pi(2e)^2 n_s}{2mc^2}\nabla \times \mathbf{A} = -\frac{4\pi(2e)^2 n_s}{2mc^2}\mathbf{B},$$

or, since $\nabla \cdot \mathbf{B} = 0$,

$$\nabla^2\mathbf{B} = \frac{4\pi(2e)^2 n_s}{(2m)c^2}\mathbf{B}. \tag{13.7.2}$$

This equation describes the decay of a magnetic field into a superconducting region at a distance z from an interface with a normal region as follows:

$$B(z) = B(0)e^{-z/l} \quad \text{with} \quad l = \sqrt{\frac{(2m)c^2}{4\pi(2e)^2 n_s}}, \tag{13.7.3}$$

where l is the penetration depth of the magnetic field into the superconductor. Since magnetic lines of flux cannot begin or end ($\nabla \cdot \mathbf{B} = 0$), they are simply expelled from the superconductor into the normal region: the Meissner effect. Typically, the magnetic field penetrates such (soft) superconductors only to a depth of $l \sim 500$ Å. (The Ginzburg–Landau theory predicts that the Meissner effect is not fully active in "hard" superconductors, admitting a higher level of magnetic flux penetration.) Finally, it should be noted that London's "perfect" conductivity equation $\frac{d\mathbf{v}_s}{dt} = \frac{e\mathbf{E}}{m}$ emerges as the time derivative of Eq. (13.7.1),

$$\frac{\partial \mathbf{J}_s}{\partial t} = -\frac{(2e)^2 n_s}{2m} \cdot \frac{1}{c}\frac{\partial \mathbf{A}}{\partial t} = \frac{(2e)^2 n_s}{2m}\mathbf{E}, \tag{13.7.4}$$

with a null scalar potential and $\mathbf{J}_s = n_s(2e)\mathbf{v}_s$ (excluding self-consistency considerations to be discussed later), again recognizing that Cooper pairs have charge $2e$ and mass $2m$. The perfect conductivity equation indicates that a supercurrent, once started, continues indefinitely after the electric field is turned off since $d\mathbf{v}_s/dt$ then vanishes and $\mathbf{v}_s$ (hence $\mathbf{J}_s$) remains unchanged thereafter.

13.8 Electromagnetic Current Response of Superconductors: Gauge Invariance and Collective Modes

The early theoretical background has provided much insight into the phenomenology of superconductivity, and is quite accurate in its description of the subject in the chosen gauge. However, it is *not* manifestly gauge invariant and that presents a serious problem in regard to number conservation. This issue was elegantly discussed by Ambegaokar and Kadanoff in Reference 21, and we start by focusing on the inclusion of electromagnetic potentials in the Gorkov equations.[21] Multiplying Eq. (13.3.22) (in the space-time matrix sense) from the right by $\overline{G}_1^{-1}$ and also eliminating $\overline{F}^{+}$ by solving Eq. (13.3.23), we have

$$\overline{G}_1^{-1}(1,1') = \overline{G}_0^{-1}(1,1') + \overline{\Delta}(1)\overline{G}_0(1',1)\overline{\Delta}^{*}(1'). \tag{13.8.1}$$

(Note that the arguments of $\overline{G}_0(1',1)$ on the right are reversed as they arise from $\widehat{\mathcal{G}_1^0}$ of Eqns. (13.6.7 and 13.6.9) or, alternatively, from $h_1^{(1)+}$ of Eq. (13.3.23)). $\overline{G}_0$ (1,1′) includes the electromagnetic potentials, which may depend on time as well as positional variables, but it is devoid of inter-particle interactions. Moreover, we recall from Eqns. (13.3.17 and 13.3.3) that

$$\overline{\Delta}(1) = -i\lambda\overline{F}(1,1) \rightarrow -\lambda < (\psi_-(1)\psi_+(1))_+ > \tag{13.8.2}$$

and

$$\overline{\Delta}^{*}(1) = i\lambda\overline{F}^{+}(1,1) \rightarrow -\lambda < (\psi_+^{+}(1)\psi_-^{+}(1))_+ > \tag{13.8.3}$$

(bear in mind that $\psi_+^{+}(1)\psi_-^{+}(1) = -\psi_-^{+}(1)\psi_+^{+}(1)$, etc., because we are dealing with Fermion operators).

Furthermore, considering a gauge transformation (Eq. (1.1.32) with $f \equiv -\Lambda$),

$$\begin{aligned} \mathbf{A}(\mathbf{x},t) &\rightarrow \mathbf{A}'(\mathbf{x},t) = \mathbf{A}(\mathbf{x},t) - \nabla\Lambda(\mathbf{x},t) \quad \text{and} \\ \phi(\mathbf{x},t) &\rightarrow \phi'(\mathbf{x},t) = \phi(\mathbf{x},t) + \frac{1}{c}\frac{\partial\Lambda(\mathbf{x},t)}{\partial t}, \end{aligned} \tag{13.8.4}$$

the quantized field variables transform as $\psi(\mathbf{x},t) \rightarrow \psi'(\mathbf{x},t) = \exp\left[-\frac{ie}{\hbar c}\Lambda(\mathbf{x},t)\right]\psi(\mathbf{x},t)$. Accordingly, the Green's functions transform as

$$\overline{G}_{\left\{{1 \atop 0}\right\}}(1,1') \rightarrow \overline{G}'_{\left\{{1 \atop 0}\right\}}(1,1') = \exp\left[-\frac{ie}{\hbar c}\left[\Lambda(1) - \Lambda(1')\right]\right]\overline{G}_{\left\{{1 \atop 0}\right\}}(1,1'), \tag{13.8.5}$$

and we also have

$$\overline{\Delta}(1) \rightarrow \overline{\Delta}'(1) = \exp\left[-\frac{2ie}{\hbar c}\Lambda(1)\right]\overline{\Delta}(1) \tag{13.8.6}$$

and

$$\overline{\Delta}^*(1) \rightarrow \overline{\Delta}'^*(1) = \exp\left[\frac{2ie}{\hbar c}\Lambda(1)\right]\overline{\Delta}^*(1). \tag{13.8.7}$$

Noting that the arguments of $\overline{G}_0$ on the far right of Eq. (13.8.1) are reversed, these relations (Eqns. 13.8.5–13.8.7) yield a gauge-transformed counterpart of Eq. (13.8.1) as

$$\overline{G}_1'^{-1}(1,1') = \overline{G}_0'^{-1}(1,1') + \exp\left[-\frac{ie}{\hbar c}\Lambda(1)\right]\overline{\Delta}(1)\overline{G}_0(1',1)\overline{\Delta}^*(1')\exp\left[\frac{ie}{\hbar c}\Lambda(1')\right], \tag{13.8.8}$$

from which it is clear that the inverse Green's functions $\overline{G}^{-1}_{\{{1 \atop 0}\}}(1,1')$ transform as

$$\overline{G}^{-1}_{\{{1 \atop 0}\}}(1,1') \rightarrow \overline{G}'^{-1}_{\{{1 \atop 0}\}}(1,1') = \exp\left[-\frac{ie}{\hbar c}\left[\Lambda(1)-\Lambda(1')\right]\right]\overline{G}^{-1}_{\{{1 \atop 0}\}}(1,1'), \tag{13.8.9}$$

just the same as the Green's functions themselves. These results will be useful in the analysis of current to follow below.

In Chapter 3 (Eq. 3.4.9), we noted that the linear current response to an impressed electromagnetic field involves matrix elements of the type $< \left[\mathbf{J}^0(\mathbf{x},t),\mathbf{J}^0(\mathbf{x}',t')\right]\cdot\mathbf{A}(\mathbf{x}',t')$, with $\mathbf{J}^0(\mathbf{x},t) = (\hbar Q/2mi)\lim_{\mathbf{x}'\rightarrow\mathbf{x}}\left[(\nabla-\nabla')\,\psi^+(\mathbf{x}',t)\psi(\mathbf{x},t)\right]$, where ψ,ψ^+ are determined in the absence of the electromagnetic field. Considering that $\lim_{\mathbf{x}'\rightarrow\mathbf{x}}$ is taken after $(\nabla-\nabla')$ is applied, this means that we need to use Green's functions that are quartic in ψ,ψ^+ jointly, with four distinct space points. For our present purpose it is convenient to generate such two-particle Green's functions in terms of variational differentiation of a non-equilibrium one-particle Green's function, $\overline{G}_1(1,1')$, with respect to a non-local instantaneous "source" potential $U(2',2)$. Recalling Eq. (9.4.4), we have ($U\rightarrow 0$ after differentiation),

$$\frac{\delta\overline{G}_1(1,1')}{\delta U(2',2)} = -\overline{G}_2(1,2;1',2'^+) + \overline{G}_1(1,1')\overline{G}_1(2,2'^+). \tag{13.8.10}$$

On the other hand, direct variational differentiation of Eq. (13.8.1) yields

$$\begin{aligned}\frac{\delta\overline{G}_1^{-1}(1,1')}{\delta U(2',2)} &= -\delta(1,2')\delta(2,1') + \overline{\Delta}(1)\frac{\delta\overline{G}_0(1',1)}{\delta U(2',2)}\overline{\Delta}^*(1')\\ &+\frac{\delta\overline{\Delta}(1)}{\delta U(2',2)}\overline{G}_0(1',1)\overline{\Delta}^*(1') + \overline{\Delta}(1)\overline{G}_0(1',1)\frac{\delta\overline{\Delta}^*(1')}{\delta U(2',2)},\end{aligned} \tag{13.8.11}$$

where we have noted that

$$\frac{\delta\overline{G}_0^{-1}(1,1')}{\delta U(2',2)} = -\delta(1,2')\delta(2,1'), \tag{13.8.12}$$

which also implies that

$$\frac{\delta \overline{G}_0(1',1)}{\delta U(2',2)} = \overline{G}_0(1,2)\overline{G}_0(1',2'). \tag{13.8.13}$$

(This result in the absence of interparticle interaction is also readily obtained by installing a Hartree–Fock approximation for $\overline{G}_2$ in Eq. (13.8.10.)) Thus Eq. (13.8.11) takes the form

$$\frac{\delta \overline{G}_1^{-1}(1,1')}{\delta U(2',2)} = -\delta(1,2')\delta(2,1') + \overline{\Delta}(1)\overline{G}_0(1,2)\overline{G}_0(1',2')\overline{\Delta}^*(1')$$
$$+\frac{\delta\overline{\Delta}(1)}{\delta U(2',2)}\overline{G}_0(1',1)\overline{\Delta}^*(1') + \overline{\Delta}(1)\overline{G}_0(1',1)\frac{\delta\overline{\Delta}^*(1')}{\delta U(2',2)}. \tag{13.8.14}$$

Operating on Eq. (13.8.14) with multiplication by $\overline{G}_1$ from both the right and the left (in the space-time matrix sense of Eq. (4.5.4)), and using Eq. (13.3.23) in integral form to identify $\overline{F}$, we obtain

$$\frac{\delta \overline{G}_1(1,1')}{\delta U(2',2)} = \overline{G}_1(1,2')\overline{G}_1(2,1') + \overline{F}(1,2)\overline{F}^+(2',1') + i\int d\overline{1}\,\overline{G}_1(1,\overline{1})\frac{\delta\overline{\Delta}(\overline{1})}{\delta U(2',2)}\overline{F}^+(\overline{1},1')$$
$$+i\int d\overline{1}\,\overline{F}(1,\overline{1})\frac{\delta\overline{\Delta}^*(\overline{1})}{\delta U(2',2)}\overline{G}_1(\overline{1},1'). \tag{13.8.15}$$

Comparing this result with Eq. (13.8.10), it is clear that the Gorkov factorization of Eqns. (13.3.5–13.3.18), $\overline{G}_2 \to \overline{FF}^+$, augmented by the Fock term, $\overline{G}_1\overline{G}_1$, is represented by the first two terms on the right of Eq. (13.8.15). The last two terms on the right represent the gap dependence on electromagnetic potentials, and are important in gauge considerations relating to charge/current conservation.

To proceed, information about $\delta\overline{\Delta}/\delta U$ and $\delta\overline{\Delta}^*/\delta U$ is needed, and may be obtained from the equations for $\overline{F}(1,1) = i\overline{\Delta}(1)/\lambda$ and $\overline{F}^+(1,1) = -i\overline{\Delta}^*(1)/\lambda$ (for example, Eq. (13.3.23) written in integral form or Eq. (13.6.7) in time-representation) with the result

$$\frac{\delta\overline{\Delta}(1)}{\delta U(2',2)} = i\lambda\overline{G}_0(1,2')\int d\overline{1}\,\overline{G}_0(2,\overline{1})\overline{G}_1(1,\overline{1})\overline{\Delta}(\overline{1}) + i\lambda\int d\overline{1}\,\overline{G}_0(1,\overline{1})\frac{\delta\overline{G}_1(1,\overline{1})}{\delta U(2',2)}\overline{\Delta}(\overline{1})$$
$$+i\lambda\int d\overline{1}\,\overline{G}_0(1,\overline{1})\overline{G}_1(1,\overline{1})\frac{\delta\overline{\Delta}(\overline{1})}{\delta U(2',2)}, \tag{13.8.16}$$

where we have used Eq. (13.8.13) for $\delta\overline{G}_0(1',1)/\delta U(2',2)$. To develop a closed set of equations for $\delta\overline{\Delta}(1)/\delta U(2',2)$ (and $\delta\overline{\Delta}^*(1)/\delta U(2',2)$), Ambegaokar and Kadanoff eliminated $\delta\overline{G}_1/\delta U$ on the right of Eq. (13.8.16) using Eq. (13.8.15) (and the equations for $\overline{G}_1$ and $\overline{F}$). They found a result that had been previously obtained by Migdal,[22]

$$\frac{\delta\overline{\Delta}(1)}{\delta U(2',2)} = 2\lambda\overline{G}_1(1,2')F(2,1) + i\lambda\int d\overline{1}\,\overline{G}_1(1,\overline{1})\frac{\delta\overline{\Delta}(\overline{1})}{\delta U(2',2)}\overline{G}_1(1,\overline{1})$$
$$+i\lambda\int d\overline{1}\overline{F}(1,\overline{1})\frac{\delta\overline{\Delta}^*(\overline{1})}{\delta U(2',2)}\overline{F}(1,\overline{1}), \tag{13.8.17}$$

as well as a similar result for $\delta\overline{\Delta}^*(1)/\delta U(2',2)$. The reader can find more detail of this derivation in Reference 21, as well as the solution of Eq. (13.8.17), etc., in terms of a spatial Fourier transform wave vector variable k. In the long wavelength limit, the result has the property

$$\overline{\Delta}\frac{\delta\overline{\Delta}^*}{\delta U}(k) + \overline{\Delta}^*\frac{\delta\overline{\Delta}}{\delta U}(k) = \frac{\delta}{\delta U}(\overline{\Delta}^*\overline{\Delta}) = 0. \tag{13.8.18}$$

This means that the magnitude of the gap parameter at low-wave number ($k < \omega_D/v_F$) does *not change* to linear order in the perturbing potential U, which can represent electromagnetic scalar or vector potentials (particularly as U is a non-local potential, and variational differentiation with respect to it introduces factors $\psi^+(\mathbf{x},t)\psi(\mathbf{x}',t')$ at different space points as needed before the action of $(\nabla - \nabla')$ to obtain the current in the subsequent limit $\mathbf{x}' \to \mathbf{x}$). Correspondingly, only the phases of $\overline{\Delta}, \overline{\Delta}^*$ are affected to linear order in electromagnetic response at long wavelengths, and we may write

$$\overline{\Delta}(1;\mathbf{A},\varphi) = \exp\left[(2ie/\hbar c)W(1;\mathbf{A},\varphi)\right]\Delta(0), \tag{13.8.19}$$

$$\overline{\Delta}^*(1;\mathbf{A},\varphi) = \exp\left[-(2ie/\hbar c)W(1;\mathbf{A},\varphi)\right]\overline{\Delta}^*(0), \tag{13.8.20}$$

where $\overline{\Delta}(0)$, $\overline{\Delta}^*(0)$ are the values in the absence of the electromagnetic potentials (zero impressed field) and $W(1;\mathbf{A},\varphi)$ is real. *However, these phase factors can be removed by a gauge transformation* (Eqns. 13.8.5–13.8.7). With the *re-gauged* electromagnetic gauge choice,

$$\Lambda(1) = -W(1;\mathbf{A},\varphi),$$

we have

$$\overline{\Delta}(1;\mathbf{A},\varphi) \to \overline{\Delta}'(1;\mathbf{A},\varphi) = \exp\left[\frac{-2ie}{\hbar c}W(1;\mathbf{A},\varphi)\right]\overline{\Delta}(1;\mathbf{A},\varphi) = \overline{\Delta}(0), \tag{13.8.21}$$

which facilitates cancellation of the phase factors involved in Eqns. (13.8.6) and (13.8.19). Also,

$$\overline{\Delta}'^*(1;\mathbf{A},\varphi) = \exp\left[\frac{2ie}{\hbar c}W(1;\mathbf{A},\varphi)\right]\overline{\Delta}^*(1;\mathbf{A},\varphi) = \overline{\Delta}^*(0). \tag{13.8.22}$$

Furthermore, we also have the re-gauged scalar potential φ' as

$$\varphi' = \varphi - \frac{1}{c}\frac{\partial\Lambda}{\partial t} = \varphi + \frac{1}{c}\frac{\partial W}{\partial t}, \tag{13.8.23}$$

and the re-gauged vector potential $\mathbf{A}'$ is given by

$$\mathbf{A}' = \mathbf{A} + \nabla\Lambda = \mathbf{A} - \nabla W. \tag{13.8.24}$$

In light of these considerations, Eq. (13.8.1) incorporates the *original* applied potentials $(\mathbf{A}, \varphi)$ in $\overline{G}_1$ and $\overline{G}_0$ in the form

$$\overline{G}_1^{-1}(1,1';\mathbf{A},\varphi) = \overline{G}_0^{-1}(1,1';\mathbf{A},\varphi) + \overline{\Delta}(1;\mathbf{A},\varphi)\overline{G}_0(1',1;\mathbf{A},\varphi)\overline{\Delta^*}(1';\mathbf{A},\varphi), \tag{13.8.25}$$

but in terms of the *re-gauged* potentials $(\mathbf{A}', \varphi')$ it has the simpler form,

$$\overline{G}_1'^{-1}(1,1';\mathbf{A}',\varphi') = \overline{G}_0'^{-1}(1,1';\mathbf{A}',\varphi') + \overline{\Delta}(0)\overline{G}_0'(1',1;\mathbf{A}',\varphi')\overline{\Delta}^*(0), \tag{13.8.26}$$

in consequence of Eqns. (13.8.21 and 13.8.22). To explore the linear response in terms of the re-gauged potentials, $\mathbf{A}', \varphi'$, we need to determine $\delta\overline{G}_1'/\delta\varphi'$ and $\delta\overline{G}_1'/\delta\mathbf{A}'$, and, to start, we apply the variational differential operator $\delta/\delta\varphi'$ to $\overline{G}_1'^{-1}$ in Eq. (13.8.26), which yields (note that $\overline{\Delta}(0)$ has *no* dependence on $\mathbf{A}'$, ϕ' since it involves no fields):

$$\begin{aligned}\frac{\delta\overline{G}_1'^{-1}}{\delta\varphi'(2)}(1,1';\mathbf{A}',\varphi') &= -\int d3\int d4\,\overline{G}_1'^{-1}(1,3)\frac{\delta\overline{G}_1'(3,4)}{\delta\varphi'(2)}\overline{G}_1'^{-1}(4,1')\\ &= \frac{\delta\overline{G}_0'^{-1}(1,1';\mathbf{A}',\varphi')}{\delta\varphi'(2)} - \overline{\Delta}(0)\int d3\int d4\,\overline{G}_0'(1',3)\frac{\delta\overline{G}_0'^{-1}(3,4)}{\delta\varphi'(2)}\overline{G}_0'(4,1)\overline{\Delta}^*(0).\end{aligned} \tag{13.8.27}$$

Recalling Eq. (9.5.5), $\delta\overline{G}_0'^{-1}(1,1';\mathbf{A}',\varphi')/\delta\varphi'(2) = -\delta(1,1')\delta(1',2)$, Eq. (13.8.27) yields

$$\begin{aligned}&-\int d3\int d4\,\overline{G}_1'^{-1}(1,3)\frac{\delta\overline{G}_1'(3,4)}{\delta\varphi'(2)}\overline{G}_1'^{-1}(4,1')\\ &\quad= -\delta(1,1')\delta(1',2) - \overline{\Delta}(0)\overline{G}_0'(1',2)\overline{G}_0'(2,1)\overline{\Delta}^*(0),\end{aligned} \tag{13.8.28}$$

and multiplying by $\overline{G}_1'$ from the right *and* from the left in the space-time matrix sense in Eqns. (13.8.27 and 13.8.28), we obtain

$$\begin{aligned}\frac{\delta\overline{G}_1'(1,1')}{\delta\varphi'(2)} &= \overline{G}_1'(1,2)\overline{G}_1'(2,1')\\ &\quad+\int d3\int d4\,\overline{G}_1'(1,3)\overline{\Delta}(0)\overline{G}_0'(3,2)\overline{G}_0'(2,4)\overline{\Delta}^*(0)\overline{G}_1'(4,1')\end{aligned} \tag{13.8.29}$$

or $(Q \to e \to 1)$

$$\frac{\delta\overline{G}_1'(1,1')}{\delta\varphi'(2)} = \left[\overline{G}_1'(1,2)\overline{G}_1'(2,1') + \overline{F}'(1,2)\overline{F}'^{+}(2,1')\right], \tag{13.8.30}$$

where we have identified $\overline{F}'$, $\overline{F}'^{+}$ using the primed (re-gauged) counterparts ($\overline{G}_1 \to \overline{G}_1'$; $\overline{F} \to \overline{F}'$) of the relations of Sections 13.3 and 13.4 (e.g., Eq. (13.3.23) in integral form), and we furthermore understand that Eq. (13.8.30) has both sides evaluated at zero-field for linear response. A similar analysis of $\delta\overline{G}_1'(1,1')/\delta\mathbf{A}'(2)$ yields

$$\frac{\delta\overline{G}_1'(1,1')}{\delta\mathbf{A}'(2)} = -\lim_{2'\to 2^+} \frac{1}{c}\frac{\nabla_2 - \nabla_{2'}}{2mi}\left[\overline{G}_1'(1,2')\overline{G}_1'(2,1') + \overline{F}'(1,2)\overline{F}'(2,1')\right], \quad (13.8.31)$$

again to be evaluated at zero-field (equilibrium) for linear response. The right-hand sides of Eqns. (13.8.30 and 13.8.31) involve a bracket having the form of a Gorkov-factorized G_2-function augmented by a Fock-term. However, these are to be evaluated at $\varphi', \mathbf{A}' \to 0$ (rather than $\varphi, \mathbf{A}$) and they participate in linear response multiplied by $\varphi', \mathbf{A}'$ (rather than $\varphi, \mathbf{A}$). Specifically, the linear perturbations of the charge density, $\rho(1)$, and current density, $\mathbf{J}(1)$, are given by

$$\rho(1) = -2ie \int d2 \left[\frac{\delta\overline{G}_1'(1,1^+)}{\delta\varphi'(2)}\varphi'(2) + \frac{\delta\overline{G}_1'(1,1^+)}{\delta\mathbf{A}'(2)}\cdot\mathbf{A}'(2)\right], \quad (13.8.32)$$

and, correspondingly,

$$\mathbf{J}(1) = -\frac{e}{m}\lim_{1'\to 1^+}\left\{(\nabla_1 - \nabla_1')\int d2\left(\frac{\delta\overline{G}_1'(1,1')}{\delta\varphi'(2)}\varphi'(2) + \frac{\delta\overline{G}_1'(1,1')}{\delta\mathbf{A}'(2)}\cdot\mathbf{A}'(2)\right)\right\}$$
$$-\frac{e^2 n}{mc}\mathbf{A}'(1). \quad (13.8.33)$$

Moreover, the evaluation of the variational derivatives, $\delta\overline{G}_1'/\delta\varphi'$ and $\delta\overline{G}_1'/\delta\mathbf{A}'$, at zero-field for linear response means that $\overline{G}_1' \to \overline{G}_1$ and $\overline{F}' \to \overline{F}$ on the right-hand sides of Eqns. (13.8.32 and 13.8.33) are to be replaced by their zero-field limits, as described in Eqns. (13.4.6 and 13.4.7). With these normal and anomalous zero-field Green's functions employed in Eqns. (13.8.30–13.8.33), one can repeat the analysis of Section 13.6 (or use direct integration) to recover Eq. (13.6.31), but with $\mathbf{A} \to \mathbf{A}'$. Moreover, as Ψ is constant at zero field, the first term on the right of Eq. (13.6.31) vanishes, yielding the London result at low wave number ($k < \overline{\Delta}(0)/v_F$) and low frequency ($\omega < 2\overline{\Delta}(0)$),

$$\mathbf{J}(1) = -\frac{2e^2 n_s}{mc}\mathbf{A}'(1) = -\frac{2e^2 n_s}{mc}(\mathbf{A}(1) - \nabla W(1)), \quad (13.8.34)$$

and the corresponding relation for $\rho(1)$ (Eq. (13.8.32)) was evaluated by Ambegaokar and Kadanoff as

$$\rho(1) = -2e^2 N(\mu)\varphi'(1) = -2e^2 N(\mu)\left(\varphi(1) + \frac{1}{c}\frac{\partial W}{\partial t}(1)\right), \quad (13.8.35)$$

where $N(\mu)$ is the density of states of one spin at the Fermi surface.

A better understanding of the physical role of $W(1;\mathbf{A},\varphi)$ can be achieved by rewriting Eq. (13.8.21) for the pair wave function in the presence of potentials, $\overline{\Delta}(1;\mathbf{A},\varphi)$, as

$$\frac{e}{c}W(1;\mathbf{A},\varphi) = \frac{\hbar}{2i}\ln\left(\frac{\overline{\Delta}(1;\mathbf{A},\varphi)}{\overline{\Delta}(0)}\right), \tag{13.8.36}$$

and applying time- and space-derivatives, with the results

$$-\frac{e}{c}\frac{\partial W(1;\mathbf{A},\varphi)}{\partial t} = \frac{1}{2}\frac{1}{\overline{\Delta}(1;\mathbf{A},\varphi)}\left(i\hbar\frac{\partial}{\partial t_1}\right)\overline{\Delta}(1;\mathbf{A},\varphi) \tag{13.8.37}$$

$$= \frac{1}{2}\frac{\overline{\Delta}^*(1;\mathbf{A},\varphi)\left(i\hbar\frac{\partial}{\partial t_1}\right)\overline{\Delta}(1;\mathbf{A},\varphi)}{\left|\overline{\Delta}(1;\mathbf{A},\varphi)\right|^2}, \tag{13.8.38}$$

and (bear in mind that $\overline{\Delta}(0)$ has neither space nor time dependence),

$$\frac{e}{c}\mathbf{\nabla}_1 W(1;\mathbf{A},\varphi) = \frac{1}{2}\frac{1}{\overline{\Delta}(1;\mathbf{A},\varphi)}\left(\frac{\hbar}{i}\nabla\right)\overline{\Delta}(1;\mathbf{A},\varphi) \tag{13.8.39}$$

$$= \frac{1}{2}\frac{\overline{\Delta}^*(1;\mathbf{A},\varphi)\left(\frac{\hbar}{i}\nabla\right)\overline{\Delta}(1;\mathbf{A},\varphi)}{\left|\overline{\Delta}(1;\mathbf{A},\varphi)\right|^2}. \tag{13.8.40}$$

The right-hand sides of Eqns. (13.8.38 and 13.8.39) represent the average energy and momentum of the superconducting state (per pair particle) at space-time point 1 in the presence of the potentials $\mathbf{A},\varphi$, as given by $-\frac{e}{c}\partial W(1;\mathbf{A},\varphi)/\partial t$ and $\frac{e}{c}\mathbf{\nabla} W(1;\mathbf{A},\varphi)$, respectively.

An explicit determination of $W(1;\mathbf{A},\varphi)$ can be accomplished by examining the equation of continuity (ensuring charge conservation),

$$\frac{\partial\rho(1)}{\partial t_1} + \mathbf{\nabla}_1\cdot\mathbf{J}(1) = 0, \tag{13.8.41}$$

and employing Eqns. (13.8.34 and 13.8.35) for $\mathbf{J}(1),\rho(1)$, respectively, with the result

$$\mathbf{\nabla}_1^2 W(1) - \frac{N(\mu)m}{n_s}\frac{\partial^2 W(1)}{\partial t_1^2} = \mathbf{\nabla}_1\cdot\mathbf{A}(1) + \frac{N(\mu)m}{n_s}c\frac{\partial\varphi(1)}{\partial t_1}. \tag{13.8.42}$$

The wave operator on the left involves the phase velocity

$$v^2 = \frac{n_s}{N(\mu)m} = \frac{v_F^2}{3}, \tag{13.8.43}$$

describing an acoustic normal mode oscillation, even in the absence of the external field driving potentials on the right of Eq. (13.8.42). Fourier-transforming with respect to space and time ($\nabla_1 \longrightarrow i\mathbf{p}$; $\partial/\partial t_1 \longrightarrow -i(\omega + i\delta)$ to ensure causality), we obtain Eq. (13.8.42) in the form

$$W(\mathbf{p}, \omega) = \frac{i(v^2\mathbf{p}\cdot\mathbf{A}(\mathbf{p}, \omega) - c\omega\varphi(\mathbf{p}, \omega))}{(\omega + i\delta)^2 - v^2p^2}. \tag{13.8.44}$$

In conjunction with Eqns. (13.8.34 and 13.8.35) this yields ρ and $\mathbf{J}$ as

$$\rho(\mathbf{p}, \omega) = \frac{2n_s e^2}{mc}\left\{\frac{cp^2\varphi(\mathbf{p}, \omega) - \omega\mathbf{p}\cdot\mathbf{A}(\mathbf{p}, \omega)}{(\omega + i\delta)^2 - v^2p^2}\right. \tag{13.8.45}$$

and

$$\mathbf{J}(\mathbf{p}, \omega) = -\frac{2n_s e^2}{mc}\left\{\mathbf{A}(\mathbf{p}, \omega) + \mathbf{p}\frac{(v^2\mathbf{p}\cdot\mathbf{A}(\mathbf{p}, \omega) - c\omega\varphi(\mathbf{p}, \omega))}{(\omega + i\delta)^2 - v^2p^2}\right\}. \tag{13.8.46}$$

These results are invariant under the gauge transformation, Λ,

$$\varphi \longrightarrow \varphi + i\frac{\omega}{c}\Lambda;\ \mathbf{A} \longrightarrow \mathbf{A} + i\mathbf{p}\Lambda. \tag{13.8.47}$$

In the d.c. limit of zero frequency, we obtain

$$\mathbf{J}(\mathbf{p}, 0) \equiv -\frac{2n_s e^2}{mc}\left\{\overleftrightarrow{\mathbf{I}} - \frac{\mathbf{pp}}{p^2}\right\}\cdot\mathbf{A}(\mathbf{p}, 0), \tag{13.8.48}$$

where $\overleftrightarrow{\mathbf{I}}$ is the unit dyadic. This is the manifestly gauge invariant generalization of the London equation, reducing to it exactly for a transverse vector potential, $\mathbf{p}\cdot\mathbf{A} = 0$. It is important to note that throughout most of this section (Section 13.8), we have discussed Green's functions for the condensate of Cooper pairs. However, electrons occupying levels deeper below $(\mu - \hbar\omega_D)$ do not participate in Cooper-pairing, nor in the supercurrent. Accordingly, those electrons are subject to the normal resistance mechanisms and their current contributions die off as a driving electric field vanishes. Therefore, the equilibrium density (n_s) and current ($\mathbf{J}$) of Eqns. (13.8.45, 13.8.46 and 13.8.48) must be understood as those of paired electrons only, to the exclusion of electrons having lower energies, as indicated by

$$n \equiv n_s \text{ and } \mathbf{J} \equiv \mathbf{J}_s.$$

The above discussion, based on Eq. (13.8.32) for ρ, lacks recognition that the density perturbation responds to the *effective* field, $V(1)$, in the medium rather than the "impressed" field, $U(1) = \varphi'(1)$, so that Eq. (13.8.35) should be supplanted by

$$\rho(1) = -2e^2N(\mu)V(1) = -2e^2N(\mu)\int d2K(1, 2)\varphi'(2), \tag{13.8.49}$$

where $K(1,2)$ is the inverse dielectric "screening" function (Section 9.4). Rewriting Eq. (9.4.11) as

$$K(1,2) = \delta(1,2) + \int d2' \int d3' \mathcal{V}(1-2') \frac{\delta\rho(2')}{\delta V(3')} K(3',2), \tag{13.8.50}$$

the polarizability, α, may be identified as

$$\alpha(1,3') = -\int d2' \mathcal{V}(1-2') \frac{\delta\rho(2')}{\delta V(3')}, \tag{13.8.51}$$

and for translationally invariant systems, space-time Fourier transformation yields Eqns. (13.8.50 and 13.8.51) in the form

$$K(\mathbf{p},\omega) = [1 + \alpha(\mathbf{p},\omega)]^{-1}, \tag{13.8.52}$$

with

$$\alpha(\mathbf{p},\omega) = -\mathcal{V}(\mathbf{p}) \frac{\delta\rho}{\delta V}(\mathbf{p},\omega). \tag{13.8.53}$$

Applying $\frac{\delta}{\delta V}$ to Eq. (13.8.49), we have

$$\frac{\delta\rho}{\delta V}(\mathbf{p},\omega) = -2e^2 N(\mu), \tag{13.8.54}$$

and for the system at hand we therefore obtain ρ as

$$\rho(\mathbf{p},\omega) = -2e^2 N(\mu) K(\mathbf{p},\omega) \varphi'(\mathbf{p},\omega) = \frac{-2e^2 N(\mu)\varphi'(\mathbf{p},\omega)}{1 + 2e^2 N(\mu)\mathcal{V}(\mathbf{p})}. \tag{13.8.55}$$

Employing this screened result in the determination of W (Eqns. (13.8.44–13.8.46)), Ambegaokar and Kadanoff used the attractive short-range potential of Eq. (13.3.11) that is constant in $(\mathbf{p},\omega)$-representation (with $\mathcal{V}(\mathbf{p}) \to -\frac{\lambda}{2}$), finding that Gorkov's decomposition of G_2 into anomalous and normal Green's functions with Hartree–Fock terms leads to results of the form of Eqns. (13.8.45 and 13.8.46), but with a modified sound velocity due to the inclusion of the effective, self-consistent potential $V(1)$, as given by

$$v_{\mathit{eff}}^2 \cong \frac{v_F^2}{3} [1 + N(\mu)\lambda], \tag{13.8.56}$$

somewhat different from the important earlier superconductor collective mode prediction of Anderson.[23]

The inclusion of long-range repulsive electron–electron Coulomb interaction, $\mathcal{V}(\mathbf{p}) = 4\pi e^2/p^2$, renders the attractive short-range interaction ($\mathcal{V}(\mathbf{p}) \to \lambda$) negligible at low wave numbers, in which case the density is essentially determined by the Poisson equation as

$$\rho(\mathbf{p},\omega) = \frac{-p^2}{4\pi}\varphi'(\mathbf{p},\omega) = \frac{-p^2}{4\pi}\left[\varphi(\mathbf{p},\omega) - \frac{i\omega}{c}W(\mathbf{p},\omega)\right], \qquad (13.8.57)$$

and the current equation remains the same, Eq. (13.8.34) (neglecting self-consistency of the vector potential),

$$\mathbf{J}(\mathbf{p},\omega) = \frac{-2e^2 n_s}{mc}\mathbf{A}'(\mathbf{p},\omega) = \frac{-2e^2 n_s}{mc}\left[\mathbf{A}(\mathbf{p},\omega) - i\mathbf{p}W(\mathbf{p},\omega)\right]. \qquad (13.8.58)$$

The determination of W follows, as before, from Eq. (13.8.41), with the result

$$W(\mathbf{p},\omega) = \frac{-i\omega c\varphi(\mathbf{p},\omega) + i\omega_p^2\mathbf{p}\cdot\mathbf{A}(\mathbf{p},\omega)/p^2}{\omega^2 - \omega_p^2}, \qquad (13.8.59)$$

and substitution of this into Eqns. (13.8.57 and 13.8.58) yields ρ and $\mathbf{J}$ as

$$\rho(\mathbf{p},\omega) = \frac{\omega_p^2}{4\pi}\left(\frac{p^2\varphi(\mathbf{p},\omega) - \omega\mathbf{p}\cdot\mathbf{A}(\mathbf{p},\omega)/c}{(\omega+i\delta)^2 - \omega_p^2}\right) \qquad (13.8.60)$$

and

$$\mathbf{J}(\mathbf{p},\omega) = \frac{2e^2 n_s\omega\mathbf{p}/m}{(\omega+i\delta)^2 - \omega_p^2}\varphi(\mathbf{p},\omega) - \frac{2e^2 n_s}{mc}\left[\overleftrightarrow{\mathbf{I}} + \frac{\omega_p^2\mathbf{p}\mathbf{p}/p^2}{(\omega+i\delta)^2 - \omega_p^2}\right]\cdot\mathbf{A}(\mathbf{p},\omega) \qquad (13.8.61)$$

(where $2n_s$ (twice the number of Cooper pairs) is the actual number density of electrons involved in the superconducting condensate). These results are, again, manifestly gauge invariant, and they exhibit a plasmon collective mode.[21]

13.9 Magnetic Flux Quantization: Type I and Type II Superconductors

Considering Eq. (13.6.28), we can write $\Psi(\mathbf{x})$ in terms of its amplitude and phase as

$$\Psi(\mathbf{x}_1) = |\Psi(\mathbf{x}_1)|\, e^{i\phi(\mathbf{x}_1)}, \qquad (13.9.1)$$

such that

$$\Psi^*\nabla\Psi - \Psi\nabla\Psi^* = 2i\,|\Psi|^2\,\nabla\phi, \qquad (13.9.2)$$

with the result (Eq. (13.6.31)):

$$\nabla\phi = \frac{2e}{c}\mathbf{A} + \frac{m\mathbf{J}_s}{e\,|\Psi|^2}. \qquad (13.9.3)$$

Line-integrating this about a closed path Γ and recalling that

$$\oint_{\Gamma} d\mathbf{x} \cdot \mathbf{A} = \int_{s} d\mathbf{S} \cdot \nabla \times \mathrm{A} = \int_{s} d\mathbf{S} \cdot \mathbf{H} = \Phi_H$$

is the flux of the magnetic field through the surface S bounded by Γ, we have the phase change of $\Delta\phi$ as

$$\Delta\phi = \frac{2e}{c}\Phi_H + \frac{m}{e}\oint_{\Gamma} \frac{d\mathbf{x} \cdot \mathbf{J}_s(\mathbf{x})}{|\Psi(\mathbf{x})|^2}. \tag{13.9.4}$$

Considering a simply connected region (containing no "swiss cheese" holes) to be the interior of a superconductor, we know that the Meissner effect mandates that all magnetic flux lines are expelled and the magnetic field becomes negligibly small in the interior, carrying the implication that $\mathbf{J}_s = (c/4\pi)\,\boldsymbol{\nabla} \times \mathbf{H}$ is also small, so that the last integral on the right of Eq. (13.9.4) is negligible, provided only that Γ is everywhere separated from the outer boundary of the superconductor by a distance larger than the penetration depth, ℓ. In this case, $\Delta\phi$, the change of phase of Ψ in integrating about Γ, vanishes. Furthermore, if there *are* interior boundaries in the form of "swiss cheese" holes, then if they are separated from the outer boundary, and from each other, by distances larger than the penetration depth, then one can choose the path of Γ to be well separated from *both* the outer and interior boundaries, and again the last integral on the right of Eq. (13.9.4) may be considered to be small. Requiring only single-valuedness of Ψ, we then have (r = integer):

$$\Delta\phi = 2\pi r = \frac{2e}{c}\Phi_H, \tag{13.9.5}$$

so that the enclosed magnetic flux is quantized in units of $\Phi_0 = 2\pi\hbar c/2e = \pi\hbar c/e$ as

$$\frac{\Phi_H}{\Phi_0} = r. \tag{13.9.6}$$

The penetration of such quantized magnetic flux structures signals the intrusion of normal regions into the superconductor as the interiors of the excluded "swiss cheese" hole cores of the multiply connected superconductor. This marks the occurrence of another type of superconductor, Type II, which admits some magnetic flux (in contrast to Type I, which admits none).

Abrikosov's analysis of such superconducting phenomenology proceeds from the Ginzburg–Landau equations (Eqns. (13.6.29 and 13.6.31) jointly with the magnetostatic equations ($\mathbf{J}_s = (c/4\pi)\,\boldsymbol{\nabla} \times \mathbf{H}$) written in dimensionless form as

$$[(i\kappa)^{-1}\boldsymbol{\nabla}' - \mathbf{A}']^2\Psi' = \Psi'(1 - |\Psi'|^2) \tag{13.9.7}$$

and

$$\boldsymbol{\nabla}' \times \mathbf{h}' = \frac{1}{2}\{\Psi'^*[(\iota\kappa)^{-1}\boldsymbol{\nabla}' - \mathbf{A}']\Psi' + \Psi'[-(\iota\kappa)^{-1}\boldsymbol{\nabla}' - \mathbf{A}']\Psi'^*\}, \tag{13.9.8}$$

where

$$\mathbf{x}' = \frac{\mathbf{x}}{l}; \ \mathbf{h}' = \frac{\mathbf{H}}{2^{\frac{1}{2}}H_c}; \ \mathbf{A}' = \frac{\mathbf{A}}{2^{\frac{1}{2}}H_c l} \ \text{and } \Psi' = \frac{\Psi}{|\Psi_\infty|}; \tag{13.9.9}$$

Ψ_∞ is the bulk equilibrium value of Ψ in the absence of a magnetic field, given by Eq. (13.6.29) as $|\Psi_\infty|^2 = n\left|\frac{T_c - T}{T_c}\right|$.

These equations couple the pair wave function $\Psi \to \Psi'$ with the actual, microscopic magnetic field $\mathbf{H} \to \mathbf{h}'$. They involve two characteristic lengths:

(i) ξ, the coherence length, which measures the characteristic length over which the pair wave function, or order parameter, varies, defined from Eq. (13.6.29) as

$$\xi^2 = \frac{\hbar^2}{4m} \cdot \frac{RT_c}{T_c - T}; \tag{13.9.10}$$

(ii) and the magnetic penetration length, l, of Eq. (13.7.3), which diverges as $T \to T_c$ (since $n_s \to |\Delta|^2$ and the gap vanishes when $T \to T_c$ (Eqns. (13.6.32) and Eqns. (13.4.12–13.4.14))), signaling the re-entry of magnetic flux into the superconductor,

$$l^2 = \frac{mc^2}{8\pi e^2 n_s}. \tag{13.9.11}$$

The ratio, $\kappa = l/\xi$, is of key importance in the analysis of the physical phenomenology embodied in the Ginzburg–Landau equations. Superconductors are classified as follows:

$$\kappa = l/\xi < 1/\sqrt{2} \quad \text{are Type I;} \tag{13.9.12}$$

$$\kappa = l/\xi > 1/\sqrt{2} \quad \text{are Type II.} \tag{13.9.13}$$

Clearly, the small penetration depth $l < \xi$ for Type I bespeaks full expulsion of the magnetic field (flux) lines (which cannot begin or end due to zero divergence). An appreciation of the significance of the Type II criterion can be found in the calculation of superconductor surface energy based on the Ginzburg–Landau equations above, jointly with surface boundary conditions of the form

$$\begin{aligned} &\Psi'(z \to -\infty) = 0; \quad h'(z \to -\infty) \to 1/\sqrt{2\pi} \quad \text{and} \\ &\Psi'(z \to +\infty) = 1; \quad h'(z \to +\infty) \to 0. \end{aligned} \tag{13.9.14}$$

The resulting surface energy[7,18] is negative for Type II superconductors, indicating that flux line penetration is energetically favorable in such a mixed state, while it is positive for Type I, mandating the complete expulsion of magnetic flux. The (lower) critical magnetic field strength at which flux line penetration into the superconductor can occur is denoted by H_{C1}.

The occurrence of several quantized magnetic flux lines in a Type II superconductor has interesting features since their associated circulatory currents can interact, and in this sense the field lines do so as well. To explore this, we apply $\nabla\times$ to both sides of Eq. (13.9.8) (equivalently Eq. (13.9.3) with $\mathbf{J}_s = (c/4\pi)\nabla \times \mathbf{H}$), obtaining

$$\nabla \times (\nabla \times \mathbf{H}) + \frac{1}{l^2}\mathbf{H} = -\nabla^2\mathbf{H} + \frac{1}{l^2}\mathbf{H} = \frac{4\pi e n_s}{mc}\nabla \times \nabla\phi. \tag{13.9.15}$$

Because ϕ is *not* single valued, $\nabla\times\nabla\phi \neq 0$ does *not* vanish identically and its flux integral is finite. Considering the cross-sectional area of a magnetic flux line to be small and in the x-y plane perpendicular to the direction of the associated field line ($\mathbf{H} = H\hat{z}$), it has been verified that we can approximately write[17,18,24]

$$\nabla \times \nabla\phi = \hat{z}\Gamma\delta^{(2)}(\mathbf{r}-\mathbf{r}'), \quad [\mathbf{r} = x\hat{\imath} + y\hat{\jmath}] \tag{13.9.16}$$

for a flux line localized at $\mathbf{r}'$ (with the 2D Dirac delta-function in the lateral plane), so that Γ is the flux integral of $\nabla \times \nabla\phi$, representing the phase change $\Delta\phi$ in a loop about the flux line:

$$\Gamma = \int d^{(2)}\mathbf{r}\hat{z}\cdot \nabla \times \nabla\phi = \oint d\mathbf{r}\cdot\nabla\phi = \Delta\phi. \tag{13.9.17}$$

The equation for $\mathbf{H}$ (Eq. (13.9.15 and 13.9.16)) written as

$$\nabla^2\mathbf{H} - \frac{1}{l^2}\mathbf{H} = -\frac{\Phi_0}{l^2}\hat{z}\delta^{(2)}(\mathbf{r}-\mathbf{r}') \tag{13.9.18}$$

is readily solved as

$$\mathbf{H}(\mathbf{r}) = \hat{z}\frac{\Phi_0}{2\pi l^2}K_0\left(\frac{|\mathbf{r}-\mathbf{r}'|}{l}\right), \tag{13.9.19}$$

where K_0 is the modified Bessel function of the third kind (Basset function) of order zero. It vanishes exponentially when $|\mathbf{r}-\mathbf{r}'| > l$, restricting the normal cross-sectional area of the flux line to be less than l^2.

Considering multiple parallel flux lines (indexed by α), Eq. (13.9.16) is supplanted by

$$\nabla \times \nabla\phi = \hat{z}\sum_{\alpha}\Gamma_\alpha\delta^{(2)}(\mathbf{r}-\mathbf{r}_\alpha),$$

and the linearity of Eq. (13.9.15) implies that $\mathbf{H} = \sum_\alpha \mathbf{H}_\alpha$, where each magnetic field line, $\mathbf{H}_\alpha$, separately satisfies an equation of the type of Eq. (13.9.18). The implications of this are best addressed through an examination of the energy, which is quadratic in the $\mathbf{H}_\alpha$ since the magnetic energy density has the form $(8\pi)^{-1}\mathbf{H}^2 = (8\pi)^{-1}\sum_{\alpha,\beta}\mathbf{H}_\alpha\cdot\mathbf{H}_\beta$ and the kinetic energy density may be written as $\frac{n_s}{2}(2m)v_s^2 = (8\pi)^{-1}l^2\sum_{\alpha,\beta}(\nabla\times\mathbf{H}_\alpha)\cdot(\nabla\times\mathbf{H}_\beta)$

(where we have used $\mathbf{v}_s = \mathbf{J}_s/2en_s$ and $\mathbf{J}_s = \frac{c}{4\pi}\nabla \times \mathbf{H}$). The equilibrium lattice structure of magnetic flux lines is determined by minimizing the energy with respect to the magnetic induction, **B**. Abrikosov's solution of the linearized Ginzburg–Landau equations consists of parallel flux lines arrayed with their cross section forming a triangular lattice in equilibrium. A thorough discussion of this is provided in References 17, 18 and 24. For magnetic fields close to H_{C1}, the flux lines are well separated, but for larger fields they start to overlap, increasing their interaction. For yet higher magnetic field strengths, the density of flux lines increases to the point where their normal cores overlap and the superconducting state is severely diminished in size, supplanted by a normal state. The associated magnetic field strength for this generally second-order phase transition is denoted as H_{C2}.

13.10 Josephson Tunneling

Josephson tunneling is concerned with charge transport between two uniform semi-infinite superconductors, "L" and "R," separated by a narrow normal barrier region, as depicted in Figure 13.2.

All tunneling transport, both normal and superconductive, is quite generally associated with the fact that a traveling electron (quasiparticle) wave function in a conductive region L does not fully terminate at a barrier that reflects most of it; some of it penetrates into the barrier region and decays exponentially as a function of the depth of penetration. Moreover, the quasiparticle wave function can recover traveling status upon leaving the barrier and entering another conductive region R. This process has come to be described by a Hamiltonian of the form

$$H = H_L + H_R + H_T, \tag{13.10.1}$$

where H_L, H_R describe the electron dynamics in regions L, R, respectively, and H_T describes a mechanism for the transfer of electrons between L and R. The determination of H_T is not trivial; it is often written in terms of the instantaneous transfer of an electron from an L eigenstate (α) to an R eigenstate (β) as

$$H_T = \sum_{\alpha,\beta} \left(T_{\alpha,\beta} \psi_\alpha^+ \phi_\beta + T_{\alpha,\beta}^* \phi_\beta^+ \psi_\alpha \right), \tag{13.10.2}$$

where ψ_α, ϕ_β and $\psi_\alpha^+, \phi_\beta^+$ annihilate and create (respectively) electrons in states α, β in L, R. This description of H_T was pioneered by Bardeen,[25] Prange,[26] Cohen, *et al.*[27] and

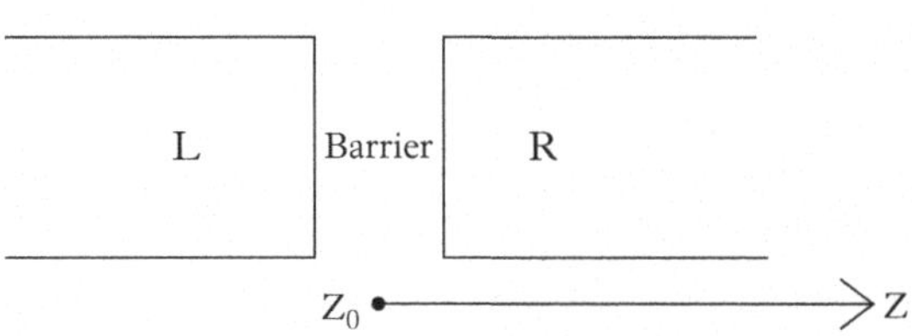

Figure 13.2: *Schematic of two superconductors L, R separated by a narrow normal barrier.*

Ambegaokar and Baratoff.[28] These authors found that the matrix elements $T_{\alpha,\beta}$ between states of approximately equal energy are given in **r**-position representation by the lateral area integral

$$T_{\alpha,\beta} = \hbar \int dx \int dy \left[\frac{\hbar}{2mi} \psi_\alpha^*(\mathbf{r}) \partial\phi_\beta(\mathbf{r})/\partial z - \phi_\beta(\mathbf{r}) \partial\psi_\alpha^*(\mathbf{r})/\partial z - \frac{e}{mc} A_\alpha \psi_\alpha^*(\mathbf{r}) \phi_\beta(\mathbf{r}) \right]_{z=z_0}, \tag{13.10.3}$$

with z_0 as a point in the barrier region and **A** is the self-consistent vector potential. (The symbols $\psi_\alpha(\mathbf{r}), \phi_\beta(\mathbf{r})$ and $\psi_\alpha^*, \phi_\beta^*(\mathbf{r})$ here represent the eigenfunctions of the H_L, H_R operators in Eq. (13.10.1)) In actual calculations, $T_{\alpha,\beta}$ is often assumed to be essentially independent of $\alpha, \beta, T_{\alpha,\beta} \to T$. While this expression for H_T describes single quasiparticle transfer from L to R, its repeated application describes the transfer of multiple quasiparticles.

In his Nobel Prize-winning prediction, Josephson applied the transfer Hamiltonian described above to second order in an iterative procedure to examine the possibility of superconductive tunneling transport of Cooper pairs. Such an iterative procedure can describe not only the transfer of single electrons and/or Bogoliubon quasiparticles, but also the transfer of multiparticle complexes between superconductors L and R in higher order iterations, and these can be sorted out to identify the tunneling of a Cooper pair from L to a Cooper pair in R, etc. Another point of view was advanced by Wallace and Stayn,[29] who invoked a canonical transformation S of the Hamiltonian H yielding its successor H' as

$$H' = e^{-S} H e^{S} = H + [H, S] + \frac{1}{2} [[H, S], S] + \ldots, \tag{13.10.4}$$

with S defined such that the first-order term in H' involving $O(H_T)$ vanishes: Writing $H_0 = H_L + H_R$, this requires

$$H_T + [H_0, S] = 0, \tag{13.10.5}$$

and introducing a complete set of H_0-eigenstates, the matrix elements of S are found to be

$$\langle n|S|m\rangle = \frac{\langle n|H_T|m\rangle}{E_m - E_n}. \tag{13.10.6}$$

The second-order term of H' arises in part from the second term on the right of Eq. (13.10.4) as $[H_T, S]$ and in part from the third term as $\frac{1}{2}$ $[[H_0, S], S]$, which combine to yield

$$\langle n|H'|m\rangle = E_0 \delta_{m,n} + \frac{1}{2} \sum_r \langle n|H_T|r\rangle \langle r|H_T|m\rangle \left\{ \frac{1}{E_m - E_r} - \frac{1}{E_r - E_n} \right\}, \tag{13.10.7}$$

where we exhibit the zero-order term and the second-order term (with the first-order term having been eliminated). The second-order term can, by itself, transfer two particles/excitations from L to R. To focus on Cooper-pair transfer, the initial and final states are essentially at the same energy $E_m \cong E_n$, so we have

$$\langle n|H'|m\rangle = E_0\delta_{m,o} + \sum_r \frac{|\langle n|H_T|r\rangle|^2}{E_n - E_r}, \tag{13.10.8}$$

which can be further adjusted to deal with Cooper pairs exclusively by requiring that the transferred pairs have equal and opposite momenta and spins. Accordingly, the appropriate tunneling Hamiltonian for Cooper pairs is written in "second-order" form as

$$H'_T = \sum_{\alpha,\beta} \left(T b_\alpha^{(L)+} b_\beta^{(R)} + T^* b_\beta^{(R)+} b_\alpha^{(L)} \right), \tag{13.10.9}$$

where

$$b_\alpha^{(L)+} = \psi_\alpha^+ \psi_{-\alpha}^+; \quad b_\alpha^{(L)} = \psi_{-\alpha}\psi_\alpha \tag{13.10.10}$$

and

$$b_\beta^{(R)+} = \psi_\beta^+ \psi_{-\beta}^+; \quad b_\beta^{(R)} = \psi_{-\beta}\psi_\beta \tag{13.10.11}$$

with $(\alpha,-\alpha)$ and $(\beta,-\beta)$ referring to equal and opposite momenta and spins in L and R, respectively.

Central to all of this is the overlap of the L and R eigenfunctions that appear, for example, in Eq. (13.10.3); in this case involving the overlap of the L and R Cooper-pair eigenfunctions of Eqns. (13.6.28 and 13.9.1) for the uniform superconductive regions L and R,

$$\Psi_{L,R} = |\Psi_{L,R}|e^{i\theta_{L,R}}. \tag{13.10.12}$$

Accordingly, Josephson tunneling devolves upon L, R wave function overlap of the form

$$\Psi_L^* \Psi_R \sim |\Psi_L||\Psi_R|e^{i(\theta_L - \theta_R)} \tag{13.10.13}$$

and since the Hamiltonian is Hermitian, the associated transfer term involves the real part of Eq. (13.10.13):

$$\langle H'_T\rangle = \langle H'_T\rangle_{max} \cos(\theta_L - \theta_R). \tag{13.10.14}$$

Impressing a uniform potential V on one of the two superconductors can be represented by the addition of a term V to its chemical potential, leading to an additional phase term $2e\,Vt$ in the argument of $\cos(\theta_L - \theta_R) \rightarrow \cos(\theta_L - \theta_R + 2e\,Vt)$, and if we also introduce the Peierls phase of Eq. (5.6.33) for an applied magnetic field, Eq. (13.10.14) becomes[30,31]

$$\langle H'_T \rangle = \langle H'_T \rangle_{max} \cos\left(\theta_L - \theta_R + 2e\, Vt/\hbar - \tfrac{2e}{\hbar c}\int_1^2 \mathbf{A}\cdot d\mathbf{l}\right)$$
$$= \langle H'_T \rangle_{max} \cos\left(\theta_L - \theta_R + 2e\, Vt/\hbar - 2e|\mathbf{A}|w/\hbar c\right),$$

where $\mathbf{A}$ is the magnetic vector potential (taken normal to the barrier surfaces and integrated across its width w). Recalling that the current operator is $\mathbf{J} = c\delta H/\delta \mathbf{A}$ (see Eq. (3.4.8)) this yields the superconductive tunneling current of Cooper pairs in the form

$$\langle \mathbf{J} \rangle = c\delta \langle H \rangle/\delta \mathbf{A} = \langle \mathbf{J} \rangle_{max} \sin(\theta_L - \theta_R + 2e\, Vt/\hbar - 2e|\mathbf{A}|w/\hbar c). \tag{13.10.15}$$

This result was obtained by Josephson[30] with the use of Green's functions; his analysis should be consulted by the reader as we have chosen a more qualitative, introductory physical discussion of the subject here. It carries the remarkable prediction that there is a non-zero tunneling supercurrent even in the complete absence of electric and magnetic fields, which is periodic in the phase difference $\theta_L - \theta_R$ of the two superconductors—and it is a steady dc current—but if a constant potential difference, V, is applied this current becomes time oscillatory.

13.11 Further Remarks and Developments

The most outstanding feature of superconductivity is the *absence of dissipation* in electrical conduction. It has facilitated the development of superconducting magnets having fields over 100,000 Gauss. Without superconductors, the production of such a field formerly involved the dissipation of millions of watts of power. On one occasion in the early 1960s, a failure of the cooling system of such a *non*-superconducting magnet resulted in an explosion at the MIT National Magnet Laboratory that vaporized the steel magnet, which then floated down a drain in gaseous state until it cooled and solidified. The maintenance of a non-dissipative equilibrium in a Type 2 superconductor is not always possible. Fluxoids (flux lines embedded in normal cross-sectional areas confining the magnetic field as described in Eq. (13.9.19)) have associated circulatory currents that interact as described below Eq. (13.9.20), repelling each other. In equilibrium, the repulsive forces on a given fluxoid due to all other fluxoids cancel, as occurs in Abrikosov's triangular array. But otherwise, with the onset of non-equilibrium, a non-vanishing force on a fluxoid results in its acquisition of a finite velocity in the plane perpendicular to the axis of the fluxoid. Such fluxoid motion bespeaks a time variation of the magnetic field at a given point, which, in turn, produces an electric field in accordance with Maxwell's equations, thus generating a mechanism of resistance with concomitant dissipation of energy. To avoid such flux flow and its undesirable consequences, "pinning" forces have been employed, involving spatial inhomogeneities due to impurities, grain boundaries, etc., that lock the vortices into fixed positions. Nevertheless, vortices can also hop among pinning sites giving rise to relatively slow flux "creep."[17,20,24]

This brief introduction to the theories of superconductivity and superfluidity as examples of the use of Green's functions and variational differential techniques cannot possibly cover the massive development of the subjects that has taken over the last half-century. The intensity of this development arises not only from the uniquely elegant conceptual structure of its mathematical description of the physical features, but also from its prominence as a macroscopic quantum phenomenon, and, in addition, it carries enormous potential for major commercial applications as materials with higher superconducting transition temperatures (with null electrical resistance) are found. The interested reader can learn more about this development from the references (as indicated in the last paragraph of Section 15.6). For example, the subject of thermodynamic fluctuations in superconductors, above and below the transition temperature, is discussed in depth in the books of Tinkham (Reference 17, Chapter 7) and Gulian and Zharkov (Reference 20, Section 7.4), wherein the role of fluctuations in electrical conduction (resistance) and specific heat is examined. In regard to superconductor tunneling phenomenology, particularly limpid discussions are provided by de Gennes (Reference 18, p.234 ff.) and Tinkham (Reference 17, Section 6.1). Also, "strong coupling" superconductors, which require more careful treatment of a strong electron–phonon interaction in generating Cooper pairs, are discussed by Scalapino in Parks' book (Reference 12b, Chapter 10).

In recent years, two two-volume treatise-like collections of review articles, much like the older books of Parks,[12b] have been published by Bennemam and Ketterson, helping to bring readers up-to-date information on superconductivity[32] and superfluidity.[33] Their two-volume book "Superconductivity" (Springer, 2008)[32] presents detailed review articles on conventional superconductors (such as the ones discussed in this chapter), including important reviews of the theoretical foundations by Pitaevskii (Chapter 2) and of the electron–phonon interaction responsible for the formation of Cooper pairs, reviewed by Marsiglio and Carbotte (Chapter 3). Furthermore, Gorkov reviews superconducting alloys, elucidating the impurity-averaging process for superconductors (Chapter 5) and vortex matter is reviewed by Blatter and Geshkenbein (Chapter 12). Fluctuation phenomena are reviewed by Larkin and Varlamov (Chapter 10) and Shafranjuk and Ketterson review principles of Josephson-function-based quantum computation (Chapter 9). There are many other interesting review articles in Volume I, including one on the coexistence of superconductivity and magnetic order by Kulic and Buzdin (Chapter 4) and another on proximity coupled systems by Chandrasekhar (Chapter 8). Moreover, nanostructures are also included in two reviews, one in Morr's article on impurity nanostructures and quantum interference in superconductors (Chapter 6) and another in Deutscher's article on nanostructured superconductors (Chapter 7). There is also an important review of tunneling spectroscopy by Zasadzinski (Chapter 15 in Volume II).

Conventionally, superconductivity was observed in materials like Hg with transition temperatures associated with liquid helium, a few degrees Kelvin. By 1980, the highest transition temperature found was 30°K in Nb_3Ge. However, in 1986 the world of physics was shocked by the discovery of Bednorz and Müller of high-temperature superconductivity in cuprate (copper oxide) materials exceeding 80°K, even up to 160°K; this seminal finding stimulated an avalanche of research aimed at even higher transition temperature materials, with room temperature commercialization as the goal.

The properties of cuprate superconductors are reviewed in an article by Schneider in Volume I (Chapter 11) and high T_c superconductivity is reviewed in Volume II by Ott (Chapter 14). Volume II also has reviews of phase-sensitive tests of pairing symmetry in cuprate superconductors by Tsuei and Kirtley (Chapter 16), and of photo-emission in high T_c superconductors by Campuzano, *et al.* (Chapter 17). Volume II is mainly focused on novel superconductors, with a holdover review (in Volume I) on unconventional superconductivity in novel materials by Maple, *et al.* (Chapter 13), and Volume II also has a review by Hayden of neutron scattering and magnetic response in superconductors (Chapter 18). Furthermore, Volume II presents reviews of heavy fermion superconductivity by Riseborough, *et al.* (Chapter 19) and of organic superconductors by Lang and Müller (Chapter 20). There is also an important review of the concepts in high T_c superconductivity by Carlson, *et al.* (Chapter 21). The spin fluctuation model for d-wave superconductivity is reviewed by Chubukov, *et al.* (Chapter 22), and Mansky, *et al.* review the electronic theory of superconductivity in high T_c cuprates (Chapter 23). Finally, Volume II includes a review of superfluid ^{3}He and the cuprate superconductors by Leggett (Chapter 24). It should also be noted that an older review of superconducting devices by Newhouse is provided in Parks' book (Reference 12, Volume 2, Chapter 22) and a related review on macroscopic quantum phenomena by Mercereau appears in Parks' Volume I (Chapter 8).

The second two-volume book of Bennemann and Ketterson, "Novel Superfluidity" (Oxford University Press, Volume I, 2013, and Volume II, 2014) also has an impressive collection of review articles. Quantized vortices in superfluid helium and atomic Bose–Einstein condensates are reviewed by Tsubota, *et al.* (Chapter 3). Bose–Einstein condensation of photons and atomic gasses are reviewed, respectively, by Klears and Weitz (Chapter 6) and Chevy and Dalibard (Chapter 7); also a theoretical overview of cold atoms is provided by Trivedi and Loh (Chapter 18). Superfluidity in exciton-polariton condensates is reviewed by Yamamoto in Chapter 9 and coherent exciton transport in semiconductors is discussed by Rontani and Sham (Chapter 15). Furthermore, there are reports on Bose-Einstein condensation of magnons at room temperature by Demokritov and Slavin (Chapter 14), and also spin superfluidity and magnon Bose–Einstein condensation are reviewed by Bunkov and Volovik (Chapter 4). There are also theoretical and experimental reviews of superfluid helium three in aerogel by Sauls (Chapter 19) and Lee and Haley (Chapter 5), respectively, and also a review of experiments on excitons in bulk semiconductors by Kuwata-Gonokami (Chapter 8). The first volume concludes with Volovnik's review "The superfluid universe" (Chapter 11). While this second two-volume book is titled for superfluids, it also contains some additional reviews of topics in superconductivity. In Volume I, Alford, *et al.* report on color superconductivity in dense quark matter (Chapter 10). Furthermore, Volume II (which has not yet appeared) is scheduled to present a review of unconventional superconductivity by Norman (Chapter 12), as well as an experimental review of iron-based superconductors by Crabtree, *et al.* (Chapter 13). Additional Volume II superfluidity reviews include one on neutron superfluidity by Gezerlis, *et al.* (Chapter 20), and another by Page, *et al.* on cooling rates and neutron superfluidity (Chapter 21). There is also a review of superfluidity and the BEC–BCS crossover in atomic Fermi gasses by Zwierlein

(Chapter 16) and another by Chin and Gemelke on the superconductor-to-Mott insulator quantum phase transition of ultra-cold bosonic atoms in optical lattices (Chapter 17). Overall, the two two-volume book series on superconductivity and superfluidity provided by Bennemann and Ketterson are an exceptionally valuable up-to-date resource, highlighted with their substantial introductory summaries/surveys incorporated in the opening chapters of the books.

13.12 Chapter 13 Problems

P.13.1. What is the relation between H_c and T_c?

P.13.2. Derive Eq. (13.8.10).

P.13.3. Verify Eqns. (13.8.45–13.8.48), proving the gauge invariance of Eq. (13.8.48).

P.13.4. Determine W, ρ and $\mathcal{J}$ for the attractive short-range potential $\mathcal{V}(\mathbf{p}) \to -\lambda/2$ in Section 13.8.

P.13.5. Prove the gauge invariance of Eqns. (13.8.60 and 13.8.61).

P.13.6. Verify Eqns. (13.9.9–13.9.11).

P.13.7. Derive Eq. (13.9.19).

P.13.8. Verify Eqns. (13.10.5–13.10.8).

P.13.9. Determine the possible eigenvalues and solutions of the linearized Ginzburg–Landau equation (Eq. 13.6.29 with $|\Psi|^2 \to 0$ and $\mathbf{A}$ taken as the vector potential of a uniform external magnetic field).

P.13.10. Having approached Ginzburg–Landau theory from the more fundamental Gorkov equations, we bypassed Landau's original variational derivation involving minimization of the Free Energy functional of the superconductor.[34,35] Read about Landau's phenomenological approach to Ginzburg–Landau theory and write a summary of it.

13.13 References for Chapter 13

R.13.1. This fundamental feature is discussed in:

(a) N.N. Bogoliubov, J. Phys. (USSR) **11**, 23 (1947);

(b) O. Penrose Phil. Mag. **42**, 1373 (1951); O. Penrose and J. Onsager, Phys. Rev. **104**, 576 (1956).

R.13.2. L.P. Kadanoff in "Statistical Physics, Phase Transitions and Superfluidity," (1966 Brandeis Summer School, Vol. I, Editors: Chretien, Gross and Deser, Gordon and Breach, p.107 ff. (1968).

R.13.3. P.C. Hohenberg and P.C. Martin, Annals of Physics (NY), **34**, 291 (1965).

R.13.4. L.D. Landau and E.M. Lifshitz, "Statistical Physics," Addison-Wesley, Section 67 (1958).

R.13.5. L.D. Landau and E.M. Lifshitz, "Fluid Mechanics," Addison-Wesley (1959).

R.13.6. L. Tisza, Nature **141**, 913 (1938).

R.13.7.

(a) A.L. Fetter and J.D. Walecka, "Quantum Theory of Many Particle Systems," McGraw-Hill, Section 55, Section 51 (1971).

(b) A.A. Abrikosov, L.P. Gorkov and I.Y. Dzyaloshinskii, "Quantum Field Theoretical Methods in Statistical Physics," Pergamon Press (1965).

R.13.8. C.N. Yang, Rev. Mod. Phys. **34**, 694 (1962).

R.13.9.

(a) P.W. Anderson, "Basic Notions of Condensed Matter Physics," Benjamin-Cumming Publ. (1984).

(b) P.W. Anderson, Rev. Mod. Phys. **38**, 298 (1966).

R.13.10. J. Bardeen, L. Cooper and J. Shriefter, Phys. Rev. **108**, 1175 (1975).

R.13.11.

(a) L.P. Kadanoff and P.C. Martin, Phys. Rev. **124**, 670 (1961) develops a Green's function description of the BCS theory of superconductivity including single-particle lifetime effects, as well as BCS electromagnetics and thermodynamics.

(b) S. Englesberg, Phys. Rev. 126, 1251 (1962) presents a variational differential formulation of the Green's function equations for superconductors.

(c) A.A. Abrikosov, L.P. Gorkov and I. Ye. Dzyaloshinskii, "Quantum Field Theoretical Methods in Statistical Physics," Pergamon Press (1965).

(d) A.L. Fetter and J.D. Walecka, "Quantum Theory of Many Particle Systems," McGraw-Hill, Section 55, Section 51 (1971).

R.13.12. Good reviews of BCS theory may be found in:

(a) Rev. Mod. Phys. **36**, (Number 1, Part 1), AIP, Jan. (1964);

(b) R.D. Parks, "Superconductivity," Vols.1,2, Dekker (1969).

R.13.13. Y. Nambu, Phys. Rev. **117**, 648 (1960).

R.13.14. D. Lurie in "Quantum Fluids," Editors: N. Wiser and D.J. Amit, Gordon and Breach, p.543 (1970).

R.13.15. V. Ambegaokar and M.R. Werthamer in "Superconductivity," Editor: R.D. Parks, Dekker (1969). Vol. 1 (Chapters 6 and 7, respectively).

R.13.16. M. Cyrot, Rep. Prog. Phys. **36**, 103 (1973).

R.13.17. M. Tinkham, "Introduction to Superconductivity," McGraw-Hill, Section 5–5.1 (1975).

R.13.18. P. deGennes, "Superconductivity of Metals and Alloys," W.A. Benjamin (1966).

R.13.19. K. Maki in "Superconductivity," Editor: R.D. Parks, Dekker (1969). Vol. 2, Chapter 18.

R.13.20. A.M. Gulian and G.F. Zharkov, "Nonequilibrium Electrons and Phonons in Superconductors," Kluwer Academic/Plenum Publ., Chapters 2 and 7 (1999).

R.13.21.

(a) V. Ambegaokar and L. Kadanoff, Il Nuovo Cimento **22**, 914 (1961);

(b) V. Ambegaokar and L. Kadanoff in "The Many-Body Problem," Editors: C. Fronsdal, Benjamin, N.Y. (1962).

R.13.22. G. Migdal, Nucl. Phys. **13**, 655 (1959).

R.13.23. P.W. Anderson, Phys. Rev. **110**, 827 (1958); ibid. **112**, 1900 (1958).

R.13.24. A.L. Fetter and P.C. Hohenberg, in "Superconductivity," Vol. 2, Editor: R.D. Parks, Chapter 14 (1969).

R.13.25. J. Bardeen, Phys. Rev. Lett. **6**, 57 (1961).

R.13.26. R. Prange, Phys. Rev. **131**, 1083 (1963).

R.13.27. M.H. Cohen, L.M. Falicov and J.C. Phillips, Phys. Rev. Lett. **8**, 316 (1962).

R.13.28. V. Ambegaokar and A. Baratoff, Phys. Rev. Lett. **10**, 486 (1963).

R.13.29. P.R. Wallace and M.J. Stavn, Canadian J. Phys. **43**, 411 (1965).

R.13.30. B. Josephson, Advan. Phys. **14**, 419 (1965); B. Josephson, Phys. Lett. **1**, 251 (1962); B. Josephson in "Superconductivity," Vol. 1, Editor: R.D. Parks, Chapt. 9 (1969).

R.13.31. P.W. Anderson in "Lectures on the Many-Body Problem," Vol. 2, Editor: E.R. Caianiello, Academic Press, p.113 (1964).

R.13.32. K.H. Bennemann and J.B. Ketterson, Editors, "Superconductivity," Vol. I (Conventional and Unconventional Superconductors) and Vol. II (Novel Superconductors), Springer (2008).

R.13.33. K.H. Bennemann and J.B. Ketterson, Editors, "Novel Superfluids," Vol. I and Vol. II, Oxford University Press (Vol. I, 2013, and Vol. II, in preparation).

R.13.34. N.R. Werthamer in Chapter 6, "Superconductivity," Editor: R.D. Parks, Vol. 1, p.321 (1969).

R.13.35. M. Cyrot, Rep. Prog. Phys. 36, 103, p.112 (1973).

Epilogue

Abstract
This epilogue is focused on providing additional references intended to help bring the reader up to date on important related subjects.

Further Developments

As explained in the Prologue, the primary purpose of this book is pedagogical: to bring the techniques of coupled quantum field theory to students at a relatively early stage of development by the use of variational calculus and an introduction to Green's functions. While real applications of these methods have been provided herein, including some involving nanostructures, Graphene and superconductors, it must be understood that a great deal more has been omitted. To assist readers interested in learning more, we provide additional references on some important related subjects, including pertinent books, reviews and papers. *Magnetism*,[1–14] (especially ferromagnetism, spin waves, and their quanta, magnons), the *Kondo Problem*[15–17] and *Heavy Fermons, Strong Correlations*[18,19] and *Bosonization*[20,21] are a few important areas to be pursued further. *Disordered Systems and Anderson Localization*[22–28] also require further attention.

The subject of transport theory is of enormous practical relevance in connection with semiconductor devices. *Formal Quantum Transport Theory*[29–42] involving the derivation of transport equations using non-equilibrium Green's functions,[29,30] alternatively, superoperators,[31] is of fundamental interest. The role of *Transport in Mesoscopic and Nanostructured Systems*[43–57] has also been thoroughly studied, as has been the subject of *Tunneling*.[58–81]

The methods discussed in this book have also been fruitfully employed in *Quantum Optics*[82,83] and *Nuclear Many-Body Theory*.[84–87] Other closely related subjects for exploration by interested readers include *Landau Fermi Liquid Theory*[88–92] and *Density Functional Theory*.[93–105]

Two of the subjects that we have chosen as examples are extremely important. Graphene (Chapter 12) and similar materials having a Dirac-type spectrum, such as *Silicene*[106–109] and *Topological Insulators*,[110–114] have the potential for enormous commercial impact as successors to silicon, holding the promise of another technological revolution in the electronic semiconductor device and computer industries. Similarly, the dissipationless electrical conduction properties of superconductors have great commercial potential if sufficiently high T_c materials can be found or developed. Moreover, superconductivity is the premier manifestation of a macroscopic quantum phenomenon. In consideration of the exceptional importance of these subjects, we have provided extensive reference lists to help bring the reader up to date on the recent developments in these areas by citing key reviews, articles and books in the pertinent chapters.

Significant advances in the quantum field theory of condensed states have emerged in some of the subjects discussed here. In particular, the appearance of a non-vanishing order parameter (in connection with long-range order[115] in a superconductor) signals that the condensed system does not have the full symmetry of the Hamiltonian of the system. Specifically, the superconducting condensate ground state, noted in Chapter 13 to be represented by a superposition of number states having the same phase, violates the conservation of number, whereas the superconducting Hamiltonian does conserve number (as the two operators commute). In this breakdown of number conservation in the condensed ground state (not being an eigenstate of definite number), there is a corresponding breakdown of local gauge symmetry.[116,117] Goldstone[118] found that such a spontaneous breakdown of a continuous symmetry of the Hamiltonian in the ground states gives rise to the existence of a scalar boson particle/excitation in the spectrum, representing a long wavelength fluctuation of the associated order parameter; provided there are no Coulomb forces. This excitation is known as a "Goldstone boson" and it has a "massless" energy–momentum dispersion relation of the form $\omega \sim cp$ (c is a constant), such as we discussed in Eq. (13.8.56). However, this is modified by the presence of the Coulomb interaction into a plasma excitation, as was shown in Eq. (13.8.61). This phenomenology is also manifested in ferromagnetism with spontaneously broken rotational symmetry and spin waves (magnons) as the associated Goldstone bosons; and also in fluids with spontaneously broken Galilean symmetry and longitudinal phonons as the Goldstone bosons; also in solids with spontaneously broken Galilean translational and rotational symmetries and longitudinal and transverse phonons as the Goldstone bosons. These, and other, broken symmetries have attracted considerable attention by quantum field theorists and a rich literature has emerged. In particular, Anderson has discussed broken symmetries at length.[116,117] Lurié[119] and Martin[120] also provide limpid discussions of Goldstone's theorem/bosons, as has Nambu.[121] Moreover, there have been concomitant developments in *Gauge Field Theory*[122] in condensed matter. Further associated developments in quantum field theory include application of the *Renormalization Group* method[123–126] and *Quantum Scaling and Critical Phenomena*.[127]

All the subjects mentioned above merit much greater attention than was provided here. This is to say that this book is just a beginning. But with this beginning the interested student can consult the literature with the ability to understand and use it fruitfully.

References for Epilogue

R.14.1. L.P. Levy, "Magnetism and Superconductivity," Springer (2000).
R.14.2. D.C. Mattis, "The Theory of Magnetism," Springer (1981).
R.14.3. S. Blundell, "Magnetism in Condensed Matter," Oxford (2001).
R.14.4. N. Majlis, "The Quantum Theory of Magnetism," World Scientific (2000).
R.14.5. P. Fazekas, "Electron Correlation and Magnetism," World Scientific (1999).
R.14.6. S. Chikazumi and C.D. Graham Jr., "Physics of Ferromagnetism," Oxford (1997).
R.14.7. A. Auerbach, "Interacting Electrons and Quantum Magnetism," Springer (1994).

R.14.8. J. Kubler, "Theory of Itinerant Electron Magnetism," Oxford (2000).
R.14.9. A. Ahroni, "Introduction to the Theory of Ferromagnetism," Oxford (1996).
R.14.10. R.M. White, "Quantum Theory of Magnetism," McGraw-Hill (1970).
R.14.11. S.V. Tyablikov, "Methods in the Quantum Theory of Magnetism," Plenum Press (1967).
R.14.12. W. Marshall, Editor, "Theory of Magnetism in Transition Metals," Academic Press (1967).
R.14.13. H.J. Zeiger and G.W. Pratt, "Magnetic Interactions in Solids," Oxford (1973).
R.14.14. A.I. Akhiezer, V.G. Baryakhtar and S.V. Peletminskii, "Spin Waves," North-Holland Publ. (1968).
R.14.15. A.C. Hewson, "The Kondo Problem to Heavy Fermions," Cambridge University Press (1993).
R.14.16. P. Fulde, J. Keller and G. Zwicknagl, "Theory of Heavy Fermion Systems," Solid State Physics *41*, 1 (1988).
R.14.17. Y. Kuramoto and Y. Kitaoka, "Dynamics of Heavy Electrons," Oxford (2000).
R.14.18. N. Nagosa, "Quantum Field Theory in Strongly Correlated Electronic Systems," Springer (1998).
R.14.19. A. Georges, G. Kotliar, W. Krauth and M.J. Rozenberg, Rev. Mod. Phys. 68, 13 (1996).
R.14.20. A.O. Gogolin, A.A. Nersesyan and A.M. Tsvelik, "Bosonization and Strongly Correlated Systems," Cambridge University Press (1998).
R.14.21. P. Kopietz, "Bosonization of Interacting Fermions in Arbitrary Dimensions," Springer (1997).
R.14.22. P.A. Lee and T.V. Ramakrishnan, Rev. Mod. Phys. 57, 287 (1985).
R.14.23. M. Janssen, "Fluctuations and Localization in Mesoscopic Electron Systems," World Scientific (2001).
R.14.24. A.L. Efros and M. Pollak, "Electron-Electron Interactions in Disordered Systems," North-Holland (1985).
R.14.25. R.J. Elliott, J.A. Krumhansl and P.L. Leath, Rev. Mod. Phys. **46**, 465 (1974).
R.14.26. J.M. Ziman, "Models of Disorder," Cambridge University Press (1979).
R.14.27. C. Papatrianafillou and E.N. Economou, Phys. Rev. B **13**, pp. 910 and 920 (1976).
R.14.28. Ping Sheng, "Introduction to Wave Scattering, Localization and Mesoscopic Phenomena," Academic Press (1995).
R.14.29. L.P. Kadanoff and G. Baym, "Quantum Statistical Mechanics," W.A. Benjamin Inc., Chapters 6–10 (1962).
R.14.30. V.G. Morozov and G. Röpke in "Progress in Nonequilibrium Green's Functions," Ed: M. Bonitz, World Scientific, p.82 (2000).
R.14.31. B. Velicky, A. Kalvolva and V. Spicka in "Progress in Nonequilib. Green's Functions III," Eds: M. Bonitz and A. Filinov, J. Phys. Conf. Series 35, p.1, (2006).
R.14.32. P. Lipavsky, V. Spicka and B. Velicky, Phys. Rev. B34, 6933 (1986), and references cited therein.

R.14.33. V. Spicka, B. Velicky and A. Kalvova, Physica E29, p.154, 175, 196, (1999), and references cited therein.
R.14.34. N. Ashby, "Applications of Quantum Field Theory to Electrical Conductivity," Ph.D thesis, Harvard University, (1961). Also, N. Ashby in "Lectures in Theoretical Physics," Boulder 3, (presented at the Theoretical Physics Institute), University of Colorado Press, Summer, (1960).
R.14.35. R. Kubo, Can. J. Phys. 31, 1274 (1956). Also R. Kubo, J. Phys. Soc. Japan 12, 570 (1957).
R.14.36. R. Zwanzig, in "Lectures in Theoretical Physics," Boulder, 3, University of Colorado Press (1960).
R.14.37. U. Fano, Phys. Rev. 131, 259 (1963); U. Fano, Reviews of Modern Physics 29, 74 (1957).
R.14.38. P.N. Argyres, in "Lectures in Theoretical Physics," Boulder, Vol. VIII-A, University of Colorado Press (1966).
R.14.39. F.A. Buot, "Rivista del Nuovo Cimento," Vol. *20* No.9, 1 (1997).
R.14.40. H.J.W. Haug and A.P. Jauho, "Quantum Kinetics in Transport and Optics of Semiconductors," 2nd Edition, Springer (2008).
R.14.41. A.P. Jauho in "Granular Nanoelectronics," Editor: R.K. Ferry, Plenum Press, p.133 (1991).
R.14.42. J.T. Devreese and V.E. van Doren, Editors, "Linear and Nonlinear Electron Transport in Solids," Plenum Press (1976).
R.14.43. D.A. Ryndyk, R. Gutierrez and G. Cuniberti, "Green Function Techniques in the Treatment of Quantum Transport at the Molecular Scale," arXiv:0805.0628v2 [cond-mat.mes-hall] (2008).
R.14.44. P. Danielewicz, Annals of Phys. **152**, 239 (1984).
R.14.45. J. Rammer, "Quantum Transport Theory," Perseus Books (1998).
R.14.46. J. Rammer, "Quantum Field Theory of Nonequilibrium States," Cambridge University Press (2007).
R.14.47. J. Rammer and H. Smith, Rev. Mod. Phys. **58**, 323 (1986).
R.14.48. J. Rammer, "Applications of Quantum Field Theory Methods in Transport Theory of Metals," Thesis at University of Copenhagen, Denmark (1984).
R.14.49. G.D. Mahan, Physics Reports **145**, 251 (1987).
R.14.50. S. Datta, J. Phys. Cond. Matter **2**, 8023 (1990).
R.14.51. S. Datta, "Electronic Transport in Mesoscopic Systems," Cambridge University Press (1995).
R.14.52. E. Akkermans and G. Montambaux, "Mesoscopic Physics of Electrons and Photons," Cambridge University Press (2007).
R.14.53. W. Magnus and W. Schoenmaker, "Quantum Transport in Submicron Devices," Springer (2002).
R.14.54. E. Scholl, "Theory of Transport Properties of Semiconductor Nanostructrures," Chapman and Hall (1998).
R.14.55. W. Schafer and M. Wegener, "Semiconductor Optics and Transport Phenomena," Springer (2002).
R.14.56. T. Ihn, "Semiconductor Nanostructures: Quantum States and Electronic Transport," Oxford University Press (2010).

R.14.57. Y.V. Nazarov and Y.M. Blanter, "Quantum Transport: Introduction to Nanoscience," Cambridge University Press (2009).
R.14.58. P. Spivak, S.V. Kravchenko, S.A. Kivelson and X. P.A. Gao, arXiv: 0905.0141vl [cond-mat.dis-nn] (2009).
R.14.59. L. Reggiani, Editor, "Hot Electron Transport in Semiconductors," Springer-Verlag (1985).
R.14.60. X.L. Lei and N.J.M. Horing, "Balance Equation Approach to Hot-Carrier Transport in Semiconductors," in "Physics of Hot Electron Transport in Semiconductors," Editor: C.S. Ting, World Scientific Press, Singapore, 1 (1992).
R.14.61. N.J.M. Horing, H.L. Cui and X.L. Lei, "Recent Developments in Hot Electron Magnetotransport Theory," in "Physics of Hot Electron Transport in Semiconductors," Editor: C.S. Ting, World Scientific Press, Singapore, 133 (1992).
R.14.62. Y. Imry, "Introduction to Mesoscopic Physics," 2nd Edition, Oxford University Press (2002).
R.14.63. H. Fukuyama and T. Ando, Editors, "Transport Phenomena in Mesoscopic Systems," Springer-Verlag (1992).
R.14.64. G. Grinstein and G. Mazenko, Editors, "Directions in Condensed Matter Physics," World Scientific (1986).
R.14.65. C.W.J. Beenaker and H. van Houten, "Quantum Transport in Semiconductor Nanostructures," Solid State Physics *44*, 1 (1991).
R.14.66. L.P. Kadanoff in "Statistical Physics, Phase Transitions of Superfluidity," Editors: M. Chretien, E.P. Gross and S. Deser, Gordon and Breach, p.116 (1968).
R.14.67. J.R. Schrieffer in "Tunneling Phenomena in Solids," Editors: E. Burstein and S. Lundquist, Plenum, Chapter 21 (1969).
R.14.68. J. Bardeen, Phys. Rev. Lett **6**, 57 (1961).
R.14.69. R. Prange, Phys. Rev. **131**, 1083 (1963).
R.14.70. T.E. Feuchtwang in "Inelastic Electron Tunneling Spectroscopy," Editor: T. Wolfram, Springer-Verlag, p.129 (1978).
R.14.71. W.A. Harrison, Phys. Rev. **123**, 85 (1961).
R.14.72. M.C. Payne, J. Phys. C: Solid State Phys. **19**, 1145 (1986).
R.14.73. T.E. Feuchtwang, Phys. Rev. B **10**, 4121 and 4135 (1974); T.E. Feuchtwang, Phys. Rev. B **12**, 3979 (1975); T.E. Feuchtwang, Phys. Rev. B **13**, 517 (1976); T.E. Feuchtwang, Phys. Rev. B **20**, 430 (1979).
R.14.74. S. Datta, "Electronic Transport in Mesoscopic Systems," Cambridge University Press, especially Chapter 12 (1995).
R.14.75. R. Lake and S. Datta, Phys. Rev. B **45**, 6670 (1992).
R.14.76. S. Datta, Phys. Rev. B **40**, 5830 (1989); S. Datta, Phys. Rev. B **46**, 9493 (1992).
R.14.77. X.J. Lu and N.J.M. Horing, Phys. Stat. Sol.(a) **172** (1992).
R.14.78. L.Y. Chen and C.S. Ting, Int. J. Mod. Phys. B **6**, 1099 (1992).

R.14.79. L.Y. Chen and P.L. Nash, Mod. Phys. Lett. B **11**, 35 (1997).
R.14.80. H. Mizuta and T. Tanoue, "The Physics and Applications of Resonant Tunneling Diodes," Cambridge University Press (1995).
R.14.81. E. Garcia-Calderon in "The Physics of Low-Dimensional Semiconductor Structures," Editors: P.N. Butcher, N.H. March and M.P. Tosi, Plenum (1993).
R.14.82. E.L. Wolf, "Principles of Electron Tunneling Spectroscopy," Oxford University Press (1985).
R.14.83. C.B. Duke, "Tunneling in Solids," Academic Press, Chapter VII (1969).
R.14.84. A.J. Bennett, C.B. Duke and S.D. Silverstein, Phys. Rev. **176**, 969 (1968).
R.14.85. H. Grabert and M.H. Devoret, "Single Charge Tunneling," Plenum Press (1992).
R.14.86. T. Dittrich, *et al.*, "Quantum Transport and Dissipation," Wiley-VCH (1998).
R.14.87. S. Takagi, "Macroscopic Quantum Tunneling," Cambridge University Press (2002).
R.14.88. M. Razavy, "Quantum Theory of Tunneling" World Scientific (2003).
R.14.89. D.K. Roy, "Quantum Mechanical Tunneling and its Applications," World Scientific (1986).
R.14.90. S.M. Barnett and P.M. Radmore, "Methods in Theoretical Quantum Optics," Oxford Science Publ. (1997).
R.14.91. H.J.W. Haug and A.P. Jauho, "Quantum Kinetics in Transport and Optics of Semiconductors," 2nd Edition, Springer (2008).
R.14.92. A.L. Fetter and J.D. Walecka, "Quantum Theory of Many–Particle Systems," McGraw-Hill (1971).
R.14.93. C. Bloch, Editor, "Many–Body Description of Nuclear Structure and Reactions," Academic Press (1966).
R.14.94. J.D. Walecka, "Theoretical Nuclear and Subnuclear Physics," Oxford (1995).
R.14.95. I.Z. Petkov and M.V. Stoitsor, "Nuclear Density Functional Theory," Oxford Sci. Publ. (1991).
R.14.96. L.P. Kadanoff and G. Baym, "Quantum Statistical Mechanics," W.A. Benjamin, Chapter 11 (1962).
R.14.97. P. Nozieres, "Interacting Fermi Systems," Benjamin (1964).
R.14.98. G. Baym and C. Pethick, "Landau Fermi–Liquid Theory," Wiley (1991).
R.14.99. H.J. Schulz, "Fermi Liquids and Non–Fermi Liquids," Lectures at Lab. de Physique des Solides, University Paris–Sud, Orsay, France.
R.14.100. N.H. March and M.P. Tosi, "Introduction to Liquid State Physics," World Scientific (2002).
R.14.101. P. Hohenberg and W. Kohn, Phys. Rev. **136**, B864 (1964).
R.14.102. W. Kohn and L. Sham, Phys. Rev. **140**, A1133 (1965).
R.14.103. J. Callaway and N.H. March, "Density Functional Methods: Theory and Applications," Solid State Physics *38*, Academic Press, 135 (1984).
R.14.104. R.O. Jones and O. Gunnarsson, "The Density Functional Formalism, its Applications and Prospects," Rev. Mod. Phys. *61*, 689 (1989).

R.14.105. R.G. Parr and W.W. Yang, "Density–Functional Theory of Atoms and Molecules," Oxford University Press (1989).
R.14.106. F. Giustino, "Materials Modeling using Density Functional Theory," Oxford University Press (2014).
R.14.107. R.F. Nalewajski, Editor, "Density Functional Theory," Vols. I, II, III, IV, Springer (1996).
R.14.108. I.Z. Petkov and M.V. Stoitsov, "Nuclear Density Functional Theory," Oxford Sci. Publ. (1991).
R.14.109. A.D. Becke, J. Chem. Phys. **140**, 18A301 (2014).
R.14.110. K. Burke, J. Chem. Phys. **136**, 150901 (2012).
R.14.111. J. Klimes and A. Michaelides, J. Chem. Phys. **137**, 120901 (2012).
R.14.112. N. Argaman and G. Makov, arXiv: physics, 9806013v2, [physics.ed-ph] (1999).
R.14.113. H.L. Neal, Am. J. Phys. **72**, 605 (2004).
R.14.114. G.G. Guzman-Verri and L.C.L. YanVoon, Phys. Rev. B **76**, 075131 (2007).
R.14.115. S. Cahangirov, *et al.*, Phys. Rev. Lett. **102**, 236804 (2009).
R.14.116. J.C. Garcia, *et al.*, J. Physical Chem.C **115**, 13242 (2011).
R.14.117. D. Jose and A. Datta, Accounts of Chemical Research **47**, 593 (2014).
R.14.118. M.Z. Hasan and C.L. Kane, Rev. Mod. Phys. **82**, 3045 (2010).
R.14.119. X.L. Qi and S.C. Zhang, Rev. Mod. Phys. **83**, 1057 (2011).
R.14.120. L. Fu and C.L. Kane, Phys. Rev. **B76**, 045302 (2007).
R.14.121. D. Hsieh, *et al.*, arXiv: 0910.2420v1 [cond–mat.mes–hall] (2009).
R.14.122. P. Di Pietro, *et al.*, Nature Nanotechnology **8**, 556 (2013).
R.14.123. R.M. White and T.H. Geballe, "Long Range Order in Solids," Academic Press (1979).
R.14.124. P.W. Anderson, "Special Effects in Superconductivity," in "Lectures on the Many Body Problem," Vol. 2, Editor: E.R. Caianiello, Academic Press, Section 4, p.132 (1964).
R.14.125. P.W. Anderson, "Basic Notions of Condensed Matter Physics," W.A. Benjamin (1984).
R.14.126. J. Goldstone, Il Nuovo Cimento **XIX**, 154 (1961).
R.14.127. D. Lurié, Chapter 22 in "Quantum Fluids," Editor: N. Wiser and D. Amit, Gordon and Breach, p.543 (1970).
R.14.128. P.C. Martin, Chapter 7 in "Superconductivity," Vol. 1, Editor: R.D. Parks, p.371 (1969).
R.14.129. Y. Nambu, Phys. Rev. **117**, 648 (1960).
R.14.130. H. Kleinert, "Gauge Fields in Condensed Matter," Vols. I and II, World Scientific (1989).
R.14.131. K. Wilson, Rev. Mod. Phys. **47**, 773 (1975).
R.14.132. D. Mattuck, "A Guide to Feynman Diagrams in the Many-Body Problem," Dover, Chapter 19, 328 (1992).
R.14.133. K. Wilson, Physics Reports **12**, 75 (1974).
R.14.134. R.J. Creswick, H.A. Farach and C.P. Poole, Jr., "Introduction to Renormalization Group Methods in Physics," Wiley (1992).
R.14.135. M.A. Continentino, "Quantum Scaling in Many-Body Systems," World Scientific (2001).

Index

A

acoustic normal mode 371, 405
Ambegaokar and Kadanoff 371, 401, 404, 407
amplitude for correlated interacting particle motions 168, 169
analytic continuation 168, 176, 206, 231
Anderson 370, 371, 380, 381, 407
Anderson Localization 421
Ando's semi-elliptic density of states 95, 113, 115
antidot lattice 340, 355
approximation procedures 206, 215, 225, 303, 323
approximations 187, 192, 195
asymptotic equivalence 150, 153

B

Bardeen–Cooper–Schrieffer energy gap 370, 382, 388
BCS 370, 382, 388
Bogoliubon excitations 371, 389
Bogoliubov 370, 372, 375, 379, 389
Bogoliubov-de Gennes energy eigenfunction equations 371, 389
Bohm–Staver phonon mode 303, 322
Born and self-consistent Born approximations 95, 109, 113, 116
Born approximation series 187, 198
Bose condensation 150, 158, 370, 371, 375, 382, 386, 387
Bose-condensed state 370, 372, 373, 379, 380
Bose–Einstein distribution function 150, 157, 175
Boson 7, 61, 63, 156, 169, 196, 219, 306, 382
Bosonization 421
bound electron–hole system 187, 199, 202
bra-ket matrix elements of operators 1, 10, 75
breakdown of local gauge symmetry 422
breakdown of number conservation 422

C

c-number coordinates and fields 61, 74, 77, 271, 370
canonical equal-time commutation relations 161, 180, 305
canonical equal-time commutation/ anticommutation relations 61, 63, 68, 188
causality 239, 290
chemical potential 150, 154, 174, 185, 193, 249, 285, 378
classical limit 238, 244–246, 288
coarse-grain average 370, 380
coherence length ξ 370, 371, 374, 410
coherent states 20, 33, 37, 38
commutation/anticommutation relations 20, 27, 28, 30, 97, 188, 312
commutator/anticommutator functions 168, 179
complete-orthonormal sets of eigenfunctions 1, 13
complex time plane 207, 232
compound nanostructures 239, 290
condensate/Cooper-pair
  anomalous bound-Cooper-pair condensate wave function 370, 382, 383
  anomalous correlation function 370, 376
  condensate ground state 422
  condensate wave function 370, 372, 373, 379, 381
  condensed state 370, 377, 382, 422
  Cooper pair 370, 382, 386, 387, 389, 393, 396–398, 406, 412, 416
  Cooper-pair density 371, 397
  Cooper-pair states 370, 383, 386
conductivity tensor 238, 273, 280
continuum of states 95, 103, 105, 121
core region 370, 375
correlation, exchange
  anomalous correlation function 370, 376
  correlated multiparticle propagation dynamics 206, 207
  correlation energies 238, 252, 255
  density-density correlation function 206, 223
  equilibrium current–current commutator correlation function 40, 52
  exchange and correlation energies 238, 252, 255
Coulombic attraction between electrons and holes 302, 306
Coulombic repulsions among electrons and among holes 302, 306
coupled field equations 61, 75
coupling strength 61, 84, 220

creation/annihilation operators
annihilation 1, 16, 32, 34, 97, 372
creation 1, 16, 35, 97, 372
creation and annihilation operators 20, 24, 25, 40, 45, 83, 169, 198, 257, 305
critical magnetic field 371, 390, 410
critical temperature for superconductivity 371, 382, 388
cumulants 87, 206, 218
current 1, 5, 33, 52, 135, 209, 271, 280

D

Darwin and Fowler 150, 151
de Haas–van Alphen effect 150, 165
Debye–Thomas–Fermi shielding law 238, 249
degenerate limit 164, 178, 238, 247, 266, 340, 344, 345, 360
Density Functional Theory 421
density matrix 40, 54, 236
diamagnetism 150, 163
dielectric tensor (inverse) 238, 276
Dirac
Dirac bra-ket notation 1, 8
Dirac delta $\delta(x)$-function 95, 96, 104, 113, 133
Dirac equation 340, 362
Dirac notation 1, 8
Dirac-like 340, 342
Dirac-type spectrum 421
Disordered Systems 421
dissipationless electrical conduction properties 421
distribution functions 150, 156, 172, 182, 234, 285
Dyson integral equation 95, 106, 107, 110–112, 114, 115

E

eigen-energy dispersion relation 340, 358
eigenstates 1, 13, 16, 20, 34
Electrodynamics 271
electromagnetic 238, 271, 288
electron 61, 80
electron propagating jointly with its induced lattice polarization 303
electron/hole/phonon interaction/propagation
correlated multiparticle propagation dynamics 206, 207
electron coupling in phonon propagation 302, 319
electron propagating jointly with its induced lattice polarization 322
electron screening of the phonon spectrum 303, 321
electron–electron interaction 302, 305
electron–electron interaction potential mediated by the one-phonon Green's function 303
electron–electron interaction potential moderated by the one-phonon Green's function 323
electron–hole interaction 302, 313
electron–hole interaction operator 303, 328, 330, 332
electron–hole–phonon system 302, 305, 323, 327, 330, 333
electron–phonon interaction 302, 306
hole 61, 80
hole–hole interaction 302, 305
hole–phonon interaction 306
one-electron propagation influenced by a non-trivial phonon field 311
particle–particle interactions 187, 191, 195, 212
phonon 61, 80
phonon source 302, 307, 323
phonon-mediated electron–electron and hole–hole interactions 335
propagation of one electron and one hole 187, 199
two-electron propagation influenced by a non-trivial phonon field 302, 312, 313
electrons in magnetic fields in 2D and 3D 116, 120
electrons in magnetic fields in both three and two-dimensions 95
elementary particle physics 61
energy eigenfunction expansion 95, 99
energy gap 370, 382, 388
energy loss of an energetic charged particle 238, 264
energy spectrum 168, 171, 340, 350, 355
Enthalpy 150, 156
entropy 150, 151
equations of motion 187–189, 191, 196
equilibrium conditions 168, 170, 176
exciton states 187, 202, 203
exciton–plasmon interaction 303, 337
exclusion of magnetic flux 371, 397, 398, 409

F

Fermi energy 150, 154
Fermi–Dirac distribution function 150, 157, 175
Fermion 7, 61, 63, 156, 169, 196, 219
ferromagnetism 421
Feynman diagrammatics 95, 107, 109
Feynman path integrals 95, 144, 145
Formal Quantum Transport Theory 421
Free Energy 340, 359
Friedel oscillations 238, 246, 248

G

gap parameter 371, 402
Gauge Field Theory 422
gauge invariance 371, 418
gauge transformations 1, 6
generalized coordinates and momenta 61, 64

Generalized Kadanoff–Baym Ansatz 207, 234
generalized shielded potential approximation 303, 331, 333
generating functional $e^{\mathcal{W}}$ 206, 212
Gibbs Free Energy 150, 155
Ginzburg–Landau equations 371, 418
Ginzburg–Landau theory 371, 391, 418
Goldstone boson 422
Goldstone's theorem/bosons 422
Gorkov equations 370, 385, 386, 388, 391, 399, 418
grand canonical ensemble average 150, 151, 168, 184
grand partition function 150, 151
grand potential 150, 151
Graphene 340, 341, 346, 352, 355, 358, 361
  Graphene antidot lattice 340, 355, 357
  Graphene Green's function 340, 346, 352, 355
  statistical thermodynamics of Graphene 340, 358
Green's functions 40, 60, 61, 85
  $G_{1<}$ 168, 173, 175, 208, 233, 344, 379
  $G_{1>}$ 168, 173, 175, 208, 233, 344, 379
  n-particle Green's function equation of motion 187, 189
  advanced Green's functions 168, 181, 233
  alternative formulation of the coupled Green's function equations 303, 323
  chain of coupled Green's function equations 302, 309, 331
  coupled-field Green's function equations 302, 307
  coupling of G_n with G_{n+1} 187, 189
  dyadic Green's functions 238, 271, 274, 291
  electron–electron interaction potential mediated by the one-phonon Green's function 303
  electron–electron interaction potential moderated by the one-phonon Green's function 323
  equivalence of Green's functions with Feynman path integrals 95, 144, 145
  factorization ansatz for G_3 187, 197
  Graphene Green's function 340, 346, 352, 358
  Green's function derivation methods 116, 123
  Green's function for one-electron propagation influenced by a non-trivial phonon field 302, 311
  Green's function for two-electron propagation influenced by a non-trivial phonon field 312, 313
  Green's function matching 95, 133, 138, 139
  infinite chain of coupled Green's functions 191, 192
  magnetic field Green's function 340, 352
  mathematically convenient non-equilibrium Green's functions 206, 207, 211, 221
  modified Green's function 168, 174
  multiparticle thermodynamic Green's functions 168, 169, 172
  non-condensate Green's function 370, 376
  non-equilibrium Green's functions 206, 207, 211, 221, 421
  one-electron Green's function 302, 311
  one-electron/one-hole Green's function 302, 311
  one-particle thermal Green's function 168, 176
  one-phonon Green's function 315
  partial Green's functions 95, 101, 103
  permanent/determinant of single-particle Green's functions 206, 215, 217
  perturbation theories for the Green's function 206, 225
  physical Green's function 206, 207, 224, 232
  product Green's functions 95, 127, 131
  retarded Green's functions 95–97, 100, 106, 124, 127, 161, 233
  single-particle Green's function equation 187, 188
  single-particle thermal Green's function 168, 169
  thermodynamic Green's functions 168, 169, 177, 238
  three-electron Green's function 312
  two-electron Green's function 302, 311
  two-electron/one-hole Green's function 312
  two-particle Green's functions 238, 258
  variational derivative of a Green's function 61, 86
Gross–Pitaevskii equation 370, 373, 378

H

Hamilton equation-of-motion method 95, 123, 124
Hamilton equations 61, 66
Hamilton Principle of Least Action 61, 63
Hamiltonian 40, 41, 45, 47, 52
Hartree approximation 187, 193, 195
Hartree equation 370, 373
Hartree–Fock approximation 187, 194, 195
Heavy Fermions 421
Heisenberg equation 40, 45, 61, 62, 68
Heisenberg picture 40, 44, 47, 62
Helmholtz Free Energy 150, 154, 160

high T_c materials 421
hole, *see 'electron/hole/phonon'*
hole–phonon interaction 302
hydrodynamic model 238, 250
Hydrogen-like spectrum 187, 202
hydrogenic atom 1–3

I

identical particles 6, 20, 22
image potential 143, 238, 258–261
imaginary frequency 168, 175
imaginary time 150, 161, 168, 174, 206, 231
impurity scattering 95, 107
infinite chain of coupled Green's functions 187
interacting many-body systems 61, 81
interaction picture 40, 47, 50
interparticle coupling interactions 61, 82
intracollisional field effect 40, 59
inverse dielectric screening function 206, 222, 238, 264
inverse dielectric screening response function 303, 333
inverse dielectric tensor 238, 276
inverse thermal energy 150, 154, 161
ion-core lattice 302, 304

J

Josephson tunneling 371, 412, 414

K

ket-bra matrix operator structures 1, 9, 10
Kohn anomaly 303, 322
Kondo Problem 421
Kramers–Krönig relations 239, 290

L

Landau Fermi Liquid Theory 421
Landau's criterion 370, 378
Langreth Algebra 207, 233
Lehmann spectral representation 168, 171
lifetime 187, 198
Lindhard 238, 244
linear and angular momentum 1, 4, 5, 20
linear current response 40, 52
linear response 40, 51
linked cluster theorem 206, 220
Liouville operators 40, 55
localized 95, 103–105
London theory 371, 397, 398
long-range order 422
longitudinal and transverse phonons 422
Lurie 422

M

macroscopic occupation 158, 370, 372
macroscopic quantum phenomenon 421
magnetic field Green's function 356
magnetic Landau quantization 340, 361
magnetic moment 150, 160, 340, 360
magnetic penetration length l 371, 410
Magnetism 421
magnons 421, 422
Martin 422
Matsubara frequencies 168, 175
Maxwell equations 271, 288
measurement symbols 20–22, 24
Meissner effect 371, 397, 398, 409
Mesoscopic and Nanostructured Systems 421
microcanonical ensemble average 150, 151
model Hamiltonian 302, 306
multiparticle energy eigenfunctions 1, 6, 7
multiple interactions 302, 307

N

Nambu 422
nanostructures 238, 271, 282, 284, 290
neutral atom 238, 256, 262, 264
non-condensate states 370, 375, 377, 385
non-conservation of number 213
non-correlation decoupling procedures 187, 192
non-equilibrium averaging 302, 319
non-equilibrium dynamics 207, 236
non-local 238, 246, 252, 263, 273
non-local dielectric function 340, 345
non-local electrostatics 95, 138, 142
non-Markovian kinetic equations 40, 58
non-relativistic many-particle physics 61, 95
non-superfluid-condensate states 370, 383
nonconservation of number 206
nonlocal plasmon spectrum 238
Nuclear Many-Body Theory 421
number conservation 370, 372, 379, 382, 383, 399, 422

O

occupation number representation 20, 25, 29
off-diagonal long-range order 370, 380, 383
one-dimensional superlattice in axial electric and magnetic fields 95, 128, 131
one-electron self-energy (Σ) 187, 203, 204
one-phonon Green's function 302
order parameter 422

P

pair wave function 371, 382, 396, 397, 405, 410
particle sources 80, 319
particle sources (η, η^{+}) 61, 206, 212, 221, 302
Pauli exclusion principle 1, 7, 20, 29
Penrose 370, 372
periodicity/antiperiodicity property 168, 174
phase change 371, 409, 411
phase change re-gauging 402, 404, 409, 411
phase difference 371, 415
phase-coherent wave packet 370, 381, 383, 386
phase-coherent wave packet superposition 370, 383, 386
Phonon, *see 'electron/hole/phonon'*
phonon-mediated electron–electron and hole–hole interactions 303
physisorption 238, 256
planar quantum well 239, 285
plasmons/collective modes
  collective mode dispersion relation 303, 335
  coupled plasmon spectra 239, 277, 289, 292, 294, 298
  dynamic, non-local plasma image potential 238, 258
  homogeneous quantum plasmas 238, 241
  hybridization of the ion and electron plasma modes 303, 321
  inhomogeneous quantum plasmas 238, 241
  metallic plasma 238, 258, 263
  non-local plasmon spectrum 244, 250
  plasma 238, 244, 256, 264, 271
  plasma dielectric response 238, 279
  plasma image potential 238, 259
  plasmon collective mode 371, 408
  surface plasmons 95, 142
polaritons 239, 279, 289, 290, 298
polarizability 238, 239, 246, 264, 303, 318, 319, 335, 336
polarizable systems 238, 256, 258
polaron 303, 322
potential sources 61, 78
power loss of an energetic charged particle 238, 264
probability amplitude 95, 97, 199, 311–313

Q

quantizing magnetic field 163, 166, 238, 340, 355
quantum dot 340, 352, 355
quantum dynamics of a multi-component system 302, 307
quantum Hamilton equations of motion 66, 302, 314
quantum mechanical ensemble averages 150, 151
quantum mechanical pictures 40, 41, 44, 47
quantum operator coordinates and fields 61, 72, 81, 87
quantum operators 51, 61, 67
Quantum Optics 421
Quantum Scaling and Critical Phenomena 422
quantum statistical mechanics 150, 156
Quantum Transport Theory 421
quantum well 239, 265, 284, 287, 295, 296
quantum wire 246, 248, 299

R

random impurity scattering 95, 108, 109, 113
Random Phase Approximation (RPA) 206, 223, 238
re-gauging 371, 402, 404
relativistic electron/positron(hole) 340, 342
relativistic quantum field theory 61, 340
Renormalization Group 422
representations 1, 8
retarded Green's functions 150
ring diagram 238, 239
RPA integral equation 206, 223

S

S-matrix 40, 51, 53
saddle potential in a time-dependent electric field 95, 124, 126
saddle-point integration 150, 154
scalar boson 422
Schrödinger equation 1, 5, 12, 13
Schrödinger picture 40, 44, 48, 62
Schwinger Action Principle 61, 64, 302
second quantization 20, 24, 31
self-energy 206, 227
semiclassical limit 238, 246
separable Hamiltonians 95, 127, 128
shielded potential perturbation theory 206, 229
Silicene 421
silicon 421
single-particle energy eigenfunctions 238, 241
single-particle spectral weight 168, 180
small-oscillation approximation 302, 304
sources 61, 78, 93, 206, 307, 311, 338
spatial translational invariance in 1D, 2D and 3D 238, 242
spatially variable effective mass 95, 133
spectral structure 168, 173
spectral weight function 168, 172
spin 1, 4, 5, 26
spin waves 421, 422
spontaneously broken Galilean translational and rotational symmetries 422
spontaneously broken rotational symmetry 422
static shielding 238, 248, 250
statistical thermodynamic information 168, 182

statistical thermodynamics 150, 151
Stern 238, 247, 266
Strong Correlations 421
Sum of Ladder Diagrams
  integral equation 187, 197
sum of ring diagrams 206, 223
sum rule 168, 180
superconductivity 370, 382, 386, 389, 391, 397, 399, 408, 415
superconductor 370, 382, 386, 389, 391, 397, 399, 408, 415, 421, 422
supercurrent 371, 382, 397, 398, 415
superfluids
  superfluid behavior 370, 378, 382
  superfluid order parameter 370, 372
  superfluidity 370, 371, 375, 382, 415
  superfluids 370, 371, 375, 382, 415
superlattice 239, 295, 297, 298
superlattice in axial electric and magnetic fields 128, 131
superoperators 40, 55, 421
symmetric formulation 206, 218

T

T-matrix 95, 106, 108, 187, 204
temperature 150, 154
tetradics 40, 55
thermal ensemble 168, 169
thermodynamic potential 150, 155
thermodynamics 150, 156
thermodynamics of superconductors 371, 389
time ordering along the contour 207, 232
time-dependent Hartree approximation 206, 222
time-development operator 95, 98, 144
time-ordered operator matrix elements 61, 82
time-ordered products 168, 175
time-translational invariance 168, 170, 176
Topological Insulators 421
transfer Hamiltonian 371, 413
transformation among representations 1, 14
transformation theory 1, 14
translation operator 1–3, 32
transport theory 421
Tunneling 421
tunneling supercurrent 371, 415
two-dimensional saddle potential in a time-dependent electric field 95, 124, 126
Type II superconductors 371, 408–410

V

van der Waals interaction 238, 256, 258
variable effective mass 95, 133
variational calculus/derivative
  variational calculus 61, 63
  variational derivative of a Green's function 61, 86
  variational derivatives with respect to sources η, η^+ 206, 214
  variational differential relations 206, 207, 216
  variational differential techniques 302, 309
  variational differentiation 302, 309, 319
  variational principle 61, 63
vertex operator 206, 228
viscosity 370, 378
von Neumann equation 40, 54
vortex 370, 375

Z

zero mass 340, 342

The manufacturer's authorised representative in the EU for product safety is Oxford University Press España S.A. of El Parque Empresarial San Fernando de Henares, Avenida de Castilla, 2 - 28830 Madrid (www.oup.es/en or product.safety@oup.com). OUP España S.A. also acts as importer into Spain of products made by the manufacturer.
Printed and bound by CPI Group (UK) Ltd, Croydon, CR0 4YY
14/09/2026
02216290-0010